Lecture Notes in Mathematics

2006

Editors:
J.-M. Morel, Cachan
F. Takens, Groningen
B. Teissier, Paris

Catherine Donati-Martin · Antoine Lejay
Alain Rouault (Eds.)

Séminaire
de Probabilités XLIII

 Springer

Editors
Catherine Donati-Martin
Laboratoire de Probabilités et Modèles
Aléatoires
Université Pierre et Marie Curie
Boite courrier 188
4, place Jussieu
75252 Paris Cedex 05, France
catherine.donati@upmc.fr

Alain Rouault
LMV
Université de Versailles-Saint-Quentin
45 avnue de Etats-Unis
78035 Versailles Cedex, France
alain.rouault@math.uvsq.fr

Antoine Lejay
IECN
Campus scientifique
BP 70239
54506 Vandoeuvre-lès-Nancy CEDEX
France
Antoine.Lejay@iecn.u-nancy.fr

ISBN: 978-3-642-15216-0   e-ISBN: 978-3-642-15217-7
DOI: 10.1007/978-3-642-15217-7
Springer Heidelberg Dordrecht London New York

Lecture Notes in Mathematics ISSN print edition: 0075-8434
ISSN electronic edition: 1617-9692

Library of Congress Control Number: 2010937110

Mathematics Subject Classification (2010): 60Gxx, 60Hxx, 60Jxx, 60Kxx, 60G22, 60G44, 60H35, 46L54

Cover design: SPi Publisher Services

Printed on acid-free paper

springer.com

Preface

The series of advanced courses, initiated in Séminaire de Probabilités XXXIII, continues with a course of Jean Picard on the representation formulae for the fractional Brownian motion. The rest of the volume covers a wide range of themes, such as stochastic calculus and stochastic differential equations, stochastic differential geometry, filtrations, analysis on Wiener space, random matrices and free probability, mathematical finance. Some of the contributions were presented at the Journées de Probabilités held in Poitiers in June 2009.

The Séminaire has now a new web site at the URL

http://portail.mathdoc.fr/SemProba/

This web site is hosted by the Cellule Math Doc funded both by the CNRS and the Université Joseph Fourier in Grenoble, France. We thank the team of the Institut de Recherche Mathématiques Avancées (IRMA) in Strasbourg for the maintenance of the former web site.

With the new web site also comes a new multicriteria research tool which improves the previous one. This tool has been developped by the Cellule MathDoc (Laurent Guillopé, Elizabeth Cherhal and Claude Goutorbe). The enormous work of indexing and commenting was started by Paul-André Meyer in 1995 with the help of other editors, with an important contribution from Michel Émery (who performed the supervision of all the work) and Marc Yor. The database covers now the contents of volumes I to XL. We expect to complete the work soon in order to provide some easy way to exploit fully the content of the Séminaire.

We remind you that the Cellule Math Doc also hosts digitized articles of many scientific journals within the NUMDAM project. All the articles of the Seminaire from Volume I in 1967 to Volume XXXVI in 2002 are freely accessible from this web site

http://www.numdam.org/numdam-bin/feuilleter?j=SPS

Finally, the Rédaction of the Séminaire is modified: Christophe Stricker and Michel Émery retired from our team after Séminaire XLII was completed. Both contributed early and continuously as authors and accepted to invest energy and time as Rédacteurs. Michel Émery was a member of the board since volume XXIX.

During all these years, the Séminaire benefited from his demanding quality requirements, be it on mathematics and on style. His meticulous reading of articles was sometimes supplemented by a rewriting suggesting notably elegant phrases instead of basic English.

While preparing this volume, we heard the sad news that Lester Dubins, professor emeritus at Berkeley University, passed away. From the early days, several talentuous mathematicians from various countries have contributed to the Séminaire and Dubins was one of them.

C. Donati-Martin
A. Lejay
A. Rouault

Contents

SPECIALIZED COURSE

Representation Formulae for the Fractional Brownian Motion 3
Jean Picard

OTHER CONTRIBUTIONS

Horizontal Diffusion in C^1 Path Space 73
Marc Arnaudon, Koléhè Abdoulaye Coulibaly,
and Anton Thalmaier

**A Stochastic Calculus Proof of the CLT for the L^2 Modulus
of Continuity of Local Time** 95
Jay Rosen

**On a Zero-One Law for the Norm Process of Transient
Random Walk** 105
Ayako Matsumoto and Kouji Yano

On Standardness and I-cosiness 127
Stéphane Laurent

On Isomorphic Probability Spaces 187
Claude Dellacherie

Cylindrical Wiener Processes 191
Markus Riedle

**A Remark on the $1/H$-Variation of the Fractional Brownian
Motion** 215
Maurizio Pratelli

Simulation of a Local Time Fractional Stable Motion 221
Matthieu Marouby

**Convergence at First and Second Order
of Some Approximations of Stochastic Integrals** 241
Blandine Bérard Bergery and Pierre Vallois

Convergence of Multi-Dimensional Quantized *SDE*'s 269
Gilles Pagès and Afef Sellami

Asymptotic Cramér's Theorem and Analysis on Wiener Space 309
Ciprian A. Tudor

Moments of the Gaussian Chaos 327
Joseph Lehec

The Lent Particle Method for Marked Point Processes 341
Nicolas Bouleau

**Ewens Measures on Compact Groups and Hypergeometric
Kernels** 351
Paul Bourgade, Ashkan Nikeghbali, and Alain Rouault

Discrete Approximation of the Free Fock Space 379
Stéphane Attal and Ion Nechita

**Convergence in the Semimartingale Topology and Constrained
Portfolios** 395
Christoph Czichowsky, Nicholas Westray, and Harry Zheng

**Closedness in the Semimartingale Topology for Spaces
of Stochastic Integrals with Constrained Integrands** 413
Christoph Czichowsky and Martin Schweizer

On Martingales with Given Marginals and the Scaling Property 437
David Baker and Marc Yor

**A Sequence of Albin Type Continuous Martingales
with Brownian Marginals and Scaling** 441
David Baker, Catherine Donati-Martin, and Marc Yor

**Constructing Self-Similar Martingales via Two Skorokhod
Embeddings** 451
Francis Hirsch, Christophe Profeta, Bernard Roynette,
and Marc Yor

Contributors

Marc Arnaudon Laboratoire de Mathématiques et Applications CNRS: UMR 6086, Université de Poitiers, Téléport 2 - BP 30179, 86962 Futuroscope Chasseneuil Cedex, France, marc.arnaudon@math.univ-poitiers.fr

Stéphane Attal Université de Lyon, Université de Lyon 1, Institut Camille Jordan, CNRS UMR 5208, 43, Boulevard du 11 novembre 1918, 69622 Villeurbanne Cedex, France, attal@math.univ-lyon1.fr

David Baker Laboratoire de Probabilités et Modèles Aléatoires, Université Pierre et Marie Curie, Université Paris 06 and CNRS, UMR 7599, 4, Place Jussieu, 75252 Paris Cedex 05, France, david.baker@etu.upmc.fr

Blandine Bérard Bergery Institut de Mathématiques Elie Cartan, Université Henri Poincaré Nancy I, B.P. 239, 54506 Vandoeuvre-lès-Nancy Cedex, France, berardb@iecn.u-nancy.fr

Nicolas Bouleau Ecole des Ponts ParisTech, Paris, France, bouleau@enpc.fr

Paul Bourgade Institut Telecom & Université Paris 6, 46 rue Barrault, 75634 Paris Cedex 13, France, paulbourgade@gmail.com

Koléhè Abdoulaye Coulibaly Laboratoire de Mathématiques et Applications CNRS: UMR 6086, Université de Poitiers, Téléport 2 - BP 30179, 86962 Futuroscope Chasseneuil Cedex, France, abdoulaye.coulibaly@math.univ-poitiers.fr and Unité de Recherche en Mathématiques, FSTC, Université du Luxembourg, 6, rue Richard Coudenhove-Kalergi, 1359 Luxembourg, Grand-Duchy of Luxembourg, abdoulaye.coulibaly@uni.lu

Christoph Czichowsky Department of Mathematics, ETH Zurich, Rämistrasse 101, CH-8092 Zurich, Switzerland, christoph.czichowsky@math.ethz.ch

Claude Dellacherie Laboratoire Raphaël Salem, C.N.R.S. et Université de Rouen Avenue de l'Université, BP.12, 76801 Saint-Étienne-du-Rouvray, France, claude.dellacherie@aliceadsl.fr

Catherine Donati-Martin Laboratoire de Probabilités et Modèles Aléatoires, Université Pierre et Marie Curie Université Paris 06 and CNRS, UMR 7599, 4, Place Jussieu, 75252 Paris Cedex 05, France, catherine.donati@upmc.fr

Francis Hirsch Laboratoire d'Analyse et Probabilités, Université d'Evry - Val d'Essonne, Boulevard F. Mitterrand, 91025 Evry Cedex, France, francis.hirsch@univ-evry.fr

Stéphane Laurent Université catholique de Louvain, Louvain-la-Neuve, Belgium Present Address: Université de Strasbourg, IRMA, UMR 7501, 7 rue René-Descartes, 67084 Strasbourg Cedex, France, laurent@math.u-strasbg.fr

Joseph Lehec CEREMADE (UMR CNRS 7534), Université Paris-Dauphine, Place du maréchal de Lattre de Tassigny, 75016 Paris, France, lehec@ceremade.dauphine.fr

Matthieu Marouby Institut de Mathématiques de Toulouse, Université Paul Sabatier, 31062 Toulouse, France, marouby@math.univ-toulouse.fr

Ayako Matsumoto T & D Financial Life Insurance Company, Japan

Ion Nechita Université de Lyon, Université de Lyon 1, Institut Camille Jordan, CNRS UMR 5208, 43, Boulevard du 11 novembre 1918, 69622 Villeurbanne Cedex, France, nechita@math.univ-lyon1.fr

Ashkan Nikeghbali Institut für Mathematik, Universität Zürich, Winterthurerstrasse 190, CH-8057 Zürich, Switzerland, ashkan.nikeghbali@math.uzh.ch

Gilles Pagès Laboratoire de Probabilités et Modèles Aléatoires, Université de Paris 6, Case 188, 4 pl. Jussieu, 75252 Paris Cedex 5, France, gilles.pages@upmc.fr

Jean Picard Laboratoire de Mathématiques, Clermont Université, Université Blaise Pascal and CNRS UMR 6620, BP 10448, 63000 Clermont-Ferrand, France, Jean.Picard@math.univ-bpclermont.fr

Maurizio Pratelli Dipartimento di Matematica, Largo B. Pontecorvo 5, 56127 Pisa, Italy, pratelli@dm.unipi.it

Christophe Profeta Institut Elie Cartan, Université Henri Poincaré, B.P. 239, 54506 Vandœuvre-lès-Nancy Cedex, France, Christophe.Profeta@iecn.u-nancy.fr

Markus Riedle School of Mathematics, The University of Manchester, Manchester M13 9PL, UK, markus.riedle@manchester.ac.uk

Jay Rosen Department of Mathematics, College of Staten Island, CUNY, Staten Island, NY 10314, USA, jrosen30@optimum.net

Alain Rouault Université Versailles-Saint Quentin, LMV, Bâtiment Fermat, 45 avenue des Etats-Unis, 78035 Versailles Cedex, France, alain.rouault@math.uvsq.fr

Bernard Roynette Institut Elie Cartan, Université Henri Poincaré, B.P. 239, 54506 Vandœuvre-lès-Nancy Cedex, France, bernard.roynette@iecn.u-nancy.fr

Martin Schweizer Department of Mathematics, ETH Zurich, Rämistrasse 101, CH-8092 Zurich, Switzerland, martin.schweizer@math.ethz.ch

Afef Sellami JP Morgan, London & Laboratoire de Probabilités et Modèles Aléatoires, Université de Paris 6, Case 188, 4 pl. Jussieu, 75252 Paris Cedex 5, France, afef.x.sellami@jpmorgan.com

Anton Thalmaier Unité de Recherche en Mathématiques, FSTC, Université du Luxembourg 6, rue Richard Coudenhove-Kalergi, 1359 Luxembourg, Grand-Duchy of Luxembourg, anton.thalmaier@uni.lu

Ciprian A. Tudor Laboratoire Paul Painlevé, Université de Lille 1, 59655 Villeneuve d'Ascq, France, tudor@math.univ-lille1.fr

Pierre Vallois Institut de Mathématiques Elie Cartan, Université Henri Poincaré Nancy I, B.P. 239, 54506 Vandoeuvre-lès-Nancy Cedex, France, Pierre.Vallois@iecn.u-nancy.fr

Nicholas Westray Department of Mathematics, Humboldt Universität Berlin, Unter den Linden 6, 10099 Berlin, Germany, westray@math.hu-berlin.de

Kouji Yano Graduate School of Science, Kobe University, Kobe, Japan, kyano@math.kobe-u.ac.jp

Marc Yor Institut Universitaire de France and Laboratoire de Probabilités et Modèles Aléatoires, Université Pierre et Marie Curie, Université Paris 06 and CNRS, UMR 7599, 4, Place Jussieu, 75252 Paris Cedex 05, France, deaproba@proba.jussieu.fr

Harry Zheng Department of Mathematics, Imperial College, London SW7 2AZ, UK, h.zheng@imperial.ac.uk

Specialized Course

Representation Formulae for the Fractional Brownian Motion

Jean Picard

Abstract We discuss the relationships between some classical representations of the fractional Brownian motion, as a stochastic integral with respect to a standard Brownian motion, or as a series of functions with independent Gaussian coefficients. The basic notions of fractional calculus which are needed for the study are introduced. As an application, we also prove some properties of the Cameron–Martin space of the fractional Brownian motion, and compare its law with the law of some of its variants. Several of the results which are given here are not new; our aim is to provide a unified treatment of some previous literature, and to give alternative proofs and additional results; we also try to be as self-contained as possible.

Keywords Fractional Brownian motion · Cameron-Martin space · Laws of Gaussian processes

1 Introduction

Consider a fractional Brownian motion $(B_t^H; t \in \mathbb{R})$ with Hurst parameter $0 < H < 1$. These processes appeared in 1940 in [24], and they generalise the case $H = 1/2$ which is the standard Brownian motion. A huge literature has been devoted to them since the late 1960s. They are often used to model systems involving Gaussian noise, but which are not correctly explained with a standard Brownian motion. Our aim here is to give a few basic results about them, and in particular to explain how all of them can be deduced from a standard Brownian motion.

The process B^H is a centred Gaussian process which has stationary increments and is H-self-similar; these two conditions can be written as

$$B_{t+t_0}^H - B_{t_0}^H \simeq B_t^H, \qquad B_{\lambda t}^H \simeq \lambda^H B_t^H \qquad (1)$$

J. Picard (✉)

Laboratoire de Mathématiques, Clermont Université, Université Blaise Pascal and CNRS UMR 6620, BP 10448, 63000 Clermont-Ferrand, France

e-mail: Jean.Picard@math.univ-bpclermont.fr

C. Donati-Martin et al. (eds.), *Séminaire de Probabilités XLIII*, Lecture Notes in Mathematics 2006, DOI 10.1007/978-3-642-15217-7_1, © Springer-Verlag Berlin Heidelberg 2011

for $t_0 \in \mathbb{R}$ and $\lambda > 0$, where the notation $Z_t^1 \simeq Z_t^2$ means that the two processes have the same finite dimensional distributions. We can deduce from (1) that B_{-t}^H and B_t^H have the same variance, that this variance is proportional to $|t|^{2H}$, and that the covariance kernel of B^H must be of the form

$$
\begin{aligned}
C(s,t) = \mathbb{E}\big[B_s^H B_t^H\big] &= \frac{1}{2}\mathbb{E}\Big[(B_s^H)^2 + (B_t^H)^2 - (B_t^H - B_s^H)^2\Big] \\
&= \frac{1}{2}\mathbb{E}\Big[(B_s^H)^2 + (B_t^H)^2 - (B_{t-s}^H)^2\Big] \\
&= \frac{\rho}{2}\left(|s|^{2H} + |t|^{2H} - |t-s|^{2H}\right)
\end{aligned}
\tag{2}
$$

for a positive parameter $\rho = \mathbb{E}[(B_1^H)^2]$ (we always assume that $\rho \neq 0$). The process B^H has a continuous modification (we always choose this modification), and its law is characterised by the two parameters ρ and H; however, the important parameter is H, and ρ is easily modified by multiplying B^H by a constant. In this article, it will be convenient to suppose $\rho = \rho(H)$ given in (51); this choice corresponds to the representation of B^H given in (6). We also consider the restriction of B^H to intervals of $\mathbb{R}$ such as $\mathbb{R}_+$, $\mathbb{R}_-$ or $[0,1]$.

Notice that the fractional Brownian motion also exists for $H = 1$ and satisfies $B_t^1 = t\,B_1^1$; this is however a very particular process which is excluded from our study (with our choice of $\rho(H)$ we have $\rho(1) = \infty$).

The standard Brownian motion $W_t = B_t^{1/2}$ is the process corresponding to $H = 1/2$ and $\rho = \rho(1/2) = 1$. It is often useful to represent B^H for $0 < H < 1$ as a linear functional of W; this means that one looks for a kernel $K^H(t,s)$ such that the Wiener-Itô integral

$$
B_t^H = \int K^H(t,s)\,dW_s
\tag{3}
$$

is a H-fractional Brownian motion. More generally, considering the family $(B^H; \ 0 < H < 1)$ defined by (3), we would like to find $K^{J,H}$ so that

$$
B_t^H = \int K^{J,H}(t,s)\,dB_s^J.
\tag{4}
$$

In this case however, we have to give a sense to the integral; the process B^J is a Gaussian process but is not a semimartingale for $J \neq 1/2$, so we cannot consider Itô integration. In order to solve this issue, we approximate B^J with smooth functions for which the Lebesgue–Stieltjes integral can be defined, and then verify that we can pass to the limit in an adequate functional space in which B^J lives almost surely. Alternatively, it is also possible to use integration by parts.

The case where $K^{J,H}$ is a Volterra kernel ($K^{J,H}(t,s) = 0$ if $s > t$) is of particular interest; in this case, the completed filtrations of B^H and of the increments of B^J satisfy $\mathcal{F}_t(B^H) \subset \mathcal{F}_t(dB^J)$, with the notation

$$
\mathcal{F}_t(X) = \sigma\,(X_s; s \leq t), \qquad \mathcal{F}_t(dX) = \sigma\,(X_s - X_u; u \leq s \leq t).
\tag{5}
$$

Notice that when the time interval is $\mathbb{R}_+$, then $\mathcal{F}_t(dB^J) = \mathcal{F}_t(B^J)$ (because $B_0^J = 0$), but this is false for $t < 0$ when the time interval is $\mathbb{R}$ or $\mathbb{R}_-$. When $\mathcal{F}_t(B^H) = \mathcal{F}_t(B^J)$, we say that the representation (4) is canonical; actually, we extend here a terminology, introduced by [25] (see [16]), which classically describes representations with respect to processes with independent increments (so here the representation (3)); such a canonical representation is in some sense unique.

Another purpose of this article is to compare B^H with two other families of processes with similar properties and which are easier to handle in some situations:

- The so-called Riemann–Liouville processes on $\mathbb{R}_+$ (they are also sometimes called type II fractional Brownian motions, see [27]), are deduced from the standard Brownian motion by applying Riemann–Liouville fractional operators, whereas, as we shall recall it, the genuine fractional Brownian motion requires a weighted fractional operator.
- We shall also consider here some processes defined by means of a Fourier-Wiener series on a finite time interval; they are easy to handle in Fourier analysis, whereas the Fourier coefficients of the genuine fractional Brownian motion do not satisfy good independence properties.

We shall prove that the Cameron–Martin spaces of all these processes are equivalent, and we shall compare their laws; more precisely, it is known from [10, 15, 16] that two Gaussian measures are either equivalent, or mutually singular, and we shall decide between these two possibilities.

Let us now describe the contents of this article. Notations and definitions which are used throughout the article are given in Sect. 2; we also give in this section a short review of fractional calculus, in particular Riemann–Liouville operators and some of their modifications which are important for our study; we introduce some functional spaces of Hölder continuous functions; much more results can be found in [35]. In Sect. 3, we give some results concerning the time inversion ($t \mapsto 1/t$) of Gaussian self-similar processes.

We enter the main topic in Sect. 4. Our first aim is to explore the relationship between two classical representations of B^H with respect to W, namely the representation of [26],

$$B_t^H = \frac{1}{\Gamma(H + 1/2)} \int_{\mathbb{R}} \left((t-s)_+^{H-1/2} - (-s)_+^{H-1/2} \right) dW_s \qquad (6)$$

on $\mathbb{R}$ (with the notation $u_+^\lambda = u^\lambda 1_{\{u>0\}}$), and the canonical representation on $\mathbb{R}_+$ obtained in [29,30], see also [8,32] (this is a representation of type (3) for a Volterra kernel K^H, and such that W and B^H generate the same filtration). Let us explain the idea by means of which this relationship can be obtained; in the canonical representation on $\mathbb{R}_+$, we want B_t^H to depend on past values W_s, $s \le t$, or equivalently, we want the infinitesimal increment dB_t^H to depend on past increments dW_s, $s \le t$. In (6), values of B_t^H for $t \ge 0$ involve values of W_s for all $-\infty \le s \le t$, so this is not convenient for a study on $\mathbb{R}_+$. However, we can reverse the time ($t \mapsto -t$) and use the backward representation

$$B_t^H = \frac{1}{\Gamma(H+1/2)} \int_0^{+\infty} \left(s^{H-1/2} - (s-t)_+^{H-1/2} \right) dW_s$$

on $\mathbb{R}_+$. Now the value of B_t^H involves the whole path of W on $\mathbb{R}_+$, but we can notice that the infinitesimal increment dB_t^H only involves future increments dW_s, $s \geq t$. Thus $dB^H(1/t)$ depends on past increments $dW(1/s)$, $s \leq t$. We can then conclude by applying the invariance of fractional Brownian motions by time inversion which has been proved in Sect. 3. This argument is justified in [29] by using the generalised processes dB_t^H/dt, but we shall avoid the explicit use of these processes here. This technique can be used to work out a general relationship of type (4) between B^H and B^J for any $0 < J, H < 1$, see Theorem 11 (such a relation was obtained by [20]).

Application of time inversion techniques also enables us to deduce in Theorem 13 a canonical representation on $\mathbb{R}_-$, and to obtain in Theorem 16 some non canonical representations of B^H with respect to itself, extending the classical case $H = 1/2$; these representations are also considered by [21].

Representations of type (3) or (4) can be applied to descriptions of the Cameron–Martin spaces $\mathcal{H}_H$ of the fractional Brownian motions B^H; these spaces are Hilbert spaces which characterise the laws of centred Gaussian processes (see Appendix C). The space $\mathcal{H}_{1/2}$ is the classical space of absolutely continuous functions h such that $h(0) = 0$ and the derivative $D^1 h$ is square integrable, and (3) implies that $\mathcal{H}_H$ is the space of functions of the form

$$t \mapsto \frac{1}{\Gamma(H+1/2)} \int_\mathbb{R} \left((t-s)_+^{H-1/2} - (-s)_+^{H-1/2} \right) f(s) ds$$

for square integrable functions f.

Sections 5 and 6 are devoted to the comparison of B^H with two processes. One of them is self-similar but has only asymptotically stationary increments in large time, and the other one has stationary increments, but is only asymptotically self-similar in small time.

In Sect. 5, we consider on $\mathbb{R}_+$ the so-called Riemann–Liouville process defined for $H > 0$ by

$$X_t^H = \frac{1}{\Gamma(H+1/2)} \int_0^t (t-s)^{H-1/2} dW_s.$$

This process is H-self-similar but does not have stationary increments; contrary to B^H, the parameter H can be larger than 1. The Cameron–Martin space $\mathcal{H}_H'$ of X^H is the space of functions

$$t \mapsto \frac{1}{\Gamma(H+1/2)} \int_0^t (t-s)^{H-1/2} f(s) ds$$

for square integrable functions f. We explain in Theorem 19 a result of [35], see [8], stating that $\mathcal{H}_H$ and $\mathcal{H}_H'$ are equivalent for $0 < H < 1$ (they are the same set with equivalent norms). We also compare the paths of B^H and X^H, and in particular study the equivalence or mutual singularity of the laws of these processes (Theorem 20); it appears that these two processes can be discriminated by looking at their behaviour in small (or large) time. As an application, we also estimate the mutual information of increments of B^H on disjoint time intervals (more results of this type can be found in [31]).

Another classical representation of the fractional Brownian motion on $\mathbb{R}$ is its spectral representation which can be written in the form

$$B_t^H = \frac{1}{\sqrt{\pi}} \int_0^{+\infty} s^{-1/2-H} \left((\cos(st) - 1) \, dW_s^1 + \sin(st) dW_s^2 \right), \qquad (7)$$

where W_t^1 and W_t^2, $t \geq 0$, are two independent standard Brownian motions; it is indeed not difficult to check that the right-hand side is Gaussian, centred, H-self-similar with stationary increments, and $1/\sqrt{\pi}$ is the constant for which this process has the same variance as (6) (see Appendix B). If now we are interested in B^H on a bounded interval, say $[0, 1]$, we look for its Fourier coefficients. Thus the aim of Sect. 6 is to study the relationship between B^H on $[0, 1]$ and some series of trigonometric functions with independent Gaussian coefficients. More precisely, the standard Brownian motion can be defined on $[0, 1]$ by series such as

$$W_t = \xi_0 t + \sqrt{2} \sum_{n \geq 1} \left(\xi_n \frac{\cos(2n\pi t) - 1}{2n\pi} + \xi_n' \frac{\sin(2n\pi t)}{2n\pi} \right), \qquad (8)$$

$$W_t = \sqrt{2} \sum_{n \geq 0} \left(\xi_n \frac{\cos((2n+1)\pi t) - 1}{(2n+1)\pi} + \xi_n' \frac{\sin((2n+1)\pi t)}{(2n+1)\pi} \right), \qquad (9)$$

or

$$W_t = \sqrt{2} \sum_{n \geq 0} \xi_n \frac{\sin((n+1/2)\pi t)}{(n+1/2)\pi}, \qquad (10)$$

where ξ_n, ξ_n' are independent standard Gaussian variables. The form (10) is the Karhunen-Loève expansion; it provides the orthonormal basis $\sqrt{2} \sin((n+1/2)\pi t)$ of $L^2([0, 1])$, such that the expansion of W_t on this basis consists of independent terms; it is a consequence of (9) which can be written on $[-1/2, 1/2]$, and of the property

$$W_t \simeq \sqrt{2} \, W_{t/2} \simeq W_{t/2} - W_{-t/2}.$$

It is not possible to write on $[0, 1]$ the analogues of these formulae for B^H, $H \neq 1/2$, but it is possible (Theorem 22) to write B^H on $[0, 1]$ as

$$B_t^H = a_0^H \xi_0 t + \sum_{n \geq 1} a_n^H \left((\cos(\pi n t) - 1) \xi_n + \sin(\pi n t) \xi_n' \right) \qquad (11)$$

with $\sum (a_n^H)^2 < \infty$. This result was proved in [18] when $H \leq 1/2$, and the case $H > 1/2$ was studied in [17] with an approximation method. Formula (11) is not completely analogous to (8), (9) or (10); contrary to these expansions of W, the σ-algebra generated by B^H in (11) is strictly smaller than the σ-algebra of the sequence (ξ_n, ξ'_n); in other words, the right hand side of (11) involves an extra information not contained in B^H, and this is a drawback for some questions. This is why we define for $H > 0$ a process

$$\widehat{B}^H_t = \xi_0 t + \sqrt{2} \sum_{n \geq 1} \left(\xi_n \frac{\cos(2\pi nt) - 1}{(2\pi n)^{H+1/2}} + \xi'_n \frac{\sin(2\pi nt)}{(2\pi n)^{H+1/2}} \right)$$

which is a direct generalisation of (8), and a similar process $\overline{B}^H_t$ which generalises (9). It appears that for $0 < H < 1$, these processes have local properties similar to B^H, and we can prove that their Cameron–Martin spaces are equivalent to $\mathcal{H}_H$ (Theorem 25). As an application, we obtain Riesz bases of $\mathcal{H}_H$, and show that functions of $\mathcal{H}_H$ can be characterised on $[0, 1]$ by means of their Fourier coefficients. We then study the equivalence or mutual singularity of the laws of B^H and $\widehat{B}^H$, $\overline{B}^H$ (Theorem 27). We also discuss the extension of (10) which has been proposed in [11]. In Theorem 29, we recover a result of [4, 37] which solves the following question: if we observe a path of a process, can we say whether it is a pure fractional Brownian motion B^J, or whether this process B^J has been corrupted by an independent fractional Brownian motion of different index H?

Technical results which are required in our study are given in the three appendices:

- A lemma about some continuous endomorphisms of the standard Cameron–Martin space (Appendix A).
- The computation of the variance of fractional Brownian motions (Appendix B).
- Results about the equivalence and mutual singularity of laws of Gaussian processes, and about their relative entropies, with in particular a short review of Cameron–Martin spaces (Appendix C).

Notice that many aspects concerning the fractional Brownian motion B^H are not considered in this work. Concerning the representations, it is possible to expand B^H on a wavelet basis; we do not consider this question to which several works have been devoted, see for instance [28]. We also do not study stochastic differential equations driven by B^H (which can be solved by means of the theory of rough paths, see [6]), or the simulation of fractional Brownian paths. On the other hand, fractional Brownian motions have applications in many scientific fields, and we do not describe any of them.

2 Fractional Calculus

Let us first give some notations. All random variables and processes are supposed to be defined on a probability space $(\Omega, \mathcal{F}, \mathbb{P})$ and the expectation is denoted by $\mathbb{E}$; processes are always supposed to be measurable functions $\Xi : (t, \omega) \mapsto \Xi_t(\omega)$, where

t is in a subset of $\mathbb{R}$ endowed with its Borel σ-algebra; the σ-algebra generated by $\varXi$ is denoted by $\sigma(\varXi)$, and for the filtrations we use the notation (5). The derivative of order n of f is denoted by $D^n f$; the function is said to be smooth if it is C^∞. The function f_1 is said to be dominated by f_2 if $|f_1| \le C f_2$. The notation $u_n \asymp v_n$ means that v_n/u_n is between two positive constants. We say that two Hilbert spaces $\mathcal{H}$ and $\mathcal{H}'$ are equivalent (and write $\mathcal{H} \sim \mathcal{H}'$) if they are the same set and

$$C_1 \|h\|_{\mathcal{H}} \le \|h\|_{\mathcal{H}'} \le C_2 \|h\|_{\mathcal{H}} \tag{12}$$

for some positive C_1 and C_2; this means that the two spaces are continuously embedded into each other. We often use the classical function $\varGamma$ defined on $\mathbb{C} \setminus \mathbb{Z}_-$, and in particular the property $\varGamma(z+1) = z\,\varGamma(z)$.

We now describe the functional spaces, fractional integrals and derivatives which are used in this work; see [35] for a much more complete study of the fractional calculus. These functional spaces are weighted Hölder spaces which are convenient for the study of the fractional Brownian motion. The results are certainly not stated in their full generality, but are adapted to our future needs.

2.1 Functional Spaces

The main property which is involved in our study is the Hölder continuity, but functions will often exhibit a different behaviour near time 0 and for large times. More precisely, on the time interval $\mathbb{R}_+^\star$, let $\mathbb{H}^{\beta,\gamma,\delta}$ for $0 < \beta < 1$ and γ, δ real, be the Banach space of real functions f such that

$$\|f\|_{\beta,\gamma,\delta} = \sup_t \frac{|f(t)|}{t^\beta t^{\gamma,\delta}} + \sup_{s<t} \frac{|f(t)-f(s)|}{(t-s)^\beta \sup_{s \le u \le t} u^{\gamma,\delta}} \tag{13}$$

is finite, with the notation

$$t^{\gamma,\delta} = t^\gamma 1_{\{t \le 1\}} + t^\delta 1_{\{t > 1\}}. \tag{14}$$

Thus functions of this space are locally Hölder continuous with index β, and parameters γ and δ make more precise the behaviour at 0 and at infinity. If $\beta + \gamma > 0$, the function f can be extended by continuity at 0 by $f(0) = \lim_0 f = 0$. If $\gamma \ge 0$ and $\delta \ge 0$ and if we consider functions f such that $\lim_0 f = 0$, then the second term of (13) dominates the first one (let s decrease to 0).

Remark 1. Define

$$\|f\|'_{\beta,\gamma,\delta} = \sup \left\{ \frac{|f(t)-f(s)|}{(2^n)^{\gamma,\delta} (t-s)^\beta},\ 2^n \le s \le t \le 2^{n+1},\ n \in \mathbb{Z} \right\}.$$

Then this semi-norm is equivalent to the second term in (13); in particular, if $\gamma \geq 0$ and $\delta \geq 0$, then $\|.\|_{\beta,\gamma,\delta}$ and $\|.\|'_{\beta,\gamma,\delta}$ are equivalent on the space of functions f such that $\lim_0 f = 0$. It is indeed easy to see that $\|.\|'_{\beta,\gamma,\delta}$ is dominated by the second term of (13). For the inverse estimation, notice that upper bounds for $|f(t) - f(s)|$ can be obtained by adding the increments of f on the dyadic intervals $[2^n, 2^{n+1}]$ intersecting $[s, t]$. More precisely, if $2^{k-1} \leq s \leq 2^k \leq 2^n \leq t \leq 2^{n+1}$, then

$$
\begin{aligned}
|f(t) - f(s)| &\leq \|f\|'_{\beta,\gamma,\delta} \sup_{k-1 \leq j \leq n} (2^j)^{\gamma,\delta} \left(\sum_{j=k}^{n-1} 2^{j\beta} + (2^k - s)^\beta + (t - 2^n)^\beta \right) \\
&\leq C \|f\|'_{\beta,\gamma,\delta} \sup_{s \leq u \leq t} u^{\gamma,\delta} \left(2^{n\beta} - 2^{k\beta} + (2^k - s)^\beta + (t - 2^n)^\beta \right) \\
&\leq 3C \|f\|'_{\beta,\gamma,\delta} \sup_{s \leq u \leq t} u^{\gamma,\delta} (t - s)^\beta
\end{aligned}
$$

because $2^{n\beta} - 2^{k\beta} \leq (2^n - 2^k)^\beta \leq (t - s)^\beta$.

In particular, one can deduce from Remark 1 that $\mathbb{H}^{\beta,\gamma,\delta}$ is continuously embedded into $\mathbb{H}^{\beta-\varepsilon,\gamma+\varepsilon,\delta+\varepsilon}$ for $0 < \varepsilon < \beta$.

Theorem 1. *The map* $(f_1, f_2) \mapsto f_1 f_2$ *is continuous from* $\mathbb{H}^{\beta,\gamma_1,\delta_1} \times \mathbb{H}^{\beta,\gamma_2,\delta_2}$ *into* $\mathbb{H}^{\beta,\beta+\gamma_1+\gamma_2,\delta+\delta_1+\delta_2}$.

Proof. This is a bilinear map, so it is sufficient to prove that the image of a bounded subset is bounded. If f_1 and f_2 are bounded in their respective Hölder spaces, it is easy to deduce that $f_1(t) f_2(t)$ is dominated by $t^{2\beta} t^{\gamma_1+\gamma_2,\delta_1+\delta_2}$. On the other hand, following Remark 1, we verify that for $2^n \leq s \leq t \leq 2^{n+1}$,

$$
\begin{aligned}
&\left| f_1(t) f_2(t) - f_1(s) f_2(s) \right| \\
&\quad \leq \left| f_1(s) \right| \left| f_2(t) - f_2(s) \right| + \left| f_2(t) \right| \left| f_1(t) - f_1(s) \right| \\
&\quad \leq C \left(s^\beta s^{\gamma_1,\delta_1} (2^n)^{\gamma_2,\delta_2} (t - s)^\beta + t^\beta t^{\gamma_2,\delta_2} (2^n)^{\gamma_1,\delta_1} (t - s)^\beta \right) \\
&\quad \leq C'(2^n)^\beta (2^n)^{\gamma_1,\delta_1} (2^n)^{\gamma_2,\delta_2} (t - s)^\beta.
\end{aligned}
$$

The theorem is therefore proved.   □

Let us define

$$
\mathbb{H}^{\beta,\gamma} = \mathbb{H}^{\beta,\gamma,0}, \qquad \mathbb{H}^\beta = \mathbb{H}^{\beta,0,0}.
$$

These spaces can be used for functions defined on a finite time interval $[0, T]$, since in this case the parameter δ is unimportant. For functions defined on $\mathbb{R}^\star_-$, we say that f is in $\mathbb{H}^{\beta,\gamma,\delta}$ if $t \mapsto f(-t)$ is in it, and for functions defined on a general interval of $\mathbb{R}$, we assume that the restrictions to $\mathbb{R}^\star_+$ and $\mathbb{R}^\star_-$ are in $\mathbb{H}^{\beta,\gamma,\delta}$. For $\gamma = 0$, the regularity at time 0 is similar to other times, so spaces $\mathbb{H}^{\beta,0,\delta}$ are invariant by the time shifts $f \mapsto f(. + t_0) - f(t_0)$. If we consider a time interval of type $[1, +\infty)$, then the parameter γ can be omitted and we denote the space by $\mathbb{H}^{\beta,.,\delta}$.

We use the notations

$$\mathbb{H}^{\beta-,\gamma,\delta+} = \bigcap_{\varepsilon>0} \mathbb{H}^{\beta-\varepsilon,\gamma,\delta+2\varepsilon}, \quad \mathbb{H}^{\beta-,\gamma} = \bigcap_{\varepsilon>0} \mathbb{H}^{\beta-\varepsilon,\gamma}, \quad \mathbb{H}^{\beta-} = \bigcap_{\varepsilon>0} \mathbb{H}^{\beta-\varepsilon}. \quad (15)$$

They are Fréchet spaces.

Example 1. If B^H is a H-fractional Brownian motion on the time interval $[0, 1]$, the probability of the event $\{B^H \in \mathbb{H}^\beta\}$ is 1 if $\beta < H$ (this follows from the Kolmogorov continuity theorem). In particular, B^H lives almost surely in $\mathbb{H}^{H-}$. We shall see in Remark 7 that this implies that on the time interval $\mathbb{R}_+$, the process B^H lives in $\mathbb{H}^{H-,0,0+}$.

The parameters γ and δ can be modified by means of some multiplication operators. More precisely, on $\mathbb{R}_+^\star$, define

$$\Pi^\alpha f(t) = t^\alpha f(t), \qquad \Pi^{\alpha_1,\alpha_2} f(t) = t^{\alpha_1}(1+t)^{\alpha_2-\alpha_1} f(t). \quad (16)$$

Theorem 2. *The operator Π^{α_1,α_2} maps continuously $\mathbb{H}^{\beta,\gamma,\delta}$ into $\mathbb{H}^{\beta,\gamma+\alpha_1,\delta+\alpha_2}$. In particular, on the time interval $(0, 1]$, the operator Π^α maps continuously $\mathbb{H}^{\beta,\gamma}$ into $\mathbb{H}^{\beta,\gamma+\alpha}$.*

Proof. The quantity $|t^\alpha - s^\alpha|(t-s)^{-\beta} t^{\beta-\alpha}$ is bounded for $2^n \le s \le t \le 2^{n+1}$, and the bound does not depend on n (use the scaling). Thus it follows from Remark 1 that the function $t \mapsto t^\alpha$ is in $\mathbb{H}^{\beta,\alpha-\beta,\alpha-\beta}$. The same property implies that $(1+t)^\alpha - (1+s)^\alpha$ is dominated by $(1+t)^{\alpha-\beta}(t-s)^\beta$ (with the same assumptions on s and t), and we can deduce that $t \mapsto (1+t)^\alpha$ is in $\mathbb{H}^{\beta,-\beta,\alpha-\beta}$ (the coefficient $-\beta$ is due to the fact that the function tends to 1 at 0). We deduce from Theorem 1 that the function $t^{\alpha_1}(1+t)^{\alpha_2-\alpha_1}$ is in $\mathbb{H}^{\beta,\alpha_1-\beta,\alpha_2-\beta}$. The operator Π^{α_1,α_2} is the multiplication by this function, and the result follows by again applying Theorem 1. □

It is then possible to deduce a density result for the spaces of (15) (the result is false with β instead of $\beta-$). Fractional polynomials are linear combinations of monomials t^α, $\alpha \in \mathbb{R}$, and these monomials are in $\mathbb{H}^{\beta,\gamma}$ on $(0, 1]$ if $\alpha \ge \beta + \gamma$.

Theorem 3. *Let $0 < \beta < 1$.*

- *On $(0, 1]$, fractional polynomials (belonging to $\mathbb{H}^{\beta-,\gamma}$) are dense in $\mathbb{H}^{\beta-,\gamma}$.*
- *On $\mathbb{R}_+^\star$, smooth functions with compact support are dense in $\mathbb{H}^{\beta-,\gamma,\delta+}$.*

Proof. Let us consider separately the two statements.

Study on $(0, 1]$. The problem can be reduced to the case $\gamma = 0$ with Theorem 2, and functions f of $\mathbb{H}^{\beta-}$ are continuous on the closed interval $[0, 1]$ with $f(0) = 0$. If f is in $\mathbb{H}^{\beta-\varepsilon}$ (for ε small), it can be approximated by classical polynomials f_n by means of the Stone-Weierstrass theorem; more precisely, if we choose the Bernstein approximations $\mathbb{E}f\left(\frac{1}{n}\sum_{j=1}^n 1_{\{U_j \le x\}}\right)$ for independent uniformly distributed variables U_j in $[0, 1]$, then f_n is bounded in $\mathbb{H}^{\beta-\varepsilon}$ and converges uniformly to f. Thus

$$\left| f_n(t) - f_n(s) - f(t) + f(s) \right|$$

$$\leq C \left(|f_n(t) - f_n(s)|^{(\beta-2\varepsilon)/(\beta-\varepsilon)} + |f(t) - f(s)|^{(\beta-2\varepsilon)/(\beta-\varepsilon)} \right)$$

$$\sup_u |f_n(u) - f(u)|^{\varepsilon/(\beta-\varepsilon)}$$

$$\leq C'(t-s)^{\beta-2\varepsilon} \sup_u |f_n(u) - f(u)|^{\varepsilon/(\beta-\varepsilon)}. \tag{17}$$

These inequalities can also be written for $s = 0$ to estimate $|f_n(t) - f(t)|$, so f_n converges to f in $\mathbb{H}^{\beta-2\varepsilon}$.

Study on $\mathbb{R}_+^\star$. The technique is similar. By means of Π^{α_1,α_2}, we can reduce the study to the case $\gamma = 0$ and $-2\beta < \delta < -\beta$. Let f be in $\mathbb{H}^{\beta-,0,\delta+}$ and let us fix a small $\varepsilon > 0$; then f is in $\mathbb{H}^{\beta-\varepsilon,0,\delta+2\varepsilon}$; in particular, it tends to 0 at 0 and at infinity. A standard procedure enables to approximate it uniformly by smooth functions f_n with compact support, such that f_n is bounded in $\mathbb{H}^{\beta-\varepsilon,0,\delta+2\varepsilon}$; to this end, we first multiply f by the function ϕ_n supported by $[2^{-n-1}, 2^{n+1}]$, taking the value 1 on $[2^{-n}, 2^n]$, and which is affine on $[2^{-n-1}, 2^{-n}]$ and on $[2^n, 2^{n+1}]$; then we take the convolution of $f \phi_n$ with $2^{n+2} \psi(2^{n+2}t)$ for a smooth function ψ supported by $[-1, 1]$ and with integral 1. By proceeding as in (17), we can see that

$$\left| f_n(t) - f_n(s) - f(t) + f(s) \right|$$

$$\leq C(t-s)^{\beta-2\varepsilon} \sup_{s \leq u \leq t} \left(u^{0,\delta+2\varepsilon} \right)^{(\beta-2\varepsilon)/(\beta-\varepsilon)} \sup_u |f_n(u) - f(u)|^{\varepsilon/(\beta-\varepsilon)}$$

so f_n converges to f in $\mathbb{H}^{\beta-2\varepsilon,0,\delta+4\varepsilon}$ because $(\delta + 2\varepsilon)(\beta - 2\varepsilon)/(\beta - \varepsilon) \leq \delta + 4\varepsilon$ for ε small enough. $\qquad\square$

2.2 Riemann–Liouville Operators

An important tool for the stochastic calculus of fractional Brownian motions is the fractional calculus obtained from the study of Riemann–Liouville operators $I_\pm^\alpha$. These operators can be defined for any real index α (and even for complex indices), but we will mainly focus on the case $|\alpha| < 1$.

2.2.1 Operators with Finite Horizon

The fractional integral operators $I_{\tau\pm}^\alpha$ (Riemann–Liouville operators) are defined for $\tau \in \mathbb{R}$ and $\alpha > 0$ by

$$I_{\tau+}^\alpha f(t) = \frac{1}{\Gamma(\alpha)} \int_\tau^t (t-s)^{\alpha-1} f(s) ds, \quad I_{\tau-}^\alpha f(t) = \frac{1}{\Gamma(\alpha)} \int_t^\tau (s-t)^{\alpha-1} f(s) ds, \tag{18}$$

respectively for $t > \tau$ and $t < \tau$. These integrals are well defined for instance if f is locally bounded on $(\tau, +\infty)$ or $(-\infty, \tau)$, and is integrable near τ. If f is integrable, they are defined almost everywhere, and $I^{\alpha}_{\tau\pm}$ is a continuous endomorphism of $L^1([\tau, T])$ or $L^1([T, \tau])$. These operators satisfy the semigroup property

$$I^{\alpha_2}_{\tau\pm} I^{\alpha_1}_{\tau\pm} = I^{\alpha_1+\alpha_2}_{\tau\pm} \tag{19}$$

which can be proved from the relation between Beta and Gamma functions recalled in (95). If α is an integer, we get iterated integrals; in particular, $I^1_{\tau\pm} f$ is $\pm$ the primitive of f taking value 0 at τ. Notice that relations (18) can also be written as

$$I^{\alpha}_{\tau+} f(t) = \frac{1}{\Gamma(\alpha)} \int_{\tau}^{t} (t-s)^{\alpha-1} \left(f(s) - f(t) \right) ds + \frac{(t-\tau)^{\alpha}}{\Gamma(\alpha+1)} f(t),$$

$$I^{\alpha}_{\tau-} f(t) = \frac{1}{\Gamma(\alpha)} \int_{t}^{\tau} (s-t)^{\alpha-1} \left(f(s) - f(t) \right) ds + \frac{(\tau-t)^{\alpha}}{\Gamma(\alpha+1)} f(t). \tag{20}$$

If f is Lipschitz with $f(\tau) = 0$, an integration by parts shows that

$$I^{\alpha}_{\tau+} f(t) = \frac{1}{\Gamma(\alpha+1)} \int_{\tau}^{t} (t-s)^{\alpha} df(s), \quad I^{\alpha}_{\tau-} f(t) = \frac{-1}{\Gamma(\alpha+1)} \int_{t}^{\tau} (s-t)^{\alpha} df(s). \tag{21}$$

For $\alpha = 0$, the operators $I^0_{\tau\pm}$ are by definition the identity (this is coherent with (21)). The study of the operators $I^{\alpha}_{\tau\pm}$ can be reduced to the study of I^{α}_{0+}, since the other cases can be deduced by means of an affine change of time.

Example 2. The value of I^{α}_{0+} on fractional polynomials can be obtained from

$$I^{\alpha}_{0+} \left(\frac{t^{\beta}}{\Gamma(\beta+1)} \right) = \frac{t^{\alpha+\beta}}{\Gamma(\alpha+\beta+1)} \tag{22}$$

which is valid for $\beta > -1$.

Riemann–Liouville operators can also be defined for negative exponents, and are called fractional derivatives. Here we restrict ourselves to $-1 < \alpha < 0$, and in this case the derivative of order $-\alpha$ is defined by

$$I^{\alpha}_{\tau+} f = D^1 I^{1+\alpha}_{\tau+} f, \quad I^{\alpha}_{\tau-} f = -D^1 I^{1+\alpha}_{\tau-} f \tag{23}$$

if $I^{1+\alpha}_{\tau\pm} f$ is absolutely continuous, for the differentiation operator D^1. The relation (22) is easily extended to negative α (with result 0 if $\alpha + \beta + 1 = 0$). Fractional derivatives operate on smooth functions, and we have the following result.

Theorem 4. *Suppose that f is smooth and integrable on $(0, 1]$. Then, for any $\alpha > -1$, $I^{\alpha}_{0+} f$ is well defined, is smooth on $(0, 1]$, and*

$$\left| D^1 I_{0+}^\alpha f(t) \right| \le C_\alpha \left(t^{\alpha-2} \int_0^{t/2} |f(s)| ds + t^{\alpha-1} \sup_{[t/2,t]} |f| \right.$$

$$\left. + t^\alpha \sup_{[t/2,t]} |D^1 f| + t^{\alpha+1} \sup_{[t/2,t]} |D^2 f| \right). \qquad (24)$$

If $D^1 f$ is integrable and $\lim_0 f = 0$, then $D^1 I_{0+}^\alpha f = I_{0+}^\alpha D^1 f$.

Proof. First suppose $\alpha > 0$. Then, for $t > u > 0$, we can write (18) in the form

$$I_{0+}^\alpha f(t) = \Gamma(\alpha)^{-1} \left(\int_0^u (t-s)^{\alpha-1} f(s) ds + \int_0^{t-u} s^{\alpha-1} f(t-s) ds \right). \qquad (25)$$

This expression is smooth, and

$$D^1 I_{0+}^\alpha f(t) = \Gamma(\alpha)^{-1} \left((\alpha-1) \int_0^u (t-s)^{\alpha-2} f(s) ds \right.$$

$$\left. + \int_0^{t-u} s^{\alpha-1} D^1 f(t-s) ds + (t-u)^{\alpha-1} f(u) \right). \qquad (26)$$

In particular, by letting $u = t/2$, we obtain (24) without the $D^2 f$ term. Moreover, if $D^1 f$ is integrable and $\lim_0 f = 0$, we see by writing

$$(t-u)^{\alpha-1} f(u) = -(\alpha-1) \int_0^u (t-s)^{\alpha-2} f(s) ds + \int_0^u (t-s)^{\alpha-1} D^1 f(s) ds$$

that

$$D^1 I_{0+}^\alpha f(t) = \Gamma(\alpha)^{-1} \left(\int_0^{t-u} s^{\alpha-1} D^1 f(t-s) ds + \int_0^u (t-s)^{\alpha-1} D^1 f(s) ds \right)$$

$$= I_{0+}^\alpha D^1 f(t)$$

(apply (25) with f replaced by $D^1 f$). Let us now consider the case $-1 < \alpha < 0$; we use the definition (23) of the fractional derivative, and in particular deduce that $I_{0+}^\alpha f$ is again smooth. Moreover, from (26),

$$D^1 I_{0+}^\alpha f(t) = D^2 I_{0+}^{\alpha+1} f(t)$$

$$= \Gamma(\alpha+1)^{-1} \left(\alpha(\alpha-1) \int_0^u (t-s)^{\alpha-2} f(s) ds + (t-u)^\alpha D^1 f(u) \right.$$

$$\left. + \int_0^{t-u} s^\alpha D^2 f(t-s) ds + \alpha(t-u)^{\alpha-1} f(u) \right).$$

We deduce (24) by letting again $u = t/2$. If $\lim_0 f = 0$ and $D^1 f$ is integrable, then

$$D^1 I_{0+}^{\alpha} f = D^2 I_{0+}^{\alpha+1} f = D^1 I_{0+}^{\alpha+1} D^1 f = I_{0+}^{\alpha} D^1 f$$

from the definition (23) and the property for $\alpha + 1$ which has already been proved.

$\square$

For $-1 < \alpha < 0$, a study of (20) shows that $I_{\tau\pm}^{\alpha} f$ is defined as soon as f is Hölder continuous with index greater than $-\alpha$, and that (20) again holds true. If f is Lipschitz and $f(\tau) = 0$, then we can write

$$I_{\tau\pm}^{\alpha} f = \pm D^1 I_{\tau\pm}^{1+\alpha} f = D^1 I_{\tau\pm}^{1+\alpha} I_{\tau\pm}^1 D^1 f = D^1 I_{\tau\pm}^1 I_{\tau\pm}^{1+\alpha} D^1 f = \pm I_{\tau\pm}^{1+\alpha} D^1 f$$

where we have used (19) in the third equality, so (21) again holds true. Thus relations (20) and (21) can be used for any $\alpha > -1$ ($\alpha \neq 0$ for (20)). By using the multiplication operators Π^{α} defined in (16), we can deduce from (20) a formula for weighted fractional operators; if f is smooth with compact support in $\mathbb{R}_+^*$, then

$$\Pi^{-\gamma} I_{0+}^{\alpha} \Pi^{\gamma} f(t) = I_{0+}^{\alpha} f(t) + \frac{1}{\Gamma(\alpha)} \int_0^t (t-s)^{\alpha-1} \left(\left(\frac{s}{t} \right)^{\gamma} - 1 \right) f(s) ds \quad (27)$$

for $\alpha > -1, \alpha \neq 0$.

Here are some results about I_{0+}^{α} related to the functional spaces of Sect. 2.1. They can easily be translated into properties of $I_{\tau\pm}^{\alpha}$, see also [32, 35].

Theorem 5. *Consider the time interval $(0, 1]$ and let $\gamma > -1$.*

- *If β and $\beta + \alpha$ are in $(0, 1)$, then the operator I_{0+}^{α} maps continuously $\mathbb{H}^{\beta,\gamma}$ into $\mathbb{H}^{\beta+\alpha,\gamma}$.*
- *The composition rule $I_{0+}^{\alpha_2} I_{0+}^{\alpha_1} = I_{0+}^{\alpha_1+\alpha_2}$ holds on $\mathbb{H}^{\beta,\gamma}$ provided β, $\beta + \alpha_1$ and $\beta + \alpha_1 + \alpha_2$ are in $(0, 1)$.*

Proof. Let us prove the first statement. Let f be in $\mathbb{H}^{\beta,\gamma}$. The property $I_{0+}^{\alpha} f(t) = O(t^{\alpha+\beta+\gamma})$ can be deduced from (20) and (22). By applying Remark 1, it is then sufficient to compare $I_{0+}^{\alpha} f$ at times s and t for $2^n \leq s \leq t \leq 2^{n+1}, n < 0$. Consider the time $v = (3s - t)/2$, so that $2^{n-1} \leq s/2 \leq v \leq s \leq 2^{n+1}$. By again applying (20), we have

$$I_{0+}^{\alpha} f(t) - I_{0+}^{\alpha} f(s)$$

$$= \frac{t^{\alpha} f(t) - s^{\alpha} f(s)}{\Gamma(\alpha+1)} + \frac{A_{v,t} - A_{v,s}}{\Gamma(\alpha)} + \frac{f(s) - f(t)}{\Gamma(\alpha)} \int_0^v (t-u)^{\alpha-1} du$$

$$+ \frac{1}{\Gamma(\alpha)} \int_0^v \left((t-u)^{\alpha-1} - (s-u)^{\alpha-1} \right) (f(u) - f(s)) du$$

with

$$A_{v,w} = \int_v^w (w - u)^{\alpha-1} \left(f(u) - f(w) \right) du = O\left((v^\gamma + w^\gamma)(w - v)^{\alpha+\beta} \right).$$

We deduce that

$$I_{0+}^\alpha f(t) - I_{0+}^\alpha f(s) = \frac{(t^\alpha - s^\alpha) f(s)}{\Gamma(\alpha + 1)} + \frac{A_{v,t} - A_{v,s}}{\Gamma(\alpha)} - \frac{f(s) - f(t)}{\Gamma(\alpha + 1)}(t - v)^\alpha$$
$$+ \frac{1}{\Gamma(\alpha)} \int_0^v \left((t - u)^{\alpha-1} - (s - u)^{\alpha-1} \right) (f(u) - f(s)) \, du.$$

$$(28)$$

The second and third terms are easily shown to be dominated by $2^{n\gamma}(t - s)^{\alpha+\beta}$. The first term is dominated by

$$\sup_{s \le u \le t} u^{\alpha-1}(t - s)s^{\beta+\gamma} \le C \, 2^{n\gamma}(t - s)^{\alpha+\beta}.$$

The last term is dominated by

$$\int_0^v \left((s - u)^{\alpha-1} - (t - u)^{\alpha-1} \right) (s - u)^\beta \, (u^\gamma + s^\gamma) \, du$$
$$\le (1 - \alpha)(t - s) \int_0^v (s - u)^{\alpha+\beta-2} \, (u^\gamma + s^\gamma) \, du$$
$$\le C(t - s) \left(2^{n\gamma}(s - v)^{\alpha+\beta-1} + \int_0^{s/2} (s - u)^{\alpha+\beta-2} \, (u^\gamma + s^\gamma) \, du \right)$$
$$\le C' 2^{n\gamma}(t - s)^{\alpha+\beta}$$

because $s - v = (t - s)/2$ and the integral on $[0, s/2]$ is proportional to $s^{\alpha+\beta+\gamma-1} \le c \, 2^{n(\alpha+\beta+\gamma-1)} \le c \, 2^{n\gamma}(t - s)^{\alpha+\beta-1}$. Thus the continuity of I_{0+}^α is proved. For the composition rule, it is easily verified for monomials $f(t) = t^\beta$ (apply (22)), and is then extended by density to the space $\mathbb{H}^{\beta-,\gamma}$ from Theorem 3. By applying this property to a slightly larger value of β, it appears that the composition rule actually holds on $\mathbb{H}^{\beta,\gamma}$.   □

Notice that fractional monomials t^κ are eigenfunctions of $\Pi^{-\alpha} I_{0+}^\alpha$ and $I_{0+}^\alpha \Pi^{-\alpha}$ when they are in the domains of definitions of these operators, so when κ is large enough. This implies that these operators commute on fractional polynomials. This property is then extended to other functions by density. In particular,

$$I_{0+}^{\alpha_2} \Pi^{-\alpha_1-\alpha_2} I_{0+}^{\alpha_1} = \Pi^{-\alpha_1} I_{0+}^{\alpha_1+\alpha_2} \Pi^{-\alpha_2}, \qquad (29)$$

see (10.6) in [35].

2.2.2 Operators with Infinite Horizon

The operators $I_{\pm}^{\alpha}$ are defined by letting $\tau \to \mp\infty$ in $I_{\tau\pm}^{\alpha}$. However, we will be more interested in the modified operators

$$\widetilde{I}_{\pm}^{\alpha} f(t) = I_{\pm}^{\alpha} f(t) - I_{\pm}^{\alpha} f(0) = \lim_{\tau \to \mp\infty} \left(I_{\tau\pm}^{\alpha} f(t) - I_{\tau\pm}^{\alpha} f(0) \right)$$

when the limit exists. For $\alpha > 0$, we can write

$$\widetilde{I}_{+}^{\alpha} f(t) = \frac{1}{\Gamma(\alpha)} \int \left((t-s)_{+}^{\alpha-1} - (-s)_{+}^{\alpha-1} \right) f(s) ds,$$

$$\widetilde{I}_{-}^{\alpha} f(t) = \frac{1}{\Gamma(\alpha)} \int \left((s-t)_{+}^{\alpha-1} - s_{+}^{\alpha-1} \right) f(s) ds \qquad (30)$$

where we use the notation $u_{+}^{\lambda} = u^{\lambda} 1_{\{u>0\}}$. These integrals are well defined if $f(t)$ is dominated by $(1 + |t|)^{\delta}$ for $\delta < 1 - \alpha$ (there are also cases where the integrals are only semi-convergent). In particular, the fractional integrals are generally not defined for large values of α, as it was the case for I_{0+}^{α}. We are going to study $\widetilde{I}_{\pm}^{\alpha}$ on the functional spaces $\mathbb{H}^{\beta,0,\delta}$.

Remark 2. The operator $\widetilde{I}_{\pm}^{\alpha}$ is a normalisation of $I_{\pm}^{\alpha}$ in the sense that it can be defined in more cases than $I_{\pm}^{\alpha} f$. For instance, for $\alpha > 0$, if we compare $I_{-}^{\alpha} f$ and $\widetilde{I}_{-}^{\alpha} f$ on $\mathbb{R}_{+}^{*}$ for $f(s) = s^{\delta}$, we see that the former one is defined for $\delta < -\alpha$, whereas the latter one is defined for $\delta < 1 - \alpha$.

Let us now consider the case $-1 < \alpha < 0$; we can let τ tend to infinity in (20) and obtain

$$\widetilde{I}_{+}^{\alpha} f(t) = \frac{1}{\Gamma(\alpha)} \int \left((t-s)_{+}^{\alpha-1} (f(s) - f(t)) - (-s)_{+}^{\alpha-1} (f(s) - f(0)) \right) ds,$$

$$\widetilde{I}_{-}^{\alpha} f(t) = \frac{1}{\Gamma(\alpha)} \int \left((s-t)_{+}^{\alpha-1} (f(s) - f(t)) - s_{+}^{\alpha-1} (f(s) - f(0)) \right) ds. \quad (31)$$

This expression is defined on $\mathbb{H}^{\beta,0,\delta}$ provided $\beta + \alpha > 0$ and $\beta + \alpha + \delta < 1$.

Let $\alpha > -1$. Suppose that f is Lipschitz and has compact support, so that f is 0 on $(-\infty, \tau]$, respectively $[\tau, +\infty)$. Then $I_{\pm}^{\alpha} f = I_{\tau\pm}^{\alpha} f$ on $[\tau, +\infty)$, respectively $(-\infty, \tau]$, so $\widetilde{I}_{\pm}^{\alpha} f(t)$ is equal to $I_{\tau\pm}^{\alpha} f(t) - I_{\tau\pm}^{\alpha} f(0)$, which can be expressed by means of (21). Thus

$$\widetilde{I}_{+}^{\alpha} f(t) = \frac{1}{\Gamma(\alpha+1)} \int \left((t-s)_{+}^{\alpha} - (-s)_{+}^{\alpha} \right) df(s),$$

$$\widetilde{I}_{-}^{\alpha} f(t) = \frac{1}{\Gamma(\alpha+1)} \int \left(s_{+}^{\alpha} - (s-t)_{+}^{\alpha} \right) df(s). \qquad (32)$$

By applying Theorem 4, we see that if f is smooth with compact support, then $\widetilde{I}^\alpha_\pm f$ is smooth and

$$D^1 \widetilde{I}^\alpha_\pm f = D^1 I^\alpha_\pm f = I^\alpha_\pm D^1 f. \tag{33}$$

Remark 3. If $f = 0$ on $\mathbb{R}_+$ and if we look for $\widetilde{I}^\alpha_+ f$ on $\mathbb{R}^*_+$, we see when $\alpha < 0$ that $f(0)$ and $f(t)$ disappear in (31), so (30) can be used on $\mathbb{R}^*_+$ for both positive and negative α, and $\widetilde{I}^\alpha_+ f$ is C^∞ on $\mathbb{R}^*_+$.

Theorem 6. *Consider the operators $\widetilde{I}^\alpha_+$ and $\widetilde{I}^\alpha_-$ on the respective time intervals $(-\infty, T]$ for $T \geq 0$, and $[T, +\infty)$ for $T \leq 0$. Let $\delta > 0$.*

- *The operator $\widetilde{I}^\alpha_\pm$ maps continuously $\mathbb{H}^{\beta,0,\delta}$ into $\mathbb{H}^{\beta+\alpha,0,\delta}$ provided β, $\beta + \alpha$ and $\beta + \alpha + \delta$ are in $(0, 1)$.*
- *The composition rule $\widetilde{I}^{\alpha_2}_\pm \widetilde{I}^{\alpha_1}_\pm = \widetilde{I}^{\alpha_1+\alpha_2}_\pm$ holds on $\mathbb{H}^{\beta,0,\delta}$ provided β, $\beta + \alpha_1$, $\beta + \alpha_1 + \alpha_2$, $\beta + \alpha_1 + \delta$ and $\beta + \alpha_1 + \alpha_2 + \delta$ are in $(0, 1)$.*

Proof. It is of course sufficient to study $\widetilde{I}^\alpha_+$. We prove separately the two statements.

Continuity of $\widetilde{I}^\alpha_+$. We want to study the continuity on the time interval $(-\infty, T]$; by means of a time shift, let us consider the time interval $(-\infty, -1]$, and let us prove that if f is in $\mathbb{H}^{\beta,\cdot,\delta}$, then the function $\lim_{\tau \to -\infty}(I^\alpha_{\tau+} f(t) - I^\alpha_{\tau+} f(-1))$ is in $\mathbb{H}^{\beta+\alpha,\cdot,\delta}$. From Remark 1, it is sufficient to estimate the increments of this function on intervals $[s, t] \subset [-2^{n+1}, -2^n]$ for $n \geq 0$. Consider the proof of Theorem 5 where I^α_{0+} is replaced by $I^\alpha_{\tau+}$, and let us estimate $I^\alpha_{\tau+} f(t) - I^\alpha_{\tau+} f(s)$ for $\tau \to -\infty$. We can write a formula similar to (28). The first term involves $(t - \tau)^\alpha - (s - \tau)^\alpha$ which tends to 0 as $\tau \to -\infty$, so this first term vanishes. The second and third terms are dealt with similarly to Theorem 5; the only difference is that the weight $2^{n\gamma}$ now becomes $2^{n\delta}$. The last term is an integral on $(-\infty, v)$ and is dominated by

$$(t - s) \int_{-\infty}^v (s - u)^{\alpha+\beta-2} |u|^\delta \, du = (t - s) \int_{(t-s)/2}^{+\infty} u^{\alpha+\beta-2} (u - s)^\delta \, du$$

$$\leq (t - s) \int_{(t-s)/2}^{+\infty} \left(u^{\alpha+\beta+\delta-2} + u^{\alpha+\beta-2} |s|^\delta \right) du$$

$$\leq C(t - s) \left((t - s)^{\alpha+\beta+\delta-1} + (t-s)^{\alpha+\beta-1} |s|^\delta \right)$$

$$\leq 2C(t - s)^{\alpha+\beta} |s|^\delta.$$

Composition rule. If f is 0 before some time τ_0, then $\widetilde{I}^{\alpha_1}_+ f(t) = I^{\alpha_1}_{\tau+} f(t) - I^{\alpha_1}_{\tau+} f(0)$ for $\tau \leq \tau_0 \wedge t$. Thus

$$\widetilde{I}^{\alpha_2}_+ \widetilde{I}^{\alpha_1}_+ f(t) = \lim_{\tau \to -\infty} \left(I^{\alpha_2}_{\tau+} \widetilde{I}^{\alpha_1}_+ f(t) - I^{\alpha_2}_{\tau+} \widetilde{I}^{\alpha_1}_+ f(0) \right)$$

with

$$I^{\alpha_2}_{\tau+}\widetilde{I}^{\alpha_1}_{+}f(t) = I^{\alpha_2}_{\tau+}I^{\alpha_1}_{\tau+}f(t) - \frac{(t-\tau)^{\alpha_2}}{\Gamma(\alpha_2+1)}I^{\alpha_1}_{\tau+}f(0) = I^{\alpha_1+\alpha_2}_{\tau+}f(t) - \frac{(t-\tau)^{\alpha_2}}{\Gamma(\alpha_2+1)}I^{\alpha_1}_{+}f(0)$$

from Theorem 5. Thus

$$\widetilde{I}^{\alpha_2}_{+}\widetilde{I}^{\alpha_1}_{+}f(t) = \widetilde{I}^{\alpha_1+\alpha_2}_{+}f(t) - \lim_{\tau\to-\infty}\frac{(t-\tau)^{\alpha_2}-(-\tau)^{\alpha_2}}{\Gamma(\alpha_2+1)}I^{\alpha_1}_{+}f(0) = \widetilde{I}^{\alpha_1+\alpha_2}_{+}f(t).$$

The case of general functions is then deduced from the density of functions with compact support in $\mathbb{H}^{\beta-,0,\delta+}$ (Theorem 3); the proof on $\mathbb{H}^{\beta,0,\delta}$ is obtained as in Theorem 5 by increasing β and decreasing δ slightly.   □

In particular, we deduce from Theorem 6 that $\widetilde{I}^{\alpha}_{\pm}$ is a homeomorphism from $\mathbb{H}^{\beta-,0,0+}$ onto $\mathbb{H}^{(\alpha+\beta)-,0,0+}$ if β and $\alpha+\beta$ are in $(0,1)$, and $\widetilde{I}^{-\alpha}_{\pm}$ is its inverse map.

2.2.3 Operators for Periodic Functions

Consider a bounded 1-periodic function f. Let $|\alpha| < 1$; if $\alpha < 0$, suppose moreover that f is in $\mathbb{H}^{\beta}$ for some $\beta > -\alpha$. Then $\widetilde{I}^{\alpha}_{+}f$ is well defined and is given by (30) or (31); moreover, this function is also 1-periodic, and is 0 at time 0; this follows from

$$I^{\alpha}_{\tau+}f(t+1) = I^{\alpha}_{(\tau-1)+}f(t)$$

so that

$$I^{\alpha}_{\tau+}f(t+1) - I^{\alpha}_{\tau+}f(0) = \left(I^{\alpha}_{(\tau-1)+}f(t) - I^{\alpha}_{(\tau-1)+}f(0)\right)$$
$$+ \left(I^{\alpha}_{(\tau-1)+}f(0) - I^{\alpha}_{\tau+}f(0)\right).$$

By letting $\tau \to -\infty$, one easily checks that the second part tends to 0, so $\widetilde{I}^{\alpha}_{+}f(t+1) = \widetilde{I}^{\alpha}_{+}f(t)$.

The following example explains the action of $\widetilde{I}^{\alpha}_{+}$ on trigonometric functions.

Example 3. Let us compute $\widetilde{I}^{\alpha}_{+}$ on the family of complex functions $\phi_r(t) = e^{irt} - 1$ for $r > 0$. Suppose $0 < \alpha < 1$. The formula

$$\Gamma(\alpha) = \int_0^\infty s^{\alpha-1}e^{-s}ds = u^\alpha \int_0^\infty s^{\alpha-1}e^{-us}ds$$

is valid for $u > 0$ and can be extended to complex numbers with positive real part. One can also write it for $u = \mp ir, r > 0$, and we obtain

$$\int_0^\infty s^{\alpha-1}e^{\pm irs}ds = e^{\pm i\alpha\pi/2}r^{-\alpha}\Gamma(\alpha) \qquad (34)$$

where the integral is only semi-convergent. Thus we obtain the classical formula (see Sect. 7 of [35])

$$
\begin{aligned}
I_+^\alpha e^{irt} &= \frac{1}{\Gamma(\alpha)} \int_{-\infty}^{t} (t-s)^{\alpha-1} e^{irs} ds \\
&= \frac{e^{irt}}{\Gamma(\alpha)} \int_{0}^{\infty} s^{\alpha-1} e^{-irs} ds = r^{-\alpha} e^{-i\alpha\pi/2} e^{irt}.
\end{aligned}
$$

We deduce that $\widetilde{I}_+^\alpha \phi_r = r^{-\alpha} e^{-i\alpha\pi/2} \phi_r$, and this relation is extended to negative α since the operators of exponents α and $-\alpha$ are the inverse of each other (Theorem 6). In particular,

$$
\begin{aligned}
\widetilde{I}_+^\alpha (1 - \cos(rt)) &= r^{-\alpha} (\cos(\alpha\pi/2) - \cos(rt - \alpha\pi/2)) \\
\widetilde{I}_+^\alpha \sin(rt) &= r^{-\alpha} (\sin(\alpha\pi/2) + \sin(rt - \alpha\pi/2)).
\end{aligned}
\tag{35}
$$

Remark 4. We can similarly study $\widetilde{I}_-^\alpha$ which multiplies ϕ_r by $r^{-\alpha} e^{i\alpha\pi/2}$; consequently, the two-sided operator $(\widetilde{I}_+^\alpha + \widetilde{I}_-^\alpha)/(2\cos(\alpha\pi/2))$ multiplies ϕ_r by $r^{-\alpha}$.

Let us now define two modifications $\widehat{I}_+^\alpha$ and $\overline{I}_+^\alpha$ of $\widetilde{I}_+^\alpha$ which will be useful for the study of the fractional Brownian motion on $[0, 1]$. Consider a bounded function f defined on the time interval $[0, 1]$ and such that $f(0) = 0$. If $\alpha < 0$, suppose again that f is in $\mathbb{H}^\beta$ for some $\beta > -\alpha$. Let $g(t)$ be the 1-periodic function coinciding on $[0, 1]$ with $f(t) - t f(1)$. We now define on $[0, 1]$

$$
\widehat{I}_+^\alpha f(t) = t f(1) + \widetilde{I}_+^\alpha g(t).
\tag{36}
$$

Thus $\widehat{I}_+^\alpha$ satisfies the formulae (35) for $r = 2n\pi$, and we decide arbitrarily that $\widehat{I}_+^\alpha t = t$. On the other hand, let h be the function with 1-antiperiodic increments, so that

$$
h(1+t) - h(1+s) = -h(t) + h(s),
$$

and coinciding with f on $[0, 1]$. We define

$$
\overline{I}_+^\alpha f(t) = \widetilde{I}_+^\alpha h(t).
\tag{37}
$$

Then $\overline{I}_+^\alpha$ satisfies (35) for $r = (2n + 1)\pi$.

It is clear that $\widehat{I}_+^{\alpha_2} \widehat{I}_+^{\alpha_1} = \widehat{I}_+^{\alpha_1+\alpha_2}$ is satisfied on $\mathbb{H}^\beta$ as soon as β, $\beta + \alpha_1$ and $\beta + \alpha_1 + \alpha_2$ are in $(0, 1)$, and the same property is valid for $\overline{I}_+^\alpha$ (apply Theorem 6). Actually, these composition rules can be used to extend the two operators to arbitrarily large values of α. Moreover, $\widehat{I}_+^\alpha$ and $\overline{I}_+^\alpha$ are homeomorphisms from $\mathbb{H}^\beta$ onto $\mathbb{H}^{\beta+\alpha}$ if β and $\beta + \alpha$ are in $(0, 1)$, and their inverse maps are $\widehat{I}_+^{-\alpha}$ and $\overline{I}_+^{-\alpha}$.

2.3 Some Other Operators

Let us describe the other operators which are used in this work. The multiplication operator Π^α, $\alpha \in \mathbb{R}$, has already been defined in (16) on $\mathbb{R}_+^*$, and let us complement it with

$$\widetilde{\Pi}^\alpha f(t) = I_{0+}^1 \Pi^\alpha D^1 f(t) = \int_0^t s^\alpha \, df(s) = t^\alpha f(t) - \alpha \int_0^t s^{\alpha-1} f(s) \, ds \quad (38)$$

for f smooth with compact support. In the last form, we see that $\widetilde{\Pi}^\alpha f$ can be defined as soon as $t^{\alpha-1} f(t)$ is integrable on any $[0, T]$, so on $\mathbb{H}^{\beta,\gamma,\delta}$ if $\alpha + \beta + \gamma > 0$.

On the other hand, let us define for $\alpha \in \mathbb{R}$ the time inversion operators T_α and T_α' on $\mathbb{R}_+^*$ by

$$T_\alpha f(t) = t^{2\alpha} f(1/t) \quad (39)$$

and

$$T_\alpha' f(t) = -I_{0+}^1 T_{\alpha-1} D^1 f(t) = -\int_0^t s^{2\alpha-2} D^1 f(1/s) \, ds = -\int_{1/t}^\infty s^{-2\alpha} \, df(s)$$

$$= t^{2\alpha} f(1/t) - 2\alpha \int_{1/t}^\infty s^{-2\alpha-1} f(s) \, ds \quad (40)$$

and the last form can be used if $t^{-2\alpha-1} f(t)$ is integrable on any $[T, \infty)$, so in particular on $\mathbb{H}^{\beta,\gamma,\delta}$ if $2\alpha > \beta + \delta$. Actually, the form of T_α and a comparison of (40) and (38) show that

$$T_\alpha = \Pi^{2\alpha} T_0 = T_0 \Pi^{-2\alpha}, \qquad T_\alpha' = \widetilde{\Pi}^{2\alpha} T_0. \quad (41)$$

Notice that T_α and T_α' are involutions, so that

$$T_\alpha T_\alpha' f(t) = f(t) - 2\alpha \, t^{2\alpha} \int_t^\infty s^{-2\alpha-1} f(s) \, ds \quad (42)$$

and

$$T_\alpha' T_\alpha f(t) = \widetilde{\Pi}^{2\alpha} \Pi^{-2\alpha} f(t) = f(t) - 2\alpha \int_0^t f(s) \frac{ds}{s} \quad (43)$$

are the inverse transformation of each other.

Theorem 7. *Let $0 < \beta < 1$ and consider the time interval $\mathbb{R}_+^*$.*

- *The operator $\widetilde{\Pi}^\alpha$ maps continuously $\mathbb{H}^{\beta,\gamma,\delta}$ into $\mathbb{H}^{\beta,\gamma+\alpha,\delta+\alpha}$ if $\beta + \gamma + \alpha > 0$ and $\beta + \delta + \alpha > 0$. It satisfies the composition rule $\widetilde{\Pi}^{\alpha_2} \widetilde{\Pi}^{\alpha_1} = \widetilde{\Pi}^{\alpha_1+\alpha_2}$ on $\mathbb{H}^{\beta,\gamma,\delta}$ if $\beta + \gamma + \alpha_1 > 0$ and $\beta + \gamma + \alpha_1 + \alpha_2 > 0$.*

- *The operator T_α maps continuously $\mathbb{H}^{\beta,\gamma,\delta}$ into $\mathbb{H}^{\beta,-\delta+2(\alpha-\beta),-\gamma+2(\alpha-\beta)}$. If moreover $2\alpha > \beta + \delta$ and $2\alpha > \beta + \gamma$, the operator T'_α satisfies the same property.*

Proof. We prove separately the two parts.

Study of $\widetilde{\Pi}^\alpha$. The continuity on $\mathbb{H}^{\beta,\gamma,\delta}$ is proved by noticing

$$\left|\widetilde{\Pi}^\alpha f(t)\right| \le \left|\Pi^\alpha f(t)\right| + C \int_0^t s^{\alpha-1+\beta} s^{\gamma,\delta} ds \le C' t^{\alpha+\beta} t^{\gamma,\delta},$$

$$\left|\widetilde{\Pi}^\alpha f(t) - \widetilde{\Pi}^\alpha f(s)\right| \le \left|\Pi^\alpha f(t) - \Pi^\alpha f(s)\right| + C \int_s^t u^{\alpha+\beta-1} u^{\gamma,\delta} du$$

$$\le \left|\Pi^\alpha f(t) - \Pi^\alpha f(s)\right| + C'(t-s)^\beta \sup_{s \le u \le t} u^\alpha u^{\gamma,\delta},$$

and by applying Theorem 2. The composition rule is evident for smooth functions (use the first equality of (38)), and can be extended by density (the parameter δ is unimportant since we only need the functions on bounded time intervals).

Study of T_α and T'_α. If f is in $\mathbb{H}^{\beta,\gamma,\delta}$, then $f(1/t)$ is dominated by $t^{-\beta} t^{-\delta,-\gamma}$, and if $2^n \le s \le t \le 2^{n+1}$,

$$\left|f(1/t) - f(1/s)\right| \le C \sup_{s \le u \le t} u^{-\delta,-\gamma} (1/s - 1/t)^\beta \le C'(2^n)^{-\delta,-\gamma} s^{-\beta} t^{-\beta} (t-s)^\beta$$

$$\le C''(2^n)^{-\delta-2\beta,-\gamma-2\beta} (t-s)^\beta,$$

so $T_0 f : t \mapsto f(1/t)$ is in $\mathbb{H}^{\beta,-2\beta-\delta,-2\beta-\gamma}$. The continuity of T_α and T'_α is then a consequence of (41) and of the continuity of $\Pi^{2\alpha}$ and $\widetilde{\Pi}^{2\alpha}$. □

Remark 5. We deduce in particular from Theorem 7 that T_α and T'_α are homeomorphisms from $\mathbb{H}^{\alpha-,0,0+}$ into itself for $0 < \alpha < 1$. We also deduce that $T_\alpha T'_\alpha$, respectively $T'_\alpha T_\alpha$, is a continuous endomorphism of $\mathbb{H}^{\beta,\gamma,\delta}$ when $2\alpha > \beta + \gamma$ and $2\alpha > \beta + \delta$, respectively when $\beta + \gamma > 0$ and $\beta + \delta > 0$; when the four conditions are satisfied, they are the inverse of each other. The form $\widetilde{\Pi}^{2\alpha} \Pi^{-2\alpha}$ of $T'_\alpha T_\alpha$ can be used on a bounded time interval $[0, T]$, and in this case we only need $\beta + \gamma > 0$.

The time inversion operator T_0 enables to write the relationship between I^α_- and I^α_{0+} on $\mathbb{R}^*_+$. If $\alpha > 0$ and if f is a smooth function with compact support in $\mathbb{R}^*_+$, we deduce from the change of variables $s \mapsto 1/s$ that

$$I^\alpha_- f(1/t) = \int_{1/t}^\infty \left(s - \frac{1}{t}\right)^{\alpha-1} f(s)ds = \int_0^t \left(\frac{1}{s} - \frac{1}{t}\right)^{\alpha-1} f(1/s)\frac{ds}{s^2}$$

so that

$$T_0 I^\alpha_- T_0 = \Pi^{1-\alpha} I^\alpha_{0+} \Pi^{-1-\alpha}. \tag{44}$$

3 Time Inversion for Self-Similar Processes

We give here time inversion properties which are valid for any H-self-similar centred Gaussian process $(\Xi_t; t > 0)$, and not only for the fractional Brownian motion. Such a process must have a covariance kernel of the form

$$C(s,t) = s^H t^H \rho(s/t) \tag{45}$$

where $\rho(u) = \rho(1/u)$ and $|\rho(u)| \leq \rho(1)$. It then follows immediately by comparing the covariance kernels that if T_H is the time inversion operator defined in (39), then one has the equality in law $T_H \Xi \simeq \Xi$. Notice that this holds even when H is not positive.

Remark 6. The Lamperti transform (see for instance [5])

$$(\Xi(t); t > 0) \mapsto \left(e^{-Ht} \Xi(e^t); t \in \mathbb{R} \right) \tag{46}$$

maps H-self-similar processes Ξ_t into stationary processes Z_t. Then $T_H \Xi \simeq \Xi$ is equivalent to the property $Z_{-t} \simeq Z_t$ which is valid for stationary Gaussian processes (invariance by time reversal).

Remark 7. We have $T_H B^H \simeq B^H$ and can deduce properties of B^H on $[1, +\infty)$ from its properties on $[0, 1]$. For instance, B^H lives in $\mathbb{H}^{H-}$ on $[0, 1]$, and we can check from Theorem 7 that T_H sends this space on $[0, 1]$ into the space $\mathbb{H}^{H-,\cdot,0+}$ on $[1, +\infty)$; thus B^H lives in $\mathbb{H}^{H-,0,0+}$ on $\mathbb{R}_+$ (notation (15)).

We now prove another time inversion property when $H > 0$ (we do not assume $H < 1$). Assume provisionally that the paths of Ξ are absolutely continuous; then its derivative $D^1 \Xi$ is $(H-1)$-self-similar, so $T_{H-1} D^1 \Xi \simeq D^1 \Xi$ and

$$T_H' \Xi = -I_{0+}^1 T_{H-1} D^1 \Xi \simeq -I_{0+}^1 D^1 \Xi = -\Xi \simeq \Xi.$$

In the general case (when Ξ is not absolutely continuous), the same property can be proved with the theory of generalised processes (as said in [29]); we here avoid this theory.

Theorem 8. *For $H > 0$, let $(\Xi_t; t \geq 0)$ be a H-self-similar centred Gaussian process, and consider the time inversion operators T_H and T_H'. Then one has the equalities in law $T_H' \Xi \simeq T_H \Xi \simeq \Xi$.*

Proof. As it has already been said in the beginning of this section, $T_H \Xi \simeq \Xi$ is obtained by comparing the covariance kernels. Since Ξ is H-self-similar, the norm of Ξ_t in $L^1(\Omega)$ is proportional to t^H, so the variable $\int_T^\infty |\Xi_t| t^{-2H-1} dt$ is in $L^1(\Omega)$ for any $T > 0$, and is therefore almost surely finite. Thus $T_H' \Xi$ is well defined. Moreover, $T_H' \Xi = T_H' T_H T_H \Xi \simeq T_H' T_H \Xi$, so let us compare the covariance kernels of Ξ and $T_H' T_H \Xi = \widetilde{\Pi}^{2H} \Pi^{-2H} \Xi$ given by (43). We have from (45) that

$$\mathbb{E}\left[\varXi_T \int_0^S \varXi_s \frac{ds}{s}\right] = T^H \int_0^S s^{H-1}\rho(s/T)ds = T^{2H} \int_0^{S/T} u^{H-1}\rho(u)du.$$

Thus

$$\mathbb{E}\left[\left(\int_0^T \varXi_t \frac{dt}{t}\right)\left(\int_0^S \varXi_s \frac{ds}{s}\right)\right]$$

$$= \int_0^T t^{2H-1} \int_0^{S/t} u^{H-1}\rho(u)du\,dt = \frac{1}{2H}\int_0^\infty \left(T \wedge \frac{S}{u}\right)^{2H} u^{H-1}\rho(u)du$$

$$= \frac{1}{2H}\left(T^{2H}\int_0^{S/T} u^{H-1}\rho(u)du + S^{2H}\int_{S/T}^\infty u^{-H-1}\rho(u)du\right)$$

$$= \frac{1}{2H}\left(T^{2H}\int_0^{S/T} u^{H-1}\rho(u)du + S^{2H}\int_0^{T/S} u^{H-1}\rho(u)du\right)$$

(we used $\rho(1/u) = \rho(u)$ in the last equality). We deduce from these two equations that

$$\mathbb{E}\left[\left(\varXi_T - 2H\int_0^T \varXi_t \frac{dt}{t}\right)\left(\varXi_S - 2H\int_0^S \varXi_s \frac{ds}{s}\right)\right] = \mathbb{E}[\varXi_T\varXi_S]$$

since the other terms cancel one another, so $T'_H T_H \varXi$ has the same covariance kernel as $\varXi$.   □

Remark 8. Theorem 8 can be applied to the fractional Brownian motion B^H. Moreover, the relations $B^H \simeq T_H B^H \simeq T'_H B^H$ can be extended to $\mathbb{R}^*$ by defining

$$T_H f(t) = |t|^{2H} f(1/t), \qquad T'_H f = \mp I_{0\pm}^1 T_{H-1} D^1 f \quad \text{on } \mathbb{R}^*_\pm.$$

Since B^H also has stationary increments, we can deduce how the law of the generalised process $D^1 B^H$ is transformed under the time transformations $t \mapsto (at + b)/(ct + d)$, see [29].

The law of the H-self-similar process $\varXi$ is therefore invariant by the transformations $T_H T'_H$ and $T'_H T_H = \widetilde{\varPi}^{2H} \varPi^{-2H}$ given by (42) and (43). We now introduce a generalisation $T_{H,L}$ of $T'_H T_H$, which was also studied in [21].

Theorem 9. *On the time interval $\mathbb{R}_+$, for $H > 0$ and $L > 0$, the operator*

$$T_{H,L} = \varPi^{H-L} T'_L T_L \varPi^{L-H} = \varPi^{H-L}\widetilde{\varPi}^{2L}\varPi^{-L-H} \tag{47}$$

is a continuous endomorphism of $\mathbb{H}^{\beta,\gamma,\delta}$ when $0 < \beta < 1$, and $\beta + \gamma$ and $\beta + \delta$ are greater than $H - L$; in particular, it is a continuous endomorphism of $\mathbb{H}^{H-,0,0+}$ if

$0 < H < 1$. *It is defined on a function f as soon as $t^{L-H-1} f(t)$ is integrable on any $[0, T]$, and it satisfies*

$$T_{H,L} f(t) = f(t) - 2L\, t^{H-L} \int_0^t f(s) s^{L-H-1} ds. \tag{48}$$

If Ξ is a H-self-similar centred Gaussian process, then $T_{H,L} \Xi$ has the same law as Ξ.

Proof. The continuity property of $T_{H,L}$ can be deduced from Theorem 7 and Remark 5. The representation (48) follows easily from (38) and the second form of $T_{H,L}$ in (47). Let Ξ be a centred Gaussian H-self-similar process; then the L^1-norm of Ξ_t is proportional to t^H, so $\int_0^T t^{L-H-1} |\Xi_t| dt$ is integrable and therefore almost surely finite for any $T > 0$. We deduce that $T_{H,L} \Xi$ is well defined; we have

$$T'_L T_L \Pi^{L-H} \Xi \simeq \Pi^{L-H} \Xi$$

because $\Pi^{L-H} \Xi$ is L-self-similar. By applying Π^{H-L} to both sides we obtain $T_{H,L} \Xi \simeq \Xi$. $\qquad\square$

Remark 9. In the non centred case, we have $T_H \Xi \simeq \Xi$ and $T'_H \Xi \simeq T_{H,L} \Xi \simeq -\Xi$.

We will resume our study of $T_{H,L}$ for self-similar processes in Sect. 4.4.

4 Representations of Fractional Brownian Motions

Starting from the classical representation of fractional Brownian motions on $\mathbb{R}$ described in Sect. 4.1, we study canonical representations on $\mathbb{R}_+$ (Sect. 4.2) and $\mathbb{R}_-$ (Sect. 4.3). In Sect. 4.4, we also consider the non canonical representations on $\mathbb{R}_+$ introduced in Theorem 9.

4.1 A Representation on $\mathbb{R}$

For $0 < H < 1$, the basic representation of a fractional Brownian motion B^H is

$$B_t^H = \kappa \int_{-\infty}^{+\infty} \left((t-s)_+^{H-1/2} - (-s)_+^{H-1/2} \right) dW_s \tag{49}$$

for a positive parameter κ, see [26]. It is not difficult to check that the integral of the right-hand side is Gaussian, centred, with stationary increments, and H-self-similar. Thus B_t^H is a fractional Brownian motion; its covariance is given by (2), and the variance ρ of B_1^H is proportional to κ^2; the precise relationship between ρ and κ

is given in Theorem 33. Subsequently, we will consider the fractional Brownian motion corresponding to

$$\kappa = \kappa(H) = 1/\Gamma(H + 1/2), \tag{50}$$

so that (following (96))

$$\rho = \rho(H) = -2\frac{\cos(\pi H)}{\pi}\Gamma(-2H), \qquad \rho(1/2) = 1. \tag{51}$$

In particular, $B^{1/2} = W$ is the standard Brownian motion. This choice of κ is due to the following result, where we use the modified Riemann–Liouville operators of Sect. 2.2.2.

Theorem 10. *The family of processes* $(B^H; 0 < H < 1)$ *defined by* (49) *with* (50) *can be written as*

$$B^H = \widetilde{I}_+^{H-1/2} W. \tag{52}$$

More generally,

$$B^H = \widetilde{I}_+^{H-J} B^J \tag{53}$$

for any $0 < J, H < 1$.

Proof. The formula (52) would hold true from (32) if W were Lipschitz with compact support; the operator $\widetilde{I}_+^{H-1/2}$ is continuous on $\mathbb{H}^{1/2-,0,0+}$ (Theorem 6) in which W lives, and Lipschitz functions with compact support are dense in this space (Theorem 3); moreover, integration by parts shows that the stochastic integral in the right-hand side of (49) can also be computed by approximating W with smooth functions with compact support, so (52) holds almost surely. Then (53) follows from the composition rules for Riemann–Liouville operators (Theorem 6). □

We deduce in particular from (53) that (52) can be reversed ($W = B^{1/2}$), and

$$W = \widetilde{I}_+^{1/2-H} B^H.$$

Thus the increments of W and B^H generate the same completed filtration, namely $\mathcal{F}_t(dB^H) = \mathcal{F}_t(dW)$ (with notation (5)).

Remark 10. Relation (53) can be written by means of (30) ($H > J$) or (31) ($H < J$). It can be written more informally as

$$B_t^H = \frac{1}{\Gamma(H - J + 1)} \int_{-\infty}^{+\infty} \left((t - s)_+^{H-J} - (-s)_+^{H-J} \right) dB_s^J,$$

where the integral is obtained by approximating B^J by Lipschitz functions with compact support, and passing to the limit.

Relations (52) or (53) can be restricted to the time interval $\mathbb{R}_-$; in order to know B^H on $\mathbb{R}_-$, we only need W on $\mathbb{R}_-$, and vice-versa. On the other hand, they cannot be used on $\mathbb{R}_+$; in order to know B^H on $\mathbb{R}_+$, we have to know W on the whole real line $\mathbb{R}$. If we want a representation on $\mathbb{R}_+$, we can reverse the time ($t \mapsto -t$) for all the processes, so that the operators $\widetilde{I}_+$ are replaced by $\widetilde{I}_-$. We obtain on $\mathbb{R}_+$ the backward representation

$$B_t^H = \widetilde{I}_-^{H-1/2} W(t) = \frac{1}{\Gamma(H+1/2)} \int_0^\infty \left(s^{H-1/2} - (s-t)_+^{H-1/2} \right) dW_s. \quad (54)$$

However, in this formula, if we want to know B^H at a single time t, we need W on the whole half-line $\mathbb{R}_+$; next section is devoted to a representation formula where we only need W on $[0,t]$.

4.2 Canonical Representation on $\mathbb{R}_+$

We shall here explain the derivation of the canonical representation of fractional Brownian motions on $\mathbb{R}_+$ which was found by [29, 30], and the general relationship between B^J and B^H which was given in [20]. More precisely, we want the various processes $(B^H; 0 < H < 1)$ to be deduced from one another, so that all of them generate the same filtration.

As explained in the introduction, we start from the relation $B^H = \widetilde{I}_-^{H-1/2} W$ of (54) and apply the time inversion $t \mapsto 1/t$ on the increments dW_t and dB_t^H; this time inversion is made by means of the operators $T'_{1/2}$ and T'_H defined in (39) (they are involutions), which preserve respectively the laws of W and B^H (Theorem 8). Thus

$$B^H \simeq \left(T'_H \widetilde{I}_-^{H-1/2} T'_{1/2} \right) W.$$

It appears that this is the canonical representation of B^H. We now make more explicit this calculation, and generalise it to the comparison of B^H and B^J for any J and H; starting from $B^H = \widetilde{I}_-^{H-J} B^J$, we can show similarly that

$$B^H \simeq \left(T'_H \widetilde{I}_-^{H-J} T'_J \right) B^J. \quad (55)$$

Theorem 11. *On the time interval $\mathbb{R}_+$, the family of fractional Brownian motions B^H, $0 < H < 1$, can be defined jointly so that $B^H = G_{0+}^{J,H} B^J$ for*

$$G_{0+}^{J,H} = \widetilde{\Pi}^{H+J-1} I_{0+}^{H-J} \widetilde{\Pi}^{1-H-J} \quad (56)$$

(see Sect. 2 for the definitions of I_{0+}^α and $\widetilde{\Pi}^\alpha$). This family of operators satisfies the composition rule $G_{0+}^{H,L} G_{0+}^{J,H} = G_{0+}^{J,L}$, and all the processes B^H generate the same

completed filtration. Moreover, the operator $G_{0+}^{J,H}$ *maps continuously* $\mathbb{H}^{J-,0,0+}$ *(where paths of* B^J *live) into* $\mathbb{H}^{H-,0,0+}$, *and can be defined by the following relation; if we define*

$$\phi^{J,H}(u) = (H - J) \int_1^u \left(v^{H+J-1} - 1 \right) (v - 1)^{H-J-1} dv + (u - 1)^{H-J} \quad (57)$$

for $0 < J, H < 1$ *and* $u > 1$, *and if*

$$K_{0+}^{J,H}(t,s) = \frac{1}{\Gamma(H - J + 1)} \phi^{J,H}\left(\frac{t}{s}\right) s^{H-J}, \quad (58)$$

then

$$G_{0+}^{J,H} f(t) = \int_0^t K_{0+}^{J,H}(t,s) df(s) \quad (59)$$

for f *Lipschitz with compact support in* $\mathbb{R}_+^\star$. *Moreover,* B^H *is given by the Itô integral*

$$B_t^H = \int_0^t K_{0+}^{1/2,H}(t,s) dW_s \quad (60)$$

for $W = B^{1/2}$.

Proof. Let us divide the proof into four steps.

Step 1: Definition of the families $G_{0+}^{J,H}$ *and* B^H. Following (55), we define

$$G_{0+}^{J,H} = T'_H \widetilde{I}_-^{H-J} T'_J, \quad B^H = G_{0+}^{1/2,H} W, \quad (61)$$

so that B^H is a H-fractional Brownian motion. The continuity of $G_{0+}^{J,H}$ from $\mathbb{H}^{J-,0,0+}$ into $\mathbb{H}^{H-,0,0+}$ is then a consequence of Theorems 6 and 7; it indeed follows from these two theorems that T'_J and T'_H are continuous endomorphisms of respectively $\mathbb{H}^{J-,0,0+}$ and $\mathbb{H}^{H-,0,0+}$, and that $\widetilde{I}_-^{H-J}$ is continuous from $\mathbb{H}^{J-,0,0+}$ into $\mathbb{H}^{H-,0,0+}$. Moreover

$$G_{0+}^{H,L} G_{0+}^{J,H} = T'_L \widetilde{I}_-^{L-H} T'_H T'_H \widetilde{I}_-^{H-J} T'_J = T'_L \widetilde{I}_-^{L-H} \widetilde{I}_-^{H-J} T'_J = T'_L \widetilde{I}_-^{L-J} T'_J = G_{0+}^{J,L}$$

and consequently

$$G_{0+}^{J,H} B^J = G_{0+}^{J,H} G_{0+}^{1/2,J} W = G_{0+}^{1/2,H} W = B^H.$$

The equality between filtrations of B^H also follows from this relation.

Step 2: Proof of (56). First assume $H > J$, and let us work on smooth functions with compact support in $\mathbb{R}_+^\star$. We deduce from (44) and the relations $T_\alpha = \Pi^{2\alpha} T_0 = T_0 \Pi^{-2\alpha}$ that

$$T_{H-1} I_-^{H-J} T_{J-1} = \Pi^{2H-2} T_0 I_-^{H-J} T_0 \Pi^{2-2J}$$
$$= \Pi^{2H-2} \Pi^{1-H+J} I_{0+}^{H-J} \Pi^{-1-H+J} \Pi^{2-2J}$$
$$= \Pi^{H+J-1} I_{0+}^{H-J} \Pi^{1-H-J}. \tag{62}$$

On the other hand, T'_α has been defined as $-I_{0+}^1 T_{\alpha-1} D^1$, and $\widetilde{I}^\alpha = I_{0+}^1 I_-^\alpha D^1$ from (33), so the definition (61) can be written as

$$G_{0+}^{J,H} = (I_{0+}^1 T_{H-1} D^1)(I_{0+}^1 I_-^{H-J} D^1)(I_{0+}^1 T_{J-1} D^1)$$
$$= I_{0+}^1 T_{H-1} I_-^{H-J} T_{J-1} D^1$$
$$= I_{0+}^1 \Pi^{H+J-1} I_{0+}^{H-J} \Pi^{1-H-J} D^1$$
$$= I_{0+}^1 \Pi^{H+J-1} I_{0+}^{H-J} D^1 I_{0+}^1 \Pi^{1-H-J} D^1$$
$$= \left(I_{0+}^1 \Pi^{H+J-1} D^1\right) I_{0+}^{H-J} \left(I_{0+}^1 \Pi^{1-H-J} D^1\right)$$
$$= \widetilde{\Pi}^{H+J-1} I_{0+}^{H-J} \widetilde{\Pi}^{1-H-J} \tag{63}$$

(we used (62) in the third equality and Theorem 4 in the fifth one). The equality can be extended to the functional space $\mathbb{H}^{J-,0,0+}$, since $G_{0+}^{J,H}$ is continuous on this space, and the right-hand side is continuous on $\mathbb{H}^{J-}$ on any interval $[0,T]$. Moreover, inverting this relation provides $G_{0+}^{H,J}$, so that this expression of $G_{0+}^{J,H}$ also holds when $H < J$.

Step 3: Proof of (59). For smooth functions f with compact support in $\mathbb{R}_+^*$, (27) yields

$$\Pi^{H+J-1} I_{0+}^{H-J} \Pi^{1-H-J} f(t)$$
$$= I_{0+}^{H-J} f(t) + \frac{1}{\Gamma(H-J)} \int_0^t \left(\left(\frac{s}{t}\right)^{1-H-J} - 1\right)(t-s)^{H-J-1} f(s) ds,$$

so (63) implies

$$G_{0+}^{J,H} f(t)$$
$$= I_{0+}^{H-J} f(t) + \frac{1}{\Gamma(H-J)} \int_0^t \left(\int_0^v \left(\left(\frac{s}{v}\right)^{1-H-J} - 1\right)(v-s)^{H-J-1} df(s)\right) dv$$
$$= \frac{1}{\Gamma(H-J+1)} \int_0^t (t-s)^{H-J} df(s)$$
$$+ \frac{H-J}{\Gamma(H-J+1)} \int_0^t \left(\int_s^t \left(\left(\frac{s}{v}\right)^{1-H-J} - 1\right)(v-s)^{H-J-1} dv\right) df(s).$$

This expression can be written as (59) for a kernel $K_{0+}^{J,H}$, and a scaling argument shows that $K_{0+}^{J,H}$ is of the form (58) for $\phi^{J,H}(u) = \Gamma(H-J+1) K_{0+}^{J,H}(u,1)$. Then (57) follows from a simple verification.

Step 4: Proof of (60). By means of an integration by parts, we write (59) for $J = 1/2$ and $H \neq 1/2$ in the form

$$
\begin{aligned}
G_{0+}^{1/2,H} f(t) &= \frac{f(t)}{t} \int_0^t K_{0+}^{1/2,H}(t,s)ds + \int_0^t K_{0+}^{1/2,H}(t,s)\left(D^1 f(s) - f(t)/t\right)ds \\
&= \frac{f(t)}{t} \int_0^t K_{0+}^{1/2,H}(t,s)ds - \int_0^t \left(f(s) - \frac{s}{t}f(t)\right) \partial_s K_{0+}^{1/2,H}(t,s)ds.
\end{aligned}
\tag{64}
$$

On the other hand,

$$
(\phi^{J,H})'(u) = (H - J)(u - 1)^{H-J-1}u^{H+J-1}
$$

so that

$$
\partial_s K_{0+}^{J,H}(t,s) = \frac{1}{\Gamma(H-J)}\left(\phi^{J,H}(\frac{t}{s})s^{H-J-1} - (t-s)^{H-J-1}\left(\frac{t}{s}\right)^{H+J}\right).
$$

An asymptotic study of (57) shows that $\phi^{1/2,H}(u)$ is $O((u-1)^{H-1/2})$ as $u \downarrow 1$ and $O(u^{2H-1} \vee 1)$ as $u \uparrow \infty$; thus $\partial_s K_{0+}^{1/2,H}(t,s)$ is $O((t-s)^{H-3/2})$ as $s \uparrow t$, and is $O(s^{-H-1/2} \vee s^{H-3/2})$ as $s \downarrow 0$. An approximation by smooth functions shows that (64) is still valid for W, and a stochastic integration by parts leads to (60).   □

Remark 11. It is also possible to write a representation $B^H = G_{T+}^{J,H} B^J$ on the time interval $[T, +\infty)$, associated to the kernel $K_{T+}^{J,H}(t,s) = K_{0+}^{J,H}(t-T, s-T)$. In [22], it is proved that letting T tend to $-\infty$, we recover at the limit (49).

Remark 12. If $H > J$, we have

$$
\phi^{J,H}(u) = (H - J)\int_1^u v^{H+J-1}(v-1)^{H-J-1}dv.
$$

If $H < J$, this integral diverges and $\phi^{J,H}(u)$ is its principal value. This function, and therefore the kernel $K_{0+}^{J,H}(t,s)$ can also be written by means of the Gauss hypergeometric function, see [8,20].

Remark 13. If $H + J = 1$, then (56) is simply written as $G_{0+}^{J,H} = I_{0+}^{H-J}$. Thus the relation between B^H and B^{1-H} is particularly simple (as it has already been noticed in [20]), but we have no intuitive explanation of this fact.

Remark 14. The expression (56) for $G_{0+}^{J,H}$ is close to the representation given in [32] for $J = 1/2$. We define

$$
Z_t^{J,H} = I_{0+}^{H-J}\widetilde{\Pi}^{1-J-H} B^J(t) = \frac{1}{\Gamma(H-J+1)}\int_0^t (t-s)^{H-J}s^{1-J-H}dB_s^J
$$

which is an Itô integral in the case $J = 1/2$, and the fractional Brownian motion B^H is given by

$$B_t^H = \widetilde{\Pi}^{H+J-1} Z^{J,H}(t) = \int_0^t s^{H+J-1} dZ_s^{J,H}$$

which can be defined by integration by parts.

Remark 15. In the case $J = 1/2$, let us compare our result with the decomposition of [8]. We look for a decomposition of $G_{0+}^{1/2,H}$ which would be valid on the classical Cameron–Martin space $\mathcal{H}_{1/2} = I_{0+}^1 L^2$ of W. To this end, we start from (63)

$$G_{0+}^{1/2,H} = I_{0+}^1 \Pi^{H-1/2} I_{0+}^{H-1/2} \Pi^{1/2-H} D^1$$

which is valid for smooth functions. When $H > 1/2$, this formula is valid on $\mathcal{H}_{1/2}$ for any finite time interval $[0, T]$ because these five operators satisfy the continuity properties

$$\mathcal{H}_{1/2} \to L^2 \to L^1 \to L^1 \to L^1 \to L^\infty$$

(use the fact that I_{0+}^α is a continuous endomorphism of L^1 for $\alpha > 0$). However, it does not make sense on $\mathcal{H}_{1/2}$ for $H < 1/2$ because $I_{0+}^{H-1/2}$ is in this case a fractional derivative, and is not defined for non continuous functions. Thus let us look for an alternative definition of the operator $G_{0+}^{1/2,H}$; in order to solve this question, we apply the property (29) of Riemann–Liouville operators and get

$$\begin{aligned}
G_{0+}^{1/2,H} &= I_{0+}^{2H} \left(I_{0+}^{1-2H} \Pi^{H-1/2} I_{0+}^{H-1/2} \right) \Pi^{1/2-H} D^1 \\
&= I_{0+}^{2H} \left(\Pi^{1/2-H} I_{0+}^{1/2-H} \Pi^{2H-1} \right) \Pi^{1/2-H} D^1 \\
&= I_{0+}^{2H} \Pi^{1/2-H} I_{0+}^{1/2-H} \Pi^{H-1/2} D^1
\end{aligned}$$

which makes sense on $\mathcal{H}_{1/2}$ if $H < 1/2$. This is the expression of [8].

Remark 16. A consequence of (60) is that we can write the conditional law of $(B_t^H; t \geq S)$ given $(B_t^H; 0 \leq t \leq S)$. This is the prediction problem, see also [13, 29].

Remark 17. Theorem 11 can also be proved by using the time inversion operators T_H rather than T_H'. If we start again from (54) and consider the process with independent increments

$$V_t^H = \int_0^t s^{H-1/2} dW_s,$$

then it appears that B_t^H depends on future values of V^H; consequently, $T_H B^H(t)$ depends on past values of $T_H V^H$. On the other hand, $T_H B^H \simeq B^H$ and

$T_H V^H \simeq V^H$ from Theorem 8, so we obtain an adapted representation of B^H with respect to V^H, and therefore with respect to W. One can verify that this is the same representation as Theorem 11; however, the composition rule for the operators $G_{0+}^{J,H}$ is less direct with this approach.

Let us give another application of Theorem 11. The process B^H has stationary increments, so a natural question is to know whether it can be written as $B_t^H = A_t^H - A_0^H$ for a stationary centred Gaussian process A^H, and to find A^H. This is clearly not possible on an infinite time interval, since the variance of B^H is unbounded. However, let us check that this is possible in an explicit way on a finite time interval, and that moreover we do not have to increase the σ-algebra of B^H. Since we are on a bounded time interval $[0, T]$, the stationarity means that $(A_{U+t}^H; \ 0 \leq t \leq T - U)$ and $(A_t^H; \ 0 \leq t \leq T - U)$ have the same law for any $0 < U < T$.

Theorem 12. *Let $T > 0$. There exists a stationary centred Gaussian process $(A_t^H; \ 0 \leq t \leq T)$ such that $B_t^H = A_t^H - A_0^H$ is a H-fractional Brownian motion on $[0, T]$, and B^H and A^H generate the same σ-algebra.*

Proof. Consider $B^H = G_{0+}^{1/2,H} W$. We look for a variable A_0^H such that $A_t^H = B_t^H + A_0^H$ is stationary; this will hold when

$$\mathbb{E}[A_t^H A_s^H] = \frac{\rho}{2}\left(t^{2H} + s^{2H} - |t - s|^{2H}\right) + \mathbb{E}[B_t^H A_0^H] + \mathbb{E}[B_s^H A_0^H] + \mathbb{E}[(A_0^H)^2]$$

is a function of $t - s$, so when

$$\mathbb{E}[B_t^H A_0^H] = -\rho t^{2H}/2.$$

By applying the operator $G_{0+}^{H,1/2}$, this condition is shown to be equivalent to

$$\mathbb{E}[W_t A_0^H] = -\frac{\rho}{2} G_{0+}^{H,1/2} t^{2H} = -\frac{\rho}{2} \frac{2H}{H + 1/2} \Gamma(H + 1/2) t^{H+1/2}$$

by using the formulae (63) and (22) for computing $G_{0+}^{H,1/2}$, and for ρ given by (51). Thus we can choose

$$A_0^H = \int_0^T \frac{d}{dt} \mathbb{E}[W_t A_0^H] dW_t = -\rho H \Gamma(H + 1/2) \int_0^T t^{H-1/2} dW_t.$$

$\square$

In particular we have $A_0^{1/2} = -W_T/2$. Of course we can add to A_0^H any independent variable; this increases the σ-algebra, but this explains the mutual compatibility of the variables A_0^H when T increases. More generally, the technique used in the proof enables to write any variable A of the Gaussian space of B^H, knowing the covariances $\mathbb{E}[A B_t^H]$.

Remark 18. We can also try to write B^H on $[0, T]$ as the increments of a process which would be stationary on $\mathbb{R}$. We shall address this question in Remark 31.

Remark 19. Another classical stationary process related to the Brownian motion is the Ornstein–Uhlenbeck process; actually there are two different fractional extensions of this process, see [5].

4.3 Canonical Representation on $\mathbb{R}_-$

In the representation (6), we have $\mathcal{F}_t(dB^H) = \mathcal{F}_t(dW)$ (with notation (5)). However, when $t < 0$, the filtration $\mathcal{F}_t(dB^H)$ is strictly included into $\mathcal{F}_t(B^H)$. We now give a representation of B^H on the time interval $\mathbb{R}_-$ for which $\mathcal{F}_t(B^H) = \mathcal{F}_t(dW)$; one can then deduce a canonical representation of B^H (see Remark 20 below). In the particular case $H = 1/2$ of a standard Brownian motion, we recover the classical representation of the Brownian bridge.

We want B_t^H, $t < 0$, to depend on past increments of W; by applying the time reversal $t \mapsto -t$, this is equivalent to wanting B_t^H, $t > 0$, to depend on future increments of W. The starting point is the operator $T_\alpha T_\alpha'$ of (42) which can be written in the form

$$T_H T_H' f(t) = -2H t^{2H} \int_t^\infty s^{-2H-1} \left(f(s) - f(t) \right) ds.$$

Thus $T_H T_H' f(t)$ depends on future increments of f, and the equality in law $B^H \simeq T_H T_H' B^H$ enables to write B^H as a process depending on future increments of another H-fractional Brownian motion. On the other hand, in the representation $B^H \simeq \widetilde{I}_-^{H-1/2} W$ of (54), future increments of B^H depend on future increments of W. Thus, in $B^H \simeq T_H T_H' \widetilde{I}_-^{H-1/2} W$, the value of B_t^H depends on future increments of W, and this answers our question. The same method can be used with W replaced by B^J.

Theorem 13. *Let B^J be a J-fractional Brownian motion on $\mathbb{R}_-$; consider the function $\phi^{J,H}$ of (57). On $\mathbb{R}_-^\star$, the operator*

$$G_+^{J,H} f(t) = \int_{-\infty}^t K_+^{J,H}(t,s) df(s)$$

for f smooth with compact support, with

$$K_+^{J,H}(t,s) = \Gamma(H - J + 1)^{-1} \phi^{J,H}(s/t)(-t)^{2H}(-s)^{-H-J}, \qquad s < t < 0,$$

can be extended to a continuous operator from $\mathbb{H}^{J-,0,0+}$ into $\mathbb{H}^{H-,0,0+}$, and $\widetilde{B}^{J,H} = G_+^{J,H} B^J$ is a H-fractional Brownian motion on $\mathbb{R}_-$. Moreover, $\mathcal{F}_t(\widetilde{B}^{J,H}) = \mathcal{F}_t(dB^J)$ (with notation (5)).

Proof. We transform the question on $\mathbb{R}_-$ into a question on $\mathbb{R}_+$ by means of the time reversal $t \mapsto -t$. Following the discussion before the theorem, we introduce on $\mathbb{R}_+^*$ the operator

$$G_-^{J,H} = T_H T_H' \widetilde{I}_-^{H-J}.$$

It follows from Theorems 6 and 7 that $G_-^{J,H}$ maps continuously $\mathbb{H}^{J-,0,0+}$ into $\mathbb{H}^{H-,0,0+}$; moreover $\widetilde{B}^{J,H} = G_-^{J,H} B^J$ is a H-fractional Brownian motion. If we compare $G_-^{J,H}$ with $G_{0+}^{J,H}$ given in (61), we see that

$$G_-^{J,H} = T_H G_{0+}^{J,H} T_J'.$$

For f smooth with compact support in $\mathbb{R}_+^*$,

$$G_{0+}^{J,H} T_J' f(t) = \int_0^t K_{0+}^{J,H}(t,s) s^{2J-2} D^1 f(1/s) ds = \int_{1/t}^\infty K_{0+}^{J,H}(t,1/s) s^{-2J} df(s)$$

so

$$G_-^{J,H} f(t) = t^{2H} \int_t^\infty K_{0+}^{J,H}(1/t,1/s) s^{-2J} df(s) = \int_t^\infty K_-^{J,H}(t,s) df(s)$$

with

$$K_-^{J,H}(t,s) = t^{2H} s^{-2J} K_{0+}^{J,H}(1/t,1/s) = \Gamma(H-J+1)^{-1} \phi^{J,H}(s/t) t^{2H} s^{-H-J}$$

(apply (58)). We still have to check that

$$\sigma\left(\widetilde{B}_s^{J,H}; s \ge t\right) = \sigma\left(B_s^J - B_u^J; s \ge u \ge t\right)$$

for $t \ge 0$. The inclusion of the left-hand side in the right-hand side follows from the discussion before the theorem. For the inverse inclusion, notice that $\widetilde{B}^{J,H} = G_-^{J,H} B^J$ can be reversed and

$$B^J = \widetilde{I}_-^{J-H} T_H' T_H \widetilde{B}^{J,H}.$$

Thus future increments of B^J depend on future increments of $T_H' T_H \widetilde{B}^{J,H}$, which depend on future values of $\widetilde{B}^{J,H}$ from (43). $\qquad\square$

Remark 20. The theorem involves $\mathcal{F}_t(dB^J)$ which is strictly smaller than $\mathcal{F}_t(B^J)$, so the representation is not really canonical on $\mathbb{R}_-$; however, $\mathcal{F}_t(dB^J)$ is also the filtration generated by (for instance) the increments of the process

$$\Upsilon_t^J = \int_{-\infty}^t (-s)^{-2J} dB_s^J = (-t)^{-2J} B_t^J + 2J \int_{-\infty}^t (-s)^{-2J-1} B_s^J ds,$$

and

$$\widetilde{B}_t^{J,H} = \int_{-\infty}^t K_+^{J,H}(t,s)(-s)^{2J} d\Upsilon_s^J. \tag{65}$$

The process Υ_t^J tends to 0 at $-\infty$, so

$$\mathcal{F}_t(\widetilde{B}^{J,H}) = \mathcal{F}_t(dB^J) = \mathcal{F}_t(d\Upsilon^J) = \mathcal{F}_t(\Upsilon^J)$$

and (65) is therefore a canonical representation on $\mathbb{R}_-$ (notice that $\Upsilon^{1/2}$ has independent increments).

Remark 21. By applying Theorem 13 with $J = 1/2$, we can predict on $\mathbb{R}_-$ future values of B^H knowing previous values; this prediction must take into account the fact $B_0^H = 0$; this can be viewed as a bridge; actually for $H = J = 1/2$, we recover the classical Brownian bridge. More precisely, $\phi^{1/2,1/2} \equiv 1$, so $K_+^{1/2,1/2}(t,s) = |t|/|s|$ on $\mathbb{R}_-$; thus $W = B^{1/2}$ and $\overline{W} = \widetilde{B}^{1/2,1/2}$ are Brownian motions on $\mathbb{R}_-$, and satisfy

$$\overline{W}_t = |t| \int_{-\infty}^t |s|^{-1} dW_s, \qquad d\overline{W}_t = -\frac{\overline{W}_t}{|t|} dt + dW_t.$$

Notice in the same vein that $B_{t-T}^H \simeq B_{T-t}^H$ on $[0,T]$ for $T > 0$, so the study on $[-T, 0]$ is related to the time reversal of B^H on $[0, T]$; some general results for this problem were obtained in [7].

4.4 Some Non Canonical Representations

Let us come back to general H-self-similar centred Gaussian processes $\Xi_t, t \geq 0$. In Theorem 9, we have proved the equality in law

$$\Xi_t \simeq T_{H,L}\Xi(t) = \Xi_t - 2L\, t^{H-L} \int_0^t s^{L-H-1} \Xi_s ds$$

for $L > 0$. When $\Xi = W$ is a standard Brownian motion so that $H = 1/2$, this is the classical Lévy family of non canonical representations of W with respect to itself. We now verify that this property of non canonical representation holds in many cases, in the sense that $\mathcal{F}_t(T_{H,L}\Xi)$ is strictly included in $\mathcal{F}_t(\Xi)$ for $t > 0$ (it is of course sufficient to consider the case $t = 1$). In the following theorem we need some notions about Cameron–Martin spaces and Wiener integrals (see a short introduction in Appendix C.1).

Theorem 14. *Let $\Xi = (\Xi_t; \, 0 \leq t \leq 1)$ be the restriction to $[0, 1]$ of a H-self-similar centred Gaussian process for $H > 0$. Let $\mathcal{W}$ be a separable Fréchet space* *of paths in which Ξ lives, and let $\mathcal{H}$ be its Cameron–Martin space. Suppose that the function $\psi(t) = t^{H+L}$ is in $\mathcal{H}$, and denote by $\langle\Xi, \psi\rangle_{\mathcal{H}}$ its Wiener integral. Then*

$$\sigma(\Xi) = \sigma(T_{H,L}\Xi) \vee \sigma(\langle\Xi, \psi\rangle_{\mathcal{H}})$$

where the two σ-algebras of the right-hand side are independent.

Proof. The operator $T_{H,L}$ operates on $\mathcal{H}$, and it is easy to check that functions proportional to ψ constitute the kernel of $T_{H,L}$. On the other hand, for any h in $\mathcal{H}$, $h \neq 0$, we can write the decomposition

$$\Xi = \langle\Xi, h\rangle_{\mathcal{H}} \frac{h}{|h|^2_{\mathcal{H}}} + \left(\Xi - \langle\Xi, h\rangle_{\mathcal{H}} \frac{h}{|h|^2_{\mathcal{H}}}\right)$$

where the two terms are independent: this is because independence and orthogonality are equivalent in Gaussian spaces, and

$$\mathbb{E}\left[\langle\Xi, h\rangle_{\mathcal{H}}\langle\Xi - \langle\Xi, h\rangle_{\mathcal{H}} \frac{h}{|h|^2_{\mathcal{H}}}, h'\rangle_{\mathcal{H}}\right] = 0$$

for any h' in $\mathcal{H}$ (apply (99)). Thus

$$T_{H,L}\Xi = \langle\Xi, h\rangle_{\mathcal{H}} \frac{T_{H,L}h}{|h|^2_{\mathcal{H}}} + \text{process independent of } \langle\Xi, h\rangle_{\mathcal{H}},$$

and $T_{H,L}\Xi$ is independent of $\langle\Xi, h\rangle_{\mathcal{H}}$ if and only if h is in the kernel of $T_{H,L}$, so if and only if h is proportional to ψ. Thus the Gaussian space of Ξ, which is generated by $\langle\Xi, h\rangle_{\mathcal{H}}$, $h \in \mathcal{H}$, is the orthogonal sum of the Gaussian space generated by $T_{H,L}\Xi$ and of the variables proportional to $\langle\Xi, \psi\rangle_{\mathcal{H}}$. We deduce the theorem. $\square$

Notice that on the other hand, the transformation $T_{H,L}$ becomes injective on the whole time interval $\mathbb{R}_+$, so $\sigma(\Xi)$ and $\sigma(T_{H,L}\Xi)$ coincide; actually, the theorem cannot be used on $\mathbb{R}_+$ because ψ is no more in $\mathcal{H}$; this can be viewed from the fact that Ξ lives in the space of functions f such that $t^{-H-1-\varepsilon} f(t)$ is integrable on $[1, \infty)$ (for $\varepsilon > 0$), so $\mathcal{H}$ is included in this space, whereas ψ does not belong to it for $\varepsilon \leq L$.

In the case where Ξ is the standard Brownian motion W, we obtain the well known property

$$\mathcal{F}_t(W) = \mathcal{F}_t(T_{1/2,L}W) \vee \sigma\left(\widetilde{\Pi}^{L-1/2}W(t)\right). \tag{66}$$

Let us prove that this property enables to write Theorem 14 in another form when Ξ has a canonical representation with respect to W, see also [21].

Theorem 15. *Consider the standard Brownian motion W on $\mathbb{R}_+$, and let*

$$\Xi_t = (AW)(t) = \int_0^t K(t, s)dW_s$$

be given by a kernel K satisfying $K(\lambda t, \lambda s) = \lambda^{H-1/2} K(t, s)$ for any $\lambda > 0$ and some $H > 0$. Suppose that $\mathcal{F}_t(\Xi) = \mathcal{F}_t(W)$ (the representation is canonical). Then Ξ is a H-self-similar process, and we have

$$T_{H,L}\Xi = AT_{1/2,L}W, \quad \mathcal{F}_t(\Xi) = \mathcal{F}_t(T_{H,L}\Xi) \vee \sigma\left(\widetilde{\Pi}^{L-1/2}W(t)\right) \qquad (67)$$

where the two σ-algebras of the right side are independent.

Proof. The scaling condition on K implies that Ξ is H-self-similar. It can be viewed for instance as a random variable in the space of functions f such that $t^{\varepsilon-1-H,-\varepsilon-H-1}f(t)$ is integrable on $\mathbb{R}_+^*$. On the other hand, notice that

$$T_{H,L} = \Pi^{H-1/2}\Pi^{1/2-L}\widetilde{\Pi}^{2L}\Pi^{-L-1/2}\Pi^{1/2-H} = \Pi^{H-1/2}T_{1/2,L}\Pi^{1/2-H} \quad (68)$$

from (47), and consider the linear functional $\Pi^{1/2-H}A$ mapping W to the $1/2$-self-similar process $\Pi^{1/2-H}\Xi$. The monomials $\psi_\beta(t) = t^\beta$, $\beta > 1/2$, generate the Cameron–Martin space $\mathcal{H}_{1/2}$ of W; we deduce from the scaling condition that they are eigenfunctions of $\Pi^{1/2-H}A$ and of $T_{1/2,L}$, so the commutativity relation

$$\Pi^{1/2-H}AT_{1/2,L} = T_{1/2,L}\Pi^{1/2-H}A \qquad (69)$$

holds on fractional polynomials, and therefore on $\mathcal{H}_{1/2}$ and on the paths of W (a linear functional of W which is zero on the Cameron–Martin space must be zero on W). We deduce from (68) and (69) that

$$T_{H,L}\Xi = \Pi^{H-1/2}T_{1/2,L}\Pi^{1/2-H}AW = \Pi^{H-1/2}\Pi^{1/2-H}AT_{1/2,L}W = AT_{1/2,L}W$$

and the first part of (67) is proved. We have moreover assumed that $\mathcal{F}_t(AW) = \mathcal{F}_t(W)$; this can be applied to the Brownian motion $T_{1/2,L}W$ so $\mathcal{F}_t(AT_{1/2,L}W) = \mathcal{F}_t(T_{1/2,L}W)$. Thus, by applying (66),

$$\mathcal{F}_t(\Xi) = \mathcal{F}_t(W) = \mathcal{F}_t(T_{1/2,L}W) \vee \sigma(\widetilde{\Pi}^{L-1/2}W(t))$$
$$= \mathcal{F}_t(AT_{1/2,L}\Xi) \vee \sigma(\widetilde{\Pi}^{L-1/2}W(t)) = \mathcal{F}_t(T_{H,L}\Xi) \vee \sigma(\widetilde{\Pi}^{L-1/2}W(t))$$

so the second part of (67) is also proved. $\square$

Remark 22. Another proof of the second part of (67) is to use directly Theorem 14; we verify that on $[0, 1]$

$$\widetilde{\Pi}^{L-1/2}W(1) = \langle W, \phi\rangle_{\mathcal{H}_{1/2}} = \langle \Xi, A\phi\rangle_{\mathcal{H}}$$

for $\phi(t) = t^{L+1/2}/(L+1/2)$, and $A\phi$ is proportional to the function $\psi(t) = t^{L+H}$ from the scaling condition.

Theorem 16. *Consider on $\mathbb{R}_+$ the family of fractional Brownian motions $B^H = G_{0+}^{1/2,H}W$, so that $B^H = G_{0+}^{J,H}B^J$. Then, for any $L > 0$, the process*

$B^{H,L} = T_{H,L}B^H$ is a H-fractional Brownian motion satisfying the relation $B^{H,L} = G_{0+}^{J,H} B^{J,L}$. Moreover, for any t,

$$\mathcal{F}_t(B^H) = \mathcal{F}_t(B^{H,L}) \vee \sigma\left(\widetilde{\Pi}^{L-1/2} W(t)\right), \tag{70}$$

and the two σ-algebras of the right-hand side are independent.

Proof. This is a direct application of Theorem 15 with $A = G_{0+}^{1/2,H}$. The first part of (67) implies that

$$B^{H,L} = G_{0+}^{1/2,H} T_{1/2,L} W,$$

and the relationship between $B^{J,L}$ and $B^{H,L}$ follows from the composition rule satisfied by the family $G_{0+}^{J,H}$. ☐

5 Riemann–Liouville Processes

In this section, we compare the fractional Brownian motion B^H with the process $X^H = I_{0+}^{H-1/2} W$.

5.1 Comparison of Processes

The processes

$$X_t^H = I_{0+}^{H-1/2} W(t) = \frac{1}{\Gamma(H-1/2)} \int_0^t (t-s)^{H-1/2} dW_s \tag{71}$$

defined on $\mathbb{R}_+$ are often called Riemann–Liouville processes. Notice that these processes can be defined for any $H > 0$. When $0 < H < 1$, these processes have paths in $\mathbb{H}^{H-}$ on bounded time intervals from Theorem 5, and can be viewed as good approximations of fractional Brownian motions B^H for large times, as it is explained in the following result.

Theorem 17. *For $0 < H < 1$, we can realise jointly the two processes (X^H, B^H) on $\mathbb{R}_+$, so that $X^H - B^H$ is C^∞ on $\mathbb{R}_+^\star$. Moreover, for $T > 0$, $S > 0$ and $1 \leq p < \infty$,*

$$\left\| \sup_{0 \leq t \leq T} \left| (X_{S+t}^H - X_S^H) - (B_{S+t}^H - B_S^H) \right| \right\|_p \leq C_p \, S^{H-1} T \tag{72}$$

(where $\|.\|_p$ denotes the $L^p(\Omega)$-norm for the probability space).

Proof. Let $(B_t^H; t \geq 0)$ be defined by $B^H = \tilde{I}_+^{H-1/2} W$ for a standard Brownian motion $(W_t; t \in \mathbb{R})$. The process W can be decomposed into the two independent processes $W_t^+ = W_t$ and $W_t^- = W_{-t}$ for $t \geq 0$, and consequently, the process B^H is decomposed into $B^H = X^H + Y^H$ where

$$X^H = \tilde{I}_+^{H-1/2} \left(W \, 1_{\mathbb{R}_+} \right) = I_{0+}^{H-1/2} W^+$$

is a Riemann–Liouville process, and $Y^H = \tilde{I}_+^{H-1/2} \left(W \, 1_{\mathbb{R}_-} \right)$ can be written by means of Remark 3; more precisely, $Y^H = I_\Delta^{H-1/2} W^-$, where

$$I_\Delta^\alpha f(t) = \frac{1}{\Gamma(\alpha)} \int_0^\infty \left((t+s)^{\alpha-1} - s^{\alpha-1} \right) f(s) ds. \tag{73}$$

We deduce from this representation that Y^H is C^∞ on $\mathbb{R}_+^\star$, so the first statement is proved. On the other hand, it follows from the scaling property that its derivative is $(H-1)$-self-similar, and is therefore of order t^{H-1} in $L^p(\Omega)$; thus the left hand side of (72) is bounded by

$$\left\| \int_S^{S+T} |D^1 Y_u^H| du \right\|_p \leq C_p \int_S^{S+T} u^{H-1} du \leq C_p \, S^{H-1} T.$$

$\square$

Remark 23. Inequality (72) says that the process $X_t^{S,H} = X_{S+t}^H - X_S^H$ is close to a fractional Brownian motion when S is large; it actually provides an upper bound for the Wasserstein distance between the laws of these two processes. A result about the total variation distance will be given later (Theorem 20).

Instead of using the representation of $B^H = \tilde{I}_+^{H-1/2} W$ on $\mathbb{R}$, we can consider the coupling based on the canonical representation of B^H on $\mathbb{R}_+$. It appears that in this case $X^H - B^H$ is not C^∞ but is still differentiable. In particular, we can deduce that the estimation (72) also holds for the coupling of Theorem 18.

Theorem 18. *Consider on $\mathbb{R}_+$ the family $B^H = G_{0+}^{1/2,H} W$ and the family X^H defined by (71). Then $X^H - B^H$ is differentiable on $\mathbb{R}_+^\star$.*

Proof. For f smooth with compact support in $\mathbb{R}_+^\star$, Theorem 4 and the expression (63) for $G_{0+}^{J,H}$ shows that $G_{0+}^{J,H} f$ and $I_{0+}^{H-J} f$ are smooth, and

$$D^1 \left(G_{0+}^{J,H} - I_{0+}^{H-J} \right) = \left(\Pi^{H+J-1} I_{0+}^{H-J} \Pi^{1-H-J} - I_{0+}^{H-J} \right) D^1.$$

We therefore deduce from (27) that

$$\frac{d}{dt}\left(G_{0+}^{J,H} - I_{0+}^{H-J}\right) f(t)$$

$$= \frac{1}{\Gamma(H-J)} \int_0^t \left(\left(\frac{t}{s}\right)^{H+J-1} - 1\right) (t-s)^{H-J-1} D^1 f(s) ds$$

$$= \frac{f(t)}{t} U(t) + \frac{1}{\Gamma(H-J)} \int_0^t \left(\left(\frac{t}{s}\right)^{H+J-1} - 1\right) (t-s)^{H-J-1}$$

$$\left(D^1 f(s) - f(t)/t\right) ds$$

$$= \frac{f(t)}{t} U(t) - \frac{1}{\Gamma(H-J)} \int_0^t \partial_s \left[\left(\left(\frac{t}{s}\right)^{H+J-1} - 1\right) (t-s)^{H-J-1}\right]$$

$$\left(f(s) - \frac{s}{t} f(t)\right) ds$$

with

$$U(t) = \frac{1}{\Gamma(H-J)} \int_0^t \left(\left(\frac{t}{s}\right)^{H+J-1} - 1\right) (t-s)^{H-J-1} ds$$

proportional to t^{H-J}. This equality can be extended to any function f of $\mathbb{H}^{J-}$, so in particular to W in the case $J = 1/2$; we deduce the differentiability announced in the theorem. $\square$

5.2 The Riemann–Liouville Cameron–Martin Space

Cameron–Martin spaces are Hilbert spaces which characterise the law of centred Gaussian variables, so in particular of centred Gaussian processes, see Appendix C.1. The Cameron–Martin spaces $\mathcal{H}_H$ of H-fractional Brownian motions are deduced from each other by means of the transforms of Theorems 10 or 11, so that

$$\mathcal{H}_H = \widetilde{I}_+^{H-J}(\mathcal{H}_J) = \widetilde{I}_-^{H-J}(\mathcal{H}_J), \qquad \mathcal{H}_H = G_{0+}^{J,H}(\mathcal{H}_J) = \widetilde{I}_-^{H-J}(\mathcal{H}_J)$$

respectively on $\mathbb{R}$ and $\mathbb{R}_+$; the space $\mathcal{H}_{1/2}$ is the classical space of absolutely continuous functions h such that $h(0) = 0$ and $D^1 h$ is in L^2. Similarly, the Cameron–Martin space of the Riemann–Liouville process X^H on $\mathbb{R}_+$ is

$$\mathcal{H}'_H = I_{0+}^{H-1/2} \mathcal{H}_{1/2} = I_{0+}^{H+1/2} L^2.$$

In particular, if f is a smooth function on $\mathbb{R}_+$ such that $f(0) = 0$, then, on the time interval $[0, T]$,

$$|f|_{\mathcal{H}'_H} = \left| D^1 I_{0+}^{1/2-H} f \right|_{L^2}$$

$$\leq C \left(\sup |D^1 f| \left(\int_0^T \left(t^{1/2-H} \right)^2 dt \right)^{1/2} \right.$$

$$\left. + \sup |D^2 f| \left(\int_0^T \left(t^{3/2-H} \right)^2 dt \right)^{1/2} \right)$$

$$\leq C' \left(T^{1-H} \sup |D^1 f| + T^{2-H} \sup |D^2 f| \right) \qquad (74)$$

from Theorem 4.

We now explain the proof of a result mentioned in [8] (Theorem 2.1) and taken from [35]. We use the equivalence of Hilbert spaces ($\mathcal{H} \sim \mathcal{H}'$) defined in (12). A probabilistic interpretation of this equivalence is given in Appendix C.1, see (100).

Theorem 19. *For $0 < H < 1$, the spaces $\mathcal{H}_H$ and $\mathcal{H}'_H$ are equivalent on $\mathbb{R}_+$.*

Proof. The proof is divided into the two inclusions; for the second one, we are going to use an analytical result proved in Appendix A. We can of course omit the case $H = 1/2$.

Proof of $\mathcal{H}'_H \subset \mathcal{H}_H$. We have seen in the proof of Theorem 17 that B^H can be written as the sum of the Riemann–Liouville process X^H and of an independent process Y^H. If we denote by $\mathcal{H}_H^\Delta$ the Cameron–Martin space of Y^H, then this decomposition implies (see (101)) that

$$\mathcal{H}_H = \mathcal{H}'_H + \mathcal{H}_H^\Delta \quad \text{with} \quad |h|_{\mathcal{H}_H} = \inf \left\{ \left(|h_1|_{\mathcal{H}'_H}^2 + |h_2|_{\mathcal{H}_H^\Delta}^2 \right)^{1/2} ; h = h_1 + h_2 \right\}. \qquad (75)$$

In particular, $\mathcal{H}'_H \subset \mathcal{H}_H$ with $|h|_{\mathcal{H}_H} \leq |h|_{\mathcal{H}'_H}$.

Proof of $\mathcal{H}_H \subset \mathcal{H}'_H$. It is sufficient from (75) to prove that $\mathcal{H}_H^\Delta$ is continuously embedded into $\mathcal{H}'_H$. Let h be in $\mathbb{H}^{1/2}$; then $|h(t)| \leq |h|_{\mathbb{H}^{1/2}} \sqrt{t}$, and we can deduce from (73) that $I_\Delta^{H-1/2} h$ is C^∞ on $\mathbb{R}_*^+$, and that the derivative of order k is dominated by $|h|_{\mathbb{H}^{1/2}} t^{H-k}$. Theorem 4 enables to deduce that $Ah = I_{0+}^{1/2-H} I_\Delta^{H-1/2} h$ is also smooth, and we have from (24) that $D^1 Ah(t)$ is dominated by $|h|_{\mathbb{H}^{1/2}}/\sqrt{t}$. Moreover, the scaling condition (93) is satisfied, so we deduce from Theorem 32 that A is a continuous endomorphism of $\mathcal{H}_{1/2}$. By composing with $I_{0+}^{H-1/2}$, we obtain that $\left| I_\Delta^{H-1/2} g \right|_{\mathcal{H}'_H}$ is dominated by $|g|_{\mathcal{H}_{1/2}}$, so

$$|h|_{\mathcal{H}_H^\Delta} = \inf \left\{ |g|_{\mathcal{H}_{1/2}} ; h = I_\Delta^{H-1/2} g \right\} \geq c |h|_{\mathcal{H}'_H}.$$

□

Remark 24. Let us give another interpretation of Theorem 19. By comparing $\mathbb{R}$ and $\mathbb{R}_+$, the fractional Brownian motion on $\mathbb{R}_+$ can be obtained as a restriction of the fractional Brownian motion on $\mathbb{R}$. This property can be extended to the Cameron–Martin spaces, and applying (101), we deduce that $\mathcal{H}_H(\mathbb{R}_+)$ consists of the restrictions to $\mathbb{R}_+$ of functions of $\mathcal{H}_H(\mathbb{R})$, and

$$|h|_{\mathcal{H}_H(\mathbb{R}_+)} = \inf\Big\{|g|_{\mathcal{H}_H(\mathbb{R})}; \ g = h \text{ on } \mathbb{R}_+\Big\},$$

so $|h|_{\mathcal{H}_H(\mathbb{R}_+)} \le |h\,1_{\mathbb{R}_+}|_{\mathcal{H}_H(\mathbb{R})}$ for h defined on $\mathbb{R}_+$. On the other hand,

$$\begin{aligned}
|h|_{\mathcal{H}'_H} &= \big|I_{0+}^{1/2-H}h\big|_{\mathcal{H}_{1/2}(\mathbb{R}_+)} = \big|(I_{0+}^{1/2-H}h)1_{\mathbb{R}_+}\big|_{\mathcal{H}_{1/2}(\mathbb{R})}\\
&= \big|\widetilde{I}_+^{\,H-1/2}\big((I_{0+}^{1/2-H}h)1_{\mathbb{R}_+}\big)\big|_{\mathcal{H}_H(\mathbb{R})} = \big|h\,1_{\mathbb{R}_+}\big|_{\mathcal{H}_H(\mathbb{R})}.
\end{aligned}$$

Thus $|h|_{\mathcal{H}_H(\mathbb{R}_+)} \le |h|_{\mathcal{H}'_H}$, and $\mathcal{H}'_H$ is continuously embedded in $\mathcal{H}_H(\mathbb{R}_+)$. The inverse inclusion means that

$$\big|h\,1_{\mathbb{R}_+}\big|_{\mathcal{H}_H(\mathbb{R})} \le C \,\inf\Big\{|g|_{\mathcal{H}_H(\mathbb{R})}; \ g = h \text{ on } \mathbb{R}_+\Big\},$$

for h defined on $\mathbb{R}_+$, and this is equivalent to

$$\big|g\,1_{\mathbb{R}_+}\big|_{\mathcal{H}_H(\mathbb{R})} \le C\,|g|_{\mathcal{H}_H(\mathbb{R})}$$

for g defined on $\mathbb{R}$; thus this means that $g \mapsto g\,1_{\mathbb{R}_+}$ is a continuous endomorphism of $\mathcal{H}_H(\mathbb{R})$. This is a known analytical result, see also Lemma 1 in [31].

Remark 25. Consider on $\mathbb{R}_+$ the even and odd parts $B_t^{H\pm} = (B_t^H \pm B_{-t}^H)/2$ of B^H. These two processes are independent (this is easily verified by computing the covariance), and $B^H 1_{\mathbb{R}_+} = B^{H+} + B^{H-}$, so their Cameron–Martin spaces $\mathcal{H}_{H\pm}$ are continuously embedded into $\mathcal{H}_H(\mathbb{R}_+)$. On the other hand

$$\begin{aligned}
|h|_{\mathcal{H}_{H\pm}} &= \inf\Big\{|g|_{\mathcal{H}_H(\mathbb{R})}; \ h(t) = \frac{1}{2}(g(t) \pm g(-t)) \text{ on } \mathbb{R}_+\Big\}\\
&\le 2\big|h\,1_{\mathbb{R}_+}\big|_{\mathcal{H}_H(\mathbb{R})} = 2|h|_{\mathcal{H}'_H} \le C|h|_{\mathcal{H}_H(\mathbb{R}_+)}
\end{aligned}$$

by means of the result of Remark 24, so the three spaces $\mathcal{H}_{H\pm}$ and $\mathcal{H}_H(\mathbb{R}_+)$ are equivalent.

Remark 26. Notice that the endomorphism of Remark 24 maps the function $h(t)$ to the function $h(t_+)$; by applying the invariance by time reversal, we deduce that the operator mapping $h(t)$ to $h(1 - (1 - t)_+)$ is also continuous, so by composing these two operators, we see that the operator mapping $h(t)$ to the function

$$h^\star(t) = \begin{cases} 0 & \text{if } t \le 0, \\ h(t) & \text{if } 0 \le t \le 1, \\ h(1) & \text{if } t \ge 1, \end{cases} \tag{76}$$

is a continuous endomorphism of $\mathcal{H}_H$. On the other hand, we have

$$|h|_{\mathcal{H}_H([0,1])} = \inf\{|g|_{\mathcal{H}_H(\mathbb{R})}; \ g = h \text{ on } [0,1]\}.$$

Thus $h \mapsto h^\star$ is continuous from $\mathcal{H}_H([0,1])$ into $\mathcal{H}_H(\mathbb{R})$.

5.3 Equivalence and Mutual Singularity of Laws

In Theorem 19, we have proved that the Cameron–Martin spaces of B^H and X^H are equivalent. It is known that the laws of two centred Gaussian processes are either equivalent, or mutually singular, see Appendix C; the equivalence of Cameron–Martin spaces is necessary for the equivalence of the laws, but is of course not sufficient (compare for instance a standard Brownian motion W_t with $2W_t$). In subsequent results, the equivalence or mutual singularity of laws of processes should be understood by considering these processes as variables with values in the space of continuous functions.

Theorem 20. *Let $0 < H < 1$. For any $S > 0$, the laws of B_t^H and $X_t^{S,H} = X_{S+t}^H - X_S^H$ are equivalent on any time interval $[0, T]$; more precisely, the relative entropies of B^H and $X^{S,H}$ with respect to each other are dominated by S^{2H-2} as $S \uparrow \infty$, and therefore tend to 0; in particular, the total variation distance between the laws of $X^{S,H}$ and B^H is dominated by S^{H-1}. In the case $S = 0$, the two laws are mutually singular as soon as $H \neq 1/2$.*

Proof. Let us consider separately the cases $S > 0$ and $S = 0$.

Equivalence for $S > 0$. Consider the coupling and notations of Theorem 17, so that the process $B_t^H = X_t^H + Y_t^H$ is written as the sum of two independent processes. This implies that $B^{S,H} = X^{S,H} + Y^{S,H}$, where $B^{S,H}$ and $Y^{S,H}$ are defined similarly to $X^{S,H}$. Theorem 19 states that the Cameron–Martin spaces of X^H and B^H are equivalent; this implies that the Cameron–Martin space of $X^{S,H}$ is equivalent to the Cameron–Martin space of $B^{S,H}$ which is $\mathcal{H}_H$, and is therefore also equivalent to $\mathcal{H}_H' = I_{0+}^{H+1/2} L^2(\mathbb{R}_+)$; thus it contains smooth functions taking value 0 at 0. But the perturbation $Y^{S,H}$ is smooth, so the equivalence of the laws of $B^{S,H}$ and $X^{S,H}$ follows from the Cameron–Martin theorem for an independent perturbation. Moreover, (103) yields an estimation of the relative entropies

$$\max\left(\mathcal{I}(B^H, X^{S,H}), \mathcal{I}(X^{S,H}, B^H)\right) \leq \frac{1}{2}\mathbb{E}|Y^{S,H}|_{\mathcal{H}_H}^2 \leq C\,\mathbb{E}|Y^{S,H}|_{\mathcal{H}_H'}^2$$

$$\leq C_T\,\mathbb{E}\left(\sup_{[0,T]}|D^1 Y^{S,H}| + \sup_{[0,T]}|D^2 Y^{S,H}|\right)^2$$

from (74). The derivative $D^k Y_t^H$ is $O(t^{H-k})$ in $L^2(\Omega)$ from the scaling property, so

$$\sup |D^1 Y_t^{S,H}| = \sup |D^1 Y_{S+t}^H| \le |D^1 Y_S^H| + \int_0^T |D^2 Y_{S+t}^H| dt = O(S^{H-1})$$

as $S \uparrow \infty$. The second derivative is even smaller (of order S^{H-2}). Thus the relative entropies are dominated by S^{2H-2}. In particular, the total variation distance is estimated from Pinsker's inequality (102).

Mutual singularity for $S = 0$. This is a consequence of Theorem 37; the two processes are self-similar, the initial σ-algebra $\mathcal{F}_{0+}(B^H)$ is almost surely trivial (Remark 46), so it is sufficient to prove that they do not have the same law. But this is evident since B^H can be written as the sum of X^H and of an independent process Y^H which is not identically zero.   □

Remark 27. In the case $S = 0$, Theorem 36 provides a criterion to decide whether a process Ξ has the law of B^H or X^H. The variances of these two processes differ (they can be computed from the calculation of Appendix B), so we can decide between them by looking at the small time behaviour of $\int_t^1 s^{-2H-1}(\Xi_s)^2 ds$. Actually, by applying the invariance by time inversion, we can also look at the behaviour in large time.

For the following result, we recall that the mutual information of two variables X_1 and X_2 is defined as the entropy of (X_1, X_2) relative to two independent copies of X_1 and X_2. We want to estimate the dependence between the increments of B^H on some interval $[S, S + T]$, $S \ge 0$, and its increments before time 0, and in particular prove that the two processes are asymptotically independent when $S \uparrow +\infty$. This result and other estimates were proved in [31] with a more analytical method; an asymptotic independence result is also given in [33].

Theorem 21. *Let $H \neq 1/2$. The joint law of the two processes $(B_t^{S,H} = B_{S+t}^H - B_S^H; \ 0 \le t \le T)$ and $(B_t^H; \ t \le 0)$ is equivalent to the product of laws as soon as $S > 0$, and the Shannon mutual information is $O(S^{2H-2})$ as $S \uparrow \infty$. If $S = 0$, the joint law and the product of laws are mutually singular.*

Proof. We consider separately the two cases.

Equivalence for $S > 0$. Let $(W_t; \ t \in \mathbb{R})$ and $(\overline{W}_t; \ t \in \mathbb{R})$ be two standard Brownian motions such that $\overline{W}_t = W_t$ for $t \ge 0$ and $(\overline{W}_t; \ t \le 0)$ is independent of W. We then consider the two fractional Brownian motions $B^H = \widetilde{I}_+^{H-1/2} W$ and $\Lambda^H = \widetilde{I}_+^{H-1/2} \overline{W}$. With the notation of Theorem 17, they can be written on $\mathbb{R}_+$ as $B^H = X^H + Y^H$ and $\Lambda^H = X^H + \overline{Y}^H$, so $\Lambda^H = B^H + \overline{Y}^H - Y^H$; by looking at the increments after time S, we have $\Lambda^{S,H} = B^{S,H} + \overline{Y}^{S,H} - Y^{S,H}$.

Conditionally on $\mathcal{F}_0(W, \overline{W}) = \mathcal{F}_0(B^H, \Lambda^H)$, the process $\overline{Y}^{S,H} - Y^{S,H}$ becomes a deterministic process which is almost surely in $\mathcal{H}_H$ (see the proof of Theorem 20), so the conditional laws of

$$(B_t^{S,H}, \, 0 \leq t \leq T; \, B_t^H, \, t \leq 0) \quad \text{and} \quad (\Lambda_t^{S,H}, \, 0 \leq t \leq T; \, B_t^H, \, t \leq 0)$$

are equivalent. We deduce that the unconditional laws are also equivalent. Moreover, the two processes of the right side are independent, and $\Lambda^{S,H} \simeq B^{S,H}$, so the equivalence of laws stated in the theorem is proved. On the other hand, the relative entropies of

$$(B_t^{S,H}, \, 0 \leq t \leq T; \, B_t^H, \, t \leq 0; \, \Lambda_t^H, \, t \leq 0)$$

and

$$(\Lambda_t^{S,H}, \, 0 \leq t \leq T; \, B_t^H, \, t \leq 0; \, \Lambda_t^H, \, t \leq 0)$$

with respect to each other are equal to

$$\frac{1}{2} \mathbb{E}|\overline{Y}^{S,H} - Y^{S,H}|^2_{\mathcal{H}_H} \leq 2\mathbb{E}|Y^{S,H}|^2_{\mathcal{H}_H} = O(S^{2H-2})$$

(proceed as in Theorem 20). If we project on the two first components, we deduce that the mutual information that we are looking for is smaller than this quantity.

Mutual singularity for $S = 0$. If we compare the law of $(B_t^H, B_{-t}^H; \, 0 \leq t \leq T)$ with the law of two independent copies of the fractional Brownian motion, we have two self-similar Gaussian processes with different laws, so the laws are mutually singular from Theorem 37. $\qquad\qquad\square$

Remark 28. As an application, we can compare B^H with its odd and even parts. Let B' and B' be two independent copies of B^H. Let $S > 0$. From Theorem 21, we have on $[0, T]$ the equivalence of laws

$$\left(B_{S+t}^H - B_S^H\right) \pm \left(B_{-S-t}^H - B_{-S}^H\right) \sim (B_{S+t} - B_S) \pm \left(B'_{-S-t} - B'_{-S}\right)$$
$$\simeq \sqrt{2}\left(B_{S+t}^H - B_S^H\right)$$
$$\simeq \sqrt{2}B_t^H.$$

Thus the law of the increments of $(B_t^H \pm B_{-t}^H)/\sqrt{2}$ on $[S, S + T]$ have a law equivalent to the law of B^H. For $S = 0$, the Cameron–Martin spaces are equivalent (Remark 25), but the laws can be proved to be mutually singular from Theorem 37.

6 Series Expansions

Let us try to write B^H on $[0, 1]$ as some series of type

$$B_t^H = \sum_n h_n(t) \xi_n$$

where h_n are deterministic functions and ξ_n are independent standard Gaussian variables. Such expansions have been described in the standard case $H = 1/2$ by [19], and actually, an expansion valid for the standard Brownian motion W can be transported to B^H by means of the operator $G_{0+}^{1/2,H}$, see [12].

If we look more precisely for a trigonometric expansion, we can apply [9] where the functions h_n are trigonometric functions, the coefficients of which are related to some Bessel function depending on H. However, we are here more interested in trigonometric functions which do not depend on H.

6.1 A Trigonometric Series

Suppose that we are interested in the Fourier series of $(B_t^H; \ 0 \le t \le 1)$. The problem is that the Fourier coefficients are not independent, since this property is already known to be false for $H = 1/2$. What is known for $H = 1/2$ is that W_t can be represented by means of (8), (9) or (10) for independent standard Gaussian variables $(\xi_n, \xi_n'; n \ge 1)$; the series converges in $L^2(\Omega)$, uniformly in t, and one easily deduces the Fourier series of W from (8). Similar representations cannot hold on $[0, 1]$ for the fractional Brownian motion as soon as $H \ne 1/2$, but it appears that one can find a representation mixing (8) and (9),

$$B_t^H \simeq a_0^H \xi_0 t + \sum_{n \ge 1} a_n^H \left((\cos(\pi n t) - 1) \xi_n + \sin(\pi n t) \xi_n' \right) \tag{77}$$

on $[0, 1]$. This question has been studied in [18] and [17] respectively for the cases $H < 1/2$ and $H > 1/2$. The sign of a_n^H is of course irrelevant so we will choose $a_n^H \ge 0$. We follow a general technique for finding series expansions of Gaussian processes from series expansions of their covariance kernels. We are going to find all the possible a_n^H for which (77) holds; it appears that a_n^H, $n \ge 1$, is unique as soon as a_0^H has been chosen in some set of possible values.

Theorem 22. *It is possible to find a sequence $(a_n^H; \ n \ge 0)$, $a_n^H \ge 0$, such that $\sum (a_n^H)^2 < \infty$ and (77) holds on $[0, 1]$ for independent standard Gaussian variables $(\xi_0, \xi_n, \xi_n'; n \ge 1)$. The convergence of the series holds uniformly in t, almost surely. If $H \le 1/2$, we have to choose a_0^H in an interval $[0, a(H)]$, $a(H) > 0$, and a_n^H is then uniquely determined; if $H > 1/2$ there is only one choice for the* *sequence. Moreover, except in the case $H = 1/2$, we must have $a_n^H \neq 0$ for all large enough n. If $H \neq 1/2$, then (77) cannot hold on $[0, T]$ for $T > 1$.*

Proof. We divide the proof into two parts.

Step 1: Study on $[0, 1]$. It is clear that the convergence of the series in (77) holds for t fixed (almost surely and in $L^2(\Omega)$); the uniform convergence comes from the Itô-Nisio theorem [19]. We have to verify that the right hand side Z has the same covariance kernel as B^H for a good choice of (a_n^H). We have

$$
\mathbb{E}[Z_s Z_t]
$$

$$
= \left(a_0^H\right)^2 st + \sum_{n \geq 1} \left(a_n^H\right)^2 ((\cos(\pi nt) - 1)(\cos(\pi ns) - 1)
$$

$$
+ \sin(\pi nt)\sin(\pi ns))
$$

$$
= \left(a_0^H\right)^2 st + \sum_{n \geq 1} \left(a_n^H\right)^2 (\cos(\pi n(t - s)) - \cos(\pi nt) - \cos(\pi ns) + 1)
$$

$$
= (f_H(t) + f_H(s) - f_H(t - s))/2
$$

with

$$
f_H(t) = \left(a_0^H\right)^2 t^2 + 2\sum_{n \geq 1} \left(a_n^H\right)^2 (1 - \cos(\pi nt)). \tag{78}
$$

If we compare this expression with (2), it appears that if f_H coincides on $[-1, 1]$ with $g_H(t) = \rho, |t|^{2H}$, then $B^H \simeq Z$ on $[0, 1]$; conversely, if $B^H \simeq Z$, then they have the same variance, so $f_H = g_H$ on $[0, 1]$ and therefore on $[-1, 1]$ (the two functions are even). Thus finding an expansion (77) on $[0, 1]$ is equivalent to finding coefficients a_n^H so that $f_H = g_H$ on $[-1, 1]$. For any choice of a_0^H, one has on $[-1, 1]$ the Fourier decomposition

$$
\rho|t|^{2H} - (a_0^H)^2 t^2 = b_0^H - 2\sum_{n \geq 1} b_n^H \cos(\pi nt).
$$

Thus the possible expansions correspond to the possible choices of a_0^H such that $b_n^H \geq 0$ for $n \geq 1$ and $\sum b_n^H < \infty$; then

$$
\rho|t|^{2H} - (a_0^H)^2 t^2 = 2\sum_{n \geq 1} b_n^H (1 - \cos(\pi nt))
$$

and we take $a_n^H = \sqrt{b_n^H}$ for $n \geq 1$. We have

$$
b_n^H = -\rho \int_0^1 t^{2H} \cos(\pi nt)dt + (a_0^H)^2 \int_0^1 t^2 \cos(\pi nt)dt
$$

$$
= \frac{2H}{\pi n}\rho \int_0^1 t^{2H-1} \sin(\pi nt)dt - \frac{2(a_0^H)^2}{\pi n} \int_0^1 t \sin(\pi nt)dt
$$

$$
= -\frac{2H(2H-1)}{\pi^2 n^2}\rho \int_0^1 t^{2H-2}\left(1 - \cos(\pi n t)\right)dt
$$

$$
+ \frac{2H}{\pi^2 n^2}\rho\left(1 - (-1)^n\right) + \frac{2(a_0^H)^2}{\pi^2 n^2}(-1)^n. \tag{79}
$$

Let us first assume $H < 1/2$; then the first term is positive, and the sum of the second and third terms is nonnegative as soon as $a_0^H \le \sqrt{2\rho H}$. Moreover

$$
cn^2 \int_0^{1/n} t^{2H}\,dt \le \int_0^1 t^{2H-2}\left(1 - \cos\left(\pi n t\right)\right)dt \le Cn^2
$$

$$
\times \int_0^{1/n} t^{2H}\,dt + 2\int_{1/n}^\infty t^{2H-2}\,dt \tag{80}
$$

so this integral is of order n^{1-2H} (actually a more precise estimate will be proved in Theorem 23), and we have $b_n^H \asymp n^{-1-2H}$. It is then not difficult to deduce that there exists a maximal $a(H) \ge \sqrt{2\rho H}$ such that if we choose a_0^H in $[0, a(H)]$, then $b_n^H \ge 0$ for any n; the value $a(H)$ is attained when one of the coefficients b_n^H becomes 0. It follows from $b_n^H \asymp n^{-1-2H}$ that $\sum b_n^H < \infty$. Let us now assume $H = 1/2$; the property $b_n^H \ge 0$ holds for $a_0^{1/2} \in [0, a(1/2)] = [0, 1]$, and $b_n^{1/2} = O(n^{-2})$. Finally, if $H > 1/2$,

$$
\begin{aligned}
b_n^H &= \frac{2H(2H-1)}{\pi^2 n^2}\rho \int_0^1 t^{2H-2}\cos(\pi n t)dt + \frac{2(a_0^H)^2 - 2\rho H}{\pi^2 n^2}(-1)^n \\
&= -\frac{2H(2H-1)(2H-2)}{\pi^3 n^3}\rho \int_0^1 t^{2H-3}\sin(\pi n t)dt + \frac{2(a_0^H)^2 - 2\rho H}{\pi^2 n^2}(-1)^n \\
&= \frac{2H(2H-1)(2H-2)(2H-3)}{\pi^4 n^4}\rho \int_0^1 t^{2H-4}\left(1 - \cos(\pi n t)\right)dt \\
&\quad - \frac{2H(2H-1)(2H-2)}{\pi^4 n^4}\rho\left(1 - (-1)^n\right) + \frac{2(a_0^H)^2 - 2\rho H}{\pi^2 n^2}(-1)^n. \tag{81}
\end{aligned}
$$

The integral of the last equality is studied like (80), and is of order n^{3-2H}, so the first term of this last equality is positive and of order n^{-1-2H}. The second term is nonnegative and smaller. If we choose $a_0^H \ne \sqrt{\rho H}$, then the third term has an alternating sign and is the dominant term, so b_n^H is not always positive. Thus we must choose $a_0^H = \sqrt{\rho H}$, and $b_n^H > 0$ for any n; we again have $b_n^H \asymp n^{-1-2H}$ so that $\sum b_n^H < \infty$. Moreover, in the two cases $H < 1/2$ and $H > 1/2$, we have $a_n^H \asymp n^{-H-1/2}$, so $a_n^H \ne 0$ for all large enough n.

Step 2: Study on larger intervals. Suppose now that (77) holds on $[0, T]$ for some $T > 1$. Then, as in previous step, we should have $f_H(t) = g_H(t) = \rho|t|^{2H}$ on $[-T, T]$. But $f_H(t) - (a_0^H)^2 t^2$ is even and 2-periodic, so

$$
f_H(1 - t) - (a_0^H)^2(1 - t)^2 = f_H(1 + t) - (a_0^H)^2(1 + t)^2.
$$

Thus

$$\rho(1-t)^{2H} - (a_0^H)^2(1-t)^2 = \rho(1+t)^{2H} - (a_0^H)^2(1+t)^2$$

for $|t| \leq \min(T-1, 1)$. By differentiating twice, it appears that this relation is false if $H \neq 1/2$. $\qquad\square$

Remark 29. For $H = 1/2$, we can choose $a_0^{1/2}$ in $[0, 1]$, and the expansion (77) is an interpolation between the decompositions containing respectively only odd terms $(a_0^{1/2} = 0)$ and only even terms $(a_0^{1/2} = 1)$, which are respectively (9) and (8).

Remark 30. Suppose that $H \leq 1/2$ with $a_0^H = 0$; the formula (77) defines a Gaussian process on the torus $\mathbb{R}/2\mathbb{Z}$ with covariance kernel

$$\mathbb{E}[B_t^H B_s^H] = \frac{\rho}{2}\left(\delta(0,t)^{2H} + \delta(0,s)^{2H} - \delta(s,t)^{2H}\right) \qquad (82)$$

for the distance δ on the torus. This is the fractional Brownian motion of [18] indexed by the torus. For $H > 1/2$, we cannot take $a_0^H = 0$; this is related to the fact proved in [18], that the fractional Brownian motion on the torus does not exist; when indeed such a process exists, we deduce from (82) that

$$\mathbb{E}[B_t^H(B_{1+t}^H - B_1^H)] = \rho\left((1-t)^{2H} - 1\right) \sim -2\rho H t$$

as $t \downarrow 0$ (use the fact $\delta(1+t,0) = 1-t$ on the torus), whereas this covariance should be dominated by t^{2H}.

Remark 31. When $H \leq 1/2$ and $a_0^H = 0$, we can write B_t^H on $[0,1]$ as $\overline{A}_t^H - \overline{A}_0^H$ for the stationary process $\overline{A}_t^H = \sum a_n^H(\cos(\pi n t)\xi_n + \sin(\pi n t)\xi_n')$. In the case $H = 1/2$, it generates the same σ-algebra as $B^{1/2}$, and this process coincides with the process $A^{1/2}$ of Theorem 12. However, a comparison of the variances of the two processes show that they are generally different when $H < 1/2$.

Remark 32. Since the two sides of (77) have stationary increments, we can replace the time intervals $[0,1]$ and $[0,T]$ of Theorem 22 by other intervals of length 1 and T containing 0.

We now study the asymptotic behaviour of the coefficients a_n^H of Theorem 22.

Theorem 23. *The expansion of Theorem 22 can be written with $a_0^H = \sqrt{\rho H}$. In this case, $a_n^H > 0$ for any n and*

$$a_n^H = (\pi n)^{-H-1/2}\left(1 + O(n^{2H-3})\right) \qquad (83)$$

for n large.

Proof. The only part which has still to be proved is (83). This will be accomplished through an asymptotic analysis of the integrals in (79) and (81). For $H = 1/2$ we have $a_n^H = (\pi n)^{-1}$ so this is trivial. If $H < 1/2$, we have

$$(1 - 2H) \int_0^1 t^{2H-2}(1 - \cos(\pi nt))dt = (1 - 2H)$$

$$\times \int_0^\infty t^{2H-2}(1 - \cos(\pi nt))dt - 1 + (1 - 2H) \int_1^\infty t^{2H-2} \cos(\pi nt)dt$$

$$= (1 - 2H)(\pi n)^{1-2H} \int_0^\infty t^{2H-2}(1 - \cos t)dt - 1$$

$$+ (1 - 2H)(\pi n)^{1-2H} \int_{\pi n}^\infty t^{2H-2} \cos t\, dt$$

$$= (\pi n)^{1-2H} \int_0^\infty t^{2H-1} \sin t\, dt - 1 + O(n^{-2}) \tag{84}$$

where we have used in the last equality

$$\left| \int_{\pi n}^\infty t^{2H-2} \cos t\, dt \right| = (2 - 2H) \left| \int_{\pi n}^\infty t^{2H-3} \sin t\, dt \right|$$

$$= (2 - 2H) \left| \sum_{k \geq n} \int_{\pi k}^{\pi(k+1)} t^{2H-3} \sin t\, dt \right|$$

$$\leq (2 - 2H) \left| \int_{\pi n}^{\pi(n+1)} t^{2H-3} \sin t\, dt \right| = O(n^{2H-3}) \tag{85}$$

(this is an alternating series). By applying (34), we deduce that

$$(1-2H)\int_0^1 t^{2H-2}(1 - \cos(\pi nt))\, dt = (\pi n)^{1-2H} \Gamma(2H) \sin(\pi H) - 1 + O\left(n^{-2}\right),$$

so (79) with $a_0^H = \sqrt{\rho H}$ implies

$$b_n^H = \rho(\pi n)^{-1-2H} \Gamma(2H + 1) \sin(\pi H) + O(n^{-4}). \tag{86}$$

Similarly, if $H > 1/2$, then (85) again holds true and

$$\int_0^1 t^{2H-2} \cos(\pi nt)dt = (\pi n)^{1-2H} \int_0^\infty t^{2H-2} \cos t\, dt + O(n^{-2})$$

$$= (\pi n)^{1-2H} \Gamma(2H - 1) \sin(\pi H) + O(n^{-2})$$

and we deduce from (81) that we again have (86). By using our choice of ρ given in (51), we obtain in both cases

$$b_n^H = -2\frac{\Gamma(-2H)\Gamma(2H+1)}{\pi^{2H+2}n^{2H+1}}\cos(\pi H)\sin(\pi H)(1 + O(n^{2H-3}))$$
$$= (\pi n)^{-2H-1}(1 + O(n^{2H-3}))$$

from (95). We deduce (83) by taking the square root. $\qquad\square$

Remark 33. Considering the expansion (77) for $a_0^H = \sqrt{\rho H}$, replacing B^H by the process

$$\check{B}_t^H = c\xi_0 t + \sum_{n\geq 1}(\pi n)^{-H-1/2}\left((\cos(\pi nt) - 1)\xi_n + \sin(\pi nt)\xi_n'\right)$$

for $c > 0$ is equivalent to multiplying ξ_0 by c/a_0^H and (ξ_n, ξ_n') by some $(1 + O(n^{2H-3}))$ which remains strictly positive. We can compare the laws of these two sequences of independent Gaussian variables by means of Kakutani's criterion (Theorem 34), and it appears that the laws of these two sequences are equivalent ($\sum n^{4H-6} < \infty$). Thus the laws of B^H and $\check{B}^H$ are equivalent on $[0, 1]$. This implies that the law of $2^{-H}\check{B}_{2t}^H$ is equivalent on $[0, 1/2]$ to the law of B_t^H; actually, we will prove in Theorem 27 that these two laws are equivalent on $[0, T]$ for any $T < 1$.

6.2 Approximate Expansions

We now consider the processes

$$\widehat{B}_t^H = \xi_0 t + \sqrt{2}\sum_{n\geq 1}\left(\xi_n\frac{\cos(2n\pi t) - 1}{(2n\pi)^{H+1/2}} + \xi_n'\frac{\sin(2n\pi t)}{(2n\pi)^{H+1/2}}\right),$$
$$\overline{B}_t^H = \sqrt{2}\sum_{n\geq 0}\left(\xi_n\frac{\cos((2n+1)\pi t) - 1}{((2n+1)\pi)^{H+1/2}} + \xi_n'\frac{\sin((2n+1)\pi t)}{((2n+1)\pi)^{H+1/2}}\right) \qquad (87)$$

on $[0, 1]$. Notice that $\widehat{B}^{1/2} \simeq \overline{B}^{1/2} \simeq W$ from (8) and (9). On the other hand, it follows from Theorem 22 that $\widehat{B}^H \not\simeq B^H$ and $\overline{B}^H \not\simeq B^H$ for $H \neq 1/2$ (because one should have $a_n^H \neq 0$ in the expansion (77) of B^H for all large enough n), but we are going to check that these two processes have a local behaviour similar to B^H. The advantage with respect to the exact expansion (77) is that the sequence of random coefficients and the process will generate the same σ-algebra. Then we will apply these approximations to some properties of the Cameron–Martin space $\mathcal{H}_H$ (Sect. 6.3), and to some equivalence of laws (Sect. 6.4). As it was the case for Riemann–Liouville processes, $\widehat{B}^H$ and $\overline{B}^H$ are not only defined for $0 < H < 1$, but also for any $H > 0$.

Let us compare $\widehat{B}^H$ and $\overline{B}^H$ with B^H for $0 < H < 1$. We use the operators $\widehat{I}_+^{\alpha}$ and $\overline{I}_+^{\alpha}$ defined in (36) and (37). By projecting on the Gaussian spaces generated by ξ_n and ξ'_n and by applying (35), we can write

$$
\widehat{I}_+^{1/2-H} \widehat{B}_t^H
$$
$$
= \xi_0 t + \sqrt{2} \sum_{n \geq 1} \left(\xi_n \frac{\cos(2\pi nt + (H-1/2)\pi/2) - \cos((H-1/2)\pi/2)}{2\pi n} \right.
$$
$$
\left. + \xi'_n \frac{\sin(2\pi nt + (H-1/2)\pi/2) - \sin((H-1/2)\pi/2)}{2\pi n} \right).
\tag{88}
$$

The two expressions (8) and (88) are related to each other by applying a rotation on the vectors (ξ_n, ξ'_n), so $\widehat{I}_+^{1/2-H} \widehat{B}^H$ and W have the same law. A similar property holds for $\overline{I}_+^{1/2-H} \overline{B}^H$, and we can therefore write

$$
\widehat{B}^H \simeq \widehat{I}_+^{H-J} \widehat{B}^J, \quad \overline{B}^H \simeq \overline{I}_+^{H-J} \overline{B}^J, \quad \widehat{B}^{1/2} \simeq \overline{B}^{1/2} \simeq W.
\tag{89}
$$

We can give an extension of Theorem 17.

Theorem 24. *It is possible to realise jointly the processes B^H, X^H, $\overline{B}^H$ and $\widehat{B}^H$ so that the differences $B^H - X^H$, $\overline{B}^H - B^H$ and $\widehat{B}^H - B^H$ are C^∞ on $(0, 1]$; moreover, the derivatives of order k of these differences are $O(t^{H-k})$ in $L^2(\Omega)$ as $t \downarrow 0$.*

Proof. We consider the coupling $B^H = \widetilde{I}_+^{H-1/2} W$, $X^H = I_{0+}^{H-1/2} W$, $\overline{B}^H = \overline{I}_+^{H-1/2} W$ and $\widehat{B}^H = \widehat{I}_+^{H-1/2} W$ for the same W on $\mathbb{R}$. The smoothness of $B^H - X^H$ is proved in Theorem 17, and the estimation of the derivatives follows by a scaling argument. On the other hand, let W_t^1 be equal to $W_t - W_1 t$ on $[0, 1]$, extend it to $\mathbb{R}$ by periodicity, and define $W_t^2 = W_{-t}^1$ for $t \geq 0$. Then, with the notation (73),

$$
\widehat{B}_t^H = W_1 t + I_{0+}^{H-1/2}(W_t - W_1 t) + I_{\triangle}^{H-1/2} W_t^2
$$
$$
= X_t^H + W_1 \left(t - \Gamma(H + 3/2)^{-1} t^{H+1/2} \right) + I_{\triangle}^{H-1/2} W_t^2
$$

The smoothness of $\widehat{B}^H - X^H$ follows; the process W_t^2 is dominated in $L^2(\Omega)$ by $\min(\sqrt{t}, 1)$, so we deduce from (73) that

$$
\left\| D^k I_{\triangle}^{H-1/2} W_t^2 \right\|_2 \leq C \int_0^\infty (t + s)^{H-k-3/2} \sqrt{s} \, ds = C' t^{H-k}
$$

for $k \geq 1$. The study of $\overline{B}^H$ is similar; let W^3 be the process W on $[0, 1]$ extended to $\mathbb{R}$ so that the increments are 1-antiperiodic, and let $W_t^4 = W_{-t}^3$; then $\overline{B}^H$ is equal to $X^H + I_\Delta^{H-1/2}W^4$; the end of the proof is identical. $\qquad\square$

6.3 Application to the Cameron–Martin Space

Let $\widehat{\mathcal{H}}_H$ and $\overline{\mathcal{H}}_H$ be the Cameron–Martin spaces of $\widehat{B}^H$ and $\overline{B}^H$ on the time interval $[0, 1]$. It follows from (89) that $\widehat{\mathcal{H}}_{1/2} = \overline{\mathcal{H}}_{1/2} = \mathcal{H}_{1/2}$, and $\widehat{\mathcal{H}}_H = \widehat{I}_+^{H-J}\widehat{\mathcal{H}}_J$ as well as $\overline{\mathcal{H}}_H = \overline{I}_+^{H-J}\overline{\mathcal{H}}_J$.

Theorem 25. *For $0 < H < 1$, the spaces $\widehat{\mathcal{H}}_H$, $\overline{\mathcal{H}}_H$ and $\mathcal{H}_H$ are equivalent on $[0, 1]$.*

Proof. We compare successively $\widehat{\mathcal{H}}_H$ and $\overline{\mathcal{H}}_H$ with $\mathcal{H}_H([0, 1])$, and use the properties of this last space described in Remark 26.

Proof of $\widehat{\mathcal{H}}_H \sim \mathcal{H}_H$. We know that $\widehat{\mathcal{H}}_H = \widehat{I}_+^{H-1/2}\mathcal{H}_{1/2}$, so it is sufficient to establish that $\widehat{I}_+^{H-1/2}$ is a homeomorphism from $\mathcal{H}_{1/2}([0, 1])$ onto $\mathcal{H}_H([0, 1])$. To this end, we are going to prove that $\widehat{I}_+^{H-J}$ is continuous from $\mathcal{H}_J([0, 1])$ into $\mathcal{H}_H([0, 1])$ for $0 < J, H < 1$. Consider a function h of $\mathcal{H}_J([0, 1])$, consider $h_0(t) = h(t) - h(1)t$, and extend it by periodicity. Then h_0 is generally not in $\mathcal{H}_J(\mathbb{R})$, but the operator $h \mapsto h_1 = h_0 1_{(-1,1]}$ is continuous from $\mathcal{H}_J([0, 1])$ into $\mathcal{H}_J(\mathbb{R})$. Moreover, the operator $h \mapsto h_2 = h_0 1_{(-\infty,-1]}$ is continuous from $\mathcal{H}_J([0, 1])$ into the space $L^\infty((-\infty, -1])$ of bounded functions supported by $(-\infty, -1]$. On the other hand, it is known that $\mathcal{H}_H = \widetilde{I}_+^{H-J}\mathcal{H}_J$ on $\mathbb{R}$, and $\widetilde{I}_+^{H-J}$ also maps continuously $L^\infty((-\infty, -1])$ into the space of smooth functions on $[0, 1]$, and therefore into $\mathcal{H}_H([0, 1])$. Thus $h \mapsto \widetilde{I}_+^{H-J}h_0 = \widetilde{I}_+^{H-J}h_1 + \widetilde{I}_+^{H-J}h_2$ is continuous from $\mathcal{H}_J([0, 1])$ into $\mathcal{H}_H([0, 1])$. If we add the operator $h \mapsto (h(1)t)$ which is also continuous, we can conclude.

Proof of $\overline{\mathcal{H}}_H \sim \mathcal{H}_H$. In this case, we let h_0 be the function h on $[0, 1]$, extended to $\mathbb{R}$ so that the increments are 1-antiperiodic. We then consider $h_1 = h_0 1_{(-2,1]}$ and $h_2 = h_0 1_{(-\infty,-2]}$. The proof is then similar, except that we do not have the term $h(1)t$ in this case. $\qquad\square$

Remark 34. In view of (7), a function h is in the space $\mathcal{H}_H(\mathbb{R})$ if its derivative D^1h (in distribution sense if $H < 1/2$) is in the homogeneous Sobolev space of order $H - 1/2$ (see for instance [31]); similarly, it follows from (87) that h is in $\widehat{\mathcal{H}}_H$ is D^1h is in the Sobolev space of order $H - 1/2$ of the torus $\mathbb{R}/\mathbb{Z}$. Thus the equivalence $\widehat{\mathcal{H}}_H \sim \mathcal{H}_H$ of Theorem 25 means that the Sobolev space on the torus is equivalent to the restriction to $[0, 1]$ of the Sobolev space on $\mathbb{R}$. This classical result is true because we deal with Sobolev spaces of order in $(-1/2, 1/2)$.

Remark 35. We have from Theorems 19 and 25 that $\mathcal{H}_H \sim \mathcal{H}'_H \sim \widehat{\mathcal{H}}_H \sim \overline{\mathcal{H}}_H$ for any $0 < H < 1$. Notice however that the comparison for instance of $\widehat{\mathcal{H}}_H$ and $\mathcal{H}'_H$ cannot be extended to the case $H > 1$; in this case indeed, functions of $\mathcal{H}'_H$ satisfy $D^1 h(0) = 0$, contrary to functions of $\widehat{\mathcal{H}}_H$.

Let us now give an immediate corollary of Theorem 25.

Theorem 26. *The sets of functions on* $[0, 1]$

$$t, \quad n^{-H-1/2}\left(1 - \cos(2n\pi t)\right), \quad n^{-H-1/2}\sin(2n\pi t),$$

and

$$n^{-H-1/2}\left(1 - \cos((2n+1)\pi t)\right), \quad n^{-H-1/2}\sin((2n+1)\pi t),$$

form two Riesz bases of $\mathcal{H}_H$. *A function h is in* $\mathcal{H}_H$ *is and only if it has the Fourier expansion*

$$h(t) - h(1)t = \sum_{n \geq 0} \alpha_n \cos(2\pi n t) + \sum_{n \geq 1} \beta_n \sin(2\pi n t)$$

with

$$\sum n^{2H+1}\left(\alpha_n^2 + \beta_n^2\right) < \infty.$$

6.4 Equivalence and Mutual Singularity of Laws

We now compare the laws of B^H, $\widehat{B}^H$ and $\overline{B}^H$ viewed as variables with values in the space of continuous functions.

Theorem 27. *Let* $H \neq 1/2$. *The laws of the processes* $\widehat{B}^H$, $\overline{B}^H$ *and* B^H *are equivalent on the time interval* $[0, T]$ *if* $T < 1$, *and are mutually singular if* $T = 1$.

Proof. We compare the laws of B^H and $\widehat{B}^H$. The study of $\overline{B}^H$ is similar.

Proof of the equivalence for $0 < T < 1$. The increments of both processes are stationary, so let us study the equivalence of $\widehat{B}_t^{S,H} = \widehat{B}_{S+t}^H - \widehat{B}_S^H$ and $B_t^{S,H} = B_{S+t}^H - B_S^H$ on $[0, T]$ for $S = 1 - T$. From Theorem 24, we can couple B^H and $\widehat{B}^H$ so that the difference is smooth on $\mathbb{R}_+^*$. Consequently, $\widehat{B}^{S,H} - B^{S,H}$ is smooth on $[0, T]$, so it lives in $\mathcal{H}_H$. Moreover, we have proved in Theorem 25 that the Cameron–Martin spaces of $\widehat{B}^H$ and B^H are equivalent, so the same is true for the Cameron–Martin spaces of $\widehat{B}^{S,H}$ and $B^{S,H}$. The equivalence of laws then follows from Theorem 35.

Proof of the mutual singularity for $T = 1$. Consider $\widehat{B}^H$ on $\mathbb{R}$. Our aim is to prove that the laws of the two processes

$$(B_t^H, B_1^H - B_{1-t}^H) \quad \text{and} \quad \left(\widehat{B}_t^H, \widehat{B}_1^H - \widehat{B}_{1-t}^H\right) = \left(\widehat{B}_t^H, -\widehat{B}_{-t}^H\right) \simeq \left(\widehat{B}_{2t}^H - \widehat{B}_t^H, \widehat{B}_t^H\right)$$

are mutually singular on the time interval $[0, 1/4]$. The law of the first process is equivalent to a couple $(B_t^{H,1}, B_t^{H,2})$ of two independent fractional Brownian motions (see Theorem 21), and $\mathcal{F}_{0+}(B^{H,1}, B^{H,2})$ is almost surely trivial. On the other hand, from the first part of this proof, the law of the second process is equivalent to the law of $(B_{2t}^H - B_t^H, B_t^H)$. We therefore obtain two self-similar processes which do not have the same law, so we deduce from Theorem 37 that the laws are mutually singular.   □

Remark 36. It follows from Remark 33 that the law of B^H is equivalent on $[0, 1]$ to the law of $(\widehat{B}^H + \overline{B}^H)/\sqrt{2}$, where $\widehat{B}^H$ and $\overline{B}^H$ are independent. We have now proved that this law is equivalent separately to the laws of $\widehat{B}^H$ and $\overline{B}^H$, but only on $[0, T]$ for $T < 1$.

Theorem 28. *Let $T > 0$. The distance in total variation between the laws of the processes $(\varepsilon^{-H}\widehat{B}_{\varepsilon t}^H; 0 \leq t \leq T)$ and $(B_t^H; 0 \leq t \leq T)$ is $O(\varepsilon^{1-H})$ as $\varepsilon \downarrow 0$. The process $\overline{B}^H$ satisfies the same property.*

Proof. As in Theorem 27, let us compare the laws of $\widehat{B}^{1/2,H}$ and $B^{1/2,H}$ on $[0, \varepsilon T]$ for $0 < \varepsilon \leq 1/(2T)$. It follows from Theorem 35 that the entropy $\mathcal{I}$ of the former process relative to the latter one satisfies

$$\mathcal{I} \leq C \, \mathbb{E} \big| \widehat{B}^{1/2,H} - B^{1/2,H} \big|_{\mathcal{H}_H([0,\varepsilon T])}^2 .$$

More precisely it is stated in Theorem 35 that the constant C involved in this domination property depends only on the constants involved in the injections of the Cameron–Martin spaces of $\widehat{B}^{1/2,H}$ and $B^{1/2,H}$ on $[0, \varepsilon T]$ into each other; but if we choose a constant which is valid for $\widehat{B}^H$ and B^H the time interval $[0, 1]$ (Theorem 25), then it is also valid for $\widehat{B}^{1/2,H}$ and $B^{1/2,H}$ on $[0, 1/2]$, and therefore on the subintervals $[0, \varepsilon T]$, $0 < \varepsilon \leq 1/(2T)$, so we can choose C not depending on ε. Thus

$$\mathcal{I} \leq C \, \mathbb{E} \big| \widehat{B}^{1/2,H} - B^{1/2,H} \big|_{\mathcal{H}'_H([0,\varepsilon T])}^2 = O(\varepsilon^{2-2H})$$

from (74). The convergence in total variation and the speed of convergence are deduced from (102). The proof for $\overline{B}^H$ is similar.   □

Remark 37. We can say that the processes $\overline{B}^H$ and $\widehat{B}^H$ are asymptotically fractional Brownian motions near time 0. The processes $\overline{B}^H$, $\widehat{B}^H$ and B^H have stationary increments, so the same local property holds at any time.

As an application, we recover a result of [4], see also [2, 37] for more general results. Notice that the equivalence stated in the following theorem may hold even when the paths of B_2^H are not in $\mathcal{H}_J$.

Theorem 29. *Let B_1^J and B_2^H be two independent fractional Brownian motions with indices $J < H$, and let $T > 0$. Then the laws of $(B_1^J + \lambda B_2^H; \lambda \geq 0)$ are pairwise equivalent on $[0, T]$ if $H > J + 1/4$. Otherwise, they are pairwise mutually singular.*

Proof. It is sufficient to prove the result for $T = 1$.

Equivalence for $H - J > 1/4$. Let us prove that the laws of B_1^J and $B_1^J + \lambda B_2^H$, are equivalent. From Theorems 22 and 23, the process B_1^J can be written as (77) for independent standard Gaussian variables (ξ_n, ξ_n') and coefficients a_n^J such that $a_n^J \neq 0$ for any n. The process B_2^H can be written similarly with coefficients a_n^H and variables (η_n, η_n'). Thus $B_1^J + \lambda B_2^H$ is the image by some functional of the sequence

$$U_n^\lambda = a_n^J (\xi_n, \xi_n') + \lambda a_n^H (\eta_n, \eta_n'),$$

and it is sufficient to prove that the laws of U_n^λ and U_n^0 are equivalent. This can be done by means of Kakutani's criterion (Theorem 34) with $\sigma_n^2 = (a_n^J)^2$ and $\bar{\sigma}_n^2 = (a_n^J)^2 + \lambda^2 (a_n^H)^2$. But

$$\sum_{n \geq 1} \left(\frac{\lambda^2 (a_n^H)^2}{(a_n^J)^2} \right)^2 \leq C \sum_{n \geq 1} n^{4(J-H)} < \infty$$

from Theorem 23.

Mutual singularity for $0 < H - J \leq 1/4$. Let us use the coupling

$$B_1^J = G_{0+}^{1/2,J} W_1, \quad B_2^H = \widetilde{I}_+^{H-1/2} W_2, \quad X_2^K = I_{0+}^{K-1/2} W_2, \quad \widehat{B}_2^K = \widehat{I}_+^{K-1/2} W_2$$

$(0 < K < 1)$, for independent W_1 on $\mathbb{R}_+$ and W_2 on $\mathbb{R}$. By applying the operator $G_{0+}^{J,1/2}$, we can write

$$
\begin{aligned}
G_{0+}^{J,1/2} \left(B_1^J + \lambda B_2^H \right) &= W_1 + \lambda G_{0+}^{J,1/2} B_2^H \\
&= W_1 + \lambda \left((G_{0+}^{J,1/2} - I_{0+}^{1/2-J}) B_2^H + I_{0+}^{1/2-J} (B_2^H - X_2^H) \right. \\
&\quad \left. + X_2^{1/2+H-J} - \widehat{B}_2^{1/2+H-J} \right) + \lambda \, \widehat{B}_2^{1/2+H-J}.
\end{aligned}
\tag{90}
$$

Let us now prove that the process inside the big parentheses lives in $\mathcal{H}_{1/2}$. We have checked in the proof of Theorem 18 that $(G_{0+}^{J,1/2} - I_{0+}^{1/2-J}) f$ is differentiable on

$\mathbb{R}_+^*$ for any f in $\mathbb{H}^{J-}$, so in particular for $f = B_2^H$; the scaling property then enables to prove that the derivative is $O(t^{H-J-1/2})$, so $(G_{0+}^{J,1/2} - I_{0+}^{1/2-J})B_2^H$ is in $\mathcal{H}_{1/2}$. Similarly, $B_2^H - X_2^H$ is smooth, so $I_{0+}^{1/2-J}(B_2^H - X_2^H)$ is also smooth, and we deduce from the same scaling property that it is in $\mathcal{H}_{1/2}$. Finally $X_2^{1/2+H-J} - \widehat{B}_2^{1/2+H-J}$ is also in $\mathcal{H}_{1/2}$ from Theorem 24. Thus we deduce that the process of (90) is obtained from $W_1 + \lambda \, \widehat{B}_2^{1/2+H-J}$ by means of a perturbation which lives in $\mathcal{H}_{1/2}$ and is independent of W_1, so the two laws are equivalent. It is then sufficient to prove that the laws of $W_1 + \lambda_i \, \widehat{B}_2^{1/2+H-J}$ for $\lambda_1 \neq \lambda_2$ are mutually singular. But these two processes can be expanded on the basis $(t, 1-\cos(2\pi n t), \sin(2\pi n t))$; the coefficients are independent with positive variance; the variance of the coefficients on $1 - \cos(2\pi n t)$ and $\sin(2\pi n t)$ is equal to $2(2\pi n)^{-2} + 2\lambda_i^2(2\pi n)^{-2(H-J+1)}$. As in the first step, we can apply Kakutani's criterion (Theorem 34) and notice that

$$\sum_{n \geq 1} \left(\frac{(\lambda_2^2 - \lambda_1^2)(2\pi n)^{-2(H-J+1)}}{(2\pi n)^{-2} + \lambda_1^2(2\pi n)^{-2(H-J+1)}} \right)^2 = \infty$$

so that the two laws are mutually singular. □

Remark 38. For $H > J$ and $\lambda > 0$, the process $B^J + \lambda B^H$ exhibits different scaling properties in finite and large time. It is locally asymptotically J-self-similar, whereas it is asymptotically H-self-similar in large time.

Another application is the comparison with B^H of a fractional analogue of the Karhunen–Loève process (10) proposed in [11].

Theorem 30. *Consider the process*

$$L_t^H = \sqrt{2} \sum_{n \geq 0} \xi_n \frac{\sin((n + 1/2)\pi t)}{((n + 1/2)\pi)^{H+1/2}}$$

for independent standard Gaussian variables ξ_n. Then the laws of $L_{S+t}^H - L_S^H$ and B^H are equivalent on $[0, T - S]$ for $0 < S < T < 1$. On the other hand, these laws are mutually singular if $S = 0$ or $T = 1$.

Proof. We deduce from Theorem 27 that the laws of $B_{t/2}^H$ and $\overline{B}_{t/2}^H$ are equivalent on $[0, 2T]$ for $T < 1$, and therefore on $[-T, T]$ (the two processes have stationary increments). Thus $(B_t^H - B_{-t}^H)/\sqrt{2}$, which has the same law as $2^{H-1/2}(B_{t/2}^H - B_{-t/2}^H)$, has a law equivalent on $[0, T]$ to the law of

$$2^{H-1/2} \left(\overline{B}_{t/2}^H - \overline{B}_{-t/2}^H \right) = 2^{H+1} \sum_{n \geq 0} \xi_n \frac{\sin((n + 1/2)\pi t)}{((2n + 1)\pi)^{H+1/2}} = L_t^H,$$

so we have the equivalence of laws

$$L_t^H \sim (B_t^H - B_{-t}^H)/\sqrt{2} \tag{91}$$

on $[0, T]$. Moreover, we deduce from Remark 28 that the increments of the right hand side of (91) on $[S, T]$ are equivalent to the increments of B^H, and this proves the first statement of the theorem. For the case $S = 0$, we have also noticed in Remark 28 that the laws of the right hand side of (91) and of B^H are mutually singular. For the case $T = 1$, we have to check that the laws of $L_1^H - L_{1-t}^H$ and of B^H are mutually singular on $[0, 1 - S]$. We have

$$\begin{aligned}
L_1^H - L_{1-t}^H &= 2^{H-1/2} \left(\overline{B}_{1/2}^H - \overline{B}_{(1-t)/2}^H - \overline{B}_{-1/2}^H + \overline{B}_{(t-1)/2}^H \right) \\
&= 2^{H-1/2} \left(2\overline{B}_{1/2}^H - \overline{B}_{(1-t)/2}^H - \overline{B}_{(1+t)/2}^H \right) \\
&\simeq 2^{H-1/2} \left(\overline{B}_{-t/2}^H + \overline{B}_{t/2}^H \right) \sim (B_t^H + B_{-t}^H)/\sqrt{2}
\end{aligned}$$

where we have used the fact that the increments of $\overline{B}^H$ are 1-antiperiodic and stationary. But the law of this process is mutually singular with the law of B^H by again applying Remark 28.   □

Appendix

We now explain some technical results which were used throughout this article.

A An Analytical Lemma

The basic result of this appendix is the following classical lemma, see Theorem 1.5 of [35].

Theorem 31. *Consider a kernel $K(t, s)$ on $\mathbb{R}_+ \times \mathbb{R}_+$ such that*

$$K(\lambda t, \lambda s) = K(t, s)/\lambda \tag{92}$$

for $\lambda > 0$, and

$$\int_0^\infty \frac{|K(1, s)|}{\sqrt{s}} ds < \infty.$$

Then $K : f \mapsto \int K(., s) f(s) ds$ defines a continuous endomorphism of L^2.

Proof. For f nonnegative, let us study

$$E(f) = \int_0^\infty \left(\int_0^\infty |K(t,s)| f(s) ds \right)^2 dt = \int_0^\infty \left(\int_0^\infty |K(1,s)| f(ts) ds \right)^2 dt$$

$$= \iiint |K(1,s)| |K(1,u)| f(ts) f(tu) ds\, du\, dt$$

from the scaling property (92) written as $K(t,s) = K(1, s/t)/t$. We have

$$\int f(ts) f(tu) dt \leq \|f\|_{L^2}^2 / \sqrt{su},$$

so

$$E(f) \leq \|f\|_{L^2}^2 \left(\int \frac{|K(1,s)|}{\sqrt{s}} ds \right)^2.$$

If now f is a real square integrable function, then $Kf(t)$ is well defined for almost any t, and

$$\int_0^\infty Kf(t)^2 dt \leq E(|f|) \leq C \|f\|_{L^2}^2.$$

$\square$

Theorem 32. *On the time interval $\mathbb{R}_+$, let*

$$A : (h(t);\ t \geq 0) \mapsto (Ah(t);\ t \geq 0)$$

be a linear operator defined on $\mathbb{H}^{1/2}$ (the space of $1/2$-Hölder continuous functions taking the value 0 at 0) such that $Ah(0) = 0$. We suppose that

$$A(h_\lambda) = (Ah)_\lambda \quad for\ h_\lambda(t) = h(\lambda t). \tag{93}$$

We also suppose that Ah is differentiable on $\mathbb{R}_+^*$ and that $h \mapsto D^1 Ah(1)$ is continuous on $\mathbb{H}^{1/2}$. Then A is a continuous endomorphism of the standard Cameron–Martin space $\mathcal{H}_{1/2} = I_{0+}^1 L^2$.

Proof. On $\mathcal{H}_{1/2}$, the linear form $h \mapsto D^1 Ah(1)$ takes the form $D^1 Ah(1) = \langle a, h \rangle_{\mathcal{H}_{1/2}}$ for some a in $\mathcal{H}_{1/2}$, so

$$D^1 Ah(t) = \frac{1}{t} D^1 (Ah)_t(1) = \frac{1}{t} D^1 Ah_t(1)$$

$$= \frac{1}{t} \langle a, h_t \rangle_{\mathcal{H}_{1/2}} = \frac{1}{t} \int D^1 a(s)\, D^1 h_t(s) ds$$

$$= \int D^1 a(s)\, D^1 h(ts) ds = \int K(t,s) D^1 h(s) ds$$

for

$$K(t,s) = D^1 a(s/t)/t.$$

Then K satisfies the scaling condition (92), and

$$\int \frac{|D^1 a(s)|}{\sqrt{s}} ds \leq \sup\left\{\langle a, h\rangle_{\mathcal{H}_{1/2}}; \ h \in \mathcal{H}_{1/2}, \ |D^1 h(s)| \leq 1/\sqrt{s}\right\}$$

$$\leq \sup\left\{D^1 Ah(1); \ h(0) = 0, \ |h(t) - h(s)| \leq 2\sqrt{t-s}\right\} < \infty$$

since $h \mapsto D^1 Ah(1)$ is continuous on $\mathbb{H}^{1/2}$. Thus we can apply Theorem 31 and deduce that $D^1 A I_{0+}^1$ is a continuous endomorphism of L^2, or, equivalently, that A is a continuous endomorphism of $\mathcal{H}_{1/2}$. □

B   Variance of Fractional Brownian Motions

We prove here a result stated in Sect. 4.1, more precisely that if B^H is given by the representation (49) with κ given by (50). then the variance ρ of B_1^H satisfies (51). We also prove that the variance of B_1^H given by the spectral representation (7) is the same.

Theorem 33. *The variance of B_1^H defined by (49) is given by*

$$\rho = \kappa^2 \frac{3/2 - H}{2H} B(2 - 2H, H + 1/2) \tag{94}$$

for the Beta function

$$B(\alpha, \beta) = \int_0^1 t^{\alpha-1}(1-t)^{\beta-1} dt, \qquad \alpha > 0, \ \beta > 0.$$

Proof. For $t > 0$, by decomposing the right-hand side of (49) into integrals on $[0, t]$ and on $\mathbb{R}_-$, we obtain

$$\mathbb{E}[(B_t^H)^2] = \kappa^2 \left(\frac{t^{2H}}{2H} + \phi(t)\right)$$

with

$$\phi(t) = \int_0^\infty \left((t+x)^{H-1/2} - x^{H-1/2}\right)^2 dx.$$

We can differentiate twice this integral and get

$$\phi'(t) = (2H - 1) \int_0^\infty \left((t+x)^{2H-2} - (t+x)^{H-3/2} x^{H-1/2}\right) dx,$$

$$\phi''(t) = (2H - 1)(2H - 2) \int_0^\infty (t+x)^{2H-3} dx$$

$$- (2H - 1)(H - 3/2) \int_0^\infty (t + x)^{H-5/2} x^{H-1/2} dx$$

$$= -(2H-1)t^{2H-2} - (2H-1)(H-3/2)t^{2H-2} \int_1^\infty y^{H-5/2}(y-1)^{H-1/2} dy$$

$$= -(2H - 1)t^{2H-2} - (2H - 1)(H - 3/2)t^{2H-2} \int_0^1 \left(\frac{1-z}{z^2}\right)^{H-1/2} dz$$

by means of the changes of variables $x = t(y - 1)$ and $y = 1/z$. Thus

$$\phi''(t) = (2H - 1)t^{2H-2} \left(-1 + (3/2 - H) B(2 - 2H, H + 1/2)\right).$$

We integrate twice this formula, and since $\phi(t)$ and $\phi'(t)$ are respectively proportional to t^{2H} and t^{2H-1}, we obtain (94) by writing $\kappa^2 (\phi(1) + 1/(2H))$. $\quad\square$

By applying properties of Beta and Gamma functions

$$B(\alpha, \beta) = \Gamma(\alpha)\Gamma(\beta)/\Gamma(\alpha + \beta),$$

$$\Gamma(z + 1) = z\,\Gamma(z), \qquad \Gamma(z)\Gamma(1 - z) = \pi/\sin(\pi z), \tag{95}$$

where Γ is defined on $\mathbb{C} \setminus \mathbb{Z}_-$, we can write equivalent forms which are used in the literature,

$$\rho = \kappa^2 \frac{3/2 - H}{2H} \frac{\Gamma(2 - 2H)\Gamma(H + 1/2)}{\Gamma(5/2 - H)}$$

$$= \kappa^2 \frac{1}{2H(1/2 - H)} \frac{\Gamma(2 - 2H)\Gamma(H + 1/2)}{\Gamma(1/2 - H)}$$

$$= \kappa^2 \frac{\cos(\pi H)}{\pi H(1 - 2H)} \Gamma(2 - 2H)\Gamma(H + 1/2)^2$$

$$= -2\kappa^2 \frac{\cos(\pi H)}{\pi} \Gamma(-2H)\Gamma(H + 1/2)^2 \tag{96}$$

where, except in the first line, we have to assume $H \neq 1/2$. Thus if we choose $\kappa = \kappa(H) = \Gamma(H + 1/2)^{-1}$ as this is done in this article, then ρ is given by (51).

If now we consider the spectral representation (7), then

$$\mathbb{E}[(B_1^H)^2] = \frac{1}{\pi} \int_0^\infty s^{-1-2H} \left((\cos s - 1)^2 + \sin^2 s\right) ds$$

$$= \frac{2}{\pi} \int_0^\infty s^{-1-2H} (1 - \cos s) ds = \frac{1}{\pi H} \int_0^\infty s^{-2H} \sin s\, ds$$

by integration by parts. If $H < 1/2$, an application of (34) shows that this variance is again given by (51); if $H > 1/2$, the same property can be proved by using another integration by parts, and the case $H = 1/2$ can be deduced from the continuity of the variance with respect to H.

Remark 39. The variance of the spectral decomposition can also be obtained as follows. The process B^H given by (7) can be written as the real part of

$$B_t^{H,\mathbb{C}} = \frac{1}{\sqrt{\pi}} \int_0^{+\infty} s^{-H-1/2} \left(e^{ist} - 1 \right) \left(dW_s^1 + i\, dW_s^2 \right)$$

$$\simeq \frac{1}{\sqrt{2\pi}} \int_{-\infty}^{+\infty} |s|^{1/2-H} \frac{e^{ist} - 1}{s} \left(dW_s^1 + i\, dW_s^2 \right).$$

The isometry property of the Fourier transform on L^2 enables to check that $B^{1/2,\mathbb{C}}$ has the same law as $W^1 + i\, W^2$, so in particular $B^{1/2}$ is a standard Brownian motion. Following Theorem 10, the general case $H \neq 1/2$ is obtained by applying $\widetilde{I}_+^{H-1/2}$ to $B^{1/2,\mathbb{C}}$ (use (34)).

C   Equivalence of Laws of Gaussian Processes

Our aim is to compare the laws of two centred Gaussian processes. It is known from [10, 15, 16] that their laws are either equivalent, or mutually singular (actually this is also true in the non centred case), and we want to decide between these two possibilities. In Sect. C.1, after a brief review of infinite dimensional Gaussian variables, we explain how the Cameron–Martin space (or reproducing kernel Hilbert space) can be used to study this question. In particular, we prove a sufficient condition for the equivalence. Then, in Sect. C.2, we describe a more computational method which can be used for self-similar processes to decide between the equivalence and mutual singularity.

C.1   Cameron–Martin Spaces

A Gaussian process can be viewed as a Gaussian variable W taking its values in an infinite-dimensional vector space $\mathcal{W}$, but the choice of $\mathcal{W}$ is not unique; in order to facilitate the study of W, it is better for $\mathcal{W}$ to have a good topological structure. This is with this purpose that the notion of abstract Wiener space was introduced by [14]; in this framework, $\mathcal{W}$ is a separable Banach space. However, more general topological vector spaces can also be considered, see for instance [3]. Here, we assume that $\mathcal{W}$ is a separable Fréchet space and we let $\mathcal{W}^\star$ be its topological dual. The space $\mathcal{W}$ is endowed with its Borel σ-algebra, which coincides with the cylindrical σ-algebra generated by the maps $w \mapsto \ell(w)$, $\ell \in \mathcal{W}^\star$. A $\mathcal{W}$-valued variable W is said to be centred Gaussian if $\ell(W)$ is centred Gaussian for any $\ell \in \mathcal{W}^\star$; the closed subspace of $L^2(\Omega)$ generated by the variables $\ell(W)$ is the Gaussian space of W. The Fernique theorem (see Theorem 2.8.5 in [3]) states that if $|.|$ is a measurable seminorm on $\mathcal{W}$ (which may take infinite values) and if $|W|$ is almost surely finite, then $\exp(\lambda|W|^2)$ is integrable for small enough positive λ.

For h in $\mathcal{W}$, define

$$|h|_{\mathcal{H}} = \sup\left\{\frac{\ell(h)}{\left\|\ell(W)\right\|_2}; \ell \in \mathcal{W}^\star\right\} \tag{97}$$

with the usual convention $0/0 = 0$. Then $\mathcal{H} = \{h; |h|_{\mathcal{H}} < \infty\}$ is a separable Hilbert space which is continuously embedded in $\mathcal{W}$ and which is called the Cameron–Martin space of W; it is dense in $\mathcal{W}$ if the topological support of the law of W is $\mathcal{W}$. It can be identified to its dual, and the adjoint of the inclusion $i : \mathcal{H} \to \mathcal{W}$ is a map $i^\star : \mathcal{W}^\star \to \mathcal{H}$ with dense image such that

$$\langle i^\star(\ell), h\rangle_{\mathcal{H}} = \ell(h), \quad \langle i^\star(\ell_1), i^\star(\ell_2)\rangle_{\mathcal{H}} = \mathbb{E}[\ell_1(W)\ell_2(W)]. \tag{98}$$

Consequently, the map $\ell \mapsto \ell(W)$ can be extended to an isometry between $\mathcal{H}$ and the Gaussian space of W, that we denote by $\langle W, h\rangle_{\mathcal{H}}$ (though W does not live in $\mathcal{H}$); thus $\ell(W) = \langle W, i^\star(\ell)\rangle_{\mathcal{H}}$ and

$$\mathbb{E}\big[\langle W, h\rangle_{\mathcal{H}} \langle W, h'\rangle_{\mathcal{H}}\big] = \langle h, h'\rangle_{\mathcal{H}}. \tag{99}$$

The variable $\langle W, h\rangle_{\mathcal{H}}$ is called the Wiener integral of h.

Example 4. When considering real continuous Gaussian processes, the space $\mathcal{W}$ can be taken to be the space of real-valued continuous functions with the topology of uniform convergence on compact subsets. The most known example is the standard Brownian motion; its Cameron–Martin space $\mathcal{H}_{1/2}$ is the space of absolutely continuous functions h such that $h(0) = 0$ and $D^1 h$ is in L^2.

Remark 40. Let $\mathcal{W}$ be the space of real-valued continuous functions. The coordinate maps $\ell_t(\omega) = \omega(t)$ are in $\mathcal{W}^\star$ and the linear subspace generated by the variables $\ell_t(W) = W_t$ is dense in the Gaussian space of W; equivalently, the space $\mathcal{H}$ is generated by the elements $i^\star(\ell_t)$. On the other hand, we deduce from (98) that

$$i^\star(\ell_t) : s \mapsto \ell_s\big(i^\star(\ell_t)\big) = \langle i^\star(\ell_s), i^\star(\ell_t)\rangle_{\mathcal{H}} = \mathbb{E}[W_s W_t].$$

Thus, if we denote by $C(s, t) = \mathbb{E}[W_s W_t]$ the covariance kernel, then $\mathcal{H}$ is the closure of the linear span of the functions $i^\star(\ell_t) = C(t, .)$ for the inner product

$$\langle C(s, .), C(t, .)\rangle_{\mathcal{H}} = C(s, t).$$

This relation is called the reproducing property, and $\mathcal{H}$ is the reproducing kernel Hilbert space of $C(., .)$. This technique can also be used for non continuous processes, see for instance [36].

Remark 41. Another viewpoint for the Wiener integrals when $W = (W_t)$ is a continuous Gaussian process is to consider the integrals $\int f(t)dW_t$ for deterministic functions f. This integral is easily defined when f is an elementary (or step) process, and we can extend by continuity this definition to more general functions. With this method, we obtain variables which are in the Gaussian space of W, but we do not necessarily obtain the whole space, see the case of the fractional Brownian motion B^H when $H > 1/2$ in [34].

Let W_1 and W_2 be two centred Gaussian variables with values in the same space $\mathcal{W}$, with Cameron–Martin spaces $\mathcal{H}_1$ and $\mathcal{H}_2$. It follows from (97) that $\mathcal{H}_1$ is continuously embedded in $\mathcal{H}_2$ if and only if

$$\left\| \ell(W_1) \right\|_2 \leq C \left\| \ell(W_2) \right\|_2 \tag{100}$$

for any $\ell \in \mathcal{W}^\star$.

Let $\mathcal{W}^1$ and $\mathcal{W}^2$ be separable Fréchet spaces, let W be a $\mathcal{W}^1$-valued centred Gaussian variable with Cameron–Martin space $\mathcal{H}^1$, and let $A : \mathcal{W}^1 \to \mathcal{W}^2$ be a measurable linear transformation which is defined on a measurable linear subspace of $\mathcal{W}^1$ supporting the law of W. Then AW is a centred Gaussian variable. If A is injective on $\mathcal{H}^1$, then the Cameron–Martin space of AW is $\mathcal{H}^2 = A(\mathcal{H}^1)$. This explains how the Cameron–Martin space $\mathcal{H}_H$ of the fractional Brownian motion B^H can be deduced from $\mathcal{H}_{1/2}$; one applies the transformations $\widetilde{I}_+^{H-1/2}$ (Theorem 10) or $G_{0+}^{1/2,H}$ (Theorem 11). On the other hand, if A is non injective, one still has $\mathcal{H}^2 = A(\mathcal{H}^1)$ and the norm is now given by

$$|h_2|_{\mathcal{H}^2} = \inf\{|h_1|_{\mathcal{H}^1}; \ A(h_1) = h_2\}. \tag{101}$$

In particular $|Ah|_{\mathcal{H}_2} \leq |h|_{\mathcal{H}_1}$. If $A = 0$ on $\mathcal{H}_1$, then $AW = 0$.

We now consider the absolute continuity of Gaussian measures with respect to one another. This notion can be studied by means of the relative entropy, or Kullback-Leibler divergence, defined for probability measures μ_1 and μ_2 by

$$\mathcal{I}(\mu_2, \mu_1) = \int \ln (d\mu_2/d\mu_1) \, d\mu_2$$

if μ_2 is absolutely continuous with respect to μ_1, and by $+\infty$ otherwise. This quantity is related to the total variation of $\mu_2 - \mu_1$ by the Pinsker inequality

$$\left(\int |d\mu_2 - d\mu_1| \right)^2 \leq 2\mathcal{I}(\mu_2, \mu_1). \tag{102}$$

The Cameron–Martin theorem enables to characterise elements of $\mathcal{H}$ amongst elements of $\mathcal{W}$. More precisely, h is in $\mathcal{H}$ if and only if the law of $W + h$ is absolutely continuous with respect to the law of W. Moreover, in this case, the density is $\exp\left(\langle W, h \rangle_{\mathcal{H}} - |h|_{\mathcal{H}}^2/2\right)$. Thus

$$\mathcal{I}(\mu', \mu) = \mathcal{I}(\mu, \mu') = |h|_{\mathcal{H}}^2/2$$

when μ and μ' are the laws of W and $W + h$.

The transformation $W \mapsto W + h$ of the Cameron–Martin space can be generalised to random h. If we add to W an independent process X taking its values in $\mathcal{H}$, it is easily seen by working conditionally on X that the laws of W and $W + X$ are again equivalent. Moreover, the law of $(W + X, X)$ is absolutely continuous with respect to the law of (W, X), with a density equal to $\exp\left(\langle W, X \rangle_{\mathcal{H}} - |X|_{\mathcal{H}}^2 / 2\right)$, and relative entropies of the two variables with respect to each other are equal to $\frac{1}{2} \mathbb{E} |X|_{\mathcal{H}}^2$. By projecting on the first component, it follows from the Jensen inequality that the relative entropy cannot increase, so

$$\max \left(\mathcal{I}(\mu', \mu), \mathcal{I}(\mu, \mu') \right) \leq \mathbb{E} |X|_{\mathcal{H}}^2 / 2 \tag{103}$$

when μ and μ' are the laws of W and $W + X$.

When $W = (W_n)$ and $\overline{W} = (\overline{W}_n)$ are two sequences consisting of independent centred Gaussian variables with positive variances, then the equivalence or mutual singularity of their laws can be decided by means of Kakutani's criterion [23]. This criterion is actually intended to general non Gaussian variables; when specialised to the Gaussian case, it leads to the following result.

Theorem 34. *Let $W = (W_n)$ and $\overline{W} = (\overline{W}_n)$ be two sequences of independent centred Gaussian variables with variances $\sigma_n^2 > 0$ and $\bar{\sigma}_n^2 > 0$. Then the laws of W and $\overline{W}$ are equivalent if and only if*

$$\sum_n \left(\frac{\bar{\sigma}_n^2}{\sigma_n^2} - 1 \right)^2 < \infty. \tag{104}$$

Returning to general Gaussian variables, we now give a sufficient condition for the equivalence of W and $W + X$ where W and X are not required to be independent. This result has been used in the proof of Theorem 27; it can be deduced from the proof of [10], but we explain its proof for completeness.

Theorem 35. *Let (W, X) be a centred Gaussian variable with values in $\mathcal{W} \times \mathcal{H}$, where $\mathcal{W}$ is a separable Fréchet space, and $\mathcal{H}$ is the Cameron–Martin space of W; thus $W + X$ is a Gaussian variable taking its values in $\mathcal{W}$; let $\mathcal{H}'$ be its Cameron–Martin space.*

- *The space $\mathcal{H}'$ is continuously embedded in $\mathcal{H}$.*
- *If moreover $\mathcal{H}$ is continuously embedded in $\mathcal{H}'$ (so that $\mathcal{H} \sim \mathcal{H}'$), then the laws of W and $W + X$ are equivalent. Moreover, the entropy of the law of $W + X$ relative to the law of W is bounded by $C \, \mathbb{E} |X|_{\mathcal{H}}^2$, where C depends only on the norms of the injections of $\mathcal{H}$ and $\mathcal{H}'$ into each other.*

Proof. We have to compare the laws of $(\ell(W); \ell \in \mathcal{W}^\star)$ and $(\ell(W + X), \ell \in \mathcal{W}^\star)$. Since $|X|_{\mathcal{H}}$ is almost surely finite, it follows from the Fernique theorem that $|X|_{\mathcal{H}}^2$ has an exponential moment and is in particular integrable, so $\ell(X) = \langle i^\star(\ell), X \rangle_{\mathcal{H}}$ is square integrable. Thus

$$\left\| \ell(W + X) \right\|_2 \leq \left\| \ell(W) \right\|_2 + C \left| i^\star(\ell) \right|_{\mathcal{H}} \leq (C + 1) \left\| \ell(W) \right\|_2$$

and the inclusion $\mathcal{H}' \subset \mathcal{H}$ follows from (100). Let us now suppose $\mathcal{H} \sim \mathcal{H}'$, so that, by again applying (100),

$$C_1 \|\ell(W)\|_2 \le \|\ell(W + X)\|_2 \le C_2 \|\ell(W)\|_2 \tag{105}$$

for positive C_1 and C_2. Let us first compare the laws of the families $(\ell(W + X); \ell \in \mathcal{W}_1^\star)$ and $(\ell(W); \ell \in \mathcal{W}_1^\star)$ for a finite-dimensional subspace $\mathcal{W}_1^\star$ of $\mathcal{W}^\star$. We have

$$\mathcal{W}_0^\star = \{\ell \in \mathcal{W}^\star; \|\ell(W)\|_2 = 0\} = \{\ell \in \mathcal{W}^\star; \|\ell(W + X)\|_2 = 0\}$$

and it is sufficient to consider the case where $\mathcal{W}_1^\star \cap \mathcal{W}_0^\star = \{0\}$. Then $|\ell| = \|\ell(W)\|_2$ and $|\ell|' = \|\ell(W + X)\|_2$ define two Euclidean structures on $\mathcal{W}_1^\star$, and it is possible to find a basis $(\ell_n; 1 \le n \le N)$ which is orthonormal for the former norm, and orthogonal for the latter norm. We have to compare the laws μ_N and μ_N' of $U_N = (\ell_n(W); 1 \le n \le N)$ and $U_N' = (\ell_n(W + X); 1 \le n \le N)$. The vectors U_N and U_N' consist of independent centred Gaussian variables; moreover, U_n has variance 1, and it follows from (105) that U_n' has a variance σ_n^2 satisfying $C_1 \le \sigma_n^2 \le C_2$. We deduce that

$$\mathcal{I}(\mu_N', \mu_N) = \frac{1}{2} \sum_{n=1}^{N} (\sigma_n^2 - 1 - \ln \sigma_n^2) \le C \sum_{n=1}^{N} (\sigma_n^2 - 1)^2.$$

But

$$\sigma_n^2 - 1 = 2\,\mathbb{E}\big[\ell_n(W)\,\ell_n(X)\big] + \mathbb{E}\big[(\ell_n(X))^2\big] \le C \left(\mathbb{E}\big[(\ell_n(X))^2\big]\right)^{1/2} \tag{106}$$

(we deduce from $\sigma_n^2 \le C_2$ that the variances of $\ell_n(X)$ are uniformly bounded), and

$$\mathcal{I}(\mu_N', \mu_N) \le C \sum_{n=1}^{N} \mathbb{E}\big[(\ell_n(X))^2\big] = C \sum_{n=1}^{N} \mathbb{E}\big[\langle i^\star(\ell_n), X\rangle_{\mathcal{H}}^2\big] \le C\,\mathbb{E}|X|_{\mathcal{H}}^2$$

because $i^\star(\ell_n)$ is from (98) an orthonormal sequence in $\mathcal{H}$. Thus the entropy of the law of $(\ell(W + X); \ell \in \mathcal{W}_1^\star)$ relative to $(\ell(W); \ell \in \mathcal{W}_1^\star)$ is bounded by an expression $C\,\mathbb{E}|X|_{\mathcal{H}}^2$ which does not depend on the choice of the finite-dimensional subspace $\mathcal{W}_1^\star$. This implies that the law in $\mathcal{W}$ of $W + X$ is absolutely continuous with respect to the law of W, and that the corresponding relative entropy is also bounded by this expression. □

Remark 42. The condition about the equivalence of Cameron–Martin spaces cannot be dropped in Theorem 35, see the counterexample of the Brownian motion $W = (W_t)$ and $X_t = -t\,W_1$.

Remark 43. If W and X are independent, then

$$\|\ell(W + X)\|_2^2 = \|\ell(W)\|_2^2 + \|\ell(X)\|_2^2 \ge \|\ell(W)\|_2^2$$

so $\mathcal{H} \subset \mathcal{H}'$ is automatically satisfied. Moreover the estimation (106) is improved and we have $\mathbb{E}\langle X, h_n\rangle_{\mathcal{H}}^2$ instead of its square root. This explains why the laws of W and $W + X$ can be equivalent even when X does not take its values in $\mathcal{H}$; when W and X consist of sequences of independent variables (and assuming again that $\mathcal{H} \sim \mathcal{H}'$), this improvement leads to the condition (104).

Remark 44. More generally, for the comparison of two centred Gaussian measures μ and μ' on a separable Fréchet space $\mathcal{W}$, a necessary condition for the equivalence of μ and μ' is the equivalence of the Cameron–Martin spaces $\mathcal{H}$ and $\mathcal{H}'$. If this condition holds, there exists a homeomorphism Q of $\mathcal{H}$ onto itself such that

$$\langle h_1, h_2\rangle_{\mathcal{H}'} = \langle h_1, Qh_2\rangle_{\mathcal{H}}.$$

Then μ and μ' are equivalent if and only if $Q - I$ is a Hilbert-Schmidt operator.

C.2 Covariance of Self-Similar Processes

Consider a square integrable H-self-similar process for $H > 0$; we now explain that if it satisfies a 0-1 law in small time, then its covariance kernel can be estimated by means of its behaviour in small time; this is a simple consequence of the Birkhoff ergodic theorem.

Theorem 36. *Let $(\varXi_t; \ t > 0)$ be a H-self-similar continuous process, and suppose that its filtration $\mathcal{F}_t(\varXi)$ is such that $\mathcal{F}_{0+}(\varXi)$ is almost surely trivial. Define*

$$\theta_r \varXi(t) = e^{Hr} \varXi(e^{-r}t), \qquad -\infty < r < +\infty.$$

Then for any measurable functional f on the space of continuous paths such that $f(\varXi)$ is integrable,

$$\lim_{T \to \infty} \frac{1}{T} \int_0^T f(\theta_r \varXi) dr = \mathbb{E}[f(\varXi)] \tag{107}$$

almost surely. In particular, if $\varXi = (\varXi^1, \ldots, \varXi^n)$ is square integrable,

$$\mathbb{E}[\varXi_u^i \varXi_v^j] = \lim_{t \to 0} \frac{1}{|\log t|} \int_t^1 \frac{\varXi_{us}^i \varXi_{vs}^j}{s^{2H+1}} ds. \tag{108}$$

Proof. One has $\theta_r \theta_{r'} = \theta_{r+r'}$, so (θ_r) is a family of shifts. Moreover, the H-self-similarity of the process $\varXi$ is equivalent to the shift invariance of its law. Events which are (θ_r)-invariant are in $\mathcal{F}_{0+}(\varXi)$ which is almost surely trivial, so the ergodic theorem enables to deduce (107). Then (108) is obtained by taking $f(\varXi) = \varXi_u^i \varXi_v^j$ and by applying the change of variable $r = \log(1/s)$ in the integral. $\qquad \square$

Remark 45. By using the Lamperti transform defined in (46), the family (θ_r) is reduced to the time translation on stationary processes.

Remark 46. In the centred Gaussian case, the law is characterised by the covariance kernel, so Theorem 36 implies that the whole law of Ξ can be deduced from its small time behaviour. The result can be applied to fractional Brownian motions of index $0 < H < 1$; by applying the canonical representation of Sect. 4, one has indeed $\mathcal{F}_{0+}(B^H) = \mathcal{F}_{0+}(W)$ and this σ-algebra is well-known to be almost surely trivial (Blumenthal 0-1 law). A simple counterexample is the fractional Brownian motion of index $H = 1$; this process (which was always excluded from our study of B^H) is given by $B_t^1 = t\,B_1$ for a Gaussian variable B_1; the assumption about $\mathcal{F}_{0+}(\Xi)$ and the conclusion of the theorem do not hold.

Remark 47. In the Gaussian case, (108) is a simple way to prove that the law of Ξ can be deduced from its small time behaviour. There are however other techniques, such as Corollary 3.1 of [1] about the law of iterated logarithm.

Theorem 37. *Let Ξ and Υ be two centred continuous H-self-similar Gaussian processes on $[0,1]$, such that $\mathcal{F}_{0+}(\Xi)$ is almost surely trivial. Then the two processes either have the same law, or have mutually singular laws.*

Proof. Gaussian measures are either equivalent, or mutually singular, so suppose that the laws of Ξ and Υ are equivalent. The process Ξ satisfies (108), so

$$\mathbb{E}[\Xi_u^i \Xi_v^j] = \lim_{t \to 0} \frac{1}{|\log t|} \int_t^1 \frac{\Upsilon_{us}^i \Upsilon_{vs}^j}{s^{2H+1}}\,ds.$$

Moreover, the right hand side is bounded in $L^p(\Omega)$ for any p, so we can take the expectation in the limit, and it follows from the self-similarity of Υ that

$$\mathbb{E}[\Xi_u^i \Xi_v^j] = \lim_{t \to 0} \frac{1}{|\log t|} \int_t^1 \frac{\mathbb{E}[\Upsilon_{us}^i \Upsilon_{vs}^j]}{s^{2H+1}}\,ds = \mathbb{E}[\Upsilon_u^i \Upsilon_v^j].$$

Thus Ξ and Υ have the same law.   □

A counterexample of this property is again the fractional Brownian motion with index $H = 1$. Processes corresponding to different variances $\rho = \mathbb{E}[(B_1)^2] > 0$ have equivalent but different laws.

References

1. Arcones, M.A.: On the law of the iterated logarithm for Gaussian processes. J. Theor. Probab. **8**(4), 877–903 (1995)
2. Baudoin, F., Nualart, D.: Equivalence of Volterra processes. Stoch. Process. Appl. **107**(2), 327–350 (2003)

3. Bogachev, V.I.: Gaussian measures. Mathematical Surveys and Monographs, vol. 62. American Mathematical Society, Providence, RI (1998)
4. Cheridito, P.: Mixed fractional Brownian motion. Bernoulli 7(6), 913–934 (2001)
5. Cheridito, P., Kawaguchi, H., Maejima, M.: Fractional Ornstein–Uhlenbeck processes. Electron. J. Probab. 8(3), 14 (2003)
6. Coutin, L., Qian, Z.: Stochastic analysis, rough path analysis and fractional Brownian motions. Probab. Theory Relat. Field. 122(1), 108–140 (2002)
7. Darses, S., Saussereau, B.: Time reversal for drifted fractional Brownian motion with Hurst index $H > 1/2$. Electron. J. Probab. 12(43), 1181–1211 (electronic) (2007)
8. Decreusefond, L., Üstünel, A.S.: Stochastic analysis of the fractional Brownian motion. Potential Anal. 10(2), 177–214 (1999)
9. Dzhaparidze, K., van Zanten, H.: A series expansion of fractional Brownian motion. Probab. Theory Relat. Field. 130(1), 39–55 (2004)
10. Feldman, J.: Equivalence and perpendicularity of Gaussian processes. Pac. J. Math. 8, 699–708 (1958)
11. Feyel, D., de La Pradelle, A.: On fractional Brownian processes. Potential Anal. 10(3), 273–288 (1999)
12. Gilsing, H., Sottinen, T.: Power series expansions for fractional Brownian motions. Theory Stoch. Process. 9(3-4), 38–49 (2003)
13. Gripenberg, G., Norros, I.: On the prediction of fractional Brownian motion. J. Appl. Probab. 33(2), 400–410 (1996)
14. Gross, L.: Abstract Wiener spaces. In: Proceedings of the Fifth Berkeley Symposium on Mathematical Statistics and Probability (Berkeley, California, 1965/66), vol. II: Contributions to Probability Theory, Part 1, pp. 31–42. University California Press, Berkeley, CA (1967)
15. Hájek, J.: On a property of normal distribution of any stochastic process. Czech. Math. J. 8(83), 610–618 (1958)
16. Hida, T., Hitsuda, M.: Gaussian processes. Translations of Mathematical Monographs, vol. 120. American Mathematical Society, Providence, RI (1993)
17. Iglói, E.: A rate-optimal trigonometric series expansion of the fractional Brownian motion. Electron. J. Probab. 10, 1381–1397 (2005)
18. Istas, J.: Spherical and hyperbolic fractional Brownian motion. Electron. Commun. Probab. 10, 254–262 (2005)
19. Itô, K., Nisio, M.: On the convergence of sums of independent Banach space valued random variables. Osaka J. Math. 5, 35–48 (1968)
20. Jost, C.: Transformation formulas for fractional Brownian motion. Stoch. Process. Appl. 116(10), 1341–1357 (2006)
21. Jost, C.: A note on ergodic transformations of self-similar Volterra Gaussian processes. Electron. Commun. Probab. 12, 259–266 (2007)
22. Jost, C.: On the connection between Molchan–Golosov and Mandelbrot–Van Ness representations of fractional Brownian motion. J. Integral Equat. Appl. 20(1), 93–119 (2008)
23. Kakutani, S.: On equivalence of infinite product measures. Ann. Math. 49(2), 214–224 (1948)
24. Kolmogorov, A.: Wienersche Spiralen und einige andere interessante Kurven im Hilbertschen Raum. Dokl. Akad. Nauk. SSSR 26, 115–118 (1940)
25. Lévy, P.: Sur une classe de courbes de l'espace de Hilbert et sur une équation intégrale non linéaire. Ann. Sci. Ecole Norm. Sup. 73(3), 121–156 (1956)
26. Mandelbrot, B., Van Ness, J.W.: Fractional Brownian motions, fractional noises and applications. SIAM Rev. 10, 422–437 (1968)
27. Marinucci, D., Robinson, P.M.: Alternative forms of fractional Brownian motion. J. Stat. Plan. Infer. 80(1–2), 111–122 (1999)
28. Meyer, Y., Sellan, F., Taqqu, M.S.: Wavelets, generalized white noise and fractional integration: the synthesis of fractional Brownian motion. J. Fourier Anal. Appl. 5(5), 465–494 (1999)
29. Molchan, G.M.: Linear problems for fractional Brownian motion: a group approach. Theor. Probab. Appl. 47(1), 69–78 (2003)
30. Molchan, G.M., Golosov, Y.I.: Gaussian stationary processes with asymptotically power spectrum (in Russian). Dokl. Akad. Nauk. SSSR 184, 546–549 (1969)

31. Norros, I., Saksman, E.: Local independence of fractional Brownian motion. Stoch. Process. Appl. **119**, 3155—3172 (2009)
32. Norros, I., Valkeila, E., Virtamo, J.: An elementary approach to a Girsanov formula and other analytical results on fractional Brownian motions. Bernoulli **5**(4), 571–587 (1999)
33. Picard, J.: A tree approach to p-variation and to integration. Ann. Probab. **36**(6), 2235–2279 (2008)
34. Pipiras, V., Taqqu, M.S.: Are classes of deterministic integrands for fractional Brownian motion on an interval complete? Bernoulli **7**(6), 873–897 (2001)
35. Samko, S.G., Kilbas, A.A., Marichev, O.I.: Fractional Integrals and Derivatives. Gordon and Breach Science Publishers, Yverdon (1993)
36. van der Vaart, A.W., van Zanten, J.H.: Reproducing kernel Hilbert spaces of Gaussian priors. In: Pushing the Limits of Contemporary Statistics: Contributions in Honor of Jayanta K. Ghosh, Inst. Math. Stat. Collect., vol. 3, pp. 200–222. Inst. Math. Statist. Beachwood, OH (2008)
37. van Zanten, H.: When is a linear combination of independent fBm's equivalent to a single fBm? Stoch. Process. Appl. **117**(1), 57–70 (2007)

Other Contributions

Horizontal Diffusion in C^1 Path Space

Marc Arnaudon, Koléhè Abdoulaye Coulibaly, and Anton Thalmaier

Abstract We define horizontal diffusion in C^1 path space over a Riemannian manifold and prove its existence. If the metric on the manifold is developing under the forward Ricci flow, horizontal diffusion along Brownian motion turns out to be length preserving. As application, we prove contraction properties in the Monge–Kantorovich minimization problem for probability measures evolving along the heat flow. For constant rank diffusions, differentiating a family of coupled diffusions gives a derivative process with a covariant derivative of finite variation. This construction provides an alternative method to filtering out redundant noise.

Keywords Brownian motion · Damped parallel transport · Horizontal diffusion · Monge–Kantorovich problem · Ricci curvature

1 Preliminaries

The main concern of this paper is to answer the following question: Given a second order differential operator L without constant term on a manifold M and a C^1 path $u \mapsto \varphi(u)$ taking values in M, is it possible to construct a one parameter family $X_t(u)$ of diffusions with generator L and starting point $X_0(u) = \varphi(u)$, such that the derivative with respect to u is locally uniformly bounded?

If the manifold is $\mathbb{R}^n$ and the generator L a constant coefficient differential operator, there is an obvious solution: the family $X_t(u) = \varphi(u) + Y_t$, where Y_t is

M. Arnaudon (✉) and K.A. Coulibaly
Laboratoire de Mathématiques et Applications CNRS: UMR 6086, Université de Poitiers,
Téléport 2 - BP 30179, 86962 Futuroscope Chasseneuil Cedex, France
e-mail: marc.arnaudon@math.univ-poitiers.fr; abdoulaye.coulibaly@math.univ-poitiers.fr

K.A. Coulibaly and A. Thalmaier
Unité de Recherche en Mathématiques, FSTC, Université du Luxembourg,
6, rue Richard Coudenhove-Kalergi, 1359 Luxembourg, Grand-Duchy of Luxembourg
e-mail: abdoulaye.coulibaly@uni.lu; anton.thalmaier@uni.lu

C. Donati-Martin et al. (eds.), *Séminaire de Probabilités XLIII*, Lecture Notes in Mathematics 73
2006, DOI 10.1007/978-3-642-15217-7_2, © Springer-Verlag Berlin Heidelberg 2011

an L-diffusion starting at 0, has the required properties. But already on $\mathbb{R}^n$ with a non-constant generator, the question becomes difficult.

In this paper we give a positive answer for elliptic operators L on general manifolds; the result also covers time-dependent elliptic generators $L = L(t)$.

It turns out that the constructed family of diffusions solves the ordinary differential equation in the space of semimartingales:

$$\partial_u X_t(u) = W(X(u))_t(\dot{\varphi}(u)), \tag{1}$$

where $W(X(u))$ is the so-called deformed parallel translation along the semimartingale $X(u)$.

The problem is similar to finding flows associated to derivative processes as studied in [7–10, 12–15]. However it is transversal in the sense that in these papers diffusions with the same starting point are deformed along a drift which vanishes at time 0. In contrast, we want to move the starting point but to keep the generator. See Stroock [22], Chap. 10, for a related construction.

Our strategy of proof consists in iterating parallel couplings for closer and closer diffusions. In the limit, the solution may be considered as an infinite number of infinitesimally coupled diffusions. We call it horizontal L-diffusion in C^1 path space.

If the generator L is degenerate, we are able to solve (1) only in the constant rank case; by parallel coupling we construct a family of diffusions satisfying (1) at $u = 0$. In particular, the derivative of $X_t(u)$ at $u = 0$ has finite variation compared to parallel transport.

Note that our construction requires only a connection on the fiber bundle generated by the "carré du champ" operator. In the previous approach of [11], a stochastic differential equation is needed and ∇ has to be the Le Jan-Watanabe connection associated to the SDE.

The construction of families of $L(t)$-diffusions $X_\cdot(u)$ with $\partial_u X_\cdot(u)$ locally uniformly bounded has a variety of applications. In Stochastic Analysis, for instance, it allows to deduce Bismut type formulas without filtering redundant noise. If only the derivative with respect to u at $u = 0$ is needed, parallel coupling as constructed in [5, 6] would be a sufficient tool. The horizontal diffusion however is much more intrinsic by yielding a flow with the deformed parallel translation as derivative, well-suited to applications in the analysis of path space. Moreover for any u, the diffusion $X_\cdot(u)$ generates the same filtration as $X_\cdot(0)$, and has the same lifetime if the manifold is complete.

In Sect. 4 we use the horizontal diffusion to establish a contraction property for the Monge–Kantorovich optimal transport between probability measures evolving under the heat flow. We only assume that the cost function is a non-decreasing function of distance. This includes all Wasserstein distances with respect to the time-dependent Riemannian metric generated by the symbol of the generator $L(t)$. For a generator which is independent of time, the proof could be achieved using simple parallel coupling. The time-dependent case however requires horizontal diffusion as a tool.

2 Horizontal Diffusion in C^1 Path Space

Let M be a complete Riemannian manifold with ρ its Riemannian distance. The Levi–Civita connection on M will be denoted by ∇.

Given a continuous semimartingale X taking values in M, we denote by $d^\nabla X = dX$ its Itô differential and by $d_m X$ the martingale part of dX. In local coordinates,

$$d^\nabla X \equiv dX = \left(dX^i + \frac{1}{2} \Gamma^i_{jk}(X) d < X^j, X^k > \right) \frac{\partial}{\partial x^i} \qquad (2)$$

where Γ^i_{jk} are the Christoffel symbols of the Levi–Civita connection on M. In addition, if

$$dX^i = dM^i + dA^i$$

where M^i is a local martingale and A^i a finite variation process, then

$$d_m X = dM^i \frac{\partial}{\partial x^i}.$$

Alternatively, if

$$P_t(X) \equiv P_t^M(X) : T_{X_0} M \to T_{X_t} M$$

denotes parallel translation along X, then

$$dX_t = P_t(X) d \left(\int_0^{\cdot} P_s(X)^{-1} \delta X_s \right)_t$$

and

$$d_m X_t = P_t(X) dN_t$$

where N_t is the martingale part of the Stratonovich integral

$$\int_0^t P(X)_s^{-1} \delta X_s.$$

If X is a diffusion with generator L, we denote by $W(X)$ the so-called deformed parallel translation along X. Recall that $W(X)_t$ is a linear map $T_{X_0} M \to T_{X_t} M$, determined by the initial condition $W(X)_0 = \mathrm{Id}_{T_{X_0} M}$ and the covariant Itô stochastic differential equation:

$$DW(X)_t = -\frac{1}{2} \mathrm{Ric}^\#(W(X)_t) dt + \nabla_{W(X)_t} Z dt. \qquad (3)$$

By definition we have

$$DW(X)_t = P_t(X) d \left(P_{\cdot}(X)^{-1} W(X) \right)_t. \qquad (4)$$

Note that the Itô differential (2) and the parallel translation require only a connection ∇ on M. For the deformed parallel translation (3) however the connection has to be adapted to a metric.

In this section the connection and the metric are independent of time. We shall see in Sect. 3 how these notions can be extended to time-dependent connections and metrics.

Theorem 2.1. *Let* $\mathbb{R} \to M$, $u \mapsto \varphi(u)$, *be a* C^1 *path in* M *and let* Z *be a vector field on* M. *Further let* X^0 *be a diffusion with generator*

$$L = \Delta/2 + Z,$$

starting at $\varphi(0)$, *and lifetime* ξ. *There exists a unique family*

$$u \mapsto (X_t(u))_{t \in [0,\xi[}$$

of diffusions with generator L, *almost surely continuous in* (t, u) *and* C^1 *in* u, *satisfying* $X(0) = X^0$, $X_0(u) = \varphi(u)$ *and*

$$\partial_u X_t(u) = W(X(u))_t(\dot{\varphi}(u)). \tag{5}$$

Furthermore, the process $X(u)$ *satisfies the Itô stochastic differential equation*

$$\mathrm{d}X_t(u) = P_{0,u}^{X_t(\cdot)} \, \mathrm{d}_m X_t^0 + Z_{X_t(u)} \, \mathrm{d}t, \tag{6}$$

where $P_{0,u}^{X_t(\cdot)} : T_{X_t^0}M \to T_{X_t(u)}M$ *denotes parallel transport along the* C^1 *curve*

$$[0, u] \to M, \quad v \mapsto X_t(v).$$

Definition 2.2. We call $t \mapsto (X_t(u))_{u \in \mathbb{R}}$ the horizontal L-diffusion in C^1 path space $C^1(\mathbb{R}, M)$ over X^0, starting at φ.

Remark 2.3. Given an elliptic generator L, we can always choose a metric g on M such that

$$L = \Delta/2 + Z$$

for some vector field Z where Δ is the Laplacian with respect to g. Assuming that M is complete with respect to this metric, the assumptions of Theorem 2.1 are fulfilled. In the non-complete case, a similar result holds with the only difference that the lifetime of $X.(u)$ then possibly depends on u.

Remark 2.4. Even if $L = \Delta/2$, the solution we are looking for is not the flow of a Cameron–Martin vector field: firstly the starting point here is not fixed and secondly the vector field would have to depend on the parameter u. Consequently one cannot apply for instance Theorem 3.2 in [15]. An adaptation of the proof of the cited result would be possible, but we prefer to give a proof using infinitesimal parallel coupling which is more adapted to our situation.

Proof (Proof of Theorem 2.1).

Without loss of generality we may restrict ourselves to the case $u \geq 0$.

A. *Existence.* Under the assumption that a solution $X_t(u)$ exists, we have for any stopping time T,

$$W_{T+t}(X(u))(\dot{\varphi}(u)) = W_t(X_{T+\cdot}(u))(\partial X_T(u)),$$

for $t \in [0, \xi(\omega) - T(\omega)[$ and $\omega \in \{T < \xi\}$. Here $\partial X_T := (\partial X)_T$ denotes the derivative process ∂X with respect to u, stopped at the random time T; note that by (5),

$$(\partial X_T)(u) = W(X(u))_T(\dot{\varphi}(u)).$$

Consequently we may localize and replace the time interval $[0, \xi[$ by $[0, \tau \wedge t_0]$ for some $t_0 > 0$, where τ is the first exit time of X from a relatively compact open subset U of M with smooth boundary.

We may also assume that U is sufficiently small and included in the domain of a local chart; moreover we can choose $u_0 \in]0, 1]$ with $\int_0^{u_0} \|\dot{\varphi}(u)\| \, du$ small enough such that the processes constructed for $u \in [0, u_0]$ stay in the domain U of the chart. At this point we use the uniform boundedness of W on $[0, \tau \wedge t_0]$.

For $\alpha > 0$, we define by induction a family of processes $(X_t^{\alpha}(u))_{t \geq 0}$ indexed by $u \geq 0$ as follows: $X^{\alpha}(0) = X^0$, $X_0^{\alpha}(u) = \varphi(u)$, and if $u \in]n\alpha, (n+1)\alpha]$ for some integer $n \geq 0$, $X^{\alpha}(u)$ satisfies the Itô equation

$$dX_t^{\alpha}(u) = P_{X_t^{\alpha}(n\alpha), X_t^{\alpha}(u)} d_m X_t^{\alpha}(n\alpha) + Z_{X_t^{\alpha}(u)} \, dt, \tag{7}$$

where $P_{x,y}$ denotes parallel translation along the minimal geodesic from x to y. We choose α sufficiently small so that all the minimizing geodesics are uniquely determined and depend smoothly of the endpoints: since $X^{\alpha}(u)$ is constructed from $X^{\alpha}(n\alpha)$ via parallel coupling (7), there exists a constant $C > 0$ such that

$$\rho(X_t^{\alpha}(u), X_t^{\alpha}(n\alpha)) \leq \rho(X_0^{\alpha}(u), X_0^{\alpha}(n\alpha)) \, e^{Ct} \leq \|\dot{\varphi}\|_{\infty} \, \alpha e^{Ct_0} \tag{8}$$

(see e.g. [16]).

The process $\partial X^{\alpha}(u)$ satisfies the covariant Itô stochastic differential equation

$$D\partial X^{\alpha}(u) = \nabla_{\partial X^{\alpha}(u)} P_{X^{\alpha}(n\alpha), \cdot} \, d_m X_t^{\alpha}(n\alpha)$$
$$+ \nabla_{\partial X^{\alpha}(u)} Z \, dt - \frac{1}{2} \text{Ric}^{\#}(\partial X^{\alpha}(u)) \, dt, \tag{9}$$

(see [3] (4.7), along with Theorem 2.2).

Step 1 We prove that if X and Y are two L-diffusions stopped at $\tau_0 := \tau \wedge t_0$ and living in U, then there exists a constant C such that

$$\mathbb{E}\left[\sup_{t \leq \tau_0} \|W(X)_t - W(Y)_t\|^2\right] \leq C \, \mathbb{E}\left[\sup_{t \leq \tau_0} \|X_t - Y_t\|^2\right]. \tag{10}$$

Here we use the Euclidean norm defined by the chart.

Write

$$L = a^{ij} \partial_{ij} + b^j \partial_j$$

with $a^{ij} = a^{ji}$ for $i, j \in \{1, \ldots, \dim M\}$.

For L-diffusions X and Y taking values in U, we denote by N^X, respectively N^Y, their martingale parts in the chart U. Then Itô's formula yields

$$\begin{aligned}
\langle (N^X)^k &- (N^Y)^k, (N^X)^k - (N^Y)^k \rangle_t \\
&= (X_t^k - Y_t^k)^2 - (X_0^k - Y_0^k)^2 \\
&\quad - 2 \int_0^t (X_s^k - Y_s^k) \, d((N_s^X)^k - (N_s^Y)^k) \\
&\quad - 2 \int_0^t (X_s^k - Y_s^k) \left(b^k(X_s) - b^k(Y_s) \right) ds.
\end{aligned}$$

Thus, for U sufficiently small, denoting by $\langle N^X - N^Y | N^X - N^Y \rangle$ the corresponding Riemannian quadratic variation, there exists a constant $C > 0$ (possibly changing from line to line) such that

$$\begin{aligned}
\mathbb{E} &\left[\langle N^X - N^Y | N^X - N^Y \rangle_{\tau_0} \right] \\
&\leq C \, \mathbb{E} \left[\sup_{t \leq \tau_0} \| X_t - Y_t \|^2 \right] + C \sum_k \mathbb{E} \left[\int_0^{\tau_0} |X_t^k - Y_t^k| \, |b^k(X_t) - b^k(Y_t)| \, dt \right] \\
&\leq C \, \mathbb{E} \left[\sup_{t \leq \tau_0} \| X_t - Y_t \|^2 \right] + C \int_0^{\tau_0} \mathbb{E} \left[\sup_{s \leq \tau_0} \| X_s - Y_s \|^2 \right] dt \\
&\leq C(1 + t_0) \, \mathbb{E} \left[\sup_{t \leq \tau_0} \| X_t - Y_t \|^2 \right].
\end{aligned}$$

Finally, again changing C, we obtain

$$\mathbb{E} \left[\langle N^X - N^Y | N^X - N^Y \rangle_{\tau_0} \right] \leq C \, \mathbb{E} \left[\sup_{t \leq \tau_0} \| X_t - Y_t \|^2 \right]. \tag{11}$$

Writing $W(X) = P(X) \left(P(X)^{-1} W(X) \right)$, a straightforward calculation shows that in the local chart

$$\begin{aligned}
dW(X) = {} & -\Gamma(X)(dX, W(X)) \\
& - \frac{1}{2}(d\Gamma)(X)(dX)(dX, W(X)) \\
& + \frac{1}{2}\Gamma(X)(dX, \Gamma(X)(dX, W(X))) \\
& - \frac{1}{2} \operatorname{Ric}^\sharp(W(X)) \, dt \\
& + \nabla_{W(X)} Z \, dt. \tag{12}
\end{aligned}$$

We are going to use (12) to evaluate the difference $W(Y) - W(X)$. Along with the already established bound (11), taking into account that $W(X)$, $W(Y)$ and the derivatives of the brackets of X and Y are bounded in U, we are able to get a bound for

$$F(t) := \mathbb{E}\left[\sup_{s \le t \wedge \tau} \|W(Y) - W(X)\|^2\right].$$

Indeed, first an estimate of the type

$$F(t) \le C_1 \mathbb{E}\left[\sup_{s \le \tau_0} \|X_s - Y_s\|^2\right] + C_2 \int_0^t F(s)\,ds, \quad 0 \le t \le t_0,$$

is derived which then by Gronwall's lemma leads to

$$F(t) \le C_1 e^{C_2 t} \mathbb{E}\left[\sup_{t \le \tau_0} \|X_t - Y_t\|^2\right]. \tag{13}$$

Letting $t = t_0$ in (13) we obtain the desired bound (10).

Step 2 We prove that there exists $C > 0$ such that for all $u \in [0, u_0]$,

$$\mathbb{E}\left[\sup_{t \le \tau_0} \rho^2\left(X_t^\alpha(u), X_t^{\alpha'}(u)\right)\right] \le C(\alpha + \alpha')^2. \tag{14}$$

From the covariant equation (9) for $\partial X_t^\alpha(v)$ and the definition of deformed parallel translation (3),

$$DW(X)_t^{-1} = \frac{1}{2}\operatorname{Ric}^\#(W(X)_t^{-1})\,dt - \nabla_{W(X)_t^{-1}}Z\,dt,$$

we have for $(t, v) \in [0, \tau_0] \times [0, u_0]$,

$$W(X^\alpha(v))_t^{-1}\,\partial X_t^\alpha(v) = \dot{\varphi}(v) + \int_0^t W(X^\alpha(v))_s^{-1}\nabla_{\partial X_s^\alpha(v)} P_{X_s^\alpha(v_\alpha),.}\,d_m X_s^\alpha(v_\alpha),$$

or equivalently,

$$\partial X_t^\alpha(v) = W(X^\alpha(v))_t\,\dot{\varphi}(v)$$

$$+ W(X^\alpha(v))_t \int_0^t W(X^\alpha(v))_s^{-1}\nabla_{\partial X_s^\alpha(v)} P_{X_s^\alpha(v_\alpha),.}\,d_m X_s^\alpha(v_\alpha) \tag{15}$$

with $v_\alpha = n\alpha$, where the integer n is determined by $n\alpha < v \le (n+1)\alpha$. Consequently, we obtain

$\rho(X_t^\alpha(u), X_t^{\alpha'}(u))$

$$= \int_0^u \left\langle d\rho, \left(\partial X_t^\alpha(v), \partial X_t^{\alpha'}(v) \right) \right\rangle dv$$

$$= \int_0^u \left\langle d\rho, \left(W(X^\alpha(v))_t \dot\varphi(v), W(X^{\alpha'}(v))_t \dot\varphi(v) \right) \right\rangle dv$$

$$+ \int_0^u \left\langle d\rho, \left(W(X^\alpha(v))_t \int_0^t W(X^\alpha(v))_s^{-1} \nabla_{\partial X_s^\alpha(v)} P_{X_s^\alpha(v_\alpha),\cdot} d_m X_s^\alpha(v_\alpha), 0 \right) \right\rangle dv$$

$$+ \int_0^u \left\langle d\rho, \left(0, W(X^{\alpha'}(v))_t \int_0^t W(X^{\alpha'}(v))_s^{-1} \nabla_{\partial X_s^{\alpha'}(v)} P_{X_s^{\alpha'}(v_{\alpha'}),\cdot} d_m X_s^{\alpha'}(v_{\alpha'}) \right) \right\rangle dv.$$

This yields, by means of boundedness of $d\rho$ and deformed parallel translation, together with (13) and the Burkholder–Davis–Gundy inequalities,

$$\mathbb{E}\left[\sup_{t \leq \tau_0} \rho^2 \left(X_t^\alpha(u), X_t^{\alpha'}(u) \right) \right] \leq C \int_0^u \mathbb{E}\left[\sup_{t \leq \tau_0} \rho^2 \left(X_t^\alpha(v), X_t^{\alpha'}(v) \right) \right] dv$$

$$+ C \int_0^u \mathbb{E}\left[\int_0^{\tau_0} \left\| \nabla_{\partial X_s^\alpha(v)} P_{X_s^\alpha(v_\alpha),\cdot} \right\|^2 ds \right] dv$$

$$+ C \int_0^u \mathbb{E}\left[\int_0^{\tau_0} \left\| \nabla_{\partial X_s^{\alpha'}(v)} P_{X_s^{\alpha'}(v_{\alpha'}),\cdot} \right\|^2 ds \right] dv.$$

From here we obtain

$$\mathbb{E}\left[\sup_{t \leq \tau_0} \rho^2 \left(X_t^\alpha(u), X_t^{\alpha'}(u) \right) \right] \leq C \int_0^u \mathbb{E}\left[\sup_{t \leq \tau_0} \rho^2 \left(X_t^\alpha(v), X_t^{\alpha'}(v) \right) \right] dv$$

$$+ C\alpha^2 \int_0^u \mathbb{E}\left[\int_0^{\tau_0} \left\| \partial X_s^\alpha(v) \right\|^2 ds \right] dv$$

$$+ C\alpha'^2 \int_0^u \mathbb{E}\left[\int_0^{\tau_0} \left\| \partial X_s^{\alpha'}(v) \right\|^2 ds \right] dv,$$

where we used the fact that for $v \in T_x M$, $\nabla_v P_{x,\cdot} = 0$, together with

$$\rho(X_s^\beta(v), X_s^\beta(v_\beta)) \leq C\beta, \quad \beta = \alpha, \alpha',$$

see estimate (8).

Now, by (9) for $D\partial X^\beta$, there exists a constant $C' > 0$ such that for all $v \in [0, u_0]$,

$$\mathbb{E}\left[\int_0^{\tau_0} \left\| \partial X_s^\beta(v) \right\|^2 ds \right] < C'.$$

Consequently,

$$\mathbb{E}\left[\sup_{t\le\tau_0}\rho^2\left(X_t^\alpha(u),X_t^{\alpha'}(u)\right)\right]\le C\int_0^u\mathbb{E}\left[\sup_{t\le\tau_0}\rho^2\left(X_t^\alpha(v),X_t^{\alpha'}(v)\right)\right]dv$$
$$+\,2CC'(\alpha+\alpha')^2,$$

which by Gronwall lemma yields

$$\mathbb{E}\left[\sup_{t\le\tau_0}\rho^2\left(X_t^\alpha(u),X_t^{\alpha'}(u)\right)\right]\le C\,(\alpha+\alpha')^2$$

for some constant $C>0$. This is the desired inequality.

Step 3 Recall that
$$L=a^{ij}\partial_{ij}+b^j\partial_j.$$

Denoting by (a_{ij}) the inverse of (a^{ij}), we let ∇' be the connection with Christoffel symbols

$$(\Gamma')_{ij}^k=-\frac{1}{2}(a_{ik}+a_{jk})b^k. \tag{16}$$

We are going to prove that all L-diffusions are ∇'-martingales:

(i) On one hand, ∇'-martingales are characterized by the fact that for any k,

$$dX^k+\frac{1}{2}(\Gamma')_{ij}^k\,d\langle X^i,X^j\rangle\quad\text{is the differential of a local martingale.} \tag{17}$$

(ii) On the other hand, L-diffusions satisfy the following two conditions:

$$dX^k-b^k(X)\,dt\quad\text{is the differential of a local martingale,} \tag{18}$$

and

$$d\langle X^i,X^j\rangle=(a^{ij}(X)+a^{ji}(X))\,dt. \tag{19}$$

From this it is clear that (16), (18) together with (19) imply (17).

From inequality (14) we deduce that there exists a limiting process

$$(X_t(u))_{0\le t\le\tau_0,\,0\le u\le u_0}$$

such that for all $u\in[0,u_0]$ and $\alpha>0$,

$$\mathbb{E}\left[\sup_{t\le\tau_0}\rho^2\left(X_t^\alpha(u),X_t(u)\right)\right]\le C\alpha^2. \tag{20}$$

In other words, for any fixed $u\in[0,u_0]$, the process $(X_t^\alpha(u))_{t\in[0,\tau_0]}$ converges to $(X_t(u))_{t\in[0,\tau_0]}$ uniformly in L^2 as α tends to 0. Since these processes are ∇'-martingales, convergence also holds in the topology of semimartingales ([4], Proposition 2.10). This implies in particular that for any $u\in[0,u_0]$, the process $(X_t(u))_{t\in[0,\tau_0]}$ is a diffusion with generator L, stopped at time τ_0.

Extracting a subsequence $(\alpha_k)_{k \geq 0}$ convergent to 0, we may assume that almost surely, for all dyadic $u \in [0, u_0]$,

$$\sup_{t \leq \tau_0} \rho \left(X_t^\alpha(u), X_t(u) \right)$$

converges to 0. Moreover we can choose $(\alpha_k)_{k \geq 0}$ of the form $\alpha_k = 2^{-n_k}$ with $(n_k)_{k \geq 0}$ an increasing sequence of positive integers. Due to (8), we can take a version of the processes $(t, u) \mapsto X_t^{\alpha_k}(u)$ such that

$$u \mapsto X_t^{\alpha_k}(u)$$

is uniformly Lipschitz in $u \in \mathbb{N} \alpha_k \cap [0, u_0]$ with a Lipschitz constant independent of k and t. Passing to the limit, we obtain that a.s for any $t \in [0, \tau_0]$, the map

$$u \mapsto X_t(u)$$

is uniformly Lipschitz in $u \in \mathscr{D} \cap [0, u_0]$ with a Lipschitz constant independent of t, where $\mathscr{D}$ is the set of dyadic numbers. Finally we can choose a version of

$$(t, u) \mapsto X_t(u)$$

which is a.s. continuous in $(t, u) \in [0, \tau_0] \times [0, u_0]$, and hence uniformly Lipschitz in $u \in [0, u_0]$.

Step 4 We prove that almost surely, $X_t(u)$ is differentiable in u with derivative

$$W(X(u))_t(\dot{\varphi}(u)).$$

More precisely, we show that in local coordinates, almost surely, for all $t \in [0, \tau_0]$, $u \in [0, u_0]$,

$$X_t(u) = X_t^0 + \int_0^u W(X(v))_t(\dot{\varphi}(v)) \, dv. \tag{21}$$

From the construction it is clear that almost surely, for all $t \in [0, \tau_0]$, $u \in [0, u_0]$,

$$X_t^{\alpha_k}(u) = X_t^0 + \int_0^u W(X^{\alpha_k}(v))_t(\dot{\varphi}(v)) \, dv$$
$$+ \int_0^u \left(W(X^{\alpha_k}(v))_t \int_0^t W(X^{\alpha_k}(v))_s^{-1} \nabla_{\partial X_s^{\alpha_k}(v)} P_{X_s^{\alpha_k}(v_{\alpha_k}),\cdot} \, d_m X_s^{\alpha_k}(v_{\alpha_k}) \right) dv.$$

This yields

$$X_t(u) - X_t^0 - \int_0^u W(X(v))_t(\dot{\varphi}(v)) \, dv$$
$$= X_t(u) - X_t^{\alpha_k}(u) + \int_0^u (W(X^{\alpha_k}(v))_t - W(X(v))_t) \, \dot{\varphi}(v) \, dv$$
$$+ \int_0^u \left(W(X^{\alpha_k}(v))_t \int_0^t W(X^{\alpha_k}(v))_s^{-1} \nabla_{\partial X_s^{\alpha_k}(v)} P_{X_s^{\alpha_k}(v_{\alpha_k}),\cdot} \, d_m X_s^{\alpha_k}(v_{\alpha_k}) \right) dv.$$

The terms of right-hand-side are easily estimated, where in the estimates the constant C may change from one line to another. First observe that

$$\mathbb{E}\left[\sup_{t\leq\tau_0}\left\|X_t(u)-X_t^{\alpha_k}(u)\right\|^2\right]\leq C\alpha_k^2.$$

Using (10) and (20) we have

$$\mathbb{E}\left[\sup_{t\leq\tau_0}\left\|\int_0^u (W(X^{\alpha_k}(v))_t - W(X(v))_t)\,dv\right\|^2\right]$$

$$\leq \mathbb{E}\left[\sup_{t\leq\tau_0}\int_0^u \|W(X^{\alpha_k}(v))_t - W(X(v))_t\|^2\,dv\right]$$

$$= \int_0^u \mathbb{E}\left[\sup_{t\leq\tau_0}\|W(X^{\alpha_k}(v))_t - W(X(v))_t\|^2\right]dv$$

$$\leq C\alpha_k^2,$$

and finally

$$\mathbb{E}\left[\sup_{t\leq\tau_0}\left\|\int_0^u W(X^{\alpha_k}(v))_t\right.\right.$$

$$\left.\left.\times\left(\int_0^t W(X^{\alpha_k}(v))_s^{-1}\nabla_{\partial X_s^{\alpha_k}(v)}P_{X_s^{\alpha_k}(v_{\alpha_k}),\bullet}\,d_m X_s^{\alpha_k}(v_{\alpha_k})\right)dv\right\|^2\right]$$

$$\leq C\int_0^u \mathbb{E}\left[\sup_{t\leq\tau_0}\left\|\int_0^t W(X^{\alpha_k}(v))_s^{-1}\nabla_{\partial X_s^{\alpha_k}(v)}P_{X_s^{\alpha_k}(v_{\alpha_k}),\bullet}\,d_m X_s^{\alpha_k}(v_{\alpha_k})\right\|^2\right]dv$$

$$\leq C\int_0^u \mathbb{E}\left[\int_0^{\tau_0}\left\|\nabla_{\partial X_s^{\alpha_k}(v)}P_{X_s^{\alpha_k}(v_{\alpha_k}),\bullet}\right\|^2 ds\right]dv \quad\text{(since } W^{-1}\text{ is bounded)}$$

$$\leq C\alpha_k^2\int_0^u \mathbb{E}\left[\int_0^{\tau_0}\|\partial X_s^{\alpha_k}(v)\|^2\,ds\right]dv$$

$$\leq C\alpha_k^2.$$

where in the last but one inequality we used $\nabla_v P_{x,\bullet} = 0$ for any $v\in T_x M$ which implies

$$\|\nabla_v P_{y,\bullet}\|^2 \leq C\,\rho(x,y)^2\|v\|^2,$$

and the last inequality is a consequence of (9).

We deduce that

$$\mathbb{E}\left[\sup_{t\leq\tau_0}\left\|X_t(u)-X_t^0-\int_0^u W(X(v))_t(\dot{\varphi}(v))\,dv\right\|^2\right]\leq C\alpha_k^2.$$

Since this is true for any α_k, using continuity in u of $X_t(u)$, we finally get almost surely for all t, u,

$$X_t(u) = X_t^0 + \int_0^u W(X(v))_t(\dot{\varphi}(v))\,\mathrm{d}v.$$

Step 5 Finally we are able to prove (6):

$$\mathrm{d}X_t(u) = P_{0,u}^{X_t(\cdot)}\,\mathrm{d}_m X_t^0 + Z_{X_t(u)}\,\mathrm{d}t.$$

Since a.s. the mapping $(t, u) \mapsto \partial X_t(u)$ is continuous, the map $u \mapsto \partial X(u)$ is continuous in the topology of uniform convergence in probability. We want to prove that $u \mapsto \partial X(u)$ is continuous in the topology of semimartingales.

Since for a given connection on a manifold, the topology of uniform convergence in probability and the topology of semimartingale coincide on the set of martingales (Proposition 2.10 of [4]), it is sufficient to find a connection on TM for which $\partial X(u)$ is a martingale for any u. Again we can localize in the domain of a chart. Recall that for all u, the process $X(u)$ is a ∇'-martingale where ∇' is defined in step 1. Then by [1], Theorem 3.3, this implies that the derivative with respect to u with values in TM, denoted here by $\partial X(u)$, is a $(\nabla')^c$-martingale with respect to the complete lift $(\nabla')^c$ of ∇'. This proves that $u \mapsto \partial X(u)$ is continuous in the topology of semimartingales.

Remark 2.5. Alternatively, one could have used that given a generator L', the topologies of uniform convergence in probability on compact sets and the topology of semimartingales coincide on the space of L'-diffusions. Since the processes $\partial X(u)$ are diffusions with the same generator, the result could be derived as well.

As a consequence, Itô integrals commute with derivatives with respect to u (see e.g. [4], Corollary 3.18 and Lemma 3.15). We write it formally as

$$D\partial X = \nabla_u \mathrm{d}X - \frac{1}{2}R(\partial X, \mathrm{d}X)\mathrm{d}X. \tag{22}$$

Since

$$\mathrm{d}X(u) \otimes \mathrm{d}X(u) = g^{-1}(X(u))\,\mathrm{d}t$$

where g is the metric tensor, (22) becomes

$$D\partial X = \nabla_u \mathrm{d}X - \frac{1}{2}\mathrm{Ric}^{\#}(\partial X)\,\mathrm{d}t.$$

On the other hand, (5) and (3) for W yield

$$D\partial X = -\frac{1}{2}\mathrm{Ric}^{\#}(\partial X)\,\mathrm{d}t + \nabla_{\partial X} Z\,\mathrm{d}t.$$

From the last two equations we obtain

$$\nabla_u \mathrm{d}X = \nabla_{\partial X} Z\,\mathrm{d}t.$$

This along with the original equation

$$dX^0 = d_m X^0 + Z_{X^0} dt$$

gives

$$dX_t(u) = P_{0,u}^{X_t(\cdot)} d_m X_t^0 + Z_{X_t(u)} dt,$$

where

$$P_{0,u}^{X_t(\cdot)}: T_{X_t} M \to T_{X_t(u)} M$$

denotes parallel transport along the C^1 curve $v \mapsto X_t(v)$.

B. *Uniqueness.* Again we may localize in the domain of a chart U. Letting $X(u)$ and $Y(u)$ be two solutions of (5), then for $(t, u) \in [0, \tau_0[\times [0, u_0]$ we find in local coordinates,

$$Y_t(u) - X_t(u) = \int_0^u \big(W(Y(v))_t - W(X(v))_t\big)(\dot{\varphi}(v)) \, dv. \tag{23}$$

On the other hand, using (10) we have

$$\mathbb{E}\left[\sup_{t \leq \tau_0} \|Y_t(u) - X_t(u)\|^2\right] \leq C \int_0^u \mathbb{E}\left[\sup_{t \leq \tau_0} \|Y_t(v) - X_t(v)\|^2\right] dv \tag{24}$$

from which we deduce that almost surely, for all $t \in [0, \tau_0]$, $X_t(u) = Y_t(u)$. Consequently, exploiting the fact that the two processes are continuous in (t, u), they must be indistinguishable. □

3 Horizontal Diffusion Along Non-Homogeneous Diffusion

In this section we assume that the elliptic generator is a C^1 function of time: $L = L(t)$ for $t \geq 0$. Let $g(t)$ be the metric on M such that

$$L(t) = \frac{1}{2}\Delta^t + Z(t)$$

where Δ^t is the $g(t)$-Laplacian and $Z(t)$ a vector field on M.

Let (X_t) be an inhomogeneous diffusion with generator $L(t)$. Parallel transport $P^t(X)_t$ along the $L(t)$-diffusion X_t is defined analogously to [2] as the linear map

$$P^t(X)_t: T_{X_0} M \to T_{X_t} M$$

which satisfies

$$D^t P^t(X)_t = -\frac{1}{2} \dot{g}^\#(P^t(X)_t) \, dt \tag{25}$$

where $\dot{g}$ denotes the derivative of g with respect to time; the covariant differential D^t is defined in local coordinates by the same formulas as D, with the only difference that Christoffel symbols now depend on t.

Alternatively, if J is a semimartingale over X, the covariant differential $D^t J$ may be defined as $\tilde{D}(0, J) = (0, D^t J)$, where $(0, J)$ is a semimartingale along (t, X_t) in $\tilde{M} = [0, T] \times M$ endowed with the connection $\tilde{\nabla}$ defined as follows: if

$$s \mapsto \tilde{\varphi}(s) = (f(s), \varphi(s))$$

is a C^1 path in $\tilde{M}$ and $s \mapsto \tilde{u}(s) = (\alpha(s), u(s)) \in T\tilde{M}$ is C^1 path over $\tilde{\varphi}$, then

$$\tilde{\nabla}\tilde{u}(s) = \left(\dot{\alpha}(s), \left(\nabla^{f(s)} u \right)(s) \right)$$

where ∇^t denotes the Levi–Civita connection associated to $g(t)$. It is proven in [2] that $P^t(X)_t$ is an isometry from $(T_{X_0} M, g(0, X_0))$ to $(T_{X_t} M, g(t, X_t))$.

The damped parallel translation $W^t(X)_t$ along X_t is the linear map

$$W^t(X)_t : T_{X_0} M \to T_{X_t} M$$

satisfying

$$D^t W^t(X)_t = \left(\nabla^t_{W^t(X)_t} Z(t, \cdot) - \frac{1}{2} (\mathrm{Ric}^t)^\sharp (W^t(X)_t) \right) dt. \tag{26}$$

If $Z \equiv 0$ and $g(t)$ is solution to the backward Ricci flow:

$$\dot{g} = \mathrm{Ric}, \tag{27}$$

then damped parallel translation coincides with the usual parallel translation:

$$P^t(X) = W^t(X),$$

(see [2], Theorem 2.3).

The Itô differential $d^\nabla Y = d^{\nabla^t} Y$ of an M-valued semimartingale Y is defined by formula (2), with the only difference that the Christoffel symbols depend on time.

Theorem 3.1. *Keeping the assumptions of this section, let*

$$\mathbb{R} \to M, \quad u \mapsto \varphi(u),$$

be a C^1 path in M and let X^0 be an $L(t)$-diffusion with starting point $\varphi(0)$ and lifetime ξ. Assume that $(M, g(t))$ is complete for every t. There exists a unique family

$$u \mapsto (X_t(u))_{t \in [0, \xi[}$$

of $L(t)$-diffusions, which is a.s. continuous in (t, u) and C^1 in u, satisfying

$$X(0) = X^0 \quad \text{and} \quad X_0(u) = \varphi(u),$$

and solving the equation

$$\partial X_t(u) = W^t(X(u))_t(\dot{\varphi}(u)). \tag{28}$$

Furthermore, $X(u)$ solves the Itô stochastic differential equation

$$d^\nabla X_t(u) = P_{0,u}^{t,X_t(\cdot)} d^{\nabla(t)} X_t + Z(t, X_t(u)) \, dt, \tag{29}$$

where

$$P_{0,u}^{t,X_t(\cdot)} : T_{X_t^0} M \to T_{X_t(u)} M$$

denotes parallel transport along the C^1 curve

$$[0, u] \to M, \quad v \mapsto X_t(v),$$

with respect to the metric $g(t)$.

   If $Z \equiv 0$ and if $g(t)$ is given as solution to the backward Ricci flow equation, then almost surely for all t,

$$\|\partial X_t(u)\|_{g(t)} = \|\dot{\varphi}(u)\|_{g(0)}. \tag{30}$$

Definition 3.2. We call

$$t \mapsto (X_t(u))_{u \in \mathbb{R}}$$

the horizontal $L(t)$-diffusion in C^1 path space $C^1(\mathbb{R}, M)$ over X^0, started at φ.

Remark 3.3. Equation (30) says that if $Z \equiv 0$ and if g is solution to the backward Ricci flow equation, then the horizontal $g(t)$-Brownian motion is length preserving (with respect to the moving metric).

Remark 3.4. Again if the manifold $(M, g(t))$ is not necessarily complete for all t, a similar result holds with the lifetime of $X_.(u)$ possibly depending on u.

Proof (Proof of Theorem 3.1). The proof is similar to the one of Theorem 2.1. We restrict ourselves to explaining the differences.

   The localization procedure carries over immediately; we work on the time interval $[0, \tau \wedge t_0]$.

   For $\alpha > 0$, we define the approximating process $X_t^\alpha(u)$ by induction as

$$X_t^\alpha(0) = X_t^0, \quad X_0^\alpha(u) = \varphi(u),$$

and if $u \in \,]n\alpha, (n+1)\alpha]$ for some integer $n \geq 0$, then $X^\alpha(u)$ solves the Itô equation

$$d^\nabla X_t^\alpha(u) = P_{X_t^\alpha(n\alpha), X_t^\alpha(u)}^t d_m X_t^\alpha(n\alpha) + Z(t, X_t(u)) \, dt \tag{31}$$

where $P_{x,y}^t$ is the parallel transport along the minimal geodesic from x to y, for the connection ∇^t.

Alternatively, letting $\tilde{X}_t^\alpha = (t, X_t^\alpha)$, we may write (31) as

$$d^{\tilde{\nabla}}\tilde{X}_t^\alpha(u) = \tilde{P}_{\tilde{X}_t^\alpha(n\alpha),\tilde{X}_t^\alpha(u)}d_m\tilde{X}_t^\alpha(n\alpha) + Z(\tilde{X}_t^\alpha(u))\,dt \tag{32}$$

where $\tilde{P}_{\tilde{x},\tilde{y}}$ denotes parallel translation along the minimal geodesic from $\tilde{x}$ to $\tilde{y}$ for the connection $\tilde{\nabla}$.

Denoting by $\rho(t, x, y)$ the distance from x to y with respect to the metric $g(t)$, Itô's formula shows that the process $\rho\left(t, X_t^\alpha(u), X_t^\alpha(n\alpha)\right)$ has locally bounded variation. Moreover since locally $\partial_t\rho(t, x, y) \le C\rho(t, x, y)$ for $x \ne y$, we find similarly to (8),

$$\rho(t, X_t^\alpha(u), X_t^\alpha(n\alpha)) \le \rho(0, X_0^\alpha(u), X_0^\alpha(n\alpha))\,e^{Ct} \le \|\dot{\varphi}\|_\infty \alpha\,e^{Ct_0}.$$

Since all Riemannian distances are locally equivalent, this implies

$$\rho(X_t^\alpha(u), X_t^\alpha(n\alpha)) \le \rho(X_0^\alpha(u), X_0^\alpha(n\alpha))\,e^{Ct} \le \|\dot{\varphi}\|_\infty \alpha\,e^{Ct_0} \tag{33}$$

where $\rho = \rho(0, \cdot, \cdot)$.

Next, differentiating (32) yields

$$\tilde{D}\partial_u\tilde{X}_t^\alpha(u) = \tilde{\nabla}_{\partial_u\tilde{X}_t^\alpha(u)}\tilde{P}_{\tilde{X}_t^\alpha(n\alpha),\cdot}d_m\tilde{X}_t^\alpha(n\alpha)$$
$$+ \tilde{\nabla}_{\partial_u\tilde{X}_t^\alpha(u)}Z\,dt - \frac{1}{2}\tilde{R}\left(\partial_u\tilde{X}_t^\alpha(u), d\,\tilde{X}_t^\alpha(u)\right)d\,\tilde{X}_t^\alpha(u).$$

Using the fact that the first component of $\tilde{X}_t^\alpha(u)$ has finite variation, a careful computation of $\tilde{R}$ leads to the equation

$$D^t\partial_u X_t^\alpha(u) = \nabla_{\partial_u X_t^\alpha(u)}^t P_{X_t^\alpha(n\alpha),\cdot}^t d_m X_t^\alpha(n\alpha)$$
$$+ \nabla_{\partial_u X_t^\alpha(u)}^t Z(t, \cdot) - \frac{1}{2}(\text{Ric}^t)^\sharp\left(\partial_u X_t^\alpha(u)\right)dt.$$

To finish the proof, it is sufficient to remark that in step 1, (11) still holds true for X and Y $g(t)$-Brownian motions living in a small open set U, and that in step 5, the map $u \mapsto \partial X(u)$ is continuous in the topology of semimartingales. This last point is due to the fact that all $\partial X(u)$ are inhomogeneous diffusions with the same generator, say L', and the fact that the topology of uniform convergence on compact sets and the topology of semimartingales coincide on L'-diffusions. □

4 Application to Optimal Transport

In this section we assume again that the elliptic generator $L(t)$ is a C^1 function of time with associated metric $g(t)$:

$$L(t) = \frac{1}{2}\Delta^t + Z(t), \quad t \in [0, T],$$

where Δ^t is the Laplacian associated to $g(t)$ and $Z(t)$ is a vector field. We assume further that for any t, the Riemannian manifold $(M, g(t))$ is metrically complete, and $L(t)$ diffusions have lifetime T.

Letting $\varphi\colon [0, T] \to \mathbb{R}_+$ be a non-decreasing function, we define a cost function

$$c(t, x, y) = \varphi(\rho(t, x, y)) \tag{34}$$

where $\rho(t, \cdot, \cdot)$ denotes distance with respect to $g(t)$.

To the cost function c we associate the Monge–Kantorovich minimization between two probability measures on M

$$\mathscr{W}_{c,t}(\mu, \nu) = \inf_{\eta \in \Pi(\mu, \nu)} \int_{M \times M} c(t, x, y) \, d\eta(x, y) \tag{35}$$

where $\Pi(\mu, \nu)$ is the set of all probability measures on $M \times M$ with marginals μ and ν. We denote

$$\mathscr{W}_{p,t}(\mu, \nu) = \left(\mathscr{W}_{\rho^p,t}(\mu, \nu) \right)^{1/p} \tag{36}$$

the Wasserstein distance associated to $p > 0$. For a probability measure μ on M, the solution of the heat flow equation associated to $L(t)$ will be denoted by μP_t.

Define a section $(\nabla^t Z)^\flat \in \Gamma(T^*M \odot T^*M)$ as follows: for any $x \in M$ and $u, v \in T_x M$,

$$(\nabla^t Z)^\flat(u, v) = \frac{1}{2} \left(g(t)(\nabla_u^t Z, v) + g(t)(u, \nabla_v^t Z) \right).$$

In case the metric is independent of t and $Z = \operatorname{grad} V$ for some C^2 function V on M, then

$$(\nabla^t Z)^\flat(u, v) = \nabla dV(u, v).$$

Theorem 4.1. *We keep notation and assumptions from above.*

(a) *Assume*

$$\operatorname{Ric}^t - \dot{g} - 2(\nabla^t Z)^\flat \geq 0, \quad t \in [0, T]. \tag{37}$$

Then the function

$$t \mapsto \mathscr{W}_{c,t}(\mu P_t, \nu P_t)$$

is non-increasing on $[0, T]$.

(b) *If for some $k \in \mathbb{R}$,*

$$\operatorname{Ric}^t - \dot{g} - 2(\nabla^t Z)^\flat \geq kg, \quad t \in [0, T], \tag{38}$$

then we have for all $p > 0$

$$\mathscr{W}_{p,t}(\mu P_t, \nu P_t) \leq e^{-kt/2} \mathscr{W}_{p,0}(\mu, \nu), \quad t \in [0, T].$$

Remark 4.2. Before turning to the proof of Theorem 4.1, let us mention that in the case $Z = 0$, g constant, $p = 2$ and $k = 0$, item (b) is due to [21] and [20]. In the case where g is a backward Ricci flow solution, $Z = 0$ and $p = 2$, statement (b) for M compact is due to Lott [18] and McCann–Topping [19]. For extensions about $\mathcal{L}$-transportation, see [23].

Proof (Proof of Theorem 4.1). (a) Assume that $\mathrm{Ric}^t - \dot{g} - 2(\nabla^t Z)^{\flat} \geq 0$. Then for any $L(t)$-diffusion (X_t), we have

$$
\begin{aligned}
d\big(g(t)&(W(X)_t, W(X)_t)\big) \\
&= \dot{g}(t)\big(W(X)_t, W(X)_t\big)\,dt + 2g(t)\big(D^t W(X)_t, W(X)_t\big) \\
&= \dot{g}(t)\big(W(X)_t, W(X)_t\big)\,dt \\
&\quad + 2g(t)\left(\nabla^t_{W(X)_t} Z(t, \cdot) - \frac{1}{2}(\mathrm{Ric}^t)^{\#}(W(X)_t), W(X)_t\right)dt \\
&= \left(\dot{g} + 2(\nabla^t Z)^{\flat} - \mathrm{Ric}^t\right)\big(W(X)_t, W(X)_t\big)\,dt \leq 0.
\end{aligned}
$$

Consequently, for any $t \geq 0$,

$$
\|W(X)_t\|_t \leq \|W(X)_0\|_0 = 1. \tag{39}
$$

For $x, y \in M$, let $u \mapsto \gamma(x, y)(u)$ be a minimal $g(0)$-geodesic from x to y in time 1: $\gamma(x, y)(0) = x$ and $\gamma(x, y)(1) = y$. Denote by $X^{x,y}(u)$ a horizontal $L(t)$-diffusion with initial condition $\gamma(x, y)$.

For $\eta \in \Pi(\mu, \nu)$, define the measure η_t on $M \times M$ by

$$
\eta_t(A \times B) = \int_{M \times M} \mathbb{P}\{X^{x,y}_t(0) \in A, \ X^{x,y}_t(1) \in B\}\,d\eta(x, y),
$$

where A and B are Borel subsets of M. Then η_t has marginals μP_t and νP_t. Consequently it is sufficient to prove that for any such η,

$$
\int_{M \times M} \mathbb{E}\big[c(t, X^{x,y}_t(0), X^{x,y}_t(1))\big]\,d\eta(x, y) \leq \int_{M \times M} c(0, x, y)\,d\eta(x, y). \tag{40}
$$

On the other hand, we have a.s.,

$$
\begin{aligned}
\rho(t, X^{x,y}_t(0), X^{x,y}_t(1)) &\leq \int_0^1 \left\|\partial_u X^{x,y}_t(u)\right\|_t du \\
&= \int_0^1 \left\|W(X^{x,y}(u))_t \, \dot{\gamma}(x, y)(u)\right\|_t du \\
&\leq \int_0^1 \left\|\dot{\gamma}(x, y)(u)\right\|_0 du \\
&= \rho(0, x, y),
\end{aligned}
$$

and this clearly implies

$$c\big(t, X_t^{x,y}(0), X_t^{x,y}(1)\big) \le c(0, x, y) \quad \text{a.s.,}$$

and then (40).

(b) Under condition (38), we have

$$\frac{\mathrm{d}}{\mathrm{d}t} g(t)\big(W(X)_t, W(X)_t\big) \le -k\, g(t)\big(W(X)_t, W(X)_t\big),$$

which implies

$$\|W(X)_t\|_t \le \mathrm{e}^{-kt/2},$$

and then

$$\rho\big(t, X_t^{x,y}(0), X_t^{x,y}(1)\big) \le \mathrm{e}^{-kt/2} \rho(0, x, y).$$

The result follows. $\qquad\qquad\square$

5 Derivative Process Along Constant Rank Diffusion

In this section we consider a generator L of constant rank: the image E of the "carré du champ" operator $\Gamma(L) \in \Gamma(TM \otimes TM)$ defines a subbundle of TM. In E we then have an intrinsic metric given by

$$g(x) = (\Gamma(L)|E(x))^{-1}, \quad x \in M.$$

Let ∇ be a connection on E with preserves g, and denote by ∇' the associated semi-connection: if $U \in \Gamma(TM)$ is a vector field, $\nabla'_v U$ is defined only if $v \in E$ and satisfies

$$\nabla'_v U = \nabla_{U_{x_0}} V + [V, U]_{x_0}$$

where $V \in \Gamma(E)$ is such that $V_{x_0} = v$ (see [11], Sect. 1.3). We denote by $Z(x)$ the drift of L with respect to the connection ∇.

For the construction of a flow of L-diffusions we will use an extension of ∇ to TM denoted by $\tilde{\nabla}$. Then the associated semi-connection ∇' is the restriction of the classical adjoint of $\tilde{\nabla}$ (see [11], Proposition 1.3.1).

Remark 5.1. It is proven in [11] that a connection ∇ always exists, for instance, we may take the Le Jan-Watanabe connection associated to a well chosen vector bundle homomorphism from a trivial bundle $M \times H$ to E where H is a Hilbert space.

If X_t is an L-diffusion, the parallel transport

$$P(X)_t : E_{X_0} \to E_{X_t}$$

along X_t (with respect to the connection $\tilde{\nabla}$) depends only on ∇. The same applies for the Itô differential $dX_t = d^\nabla X_t$. We still denote by $d_m X_t$ its martingale part.

We denote by

$$\hat{P}'(X)_t : T_{X_0}M \to T_{X_t}M$$

the parallel transport along X_t for the adjoint connection $(\tilde{\nabla})'$, and by $\tilde{D}'J$ the covariant differential (with respect to $(\tilde{\nabla})'$) of a semimartingale $J \in TM$ above X; compare (4) for the definition.

Theorem 5.2. *We keep the notation and assumptions from above. Let x_0 be a fixed point in M and $X_t(x_0)$ an L-diffusion starting at x_0. For $x \in M$ close to x_0, we define the L-diffusion $X_t(x)$, started at x, by*

$$dX_t(x) = \hat{P}_{X_t(x_0),X_t(x)} \, d_m X_t(x_0) + Z(X_t(x)) \, dt \qquad (41)$$

where $\hat{P}_{x,y}$ denotes parallel transport (with respect to $\tilde{\nabla}$) along the unique $\tilde{\nabla}$-geodesic from x to y. Then

$$\tilde{D}' T_{x_0} X = \tilde{\nabla}_{T_{x_0}X} Z \, dt - \frac{1}{2} \operatorname{Ric}^\#(T_{x_0}X) \, dt \qquad (42)$$

where

$$\operatorname{Ric}^\#(u) = \sum_{i=1}^d \tilde{R}(u, e_i)e_i, \quad u \in T_xM,$$

and $(e_i)_{i=1,\ldots,d}$ an orthonormal basis of E_x for the metric g.

Under the additional assumption that $Z \in \Gamma(E)$, the differential $\tilde{D}' T_{x_0}X$ does not depend on the extension $\tilde{\nabla}$, and we have

$$\tilde{D}' T_{x_0} X = \nabla_{T_{x_0}X} Z \, dt - \frac{1}{2} \operatorname{Ric}^\#(T_{x_0}X) \, dt. \qquad (43)$$

Proof. From [3, eq. (7.4)] we have

$$
\begin{aligned}
\tilde{D}' T_{x_0} X = & \ \tilde{\nabla}_{T_{x_0}X} \hat{P}_{X_t(x_0),\,.} \, d_m X_t(x_0) + \tilde{\nabla}_{T_{x_0}X} Z \, dt \\
& - \frac{1}{2}\left(\tilde{R}'(T_{x_0}X, dX(x_0)) \, dX(x_0) + \tilde{\nabla}' \tilde{T}'(dX(x_0), T_{x_0}X, dX(x_0)) \right) \\
& - \frac{1}{2} \tilde{T}'(\tilde{D}' T_{x_0}X, dX)
\end{aligned}
$$

where $\tilde{T}'$ denotes the torsion tensor of $\tilde{\nabla}'$. Since for all $x \in M$, $\tilde{\nabla}_v \hat{P}_{x,.} = 0$ if $v \in T_xM$, the first term in the right vanishes. As a consequence, $\tilde{D}' T_{x_0}X$ has finite variation, and $T'(\tilde{D}' T_{x_0}X, dX) = 0$. Then using the identity

$$\tilde{R}'(v, u)u + \tilde{\nabla}' \tilde{T}'(u, v, u) = \tilde{R}(v, u)u, \quad u, v \in T_xM,$$

which is a particular case of identity (C.17) in [11], we obtain

$$\tilde{D}' T_{x_0} X = \tilde{\nabla}_{T_{x_0} X} Z \, dt - \frac{1}{2} \tilde{R}(T_{x_0} X, dX(x_0)) dX(x_0).$$

Finally writing

$$\tilde{R}(T_{x_0} X, dX(x_0)) dX(x_0) = \mathrm{Ric}^\sharp (T_{x_0} X) \, dt$$

yields the result. □

Remark 5.3. In the non-degenerate case, ∇ is the Levi–Civita connection associated to the metric generated by L, and we are in the situation of Sect. 2. In the degenerate case, in general, ∇ does not extend to a metric connection on M. However conditions are given in [11] (1.3.C) under which $P'(X)$ is adapted to some metric, and in this case $T_{x_0} X$ is bounded with respect to the metric.

One would like to extend Theorem 2.1 to degenerate diffusions of constant rank, by solving the equation

$$\partial_u X(u) = \tilde{\nabla}_{\partial_u X(u)} Z \, dt - \frac{1}{2} \mathrm{Ric}^\sharp (\partial_u X(u)) \, dt.$$

Our proof does not work in this situation for two reasons. The first one is that in general $\tilde{P}'(X)$ is not adapted to a metric. The second one is the lack of an inequality of the type (8) since ∇ does not have an extension $\tilde{\nabla}$ which is the Levi–Civita connection of some metric.

Remark 5.4. When M is a Lie group and L is left invariant, then $\tilde{\nabla}$ can be chosen as the left invariant connection. In this case $(\tilde{\nabla})'$ is the right invariant connection, which is metric.

Acknowledgements The first named author wishes to thank the University of Luxembourg for support.

Note added in proof Using recent results of Kuwada and Philipowski [17], the condition at the beginning of Sect. 4 that $L(t)$ diffusions have lifetime T is automatically satisfied in the case of a family of metrics $g(t)$ evolving by backward Ricci flow on a $g(0)$-complete manifold M. Thus our Theorem 4.1 extends in particular the result of McCann–Topping [19] and Topping [23] from compact to complete manifolds.

References

1. Arnaudon, M.: Differentiable and analytic families of continuous martingales in manifolds with connections. Probab. Theor. Relat. Field **108**(3), 219–257 (1997)
2. Arnaudon, M., Coulibaly, K.A., Thalmaier, A.: Brownian motion with respect to a metric depending on time; definition, existence and applications to Ricci flow. C. R. Math. Acad. Sci. Paris **346**(13–14), 773–778 (2008)

3. Arnaudon, M., Thalmaier, A.: Horizontal martingales in vector bundles, Séminaire de Probabilités XXXVI. Lecture Notes in Mathematics, vol. 1801, pp. 419–456. Springer, Berlin (2003)

4. Arnaudon, M., Thalmaier, A.: Stability of stochastic differential equations in manifolds, Séminaire de Probabilités XXXII. Lecture Notes in Mathematics, vol. 1686, pp. 188–214. Springer, Berlin (1998)

5. Arnaudon, M., Thalmaier, A., Wang, F.-Y.: Gradient estimate and Harnack inequality on noncompact Riemannian manifolds. Stoch. Process. Appl. **119**(10), 3653–3670 (2009)

6. Arnaudon, M., Thalmaier, A., Wang, F.-Y.: Harnack inequality and heat kernel estimates on manifolds with curvature unbounded below. Bull. Sci. Math. **130**, 223–233 (2006)

7. Cipriano, F., Cruzeiro, A.B.: Flows associated with irregular $\mathbb{R}^d$ vector fields. J. Diff. Equ. **210**(1), 183–201 (2005)

8. Cipriano, F., Cruzeiro, A.B.: Flows associated to tangent processes on the Wiener space. J. Funct. Anal. **166**(2), 310–331 (1999)

9. Cruzeiro, A.B.: Equations différentielles sur l'espace de Wiener et formules de Cameron-Martin non linéaires. J. Funct. Anal. **54**(2), 206–227 (1983)

10. Driver, B.K.: A Cameron-Martin type quasi-invariance theorem for Brownian motion on a compact manifold. J. Funct. Anal. **110**, 272–376 (1992)

11. Elworthy, K.D., Le Jan, Y., Li, X.-M.: On the geometry of diffusion operators and stochastic flows. Lecture Notes in Mathematics, vol. 1720, p. 118. Springer, Berlin (1999)

12. Fang, S., Luo, D.: Quasi-invariant flows associated to tangent processes on the Wiener space, Bull. Sci. Math. **134**(3), 295–328 (2010)

13. Gong, F., Zhang, J.: Flows associated to adapted vector fields on the Wiener space, J. Funct. Anal. **253**(2), 647–674 (2007)

14. Hsu, E.P.: Quasi-invariance of the Wiener measure on path spaces: noncompact case, J. Funct. Anal. **193**(2), 278–290 (2002)

15. Hsu, E.P.: Quasi-invariance of the Wiener measure on the path space over a compact Riemannian manifold. J. Funct. Anal. **134**(2), 417–450 (1995)

16. Kendall, W.: Nonnegative Ricci curvature and the Brownian coupling property. Stochastics **19**(1–2), 111–129 (1986)

17. Kuwada, K., Philipowski, R.: Non-explosion of diffusion processes on manifolds with time-dependent metric, To appear in Math. Z. arXiv:0910.1730

18. Lott, J.: Optimal transport and Ricci curvature for metric-measure spaces. Surveys in differential geometry, vol. 11, pp. 229–257. International Press, Somerville, MA (2007)

19. McCann, R.J., Topping, P.M.: Ricci flow, entropy and optimal transportation. Amer. J. Math. **132**(3), 711–730 (2010)

20. Otto, F., Westdickenberg, M.: Eulerian calculus for the contraction in the Wasserstein distance, SIAM J. Math. Anal. **37**(4), 1227–1255 (electronic) (2005)

21. von Renesse, M.K., Sturm, K.T.: Transport inequalities, gradient estimates, entropy, and Ricci curvature, Comm. Pure Appl. Math. **58**(7), 923–940 (2005)

22. Stroock, D.W.: An introduction to the analysis of paths on a Riemannian manifold, Mathematical Surveys and Monographs, vol. 74. American Mathematical Society, Providence, RI (2000)

23. Topping, P.: $\mathcal{L}$-optimal transportation for Ricci flow, J. Reine Angew. Math. **636**:93–122 (2009)

A Stochastic Calculus Proof of the CLT
for the L^2 Modulus of Continuity of Local Time

Jay Rosen

Abstract We give a stochastic calculus proof of the Central Limit Theorem

$$\frac{\int (L_t^{x+h} - L_t^x)^2 \, dx - 4ht}{h^{3/2}} \overset{\mathcal{L}}{\Longrightarrow} c \left(\int (L_t^x)^2 \, dx \right)^{1/2} \eta$$

as $h \to 0$ for Brownian local time L_t^x. Here η is an independent normal random variable with mean zero and variance one.

Keywords Central limit theorem · Moduli of continuity · Local time · Brownian motion

MSC 2000: Primary 60F05, 60J55, 60J65.

1 Introduction

In [3] we obtain almost sure limits for the L^p moduli of continuity of local times of a very wide class of symmetric Lévy processes. More specifically, if $\{L_t^x \, ; \, (x,t) \in R^1 \times R_+^1\}$ denotes Brownian local time then for all $p \geq 1$, and all $t \in R_+$,

$$\lim_{h \downarrow 0} \int_a^b \left| \frac{L_t^{x+h} - L_t^x}{\sqrt{h}} \right|^p dx = 2^p E(|\eta|^p) \int_a^b |L_t^x|^{p/2} \, dx$$

for all a, b in the extended real line almost surely, and also in L^m, $m \geq 1$. (Here η is normal random variable with mean zero and variance one.) In particular when $p = 2$ we have

$$\lim_{h \downarrow 0} \int \frac{(L_t^{x+h} - L_t^x)^2}{h} \, dx = 4t, \qquad \text{almost surely.} \tag{1}$$

J. Rosen (✉)
Department of Mathematics, College of Staten Island, CUNY, Staten Island, NY 10314, USA
e-mail: jrosen30@optimum.net

C. Donati-Martin et al. (eds.), *Séminaire de Probabilités XLIII*, Lecture Notes in Mathematics 95
2006, DOI 10.1007/978-3-642-15217-7_3, © Springer-Verlag Berlin Heidelberg 2011

We refer to $\int (L_t^{x+h} - L_t^x)^2 \, dx$ as the L^2 modulus of continuity of Brownian local time.

In our recent paper [1] we obtain the central limit theorem corresponding to (1).

Theorem 1. *For each fixed t*

$$\frac{\int (L_t^{x+h} - L_t^x)^2 \, dx - 4ht}{h^{3/2}} \overset{\mathcal{L}}{\Longrightarrow} c \left(\int (L_t^x)^2 \, dx \right)^{1/2} \eta \tag{2}$$

as $h \to 0$, *with* $c = \left(\frac{64}{3} \right)^{1/2}$. *Equivalently*

$$\frac{\int (L_t^{x+1} - L_t^x)^2 \, dx - 4t}{t^{3/4}} \overset{\mathcal{L}}{\Longrightarrow} c \left(\int (L_1^x)^2 \, dx \right)^{1/2} \eta \tag{3}$$

as $t \to \infty$. *Here* η *is an independent normal random variable with mean zero and variance one.*

It can be shown that

$$E \left(\int (L_t^{x+1} - L_t^x)^2 \, dx \right) = 4 \left(t - \frac{2t^{1/2}}{\sqrt{2\pi}} \right) + O(1).$$

so that (3) can be written as

$$\frac{\int (L_t^{x+1} - L_t^x)^2 \, dx - E \left(\int (L_t^{x+1} - L_t^x)^2 \, dx \right)}{t^{3/4}} \overset{\mathcal{L}}{\Longrightarrow} c \left(\int (L_1^x)^2 \, dx \right)^{1/2} \eta$$

with a similar statement for (2).

Our proof of Theorem 1 in [1] is rather long and involved. We use the method of moments, but rather than study the asymptotics of the moments of (2), which seem intractable, we study the moments of the analogous expression where the fixed time t is replaced by an independent exponential time of mean $1/\lambda$. An important part of the proof is then to "invert the Laplace transform" to obtain the asymptotics of the moments for fixed t.

The purpose of this paper is to give a new and shorter proof of Theorem 1 using stochastic integrals, following the approach of [8,9]. Our proof makes use of certain differentiability properties of the double and triple intersection local time, $\alpha_{2,t}(x)$ and $\alpha_{3,t}(x, y)$, which are formally given by

$$\alpha_{2,t}(x) = \int_0^t \int_0^s \delta(W_s - W_r - x) \, dr \, ds$$

and

$$\alpha_{3,t}(x, y) = \int_0^t \int_0^s \int_0^r \delta(W_r - W_{r'} - x)\delta(W_s - W_r - y) \, dr' \, dr \, ds.$$

More precisely, let $f(x)$ be a smooth positive symmetric function with compact support and $\int f(x)\,dx = 1$. Set $f_\epsilon(x) = \frac{1}{\epsilon} f(x/\epsilon)$. Then

$$\alpha_{2,t}(x) = \lim_{\epsilon \to 0} \int_0^t \int_0^s f_\epsilon(W_s - W_r - x)\,dr\,ds$$

and

$$\alpha_{3,t}(x, y) = \lim_{\epsilon \to 0} \int_0^t \int_0^s \int_0^r f_\epsilon(W_r - W_{r'} - x) f_\epsilon(W_s - W_r - y)\,dr'\,dr\,ds$$

exist almost surely and in all L^p, are independent of the particular choice of f, and are continuous in (x, y, t) almost surely, [6]. It is easy to show, see [7, Theorem 2], that for any measurable $\phi(x)$

$$\int_0^t \int_0^s \phi(W_s - W_r)\,dr\,ds = \int \phi(x)\alpha_{2,t}(x)\,dx \tag{4}$$

and for any measurable $\phi(x, y)$

$$\int_0^t \int_0^s \int_0^r \phi(W_r - W_{r'}, W_s - W_r)\,dr'\,dr\,ds = \int \phi(x, y)\alpha_{3,t}(x, y)\,dx\,dy. \tag{5}$$

To express the differentiability properties of $\alpha_{2,t}(x)$ and $\alpha_{3,t}(x, y)$ which we need, let us set

$$v(x) = \int_0^\infty e^{-s/2} p_s(x)\,ds = e^{-|x|}.$$

The following result is [7, Theorem 1].

Theorem 2. *It holds that*

$$\gamma_{2,t}(x) =: \alpha_{2,t}(x) - tv(x)$$

and

$$\gamma_{3,t}(x, y) =: \alpha_{3,t}(x, y) - \gamma_{2,t}(x)v(y) - \gamma_{2,t}(y)v(x) - tv(x)v(y)$$

are C^1 in (x, y) and $\nabla\gamma_{2,t}(x), \nabla\gamma_{3,t}(x, y)$ are continuous in (x, y, t).

Our new proof of Theorem 1 is given in Sect. 2.

Our original motivation for studying the asymptotics of $\int (L_t^{x+h} - L_t^x)^2\,dx$ comes from our interest in the Hamiltonian

$$H_n = \sum_{i,j=1, i \neq j}^n 1_{\{S_i = S_j\}} - \frac{1}{2} \sum_{i,j=1, i \neq j}^n 1_{\{|S_i - S_j| = 1\}},$$

for the critical attractive random polymer in dimension one, [2], where $\{S_n \; ; \; n = 0, 1, 2, \ldots\}$ is a simple random walk on Z^1. Note that $H_n = \sum_{x \in Z^1} \left(l_n^x - l_n^{x+1} \right)^2$, where $l_n^x = \sum_{i=1}^{n} 1_{\{S_i = x\}}$ is the local time for the random walk S_n.

2   A Stochastic Calculus Approach

By [4, Lemma 2.4.1] we have that

$$L_t^x = \lim_{\epsilon \to 0} \int_0^t f_\epsilon(W_s - x) \, ds$$

almost surely, with convergence locally uniform in x. Hence

$$\int L_t^{x+h} L_t^x \, dx$$

$$= \int \lim_{\epsilon \to 0} \left(\int_0^t f_\epsilon(W_s - (x+h)) \, ds \right) \left(\int_0^t f_\epsilon(W_r - x) \, dr \right) dx$$

$$= \lim_{\epsilon \to 0} \int \left(\int_0^t f_\epsilon(W_s - (x+h)) \, ds \right) \left(\int_0^t f_\epsilon(W_r - x) \, dr \right) dx$$

$$= \lim_{\epsilon \to 0} \int_0^t \int_0^t f_\epsilon * f_\epsilon(W_s - W_r - h) \, dr \, ds$$

$$= \lim_{\epsilon \to 0} \int_0^t \int_0^s f_\epsilon * f_\epsilon(W_s - W_r - h) \, dr \, ds$$

$$+ \lim_{\epsilon \to 0} \int_0^t \int_0^r f_\epsilon * f_\epsilon(W_r - W_s + h) \, ds \, dr$$

$$= \alpha_{2,t}(h) + \alpha_{2,t}(-h). \tag{6}$$

Note that

$$\int (L_t^{x+h} - L_t^x)^2 \, dx = 2 \left(\int (L_t^x)^2 \, dx - \int L_t^{x+h} L_t^x \, dx \right)$$

and thus

$$\int (L_t^{x+h} - L_t^x)^2 \, dx = 2 \left(2\alpha_{2,t}(0) - \alpha_{2,t}(h) - \alpha_{2,t}(-h) \right).$$

Hence we can prove Theorem 1 by showing that for each fixed t

$$\frac{2 \left(2\alpha_{2,t}(0) - \alpha_{2,t}(h) - \alpha_{2,t}(-h) \right) - 4ht}{h^{3/2}} \overset{\mathcal{L}}{\Longrightarrow} c \sqrt{\alpha_{2,t}(0)} \, \eta \tag{7}$$

as $h \to 0$, with $c = \left(\frac{128}{3}\right)^{1/2}$. Here we used the fact, which follows from (6), that $\int (L_1^x)^2 \, dx = 2 \, \alpha_{2,t}(0)$.

In proving (7) we will need the following Lemma. Compare Tanaka's formula, [5, Chap. VI, Theorem 1.2].

Lemma 1. *For any $a \in R^1$,*

$$\alpha_{2,t}(a) = 2 \int_0^t (W_t - W_s - a)^+ \, ds - 2(-a)^+ t - 2 \int_0^t \int_0^s 1_{\{W_s - W_r > a\}} \, dr \, dW_s.$$

Proof. Set

$$g_\epsilon(x) = \int_0^\infty y f_\epsilon(x - y) \, dy$$

so that

$$g_\epsilon'(x) = \int_0^\infty y f_\epsilon'(x - y) \, dy = \int_0^\infty f_\epsilon(x - y) \, dy \qquad (8)$$

and consequently

$$g_\epsilon''(x) = f_\epsilon(x). \qquad (9)$$

Let

$$F_a(t, x) = \int_0^t g_\epsilon(x - W_s - a) \, ds.$$

Then by Ito's formula, [5, Chap. IV, (3.12)], applied to the non-anticipating functional $F_a(t, x)$ we have

$$\int_0^t g_\epsilon(W_t - W_s - a) \, ds$$

$$= \int_0^t g_\epsilon(-a) \, ds + \int_0^t \int_0^s g_\epsilon'(W_s - W_r - a) \, dr \, dW_s$$

$$+ \frac{1}{2} \int_0^t \int_0^s g_\epsilon''(W_s - W_r - a) \, dr \, ds.$$

It is easy to check that locally uniformly

$$\lim_{\epsilon \to 0} g_\epsilon(x) = x^+$$

and hence using (9) we obtain

$$\alpha_{2,t}(a)$$

$$= 2 \int_0^t (W_t - W_s - a)^+ \, ds - 2(-a)^+ t - 2 \lim_{\epsilon \to 0} \int_0^t \int_0^s g_\epsilon'(W_s - W_r - a) \, dr \, dW_s.$$

From (8) we can see that $\sup_x |g'_\epsilon(x)| \leq 1$ and

$$\lim_{\epsilon \to 0} g'_\epsilon(x) = 1_{\{x>0\}} + \frac{1}{2} 1_{\{x=0\}}.$$

Thus by the dominated convergence theorem

$$\lim_{\epsilon \to 0} \int_0^t E\left(\left(\int_0^s \{g'_\epsilon(W_s - W_r - a) - 1_{\{W_s - W_r > a\}}\} \, dr\right)^2\right) ds = 0$$

which completes the proof of our Lemma.   □

Proof (Proof of Theorem 1). If we now set

$$J_h(x) = 2x^+ - (x - h)^+ - (x + h)^+ \begin{cases} -x - h & \text{if } -h \leq x \leq 0, \\ x - h & \text{if } 0 \leq x \leq h. \end{cases} \tag{10}$$

and

$$K_h(x) = 21_{\{x>0\}} - 1_{\{x>h\}} - 1_{\{x>-h\}} = 1_{\{0<x\leq h\}} - 1_{\{-h<x\leq 0\}}$$

we see from Lemma 1 that

$$2\{2\alpha_t(0) - \alpha_t(h) - \alpha_t(-h)\} - 4ht$$
$$= 4 \int_0^t J_h(W_t - W_s) \, ds - 4 \int_0^t \int_0^s K_h(W_s - W_r) \, dr \, dW_s. \tag{11}$$

By (10),

$$\int_0^t J_h(W_t - W_s) \, ds = \int J_h(W_t - x) L_t^x \, dx = O(h^2 \sup_x L_t^x).$$

Hence to prove (7) it suffices to show that for each fixed t

$$\frac{\int_0^t \int_0^s K_h(W_s - W_r) \, dr \, dW_s}{h^{3/2}} \overset{\mathcal{L}}{\Longrightarrow} \left(\frac{8}{3}\right)^{1/2} \sqrt{\alpha_{2,t}(0)} \, \eta \tag{12}$$

as $h \to 0$. Let
$$M_t^h = h^{-3/2} \int_0^t \int_0^s K_h(W_s - W_r) \, dr \, dW_s.$$

It follows from the proof of Theorem 2.6 in [5, Chap. XIII], (the Theorem of Papanicolaou, Stroock, and Varadhan) that to establish (12) it suffices to show that

$$\lim_{h \to 0} \langle M^h, W \rangle_t = 0 \tag{13}$$

and

$$\lim_{h \to 0} \langle M^h, M^h \rangle_t = \frac{8}{3} \alpha_{2,t}(0) \tag{14}$$

uniformly in t on compact intervals.

By (4), and using the fact that $K_h(x) = K_1(x/h)$, we have that

$$\langle M^h, W \rangle_t = h^{-3/2} \int_0^t \int_0^s K_h(W_s - W_r) \, dr \, ds$$

$$= h^{-3/2} \int K_h(x) \alpha_{2,t}(x) \, dx$$

$$= h^{-1/2} \int K_1(x) \alpha_{2,t}(hx) \, dx$$

$$= \int_0^1 \frac{\alpha_{2,t}(hx) - \alpha_{2,t}(-hx)}{h^{1/2}} \, dx.$$

But $v(hx) = v(-hx)$, so by Lemma 2 we have that

$$\alpha_{2,t}(hx) - \alpha_{2,t}(-hx) = \gamma_{2,t}(hx) - \gamma_{2,t}(-hx) = O(h)$$

which completes the proof of (13).

We next analyze

$$\langle M^h, M^h \rangle_t = h^{-3} \int_0^t \left(\int_0^s K_h(W_s - W_r) \, dr \right)^2 \, ds$$

$$= h^{-3} \int_0^t \left(\int_0^s K_h(W_s - W_r) \, dr \right) \left(\int_0^s K_h(W_s - W_{r'}) \, dr' \right) \, ds$$

$$= h^{-3} \int_0^t \left(\int_0^s \int_0^r K_h(W_s - W_{r'}) K_h(W_s - W_r) \, dr' \, dr \right) \, ds$$

$$+ h^{-3} \int_0^t \left(\int_0^s \int_0^{r'} K_h(W_s - W_r) K_h(W_s - W_{r'}) \, dr \, dr' \right) \, ds. \tag{15}$$

By (5) we have that

$$\int_0^t \int_0^s \int_0^r K_h(W_s - W_{r'}) K_h(W_s - W_r) \, dr' \, dr \, ds$$

$$= \int_0^t \int_0^s \int_0^r K_h(W_s - W_r + W_r - W_{r'}) K_h(W_s - W_r) \, dr' \, dr \, ds$$

$$= \int \int K_h(x + y) K_h(y) \alpha_{3,t}(x, y) \, dx \, dy.$$

Using $K_h(x) = K_1(x/h)$ we have

$$h^{-3} \int_0^t \int_0^s \int_0^r K_h(W_s - W_{r'}) K_h(W_s - W_r) \, dr' \, dr \, ds$$

$$= h^{-3} \int \int K_h(x+y) K_h(y) \alpha_{3,t}(x,y) \, dx \, dy$$

$$= h^{-1} \int \int K_1(x+y) K_1(y) \alpha_{3,t}(hx, hy) \, dx \, dy$$

$$= h^{-1} \int \int K_1(x) K_1(y) \alpha_{3,t}(h(x-y), hy) \, dx \, dy$$

$$= h^{-1} \int_0^1 \int_0^1 A_{3,t}(h, x, y) \, dx \, dy$$

where

$$A_{3,t}(h, x, y) = \alpha_{3,t}(h(x-y), hy) - \alpha_{3,t}(h(-x-y), hy)$$
$$-\alpha_{3,t}(h(x+y), -hy) + \alpha_{3,t}(-h(x-y), -hy).$$

It remains to consider

$$\lim_{h \to 0} \frac{A_{3,t}(h, x, y)}{h}.$$

We now use Lemma 2. Using the fact that $\gamma_{3,t}(x, y)$, $\gamma_{2,t}(x)$ are continuously differentiable

$$\gamma_{3,t}(h(x-y), hy) - \gamma_{3,t}(h(-x-y), hy)$$

$$= h(x-y) \frac{\partial}{\partial x} \gamma_{3,t}(0, hy) - h(-x-y) \frac{\partial}{\partial x} \gamma_{3,t}(0, hy) + o(h)$$

$$= 2hx \frac{\partial}{\partial x} \gamma_{3,t}(0, 0) + o(h)$$

and similarly

$$\gamma_{3,t}(-h(x-y), -hy) - \gamma_{3,t}(h(x+y), -hy)$$

$$= -h(x-y) \frac{\partial}{\partial x} \gamma_{3,t}(0, -hy) - h(x+y) \frac{\partial}{\partial x} \gamma_{3,t}(0, -hy) + o(h)$$

$$= -2hx \frac{\partial}{\partial x} \gamma_{3,t}(0, 0) + o(h)$$

and these two terms cancel up to $o(h)$.

Next,

$$\gamma_{2,t}(h(x-y))v(hy) - \gamma_{2,t}(h(-x-y))v(hy)$$
$$+ \gamma_{2,t}(-h(x-y))v(-hy) - \gamma_{2,t}(h(x+y))v(-hy)$$
$$= h(x-y)\gamma'_{2,t}(0)v(0) - h(-x-y)\gamma'_{2,t}(0)v(0)$$

$$-h(x-y)\gamma'_{2,t}(0)v(0) - h(x+y)\gamma'_{2,t}(0)v(0) + o(h)$$
$$= o(h).$$

On the other hand, using $v(x) = e^{-|x|} = 1 - |x| + O(x^2)$ we have

$$v(h(x-y))\gamma_{2,t}(hy) - v(h(-x-y))\gamma_{2,t}(hy)$$
$$+ v(-h(x-y))\gamma_{2,t}(-hy) - v(h(x+y))\gamma_{2,t}(-hy)$$
$$= -|h(x-y)|\gamma_{2,t}(0) + |h(-x-y)|\gamma_{2,t}(0)$$
$$-|h(x-y)|\gamma_{2,t}(0) + |h(x+y)|\gamma_{2,t}(0) + o(h)$$
$$= 2h(|x+y| - |x-y|)\gamma_{2,t}(0) + o(h).$$

and similarly

$$v(h(x-y))v(hy) - v(h(-x-y))v(hy)$$
$$+ v(-h(x-y))v(-hy) - v(h(x+y))v(-hy)$$
$$= -|h(x-y)|v(0) + |h(-x-y)|v(0)$$
$$-|h(x-y)|v(0) + |h(x+y)|v(0) + O(h^2)$$
$$= 2h(|x+y| - |x-y|)v(0) + O(h^2).$$

Putting this all together and using the fact that $\alpha_{2,t}(0) = \gamma_{2,t}(0) + tv(0)$ we see that

$$\int_0^1 \int_0^1 A_{3,t}(h,x,y)\,dx\,dy = 2h\alpha_{2,t}(0) \int_0^1 \int_0^1 (|x+y| - |x-y|))\,dx\,dy + o(h).$$

Of course

$$\int_0^1 \int_0^1 (|x+y| - |x-y|))\,dx\,dy = \int_0^1 \int_0^x 2y\,dy\,dx + \int_0^1 \int_0^y 2x\,dx\,dy = \frac{2}{3}$$

so that

$$\lim_{h \to 0} \frac{\int_0^1 \int_0^1 A_{3,t}(h,x,y)\,dx\,dy}{h} = \frac{4}{3}\alpha_{2,t}(0).$$

By (15) this gives (14). $\qquad\qquad\qquad\qquad\qquad\qquad\qquad\qquad\qquad\qquad\qquad\qquad\square$

Acknowledgements This research was supported, in part, by grants from the National Science Foundation and PSC-CUNY.

References

1. Chen, X., Li, W., Marcus, M., Rosen, J.: A CLT for the L^2 modulus of continuity of local times of Brownian motion. Ann. Probab. **38**(1), 396–438 (2010) arxiv:0901.1102
2. van der Hofstad, R., Klenke, A., Konig, W.: The critical attractive random polymer in dimension one. J. Stat. Phys. **106**(3–4), 477–520 (2002)

3. Marcus, M.B., Rosen, J.: L^p moduli of continuity of Gaussian processes and local times of symmetric Lévy processes. Ann. Probab. **36**, 594–622 (2008)
4. Marcus, M.B., Rosen, J.: Markov processes, Gaussian processes and local times. Cambridge studies in advanced mathematics, vol. 100. Cambridge University Press, Cambridge (2006)
5. Revuz, D., Yor, M.: Continuous martingales and Brownian motion, 3rd edn. Springer, Berlin (1999)
6. Rosen, J.: Joint continuity of renormalized intersection local times. Ann. Inst. Henri Poincare. **32** 671–700 (1996)
7. Rosen, J.: Continuous differentiability of renormalized intersection local times in R^1. Ann. Inst. Henri Poincare, to appear. arxiv:0910.2919
8. Weinryb, S., Yor, M.: Le mouvement brownien de Lévy indexé par R^3 comme limite centrale des temps locaux d'intersection. In: Séminaire de Probabilités XXII. Lecture Notes in Mathematics vol. 1321, pp. 225–248. Springer, New York (1988). To appear
9. Yor, M.: Le drap brownien comme limite en lois des temps locaux linéaires. Séminaire de Probabilités XVII, Lecture Notes in Mathematics vol. 986, pp. 89–105. Springer, New York (1983)

On a Zero-One Law for the Norm Process
of Transient Random Walk

Ayako Matsumoto and Kouji Yano

Abstract A zero-one law of Engelbert–Schmidt type is proven for the norm process of a transient random walk. An invariance principle for random walk local times and a limit version of Jeulin's lemma play key roles.

Keywords Zero-one law · Random walk · Local time · Jeulin's lemma

AMS 2000 Subject class: Primary 60G50; Secondary 60F20, 60J55

1 Introduction

Let $S = (S_n : n \in \mathbb{Z}_{\geq 0})$ be a random walk in $\mathbb{Z}^d$ starting from the origin. Let $\|\cdot\|$ be a norm on $\mathbb{R}^d$ taking integer values on the integer lattice $\mathbb{Z}^d$. The norm $\|\cdot\|$ cannot be the Euclidean norm denoted by $|x| = \sqrt{|x^1|^2 + \cdots + |x^d|^2}$. By the *norm process of the random walk S*, we mean the process $\|S\| = (\|S_n\| : n \in \mathbb{Z}_{\geq 0})$. The purpose of the present paper is to study summability of $f(\|S_n\|)$ for a non-negative function f on $\mathbb{Z}$.

Set $X_n = S_n - S_{n-1}$ for $n \in \mathbb{Z}_{\geq 1}$. Then X_n's are independent identically-distributed random vectors taking values in $\mathbb{Z}^d$. We suppose that $E[X_1^i] = 0$ and $E[(X_1^i)^2] < \infty$, $i = 1, 2, \ldots, d$. Let Q denote the covariance matrix of X_1, i.e., $Q = (E[X_1^i X_1^j])_{i,j}$. We introduce the following assumption:

(A0) $Q = \sigma^2 I$ for some constant $\sigma > 0$, where I stands for the identity matrix.

We write

$$B(0; r) = \{x \in \mathbb{R}^d : \|x\| \leq r\}, \tag{1}$$

$$\partial B(0; r) = \{x \in \mathbb{R}^d : \|x\| = r\}. \tag{2}$$

A. Matsumoto
T & D Financial Life Insurance Company, Japan

K. Yano (✉)
Graduate School of Science, Kobe University, Kobe, Japan
e-mail: kyano@math.kobe-u.ac.jp

C. Donati-Martin et al. (eds.), *Séminaire de Probabilités XLIII*, Lecture Notes in Mathematics 105
2006, DOI 10.1007/978-3-642-15217-7_4, © Springer-Verlag Berlin Heidelberg 2011

For $k \in \mathbb{Z}_{\geq 0}$, we set

$$N(k) = \sharp(\partial B(0; k) \cap \mathbb{Z}^d) = \sharp\left\{ x \in \mathbb{Z}^d : \|x\| = k \right\}. \tag{3}$$

We call B a d-*polytope* if B is a bounded convex region in a d-dimensional space enclosed by a finite number of $(d-1)$-dimensional hyperplanes. The part of the polytope B which lies in one of the hyperplanes is called a *cell*. (See, e.g., [4] for this terminology.) We introduce the following assumptions:

(A1) $\|x\| \in \mathbb{Z}_{\geq 0}$ for any $x \in \mathbb{Z}^d$.
(A2) For each $k \in \mathbb{Z}_{\geq 1}$, the set $B(0; k)$ is a d-polytope whose vertices are contained in $\mathbb{Z}^d$. Consequently, its boundary $\partial B(0; k)$ is the union of all cells of the d-polytope $B(0; k)$.
(A3) For any $k \in \mathbb{Z}_{\geq 1}$, there exists a finite partition of $\partial B(0; 1)$, which is denoted by $\{U_j^{(k)} : j = 1, \ldots, M(k)\}$, such that the following statements hold:

    (i) $M(k) \leq N(k)$ and $M(k)/N(k) \to 1$ as $k \to \infty$.
    (ii) Each $U_j^{(k)}$ contains at least one point of $\partial B(0; 1) \cap (k^{-1}\mathbb{Z}^d)$.
    (iii) The $U_j^{(k)}$'s for $j = 1, \ldots, M(k)$ have a common area.
    (iv) $\max_j \max\{\|x - y\| : x, y \in U_j^{(k)}\} \to 0$ as $k \to \infty$.

Note that these assumptions (A0)–(A3) imply that $N(k) \to \infty$ as $k \to \infty$. Our main theorem is the following:

Theorem 1.1. *Suppose that $d \geq 3$ and that* (A0)–(A3) *hold. Then, for any nonnegative function f on $\mathbb{Z}_{\geq 0}$, the following conditions are equivalent:*

(I) $P\left(\sum_{n=1}^{\infty} f(\|S_n\|) < \infty\right) > 0.$

(II) $P\left(\sum_{n=1}^{\infty} f(\|S_n\|) < \infty\right) = 1.$

(III) $E\left[\sum_{n=1}^{\infty} f(\|S_n\|)\right] < \infty.$

(IV) $\sum_{k=1}^{\infty} k^{2-d} N(k) f(k) < \infty.$

Suppose, moreover, that
(A4) *There exists $k_0 \in \mathbb{Z}_{\geq 1}$ such that $N(k)$ is non-decreasing in $k \geq k_0$.*
Then the above conditions are equivalent to

(V) $\sum_{k=1}^{\infty} k f(k) < \infty.$

We will prove, in Sect. 5, that (III) and (IV) are equivalent, by virtue of the asymptotic behavior of the Green function due to Spitzer [28] (see Theorem 5.1). We will prove, in Sect. 6, that (I) implies (IV), where a key role is played by a limit version of *Jeulin's lemma* (see Proposition 3.2). Note that (III) trivially implies (II) and that (II) trivially implies (I).

Sect. 5

(III) $\Longleftrightarrow$ (IV)

trivial $\Downarrow$ $\Uparrow$ Sect. 6

(II) $\Longrightarrow$ (I)

trivial

The equivalence between (I) and (II) may be considered to be a *zero-one law of Engelbert–Schmidt type*; see Sect. 2. However, we remark that this equivalence follows also from the *Hewitt–Savage zero-one law* (see, e.g. [2, Theorem 7.36.5]). In fact, the event $\{\sum f(\|S_n\|) < \infty\}$ is *exchangeable*, i.e., invariant under permutation of any finite number of the sequence (X_n).

If $d = 1$ or 2, the random walk S is recurrent, and hence it is obvious that the conditions (I)–(III) are equivalent to stating that $f(k) \equiv 0$. This is why we confine ourselves to the case $d \geq 3$, where the random walk S is transient so that $\|S_n\|$ diverges as $n \to \infty$. In the case $d \geq 3$, the summability of $f(\|S_n\|)$ depends upon how rapidly the function $f(k)$ vanishes as $k \to \infty$. Theorem 1.1 gives a criterion for the summability of $f(\|S_n\|)$ in terms of summability of $k f(k)$.

Consider the *max norm*

$$\|x\|_\infty^{(d)} = \max_{i=1,\ldots,d} |x^i|, \quad x = (x^1, \ldots, x^d) \in \mathbb{R}^d \tag{4}$$

and the ℓ^1-*norm*

$$\|x\|_1^{(d)} = \sum_{i=1}^d |x^i|, \quad x = (x^1, \ldots, x^d) \in \mathbb{R}^d. \tag{5}$$

We will show in Sect. 4 that these norms satisfy (A1)–(A4). Thus we obtain the following corollary:

Corollary 1.2. *Let S be a simple random walk of dimension $d \geq 3$ and take $\|\cdot\|$ as the max norm or the ℓ^1-norm. Then, for any non-negative function f on $\mathbb{Z}_{\geq 0}$, the conditions (I)–(V) are equivalent.*

The organization of this paper is as follows. In Sect. 2, we give a brief summary of known results of zero-one laws of Engelbert–Schmidt type. In Sect. 3, we recall Jeulin's lemma. We also state and prove its limit version in discrete time. In Sect. 4, we present some examples of norms which satisfy (A1)–(A4). Sections 5 and 6 are devoted to the proof of Theorem 1.1. In Sect. 7, we present some results about Jeulin's lemma obtained by Shiga (Shiga, unpublished).

2 Zero-One Laws of Engelbert–Schmidt Type

Let us give a brief summary of known results of zero-one laws concerning finiteness of certain integrals, which we call *zero-one laws of Engelbert–Schmidt type*.

$1°$). Let $(B_t : t \geq 0)$ be a one-dimensional Brownian motion starting from the origin. The following theorem, which originates from Shepp–Klauder–Ezawa [27] with motivation in quantum theory, is due to Engelbert–Schmidt [5, Theorem 1] with motivation in construction of a weak solution of a certain stochastic differential equation by means of time-change method.

Theorem 2.1 ([5, 27]). *Let f be a non-negative Borel function on $\mathbb{R}$. Then the following conditions are equivalent:*

(B1) $P\left(\int_0^t f(B_s)ds < \infty \text{ for every } t \geq 0\right) > 0.$

(B2) $P\left(\int_0^t f(B_s)ds < \infty \text{ for every } t \geq 0\right) = 1.$

(B3) $f(x)$ *is integrable on all compact subsets of* $\mathbb{R}$.

The proof of Theorem 2.1 was based on the formula

$$\int_0^t f(B_s)ds = \int_{\mathbb{R}} f(x)L_t^B(x)dx \tag{6}$$

where $L_t^B(x)$ stands for the local time at level x by time t (see [15]).

Engelbert–Schmidt [6, Theorem 1] proved that a similar result holds for a Bessel process of dimension $d \geq 2$ starting from a positive number.

$2°$). Let $(R_t : t \geq 0)$ be a Bessel process of dimension $d > 0$ starting from the origin, i.e., $R_t = \sqrt{Z_t}$ where Z_t is the unique non-negative strong solution of

$$Z_t = td + 2\int_0^t \sqrt{|Z_s|}dB_s. \tag{7}$$

The following theorem is due to Pitman–Yor [24, Proposition 1] and Xue [29, Proposition 2].

Theorem 2.2 ([24, 29]). *Suppose that $d \geq 2$. Let f be a non-negative Borel function on $[0, \infty)$. Then the following conditions are equivalent:*

(R1) $P\left(\int_0^t f(R_s)ds < \infty \text{ for every } t \geq 0\right) > 0;$

(R2) $P\left(\int_0^t f(R_s)ds < \infty \text{ for every } t \geq 0\right) = 1;$

(R3) $f(r)$ *is integrable on all compact subsets of* $(0, \infty)$ *and*

(R3a) $\int_0^c f(r)r(\log \frac{1}{r})_+ dr < \infty$ *if $d = 2$;*

(R3b) $\int_0^c f(r)r dr < \infty$ *if $d > 2$*

where c is an arbitrary positive number.

The proof of Theorem 2.2 was done by applying Jeulin's lemma (see Theorem 3.1 below) to the total local time, where the assumption of Jeulin's lemma was assured by the *Ray–Knight theorem* (see Le Gall [19, pp. 299]).

3°). Xue [29, Corollary 4] generalized Engelbert–Schmidt [6, Corollary on pp. 227] and proved the following theorem.

Theorem 2.3 ([29]). *Suppose that $d > 2$. Let f be a non-negative Borel function on $[0, \infty)$. Then the following conditions are equivalent:*

(RI) $P\left(\int_0^\infty f(R_t)dt < \infty\right) > 0.$

(RII) $P\left(\int_0^\infty f(R_t)dt < \infty\right) = 1.$

(RIII) $E\left[\int_0^\infty f(R_t)dt\right] < \infty.$

(RIV) $\int_0^\infty rf(r)dr < \infty.$

The proof of Theorem 2.3 was based on Jeulin's lemma and the Ray–Knight theorem. Our results (Theorem 1.1 and Corollary 1.2) may be considered to be random walk versions of Theorem 2.3. Note that, in Theorem 2.3, the condition (RIII), which is obviously stronger than (RII), is in fact equivalent to (RII). We remark that, in Theorem 2.3, we consider the perpetual integral $\int_0^\infty f(R_t)dt$ instead of the integrals on compact intervals.

4°). Höhnle–Sturm [13, 14] obtained a zero-one law about the event

$$\left\{\int_0^t f(X_s)ds < \infty \text{ for every } t \geq 0\right\} \tag{8}$$

where $(X_t : t \geq 0)$ is a symmetric Markov process which takes values in a Lusin space and which has a strictly positive density. Their proof was based on excessive functions. As an application, they obtained the following theorem ([14, pp. 411]).

Theorem 2.4 ([14]). *Suppose that $0 < d < 2$. Let f be a non-negative Borel function on $[0, \infty)$. Then the following conditions are equivalent:*

(Ri) $P\left(\int_0^t f(R_s)ds < \infty \text{ for every } t \geq 0\right) > 0.$

(Rii) $P\left(\int_0^t f(R_s)ds < \infty \text{ for every } t \geq 0\right) = 1.$

(Riii) $f(x)$ is integrable on all compact subsets of $[0, \infty)$ and $\int_0^1 f(x)x^{d-1}dx < \infty.$

See also Cherny [3, Corollary 2.1] for another approach.

5°). Engelbert–Senf [7] studied integrability of $\int_0^\infty f(Y_s)ds$ where $(Y_t : t \geq 0)$ is a Brownian motion with constant drift. See Salminen–Yor [26] for a generalization of this direction. See also Khoshnevisan–Salminen–Yor [18] for a generalization of the case where $(Y_t : t \geq 0)$ is a certain one-dimensional diffusion process.

3 Jeulin's Lemma and its Limit Version in Discrete Time

3.1 Jeulin's Lemma

Jeulin [16, Lemma 3.22] gave quite a general theorem about integrability of a function of a stochastic process. He gave detailed discussions in [17] about his lemma. Among the applications presented in [17], let us focus on the following theorem:

Theorem 3.1 ([16, 17]). *Let $(X(t) : 0 < t \leq 1)$ be a non-negative measurable process and φ a positive function on $(0, 1]$. Suppose that there exists a random variable X with*

$$E[X] < \infty \quad and \quad P(X > 0) = 1 \tag{9}$$

such that

$$\frac{X(t)}{\varphi(t)} \overset{law}{=} X \quad holds \ for \ each \ fixed \ 0 < t \leq 1. \tag{10}$$

Then, for any non-negative Borel measure μ on $(0, 1]$, the following conditions are equivalent:

(JI) $P\left(\int_0^1 X(t)\mu(dt) < \infty\right) > 0.$

(JII) $P\left(\int_0^1 X(t)\mu(dt) < \infty\right) = 1.$

(JIII) $E\left[\int_0^1 X(t)\mu(dt)\right] < \infty.$

(JIV) $\int_0^1 \varphi(t)\mu(dt) < \infty.$

A good elementary proof of Theorem 3.1 can be found in Xue [29, Lemma 2].

For several applications of Jeulin's lemma (Theorem 3.1), see Yor [30], Pitman–Yor [23, 24], Xue [29], Peccati–Yor [22], Funaki–Hariya–Yor [11, 12], and Fitzsimmons–Yano [9].

We cannot remove the assumption $E[X] < \infty$ from Theorem 3.1; see Proposition 7.1.

3.2 A Limit Version of Jeulin's Lemma in Discrete Time

For our purpose, we would like to replace the assumption (10) which requires identity in law by a weaker assumption which requires convergence in law. The following proposition plays a key role in our purpose (see also Corollary 7.3).

Proposition 3.2. *Let $(V(k) : k \in \mathbb{Z}_{\geq 1})$ be a non-negative measurable process and Φ a positive function on $\mathbb{Z}_{\geq 1}$. Suppose that there exists a random variable X with*

$$P(X > 0) = 1 \tag{11}$$

such that

$$\frac{V(k)}{\Phi(k)} \xrightarrow{\text{law}} X \quad as \ k \to \infty. \tag{12}$$

Then, for any non-negative function f on $\mathbb{Z}_{\geq 1}$, it holds that

$$P\left(\sum_{k=1}^{\infty} f(k)V(k) < \infty\right) > 0 \quad implies \quad \sum_{k=1}^{\infty} f(k)\Phi(k) < \infty. \tag{13}$$

The following proof of Proposition 3.2 is a slight modification of that of [29, Lemma 2].

Proof. Suppose that $P(\sum f(k)V(k) < \infty) > 0$. Then there exists a number C such that the event

$$B = \left\{\sum_{k=1}^{\infty} f(k)V(k) \leq C\right\} \tag{14}$$

has positive probability. Since $P(X \leq 0) = 0$, there exists a positive number u_0 such that $P(X \leq u_0) < P(B)/4$. By assumption (12), we see that there exists u_1 with $0 < u_1 < u_0$ such that

$$P(V(k)/\Phi(k) \leq u_1) \xrightarrow{k \to \infty} P(X \leq u_1) < \frac{1}{4}P(B). \tag{15}$$

Then, for some large number k_0, we have

$$P(V(k)/\Phi(k) \leq u_1) \leq \frac{1}{2}P(B), \quad k \geq k_0. \tag{16}$$

Now we obtain

$$C \geq E\left[1_B \sum_{k=1}^{\infty} f(k)V(k)\right] \tag{17}$$

$$= \sum_{k=1}^{\infty} f(k)\Phi(k)E\left[1_B \cdot \frac{V(k)}{\Phi(k)}\right] \tag{18}$$

$$= \sum_{k=1}^{\infty} f(k)\Phi(k) \int_0^{\infty} P(B \cap \{V(k)/\Phi(k) > u\})du \tag{19}$$

$$\geq \sum_{k=k_0}^{\infty} f(k)\Phi(k) \int_0^{u_1} [P(B) - P(V(k)/\Phi(k) \leq u)]_+)du \tag{20}$$

$$\geq \frac{1}{2} P(B) u_1 \sum_{k=k_0}^{\infty} f(k) \Phi(k). \tag{21}$$

Since $P(B) u_1 > 0$, we conclude that $\sum f(k) \Phi(k) < \infty$.

4 Examples of Norms

Let us introduce several notations. For an index set A (we shall take $A = \mathbb{Z}_{\geq 0}$ or $\mathbb{Z}^d \setminus \{0\}$ later), we denote $\mathcal{M}(A)$ the set of all non-negative functions on A. For three functions $f, g, h \in \mathcal{M}(A)$, we say that

$$f(a) \sim g(a) \quad \text{as } h(a) \to \infty \tag{22}$$

if $f(a)/g(a) \to 1$ as $h(a) \to \infty$. For two functions $f, g \in \mathcal{M}(A)$, we say that

$$f(a) \asymp g(a) \quad \text{for } a \in A \tag{23}$$

if there exist positive constants c_1, c_2 such that

$$c_1 f(a) \leq g(a) \leq c_2 f(a) \quad \text{for } a \in A. \tag{24}$$

For two functionals F, G on $\mathcal{M}(A)$, we say that

$$F(f) \asymp G(f) \quad \text{for } f \in \mathcal{M}(A) \tag{25}$$

if there exist positive constants c_1, c_2 such that

$$c_1 F(f) \leq G(f) \leq c_2 F(f) \quad \text{for } f \in \mathcal{M}(A). \tag{26}$$

Now let us present several examples of norms which satisfy (A1)–(A4).

Example 4.1 (Max norms). Consider $\|x\|_{\infty}^{(d)} = \max_i |x^i|$. It is obvious that the conditions (A1)–(A3) are satisfied. In fact, the partition of $\partial B(0; 1)$ in (A3) can be obtained by separating $\partial B(0; 1)$ by hyperplanes $\{x \in \mathbb{R}^d : x^i = j/k\}$ for $i = 1, \ldots, d$ and $j = -k, \ldots, k$. Let us study $N(k) = N_{\infty}^{(d)}(k)$ and its asymptotic behavior. For $k \in \mathbb{Z}_{\geq 1}$, we have

$$N_{\infty}^{(d)}(k) = \sharp\{x \in \mathbb{Z}^d : \|x\| \leq k\} - \sharp\{x \in \mathbb{Z}^d : \|x\| \leq k - 1\} \tag{27}$$

$$= (2k + 1)^d - (2k - 1)^d. \tag{28}$$

Now we obtain

$$N_{\infty}^{(d)}(k) \sim d 2^d k^{d-1} \quad \text{as } k \to \infty. \tag{29}$$

Example 4.2 (ℓ^1-norms). Consider

$$\|x\|_1^{(d)} = \sum_{i=1}^{d} |x^i|, \quad x \in \mathbb{R}^d. \tag{30}$$

It is obvious that the conditions (A1)–(A3) are satisfied. In this case,

$$N(k) = N_1^{(d)}(k) = \#\{x \in \mathbb{Z}^d : \|x\|_1^{(d)} = k\} \tag{31}$$

satisfies the recursive relation

$$N_1^{(d)}(k) = \sum_{j=0}^{k} N_1^{(1)}(j) N_1^{(d-1)}(k-j), \quad d \geq 2, \, k \geq 0 \tag{32}$$

with initial condition

$$N_1^{(1)}(k) = \begin{cases} 1 & \text{if } k = 0, \\ 2 & \text{if } k \geq 1. \end{cases} \tag{33}$$

Since the moment generating function may be computed as

$$\sum_{k=0}^{\infty} s^k N_1^{(d)}(k) = \left(\frac{1+s}{1-s} \right)^d, \quad 0 < s < 1, \tag{34}$$

we see, by Tauberian theorem (see, e.g., [8, Theorem XIII.5.5]), that

$$N_1^{(d)}(k) \sim \frac{2^d}{(d-1)!} k^{d-1} \quad \text{as } k \to \infty. \tag{35}$$

Example 4.3 (Weighted ℓ^1-norms). Consider

$$\|x\|_{w1}^{(d)} = \sum_{i=1}^{d} i |x^i|, \quad x \in \mathbb{R}^d. \tag{36}$$

The conditions (A1)–(A3) are obviously satisfied.

Now let us discuss the asymptotic behavior of $N_{w1}^{(d)}(k)$. Note that

$$N(k) = N_{w1}^{(d)}(k) = \#\{x \in \mathbb{Z}^d : \|x\|_{w1}^{(d)} = k\} \tag{37}$$

satisfies the recursive relation

$$N_{w1}^{(d)}(k) = \sum_{j \in \mathbb{Z}_{\geq 0} : k - dj \geq 0} N_1^{(1)}(j) N_{w1}^{(d-1)}(k-dj), \quad d \geq 2, \, k \geq 0 \tag{38}$$

with initial condition $N_{w1}^{(1)}(k) \equiv N_1^{(1)}(k)$. Then, by induction, we can easily see that

$$|N_{w1}^{(d)}(k) - a^{(d)}k^{d-1}| \leq b^{(d)}k^{d-2}, \quad k \in \mathbb{Z}_{\geq 1}, \ d \geq 2 \tag{39}$$

for some positive constants $a^{(d)}, b^{(d)}$ where $a^{(d)}$ is defined recursively as

$$a^{(1)} = 2, \quad a^{(d)} = \frac{2}{d(d-1)}a^{(d-1)} \ (d \geq 2). \tag{40}$$

In particular, we see that $N_{w1}^{(d)}(k) \sim a^{(d)}k^{d-1}$ as $k \to \infty$. For instance, by easy computations, we obtain

$$N_{w1}^{(2)}(k) = \begin{cases} 1 & \text{if } k = 0, \\ 2k & \text{if } k \geq 1 \end{cases} \tag{41}$$

and

$$N_{w1}^{(3)}(k) = \begin{cases} 1 & \text{if } k = 0, \\ \frac{2}{3}k^2 + 2 & \text{if } k \equiv 0 \text{ modulo } 3, \ k \neq 0, \\ \frac{2}{3}k^2 + \frac{4}{3} & \text{if } k \equiv 1, 2 \text{ modulo } 3. \end{cases} \tag{42}$$

Example 4.4 (Transformation by unimodular matrices). Let A be a unimodular $d \times d$ matrix, i.e. A is a $d \times d$ matrix whose entries are integers and whose determinant is 1 or -1. Let $\|\cdot\|_0$ be a norm on $\mathbb{R}^d$ satisfying (A1)–(A3). Then the norm $\|x\| = \|Ax\|_0$ also satisfies (A1)–(A3). Note that

$$\sharp\{x \in \mathbb{Z}^d : \|x\| = k\} = \sharp\{x \in \mathbb{Z}^d : \|x\|_0 = k\}, \quad k \in \mathbb{Z}_{\geq 0}. \tag{43}$$

For example, the norm on $\mathbb{R}^3$ defined as

$$\|(x^1, x^2, x^3)\| = |x^1 - x^2| + |x^2 - x^3| + |x^1 - x^2 + x^3| \tag{44}$$

satisfies (A1)–(A3).

Remark 4.5. Let us consider the norm $2\|x\|_\infty^{(d)}$. Then the conditions (A1)–(A2) are satisfied, but neither of (A3) nor (A4) is; in fact,

$$N(k) = \begin{cases} N_\infty^{(d)}(k/2) & \text{if } k \text{ is even}, \\ 0 & \text{if } k \text{ is odd}. \end{cases} \tag{45}$$

Nevertheless, we see that the conditions (I)–(IV) are equivalent to each other and also to

$$\sum_{k=1}^{\infty} k f(2k) < \infty, \tag{46}$$

which is strictly weaker than (V) because there is no restriction on the values of $f(2k+1)$.

5 Equivalence Between (III) and (IV)

Let us introduce the *random walk local times*:

$$L_n^S(x) = \sharp \{m = 1, 2, \ldots, n : S_m = x\}, \quad x \in \mathbb{Z}^d, \tag{47}$$
$$L_n^{\|S\|}(k) = \sharp \{m = 1, 2, \ldots, n : \|S_m\| = k\}, \quad k \in \mathbb{Z}_{\geq 0}. \tag{48}$$

Then, for any non-negative function g on $\mathbb{Z}^d$, we have

$$\sum_{n=1}^{\infty} g(S_n) = \sum_{x \in \mathbb{Z}^d} g(x) L_\infty^S(x). \tag{49}$$

Taking the expectations of both sides, we have

$$E\left[\sum_{n=1}^{\infty} g(S_n)\right] = \sum_{x \in \mathbb{Z}^d} g(x) E\left[L_\infty^S(x)\right]. \tag{50}$$

It is obvious by definition that

$$E\left[L_\infty^S(x)\right] = \sum_{n=1}^{\infty} P(S_n = x) = G(0, x) \tag{51}$$

where $G(x, y)$ is the *Green function* given as

$$G(x, y) = \sum_{n=1}^{\infty} P_x(S_n = y). \tag{52}$$

Let $|\cdot|$ denote the Euclidean norm of $\mathbb{R}^d$, i.e., $|x|^2 = \sum_{i=1}^{d} (x^i)^2$. We recall the following asymptotic behavior of the Green function:

Theorem 5.1 ([28]). *It holds that*

$$G(0, x) \sim \frac{\Gamma(d/2 - 1)}{2\pi^{d/2}} |\det Q|^{-1/2} (x, Q^{-1}x)^{1-d/2} \quad \text{as } |x| \to \infty. \tag{53}$$

In particular, if $Q = \sigma^2 I$, then

$$|x|^{d-2}G(0, x) \rightarrow \frac{\Gamma(d/2 - 1)}{2\pi^{d/2}}\sigma^{-2} \quad as \; |x| \rightarrow \infty. \tag{54}$$

We can prove Theorem 5.1 in the same way as in Spitzer [28, P26.1], so we omit the proof.

Proposition 5.2. *It holds that*

$$E\left[\sum_{n=1}^{\infty} g(S_n)\right] \asymp g(0) + \sum_{x \in \mathbb{Z}^d \setminus \{0\}} g(x)\|x\|^{2-d} \quad for \; g \in \mathcal{M}(\mathbb{Z}^d). \tag{55}$$

Proof. Since $\|x\| \asymp |x|$ for $x \in \mathbb{Z}^d$, it follows from Theorem 5.1 that

$$G(0, x) \asymp \|x\|^{2-d} \quad for \; x \in \mathbb{Z}^d \setminus \{0\}. \tag{56}$$

Combining it with (50), we obtain the desired result.

Remark 5.3. It is now obvious from Proposition 5.2 that

$$\sum_{x \in \mathbb{Z}^d} g(x)\|x\|^{2-d} < \infty \quad \text{implies} \quad P\left(\sum_{n=1}^{\infty} g(S_n) < \infty\right) = 1. \tag{57}$$

But we do not know whether the converse is true or not.

The following proposition proves part of Theorem 1.1.

Proposition 5.4. *Suppose that the condition* (A1) *is satisfied. Then it holds that*

$$E\left[\sum_{n=1}^{\infty} f(\|S_n\|)\right] \asymp f(0) + \sum_{k=1}^{\infty} f(k)k^{2-d} N(k) \quad for \; f \in \mathcal{M}(\mathbb{Z}_{\geq 0}) \tag{58}$$

and, in particular, that (III) *and* (IV) *are equivalent. If, moreover, the condition* (A4) *is satisfied, then it holds that*

$$E\left[\sum_{n=1}^{\infty} f(\|S_n\|)\right] \asymp f(0) + \sum_{k=1}^{\infty} kf(k) \quad for \; f \in \mathcal{M}(\mathbb{Z}_{\geq 0}) \tag{59}$$

and, in particular, that (IV) *and* (V) *are equivalent.*

Proof. The former half of Proposition 5.4 is immediate from Propositions 5.2 and 5.7 for $g(x) = f(\|x\|)$. The latter half is immediate from Proposition 5.7 below.

Remark 5.5. Let $p(x)$ denote the probability that the process visits x at least once:

$$p(x) = P(L_\infty^S(x) \geq 1) = P(T_x < \infty), \quad x \in \mathbb{Z}^d \tag{60}$$

where $T_x = \inf\{n \geq 1 : S_n = x\}$ is the first hitting time of x. Since $L_\infty^S(x) = L_\infty^S(x) \circ \theta_{T_x} + 1$ and by translation invariance, we may compute the distribution of the total local time $L_\infty^S(x)$ as

$$P(L_\infty^S(x) \geq n) = p(x)p(0)^{n-1}, \quad x \in \mathbb{Z}^d, \ n = 1, 2, \ldots \tag{61}$$

See [20] for some general discussions for symmetric Markov processes. Note that the Green function $G(0, x)$ may be expressed as

$$G(0, x) = E\left[L_\infty^S(x)\right] = \sum_{n=1}^{\infty} P(L_\infty^S(x) \geq n) = \frac{p(x)}{1 - p(0)}. \tag{62}$$

Remark 5.6. We do not know any explicit result about the law of the total local time $L_\infty^{\|S\|}(k)$ for the norm process $\|S\|$.

Proposition 5.7. *Let $\|\cdot\|$ be a norm on $\mathbb{R}^d$. Suppose that the condition (A4) is satisfied. Then there exists $k_1 \in \mathbb{Z}_{\geq 1}$ such that $N(k) \asymp k^{d-1}$ for $k \geq k_1$.*

Proof. By (28), we have

$$\sharp \left\{x \in \mathbb{Z}^d : \|x\|_\infty^{(d)} \leq k\right\} \asymp k^d \quad \text{for } k \in \mathbb{Z}_{\geq 1}. \tag{63}$$

Note that $\|x\| \asymp \|x\|_\infty^{(d)}$ for $x \in \mathbb{Z}^d$; in fact, any two norms on $\mathbb{R}^d$ are mutually equivalent. This immediately implies that

$$\sum_{j=0}^{k} N(j) = \sharp \left\{x \in \mathbb{Z}^d : \|x\| \leq k\right\} \asymp k^d \quad \text{for } k \in \mathbb{Z}_{\geq 1}. \tag{64}$$

Hence there exist constants c_1, c_2 such that

$$c_1 k^d \leq \sum_{j=0}^{k} N(j) \leq c_2 k^d \quad \text{for } k \in \mathbb{Z}_{\geq 1}. \tag{65}$$

By the condition (A4), we have

$$kN(k) = \sum_{j=k+1}^{2k} N(k) \leq \sum_{j=k+1}^{2k} N(j) \leq c_2(2k)^d \quad \text{for } k \in \mathbb{Z}_{\geq 1}. \tag{66}$$

Now we obtain $N(k) \leq c_3 k^{d-1}$ with $c_3 = c_2 2^d$. Again by the condition (A4), we have

$$kN(k) = \sum_{j=1}^{k} N(k) \geq \sum_{j=0}^{k} N(j) \geq c_1 k^d \quad \text{for } k \in \mathbb{Z}_{\geq 1}. \tag{67}$$

Now we obtain $N(k) \geq c_1 k^{d-1}$. This completes the proof.

6 Proving that (I) implies (IV)

By the assumption (A2), we may identify each cell of $B(0; r)$ with a subset of $\mathbb{R}^{d-1}$. So we may introduce the area measure λ on $\partial B(0; 1)$. We define $\mu(\cdot) = \lambda(\cdot)/\lambda(\partial B(0; 1))$ and call it the *uniform measure* on $\partial B(0; 1)$.

For $k \in \mathbb{Z}_{\geq 1}$, we define a probability measure on $\mathbb{R}^d$ by

$$\mu_k(A) = \frac{1}{N(k)} \sharp \left\{ x \in k^{-1}\mathbb{Z}^d \cap A : \|x\| = 1 \right\}, \quad A \in \mathcal{B}(\mathbb{R}^d). \tag{68}$$

Proposition 6.1. *Suppose that (A1)–(A3) are satified. Then, as $k \to \infty$, the measure μ_k converges weakly to μ.*

Proof. Let $\{U_j^{(k)} : j = 1, \ldots, M(k)\}$ be such as in the assumption (A3). Then we see that $\mu(U_j^{(k)}) = M(k)^{-1}$ for any j and any k. For $j = 1, \ldots, M(k)$, choose $x_j^{(k)} \in U_j^{(k)} \cap (k^{-1}\mathbb{Z}^d)$. We may choose $\{x_j^{(k)} : j = M(k) + 1, \ldots, N(k)\}$ so that $\{x_j^{(k)} : j = 1, \ldots, N(k)\}$ is an enumeration of the points of $\{x \in k^{-1}\mathbb{Z}^d : \|x\| = 1\}$.

Let $f : \mathbb{R}^d \to \mathbb{R}$ be a continuous function with compact support. It suffices to prove that

$$\int_{\mathbb{R}^d} f(x)\mu_k(dx) \overset{k \to \infty}{\longrightarrow} \int_{\partial B(0;1)} f(x)\mu(dx). \tag{69}$$

Note that

$$\int_{\mathbb{R}^d} f(x)\mu_k(dx) = \frac{1}{N(k)} \sum_{j=1}^{N(k)} f(x_j^{(k)}). \tag{70}$$

Since $M(k)/N(k) \to 1$ as $k \to \infty$, it suffices to prove that

$$\frac{1}{M(k)} \sum_{j=1}^{M(k)} f(x_j^{(k)}) \overset{k \to \infty}{\longrightarrow} \int_{\partial B(0;1)} f(x)\mu(dx). \tag{71}$$

Since $\partial B(0;1) = \cup_j U_j^{(k)}$ and $\mu(U_j^{(k)}) = M(k)^{-1}$, we obtain

$$\left| \frac{1}{M(k)} \sum_{j=1}^{M(k)} f\left(x_j^{(k)}\right) - \int_{\partial B(0;1)} f(x)\mu(dx) \right| \tag{72}$$

$$\leq \frac{1}{M(k)} \sum_{j=1}^{M(k)} \int_{U_j^{(k)}} \left| f\left(x_j^{(k)}\right) - f(x) \right| \mu(dx) \tag{73}$$

$$\leq \max_{1 \leq j \leq M(k)} \max_{x,y \in U_j^{(k)}} |f(y) - f(x)|. \tag{74}$$

By uniform continuity of f and by the assumption (A3), the quantity (74) converges to 0 as $k \to \infty$. Therefore the proof is complete.

Let (B_t) denote a standard Brownian motion of dimension $d \geq 3$ starting from the origin. Set

$$g(x, y) = \int_0^\infty \frac{ds}{(2\pi s)^{d/2}} \exp\left(-\frac{|x-y|^2}{2s}\right), \quad x, y \in \mathbb{R}^d. \tag{75}$$

For the uniform measure μ on $\partial B(0;1)$, we define

$$g\mu(x) = \int_{\mathbb{R}^d} g(x, y)\mu(dy), \quad x \in \mathbb{R}^d. \tag{76}$$

Then it is well-known (see [21]; see also [10, Theorem 5.2.5]) that there exists a unique positive continuous additive functional (L_t^μ) such that

$$g\mu(\sigma B_t) - g\mu(\sigma B_0) + L_t^\mu \tag{77}$$

is a martingale with zero mean. The process (L_t^μ) is called the *local time process on the union of cells* $\partial B(0;1)$ *for* (σB_t). The relation between the measure μ and the positive continuous additive functional (L_t^μ) is called the *Revuz correspondence* (see [25]).

The following theorem is an invariance principle for the random walk local time of the norm process.

Theorem 6.2. *Suppose that (A0)–(A3) are satisfied. Then it holds that*

$$\frac{L_\infty^{\|S\|}(k)}{k^{2-d} N(k)} \xrightarrow{\text{law}} L_\infty^\mu \quad \text{as } k \to \infty. \tag{78}$$

Proof. Note that

$$\frac{L_\infty^{\|S\|}(k)}{k^{2-d}N(k)} = k^{d-2}\sum_{n=1}^{\infty}\mu_k\left(\left\{\frac{S_n}{k}\right\}\right).\tag{79}$$

Hence we obtain the desired result as an immediate consequence of Proposition 6.1 and Bass–Khoshnevisan [1, Proposition 6.3].

Now we are in a position to prove that (I) implies (IV) in Theorem 1.1.

Proof that (I) *implies* (IV) *in Theorem 1.1.* Let us check that the assumptions of Proposition 3.2 are satisfied for $V(k) = L_\infty^{\|S\|}(k)$, $\Phi(k) = k^{2-d}N(k)$ and $X = L_\infty^\mu$.

By Theorem 6.2, assumption (12) is satisfied.

Let us show that $P(L_\infty^\mu \le 0) = 0$. The first hitting place of the union of cells $\partial B(0; 1)$ for the Brownian motion is almost surely an interior point of some cell of the d-polytope $B(0; 1)$ by assumption (A2). Hence it holds that, starting afresh at the first hitting time, the local time on the union of cells $\partial B(0; 1)$ is locally equal to the local time on the hyperplane which contains the cell. Since the local time at the origin for one-dimensional Brownian motion is positive almost surely at any positive time, we see that L_∞^μ is positive almost surely.

Thus we may apply Proposition 3.2 (or Corollary 7.3) and we see that (I) implies (IV). The proof is now complete.   □

7 A Remark on Jeulin's Lemma

The results of this section are mainly due to Tokuzo Shiga (2007, unpublished).

7.1 *Counterexample to Jeulin's Lemma without $E[X] < \infty$*

The following proposition gives a counterexample to Jeulin's lemma (Theorem 3.1) without $E[X] < \infty$.

Proposition 7.1. *(Shiga, unpublished) There exist a non-negative measurable process $(X(t) : 0 < t \le 1)$, a positive function φ on $(0, 1]$, a random variable X, and a non-negative Borel measure μ on $(0, 1]$ such that*

$$E[X] = \infty \quad and \quad P(X > 0) = 1,\tag{80}$$

$$\frac{X(t)}{\varphi(t)} \overset{\text{law}}{=} X \quad holds\ for\ each\ fixed\ 0 < t \le 1,\tag{81}$$

$$\int_0^1 \varphi(t)\mu(\mathrm{d}t) < \infty\tag{82}$$

but

$$P\left(\int_\varepsilon^1 X(t)\mu(dt) < \infty \ (\forall \varepsilon > 0), \quad \int_0^1 X(t)\mu(dt) = \infty\right) = 1. \qquad (83)$$

Proof. Let $(X(t))$ be an α-stable subordinator with $0 < \alpha \leq 1/2$. Then we have (80) and (81) for $\varphi(t) \equiv t^{1/\alpha}$. Set

$$\mu(dt) = t^{-1-1/\alpha}(\log 1/t)^{-1/\alpha}dt \qquad (84)$$

so that $\mu((t, 1])^\alpha \sim Ct^{-1}(\log 1/t)^{-1}$ as $t \to 0+$ for some positive constant C. Thus we obtain (82). Since we have

$$E\left[\exp - \int_0^1 X(t)\mu(dt)\right] = \exp - \int_0^1 \mu((t, 1])^\alpha dt = 0, \qquad (85)$$

we obtain (83).

7.2 A Limit Version of Jeulin's Lemma

Theorem 7.2 (Shiga, unpublished). *Let $(X(t) : 0 < t \leq 1)$ be a non-negative measurable process, φ a positive function defined on $(0, 1]$, and μ a non-negative Borel measure on $(0, 1]$. Suppose that there exists a random variable X with $P(X > 0) > 0$ such that*

$$\frac{X(t)}{\varphi(t)} \xrightarrow{\text{law}} X \quad as \ t \to 0+. \qquad (86)$$

Suppose, moreover, that

$$\int_\varepsilon^1 \varphi(t)\mu(dt) < \infty \quad for \ every \ 0 < \varepsilon < 1. \qquad (87)$$

Then it holds that

$$P\left(\int_0^1 X(t)\mu(dt) < \infty\right) = 1 \quad implies \quad \int_0^1 \varphi(t)\mu(dt) < \infty. \qquad (88)$$

Proof. Suppose that

$$P\left(\int_0^1 X(t)\mu(dt) < \infty\right) = 1 \qquad (89)$$

but that $\int_0^1 \varphi(s)\mu(\mathrm{d}s) = \infty$. For each $\varepsilon > 0$, we define a probability measure μ_ε by

$$\mu_\varepsilon(\mathrm{d}t) = C_\varepsilon^{-1} 1_{(\varepsilon,1]}(t)\varphi(t)\mu(\mathrm{d}t) \quad \text{with} \quad C_\varepsilon = \int_\varepsilon^1 \varphi(t)\mu(\mathrm{d}t) \qquad (90)$$

where C_ε is finite by the assumption (87). Then $C_\varepsilon \to \infty$ and $\mu_\varepsilon \xrightarrow{\mathrm{d}} \delta_0$ as $\varepsilon \to 0+$, where δ_0 stands for the unit point mass at 0. Using Jensen's inequality and changing the order of integration, we have

$$E\left[\exp -C_\varepsilon^{-1}\int_\varepsilon^1 X(t)\mu(\mathrm{d}t)\right] = E\left[\exp -\int_\varepsilon^1 \frac{X(t)}{\varphi(t)}\mu_\varepsilon(\mathrm{d}t)\right] \qquad (91)$$

$$\leq \int_\varepsilon^1 E\left[\exp -\frac{X(t)}{\varphi(t)}\right]\mu_\varepsilon(\mathrm{d}t). \qquad (92)$$

Hence it follows from (89) and (86) that

$$1 \leq \lim_{t\to 0+} E\left[\exp -\frac{X(t)}{\varphi(t)}\right] = E\left[e^{-X}\right], \qquad (93)$$

which implies $P(X = 0) = 1$. This is a contradiction to the assumption that $P(X > 0) > 0$.

From Theorem 7.2, we obtain another version of Jeulin's lemma in discrete time.

Corollary 7.3. *Let $(V(k) : k \in \mathbb{Z}_{\geq 1})$ be a non-negative measurable process and Φ a positive function on $\mathbb{Z}_{\geq 1}$. Suppose that there exists a random variable X with*

$$P(X > 0) > 0 \qquad (94)$$

such that

$$\frac{V(k)}{\Phi(k)} \xrightarrow{\text{law}} X \quad \text{as } k \to \infty. \qquad (95)$$

Then, for any non-negative function f on $\mathbb{Z}_{\geq 1}$, it holds that

$$P\left(\sum_{k=1}^\infty f(k)V(k) < \infty\right) = 1 \quad \text{implies} \quad \sum_{k=1}^\infty f(k)\Phi(k) < \infty. \qquad (96)$$

Proof. Take

$$X(t) = V([1/t]), \quad \varphi(t) = \Phi([1/t]) \qquad (97)$$

where $[x]$ stands for the smallest integer which does not exceed x and

$$\mu = \sum_{k=1}^{\infty} f(k)\delta_{1/k}. \tag{98}$$

Then the desired result is immediate from Theorem 7.2.

Proposition 3.2 and Corollary 7.3 cannot be unified in the following sense:

Proposition 7.4. *There exist a non-negative measurable process $(V(k) : k \in \mathbb{Z}_{\geq 1})$, a positive function Φ on $\mathbb{Z}_{\geq 1}$, a random variable X, and a non-negative function f on $\mathbb{Z}_{\geq 1}$ such that*

$$P(X > 0) > 0, \quad \frac{V(k)}{\Phi(k)} \xrightarrow{\text{law}} X \text{ as } k \to \infty, \tag{99}$$

and

$$P\left(\sum_{k=1}^{\infty} f(k)V(k) < \infty\right) > 0 \tag{100}$$

but

$$\sum_{k=1}^{\infty} f(k)\Phi(k) = \infty. \tag{101}$$

Proof. Let X be such that

$$P(X = 0) = P(X = 1) = \frac{1}{2} \tag{102}$$

and set $V(k) = X$ for $k \in \mathbb{Z}_{\geq 1}$. Then we have (99)–(101) for $\Phi(k) \equiv 1$ and $f(k) \equiv 1$.

7.3 A Counterexample

We give a counterexample to the converse of (96) where the assumptions of Corollary 7.3 are satisfied.

Proposition 7.5. *(Shiga, unpublished) There exist a non-negative measurable process $(V(k) : k \in \mathbb{Z}_{\geq 1})$, a positive function Φ on $\mathbb{Z}_{\geq 1}$, and a non-negative function f on $\mathbb{Z}_{\geq 1}$ such that*

$$\frac{V(k)}{\Phi(k)} \xrightarrow{\text{law}} 1 \quad as \ k \to \infty \tag{103}$$

and

$$\sum_{k=1}^{\infty} f(k)\Phi(k) < \infty. \tag{104}$$

but

$$P\left(\sum_{k=1}^{\infty} f(k)V(k) = \infty\right) = 1. \tag{105}$$

Proof. Let $0 < \alpha < 1/2$. Let $(V_0(k))$ be a sequence of i.i.d. random variables such that

$$E[e^{-\lambda V_0(k)}] = e^{-\lambda^{\alpha}}, \quad \lambda > 0, \, k \in \mathbb{Z}_{\geq 1}. \tag{106}$$

Set $\Phi(k) \equiv k$ and $f(k) \equiv k^{-1/\alpha}$. Then (104) holds and we have

$$\frac{V_0(k)}{\Phi(k)} \xrightarrow{\text{law}} 0 \quad \text{as } k \to \infty. \tag{107}$$

Since we have

$$E\left[\exp - \sum_{k=1}^{\infty} f(k)V_0(k)\right] = \prod_{k=1}^{\infty} E\left[e^{-f(k)V_0(k)}\right] = \exp - \sum_{k=1}^{\infty} k^{-1} = 0, \tag{108}$$

we obtain

$$P\left(\sum_{k=1}^{\infty} f(k)V_0(k) = \infty\right) = 1. \tag{109}$$

Since we may take $V(k) = k + V_0(k)$, we also obtain (103) and (105). The proof is now complete.

Acknowledgements The authors would like to thank Professor Tokuzo Shiga who kindly allowed them to append to this paper his detailed study (Shiga, unpublished) about Jeulin's lemma. They also thank Professors Marc Yor, Katsushi Fukuyama and Patrick J. Fitzsimmons for valuable comments. They are thankful to the referee for pointing out several errors in the earlier version. The first author, Ayako Matsumoto, expresses her sincerest gratitudes to Professors Yasunari Higuchi and Taizo Chiyonobu for their encouraging guidance in her study of mathematics. The research of the second author, Kouji Yano, was supported by KAKENHI (20740060).

References

1. Bass, R.F., Khoshnevisan, D.: Local times on curves and uniform invariance principles. Probab. Theory Relat. Field. **92**(4), 465–492 (1992)
2. Billingsley, P.: Probability and Measure. Wiley Series in Probability and Mathematical Statistics, 3rd edn. Wiley, New York, (1995)
3. Cherny, A.S.: Convergence of some integrals associated with Bessel processes. Theor. Probab. Appl. **45**(2), 195–209 (2001) Translated from Russian original, Teor. Veroyatnost. i Primenen. **45**(2), 251–267 (2000)
4. Coxeter, H.S.M.: Regular Polytopes, 3rd edn. Dover Publications Inc., New York (1973)
5. Engelbert, H.J., Schmidt, W.: On the behaviour of certain functionals of the Wiener process and applications to stochastic differential equations. In: Stochastic Differential Systems (Visegrád, 1980). Lecture Notes in Control and Information Science, vol. 36, pp. 47–55. Springer, Berlin (1981)
6. Engelbert, H.J., Schmidt, W.: On the behaviour of certain Bessel functionals. An application to a class of stochastic differential equations. Math. Nachr. **131**, 219–234 (1987)
7. Engelbert, H.J., Senf, T.: On functionals of a Wiener process with drift and exponential local martingales. In: Stochastic Processes and Related Topics (Georgenthal, 1990). Mathematics Research, vol. 61, pp. 45–58. Akademie-Verlag, Berlin (1991)
8. Feller, W.: An Introduction to Probability Theory and Its Applications, vol. II, 2nd edn. Wiley, New York (1971)
9. Fitzsimmons, P.J., Yano, K.: Time change approach to generalized excursion measures, and its application to limit theorems. J. Theor. Probab. **21**(1), 246–265 (2008)
10. Fukushima, M., Ōshima, Y., Takeda, M.: Dirichlet forms and symmetric Markov processes. de Gruyter Studies in Mathematics, vol. 19. Walter de Gruyter & Co., Berlin (1994)
11. Funaki, T., Hariya, Y., Yor, M.: Wiener integrals for centered Bessel and related processes. II. ALEA Lat. Am. J. Probab. Math. Stat. **1**, 225–240 (2006) (electronic)
12. Funaki, T., Hariya, Y., Yor, M.: Wiener integrals for centered powers of Bessel processes. I. Markov Process Related Fields, **13**(1), 21–56 (2007)
13. Höhnle, R., Sturm, K.-Th.: A multidimensional analogue to the 0-1-law of Engelbert and Schmidt. Stoch. Stoch. Rep. **44**(1–2), 27–41 (1993)
14. Höhnle, R., Sturm, K.-Th.: Some zero-one laws for additive functionals of Markov processes. Probab. Theory Relat. Field. **100**(4), 407–416 (1994)
15. Itô, K., McKean, H. P. Jr. Diffusion processes and their sample paths. Die Grundlehren der Mathematischen Wissenschaften, Band 125. Academic, New York, 1965.
16. Jeulin, Th.: Semi-martingales et grossissement d'une filtration. Lecture Notes in Mathematics, vol. 833. Springer, Berlin (1980)
17. Jeulin, Th.: Sur la convergence absolue de certaines intégrales. In: Séminaire de Probabilités, XVI. Lecture Notes in Mathematics, vol. 920, pp. 248–256. Springer, Berlin (1982)
18. Khoshnevisan, D., Salminen, P., Yor, M.: A note on a.s. finiteness of perpetual integral functionals of diffusions. Electron. Commun. Probab. **11**, 108–117 (2006) (electronic)
19. Le Gall, J.-F.: Sur la mesure de Hausdorff de la courbe brownienne. In: Séminaire de Probabilités, XIX, 1983/84. Lecture Notes in Mathematics, vol. 1123, pp 297–313. Springer, Berlin (1985)
20. Marcus, M.B., Rosen, J.: Moment generating functions for local times of symmetric Markov processes and random walks. In: Probability in Banach spaces, 8 (Brunswick, ME, 1991), Progress in Probability, vol. 30, pp. 364–376. Birkhäuser Boston, Boston, MA (1992)
21. Meyer, P.-A.: La formule d'Itô pour le mouvement brownien d'après G. Brosamler. In: Séminaire de Probabilités, XII (Univ. Strasbourg, Strasbourg, 1976/1977). Lecture Notes in Mathematics, vol. 649, pp. 763–769. Springer, Berlin (1978)
22. Peccati, G., Yor, M.: Hardy's inequality in $L^2([0, 1])$ and principal values of Brownian local times. In: Asymptotic methods in stochastics. Fields Institute Communications, vol. 44, pp. 49–74. American Mathematical Society, Providence, RI (2004)

23. Pitman, J., Yor, M.: A decomposition of Bessel bridges. Z. Wahrsch. Verw. Gebiete. **59**(4), 425–457 (1982)
24. Pitman, J.W., Yor, M.: Some divergent integrals of Brownian motion. Adv. Appl. Probab. (Suppl.) 109–116 (1986)
25. Revuz, D.: Mesures associées aux fonctionnelles additives de Markov. I. Trans. Am. Math. Soc. **148**, 501–531 (1970)
26. Salminen, P., Yor, M.: Properties of perpetual integral functionals of Brownian motion with drift. Ann. Inst. H. Poincaré Probab. Stat. **41**(3), 335–347 (2005)
27. Shepp, L.A., Klauder, J.R., Ezawa, H.: On the divergence of certain integrals of the Wiener process. Ann. Inst. Fourier (Grenoble), **24**(2), vi, 189–193 (1974) Colloque International sur les Processus Gaussiens et les Distributions Aléatoires (Colloque Internat. du CNRS, No. 222, Strasbourg, 1973)
28. Spitzer, F.: Principles of random walks, 2nd edn. Graduate Texts in Mathematics, vol. 34. Springer, New York (1976)
29. Xue, X.X.: A zero-one law for integral functionals of the Bessel process. In: Séminaire de Probabilités, XXIV, 1988/89. Lecture Notes in Mathematics, vol. 1426, pp. 137–153. Springer, Berlin (1990)
30. Yor, M.: Application d'un lemme de T. Jeulin au grossissement de la filtration brownienne. In: Séminaire de Probabilités, XIV (Paris, 1978/1979) (French), Lecture Notes in Mathematics, vol. 784, pp. 189–199, Springer, Berlin (1980)

On Standardness and I-cosiness

Stéphane Laurent

Abstract The object of study of this work is the invariant characteristics of filtrations in discrete, negative time, pioneered by Vershik. We prove the equivalence between I-cosiness and standardness without using Vershik's standardness criterion. The equivalence between I-cosiness and productness for homogeneous filtrations is further investigated by showing that the I-cosiness criterion is equivalent to Vershik's first level criterion separately for each random variable. We also aim to derive the elementary properties of both these criteria, and to give a survey and some complements on the published and unpublished literature.

Keywords Standard filtration · Cosy filtration · Self-joining of a filtration

1 Introduction

A filtration $\mathcal{F} = (\mathcal{F}_n)_{n \leq 0}$ in discrete, negative time, is said to be of *local product type* if there exists a sequence $(V_n)_{n \leq 0}$ of independent random variables such that for each $n \leq 0$, one has $\mathcal{F}_n = \mathcal{F}_{n-1} \vee \sigma(V_n)$ and V_n is independent of $\mathcal{F}_{n-1}$. Such random variables V_n are called *innovations* of $\mathcal{F}$. A typical example is the case of a filtration generated by a sequence of independent random variables, termed as *filtration of product type*.

Originally, the theory of decreasing sequences of measurable partitions investigated by Vershik [35–38, 40] was mainly oriented towards characterizing productness for *homogeneous* filtrations of local product type, that is, those for which each innovation V_n has either a uniform distribution on a finite set or a diffuse law. The *standardness criterion* introduced by Vershik provides such a characterization under the assumption that the final σ-field $\mathcal{F}_0$ of the filtration is essentially separable (in

S. Laurent (✉)

Université catholique de Louvain, Louvain-la-Neuve, Belgium

Present Address: Université de Strasbourg, IRMA, UMR 7501, 7 rue René-Descartes, 67084 Strasbourg Cedex, France

e-mail: laurent@math.u-strasbg.fr

C. Donati-Martin et al. (eds.), *Séminaire de Probabilités XLIII*, Lecture Notes in Mathematics 2006, DOI 10.1007/978-3-642-15217-7_5, © Springer-Verlag Berlin Heidelberg 2011

other words, it is countably generated up to negligible sets, and we also say that the filtration is *essentially separable*).

Vershik's standardness criterion makes sense not only in the context of filtrations of local product type, and it characterizes essentially separable filtrations $\mathcal{F} = (\mathcal{F}_n)_{n \leq 0}$ having an *extension* of product type, hereafter called *standard filtrations*.

Vershik's theory of filtrations in discrete, negative time remained unknown to the western probabilistic culture for about 25 years, until Dubins, Feldman, Smorodinsky and Tsirelson used *Vershik's standardness criterion* in [10]. Later, in [14], Émery and Schachermayer partially translated Vershik's theory into the language of stochastic processes, and introduced the *I-cosiness criterion*, inspired by the notion of cosiness which Tsirelson devised in [34] and by Smorodinsky's proof in [32] that the filtration of a *split-word process* is not standard. In the context of essentially separable filtrations, the results of Vershik's theory of filtrations are summarized in Fig. 1.

Among the contents of this paper is a proof of the following theorem.

Theorem A. *A homogeneous filtration $\mathcal{F} = (\mathcal{F}_n)_{n \leq 0}$ with an essentially separable final σ-field $\mathcal{F}_0$ is I-cosy if and only if it is generated by a sequence of independent random variables.*

The proof of this theorem is incomplete in [14], for only the case of homogeneous filtrations with diffuse innovations is considered there. Moreover, Vershik's standardness criterion is used to establish this result, whereas we give a more direct proof without using this criterion, which actually is not even stated in the present paper.

The proofs given in the literature [14,15,40] of Theorem A, or of the equivalence between Vershik's standardness criterion and productness for a homogeneous filtration, use *Vershik's first level criterion* as a key step, without naming it. Vershik's

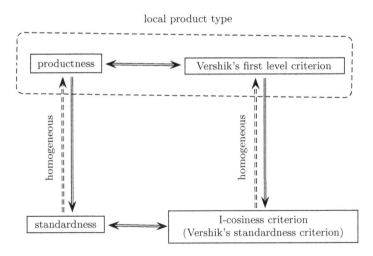

Fig. 1 Theorems for essentially separable filtrations

first level criterion is known to be equivalent to productness when the filtration is essentially separable. Roughly speaking, this criterion says that any random variable measurable with respect to the final σ-field can be approximated by a function of finitely many innovations.

Thus, Theorem A derives from the equivalence between I-cosiness and Vershik's first level criterion for an essentially separable homogeneous filtration. We will see that this equivalence is still valid without assuming the filtration to be essentially separable, and thus we will deduce Theorem A from Theorem A' below.

Theorem A'. *A homogeneous filtration $\mathcal{F} = (\mathcal{F}_n)_{n \leqslant 0}$ is I-cosy if and only if it satisfies Vershik's first level criterion.*

Actually all our results will be stated under a weaker assumption than essential separability of the filtration. Namely, in this paper, the standing assumption on filtrations is *local separability*; we say that a filtration is *locally separable* if it admits essentially separable increments, with a final σ-field which is not necessarily essentially separable. All these results are summarized in Fig. 2.

For example, Theorem B below admits Theorem B' as its analogue for locally separable filtration. Theorems B and B' are elementarily deduced from Theorems A and A' respectively.

Theorem B. *An essentially separable filtration $\mathcal{F} = (\mathcal{F}_n)_{n \leqslant 0}$ is I-cosy if and only if it is standard.*

Theorem B'. *A locally separable filtration $\mathcal{F} = (\mathcal{F}_n)_{n \leqslant 0}$ is I-cosy if and only if it is weakly standard.*

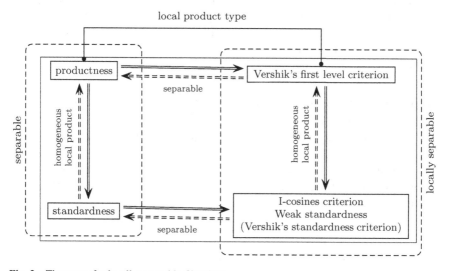

Fig. 2 Theorems for locally separable filtrations

The definition of *weak standardness* is analogous to that of standardness, with productness replaced by Vershik's first level criterion: whereas a filtration is standard if it admits an extension of product type, a filtration is weakly standard if it admits an extension satisfying Vershik's first level criterion.

Our hypothesis of locally separability, less stringent than requiring the final σ-field to be essentially separable, has no practical interest; but it requires no additional efforts, and it sometimes provides a better understanding of the results.

Actually, our additional efforts are oriented towards investigating the I-cosiness criterion *for a random variable with respect to a filtration*, and not only for the whole filtration, in the following sense. The definition of the I-cosiness criterion for a filtration $\mathcal{F} = (\mathcal{F}_n)_{n \leq 0}$ requires a certain property, say $I(X)$, to hold for each 'test' random variable X measurable with respect to the final σ-field $\mathcal{F}_0$. This property $I(X)$ will be called *I-cosiness of the random variable X (with respect to $\mathcal{F}$)*. Shortly:

$$\underbrace{\forall X, \overbrace{I(X)}^{\text{I-cosiness of } X}.}_{\text{I-cosiness of } \mathcal{F}}$$

Vershik's first level criterion has the same structure, and we will similarly define *Vershik's first level criterion for a random variable*. Then Theorem A' will be an immediate consequence of Theorem A" below.

Theorem A". *Let $\mathcal{F} = (\mathcal{F}_n)_{n \leq 0}$ be a homogeneous filtration. Then a random variable is I-cosy with respect to $\mathcal{F}$ if and only if it satisfies Vershik's first level criterion with respect to $\mathcal{F}$.*

This theorem is more interesting than Theorem A, and its proof is not simplified when $\mathcal{F}$ is essentially separable. Thus, our generalization to locally separable filtrations is only a by-product of our investigations of the I-cosiness criterion and Vershik's first level criterion at the level of random variables. We will also obtain the following characterization of I-cosiness for a random variable with respect to a general locally separable filtration.

Theorem. *Let $\mathcal{F} = (\mathcal{F}_n)_{n \leq 0}$ be a locally separable filtration. Then a random variable is I-cosy with respect to $\mathcal{F}$ if and only if it satisfies Vershik's first level criterion with respect to a homogeneous extension of $\mathcal{F}$ with diffuse innovations.*

In the same spirit, Vershik's standardness criterion, which is not stated in this paper, is investigated "random variable by random variable" in [21], where we show it to be equivalent to the I-cosiness criterion under the local separability assumption. The proof is self-contained and no familiarity with the subject is needed. Throughout this paper, we will sometimes announce results from [21].

1.1 Main Notations and Conventions

By a probability space, we always mean a triple $(\Omega, \mathcal{A}, \mathbb{P})$ where the σ-field $\mathcal{A}$ is $\mathbb{P}$-complete. By a σ-field $\mathcal{C} \subset \mathcal{A}$ we always mean an $(\mathcal{A}, \mathbb{P})$-complete σ-field. By a random variable on $(\Omega, \mathcal{A}, \mathbb{P})$, we mean a $\mathbb{P}$-equivalence class of measurable maps from Ω to a separable metric space. By convention, the σ-field generated by an empty family of random variables equals the trivial σ-field $\{\varnothing, \Omega\}$ up to negligible sets. A σ-field $\mathcal{C}$ is *essentially separable* if it is countably generated up to negligible sets. Thus a random variable X generates an essentially separable σ-field $\sigma(X)$; equivalently, an essentially separable σ-field is a σ-field generated by a real-valued random variable. We will extensively use the following elementary lemma, which is often implicit in the probabilistic literature.

Lemma 1.1. *On $(\Omega, \mathcal{A}, \mathbb{P})$, let $\mathcal{B}$ and $\mathcal{C}$ be two σ-fields. For any random variable X measurable with respect to $\mathcal{B} \vee \mathcal{C}$, there exist a $\mathcal{B}$-measurable random variable B and a $\mathcal{C}$-measurable random variable C such that $\sigma(X) \subset \sigma(B, C)$.*

This lemma derives from the equality $\mathcal{B} \vee \mathcal{C} = \bigcup_{B,C} \sigma(B, C)$, where B and C range over all $\mathcal{B}$-measurable r.v. and all $\mathcal{C}$-measurable r.v. respectively. Of course we can also take bounded random variables B and C in this lemma.

We use the notation $L^0(\mathcal{C}; (E, \rho))$ or, shorter $L^0(\mathcal{C}; E)$, to denote the metrizable topological space of all $\mathcal{C}$-measurable random variables taking their values in a separable metric space (E, ρ); the space $L^0(\mathcal{C}; (E, \rho))$ is endowed with the topology of convergence in probability; when $E = \mathbb{R}$ we just write $L^0(\mathcal{C})$. Similarly, the space $L^1(\mathcal{C}; E)$ is the set of all $\mathcal{C}$-measurable random variables X taking their values in E such that $\mathbb{E}[\rho(X, x)]$ is finite for some ($\Leftrightarrow$ for all) $x \in E$; the space $L^1(\mathcal{C}; E)$ is endowed with the metric $(X, Y) \mapsto \mathbb{E}[\rho(X, Y)]$. It is well-known that $L^0(\mathcal{C}; (E, \rho)) = L^1(\mathcal{C}; (E, \rho \wedge 1))$. The set of all simple, E-valued, $\mathcal{C}$-measurable random variables is a dense subset of $L^1(\mathcal{C}; E)$. If F is a finite set, we denote by $L(\mathcal{C}; F)$ the set of all $\mathcal{C}$-measurable random variables taking their values in F, considered as a metric space with the metric $(S, T) \mapsto \mathbb{P}[S \neq T]$. Thus $L(\mathcal{C}; F) = L^1(\mathcal{C}; F)$ where F is equipped with the $0 - 1$ distance. The Borel σ-field on a separable metric space E is denoted by $\mathfrak{B}_E$.

A *Polish metric space* is a complete separable metric space. A *Polish space* is a topological space that admits a separable and complete metrization. A *Polish probability space* is (the completion of) a probability space on a Polish space with its Borel σ-field. Any Polish space F has the *Doob property*: for any measurable space $(\Omega, \mathcal{A})$, if $X: \Omega \to T$ is a measurable function taking its values in a measurable space T and $Y: \Omega \to E$ is a $\sigma(X)$-measurable function taking its values in a Polish space E, then there exists a measurable function $f: T \to E$ such that $Y = f(X)$ (see for instance [8]). We will sometimes use the Doob property without invoking its name, or we will also call it *Doob's functional representation theorem*.

When X is a random variable taking values in a Polish space, the existence of the *conditional law* of X given any σ-field $\mathcal{C}$ is guaranteed (see [11]); we denote it by $\mathcal{L}(X \mid \mathcal{C})$. It is itself a random variable in a Polish space. Some details on conditional laws are provided in Annex 5.1.

1.2 Lebesgue Isomorphisms

A *Lebesgue isomorphism* between two probability spaces $(E, \mathcal{B}, \mu)$ and $(F, \mathcal{L}, \nu)$ is a bimeasurable bijection T from a set $E_0 \in \mathcal{B}$ of full μ-measure into a set $F_0 \in \mathcal{L}$ of full ν-measure, and satisfying $T(\mu) = \nu$. Any Polish probability space is Lebesgue isomorphic to the completion of the Borel space $\mathbb{R}$ equipped with some probability measure (see [6, 25, 29]).

1.3 Filtrations in Discrete, Negative Time

On an underlying probability space $(\Omega, \mathcal{A}, \mathbb{P})$, a filtration is an increasing sequence of sub-σ-fields of $\mathcal{A}$ indexed by a time-axis. Most filtrations considered in this paper are indexed by the time axis $-\mathbb{N} = \{\ldots, -2, -1, 0\}$. If the time axis of a filtration $\mathcal{F}$ is not specified, it will be understood that $\mathcal{F} = (\mathcal{F}_n)_{n \leq 0}$ is a filtration in discrete, negative time. We say that a filtration $\mathcal{F}$ is *essentially separable* if the final σ-field $\mathcal{F}_0$ is essentially separable; equivalently, each σ-field $\mathcal{F}_n$ is essentially separable. We say that a filtration $\mathcal{F}$ is *Kolmogorovian* if the tail σ-field $\mathcal{F}_{-\infty} := \bigcap_{n \leq 0} \mathcal{F}_n$ equals the trivial σ-field $\{\emptyset, \Omega\}$ up to negligible sets. A filtration $\mathcal{F}$ is *included* in a filtration $\mathcal{G}$, and this is denoted by $\mathcal{F} \subset \mathcal{G}$, if $\mathcal{F}_n \subset \mathcal{G}_n$ for each $n \leq 0$. The *supremum* $\mathcal{F} \vee \mathcal{G}$ of two filtrations $\mathcal{F}$ and $\mathcal{G}$ is the smallest filtration containing both $\mathcal{F}$ and $\mathcal{G}$; it is given by $(\mathcal{F} \vee \mathcal{G})_n = \mathcal{F}_n \vee \mathcal{G}_n$. The independent product of two filtrations $\mathcal{F} = (\mathcal{F}_n)_{n \leq 0}$ and $\mathcal{G} = (\mathcal{G}_n)_{n \leq 0}$ respectively defined on two probability spaces $(\Omega, \mathcal{A}, \mathbb{P})$ and $(\Omega^*, \mathcal{A}^*, \mathbb{P}^*)$ is the filtration $\mathcal{F} \otimes \mathcal{G}$ defined on the product probability space $(\Omega, \mathcal{A}, \mathbb{P}) \otimes (\Omega^*, \mathcal{A}^*, \mathbb{P}^*)$ by $(\mathcal{F} \otimes \mathcal{G})_n = \mathcal{F}_n \otimes \mathcal{G}_n$.

1.4 Random Variables, Processes

It is understood, if not otherwise specified, that a random variable takes its values in a separable metric space or in $\mathbb{R}$ if this is clear from the context. By a process, we mean a sequence of random variables (each taking its values in a separable metric space if nothing else is specified). Most processes considered in this paper are indexed by the time-axis $-\mathbb{N}$. Such a process $(X_n)_{n \leq 0}$ generates a filtration $\mathcal{F} = (\mathcal{F}_n)_{n \leq 0}$ defined by $\mathcal{F}_n = \sigma(X_m; m \leq n)$. The process $(X_n)_{n \leq 0}$ is *Markovian* if for each $n \leq 0$ the σ-field $\sigma(X_n)$ is conditionally independent of $\mathcal{F}_{n-1}$ given $\sigma(X_{n-1})$. Given a filtration $\mathcal{G} \supset \mathcal{F}$, the process $(X_n)_{n \leq 0}$ is *Markovian with respect to* $\mathcal{G}$ if for each $n \leq 0$ the σ-field $\sigma(X_n)$ is conditionally independent of $\mathcal{G}_{n-1}$ given $\sigma(X_{n-1})$. Equivalently, the process is Markovian and its filtration $\mathcal{F}$ is *immersed* in $\mathcal{G}$, as we shall see below.

1.5 Preliminary Notion: Immersion

The notion of immersion will be used throughout all this paper. We say that a filtration $\mathcal{F}$ is *immersed* in a filtration $\mathcal{G}$ if $\mathcal{F} \subset \mathcal{G}$ and if every $\mathcal{F}$-martingale is a $\mathcal{G}$-martingale; the notation $\mathcal{F} \overset{m}{\subset} \mathcal{G}$ means that $\mathcal{F}$ is immersed in $\mathcal{G}$. Obviously, the binary relation $\overset{m}{\subset}$ defines a partial order on the set of filtrations on $(\Omega, \mathcal{A}, \mathbb{P})$. A typical example is provided by Lemma 1.2 and some usual characterizations of immersion are given in Lemma 1.3.

Lemma 1.2. *Let $\mathcal{F}$ and $\mathcal{G}$ be two independent filtrations. Then both $\mathcal{F}$ and $\mathcal{G}$ are immersed in $\mathcal{F} \vee \mathcal{G}$.*

Two filtrations $\mathcal{F}$ and $\mathcal{G}$ both immersed in $\mathcal{F} \vee \mathcal{G}$ are said to be *jointly immersed*; it suffices that $\mathcal{F} \overset{m}{\subset} \mathcal{H}$ and $\mathcal{G} \overset{m}{\subset} \mathcal{H}$ for some filtration $\mathcal{H}$.

Lemma 1.3. *Let $\mathcal{F} = (\mathcal{F}_n)_{n \leq 0}$ and $\mathcal{G} = (\mathcal{G}_n)_{n \leq 0}$ be two filtrations on a probability space $(\Omega, \mathcal{A}, \mathbb{P})$. The following conditions are equivalent:*

(i) $\mathcal{F}$ *is immersed in* $\mathcal{G}$.

(ii) $\mathcal{F}$ *is included in* $\mathcal{G}$ *and the σ-field $\mathcal{F}_0$ is conditionally independent of $\mathcal{G}_n$ given $\mathcal{F}_n$ for each $n \leq 0$.*

(iii) *For every random variable $X \in L^1(\mathcal{F}_0)$, one has $\mathbb{E}[X \mid \mathcal{G}_n] = \mathbb{E}[X \mid \mathcal{F}_n]$ for each $n \leq 0$.*

(iv) *For every $\mathcal{F}_0$-measurable random variable Y taking its values in a Polish space, one has $\mathcal{L}(Y \mid \mathcal{G}_n) = \mathcal{L}(Y \mid \mathcal{F}_n)$ for each $n \leq 0$.*

Note also that immersion of $\mathcal{F}$ in $\mathcal{G}$ implies $\mathcal{F}_n = \mathcal{F}_0 \cap \mathcal{G}_n$ for all $n \leq 0$. Proofs of the preceding two lemmas are left as an exercise to the reader, as well as those of the next three lemmas, which will frequently be used in this paper. The third one is a straightforward consequence of the first two ones.

Lemma 1.4. *A filtration $\mathcal{F}$ is immersed in a filtration $\mathcal{G}$ if and only if $\mathcal{F} \subset \mathcal{G}$ and for every integer $n < 0$, the σ-field $\mathcal{F}_{n+1}$ is conditionally independent of $\mathcal{G}_n$ given $\mathcal{F}_n$.*

Lemma 1.5. *If $\mathcal{B}$, $\mathcal{C}$ and $\mathcal{D}$ are three σ-fields such that $\mathcal{B}$ and $\mathcal{C}$ are conditionally independent given $\mathcal{D}$, then $\mathcal{D} \vee \mathcal{B}$ and $\mathcal{C}$ are also conditionally independent given $\mathcal{D}$.*

Lemma 1.6. *Let $\mathcal{F} = (\mathcal{F}_n)_{n \leq 0}$ and $\mathcal{G} = (\mathcal{G}_n)_{n \leq 0}$ be two filtrations such that $\mathcal{F} \subset \mathcal{G}$. Let $(V_n)_{n \leq 0}$ be a process such that $\mathcal{F}_n \subset \mathcal{F}_{n-1} \vee \sigma(V_n)$ for every $n \leq 0$. If V_n is conditionally independent of $\mathcal{G}_{n-1}$ given $\mathcal{F}_{n-1}$ for every $n \leq 0$, then $\mathcal{F}$ is immersed in $\mathcal{G}$.*

Here are two straightforward applications of Lemma 1.6. First, the filtration $\mathcal{F}$ generated by an independent sequence $(V_n)_{n \leq 0}$ of random variables is immersed in a filtration $\mathcal{G}$ if and only if $\mathcal{F} \subset \mathcal{G}$ and V_n is independent of $\mathcal{G}_{n-1}$ for every $n \leq 0$. Second, a Markov process $(V_n)_{n \leq 0}$ is Markovian with respect to a filtration $\mathcal{G}$ if and only if its generated filtration $\mathcal{F}$ is immersed in $\mathcal{G}$.

1.6 Preliminary Notion: Isomorphic σ-fields and Filtrations

An *embedding* Ψ between two probability spaces $(\Omega, \mathcal{B}, \mathbb{P})$ and $(\Omega', \mathcal{A}', \mathbb{P}')$ is (necessarily injective) map from the quotient σ-field $\mathcal{B}/\mathbb{P}$ to the quotient σ-field $\mathcal{A}'/\mathbb{P}'$ that preserves the σ-field structures and the probabilities. We shortly write $\Psi \colon \mathcal{B} \to \mathcal{A}'$. Is is called an *isomorphism* if moreover it is onto. Up to isomorphism, an essentially separable σ-field is characterized by the descending sequence (possibly empty, finite, or denumerable) of the masses of its atoms. An embedding Ψ extends uniquely to random variables taking their values in a Polish space, and we call $\Psi(X)$ the *copy* of such a random variable X. Details are provided in Annex 5.1. However this Annex can be skipped since there is no risk when naively using isomorphisms: any expected property such as $\Psi(f(X)) = f(\Psi(X))$, $\Psi(X, Y) = (\Psi(X), \Psi(Y))$, $\sigma(\Psi(X)) = \Psi(\sigma(X))$, $\Psi(\mathbb{E}[X \mid \mathcal{C}]) = \mathbb{E}'[\Psi(X) \mid \Psi(\mathcal{C})]$, is true.

The definition of isomorphic σ-fields extends naturally to filtrations as follows. Two filtrations $\mathcal{F} = (\mathcal{F}_n)_{n \leq 0}$ and $\mathcal{F}' = (\mathcal{F}'_n)_{n \leq 0}$, defined on possibly different probability spaces, are *isomorphic* if there is an isomorphism $\Psi \colon \mathcal{F}_0 \to \mathcal{F}'_0$ such that $\Psi(\mathcal{F}_n) = \mathcal{F}'_n$ for every $n \leq 0$. We say that $\Psi \colon \mathcal{F} \to \mathcal{F}'$ is an isomorphism. We denote by $\Psi(\mathcal{F})$ the filtration $(\mathcal{F}'_n)_{n \leq 0} = (\Psi(\mathcal{F}_n))_{n \leq 0}$ and we call it the *copy* of the filtration $\mathcal{F}$ by the isomorphism Ψ.

A typical example of isomorphic filtrations is the case when $\mathcal{F}$ and $\mathcal{F}'$ are respectively generated by two processes $(X_n)_{n \leq 0}$ and $(X'_n)_{n \leq 0}$ having the same law. In the case when the X_n (hence the X'_n) take their values in Polish spaces, there exists a unique isomorphism $\Psi \colon \mathcal{F} \to \mathcal{F}'$ that sends X_n to X'_n for each $n \leq 0$. This stems from Lemma 5.7. Another typical example of isomorphic filtrations is provided by the following lemma.

Lemma 1.7. *Let $\mathcal{F}$ and $\mathcal{G}$ be two independent filtrations. Then $\mathcal{F} \vee \mathcal{G}$ is isomorphic to the independent product $\mathcal{F} \otimes \mathcal{G}$ of $\mathcal{F}$ and $\mathcal{G}$.*

Proof. By Proposition 5.11, there exists a unique isomorphism extending the canonical embeddings $\iota_1 \colon \mathcal{F}_0 \to \mathcal{F}_0 \otimes \mathcal{G}_0$ and $\iota_2 \colon \mathcal{G}_0 \to \mathcal{F}_0 \otimes \mathcal{G}_0$ (defined in Example 5.2). $\square$

2   Vershik's First Level Criterion

This section deals with filtrations of *product type* and *Vershik's first level criterion*.

Definition 2.1. A filtration is of *product type* if it is generated by a sequence of independent random variables.

As we shall see, productness is equivalent to Vershik's first level criterion for an essentially separable filtration of *local product type* (Theorem 2.25). This result is far from new: Vershik's first level criterion appears, but without a name, in [14, 15, 40]. Corollary 2.46 shows that the assumption of essential separability cannot be waived: there exist some filtrations of local product type satisfying

Vershik's first level criterion but which are not essentially separable, hence not of product type.

As said in the introduction, the important Theorem A is deduced from Theorem A' and from the equivalence between Vershik's first level criterion and productness for essentially separable filtrations of local product type. Vershik's first level criterion will also be used in Sect. 3 to extend the notion of standardness to the notion of weak standardness (Definition 3.21).

Theorem A' stated in the introduction is directly deduced from Theorem A"; actually the latter will be proved (in Sect. 4) with the help of the equivalent "self-joining version" of Vershik's first level criterion, which we study in Sect. 2.2 and call *Vershik's self-joining criterion*.

In Sect. 2.3 we introduce the filtrations of *split-word processes*. We state the theorems on productness for these filtrations which are found in the literature, and we initiate the proofs of these theorems assuming some intermediate key results. At this stage, we will not have at our disposal the tools for finishing these proofs; they will be pursued at the end of each following section, illustrating the new tools we shall acquire.

2.1   Productness and Vershik's First Level Criterion

In this section, we define Vershik's first level criterion and prove its equivalence (Theorem 2.25) with productness for an essentially separable filtration of local product type (Definition 2.3). We will derive this theorem from Theorem 2.23 which gives a characterization of Vershik's first level criterion for filtrations of local product type that are not necessarily essentially separable.

With the terminology of Definition 2.2 below, a filtration of product type is a filtration for which there exists a generating innovation.

Definition 2.2. Given two σ-fields $\mathcal{B}$ and $\mathcal{C}$ such that $\mathcal{C} \subset \mathcal{B}$, an *independent complement* of $\mathcal{C}$ in $\mathcal{B}$ is a random variable V taking its values in a Polish space, independent of $\mathcal{C}$ and such that $\mathcal{B} = \mathcal{C} \vee \sigma(V)$. An *innovation*, or a *global innovation*, of a filtration $\mathcal{F} = (\mathcal{F}_n)_{n \leq 0}$ is a process $(V_n)_{n \leq 0}$ such that for each $n \leq 0$, the random variable V_n is an independent complement of $\mathcal{F}_{n-1}$ in $\mathcal{F}_n$. An innovation $(V_n)_{n \leq 0}$ is called *generating* if $\mathcal{F}$ is generated by the process $(V_n)_{n \leq 0}$. For two given integers n_0 and m_0 such that $n_0 < m_0 \leq 0$, an *innovation*, or a *local innovation* of $\mathcal{F}$ from n_0 to m_0 is a sequence of random variables $(V_{n_0+1}, \ldots, V_{m_0})$ such that V_n is an independent complement of $\mathcal{F}_{n-1}$ in $\mathcal{F}_n$ for each $n \in \{n_0 + 1, \ldots, m_0\}$.

The random variables V_n appearing in a global or a local innovation of a filtration $\mathcal{F}$ are themselves called innovations of $\mathcal{F}$.

Definition 2.3. A filtration $\mathcal{F} = (\mathcal{F}_n)_{n \leq 0}$ is of *local product type* if there exists a global innovation of $\mathcal{F}$.

Innovations are not unique in general. They are described by Lemma 2.4 below. The notion of *Lebesgue isomorphism* has been recalled in Sect. 1.

Lemma 2.4. *Let $(\Omega, \mathcal{A}, \mathbb{P})$ be a probability space, $\mathcal{C}$ and $\mathcal{B}$ two sub-σ-fields of $\mathcal{A}$, and V an independent complement of $\mathcal{C}$ in $\mathcal{B}$ taking values in a Polish space E. Let V' be a random variable taking values in a Polish space E'. Then V' is an independent complement of $\mathcal{C}$ in $\mathcal{B}$ if and only if there exist a $\mathcal{C}$-measurable random variable C and a measurable function $\phi: \mathbb{R} \times E \to E'$ such that $V' = \phi(C, V)$ and, almost surely, the random map $T_C: v \mapsto \phi(C, v)$ is a Lebesgue isomorphism from the probability space induced by V into the probability space induced by V'. In particular, the σ-fields $\sigma(V)$ and $\sigma(V')$ are isomorphic.*

Proof. The 'if' part is easy to verify. To show the 'only if' part, assume V' to be an independent complement of $\mathcal{C}$ in $\mathcal{B}$. There exist (Lemma 1.1) two $\mathcal{C}$-measurable random variables C_1 and C_2 such that $\sigma(V') \subset \sigma(C_1, V)$ and $\sigma(V) \subset \sigma(C_2, V')$. We introduce a real-valued $\mathcal{C}$-measurable random variable C such that $\sigma(C_1, C_2) \subset \sigma(C)$ and we denote its law by $\mathbb{P}_C$. By Doob's functional representation theorem, there exist two measurable functions ϕ and ψ such that $V' = \phi(C, V)$ and $V = \psi(C, V')$. Considering the conditional laws given $\mathcal{C}$, we see that $\phi(c, V)$ has the same law as V' and $\psi(c, V')$ has the same law as V for $\mathbb{P}_C$-almost every c. Moreover, since

$$1 = \mathbb{P}\left[V' = \phi\left(C, \psi(C, V')\right)\right] = \int \mathbb{P}\left[V' = \phi\left(c, \psi(c, V')\right)\right] d\mathbb{P}_C(c),$$

one has $V' = \phi(c, \psi(c, V'))$ almost surely for $\mathbb{P}_C$-almost every c. In the same way one has $V = \psi(c, \phi(c, V))$ almost surely for $\mathbb{P}_C$-almost every c. Hence, for $\mathbb{P}_C$-almost every c, the Borel subset $E_0 := \{v \in E \mid v = \psi(c, \phi(c, v))\}$ of E has full $\mathbb{P}_V$-measure, the Borel subset $E'_0 := \{v' \in E' \mid v' = \phi(c, \psi(c, v'))\}$ of E' has full $\mathbb{P}_{V'}$-measure, and the maps $v \mapsto \phi(c, v)$ and $v' \mapsto \psi(c, v')$ define mutual inverse bijections between E_0 and E'_0. Finally $v \mapsto \phi(c, v)$ defines a Lebesgue isomorphism from the probability space induced by V into the probability space induced by V'. Consequently, the σ-fields $\sigma(V)$ and $\sigma(V')$ have the same descending sequences of masses of their atoms, and hence are isomorphic. $\square$

A straightforward application of Lemma 1.6 gives the following lemma:

Lemma 2.5. *Let $\mathcal{F} = (\mathcal{F}_n)_{n \leq 0}$ be a filtration of local product type and $(V_n)_{n \leq 0}$ an innovation of $\mathcal{F}$. Then the filtration generated by $(V_n)_{n \leq 0}$ is immersed in $\mathcal{F}$. Consequently one has $\sigma(V_m; m \leq n) = \mathcal{F}_n \cap \sigma(V_m; m \leq 0)$ for each $n \leq 0$.*

Now we turn to Vershik's first level criterion. We shall see at the end of this section that this criterion is equivalent to productness for an essentially separable filtration of local product type (Theorem 2.25).

Definition 2.6. On $(\Omega, \mathcal{A}, \mathbb{P})$, let $\mathcal{F} = (\mathcal{F}_n)_{n \leq 0}$ be a filtration of local product type.

- Let (E, ρ) be a separable metric space and $X \in L^1(\mathcal{F}_0; E)$. The random variable X satisfies *Vershik's first level criterion* (with respect to $\mathcal{F}$) if for every $\delta > 0$, there exist an integer $n_0 < 0$, an innovation $(V_{n_0+1}, \ldots, V_0)$ of $\mathcal{F}$ from n_0 to 0, and a random variable $S \in L^1(\sigma(V_{n_0+1}, \ldots, V_0); E)$ such that $\mathbb{E}[\rho(X, S)] < \delta$.
- A σ-field $\mathcal{E}_0 \subset \mathcal{F}_0$ satisfies *Vershik's first level criterion* (with respect to $\mathcal{F}$) if every random variable $X \in L^1(\mathcal{E}_0; \mathbb{R})$ satisfies Vershik's first level criterion with respect to $\mathcal{F}$.
- The filtration $\mathcal{F} = (\mathcal{F}_n)_{n \leq 0}$ satisfies *Vershik's first level criterion* if the σ-field $\mathcal{F}_0$ satisfies Vershik's first level criterion with respect to $\mathcal{F}$.

When there is no ambiguity, we will omit the specification *with respect to $\mathcal{F}$* in this definition. We will see (Proposition 2.17) that Vershik's first level criterion for a random variable X is equivalent to Vershik's first level criterion for the σ-field $\sigma(X)$. It is clear that Vershik's first level criterion is preserved by isomorphism. The following proposition is easily established from the definition; its proof is left to the reader.

Proposition 2.7. *Let $\mathcal{F} = (\mathcal{F}_n)_{n \leq 0}$ be a filtration of local product type and let (E, ρ) be a separable metric space. The set of random variables $X \in L^1(\mathcal{F}_0; E)$ satisfying Vershik's first level criterion is closed in $L^1(\mathcal{F}_0; E)$.*

To establish other properties of Vershik's first level criterion, it will be convenient to rephrase it with the help of the following notion:

Definition 2.8. Let $(\Omega, \mathcal{A}, \mathbb{P})$ be a probability space and $\mathcal{B} \subset \mathcal{A}$ be a σ-field. A family C of sub-σ-fields of $\mathcal{A}$ is *substantial* in $\mathcal{B}$ if the L^1-closure of $\bigcup_{\mathcal{C} \in C} L^1(\Omega, \mathcal{C}, \mathbb{P})$ contains $L^1(\Omega, \mathcal{B}, \mathbb{P})$.

Thus, we can restate Definition 2.6 of Vershik's first level criterion for a σ-field as follows.

Definition 2.9. Let $\mathcal{F} = (\mathcal{F}_n)_{n \leq 0}$ be a filtration of local product type. Call C^{loc} the family of all σ-fields $\sigma(V_{n_0+1}, \ldots, V_0)$ generated by local innovations $(V_{n_0+1}, \ldots, V_0)$ from n_0 to 0, for all $n_0 < 0$. A σ-field $\mathcal{E}_0 \subset \mathcal{F}_0$ satisfies *Vershik's first level criterion* if C^{loc} is substantial in $\mathcal{E}_0$.

The notion of substantial family of σ-fields appears in [13] in a slightly different form. Lemma below is a duplicate of Lemma 2 in [13], which could be proved identically in spite of this difference between the two notions of substantialness.

Lemma 2.10. *Let $(\Omega, \mathcal{A}, \mathbb{P})$ be a probability space, $\mathcal{B} \subset \mathcal{A}$ a σ-field, and C a family of sub-σ-fields of $\mathcal{A}$. The following three conditions are equivalent:*

(i) C is substantial in $\mathcal{B}$.

(ii) For each finite set F, the closure of $\bigcup_{\mathcal{C} \in C} L(\mathcal{C}; F)$ in $L(\mathcal{A}; F)$ contains $L(\mathcal{B}; F)$.

(iii) For each separable metric space E, the closure of $\bigcup_{\mathcal{C} \in C} L^1(\mathcal{C}; E)$ in $L^1(\mathcal{A}; E)$ contains $L^1(\mathcal{B}; E)$.

Proposition 2.11. *Let $\mathcal{F} = (\mathcal{F}_n)_{n \leqslant 0}$ be a filtration of local product type. If a σ-field $\mathcal{E}_0 \subset \mathcal{F}_0$ satisfies Vershik's first level criterion, then for any separable metric space E, every random variable $X \in L^1(\mathcal{E}_0; E)$ satisfies Vershik's first level criterion.*

Proof. Left to the reader as an easy application of the definitions and the previous lemma.   □

The following lemma provides a typical example of substantialness.

Lemma 2.12. *Let $(\Omega, \mathcal{A}, \mathbb{P})$ be a probability space and $(\mathcal{B}_m)_{m \in \mathbb{N}}$ an increasing sequence of sub-σ-fields of $\mathcal{A}$. Then the family of σ-fields $\{\mathcal{B}_m; m \in \mathbb{N}\}$ is substantial in $\bigvee_m \mathcal{B}_m$.*

Proof. A classical result says that for any set $B \in \bigvee_m \mathcal{B}_m$, there exist some $B_1, B_2, \ldots \in \bigcup_m \mathcal{B}_m$ such that $\mathbb{P}[B \vartriangle B_m] \to 0$ (this is easily established by a monotone class argument). With the help of this, the lemma follows from Lemma 2.10.(ii).   □

Proposition 2.13. *Let $\mathcal{F} = (\mathcal{F}_n)_{n \leqslant 0}$ be a filtration of local product type and $(\mathcal{B}_m)_{m \geqslant 0}$ an increasing sequence of sub-σ-fields of $\mathcal{F}_0$. If each $\mathcal{B}_m$ satisfies Vershik's first level criterion, then so does $\bigvee_m \mathcal{B}_m$.*

Proof. Straightforward from the previous lemma and Proposition 2.7.   □

Corollary 2.14. *Let $\mathcal{F} = (\mathcal{F}_n)_{n \leqslant 0}$ be a filtration of local product type and $(V_n)_{n \leqslant 0}$ an innovation of $\mathcal{F}$. The σ-field $\sigma(V_n; n \leqslant 0)$ satisfies Vershik's first level criterion. Consequently, a filtration of product type satisfies Vershik's first level criterion.*

Proof. Obviously, the σ-field $\mathcal{B}_m := \sigma(V_n; -m \leqslant n \leqslant 0)$ satisfies Vershik's first level criterion for every $m \in \mathbb{N}$. Hence, the result derives from Proposition 2.13.   □

This corollary contains the easy implications of the equivalences stated in Theorems 2.23 and 2.25 towards which we orient the rest of this section.

Lemma 2.16 is the key lemma to prove the equivalence between Vershik's first level criterion for a random variable X and Vershik's first level criterion for the σ-field $\sigma(X)$. It characterizes substantialness of a family of σ-fields in an essentially separable σ-field $\mathcal{B}$ by a property on a random variable X generating $\mathcal{B}$. It will be proved with the help of the following lemma, which we shall also use several times in the next sections.

Lemma 2.15. *Let $(\Omega, \mathcal{A}, \mathbb{P})$ be a probability space and $X \in L^1(\mathcal{A}; E)$ where (E, ρ) is a separable metric space. The set of all random variables of the form $f(X)$ where $f : E \to \mathbb{R}$ is Lipschitz function, is a dense subset of $L^1(\sigma(X))$.*

Proof. Let us denote by $\ell(X)$ this set of random variables. For every open set $O \subset \mathbb{R}$, the sequence of random variables $X_m := (m\rho(X, O^c)) \wedge 1$ converges almost surely to $\mathbb{1}_{X \in O}$, and $x \mapsto (m\rho(x, O^c)) \wedge 1$ is a Lipschitz function. It follows that the L^1-closure of $\ell(X)$ contains all linear combinations of indicator random variables $\mathbb{1}_{\{X \in O_i\}}$ where O_i is an open set, and therefore is dense in $L^1(\sigma(X))$.   □

Lemma 2.16. *Let $(\Omega, \mathcal{A}, \mathbb{P})$ be a probability space, (E, ρ) a separable metric space, $X \in L^1(\mathcal{A}; E)$, and $\mathbf{C}$ a family of sub-σ-fields of $\mathcal{A}$. Then $\mathbf{C}$ is substantial in $\sigma(X)$ if and only if for every $\delta > 0$, there exist $\mathcal{C} \in \mathbf{C}$ and a random variable $C \in L^1(\mathcal{C}; E)$ such that $\mathbb{E}[\rho(X, C)] < \delta$.*

Proof. This easily results from Lemma 2.10.(iii) and Lemma 2.15.   □

Proposition 2.17. *Let $\mathcal{F} = (\mathcal{F}_n)_{n \leq 0}$ be a filtration of local product type. Let (E, ρ) be a separable metric space and $X \in L^0(\mathcal{F}_0; E)$. The following conditions are equivalent.*

(i) The σ-field $\sigma(X)$ satisfies Vershik's first level criterion.
(ii) For every $\delta > 0$, there exist an integer $n_0 < 0$, an innovation $(V_{n_0+1}, \ldots, V_0)$ of $\mathcal{F}$ from n_0 to 0, and a random variable $S \in L^0(\sigma(V_{n_0+1}, \ldots, V_0); E)$ such that $\mathbb{P}[\rho(X, S) > \delta] < \delta$.

If $X \in L^1(\mathcal{F}_0; E)$, these conditions are also equivalent to

1. X satisfies Vershik's first level criterion.

Proof. We know (Definition 2.9) that (i) is equivalent to $\mathbf{C}^{\mathrm{loc}}$ being substantial in $\sigma(X)$. Thus, Lemma 2.16 directly shows (i) $\Longleftrightarrow$ (iii), and (i) $\Longleftrightarrow$ (ii) derives from the same lemma by replacing ρ with $\rho \wedge 1$.   □

Most of the results in the sequel of this section will be established with the help of the following elementary Lemmas 2.18 and 2.19. Lemma 2.18 is a duplicate of Lemma 3 in [13], which could be proved identically in spite of the difference between our notion of substantialness and the one given in [13].

Lemma 2.18. *Let $(\Omega, \mathcal{A}, \mathbb{P})$ be a probability space, $\mathcal{B}$ and $\mathcal{D}$ two sub-σ-fields of $\mathcal{A}$, and $\mathbf{C}$ a family of sub-σ-fields of $\mathcal{A}$. If $\mathbf{C}$ is substantial in $\mathcal{B}$, then the family of σ-fields $\{\mathcal{C} \vee \mathcal{D} \mid \mathcal{C} \in \mathbf{C}\}$ is substantial in $\mathcal{B} \vee \mathcal{D}$.*

Lemma 2.19. *Let $(\Omega, \mathcal{A}, \mathbb{P})$ be a probability space, $\mathcal{E} \subset \mathcal{A}$ a σ-field, $\mathbf{C}$ a family of sub-σ-fields of $\mathcal{E}$, and $\mathbf{D}$ a family of sub-σ-fields of $\mathcal{A}$ such that each $\mathcal{D} \in \mathbf{D}$ is independent of $\mathcal{E}$. If the family of σ-fields $\{\mathcal{C} \vee \mathcal{D} \mid \mathcal{C} \in \mathbf{C}, \mathcal{D} \in \mathbf{D}\}$ is substantial in a σ-field $\mathcal{B} \subset \mathcal{E}$, then $\mathbf{C}$ is substantial in $\mathcal{B}$.*

Proof. Let $X \in L^1(\mathcal{B})$ and $\delta > 0$. Assuming that $\{\mathcal{C} \vee \mathcal{D} \mid \mathcal{C} \in \mathbf{C}, \mathcal{D} \in \mathbf{D}\}$ is substantial in $\mathcal{B}$, there exist $\mathcal{C} \in \mathbf{C}$, $\mathcal{D} \in \mathbf{D}$, and a $\mathcal{C} \vee \mathcal{D}$-measurable random variable S such that $\mathbb{E}[|X - S|] < \delta$. One can write (Lemma 1.1) $S = f(C, D)$ where C and D are random variables measurable with respect to $\mathcal{C}$ and $\mathcal{D}$ respectively and f is measurable, thus $\mathbb{E}[S \mid \mathcal{E}]$ is measurable with respect to $\mathcal{C}$ because $\mathbb{E}[S \mid \mathcal{E}] = h(C)$ where $h(c) = \mathbb{E}[f(c, D)]$. Using the L^1-contractivity of conditional expectations, we get $\mathbb{E}[|X - \mathbb{E}[S \mid \mathcal{C}]|] = \mathbb{E}[|\mathbb{E}(X - S \mid \mathcal{E})|] < \mathbb{E}[|X - S|] < \delta$, which shows that $\mathbf{C}$ is substantial in $\mathcal{B}$.   □

Proposition 2.21 highlights the asymptotic nature of Vershik's first level criterion. It is proved with the help of the following lemma.

Lemma 2.20. *Let $\mathcal{F} = (\mathcal{F}_n)_{n \leq 0}$ be a filtration of local product type and $N \leq 0$ an integer. If a σ-field $\mathcal{E}_N \subset \mathcal{F}_N$ satisfies Vershik's first level criterion with respect to $\mathcal{F}$, then it satisfies Vershik's first level criterion with respect to the truncated filtration $(\mathcal{F}_{N+n})_{n \leq 0}$.*

Proof. This results from Lemma 2.19 by taking $\mathcal{E} = \mathcal{F}_N$, $\mathcal{B} = \mathcal{E}_N$, and by considering the family C consisting of all the σ-fields $\sigma(V_{n_0+1}, \ldots, V_N)$ generated by local innovations $(V_{n_0+1}, \ldots, V_N)$ of $\mathcal{F}$ from n_0 to N for all $n_0 < N$, and the family D consisting of all the σ-fields $\sigma(V_{N+1}, \ldots, V_0)$ generated by local innovations $(V_{N+1}, \ldots, V_0)$ of $\mathcal{F}$ from N to 0.   □

Proposition 2.21. *Let $\mathcal{F} = (\mathcal{F}_n)_{n \leq 0}$ be a filtration of local product type. The following conditions are equivalent:*

  (i) *$\mathcal{F}$ satisfies Vershik's first level criterion.*
 (ii) *For every $N \in -\mathbb{N}$, the truncated filtration $(\mathcal{F}_{N+n})_{n \leq 0}$ satisfies Vershik's first level criterion.*
(iii) *There exists $N \in -\mathbb{N}$ such that the truncated filtration $(\mathcal{F}_{N+n})_{n \leq 0}$ satisfies Vershik's first level criterion.*

Proof. Let C be the family of σ-fields $\sigma(V_{n_0+1}, \ldots, V_N)$ generated by all local innovations $(V_{n_0+1}, \ldots, V_N)$ from n_0 to N for all $n_0 < N$, let $D = \sigma(V_{N+1}, \ldots, V_0)$ where $(V_{N+1}, \ldots, V_0)$ is an innovation of $\mathcal{F}$ from N to 0, and let $\mathcal{B} = \mathcal{F}_N$. Lemma 2.18 applied with these notations shows that (iii) $\Longrightarrow$ (i); Lemma 2.20 shows that (i) $\Longrightarrow$ (ii); finally, (ii) $\Longrightarrow$ (iii) is trivially true.   □

The following proposition will help in Sect. 2.3.

Proposition 2.22. *Let $\mathcal{F} = (\mathcal{F}_n)_{n \leq 0}$ be a filtration of local product type, $(V_n)_{n \leq 0}$ an innovation of $\mathcal{F}$, and $(\mathcal{C}_n)_{n \leq 0}$ a sequence of σ-fields such that $\mathcal{C}_n \subset \mathcal{F}_n$ for each $n \leq 0$ and such that $(\mathcal{C}_n \vee \sigma(V_{n+1}, \ldots, V_0))_{n \leq 0}$ is an increasing sequence of σ-fields. Define $\mathcal{C}_\infty = \bigvee_n \mathcal{C}_n$. If $\mathcal{C}_n$ satisfies Vershik's first level criterion for every $n \leq 0$, then $\mathcal{C}_\infty \vee \sigma(V_n; n \leq 0)$ satisfies Vershik's first level criterion.*

Proof. Thanks to Proposition 2.13, it suffices to show that each σ-field $\mathcal{C}_n \vee \sigma(V_{n+1}, \ldots, V_0)$ satisfies Vershik's first level criterion. Let C be the family consisting of all the σ-fields $\sigma(V'_{n_0+1}, \ldots, V'_n)$ generated by some local innovation $(V'_{n_0+1}, \ldots, V'_n)$ of $\mathcal{F}$ from n_0 to n for some $n_0 < n$. By Lemma 2.20, C is substantial in $\mathcal{C}_n$. Let $D = \sigma(V_{n+1}, \ldots, V_0)$. Then apply Lemma 2.18.   □

Theorem 2.23. *Let $\mathcal{F} = (\mathcal{F}_n)_{n \leq 0}$ be a filtration of local product type on $(\Omega, \mathcal{A}, \mathbb{P})$. Then $\mathcal{F}$ satisfies Vershik's first level criterion if and only if for every separable metric space E and every random variable $X \in L^1(\mathcal{F}_0; E)$, there exists an innovation $(V_n)_{n \leq 0}$ of $\mathcal{F}$ such that X is measurable with respect to $\sigma(V_n; n \leq 0)$.*

Proof. The 'if' part follows from Corollary 2.14 and Proposition 2.11. The 'only if' part is proved as follows. Assume that $\mathcal{F}$ satisfies Vershik's first level criterion and let $X \in L^1(\mathcal{F}_0; E)$. Consider a random variable $Y \in L^1(\mathcal{F}_0; \mathbb{R})$ such that $\sigma(X) = \sigma(Y)$. Let $(\delta_k)_{k \leq 0}$ be a sequence of positive numbers such that $\delta_k \to 0$.

Consider the following construction at rank k: we have an integer $n_k < 0$, an innovation $(V_n; n_k < n \leq 0)$ of $\mathcal{F}$ from n_k to 0 and a random variable $X_k \in L^1(\sigma(V_n; n_k < n \leq 0))$ such that $\mathbb{E}[|Y - X_k|] < \delta_k$. We firstly apply the Vershik first level property of Y with respect to $\mathcal{F}$ to obtain this construction for $k = 0$. When the construction is performed at rank k, we perform it at rank $k-1$ by exhibiting an innovation $(V_{n_{k-1}+1}, \ldots, V_{n_k})$ from an integer $n_{k-1} < n_k$ to n_k, and a random variable $X_{k-1} \in L^1(\sigma(V_n; n_{k-1} < n \leq 0))$ such that $\mathbb{E}[|Y - X_{k-1}|] < \delta_{k-1}$. To do so, we apply Proposition 2.21 to get Vershik's first level criterion of the truncated filtration $(\mathcal{F}_{n_k+n})_{n \leq 0}$ and then we use the fact, due to Lemma 2.18, that the family of σ-fields of the form $\sigma(V_{m+1}, \ldots, V_{n_k}, V_{n_k+1}, \ldots, V_0)$ where $(V_{m+1}, \ldots, V_{n_k})$ is an innovation of $\mathcal{F}$ from some $m < n_k$ to n_k, is substantial in $\mathcal{F}_0$.

Continuing so, we obtain a global innovation $(V_n)_{n \leq 0}$ of $\mathcal{F}$ and a sequence of random variables $(X_k)_{k \leq 0}$ in $L^1(\sigma(V_n; n \leq 0))$ converging in L^1 to Y. □

Remark 2.24. We do not know if Theorem 2.23 is true "random variable by random variable". More precisely, we do not know if each random variable satisfying Vershik's first level criterion is measurable with respect to the σ-field generated by some global innovation of the filtration.

Theorem 2.25 (Vershik's First Level Criterion). *Let $\mathcal{F}$ be an essentially separable filtration of local product type. Then $\mathcal{F}$ satisfies Vershik's first level criterion if and only if $\mathcal{F}$ is of product type.*

Proof. The 'if' part is given in Corollary 2.14. To show the converse, apply Theorem 2.23 with a random variable X generating $\mathcal{F}_0$. This yields a global innovation $(V_n)_{n \leq 0}$ such that $\mathcal{F}_0 = \sigma(V_n; n \leq 0)$. Therefore $\mathcal{F}_n \cap \sigma(V_n; n \leq 0) = \mathcal{F}_n$ for every $n \leq 0$. Consequently $\mathcal{F}_n = \sigma(V_m; m \leq n)$ for every $n \leq 0$ because the filtration generated by $(V_n)_{n \leq 0}$ is immersed in $\mathcal{F}$ (Lemma 2.5). □

Obviously, a filtration of local product type which is not essentially separable cannot be of product type. However we will see in Sect. 3 that it is possible that such a filtration satisfies Vershik's first level criterion (Corollary 2.46). It is then interesting to notice that a filtration satisfying Vershik's first level criterion is Kolmogorovian, even if it is not of product type.

Corollary 2.26. *A filtration of local product type satisfying Vershik's first criterion is Kolmogorovian.*

Proof. Let $\mathcal{F}$ be such a filtration and $A \in \mathcal{F}_{-\infty}$. Thanks to Theorem 2.23, there exists a global innovation $(V_n)_{n \leq 0}$ such that $\mathbb{1}_A$ is measurable with respect to $\sigma(V_n; n \leq 0)$. As the filtration generated by $(V_n)_{n \leq 0}$ is immersed in $\mathcal{F}$ (Lemma 2.5), it follows that $\mathbb{1}_A$ is measurable with respect to the trivial σ-field $\cap_n \sigma(V_m; m \leq n)$. □

Below is another corollary of Theorem 2.23 which will be used in Sect. 3 to prove a result on weak standardness (Proposition 3.23).

Corollary 2.27. *Let $\mathcal{F}$ and $\mathcal{G}$ be two independent filtrations of local product type satisfying Vershik's first level criterion. Then $\mathcal{F} \vee \mathcal{G}$ is a filtration of local product type satisfying Vershik's first level criterion.*

Proof. Let $R \in L^1\,(\mathcal{F}_0 \vee \mathcal{G}_0; [0,1])$. We can write $R = f(X,Y)$ where X and Y are random variables measurable with respect to $\mathcal{F}_0$ and $\mathcal{G}_0$ respectively and f is a Borelian function. By Theorem 2.23, there exist a global innovation $(V_n)_{n \leqslant 0}$ of $\mathcal{F}$ and a global innovation $(W_n)_{n \leqslant 0}$ of $\mathcal{G}$ such that X is measurable with respect to $\sigma(V_n; n \leqslant 0)$ and Y is measurable with respect to $\sigma(W_n; n \leqslant 0)$. Setting $Z_n = (V_n, W_n)$, then $(Z_n)_{n \leqslant 0}$ obviously is an innovation of $\mathcal{F} \vee \mathcal{G}$ and R is measurable with respect to $\sigma(Z_n; n \leqslant 0)$. Thus $\mathcal{F} \vee \mathcal{G}$ satisfies Vershik's first level criterion due to Theorem 2.23.   □

The converse of Corollary 2.27 holds true; actually we could prove the following stronger result, but we will not need it: *Let $\mathcal{F}$ and $\mathcal{G}$ be two independent filtrations of local product type. Let E_1 and E_2 be Polish spaces, $X \in L^1(\mathcal{F}_0; E_1)$ and $Y \in L^1(\mathcal{G}_0; E_2)$. Then X and Y satisfy Vershik's first level criterion with respect to $\mathcal{F}$ and $\mathcal{G}$ respectively, if and only if (X,Y) satisfies Vershik's first level criterion with respect to $\mathcal{F} \vee \mathcal{G}$.*

Remark 2.28. If $\mathcal{F}$ is a filtration of local product type generated by a martingale $(M_n)_{n \leqslant 0}$, then it is possible to show that $\mathcal{F}$ satisfies Vershik's first level criterion if and only if the random variable M_0 satisfies Vershik's first level criterion. We will not show this fact as we shall see in the next section that Vershik's first level criterion is equivalent to *Vershik's self-joining criterion*, and a result in [21] says that the same fact holds for Vershik's self-joining criterion. Actually this result says that the same fact holds more generally for any *self-joining criterion*, a notion defined in [21] that includes Vershik's self-joining criterion and the I-cosiness criterion as particular cases (see also Remark 3.44 and the first paragraph of Sect. 3.5).

2.2 Rosenblatt's and Vershik's Self-Joining Criteria

Given a filtration $\mathcal{F}$ of local product type, Proposition 2.33 gives a "self-joining criterion" for a global innovation of $\mathcal{F}$ to be generating (*Rosenblatt's self-joining criterion*), and Theorem 2.38 (*Vershik's self-joining criterion*) gives a "self-joining criterion" for $\mathcal{F}$ to satisfy Vershik's first level criterion. More precisely, these criteria are stated "random variable by random variable". The terminologies are discussed at the end of the section. Both these criteria are a particular form of the I-cosiness criterion (Definition 3.29). Rosenblatt's self-joining criterion will be illustrated by the example given in Sect. 3.1. Vershik's self-joining criterion will be used to establish Theorem A" stated in the introduction (and restated in Theorem 4.4).

2.2.1 Joinings

As also does the I-cosiness criterion, Rosenblatt's self-joining criterion and Vershik's self-joining criterion involve *joinings* of filtrations, defined below.

Definition 2.29. Let $\mathcal{F}$ be a filtration.

1. A *joining* of $\mathcal{F}$ is a pair $(\mathcal{F}', \mathcal{F}'')$ of two filtrations $\mathcal{F}'$ and $\mathcal{F}''$ defined on the same probability space which are both isomorphic to $\mathcal{F}$ and *jointly immersed*, that is, $\mathcal{F}'$ and $\mathcal{F}''$ are both immersed in $\mathcal{F}' \vee \mathcal{F}''$ (or, equivalently, in a same filtration).
2. A joining $(\mathcal{F}', \mathcal{F}'')$ of $\mathcal{F}$ is *independent in small time* if the σ-fields $\mathcal{F}'_{n_0}$ and $\mathcal{F}''_{n_0}$ are independent for some integer $n_0 \leq 0$. We also say that $(\mathcal{F}', \mathcal{F}'')$ is a joining of $\mathcal{F}$ *independent up to n_0* to specify this integer.

A typical example of joining $(\mathcal{F}', \mathcal{F}'')$ is the case where $\mathcal{F}'$ and $\mathcal{F}''$ are two independent copies of $\mathcal{F}$ (Lemma 1.2).

Rigorously, a joining is the pair $(\Psi'(\mathcal{F}), \Psi''(\mathcal{F}))$ given by a probability space $(\overline{\Omega}, \overline{\mathcal{A}}, \overline{\mathbb{P}})$ and two embeddings $\Psi' \colon \mathcal{F}_0 \to \overline{\mathcal{A}}$ and $\Psi'' \colon \mathcal{F}_0 \to \overline{\mathcal{A}}$, with the additional property of joint immersion. Considering a joining $(\mathcal{F}', \mathcal{F}'')$ of a filtration $\mathcal{F} = (\mathcal{F}_n)_{n \leq 0}$, and given a $\mathcal{F}_0$-measurable random variable X valued in a Polish space, we will traditionally denote by X' and X'' the respective copies of X given by the two underlying embeddings Ψ' and Ψ''. Of course, Y' and Y'' will denote the copies of a $\mathcal{F}_0$-measurable random variable Y, and so on. In the same way, the two copies of a σ-field $\mathcal{B} \subset \mathcal{F}_0$ will be respectively denoted by $\mathcal{B}'$ and $\mathcal{B}''$, and the two copies of a filtration $\mathcal{E} \subset \mathcal{F}$ will be respectively denoted by $\mathcal{E}'$ and $\mathcal{E}''$. Note that, given a filtration $\mathcal{E}$ immersed in $\mathcal{F}$, a joining $(\mathcal{F}', \mathcal{F}'')$ of $\mathcal{F}$ induces a joining $(\mathcal{E}', \mathcal{E}'')$ of $\mathcal{E}$.

We shall need the following lemma in the proof of Theorem 2.38. Its easy proof is left to the reader.

Lemma 2.30. *Let $\mathcal{F} = (\mathcal{F}_n)_{n \leq 0}$ be a filtration and $n_0 \leq 0$ an integer. Let $(\mathcal{F}', \mathcal{F}'')$ be a joining of $\mathcal{F}$ independent up to n_0. Then the σ-fields $\mathcal{F}'_0$ and $\mathcal{F}''_{n_0}$ are independent.*

The next lemma is obvious from the definitions; it will be used to construct joinings of a filtration of local product type with the help of innovations.

Lemma 2.31. *On $(\Omega, \mathcal{A}, \mathbb{P})$, let $\mathcal{F} = (\mathcal{F}_n)_{n \leq 0}$ be a filtration and V_0 be an independent complement of $\mathcal{F}_{-1}$ in $\mathcal{F}_0$. On $(\overline{\Omega}, \overline{\mathcal{A}}, \overline{\mathbb{P}})$, let $(\mathcal{F}'_n)_{n \leq -1}$ and $(\mathcal{F}''_n)_{n \leq -1}$ be two jointly immersed isomorphic copies of $(\mathcal{F}_n)_{n \leq -1}$, given by two isomorphisms $\Psi'_{-1} \colon \mathcal{F}_{-1} \to \mathcal{F}'_{-1}$ and $\Psi''_{-1} \colon \mathcal{F}_{-1} \to \mathcal{F}''_{-1}$, and let V'_0 and V''_0 be two random variables each having the same law as V_0 and independent of $\mathcal{F}'_{-1} \vee \mathcal{F}''_{-1}$. We put $\mathcal{F}'_0 = \mathcal{F}'_{-1} \vee \sigma(V'_0)$ and $\mathcal{F}''_0 = \mathcal{F}''_{-1} \vee \sigma(V''_0)$. Then $(\mathcal{F}', \mathcal{F}'')$ is a joining of $\mathcal{F}$, given by two unique isomorphisms Ψ' and Ψ'' that respectively extend Ψ'_{-1} and Ψ''_{-1} and respectively send V_0 to V'_0 and V''_0.*

Proof. The two isomorphisms Ψ' and Ψ'' are given by Corollary 5.12. If $(\mathcal{F}'_n)_{n \leq -1}$ is immersed in $(\mathcal{F}'_n \vee \mathcal{F}''_n)_{n \leq -1}$, then we can see by Lemma 1.4 that $\mathcal{F}'$ is immersed in $\mathcal{F}' \vee \mathcal{F}''$ if and only if $\mathcal{F}'_0$ and $\mathcal{F}'_{-1}$ are conditionally independent given $\mathcal{F}'_{-1} \vee \mathcal{F}''_{-1}$. By Lemma 1.5, this conditional independence holds if (and only if) V'_0 is independent of $\mathcal{F}'_{-1} \vee \mathcal{F}''_{-1}$. Thus $\mathcal{F}'$ is immersed in $\mathcal{F}' \vee \mathcal{F}''$ and the same fact obviously holds for $\mathcal{F}''$. □

2.2.2 Rosenblatt's Self-Joining Criterion (for Generatingness)

Let $\mathcal{F} = (\mathcal{F}_n)_{n \leq 0}$ be a filtration of local product type and $(V_n)_{n \leq 0}$ a global innovation of $\mathcal{F}$. Proposition 2.33 below gives a necessary and sufficient condition for a random variable to be measurable with respect to the σ-field generated by $(V_n)_{n \leq 0}$. The proposition involves joinings $(\mathcal{F}', \mathcal{F}'')$ constructed as follows.

Given an integer $n_0 < 0$, we consider, on some probability space $(\overline{\Omega}, \overline{\mathcal{A}}, \overline{\mathbb{P}})$, two copies $(\mathcal{F}'_n)_{n \leq n_0}$ and $(\mathcal{F}''_n)_{n \leq n_0}$ of the filtration $(\mathcal{F}_n)_{n \leq n_0}$ and a random vector $(V'_{n_0+1}, \ldots, V'_0)$ having the same law as $(V_{n_0+1}, \ldots, V_0)$ and independent of $\mathcal{F}'_{n_0} \vee \mathcal{F}''_{n_0}$. We complete the filtrations $(\mathcal{F}'_n)_{n \leq n_0}$ and $(\mathcal{F}''_n)_{n \leq n_0}$ up to time 0 by putting

$$\mathcal{F}'_n = \mathcal{F}'_{n_0} \vee \sigma(V'_{n_0+1}, \ldots, V'_n) \quad \text{and} \quad \mathcal{F}''_n = \mathcal{F}''_{n_0} \vee \sigma(V'_{n_0+1}, \ldots, V'_n)$$

for each $n \in \{n_0 + 1, \ldots, 0\}$. Assuming that $(\mathcal{F}'_n)_{n \leq n_0}$ and $(\mathcal{F}''_n)_{n \leq n_0}$ are jointly immersed, one easily checks with the help of Lemma 2.31 that $(\mathcal{F}', \mathcal{F}'')$ is a joining of $\mathcal{F}$, given by two isomorphisms $\Psi' : \mathcal{F} \to \mathcal{F}'$ and $\Psi'' : \mathcal{F} \to \mathcal{F}''$, both of them sending V_n to V'_n for each $n \in \{n_0 + 1, \ldots, 0\}$. For a time $n \geq n_0$, the joining can be pictured as follows:

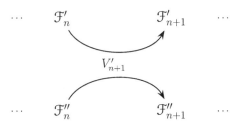

Such a joining $(\mathcal{F}', \mathcal{F}'')$ is characterized by the fact that $V'_n = V''_n$ for every $n \in \{n_0 + 1, \ldots, 0\}$.

Definition 2.32 (Rosenblatt's Self-Joining Criterion). Let $\mathcal{F} = (\mathcal{F}_n)_{n \leq 0}$ be a filtration of local product type and $(V_n)_{n \leq 0}$ an innovation of $\mathcal{F}$. Let (E, ρ) be a Polish metric space and $X \in L^1(\mathcal{F}_0; E)$. We say that X satisfies *Rosenblatt's self-joining criterion with* $(V_n)_{n \leq 0}$ if for each real number $\delta > 0$, there exists, on some probability space $(\overline{\Omega}, \overline{\mathcal{A}}, \overline{\mathbb{P}})$, a joining $(\mathcal{F}', \mathcal{F}'')$ of $\mathcal{F}$ independent up to an integer $n_0 \leq 0$, such that $V'_n = V''_n$ for every $n \in \{n_0 + 1, \ldots, 0\}$, and for which one has $\overline{\mathbb{E}}[\rho(X', X'')] < \delta$, where X' and X'' are the respective copies of X in $\mathcal{F}'$ and in $\mathcal{F}''$.

Proposition 2.33 (Rosenblatt's Self-Joining Criterion). *Let* $\mathcal{F} = (\mathcal{F}_n)_{n \leq 0}$ *be a filtration of local product type and* $(V_n)_{n \leq 0}$ *an innovation of* $\mathcal{F}$. *Let* (E, ρ) *be a Polish metric space and* $X \in L^1(\mathcal{F}_0; E)$. *Then* X *is measurable with respect to* $\sigma(V_n; n \leq 0)$ *if and only if* X *satisfies Rosenblatt's self-joining criterion with* $(V_n)_{n \leq 0}$.

Proof. If X is measurable with respect to $\sigma(V_n; n \leq 0)$ then, by Lemma 2.12 and Lemma 2.10, for any $\delta > 0$ there exist an integer $n_0 \leq 0$ and a random variable $S \in L^1\big(\sigma(V_{n_0+1}, \ldots, V_0); E\big)$ such that $\mathbb{E}[\rho(X, S)] < \delta/2$. Considering a joining $(\mathcal{F}', \mathcal{F}'')$ of $\mathcal{F}$ independent up to n_0 as defined in the proposition, one has $\overline{\mathbb{E}}[\rho(X', S')] = \overline{\mathbb{E}}[\rho(X'', S'')] = \mathbb{E}[\rho(X, S)]$ due to isomorphisms, and $S' = S''$ because S is measurable with respect to $\sigma(V_{n_0+1}, \ldots, V_0)$. This gives $\overline{\mathbb{E}}[\rho(X', X'')] < \delta$ by the triangular inequality, thereby showing that X satisfies Rosenblatt's self-joining criterion.

Conversely, assume that Rosenblatt's self-joining criterion holds for X. Then, considering two independent copies $\mathcal{F}'$ and $\mathcal{F}^*$ of $\mathcal{F}$, we see that the family of σ-fields $\big\{\mathcal{F}_0^* \vee \sigma(V_m'; n < m \leq 0) \mid n < 0\big\}$ is substantial in $\sigma(X')$ (Definition 2.8 and Lemma 2.16). As $\mathcal{F}_0^*$ is independent of $\mathcal{F}_0'$, Lemma 2.19 shows that the family of σ-fields $\{\sigma(V_m'; n < m \leq 0) \mid n < 0\}$ is substantial in $\sigma(X')$. Consequently X' is measurable with respect to $\sigma(V_n'; n \leq 0)$, and therefore, due to isomorphism, X is measurable with respect to $\sigma(V_n; n \leq 0)$. $\qquad\square$

Corollary 2.34. *Let* $\mathcal{F} = (\mathcal{F}_n)_{n \leq 0}$ *be an essentially separable filtration of local product type and* $(V_n)_{n \leq 0}$ *an innovation of* $\mathcal{F}$. *Then the following conditions are equivalent:*

(i) $\mathcal{F}$ *is the filtration of product type generated by* $(V_n)_{n \leq 0}$.

(ii) *For every Polish space* E, *every random variable* $X \in L^1(\mathcal{F}_0; E)$ *satisfies Rosenblatt's self-joining criterion with* $(V_n)_{n \leq 0}$.

(iii) *For some Polish space* E, *there exists a random variable* $X \in L^1(\mathcal{F}_0; E)$ *generating* $\mathcal{F}_0$ *and satisfying Rosenblatt's self-joining criterion with* $(V_n)_{n \leq 0}$.

(iv) *Every random variable* $X \in L^1(\mathcal{F}_0; \mathbb{R})$ *satisfy Rosenblatt's self-joining criterion with* $(V_n)_{n \leq 0}$.

Proof. The previous proposition shows that (i) $\implies$ (ii), (iii), (iv), and that each of (ii), (iii) and (iv) implies that $\mathcal{F}_0 = \sigma(V_n; n \leq 0)$. As the filtration generated by $(V_n)_{n \leq 0}$ is immersed in $\mathcal{F}$ (Lemma 2.5), this yields $\sigma(V_m; m \leq n) = \mathcal{F}_n$ for each $n \leq 0$. $\qquad\square$

2.2.3 Vershik's Self-Joining Criterion

Vershik's self-joining criterion is the "self-joining version" of Vershik's first level criterion (Definition 2.6). Given a filtration $\mathcal{F}$ of local product type and an innovation $(V_n)_{n \leq 0}$ of $\mathcal{F}$, its statement involves the joinings $(\mathcal{F}', \mathcal{F}'')$ of $\mathcal{F}$ defined as follows and including as particular cases the joinings involved in Rosenblatt's self-joining criterion.

Given an integer $n_0 \leq 0$, we consider, on some probability space $(\overline{\Omega}, \overline{\mathcal{A}}, \overline{\mathbb{P}})$, two copies $(\mathcal{F}'_n)_{n \leq n_0}$ and $(\mathcal{F}''_n)_{n \leq n_0}$ of the filtration $(\mathcal{F}_n)_{n \leq n_0}$ and a random vector $(V'_{n_0+1}, \ldots, V'_0)$ having the same law as $(V_{n_0+1}, \ldots, V_0)$ and independent of $\mathcal{F}'_{n_0} \vee \mathcal{F}''_{n_0}$. We complete the filtrations $(\mathcal{F}'_n)_{n \leq n_0}$ and $(\mathcal{F}''_n)_{n \leq n_0}$ up to time 0 by defining $\mathcal{F}'_n$ and $\mathcal{F}''_n$ for each $n \in \{n_0 + 1, \ldots, 0\}$ by

$$\mathcal{F}'_n = \mathcal{F}'_{n_0} \vee \sigma(V'_{n_0+1}, \ldots, V'_n) \quad \text{and} \quad \mathcal{F}''_n = \mathcal{F}''_{n_0} \vee \sigma(V''_{n_0+1}, \ldots, V''_n),$$

where the random vector $(V''_{n_0+1}, \ldots, V''_0)$ is constructed as follows. At each step $n \in \{n_0 + 1, \ldots, 0\}$, calling E_n the Polish state space of V_n, we consider a measurable function $\phi^n : \mathbb{R} \times E_n \to E_n$ such that for each fixed $x \in \mathbb{R}$, the function $v \mapsto \phi^n(x, v)$ is a Lebesgue automorphism of the probability space induced by V_n, and then we put $V''_n = \phi^n(\bar{H}_{n-1}, V'_n)$ where $\bar{H}_{n-1}$ is some random variable measurable with respect to $\mathcal{F}'_{n-1} \vee \mathcal{F}''_{n-1}$. This construction can be pictured as follows:

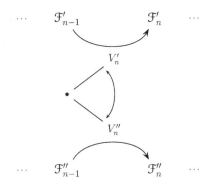

It is clear from this construction that V''_n has the same law as V'_n and that each of V'_n and V''_n is independent of $\mathcal{F}'_{n-1} \vee \mathcal{F}''_{n-1}$. Therefore, assuming that $(\mathcal{F}'_n)_{n \leq n_0}$ and $(\mathcal{F}''_n)_{n \leq n_0}$ are jointly immersed, Lemma 2.31 ensures that $(\mathcal{F}', \mathcal{F}'')$ is a joining of $\mathcal{F}$, given by two isomorphisms $\Psi' : \mathcal{F} \to \mathcal{F}'$ and $\Psi'' : \mathcal{F} \to \mathcal{F}''$ that respectively send V_n to V'_n and V''_n for each $n \in \{n_0 + 1, \ldots, 0\}$. Such joinings will be given a name in the following definition.

Definition 2.35. Let $\mathcal{F} = (\mathcal{F}_n)_{n \leq 0}$ be a filtration of local product type and $(V_n)_{n \leq 0}$ an innovation of $\mathcal{F}$. A joining $(\mathcal{F}', \mathcal{F}'')$ of $\mathcal{F}$ is *permutational after n_0* for an integer $n_0 \leq 0$ if for each $n \in \{n_0 + 1, \ldots, 0\}$ we have $V''_n = T_n(V'_n)$ where T_n is a random Lebesgue automorphism of the probability space induced by V_n, defined by $T_n(\cdot) = \phi^n(\bar{H}_{n-1}, \cdot)$ where ϕ^n is a measurable function and $\bar{H}_{n-1}$ is some random variable measurable with respect to $\mathcal{F}'_{n-1} \vee \mathcal{F}''_{n-1}$.

The joinings featuring in Rosenblatt's self-joining criterion appear as the particular case when the T_n are almost surely equal to identity. Note that the definition does not depend on the choice of the innovation $(V_n)_{n \leq 0}$ in view of Lemma 2.4. Actually we can see by this lemma that a joining $(\mathcal{F}', \mathcal{F}'')$ permutational after n_0 is characterized by the fact that $\mathcal{F}'_n \vee \mathcal{F}''_n = \mathcal{F}'_n \vee \mathcal{F}''_{n_0}$ for each $n \in \{n_0 + 1, \ldots, 0\}$, and this characterization does not involve any innovation of $\mathcal{F}$.

The following easy lemma will be used in the proof of Theorem 2.38 and Lemma 4.2

Lemma 2.36. *In the context of the above definition, and, in addition, given an $\mathcal{F}_0$-measurable random variable X, there exist two $\mathcal{F}_{n_0}$-measurable random variables C_{n_0} and D_{n_0} such that $\sigma(X) \subset \sigma(D_{n_0}, V_{n_0+1}, \ldots, V_0)$ and it is possible to write*

$$T_n(\cdot) = \psi^n(C'_{n_0}, D''_{n_0}, V'_{n_0+1}, \ldots, V'_{n-1}, \cdot)$$

for every $n \in \{n_0 + 1, \ldots, 0\}$, where ψ^n is a measurable function.

Proof. The proof is a successive application of Lemma 1.1. We use the notations of Definition 2.35, thus one has $V''_n = T_n(V'_n)$ where the random transformations T_n are written in form $T_n(\cdot) = \phi^n(\bar{H}_{n-1}, \cdot)$. To show the lemma, it suffices to find C_{n_0} and D_{n_0} such that $\sigma(X'') \subset \sigma(D''_{n_0}, V''_{n_0+1}, \ldots, V''_0)$ and $\sigma(\bar{H}_{n-1}) \subset \sigma(C'_{n_0}, D''_{n_0}, V'_{n_0+1}, \ldots, V'_{n-1})$ for each $n \in \{n_0 + 1, \ldots, 0\}$. Note that $(V'_{n_0+1}, \ldots, V'_0)$ is a local innovation of $\mathcal{F}' \vee \mathcal{F}''$ from n_0 to 0.

For each $n \in \{n_0 + 1, \ldots, 0\}$, take (Lemma 1.1) an $(\mathcal{F}'_{n_0} \vee \mathcal{F}''_{n_0})$-measurable r.v. $\bar{S}^n_{n_0}$ such that $\sigma(\bar{H}_{n-1}) \subset \sigma(\bar{S}^n_{n_0}, V'_{n_0+1}, \ldots, V'_{n-1})$. Then, take a r.v. $\bar{R}_{n_0}$ such that $\sigma(\bar{R}_{n_0}) = \sigma(\bar{S}^n_{n_0}; n \in \{n_0 + 1, \ldots, 0\})$, so that one has $\sigma(\bar{H}_{n-1}) \subset \sigma(\bar{R}_{n_0}, V'_{n_0+1}, \ldots, V'_{n-1})$ for each $n \in \{n_0 + 1, \ldots, 0\}$. Then, take (Lemma 1.1) an $\mathcal{F}'_{n_0}$-measurable r.v. C'_{n_0} and an $\mathcal{F}''_{n_0}$-measurable r.v. B''_{n_0} such that $\sigma(\bar{R}_{n_0}) \subset \sigma(C'_{n_0}, B''_{n_0})$, and, finally, take (Lemma 1.1) an $\mathcal{F}''_{n_0}$-measurable r.v. X''_{n_0} such that $\sigma(X'') \subset \sigma(X''_{n_0}, V''_{n_0+1}, \ldots, V''_0)$ and take a r.v. D''_{n_0} such that $\sigma(D''_{n_0}) = \sigma(B''_{n_0}, X''_{n_0})$. $\square$

Definition 2.37. Let $\mathcal{F} = (\mathcal{F}_n)_{n \leq 0}$ be a filtration of local product type, (E, ρ) a Polish metric space, and let $X \in L^1(\mathcal{F}_0; E)$. We say that X satisfies *Vershik's self-joining criterion* if for each real number $\delta > 0$, there exist an integer $n_0 \leq 0$ and, on some probability space $(\bar{\Omega}, \bar{\mathcal{A}}, \bar{\mathbb{P}})$, a joining $(\mathcal{F}', \mathcal{F}'')$ of $\mathcal{F}$ independent up to n_0 and permutational after n_0 such that one has $\bar{\mathbb{E}}[\rho(X', X'')] < \delta$, where X' and X'' are the respective copies of X in $\mathcal{F}'$ and in $\mathcal{F}''$.

Theorem 2.38. *Let $\mathcal{F} = (\mathcal{F}_n)_{n \leq 0}$ be a filtration of local product type. Let (E, ρ) be a Polish metric space and $X \in L^1(\mathcal{F}_0; E)$. Then X satisfies Vershik's self-joining criterion if and only if X satisfies Vershik's first level criterion.*

Proof. The proof of the 'only if' part is similar the proof of the 'only if' part of Proposition 2.33. Indeed, assuming that X satisfies Vershik's first level criterion, then for any $\delta > 0$ there exist $n_0 \leq 0$, a local innovation $(\tilde{V}_{n_0+1}, \ldots, \tilde{V}_0)$ from n_0 to 0, and a random variable $S \in L^1(\sigma(\tilde{V}_{n_0+1}, \ldots, \tilde{V}_0); E)$ such that $\mathbb{E}[\rho(X, \tilde{S})] < \delta/2$. Thus Vershik's self-joining criterion for X can be proved in the same way as we have proved Rosenblatt's self-joining criterion for X in the first part of the proof of Proposition 2.33, by replacing $(V_{n_0+1}, \ldots, V_0)$ with $(\tilde{V}_{n_0+1}, \ldots, \tilde{V}_0)$.

Conversely, assume Vershik's self-joining criterion holds for X. Fix $\delta > 0$, $n_0 \leqslant 0$ and $(\mathcal{F}', \mathcal{F}'')$ as in Definition 2.37. By Lemma 2.36, there are two $\mathcal{F}_{n_0}$-measurable random variables C_{n_0} and D_{n_0} such that $\sigma(X) \subset \sigma(D_{n_0}, V_{n_0+1}, \ldots, V_0)$ and

$$T_n(\cdot) = \psi^n(C'_{n_0}, D''_{n_0}, V'_{n_0+1}, \ldots, V'_{n-1}, \cdot)$$

for every $n \in \{n_0 + 1, \ldots, 0\}$. By taking a Borelian function f such that $X = f(D_{n_0}, V_{n_0+1}, \ldots, V_0)$, one has $X'' = h_{D''_{n_0}}(C'_{n_0}, V'_{n_0+1}, \ldots, V'_0)$ where, for a given value y of D''_{n_0},

$$h_y(C'_{n_0}, V'_{n_0+1}, \ldots, V'_0) = f\left(y, \widetilde{V}_{n_0+1}, \ldots, \widetilde{V}_0\right),$$

with

$$\widetilde{V}_n = \psi^n(C'_{n_0}, y, V'_{n_0+1}, \ldots, V'_{n-1}, V'_n),$$

for each $n \in \{n_0 + 1, \ldots, 0\}$, and thus $\widetilde{V}_n$ is an independent complement of $\mathcal{F}'_{n-1}$ in $\mathcal{F}'_n$ by Lemma 2.4. In view of Lemma 2.30, the random variable D''_{n_0} is independent of $\mathcal{F}'_0$. The assumption that $\overline{\mathbb{E}}\left[\rho(X', X'')\right] < \delta$ therefore implies that for some y, we have $\overline{\mathbb{E}}\left[\rho\left(X', \widetilde{S}\right)\right] < \delta$ where $\widetilde{S} = h_y(C'_{n_0}, V'_{n_0+1}, \ldots, V'_0)$. Thus X' satisfies Vershik's first level criterion with respect to $\mathcal{F}'$ because the random variable $\widetilde{S}$ is measurable with respect to $\sigma(\widetilde{V}_{n_0+1}, \ldots, \widetilde{V}_0)$. Obviously, Vershik's first level criterion with respect to $\mathcal{F}$ is satisfied for X due to isomorphism. □

2.2.4 On the Terminology

We have called the self-joining criterion of Proposition 2.33 *Rosenblatt's self-joining criterion* because this result was often used in the works of Rosenblatt [26–28] and their further developments [5, 17]. The self-joining criterion in Definition 2.37 is called *Vershik's self-joining criterion* because it is close to the "combinatorial" standardness criterion stated in Vershik's works in the particular case of homogeneous filtrations (defined in the introduction and in Definition 4.1) with atomic innovations.

2.3 Example: Split-Word Processes

We shall define the split-word processes. Their filtrations are known to be of product type or not according to some conditions on the parameters defining these processes. Admitting the key results found in the literature, we will start proving the main theorem on productness for these filtrations (Theorem 2.39). At this stage we do not yet have at our disposal the tools needed to entirely prove this theorem; its proof will be continued at the end of each section.

To define a *split-word process*, the first ingredient is a probability space $(A, \mathfrak{A}, \mu)$ called the *alphabet*. A *word* on A is an element $w \in A^\ell$ for some integer $\ell \geq 1$, called the *length* of w, and $w(1), \ldots, w(\ell)$ are the *letters* of w. Given any function f from A to a set B we naturally define $f(w) \in B^\ell$ as the word on B with letters $f(w(1)), \ldots, f(w(\ell))$ in this order.

The second ingredient is the *splitting sequence* $(r_n)_{n \leq 0}$, consisting of integers $r_n \geq 2$. Given this sequence, we define the sequence $(\ell_n)_{n \leq 0}$ of *lengths* by $\ell_n = \prod_{k=n+1}^{0} r_k$ for all $n \leq 0$; in other words, the sequence $(\ell_n)_{n \leq 0}$ is recursively defined by $\ell_0 = 1$ and $\ell_{n-1} = r_n \ell_n$.

Then we define the *split-word process* with alphabet $(A, \mathfrak{A}, \mu)$ and splitting sequence $(r_n)_{n \leq 0}$ to be the (non time-homogeneous) Markov process $(W_n, \eta_n)_{n \leq 0}$ whose law is characterized by the following two conditions:

⋄ For each $n \leq 0$, W_n is a random word on A of length ℓ_n, whose letters are i.i.d. random variables with law μ, and η_n is independent of W_n and has the uniform law on the set $\{1, \ldots, r_n\}$.

⋄ The transition from $n - 1$ to n is obtained by taking η_n independent of (W_{n-1}, η_{n-1}), and then by choosing W_n as the η_n-th subword of W_{n-1} considered as a concatenation of r_n subwords of equal length ℓ_n.

We denote by $\mathcal{F}$ the filtration generated by $(W_n, \eta_n)_{n \leq 0}$. Of course the process $(\eta_n)_{n \leq 0}$ is an innovation of $\mathcal{F}$. It can be shown that $\mathcal{F}_{-\infty}$ is degenerate whatever the alphabet $(A, \mathfrak{A}, \mu)$ and the splitting sequence $(r_n)_{n \leq 0}$; the proof is the same as in [14] where the particular case that $r_n \equiv 2$ and μ is uniform on a finite alphabet A is treated. We will derive the following theorem using results from the literature.

Theorem 2.39. *Call* (Δ) *the condition on the splitting sequence* $(r_n)_{n \leq 0}$:

$$\sum_{k=-\infty}^{0} \frac{\log(r_k)}{\ell_k} < \infty. \tag{Δ}$$

Then:

(a) If (Δ) *holds, then* $\mathcal{F}$ *is not of product type unless* $(A, \mathfrak{A}, \mu)$ *is degenerate.*

(b) If (Δ) *does not hold and if* $(A, \mathfrak{A}, \mu)$ *is Polish, then* $\mathcal{F}$ *is of product type.*

Remark 2.40. As we have seen in this section, part (a) of Theorem 2.39 means that under condition (Δ), Vershik's first level criterion, or Vershik's self-joining criterion, fails to be true for some random variables. It would be interesting to have more information on those random variables. For example, does the final letter W_0 of the split-word process $(W_n, \eta_n)_{n \leq 0}$ never satisfy Vershik's first level criterion under condition (Δ) on the splitting sequence, whatever the non-trivial alphabet?

Remark 2.41 (Vershik's Example 1). The similar theorem holds for the so-called *Example 1* in [40]. Part (a) of Theorem 2.39 for Vershik's Example 1 is shown in [40] with the help of its notion of entropy of filtrations. Part (b) is shown by Heicklen in [18] in the particular case when A is finite and μ is the uniform probability on A

(thus Heicklen shows the analogue of Ceillier's result 2.44 for Vershik's Example 1). This example deals with filtrations defined as sequences of the invariant σ-fields of the actions of a decreasing sequence of groups on a Lebesgue probability space. Unfortunately we have failed to check whether or not Vershik's Example 1 is exactly the same as the filtrations of the split-word processes, but similar mathematics appear in the proof of the two theorems. In fact, we know that the following coincidence holds: Vershik's self-joining criterion of the random variable W_0 with respect to the filtration of the split-word process $(W_n, \eta_n)_{n \leq 0}$ can be expressed in a problem of purely combinatorial nature, and this problem is also equivalent to Vershik's self-joining criterion of a certain random variable W_0^* with respect to the corresponding filtration of Vershik's Example 1.[1] Other examples of filtrations for which Vershik's self-joining criterion of a certain random variable is equivalent to Vershik's self-joining criterion of W_0, are given in [40].

Remark 2.42 (The scale of an automorphism). Theorem 2.39 is closely related to the *scale* of Bernoulli automorphisms. The notion of *scale of an automorphism* has been introduced by Vershik in [39]. The scale is a set of sequences of integers $(r_n)_{n \leq 0}$. Vershik asserts in [39], without giving a proof, that $(r_n)_{n \leq 0}$ does not belong to the scale of Bernoulli automorphisms under condition (Δ), and he proves that it belongs to the scale of Bernoulli automorphisms under a stronger condition than $\neg(\Delta)$. We have checked, by using the first definition of the scale given by Vershik, that a sequence $(r_n)_{n \leq 0}$ belongs to the scale of a Bernoulli automorphism if and only if Vershik's self-joining criterion holds for a certain random variable with respect to the filtration $\mathcal{F}$ of a split-word process with splitting sequence $(r_n)_{n \leq 0}$. Thus, part (b) of Theorem 2.39 shows that $(r_n)_{n \leq 0}$ *belongs to the scale of Bernoulli automorphisms under condition* $\neg(\Delta)$, thereby improving the proposition in [39]. In [21], we describe this random variable and we argue that Vershik's self-joining criterion of this random variable is actually equivalent to productness of $\mathcal{F}$. Finally Theorem 2.39 then shows that *the scale of a Bernoulli automorphism is the set of sequences* $(r_n)_{n \leq 0}$ *satisfying* $\neg(\Delta)$.

We shall derive Theorem 2.39 from the following two facts, which will be admitted:

Result 2.43. *If* (Δ) *holds and A is finite, then $\mathcal{F}$ is not of product type, unless μ is degenerate.*

If $r_n \equiv 2$, condition (Δ) holds; in this case, Smorodinsky [32] has shown that $\mathcal{F}$ is not of product type. His proof is copied in [14] with the language of I-cosiness. Result 2.43 is shown in [20] with the help of Vershik's self-joining criterion (Theorem 2.38) and by generalizing some lemmas given in [14] for the

---

[1] With the terminology of [21], the respective sequences of *Vershik's progressive predictions* $(\pi_n W_0)_{n \leq 0}$ and $(\pi_n W_0^*)_{n \leq 0}$ of W_0 and W_0^* are two processes with the same law; that shows that the Vershik property, or the I-cosiness, or Vershik's self-joining criterion, are the same for W_0 and W_0^*.

particular case $r_n \equiv 2$. Result 2.43 is also shown in [7] by means of the I-cosiness criterion, by a more direct generalization of the proof given in [14].

Result 2.44. *If A is finite and μ is the uniform probability on A, then $\mathcal{F}$ is of product type if (Δ) does not hold.*

Result 2.44 is due to Ceillier [7]. This is the most recent and from some point of view the most difficult part of Theorem 2.39.

We do not yet have the material needed to prove Theorem 2.39 from these two results. We will only give, in Result 2.45 below, the first step towards the derivation of part (b) of the theorem from Result 2.44. Result 2.43 will be discussed in Sect. 3 where we shall demonstrate how to derive part (a) of Theorem 2.39 from this result. Finally we shall demonstrate in Sect. 4 how to derive part (b) of Theorem 2.39 from Result 2.45.

Result 2.45. *If $A = [0, 1]$ and μ is the Lebesgue measure, then $\mathcal{F}$ is of product type if (Δ) does not hold.*

Proof. Consider the split-word process $(W_n, \eta_n)_{n \leq 0}$ on the alphabet $A = [0, 1]$ equipped with the Lebesgue measure, and its generated filtration $\mathcal{F}$. Assuming Result 2.43, we shall see that $\mathcal{F}$ satisfies Vershik's first level criterion if (Δ) does not hold. This will prove that $\mathcal{F}$ is of product type thanks to Vershik's first level theorem (Theorem 2.25). By Proposition 2.22, for $\mathcal{F}$ to satisfy Vershik's first level criterion, it suffices that each σ-field $\sigma(W_n)$ satisfies Vershik's first level criterion. In turn, this is proved as follows. For each $k \in \mathbb{N}$, we define the approximation of identity $f^k : [0, 1] \to [0, 1]$ by

$$f^k(u) = \sum_{i=0}^{2^k-1} \frac{i}{2^k} \mathbb{1}_{\left\{\frac{i}{2^k} < u \leq \frac{i+1}{2^k}\right\}}.$$

By result 2.44, we can see that $\left(f^k(W_n), \eta_n\right)_{n \leq 0}$ is a split-word process which generates a filtration $\mathcal{F}^k$ of product type. Consequently, each filtration $\mathcal{F}^k$ satisfies Vershik's first level criterion (Corollary 2.14). Lemma 2.4 shows that any innovation of a filtration $\mathcal{F}^k$ is also an innovation of $\mathcal{F}$. Therefore, all random variables $f^k(W_n)$ satisfy Vershik's first level criterion with respect to $\mathcal{F}$. Hence, due to Proposition 2.7, the random variable W_n satisfies Vershik's first level criterion with respect to $\mathcal{F}$. This amounts to saying that the σ-field $\sigma(W_n)$ satisfies Vershik's first level criterion (Proposition 2.17). $\qquad \square$

Before turning to the next section, we give a corollary of part (b) of Theorem 2.39 showing that, as announced above Corollary 2.26, there exist some filtrations of local product type satisfying Vershik's first level criterion although they are not of product type. Note that the existence of split-word processes with a non-Polish alphabet is guaranteed by Ionescu–Tulcea's theorem.

Corollary 2.46. *In the context of Theorem 2.39, if (Δ) does not hold, then $\mathcal{F}$ satisfies Vershik's first level criterion, whatever $(A, \mathfrak{A}, \mu)$. However, $\mathcal{F}$ is not of product type if $(A, \mathfrak{A}, \mu)$ is not essentially separable.*

Proof. By Proposition 2.22, it suffices to show that each σ-field $\sigma(W_n)$ satisfies Vershik's first level criterion. Thus, considering a measurable function $f: A^{\ell_n} \to \mathbb{R}$ such that the random variable $f(W_n)$ is integrable, we have to show that this random variable satisfies Vershik's first level criterion. As f is measurable with respect to $\mathfrak{A}^{\otimes \ell_n}$, there exist (Lemma 1.1) some essentially separable σ-fields $\mathfrak{C}_1 \subset \mathfrak{A}, \ldots,$ $\mathfrak{C}_{\ell_n} \subset \mathfrak{A}$ such that f is measurable with respect to $\mathfrak{C}_1 \otimes \cdots \otimes \mathfrak{C}_{\ell_n}$. Introduce the essentially separable σ-field $\mathfrak{B} = \mathfrak{C}_1 \vee \cdots \vee \mathfrak{C}_{\ell_n}$; then f is measurable with respect to $\mathfrak{B}^{\otimes \ell_n}$. Thus, considering a measurable function $G: A \to \mathbb{R}$ such that $\sigma(G) = \mathfrak{B}$, the random variable $f(W_n)$ is measurable with respect to the σ-field generated by the random word $X_n := G\big(W_n(1)\big) \ldots G\big(W_n(\ell_n)\big)$. The process $(X_n, \eta_n)_{n \leqslant 0}$ is a split-word process on a Polish alphabet, and thus we know from Theorem 2.39 that it generates a filtration of product type, so the σ-field $\sigma(X_n)$ satisfies Vershik's first level criterion with respect to the filtration generated by $(X_n, \eta_n)_{n \leqslant 0}$. By Lemma 2.4, any innovation of this filtration is also an innovation of $\mathcal{F}$, hence the σ-field $\sigma(X_n)$ satisfies Vershik's first level criterion with respect to $\mathcal{F}$, and so does also the random variable $f(W_n)$. □

3 Standardness and I-cosiness

Section 3.1 deals with *standard filtrations*, defined as filtrations that are *immersible* in a filtration (of product type) generated by independent random variables having a diffuse law. We shall see that any filtration of product type is standard and that being standard is equivalent to being immersible in a filtration of product type.

In Sect. 3.2, as an example of a sufficient condition for standardness, we provide an unpublished result from Tsirelson on the existence of a *generating parameterization* for a filtration under a certain ergodicity condition on a Markov process generating this filtration. The proof we give makes use of Rosenblatt's self-joining criterion (Proposition 2.33). As we shall see, the existence of a generating parameterization for a filtration obviously implies standardness, but we disagree with a result in a literature asserting that the converse is true. This point is discussed in Sect. 3.3.

In Section 3.4, the notion of standardness is extended to its analogue for *locally separable filtrations* (Definition 3.17), namely *weak standardness*, defined similarly to standardness but with productness replaced by Vershik's first level criterion.

Next, the I-cosiness criterion is defined in Sect. 3.5 and its basic properties are given. The implications productness $\Rightarrow$ I-cosiness and Vershik's first level criterion $\Rightarrow$ I-cosiness will be directly deduced from Rosenblatt's self-joining criterion and Vershik's self-joining criterion, defined in the previous section. I-cosiness of standard or weakly standard filtrations will follow as an obvious consequence of

I-cosiness being inherited by immersion. As an illustration of the I-cosiness criterion, we give a sufficient condition for a stationary Markov process to generate an I-cosy filtration.

Finally, in Sect. 3.6, we pursue the proof of Theorem 2.39 (productness of the filtrations of split-words processes).

3.1 Standardness, Superinnovations, Parameterizations

Two notions must preliminarily be defined before standardness: the notion of a *standard conditionally non-atomic filtration* and the notion of an *extension* of a filtration.

Definition 3.1. A *conditionally non-atomic* filtration is a filtration of local product type admitting a global innovation $(U_n)_{n \leq 0}$ such that each U_n is uniformly distributed on the interval $[0, 1]$. A *standard conditionally non-atomic* filtration is a filtration generated by a sequence $(U_n)_{n \leq 0}$ of independent random variables uniformly distributed on the interval $[0, 1]$.

Observe that all standard conditionally non-atomic filtrations are isomorphic to each other. Remark also that *"uniformly distributed on the interval $[0, 1]$"* in this definition can equivalently be replaced with *"having a diffuse law"*. The two lemmas below respectively characterize standard conditionally non-atomic filtrations and conditionally non-atomic filtrations.

Lemma 3.2. *A filtration is standard conditionally non-atomic if and only if it is of product type and conditionally non-atomic.*

Proof. The 'only if' part is obvious. The 'if' part follows from the fact that for a conditionally non-atomic filtration $\mathcal{F}$, any independent complement of $\mathcal{F}_{n-1}$ in $\mathcal{F}_n$ necessarily has a diffuse law, by virtue of Lemma 2.4.   □

Lemma 3.3. *A filtration $\mathcal{F} = (\mathcal{F}_n)_{n \leq 0}$ is conditionally non-atomic if and only if for every $n \leq 0$, there exists an $\mathcal{F}_n$-measurable random variable V_n such that $\mathcal{F}_{n-1} \vee \sigma(V_n) = \mathcal{F}_n$ and such that the conditional law $\mathcal{L}(V_n \mid \mathcal{F}_{n-1})$ is almost surely diffuse.*

Proof. The 'only if' part is trivially true. For the 'if' part, consider V_n as in the lemma and let $F(\cdot \mid \mathcal{F}_{n-1})$ be the conditional cumulative distribution function of V_n given $\mathcal{F}_{n-1}$. One easily checks that the random variable $F(V_n \mid \mathcal{F}_{n-1})$ is an innovation from $\mathcal{F}_{n-1}$ into $\mathcal{F}_n$ having uniform law on $[0, 1]$.   □

To define standardness, we need one more notion, an *extension* of a filtration. Roughly speaking, an extension of $\mathcal{F}$ is a filtration in which $\mathcal{F}$ "can be immersed".

Definition 3.4. Let $\mathcal{F}$ and $\mathcal{G}'$ be two filtrations defined on possibly different probability spaces. The filtration $\mathcal{F}$ is *immersible* in the filtration $\mathcal{G}'$, and $\mathcal{G}'$ is an *extension* of $\mathcal{F}$, if $\mathcal{F}$ is isomorphic to some filtration $\mathcal{F}'$ immersed in $\mathcal{G}'$.

Lemma 3.5. *Let $\mathcal{F} = (\mathcal{F}_n)_{n \leq 0}$ and $\mathcal{G} = (\mathcal{G}_n)_{n \leq 0}$ be two filtrations defined on possibly different probability spaces. Then both $\mathcal{F}$ and $\mathcal{G}$ are immersible in the product filtration $\mathcal{F} \otimes \mathcal{G}$.*

Proof. Let $\iota_1 \colon \mathcal{F} \to \mathcal{F} \otimes \mathcal{G}$ and $\iota_2 \colon \mathcal{G} \to \mathcal{F} \otimes \mathcal{G}$ be the identifications with the first factor and the second factor respectively (see Example 5.2). Then we know from Lemma 1.2 that the two independent filtrations $\iota_1(\mathcal{F})$ and $\iota_2(\mathcal{G})$ are both immersed in $\iota_1(\mathcal{F}) \vee \iota_2(\mathcal{G}) = \mathcal{F} \otimes \mathcal{G}$.   □

Now we turn on to the notion of standard filtrations.

Definition 3.6. A *standard* filtration is a filtration immersible in a standard conditionally non-atomic filtration.

As obvious facts on standardness, we note:

- Standardness is preserved by isomorphism.
- A standard conditionally non-atomic filtration in the sense of Definition 3.1 is standard and is conditionally non-atomic. But at this stage we are not yet able to prove the converse; this will be done in Sect. 4 (Corollary 4.7).
- The standardness property for a filtration is inherited by immersion, i.e., any filtration immersible in a standard filtration is itself standard.
- A standard filtration is essentially separable.

Proposition 3.7. *Any filtration of product type is standard, and a filtration is standard if and only if it is immersible in a filtration of product type.*

Proof. The independent product of a filtration $\mathcal{F}$ with a standard non-atomic filtration is an extension of $\mathcal{F}$ (Lemma 3.5). Obviously, this product filtration is itself standard non-atomic if $\mathcal{F}$ is of product type, and hence $\mathcal{F}$ is standard. Consequently, a filtration is standard if it is immersible in a filtration of product type because standardness is inherited by immersion. The converse is obvious from the Definition of standardness.   □

As we know from the pioneering works of Vershik, there exist some Kolmogorovian essentially separable filtrations that are not standard, i.e., which cannot be immersed in a standard conditionally non-atomic filtration. However Theorem 3.9 below shows that *any essentially separable filtration can be immersed in the supremum of two jointly immersed standard conditionally non-atomic filtrations*. This theorem firstly says that an essentially separable conditionally non-atomic filtration *equals* the supremum of two jointly immersed standard conditionally non-atomic filtrations. This is the main assertion and it is due to Parry [24]. The second assertion of this theorem follows from Lemma 3.8 below. An interesting consequence of Theorem 3.9 is given in [21] and is invoked in our Remark 3.34. Except for this remark and for the remark below Proposition 3.46, we will never use this theorem.

Lemma 3.8. *Any essentially separable filtration admits an essentially separable conditionally non-atomic extension.*

Proof. Using Lemma 3.3, it is easy to see that the independent product of an essentially separable filtration and a standard conditionally non-atomic filtration is an essentially separable conditionally non-atomic filtration. Then the lemma immediately follows from Lemma 3.5.   □

Theorem 3.9. *Let $\mathcal{F} = (\mathcal{F}_n)_{n\leq 0}$ be an essentially separable conditionally non-atomic filtration. Then $\mathcal{F} = \mathcal{H}_1 \vee \mathcal{H}_2$ where $\mathcal{H}_1$ and $\mathcal{H}_2$ are two jointly immersed standard conditionally non-atomic filtrations. Consequently, any essentially separable filtration is immersible in such a filtration $\mathcal{H}_1 \vee \mathcal{H}_2$.*

Proof. The consequence follows from Lemma 3.8. To prove the first assertion, we strictly follow [24]. Consider an essentially separable conditionally non-atomic filtration $\mathcal{F} = (\mathcal{F}_n)_{n\leq 0}$. Let $(U_n)_{n\leq 0}$ be an innovation of $\mathcal{F}$ such that each U_n is uniformly distributed on the interval $[0, 1]$. For every $n \leq 0$, let X_n be a random variable generating $\mathcal{F}_n$. Then X_n necessarily has a diffuse law, and we assume without loss of generality that this is the uniform law on $[0, 1]$. Define $V_n = X_{n-1} + U_n$ (mod 1) for every $n \leq 0$. By Lemma 2.4, $(V_n)_{n\leq 0}$ is an innovation of $\mathcal{F}$. It suffices to define $\mathcal{H}_1$ as the filtration generated by $(U_n)_{n\leq 0}$ and $\mathcal{H}_2$ as the filtration generated by $(V_n)_{n\leq 0}$. Then we easily see that $\mathcal{F} = \mathcal{H}_1 \vee \mathcal{H}_2$, and the joint immersion of $\mathcal{H}_1$ and $\mathcal{H}_2$ is a consequence of Lemma 1.6.   □

A typical example of a standard conditionally non-atomic extension of a filtration $\mathcal{F}$ is the filtration generated by a generating parameterization of $\mathcal{F}$, defined below.

Definition 3.10. A (global) *superinnovation* of a filtration $\mathcal{F} = (\mathcal{F}_n)_{n\leq 0}$ is a process $(V_n)_{n\leq 0}$ such that for each $n \leq 0$, the random variable V_n takes its values in a Polish space, is independent of $\mathcal{F}_{n-1} \vee \sigma(V_m; m \leq n - 1)$, and satisfies $\mathcal{F}_n \subset \mathcal{F}_{n-1} \vee \sigma(V_n)$. The superinnovation $(V_n)_{n\leq 0}$ is a *generating superinnovation* if moreover $\mathcal{F}$ is contained in the filtration generated by $(V_n)_{n\leq 0}$. A filtration $\mathcal{F}$ *admits a superinnovation* if there exists a superinnovation of a filtration isomorphic to $\mathcal{F}$. A superinnovation $(V_n)_{n\leq 0}$ of $\mathcal{F}$ is called a *parameterization* if V_n has the uniform law on $[0, 1]$ for every $n \leq 0$.

Obviously, a filtration admits a parameterization if and only if it admits a superinnovation $(V_n)_{n\leq 0}$ such that each V_n has a diffuse law. Proposition below shows that a filtration admitting a generating superinnovation is actually immersed in the filtration generated by this superinnovation.

Proposition 3.11. *On $(\Omega, \mathcal{A}, \mathbb{P})$, let $\mathcal{F} = (\mathcal{F}_n)_{n\leq 0}$ be a filtration and $(V_n)_{n\leq 0}$ a process such that each V_n takes its values in a Polish space. Let $\mathcal{V} = (\mathcal{V}_n)_{n\leq 0}$ be the filtration generated by $(V_n)_{n\leq 0}$. The following conditions are equivalent:*

(i) *$(V_n)_{n\leq 0}$ is a superinnovation of $\mathcal{F}$.*
(ii) *$(V_n)_{n\leq 0}$ is a sequence of independent random variables, one has $\mathcal{F}_n \subset \mathcal{F}_{n-1} \vee \sigma(V_n)$ for each $n \leq 0$, and $\mathcal{V}$ is immersed in $\mathcal{F} \vee \mathcal{V}$.*
(iii) *$(V_n)_{n\leq 0}$ is a sequence of independent random variables, one has $\mathcal{F}_n \subset \mathcal{F}_{n-1} \vee \sigma(V_n)$ for each $n \leq 0$, and $\mathcal{F}$ and $\mathcal{V}$ are jointly immersed.*

Consequently, if $(V_n)_{n \leqslant 0}$ *is a generating superinnovation for* $\mathcal{F}$*, then* $\mathcal{F}$ *is immersed in* $\mathcal{V}$*, and hence* $\mathcal{F}$ *is standard.*

Proof. If $(V_n)_{n \leqslant 0}$ is a sequence of independent random variables, then Lemma 1.6 shows that $\mathcal{V}$ is immersed in $\mathcal{F} \vee \mathcal{V}$ if and only if V_n is independent of $\mathcal{F}_{n-1} \vee \sigma(V_m; m \leqslant n - 1)$ for each $n \leqslant 0$. It follows that (i) $\Longleftrightarrow$ (ii). If $(V_n)_{n \leqslant 0}$ is a superinnovation of $\mathcal{F}$, then $\mathcal{F}$ is immersed in $\mathcal{F} \vee \mathcal{V}$ as an easy consequence of Lemma 1.6. That finally shows that (i) $\Longleftrightarrow$ (ii) $\Longleftrightarrow$ (iii). Consequently, if $(V_n)_{n \leqslant 0}$ is a generating superinnovation, then $\mathcal{F}$ is standard by Proposition 3.7 since it is immersed in the product type filtration $\mathcal{V} = \mathcal{F} \vee \mathcal{V}$.   □

Remark 3.12. It is claimed in the literature that any standard filtration admits a generating parameterization. Actually some confusion occurred, and to our knowledge there exists no valid proof of this assertion. This will be discussed in Sect. 3.3.

3.2 An Example from Tsirelson

Theorem 3.15 below gives a sufficient condition on a Markov process for its generated filtration to admit a generating parameterization, and hence to be standard (Proposition 3.11). This result is borrowed from Tsirelson (About Yor's Problem, Tel Aviv University, unpublished preprint). Lemma 3.14 is the key point of the proof. The last part of the proof we give illustrates Rosenblatt's self-joining criterion (Proposition 2.33).

The following lemma will be used in the proof of Lemma 3.14 (and in the statement of Lemma 3.24 and the proof of Lemma 3.27). This lemma is a verbatim copy of Lemma 6.4.6 in [6].

Lemma 3.13. *Let* $(\Omega, \mathcal{A})$ *be a measurable space and* $f : \Omega \times \mathbb{R} \to \mathbb{R}$ *a function satisfying the following conditions: for every fixed* $t \in \mathbb{R}$*, the function* $\omega \mapsto f(\omega, t)$ *is* $\mathcal{A}$*-measurable, and for every fixed* $\omega \in \Omega$*, the function* $t \mapsto f(\omega, t)$ *is right-continuous. Then the function* f *is measurable with respect to* $\mathcal{A} \otimes \mathfrak{B}_{\mathbb{R}}$*.*

The statement and the proof of Lemma 3.14 and Theorem 3.15 involve infimum of measures. Given two measures ν_1 and ν_2 on a measurable space, we denote by $\nu_1 \wedge \nu_2$ the infimum of ν_1 and ν_2. The existence of this measure is guaranteed; more generally, the infimum of an infinite family of measures always exists (see [33], Theorem 7.1, or [9], Appendices III & IV.) When $\nu_1 = f_1 \cdot \mu$ and $\nu_2 = f_2 \cdot \mu$ (Radon-Nikodým derivatives), $\nu_1 \wedge \nu_2 = (f_1 \wedge f_2) \cdot \mu$.

Lemma 3.14. *Let* X *and* Y *be two random variables taking values in some Polish spaces* E *and* F *respectively, and* $\{\nu_x\}_{x \in E}$ *a regular version of the conditional law of* Y *given* X*. Denote by* μ_1 *and* μ_2 *the respective laws of* X *and* Y *and by Leb the Lebesgue probability measure on* $[0, 1]$*. There exists a measurable function* $p : E \times F \to [0, 1]$ *such that for* μ_1*-almost all* $x \in E$*, the function* $y \mapsto p(x, y)$ *is a Radon-Nikodým derivative of* $\nu_x \wedge \mu_2$ *with respect to* μ_2*, and there exists a* *measurable function* $\alpha\colon E \times F \times [0, 1] \to \mathbb{R}$ *such that one has* $\alpha(x, \mu_2 \times \mathrm{Leb}) = \nu_x$ *for every* x *and* $\alpha(x, y, u) = y$ *for every* x, y, u *satisfying* $u \leqslant p(x, y)$.

Proof. Since any Polish probability space is Lebesgue isomorphic to a probability space on $\mathbb{R}$, it suffices to do the proof for $F = \mathbb{R}$. Let f be a Radon–Nikodým derivative of the absolutely continuous part in the Lebesgue decomposition of the joint distribution of (X, Y) with respect to $\mu_1 \otimes \mu_2$. The set of values of $x \in E$ such that $\int f(x, y) d\mu_2(y) = 0$, is μ_1-negligible, and for those values of x for which $\int f(x, y) d\mu_2(y) \neq 0$, it is not difficult to check that the function $p_x\colon y \mapsto \min\{f(x, y)/\int f(x, z) d\mu_2(z), 1\}$ is a Radon–Nikodým derivative of the measure $\nu_x \wedge \mu_2$ with respect to μ_2. We define p by $p(x, y) = p_x(y)$. Now we are going to construct α. The function $t \mapsto \nu_x(]-\infty, t]) - (\nu_x \wedge \mu_2)(]-\infty, t])$ is right-continuous and increasing, and takes its values in $[0, 1 - m_x]$ where m_x is the total mass of $\nu_x \wedge \mu_2$. Call g_x the right-continuous inverse of this function. The function $(x, v) \mapsto g_x(v)$ is measurable by virtue of Lemma 3.13. Then put $\alpha(x, y, u) = g_x\big(\frac{1-m_x}{1-p_x(y)}(1 - u)\big)$ for $u \in\,]p_x(y), 1]$. One checks without difficulty that $\alpha(x, \mu_2 \times \mathrm{Leb}) = \nu_x$. $\qquad\square$

Theorem 3.15. *Consider a Markov process* $(X_n)_{n \leqslant 0}$ *where each* X_n *takes its values in a Polish space. Denote by* μ_n *the law of* X_n *and by* $\nu^n_{x_{n-1}}$ *the conditional law* $\mathcal{L}(X_n \mid X_{n-1} = x_{n-1})$. *Let* m_n *be the* $\mu_{n-1} \otimes \mu_{n-1}$-*essential infimum over* x'_{n-1}, x''_{n-1} *of the total masses of the measures* $\nu^n_{x'_{n-1}} \wedge \nu^n_{x''_{n-1}} \wedge \mu_n$. *If* $\sum m_n = +\infty$, *then the filtration generated by* $(X_n)_{n \leqslant 0}$ *admits a generating parameterization.*

As an application of this theorem, we can see that the filtration generated by a stationary random walk on the vertices of a triangle admits a generating parameterization. Note that the number m_n defined in Theorem 3.15 satisfies $m_n \geqslant \beta(\nu^n)$ where $\beta(\nu^n)$ is the total mass of the essential infimum over all x of the measures ν^n_x. Thus the condition $\sum \beta(\nu^n) = +\infty$ guarantees the existence of a generating parameterization. In the stationary case, this condition is equivalent to the existence of a positive non-null measure that minorizes the probability measures $\mathcal{L}(X_n \mid X_{n-1} = x)$ for almost all x. A weaker minorization condition given in [17] guarantees the existence of a generating parameterization in the stationary case.

Proof (Proof of Theorem 3.15). Let p_n and α_n be the functions p and α obtained from Lemma 3.14 applied with $X = X_{n-1}$ and $Y = X_n$. We write $p^n_x(y) = p_n(x, y)$. Now consider the Markov process $\big(X'_n, (Y'_n, U'_n)\big)_{n \leqslant 0}$ defined as follows:

- For each $n \leqslant 0$, (Y'_n, U'_n) is independent of the past up to $n - 1$, Y'_n has law μ_n, U'_n has the uniform law on $[0, 1]$, Y'_n and U'_n are independent.
- $X'_n = \alpha_n(X'_{n-1}, Y'_n, U'_n)$ for each $n \leqslant 0$.

Then we know from Lemma 3.14 that $(X'_n)_{n \leqslant 0}$ has the same distribution as the Markov process $(X_n)_{n \leqslant 0}$, and that $X'_n = Y'_n$ if $U'_n \leqslant p^n_{X'_{n-1}}(Y'_n)$. Let $\mathcal{F}'$ be the filtration generated by $(X'_n)_{n \leqslant 0}$. Obviously, the process $(Y'_n, U'_n)_{n \leqslant 0}$ is a superinnovation of $\mathcal{F}'$ (Definition 3.10), and we shall show with the help of Rosenblatt's self-joining criterion (Proposition 2.33) that this superinnovation is generating if $\sum m_n = +\infty$.

Let $(X_n^*)_{n \leq 0}$ be a copy of $(X_n)_{n \leq 0}$ lying on the same probability space $(\overline{\Omega}, \overline{\mathcal{A}}, \overline{\mathbb{P}})$ as $(X_n', (Y_n', U_n'))_{n \leq 0}$ and independent of $(X_n', (Y_n', U_n'))_{n \leq 0}$. For a given integer $n_0 < 0$, we define another copy $(X_n'')_{n \leq 0}$ of $(X_n)_{n \leq 0}$ by setting $X_n'' = X_n^*$ for $n \leq n_0$ and $X_{n+1}'' = \alpha_n(X_n'', Y_{n+1}', U_{n+1}')$ for n going from n_0 to -1. By Proposition 2.33, it suffices to show that $\overline{\mathbb{P}}[X_n' \neq X_n''] \to 0$ as n_0 goes to $-\infty$ for each $n \leq 0$. For $n \geq n_0$ one has $(X_n' = X_n'') \implies (X_{n+1}' = X_{n+1}'')$, so it suffices to show that the inequality $\overline{\mathbb{P}}[X_n' \neq X_n'' \mid X_{n-1}', X_{n-1}''] \leq 1 - m_n$ almost surely holds for every $n \in \{n_0 + 1, \ldots, 0\}$, because this yields the inequality $\overline{\mathbb{P}}[X_n' \neq X_n''] \leq \prod_{k=n_0+1}^{n}(1 - m_k)$. Now one has

$$\overline{\mathbb{P}}\left[\alpha_n(x_{n-1}', Y_n', U_n') = \alpha_n(x_{n-1}'', Y_n', U_n')\right] \leq \overline{\mathbb{P}}\left[U_n' \leq p_{x_{n-1}'}^n(Y_n') \wedge p_{x_{n-1}''}^n(Y_n')\right].$$

But we can see that

$$\overline{\mathbb{P}}\left[U_n' \leq p_{x_{n-1}'}^n(Y_n') \wedge p_{x_{n-1}''}^n(Y_n')\right] = \int d\mu_n(y_n)(p_{x_{n-1}'}^n(y_n) \wedge p_{x_{n-1}''}^n(y_n) \wedge 1)$$

is nothing but the total mass of $\nu_{x_{n-1}'}^n \wedge \nu_{x_{n-1}''}^n \wedge \mu_n$. The proof is over. $\qquad \square$

Remark 3.16. It is known that the total variation $\|\nu_1 - \nu_2\| \in [0, 2]$ between two probability measures ν_1 and ν_2 on an arbitrary measurable space E satisfies $\|\nu_1 - \nu_2\| = 2(1 - (\nu_1 \wedge \nu_2)(E))$ (see [33]). Therefore the number m_n defined in Theorem 3.15 satisfies $2m_n \leq 2 - \alpha(\nu^n)$ where $\alpha(\nu^n) = \text{ess sup}_{x', x''} \|\nu_{x'}^n - \nu_{x''}^n\|$. It would be interesting to know if the condition $\sum_n (2 - \alpha(\nu^n)) = +\infty$ guarantees standardness of the filtration.

3.3 Erratum on Generating Parameterizations

Some confusion occurred in the articles [15, 30, 32]. It was erroneously claimed that, for a discrete negative-time filtration, every standard conditionally non-atomic extension is *obviously* induced by a generating parameterization (Definition 3.10). Schachermayer gave a counter-example in [31]. As this (false) claim was considered as a proof that every standard filtration admits a generating parameterization, a new proof of the latter fact was needed. Feldman and Smorodinsky claimed in [16] that this fact is nonetheless true, and gave a proof using results from the literature. Unfortunately, a confusion of the same kind occurred again in the proof proposed in [16]. Thus, to our knowledge, *there does not exist any proof of this assertion.* The proof given in [20] contains an error too.

The reiterated confusion in the proof proposed in [16] lies on page 1086 of [15], where it is claimed that if the independent product of a filtration $\mathcal{F}$ with a standard non-atomic filtration is itself standard non-atomic, then any generating innovation of the latter is a generating parameterization of $\mathcal{F}$. This is false, as we shall see, and this is unfortunately an ingredient in the proof proposed in [16]. This confusion is of the same kind as the one which was pointed out in [31]: a filtration is immersed in its independent product with another filtration, but a sequence that generates the product may not be a parameterization of $\mathcal{F}$, even though it generates a standard conditionally non-atomic extension of $\mathcal{F}$.

Here is a counter-example. It is of the same spirit as the counter-example given in [31]. On a probability space $(\Omega, \mathcal{A}, \mathbb{P})$, consider a sequence $(U_n)_{n \leq 0}$ of independent random variables uniformly distributed in $[0, 1]$, and a random variable U_0^* uniformly distributed in $[0, 1]$ and independent of $(U_n)_{n \leq 0}$. Define $X_0 = U_{-1} + U_0$ (mod 1). Let $\mathcal{F}$ be the filtration defined by $\mathcal{F}_n = \{\varnothing, \Omega\}$ for $n \leq -1$ and $\mathcal{F}_0 = \sigma(X_0)$. Let $\mathcal{G}$ be the filtration defined by $\mathcal{G}_n = \sigma(U_m, m \leq n)$ for $n \leq -1$ and $\mathcal{G}_0 = \mathcal{G}_{-1} \vee \sigma(U_0^*)$. Then $\mathcal{G}$ is a standard conditionally non-atomic filtration independent of $\mathcal{F}$. Consider any random variable V_0 uniformly distributed on $[0, 1]$ and such that $\sigma(V_0) = \sigma(U_0, U_0^*)$. One easily verifies that $\mathcal{F} \vee \mathcal{G}$ is generated by the sequence of independent random variables $(\ldots, U_{-2}, U_{-1}, V_0)$. However, $(\ldots, U_{-2}, U_{-1}, V_0)$ is not a parameterization for $\mathcal{F}$, because the inclusion $\mathcal{F}_0 \subset \mathcal{F}_{-1} \vee \sigma(V_0)$ does not hold.

So we consider the statement S_1: "*Standardness is equivalent to the existence of a generating parametrization*" as an open question. Note that S_1 is equivalent to S_2: "*A filtration immersed in a filtration which admits a generating parameterization, admits itself a generating parameterization*". Indeed, if every standard filtration admits a generating parameterization, then S_2 is true owing to the fact that standardness is inherited by immersion. Conversely, if S_2 is true, then every standard filtration admits a generating parameterization because a standard conditionally non-atomic filtration obviously admits a generating parameterization.

3.4 Weak Standardness. Locally Separable Filtrations

Obviously, standard filtrations must be essentially separable. For essentially separable filtrations, it is already known [14] that standardness is equivalent to I-cosiness (Definition 3.29), and also to Vershik's standardness criterion (which is not stated in this paper; see [14,21]). However the I-cosiness criterion and Vershik's standardness criterion could a priori be satisfied for a filtration which is not essentially separable. We shall soon define *weak standardness*, and we shall see in Sect. 4 that weak standardness is equivalent to I-cosiness for a *locally separable filtration*,[2] defined as follows.

Definition 3.17. A filtration $\mathcal{F} = (\mathcal{F}_n)_{n \leq 0}$ is *locally separable* if for each $n \leq 0$, there exists a random variable V_n such that $\mathcal{F}_n = \mathcal{F}_{n-1} \vee \sigma(V_n)$.

---

[2] It is also proved in [21] that the equivalence between I-cosiness and Vershik's standardness criterion remains true for locally separable filtrations.

Thus, any essentially separable filtration is locally separable, and filtrations of local product type (Definition 2.3), obviously are locally separable filtrations. We take the opportunity of Definition 3.17 to state a conjecture about local separability.

Conjecture 3.18. A filtration immersible in a locally separable filtration is itself locally separable.

We will later see (Corollary 3.28) that locally separable filtrations are precisely filtrations that admit a superinnovation or, equivalently, a parameterization (Definition 3.10).

The notion of *weak standardness* defined below invokes Vershik's first level criterion (Definition 2.6, Definition 2.9, Proposition 2.17). The generalization from standardness to weak standardness is based on the fact that Vershik's first level criterion is equivalent to productness for an essentially separable filtration (Theorem 2.25). Inspired by Lemma 3.2, we first define a *weakly standard conditionally non-atomic filtration* as the following generalization of a standard conditionally non-atomic filtration.

Definition 3.19. A *weakly standard conditionally non-atomic* filtration is a conditionally non-atomic filtration satisfying Vershik's first level criterion.

Lemma 3.20. *A filtration is standard non-atomic if and only if it is weakly standard conditionally non-atomic and essentially separable.*

Proof. This follows from Vershik's first level criterion (Theorem 2.25) and Lemma 3.2.   □

Then, the notion of weak standardness is defined analogously to the notion of standardness (Definition 3.6).

Definition 3.21. A filtration is *weakly standard* if it is immersible in a weakly standard conditionally non-atomic filtration.

As obvious remarks, we note:

- Weak standardness is preserved by isomorphism.
- If Conjecture 3.18 is true, then every weakly standard filtration is locally separable.
- Weak standardness is hereditary for immersion: a filtration immersible in a weakly standard filtration is itself weakly standard.
- As a product type filtration satisfies Vershik's first level criterion (Theorem 2.25), standardness implies weak standardness. Thus a standard filtration is weakly standard and essentially separable. However we are not yet able to show the converse; it will be proved in Sect. 4 (Corollary 4.10).
- Obviously, a weakly standard conditionally non-atomic filtration is weakly standard and conditionally non-atomic. We will see in Sect. 4 that the converse is true (Corollary 4.6).

Proposition 3.23 below is analogous to Proposition 3.7. Its proof invokes the following lemma.

Lemma 3.22. *The independent product of a conditionally non-atomic filtration and a locally separable filtration is conditionally non-atomic.*

Proof. Left to the reader as an easy application of Lemma 3.3.   □

Proposition 3.23. *Any filtration satisfying Vershik's first level criterion is weakly standard; and a filtration is weakly standard if and only if it is immersible in a filtration satisfying Vershik's first level criterion.*

Proof. The independent product of a filtration $\mathcal{F}$ with a standard conditionally non-atomic filtration is an extension of $\mathcal{F}$ (Lemma 3.5), and this product filtration is conditionally non-atomic by Lemma 3.22. Moreover, a standard conditionally non-atomic filtration satisfies Vershik's first level criterion (Corollary 2.14), hence if $\mathcal{F}$ satisfies Vershik's first level criterion, then so does this product filtration in view of Corollary 2.27. Thus we have proved that every filtration satisfying Vershik's first level criterion is weakly standard. Consequently, a filtration is weakly standard if it is immersible in a filtration satisfying Vershik's first level criterion because weak standardness is hereditary for immersion. The converse is obviously true from the definition of weak standardness.   □

Before turning to the next section devoted to I-cosiness, we are going to prove (Corollary 3.28) that a filtration is locally separable if and only if it admits a global parameterization (Definition 3.10). We will make use of the following lemma. The construction appearing in this lemma is the general *conditional quantile transformation* (see [23]). In the statement of this lemma, we implicitly use Lemma 3.13 to justify the measurability of $F_{\mathcal{C}}$ and $F_{\mathcal{C}}^-$ with respect to $\mathcal{C} \otimes \mathcal{B}_{\mathbb{R}}$ and the measurability of $G_{\mathcal{C}}$ with respect to $\mathcal{C} \otimes \mathcal{B}_{[0,1]}$.

Lemma 3.24. *Let X be a real random variable on a probability space $(\Omega, \mathcal{A}, \mathbb{P})$ and $\mathcal{C} \subset \mathcal{A}$ be a σ-field. Let $F_{\mathcal{C}}$ be the cumulative distribution function of the conditional law of X given $\mathcal{C}$, and let $F_{\mathcal{C}}^-(x) = \lim_{x' \to x^-} F_{\mathcal{C}}(x')$ be the left limit of $F_{\mathcal{C}}(x')$ as x' approaches x. Let ξ be a random variable with uniform law on $[0,1]$ and independent of $\mathcal{C} \vee \sigma(X)$. We put*

$$U = F_{\mathcal{C}}^-(X) + \xi \left(F_{\mathcal{C}}(X) - F_{\mathcal{C}}^-(X) \right).$$

Then U is a random variable independent of $\mathcal{C}$, uniformly distributed on $[0,1]$, and one has $X = G_{\mathcal{C}}(U)$ where $G_{\mathcal{C}}$ is the right-continuous inverse function of $F_{\mathcal{C}}$, defined by $G_{\mathcal{C}}(u) = \inf \{x \mid F_{\mathcal{C}}(x) > u\}$.

Proof. It suffices to show the lemma in the case when $\mathcal{C}$ is degenerate. We write F, F^- and G instead of $F_{\mathcal{C}}$, $F_{\mathcal{C}}^-$ and $G_{\mathcal{C}}$ respectively. Denote by $S = \{x_1, x_2, \ldots\}$ the denumerable set of atoms of X. Conditionally on the event $X = x_i$, the random variable U has the uniform law on $(F^-(x_i), F(x_i))$ and one has $X = G(U)$; and conditionally on the event $X \notin S$, one has $U = F(X)$, so the distribution of U is uniform on $[0, 1] \setminus \bigcup_i [F^-(x_i), F(x_i)]$, and one has $X = G(U)$. Finally the distribution of U is the uniform law on $[0, 1]$ and one has $X = G(U)$. $\quad\square$

Proposition 3.25. *Any locally separable filtration admits a global parameterization.*

Proof. Let $\mathcal{F} = (\mathcal{F}_n)_{n \leq 0}$ be a locally separable filtration on $(\Omega, \mathcal{A}, \mathbb{P})$. For each $n \leq 0$, let V_n be a random variable such that $\mathcal{F}_n = \mathcal{F}_{n-1} \vee \sigma(V_n)$. On some probability space $(\Omega^*, \mathcal{C}^*, \mathbb{P}^*)$, consider a sequence $(\xi_n)_{n \leq 0}$ of independent random variables having uniform law on $[0, 1]$. We work on the product probability space $(\widehat{\Omega}, \widehat{\mathcal{A}}, \widehat{\mathbb{P}}) := (\Omega, \mathcal{F}_0, \mathbb{P}) \otimes (\Omega^*, \mathcal{C}^*, \mathbb{P}^*)$ and we identify $\mathcal{F}$ and $(\xi_n)_{n \leq 0}$ with their image under the canonical embedding (see Example 5.2) from $(\Omega, \mathcal{F}_0, \mathbb{P})$ to $(\widehat{\Omega}, \widehat{\mathcal{A}}, \widehat{\mathbb{P}})$ and the canonical embedding from $(\Omega^*, \mathcal{C}^*, \mathbb{P}^*)$ to $(\widehat{\Omega}, \widehat{\mathcal{A}}, \widehat{\mathbb{P}})$ respectively, so the sequence $(\xi_n)_{n \leq 0}$ is independent of $\mathcal{F}_0$. We denote by $\mathcal{D} = (\mathcal{D}_n)_{n \leq 0}$ the filtration generated by $(\xi_n)_{n \leq 0}$. For each $n \leq 0$, the random variable ξ_n is independent of $\mathcal{F}_0 \vee \mathcal{D}_{n-1}$ and, since $\mathcal{F}$ and $\mathcal{D}$ are jointly immersed (Lemma 1.2), one has $\mathcal{L}(V_n \mid \mathcal{F}_{n-1} \vee \mathcal{D}_{n-1}) = \mathcal{L}(V_n \mid \mathcal{F}_{n-1})$. For each $n \leq 0$, let U_n be the random variable called U in Lemma 3.24 when this lemma is applied with $X = V_n$, $\mathcal{C} = \mathcal{F}_{n-1} \vee \mathcal{D}_{n-1}$ and $\xi = \xi_n$. One easily checks that $(U_n)_{n \leq 0}$ is a parameterization of $\mathcal{F}$. $\quad\square$

Remark 3.26. Let $\mathcal{U} = (\mathcal{U}_n)_{n \leq 0}$ be the filtration generated by the parameterization $(U_n)_{n \leq 0}$ of $\mathcal{F}$ in the preceding proof. It follows from Lemma 1.6 that $\mathcal{F} \vee \mathcal{U}$ is immersed in $\mathcal{F} \vee \mathcal{D}$. Therefore, if $\mathcal{F}$ is standard, then $\mathcal{F} \vee \mathcal{U}$ is also standard because $\mathcal{F} \vee \mathcal{D}$ is standard (Proposition 3.7). It is shown in [21], with the help of Vershik's standardness criterion, that this is actually true for an arbitrary parameterization $(U_n)_{n \leq 0}$ of $\mathcal{F}$.

The converse of Proposition 3.25 is an easy consequence of (ii) $\Longrightarrow$ (i) in the following lemma.

Lemma 3.27. *On $(\Omega, \mathcal{A}, \mathbb{P})$, let $\mathcal{C}$ and $\mathcal{B}$ be two σ-fields such that $\mathcal{C} \subset \mathcal{B}$. The following conditions are equivalent:*

(i) There exists a random variable V such that $\mathcal{B} = \mathcal{C} \vee \sigma(V)$.

(ii) There exists a random variable W such that $\mathcal{B} \subset \mathcal{C} \vee \sigma(W)$.

(iii) There exist a probability space $(\Omega', \mathcal{A}', \mathbb{P}')$ and an embedding $\Psi : \mathcal{B} \to \mathcal{A}'$ such that $\Psi(\mathcal{B}) \subset \Psi(\mathcal{C}) \vee \sigma(U')$ where U' is a random variable uniformly distributed on $[0, 1]$ and independent of $\Psi(\mathcal{C})$.

Proof. Obviously, (i) $\Longrightarrow$ (ii) is true. To prove (ii) $\Longrightarrow$ (i) we make the non restrictive assumption that W is valued in a Polish space. Then (ii) $\Longrightarrow$ (i) is obtained by putting $V = \mathcal{L}(W \mid \mathcal{B})$. Indeed, any bounded random variable X measurable with respect to $\mathcal{C} \vee \sigma(W)$ can be written $X = f(C, W)$ where C is a $\mathcal{C}$-measurable random variable and f is a bounded measurable function (Lemma 1.1); thus $\mathbb{E}[X \mid \mathcal{B}] = \int f(C, w) \mathcal{L}(W \mid \mathcal{B})(dw)$ is measurable with respect to $\mathcal{C} \vee \sigma(V)$, and hence $\mathcal{B} \subset \mathcal{C} \vee \sigma(V)$. To prove (i) $\Longrightarrow$ (iii) we make the non restrictive assumption that V is real-valued. Consider the product probability space of $(\Omega, \mathcal{A}, \mathbb{P})$ with a probability space on which is defined a random variable U^* having the uniform law on $[0, 1]$. Let $\mathcal{C}' = \iota_1(\mathcal{C})$ and $U' = \iota_2(U^*)$ respectively be the copies of $\mathcal{C}$ and U^* on the product probability space with the canonical embeddings $\iota_1 \colon \mathcal{C} \to \mathcal{C} \otimes \sigma(U^*)$ and $\iota_2 \colon \sigma(U^*) \to \mathcal{C} \otimes \sigma(U^*)$ (see Example 5.2). We introduce the right-continuous inverse $G(\cdot \mid \mathcal{C}')$ of the cumulative distribution function of the copy of the conditional law $\mathcal{L}(V \mid \mathcal{C})$ with the first embedding ι_1, and then we put $V' = G(U' \mid \mathcal{C}')$. Then V' is a well-defined random variable by virtue of Lemma 3.13, and, with the help of Lemma 5.8, it is easy to check that the conditional law $\mathcal{L}(V' \mid \mathcal{C}')$ is the copy of $\mathcal{L}(V \mid \mathcal{C})$ with ι_1, therefore the embedding Ψ is given by Corollary 5.12. Finally, (iii) $\Longrightarrow$ (i) is a consequence of (ii) $\Longrightarrow$ (i). $\qquad\square$

Corollary 3.28. *Let $\mathcal{F} = (\mathcal{F}_n)_{n \leq 0}$ be a filtration. The following conditions are equivalent:*

(i) $\mathcal{F}$ is locally separable.
(ii) For each $n \leq 0$, one has $\mathcal{F}_n \subset \mathcal{F}_{n-1} \vee \sigma(W_n)$ for some random variable W_n.
(iii) $\mathcal{F}$ admits a global parameterization.
(iv) $\mathcal{F}$ admits a global superinnovation.

Proof. This stems from Lemma 3.27 and Proposition 3.25. $\qquad\square$

3.5 I-cosiness

The *I-cosiness criterion* introduced in [14] was inspired from two sources: it is a variant of the notion of *cosiness* introduced by Tsirelson in [34] in the framework of continuous time, and the authors of [14] noticed that I-cosiness is used (but not named) in [32] to prove Result 2.43 about the split-word filtrations. In fact, as was pointed out in [3], "there is a whole range of possible variations" on the definition of cosiness introduced in [34], and the main underlying idea, due to Tsirelson, is what [21] calls a self-joining criterion, which comprises these possible variants of cosiness and in particular the I-cosiness criterion, as well as Rosenblatt's self-joining criterion and Vershik's self-joining criterion introduced in Sect. 2.2. Actually many elementary results we give on I-cosiness remain valid for any self-joining criterion, as defined in [21]. However we prefer not to introduce this notion here: there are already too many definitions.

It is shown in [14] that standardness and I-cosiness are equivalent properties for an essentially separable filtration. In the next section, we will give another proof of this fact and extend it to locally separable filtrations by showing that I-cosiness is equivalent to weak standardness.

In this section, we define I-cosiness and give more or less elementary results concerning it. We shall see that, for a filtration of product type and a filtration satisfying Vershik's first level criterion, I-cosiness is straightforward from Rosenblatt's self-joining criterion and Vershik's self-joining criterion respectively (Sect. 2.2). That a standard or a weakly standard filtration is I-cosy follows from the fact that I-cosiness, as standardness and weak standardness, is inherited by immersion. We end this section by giving an example of stationary Markov processes whose filtrations are I-cosy.

Definition 3.29. Let $\mathcal{F} = (\mathcal{F}_n)_{n \leqslant 0}$ be a filtration.

1. Let (E, ρ) be a Polish metric space and $X \in L^1(\mathcal{F}_0; E)$ We say that the random variable X is *I-cosy (with respect to $\mathcal{F}$)* if for each real number $\delta > 0$, there exist two filtrations $\mathcal{F}'$ and $\mathcal{F}''$ defined on a probability space $(\overline{\Omega}, \overline{\mathcal{A}}, \overline{\mathbb{P}})$ such that:

   (i) $(\mathcal{F}', \mathcal{F}'')$ is a joining of $\mathcal{F}$ independent in small time (Definition 2.29).
   (ii) $\overline{\mathbb{E}}[\rho(X', X'')] < \delta$, where X' and X'' are the respective copies of X in $\mathcal{F}'$ and in $\mathcal{F}''$.

2. We say that a σ-field $\mathcal{E}_0 \subset \mathcal{F}_0$ is *I-cosy (with respect to $\mathcal{F}$)* if every random variable $X \in L^1(\mathcal{E}_0)$ is I-cosy with respect to $\mathcal{F}$.
3. We say that the filtration $\mathcal{F}$ is *I-cosy* if the final σ-field $\mathcal{F}_0$ is I-cosy with respect to $\mathcal{F}$.

As with Vershik's first level criterion, we will sometimes omit to specify *with respect to $\mathcal{F}$* when no ambiguity is possible. We will see in Proposition 3.36 that I-cosiness of a random variable X is equivalent to the σ-field $\sigma(X)$ being I-cosy. It is clear that I-cosiness is preserved by isomorphism. Another obvious property of I-cosiness is hereditability by immersion, stated in the next lemma.

Lemma 3.30. *Let $\mathcal{F} = (\mathcal{F}_n)_{n \leqslant 0}$ be a filtration, E a Polish space, $X \in L^1(\mathcal{F}_0; E)$, and $\mathcal{E}$ a filtration immersed in $\mathcal{F}$. If X is I-cosy with respect to $\mathcal{F}$ and is $\mathcal{E}_0$-measurable, then X is I-cosy with respect to $\mathcal{E}$. Consequently, if $\mathcal{F}$ is I-cosy, so is also $\mathcal{E}$.*

Proof. This is a straightforward consequence of the definition of I-cosiness and the transitivity property of immersion. □

Proposition 3.31. *A filtration of product type is I-cosy. A filtration satisfying Vershik's first level criterion is I-cosy. More precisely, a random variable, or a σ-field, satisfying Vershik's first level criterion with respect to a filtration of local product type, is I-cosy with respect to this filtration.*

Proof. This follows from Rosenblatt's self-joining criterion (Proposition 2.33) and Vershik's self-joining criterion (Theorem 2.38). Indeed, each of these criteria is a particular case of the I-cosiness criterion. □

Corollary 3.32. *Any standard (Definition 3.6) or weakly standard (Definition 3.21) filtration is I-cosy.*

Proof. This follows from Lemma 3.30 and Proposition 3.31. □

Below we shall list some elementary properties of I-cosiness.

Lemma 3.33. *Let $\mathcal{F} = (\mathcal{F}_n)_{n \leqslant 0}$ be a filtration and E a Polish metric space. The random variables $X \in L^1(\mathcal{F}_0; E)$ which are I-cosy form a closed subset of $L^1(\mathcal{F}_0; E)$.*

Proof. Take $R \in L^1(\mathcal{F}_0; E)$ in the L^1-closure of the set of I-cosy random variables $X \in L^1(\mathcal{F}_0; E)$. Given $\delta > 0$, there exists an I-cosy $X \in L^1(\mathcal{F}_0; E)$ such that $\mathbb{E}[\rho(R, X)] < \delta/3$. By I-cosiness, there exists a joining $(\mathcal{F}', \mathcal{F}'')$ such that $\overline{\mathbb{E}}[\rho(X', X'')] < \delta/3$. By isomorphisms, we have $\overline{\mathbb{E}}[\rho(X', R')] = \overline{\mathbb{E}}[\rho(X'', R'')] < \delta/3$; so the triangular inequality gives $\overline{\mathbb{E}}[\rho(R', R'')] < \delta$. $\qquad\square$

Remark 3.34. If X and Y are two I-cosy random variables, it is not true in general that (X, Y) is also I-cosy. This is shown in [21] with the help of Theorem 3.9 and Vershik's standardness criterion.

The following lemma plays the same role for I-cosiness as Lemma 2.10 for Vershik's first level criterion. In the second condition, we consider $L(\mathcal{E}_0; F)$ as the space $L^1(\mathcal{E}_0; (F, \rho))$ where the metric ρ is the $0 - 1$ distance; so $\overline{\mathbb{E}}[\rho(X', X'')] = \overline{\mathbb{P}}[X' \neq X'']$.

Lemma 3.35. *Let $\mathcal{F} = (\mathcal{F}_n)_{n \leq 0}$ be a filtration on $(\Omega, \mathcal{A}, \mathbb{P})$ and $\mathcal{E}_0 \subset \mathcal{F}_0$ a σ-field. The following conditions are equivalent:*

(i) *The σ-field $\mathcal{E}_0$ is I-cosy with respect to $\mathcal{F}$.*
(ii) *For any finite set $F \subset \mathbb{R}$, every random variable $X \in L(\mathcal{E}_0; F)$ is I-cosy with respect to $\mathcal{F}$.*
(iii) *For any Polish space E, every random variable $X \in L^1(\mathcal{E}_0; E)$ is I-cosy with respect to $\mathcal{F}$.*

Proof. (iii) $\Longrightarrow$ (i) is trivial.

(i) $\Longrightarrow$ (ii): Fix F finite, $R \in L(\mathcal{E}_0; F)$, and $\delta > 0$. Let a be the minimum distance $|s - t|$ between two distinct elements s, t of F. Applying hypothesis (i), one obtains a joining $(\mathcal{F}', \mathcal{F}'')$ such that $\overline{\mathbb{E}}[|R' - R''|] < \epsilon\delta$ where $\epsilon = \min\{a, \delta\}$; hence we have $\overline{\mathbb{P}}(|R' - R''| \geq \epsilon) < \delta$ and therefore $\overline{\mathbb{P}}[R' \neq R''] < \delta$.

(ii) $\Longrightarrow$ (iii): Fix $X \in L^1(\mathcal{E}_0; (E, \rho))$ and $\delta > 0$. There exist some finite subset F of E and some $R \in L(\mathcal{E}_0; F)$ such that $\mathbb{E}[\rho(X, R)] < \delta/3$. Call d the diameter of F. Given a measurable injection $\phi: F \to \mathbb{R}$ and applying hypothesis (ii) to the random variable $\phi(R)$, one obtains a joining $(\mathcal{F}', \mathcal{F}'')$ such that $\overline{\mathbb{P}}[R' \neq R''] < \delta/(3d)$; so $\overline{\mathbb{E}}[\rho(R', R'')] < \delta/3$. Now the isomorphisms give $\overline{\mathbb{E}}[\rho(X', R')] = \overline{\mathbb{E}}[\rho(X'', R'')] < \delta/3$, wherefrom $\overline{\mathbb{E}}[\rho(X', X'')] < \delta$ by the triangular inequality. $\qquad\square$

Proposition 3.36. *Let $\mathcal{F} = (\mathcal{F}_n)_{n \leq 0}$ be a filtration. Let (E, ρ) be a Polish metric space and $X \in L^0(\mathcal{F}_0; E)$. The following conditions are equivalent.*

(i) *The σ-field $\sigma(X)$ is I-cosy.*
(ii) *For every $\delta > 0$, there exists, on some probability space $(\overline{\Omega}, \overline{\mathcal{A}}, \overline{\mathbb{P}})$, a joining $(\mathcal{F}', \mathcal{F}'')$ of $\mathcal{F}$ independent in small time such that $\overline{\mathbb{P}}[\rho(X', X'') > \delta] < \delta$, where X' and X'' are the respective copies of X in $\mathcal{F}'$ and in $\mathcal{F}''$.*

If $X \in L^1(\mathcal{F}_0; E)$, these conditions are also equivalent to:
(iii) *X is I-cosy.*

Proof. We first assume that $X \in L^1(\mathcal{F}_0; E)$ and show (i) $\Longleftrightarrow$ (iii). If the σ-field $\sigma(X)$ is I-cosy, then X is I-cosy by Lemma 3.35. Conversely, assume X to be I-cosy. We know from Lemma 2.15 that the set of random variables of the form $f(X)$ with $f: E \to \mathbb{R}$ Lipschitz, is a dense subset of $L^1(\sigma(X))$. It is easy to see that such a random variable $f(X)$ is I-cosy. Therefore the σ-field $\sigma(X)$ is I-cosy as a consequence of Lemma 3.33. The proof that (i) $\Longleftrightarrow$ (iii). It is then not hard to derive (i) $\Longleftrightarrow$ (ii) by replacing ρ with $\rho \wedge 1$.   □

We will use the following lemma to prove the asymptotic character of I-cosiness (Proposition 3.38) and to prove Proposition 3.43. This lemma involves I-cosiness of a filtration $(\mathcal{F}_n)_{n \leqslant N}$ with time-axis $-\mathbb{N} \cap]-\infty, N]$ for some integer $N \leqslant 0$, whereas I-cosiness is defined for a filtration indexed by $-\mathbb{N}$; but it is clear how to adapt the definition to this time-axis and obviously I-cosiness of $(\mathcal{F}_n)_{n \leqslant N}$ is equivalent to I-cosiness of $(\mathcal{F}_{N+n})_{n \leqslant 0}$.

Lemma 3.37. *On $(\Omega, \mathcal{A}, \mathbb{P})$, let $\mathcal{F} = (\mathcal{F}_n)_{n \leqslant 0}$ be a filtration and $(V_n)_{n \leqslant 0}$ a superinnovation of $\mathcal{F}$. Let $N < 0$ be an integer and $\mathcal{E}_N \subset \mathcal{F}_N$ a σ-field. If $\mathcal{E}_N$ is I-cosy with respect to the truncated filtration $(\mathcal{F}_n)_{n \leqslant N}$, then the σ-field $\mathcal{E}_0 := (\mathcal{E}_N \vee \sigma(V_{N+1}, \ldots, V_0)) \cap \mathcal{F}_0$ is I-cosy with respect to $\mathcal{F}$.*

Proof. We introduce the filtration $\mathcal{G} = (\mathcal{G}_n)_{n \leqslant 0}$ defined to be equal to $\mathcal{F}$ up to time N and for which $(V_{N+1}, \ldots, V_0)$ is an innovation from N to 0; precisely, we put

$$\mathcal{G}_n = \begin{cases} \mathcal{F}_n & \text{if } n \leqslant N; \\ \mathcal{F}_N \vee \sigma(V_{N+1}, \ldots, V_n) & \text{if } n \in \{N+1, \ldots, 0\}. \end{cases}$$

We can see by Lemma 1.6 that $\mathcal{F}$ is immersed in $\mathcal{G}$. Now we consider a random variable $X \in L^1(\mathcal{E}_0)$, and we shall see that X is I-cosy. Take $\delta > 0$ and take (with Lemma 1.1) a random variable $Y_N \in L^1(\mathcal{E}_N)$ such that $\sigma(X) \subset \sigma(Y_N, V_{N+1}, \ldots, V_0)$. We put $k = |N| + 1$ and equip $\mathbb{R}^k$ with the ℓ^1-metric. By Lemma 2.15, there exists a Lipschitz function $f: \mathbb{R}^k \to \mathbb{R}$ such that $\mathbb{E}[|X - R|] < \delta$ where $R = f(Y_N, V_{N+1}, \ldots, V_0)$. Let c be a Lipschitz constant for f. As Y_N is I-cosy with respect to the truncated filtration $(\mathcal{F}_n)_{n \leqslant N}$, there exist, on some probability space $(\overline{\Omega}, \overline{\mathcal{A}}, \overline{\mathbb{P}})$, two jointly immersed isomorphic copies $(\mathcal{F}'_n)_{n \leqslant N}$ and $(\mathcal{F}''_n)_{n \leqslant N}$ of $(\mathcal{F}_n)_{n \leqslant N}$ such that $\overline{\mathbb{E}}[|Y'_N - Y''_N|] < \delta/c$ where Y'_N and Y''_N are the respective copies of Y_N. We introduce the product probability space

$$(\widehat{\Omega}, \widehat{\mathcal{A}}, \widehat{\mathbb{P}}) = (\overline{\Omega}, \overline{\mathcal{A}}, \overline{\mathbb{P}}) \otimes (\Omega, \sigma(V_{N+1}, \ldots, V_0), \mathbb{P}),$$

and the canonical embeddings $\iota_1: \overline{\mathcal{A}} \to \widehat{\mathcal{A}}$ and $\iota_2: \sigma(V_{N+1}, \ldots, V_0) \to \widehat{\mathcal{A}}$ (see Example 5.2). We use a "hat" to identify random variables and σ-fields through these embeddings: for example we put $\widehat{X}_1 = \iota_1(X_1)$ and $\widehat{X}_2 = \iota_2(X_2)$ for any random variables X_1 and X_2 measurable with respect to $\overline{\mathcal{A}}$ and $\sigma(V_{N+1}, \ldots, V_0)$ respectively. Then, for each $n \leqslant N$ we define $\widehat{\mathcal{G}}'_n = \widehat{\mathcal{F}}'_n$ and $\widehat{\mathcal{G}}'_n = \widehat{\mathcal{F}}'_n$, and for each $n \in \{N+1, \ldots, 0\}$ we define the σ-fields

$$\widehat{\mathcal{G}}'_n = \widehat{\mathcal{F}}'_N \vee \sigma(\widehat{V}_{N+1}, \ldots, \widehat{V}_n) \quad \text{and} \quad \widehat{\mathcal{G}}''_n = \widehat{\mathcal{F}}''_N \vee \sigma(\widehat{V}_{N+1}, \ldots, \widehat{V}_n).$$

Using Lemma 2.31, it is a child's play to verify that $\widehat{\mathcal{G}}' := (\widehat{\mathcal{G}}'_n)_{n\leq 0}$ and $\widehat{\mathcal{G}}'' := (\widehat{\mathcal{G}}''_n)_{n\leq 0}$ are two jointly immersed isomorphic copies of $\mathcal{G}$. The respective copies of R are $\widehat{R}' = f(\widehat{Y}'_N, \widehat{V}_{N+1}, \ldots, \widehat{V}_0)$ and $\widehat{R}'' = f(\widehat{Y}''_N, \widehat{V}_{N+1}, \ldots, \widehat{V}_0)$. Thus we have $\widehat{\mathbb{E}}[|\widehat{R}' - \widehat{R}''|] \leq c\,\widehat{\mathbb{E}}[|\widehat{Y}'_N - \widehat{Y}''_N|]$ because f is c-Lipschitz. Due to isomorphism, we have $\widehat{\mathbb{E}}[|\widehat{Y}'_N - \widehat{Y}''_N|] = \overline{\mathbb{E}}[|Y'_N - Y''_N|]$ and $\widehat{\mathbb{E}}[|\widehat{X}' - \widehat{R}'|] = \widehat{\mathbb{E}}[|\widehat{X}'' - \widehat{R}''|] = \mathbb{E}[|X - R|]$ where $\widehat{X}'$ and $\widehat{X}''$ are the respective copies of X; consequently, $\widehat{\mathbb{E}}[|\widehat{X}' - \widehat{X}''|] < 3\delta$, thereby showing that X is I-cosy with respect to $\mathcal{G}$. As $\mathcal{F}$ is immersed in $\mathcal{G}$, we see that X is I-cosy with respect to $\mathcal{F}$ (Lemma 3.30). □

Proposition 3.38. *Let $\mathcal{F} = (\mathcal{F}_n)_{n\leq 0}$ be a locally separable filtration. The following conditions are equivalent:*

(i) *$\mathcal{F}$ is I-cosy.*
(ii) *For every $N \in -\mathbb{N}$, the truncated filtration $(\mathcal{F}_n)_{n\leq N}$ is I-cosy.*
(iii) *There exists $N \in -\mathbb{N}$ such that the truncated filtration $(\mathcal{F}_n)_{n\leq N}$ is I-cosy.*

Proof. It is easy to convince oneself that (i) $\Longrightarrow$ (ii). Obviously, (ii) $\Longrightarrow$ (iii) is true. We now show that (iii) $\Longrightarrow$ (i). Let $(V_n)_{n\leq 0}$ be a superinnovation of $\mathcal{F}$, whose existence is provided by Corollary 3.28. We assume that the truncated filtration $(\mathcal{F}_n)_{n\leq N}$ is I-cosy for some N. Then we know from Lemma 3.37 that $\mathcal{F}$ is I-cosy because of $\mathcal{F}_0 \subset \mathcal{F}_N \vee \sigma(V_{N+1}, \ldots, V_0)$. □

Remark 3.39. When $\mathcal{F}$ is essentially separable, a result in [12] states that Proposition 3.38 is more generally true for a truncation with an $\mathcal{F}$-stopping time N. We have not attempted to generalize this result to locally separable filtrations.

As another application of Lemma 3.37, we shall give a result on I-cosiness for the filtration generated by processes enjoying the properties of the following definition.

Definition 3.40. Let $(X_n)_{n\leq 0}$ be a process, and let $\phi: -\mathbb{N} \to -\mathbb{N}$ be a strictly increasing map with $\phi(0) = 0$.

1. We say that ϕ is a *sequence of memory-loss times of type I* for $(X_n)_{n\leq 0}$ if X_n is conditionally independent of $\sigma(X_m; m < n)$ given $(X_{\phi(k-1)}, \ldots, X_{n-1})$ for every $k, n \in -\mathbb{N}$ satisfying $\phi(k-1) < n \leq \phi(k)$.
2. Let $\mathcal{F}$ be the filtration generated by $(X_n)_{n\leq 0}$. We say that ϕ is a *sequence of memory-loss times of type II* for $(X_n)_{n\leq 0}$ if there exist a probability space $(\overline{\Omega}, \overline{\mathcal{A}}, \overline{\mathbb{P}})$, an embedding $\Psi: \mathcal{F}_0 \to \overline{\mathcal{A}}$, and a parameterization $(U'_n)_{n\leq 0}$ of the filtration $\mathcal{F}' := \Psi(\mathcal{F})$ such that

$$\Psi\big(\sigma(X_n)\big) \subset \Psi\big(\sigma(X_{\phi(k-1)})\big) \vee \sigma(U'_{\phi(k-1)+1}, \ldots, U'_n)$$

for every $k, n \in -\mathbb{N}$ satisfying $\phi(k-1) < n \leq \phi(k)$.

Obviously, a process is Markovian if and only if it admits the identity map $\phi: -\mathbb{N} \to -\mathbb{N}$ as a sequence of memory-loss times of type I.

Lemma 3.41. *Let* $(X_n)_{n \leq 0}$ *be a process, and let* $\phi: -\mathbb{N} \to -\mathbb{N}$ *be a strictly increasing map with* $\phi(0) = 0$. *If* ϕ *is a sequence of memory-loss times of type I for* $(X_n)_{n \leq 0}$, *then* ϕ *is a sequence of memory-loss times of type II for* $(X_n)_{n \leq 0}$.

In the particular case when $(X_n)_{n \leq 0}$ is a Markov process, this lemma shows that, up to isomorphism, there is a parameterization $(U_n)_{n \leq 0}$ of the filtration $\mathcal{F}$ generated by $(X_n)_{n \leq 0}$ for which $\sigma(X_n) \subset \sigma(X_{n-1}, U_n)$ for every $n \leq 0$. We shall use the following lemma in the proof of Lemma 3.41, which is copied from Lemma 2.22 in [19].

Lemma 3.42. *Let* $\Lambda = \{\Lambda_s\}_{s \in S}$ *be a probability kernel from a measurable space* S *to a Polish space* E. *Then there exists a measurable function* $(s, u) \mapsto \Delta_s(u)$ *from* $S \times [0, 1]$ *to* E *such that for all* $s \in S$, *the probability* Λ_s *is the image of the Lebesgue measure on* $[0, 1]$ *under the mapping* $\Delta_s: [0, 1] \to E$.

Proof of Lemma 3.41. We can assume that each X_n takes its values in a Polish space, by replacing, if needed, X_n with a real-valued random variable generating the same σ-field. Let $k, n \in -\mathbb{N}$ such that $n \in \{\phi(k-1) + 1, \ldots, \phi(k)\}$. By Lemma 3.42, there exists a measurable function $G_{k,n}(x_{\phi(k-1)}, \ldots, x_{n-1}, u)$ such that the function $u \mapsto G_{k,n}(X_{\phi(k-1)}, \ldots, X_{n-1}, u)$ almost surely carries the Lebesgue measure to the conditional law $\mathcal{L}(X_n \mid X_{\phi(k-1)}, \ldots, X_{n-1})$. Consider a process $(X'_n, U'_n)_{n \leq 0}$ defined by the following conditions:

- For each time $n \leq 0$, the random variable U_n is independent of the past and has the uniform law $[0, 1]$.
- For each $k \leq 0$, the random variable $X'_{\phi(k)}$ has the same law as $X_{\phi(k)}$.
- We have $X'_n = G_{k,n}(X'_{\phi(k-1)}, \ldots, X'_{n-1}, U'_n)$ for each $k, n \in -\mathbb{N}$ satisfying $\phi(k-1) < n \leq \phi(k)$.

Assuming that ϕ is a sequence of memory-loss times of type I for $(X_n)_{n \leq 0}$, it is easy to check that these two conditions uniquely define the law of $(X'_n, U'_n)_{n \leq 0}$, that $(X'_n)_{n \leq 0}$ is a copy of $(X_n)_{n \leq 0}$, and then that ϕ is a sequence of memory-loss times of type II for $(X_n)_{n \leq 0}$. $\square$

Proposition 3.43. *Let* $(X_n)_{n \leq 0}$ *be a process,* $\mathcal{F}$ *the filtration it generates, and* $\phi: -\mathbb{N} \to -\mathbb{N}$ *a strictly increasing map with* $\phi(0) = 0$, *assumed to be a sequence of memory-loss times of type II for* $(X_n)_{n \leq 0}$. *Then the following conditions are equivalent:*

(i) $\mathcal{F}$ *is I-cosy.*

(ii) *For each* $n \leq 0$, *the* σ-*field* $\sigma(X_{\phi(n)})$ *is I-cosy with respect to* $\mathcal{F}$.

(iii) *For each* $n \leq 0$, *the* σ-*field* $\sigma(X_{\phi(n)})$ *is I-cosy with respect to the truncated filtration* $(\mathcal{F}_m)_{m \leq \phi(n)}$.

Proof. One obviously has (i) $\implies$ (ii) $\implies$ (iii). Assuming that (iii) holds for some $n \leq 0$, then it is easy to show with the help of Lemma 3.37 that $\sigma(X_{\phi(n)}, X_{\phi(n)+1}, \ldots, X_0)$ is I-cosy with respect to $\mathcal{F}$ when ϕ is a sequence of memory-loss times of type II for $(X_n)_{n \leq 0}$. Hence, (iii) $\implies$ (i) follows from the L^1-closure of the set of I-cosy random variables (Lemma 3.33) and Lemma 2.12. $\square$

Remark 3.44. If $\mathcal{F}$ is the filtration generated by a martingale $(M_n)_{n \leq 0}$, it is possible to show that $\mathcal{F}$ is I-cosy if and only if the random variable M_0 is I-cosy. This is deduced from the same result stated in [21] for a general self-joining criterion (see also Remark 2.28).

Proposition 3.46 below will be used in the proof of Theorem 4.9. Its proof invokes the following lemma, copied verbatim from [14], to which we refer for a proof.

Lemma 3.45. *On $(\Omega, \mathcal{A}, \mathbb{P})$, let $\mathcal{F}, \mathcal{G}, \mathcal{H}, \mathcal{K}$ be four filtrations such that $\mathcal{F}$ is immersed in $\mathcal{H}$ and $\mathcal{G}$ is immersed in $\mathcal{K}$. If $\mathcal{H}$ and $\mathcal{K}$ are independent, then $\mathcal{F} \vee \mathcal{G}$ is immersed in $\mathcal{H} \vee \mathcal{K}$.*

Proposition 3.46. *Let $\mathcal{F} = (\mathcal{F}_n)_{n \leq 0}$ and $\mathcal{G} = (\mathcal{G}_n)_{n \leq 0}$ be two filtrations on some possibly different probability spaces. Let (E_1, ρ_1) and (E_2, ρ_2) be two Polish metric spaces, $X \in L^1(\mathcal{F}_0; E_1)$ and $Y \in L^1(\mathcal{G}_0; E_2)$. If X is I-cosy with respect to $\mathcal{F}$ and Y is I-cosy with respect to $\mathcal{G}$, then (X, Y) is I-cosy with respect to $\mathcal{F} \otimes \mathcal{G}$. As a consequence, the supremum of two independent filtrations is I-cosy if and only if each of these two filtrations is I-cosy.*

However, the supremum of two jointly immersed I-cosy filtrations is not I-cosy in general; this clearly results from Theorem 3.9.

Proof (Proof of Proposition 3.46). The proof of the consequence is left to the reader (the 'only if' part obviously follows from Lemma 1.2 and Lemma 3.30). Now we prove the first part of the proposition. Assume I-cosiness of both X and Y. The random pair (X, Y) takes its values in the Polish metric space $E_1 \times E_2$ equipped with the metric $\rho = \rho_1 + \rho_2$. Let $\delta > 0$. Let $(\mathcal{F}', \mathcal{F}'')$ be a joining of $\mathcal{F}$ on $(\overline{\Omega}, \overline{\mathcal{A}}, \overline{\mathbb{P}})$ independent in small time and such that $\overline{\mathbb{E}}[\rho_1(X', X'')] < \delta/2$, and let $(\widetilde{\mathcal{G}}', \widetilde{\mathcal{G}}'')$ be a joining of $\mathcal{G}$ on $(\widetilde{\Omega}, \widetilde{\mathcal{A}}, \widetilde{\mathbb{P}})$ independent in small time and such that $\widetilde{\mathbb{E}}[\rho_2(\widetilde{Y}', \widetilde{Y}'')] < \delta/2$. Let $\widehat{\mathcal{E}}'$ and $\widehat{\mathcal{E}}''$ be the filtrations defined on $(\widehat{\Omega}, \widehat{\mathcal{A}}, \widehat{\mathbb{P}}) := (\overline{\Omega}, \overline{\mathcal{A}}, \overline{\mathbb{P}}) \otimes (\widetilde{\Omega}, \widetilde{\mathcal{A}}, \widetilde{\mathbb{P}})$ by $\widehat{\mathcal{E}}'_n = \mathcal{F}'_n \otimes \widetilde{\mathcal{G}}'_n$ and $\widehat{\mathcal{E}}''_n = \mathcal{F}''_n \otimes \widetilde{\mathcal{G}}''_n$. It follows from Lemma 3.45 that $(\widehat{\mathcal{E}}', \widehat{\mathcal{E}}'')$ is a joining of $\mathcal{F} \otimes \mathcal{G}$, and clearly it is independent in small time. The copies $(\widehat{X}', \widehat{Y}')$ and $(\widehat{X}'', \widehat{Y}'')$ of (X, Y) in $(\widehat{\mathcal{E}}', \widehat{\mathcal{E}}'')$ satisfy $\widehat{\mathbb{E}}[\rho_1(\widehat{X}', \widehat{X}'')] = \overline{\mathbb{E}}[\rho_1(X', X'')]$ and $\widehat{\mathbb{E}}[\rho_2(\widehat{Y}', \widehat{Y}'')] = \widetilde{\mathbb{E}}[\rho_2(\widetilde{Y}', \widetilde{Y}'')]$, therefore $\widehat{\mathbb{E}}[\rho((\widehat{X}', \widehat{Y}'), (\widehat{X}'', \widehat{Y}''))] < \delta$. $\square$

Below is a corollary of Proposition 3.46. We commit a slight abuse of language in assertions (ii) and (iii). Both these assertions say that the σ-field $\mathcal{E}_0$ is I-cosy with respect to an extension of the filtration $\mathcal{F}$ (it is understood in assertion (ii) that the independent product of $\mathcal{F}$ with a filtration is an extension of $\mathcal{F}$, which is the content of Lemma 3.5). This more rigorously means that I-cosiness holds for the image of $\mathcal{E}_0$ under the underlying embedding from $\mathcal{F}$ to this extension of $\mathcal{F}$.

Corollary 3.47. *Let $\mathcal{F}$ be a locally separable filtration and $\mathcal{E}_0 \subset \mathcal{F}_0$ a σ-field. Then the following facts are equivalent.*

(i) $\mathcal{E}_0$ is I-cosy with respect to $\mathcal{F}$.
(ii) $\mathcal{E}_0$ is I-cosy with respect to the independent product of $\mathcal{F}$ with a standard conditionally non-atomic filtration.
(iii) $\mathcal{E}_0$ is I-cosy with respect to some conditionally non-atomic extension of $\mathcal{F}$.

Consequently, letting E be a Polish space and $X \in L^1(\mathcal{F}_0; E)$, the analogous three statements with X instead of $\mathcal{E}_0$ also are equivalent.

Proof. The consequence follows from Proposition 3.36. Inheritability of I-cosiness (Lemma 3.30) gives (iii) $\implies$ (i). Let $\mathcal{G}$ be a standard conditionally non-atomic filtration. By Lemma 3.22, the product filtration $\mathcal{F} \otimes \mathcal{G}$ is conditionally non-atomic, which shows that (ii) $\implies$ (iii). It remains to show that (i) $\implies$ (ii). Assume $\mathcal{E}_0$ is I-cosy with respect to $\mathcal{F}$. As $\mathcal{G}$ is I-cosy (Proposition 3.31), the σ-field $\mathcal{E}_0 \otimes \mathcal{G}_0$ is I-cosy with respect to $\mathcal{F} \otimes \mathcal{G}$ by Proposition 3.46. In particular, the image of $\mathcal{E}_0$ under the identification with the first factor is I-cosy with respect to $\mathcal{F} \otimes \mathcal{G}$. □

Remark 3.48. A stronger result is derived in [21] from the equivalence between I-cosiness and Vershik's standardness criterion: *in the same context of the above corollary, if $\mathcal{G}$ is a locally separable extension of $\mathcal{F}$, then $\mathcal{E}_0$ is I-cosy with respect to $\mathcal{F}$ if and only if $\mathcal{E}_0$ is I-cosy with respect to $\mathcal{G}$.* We do not know how to prove this result without using Vershik's standardness criterion.

3.5.1 Example: Random Dynamical Systems

We will give in Theorem 3.53 a sufficient condition for a stationary Markov process $(X_n)_{n \leq 0}$ to generate an I-cosy filtration. In the proof we give, we will firstly argument that, due to Proposition 3.43 and stationarity, it suffices to show I-cosiness of X_0. Theorem 3.53 firstly requires the stationary process to be *couplable* (Definition 3.49). Assuming this condition will allow us to construct some copies $(X'_n)_{n \leq 0}$ and $(X''_n)_{n \leq 0}$ of $(X_n)_{n \leq 0}$ generating jointly immersed filtrations and independent up to some time n_0 sufficiently small for the random variables X'_T and X''_T to be arbitrarily close with high probability for some random time $T \in \{n_0 + 1, \dots, 0\}$. The second condition is the *stochastic self-contractivity* (Definition 3.52) of the Markovian kernel. This condition will allow us to maintain the distance between X'_n and X''_n for n going from T to 0.

Definition 3.49. Let (E, ρ) be a Polish metric space. Let $(X_n)_{n \in \mathbb{Z}}$ be a stationary Markov process in E. This process is *couplable* if for every $\delta > 0$, there exist, on some probability space $(\overline{\Omega}, \overline{\mathcal{A}}, \overline{\mathbb{P}})$, two jointly immersed copies $(\mathcal{F}'_n)_{n \geq 0}$ and

$(\mathcal{F}_n'')_{n\geq 0}$ of the filtration generated by $(X_n)_{n\geq 0}$ such that $\mathcal{F}_0'$ is independent of $\mathcal{F}_0''$ and the stopping time $\inf\{n \geq 0 \mid \rho(X_n', X_n'') < \delta\}$ is almost surely finite.

Example 3.50 (Stationary Markov chain on a denumerable space). Let $(X_n)_{n\in\mathbb{Z}}$ be a stationary Markov process on a denumerable state space, equipped with the $0-1$ distance. Considering two independent copies $(X_n')_{n\geq 0}$ and $(X_n'')_{n\geq 0}$ of $(X_n)_{n\geq 0}$, it is known that the so-called product chain $(X_n', X_n'')_{n\geq 0}$ is recurrent under the assumption that the Markov kernel of $(X_n)_{n\geq 0}$ is *positive recurrent, irreducible, and aperiodic* (see [33]). In particular the product chain almost surely visits the diagonal, and hence $(X_n)_{n\in\mathbb{Z}}$ is couplable under this assumption. In fact, we can see that if $(X_n)_{n\in\mathbb{Z}}$ generates a Kolmogorovian filtration, then the Markov kernel is positive recurrent, irreducible, and aperiodic. Indeed, it is known (see [19]) that an irreducible Markov kernel is positive recurrent whenever it admits an invariant probability measure, and it is not difficult to check that the Markov kernel is irreducible and aperiodic if $(X_n)_{n\in\mathbb{Z}}$ generates a Kolmogorovian filtration.

Example 3.51 (Random walk on the circle). Let $(X_n)_{n\in\mathbb{Z}}$ be the stationary Markov process on the one-dimensional torus $\mathbb{R}/\mathbb{Z}$ defined as follows. For a given $\alpha \in \mathbb{R}/\mathbb{Z}$:

- X_n has the uniform distribution on the one-dimensional torus $\mathbb{R}/\mathbb{Z}$.
- Given $X_n = x$, the random variable X_{n+1} takes as possible values $x \pm \alpha$ with equal probability.

When the "step" α is irrational, then the set $\{x + m\alpha \in \mathbb{R}/\mathbb{Z} \mid m \in \mathbb{N}\}$ is a dense subset of the circle $\mathbb{R}/\mathbb{Z}$. Thus, considering two independent copies $(X_n')_{n\geq 0}$ and $(X_n'')_{n\geq 0}$ of $(X_n)_{n\geq 0}$, the product chain $(X_n', X_n'')_{n\geq 0}$ almost surely visits any open set because of the property of recurrence of a random walk on $\mathbb{Z}^2$. Hence $(X_n)_{n\in\mathbb{Z}}$ is couplable when the step is irrational.

Before defining stochastic self-contractivity, we need to introduce the following decomposition of probability kernels, which is usual in the theory of random dynamical systems. Let $(X_n)_{n\in\mathbb{Z}}$ be a stationary Markov process in a Polish space E, with transition kernel $\{P_x\}$. According to Lemma 3.42, it is always possible to write the kernel as

$$P_x(f) = \int f \circ \nabla_v(x) d\gamma(v) \tag{1}$$

where γ is a probability measure on some Polish space and $(x, v) \mapsto \nabla_v(x)$ is measurable. Given a kernel written in form (1), we can consider a stationary Markov processes $(X_n', V_n')_{n\leq 0}$ with probability transition kernel $\{Q_{x,v}\}$ defined by $Q_{x,v}(h) = \int h(\nabla_t(x), t) d\gamma(t)$ and instantaneous law ν defined by $\nu(h) = \int\int h(\nabla_t(x), t) d\gamma(t)d\mu(x)$ where μ is the instantaneous law of $(X_n)_{n\leq 0}$. Thus $(X_n')_{n\leq 0}$ has the same law as $(X_n)_{n\leq 0}$, the random variable V_n' is distributed according to γ for every n, one has $X_{n+1}' = \nabla_{V_{n+1}'}(X_n')$ and the process $(V_n')_{n\leq 0}$ is a superinnovation of the filtration $\mathcal{F}'$ generated by $(X_n')_{n\leq 0}$ (this is actually our proof of Lemma 3.41 in the particular case where ϕ is the identity map).

Definition 3.52. Let (E, ρ) be a Polish metric space. Let $X = (X_n)_{n \in \mathbb{Z}}$ be a stationary Markov process in E with instantaneous distribution μ. If the transition kernel $\{P_x\}$ can be written as in (1) with the property that there exists a probability kernel $\{\Lambda_{x,y}\}$ such that $\Lambda_{x,y}$ is a joining of γ (i.e. the margins are both γ) for all (x, y) and

$$\int \rho\left(\nabla_v(x), \nabla_{v'}(y)\right) \mathrm{d}\Lambda_{x,y}(v, v') \leqslant \rho(x, y), \tag{2}$$

then X is said to be *stochastically self-contractive*.

Example 3.50 (continued). Let $(X_n)_{n \in \mathbb{Z}}$ be a stationary Markov process on a denumerable state space. Given any decomposition (1), and setting $\Lambda_{x,y}$ to be the joining of γ supported by the diagonal for any x and y, then we see that $(X_n)_{n \in \mathbb{Z}}$ is stochastically self-contractive when we consider the discrete $0 - 1$ distance on the denumerable state space.

Example 3.51 (continued). Let $(X_n)_{n \in \mathbb{Z}}$ be the random walk on the circle. The natural decomposition (1) consists in taking $\nabla_v(x) = x + v\alpha$ and γ the law of $\varepsilon_n := \mathbb{1}_{\{X_n = X_{n-1} + \alpha\}}$. Setting $\Lambda_{x,y}$ to be the joining of γ supported by the diagonal of $\{0, 1\}^2$ for any x and y, we see that $(X_n)_{n \in \mathbb{Z}}$ is stochastically self-contractive.

Theorem 3.53. *Let $(X_n)_{n \in \mathbb{Z}}$ be a stationary Markov process in a Polish bounded metric space (E, ρ). If $(X_n)_{n \in \mathbb{Z}}$ is couplable and stochastically self-contractive, then the filtration generated by $(X_n)_{n \leqslant 0}$ is I-cosy.*

Thus, in view of Examples 3.50 and 3.51, we know from this theorem that a Markov chain on a denumerable state space generates an I-cosy filtration whenever this filtration is Kolmogorovian, and we know that the random walk on the circle with an irrational step generates an I-cosy filtration (we still give a remark on this example at the end of this section).

Lemma 3.54. *Let $(X_n, V_n)_{n \leqslant 0}$ be a Markov process such that each X_n and each V_n takes its values in a Polish space and such that the process $(V_n)_{n \leqslant 0}$ is a superinnovation (Definition 3.10) of the filtration $\mathcal{F}$ generated by $(X_n)_{n \leqslant 0}$ satisfying in addition $\sigma(X_{n+1}) \subset \sigma(X_n) \vee \sigma(V_{n+1})$ for every $n < 0$. We consider a measurable function f_n such that $X_{n+1} = f_n(X_n, V_{n+1})$ for every $n < 0$.*

On $(\overline{\Omega}, \overline{\mathcal{A}}, \overline{\mathbb{P}})$, let $\mathcal{F}'$ be a copy of $\mathcal{F}$ and $(V_n'')_{n \leqslant 0}$ a sequence of independent random variables having the same law as $(V_n)_{n \leqslant 0}$. We suppose that the filtration generated by $(V_n'')_{n \leqslant 0}$ and the filtration $\mathcal{F}'$ are jointly immersed in some filtration $\overline{\mathcal{H}}$.

Let T be a $\overline{\mathcal{H}}$-stopping time in $-\mathbb{N} \cup \{+\infty\}$. Let $(X_n'')_{n \leqslant 0}$ be the process defined by

$$\begin{cases} X_n'' = X_n' & \text{if } n \leqslant T; \\ X_{n+1}'' = f_n(X_n'', V_{n+1}'') & \text{for } n \text{ from } T \text{ to } -1. \end{cases}$$

Then $(X_n'')_{n \leqslant 0}$ is a copy of $(X_n)_{n \leqslant 0}$ and the filtration it generates is immersed in $\overline{\mathcal{H}}$.

Proof. Let $n \in -\mathbb{N}^*$. We denote by $\{P_x^n\}$ a regular version of the conditional law of X_{n+1} given X_n. Let g be a bounded Borelian function. One easily checks that

$$\mathbb{1}_{T>n}\,\overline{\mathbb{E}}\left[g(X''_{n+1})\,|\,\overline{\mathcal{H}}_n\right] = \mathbb{1}_{T>n}\,P^n_{X''_n}(g).$$

On the other hand, as V''_{n+1} is independent of $\overline{\mathcal{H}}_n$, we have

$$\mathbb{1}_{T\leqslant n}\,\overline{\mathbb{E}}\left[g(X''_{n+1})\,|\,\overline{\mathcal{H}}_n\right] = \mathbb{1}_{T\leqslant n}\,\overline{\mathbb{E}}\left[g\circ f_n(X''_n, V''_{n+1})\,|\,\overline{\mathcal{H}}_n\right] = \mathbb{1}_{T\leqslant n}\,P^n_{X''_n}(g).$$

Hence, the process $(X''_n)_{n\leqslant 0}$ is Markovian with respect to $\overline{\mathcal{H}}$ and has the same Markov kernels $\{P^n_x\}$ as the Markov process $(X_n)_{n\leqslant 0}$.

It remains to check that X''_n has the same law as X_n for all $n \leqslant 0$. For $m \leqslant n$, one has

$$\overline{\mathbb{E}}\left[g(X''_n)\,\mathbb{1}_{T=m}\,|\,\overline{\mathcal{H}}_m\right] = \mathbb{1}_{T=m}\,\overline{\mathbb{E}}\left[h(X'_m, V''_{m+1}, \ldots, V''_n)\,|\,\overline{\mathcal{H}}_m\right]$$

where

$$h(\cdot, v_{m+1}, \ldots, v_n) = g\circ f_{n-1}(\cdot, v_n)\circ\cdots\circ f_{m+1}(\cdot, v_{m+2})\circ f_m(\cdot, v_{m+1}).$$

But, because $(V''_{m+1}, \ldots, V''_n)$ is independent of $\overline{\mathcal{H}}_m$, we see that

$$\overline{\mathbb{E}}\left[h(X'_m, V''_{m+1}, \ldots, V''_n)\,|\,\overline{\mathcal{H}}_m\right] = Q_{X'_m}(g),$$

where $\{Q_x\}$ a regular version of the conditional law of X_n given X_m, and thus we have

$$\overline{\mathbb{E}}\left[h(X'_m, V''_{m+1}, \ldots, V''_n)\,|\,\overline{\mathcal{H}}_m\right] = \overline{\mathbb{E}}\left[g(X'_n)\,|\,\overline{\mathcal{H}}_m\right].$$

Hence we obtain $\overline{\mathbb{E}}\left[g(X''_n)\,\mathbb{1}_{T=m}\right] = \overline{\mathbb{E}}\left[g(X'_n)\,\mathbb{1}_{T=m}\right]$. As we obviously have $\overline{\mathbb{E}}\left[g(X''_n)\,\mathbb{1}_{T\geqslant n}\right] = \overline{\mathbb{E}}\left[g(X'_n)\,\mathbb{1}_{T\geqslant n}\right]$, we finally obtain $\overline{\mathbb{E}}\left[g(X''_n)\right] = \overline{\mathbb{E}}\left[g(X'_n)\right]$. $\square$

Proof of Theorem 3.53. Let $\mathcal{F}$ be the filtration generated by the stationary Markov process $(X_n)_{n\leqslant 0}$. To show that $\mathcal{F}$ is I-cosy, it suffices, thanks to Proposition 3.43, to prove that for each $n \leqslant 0$, the σ-field $\sigma(X_n)$, or equivalently (Proposition 3.36) the random variable X_n, is I-cosy with respect to the truncated filtration $(\mathcal{F}_m)_{m\leqslant n}$. We prove this for $n = 0$ only as our construction will obviously adapt to an arbitrary n due to stationarity.

Set $\delta > 0$ and define $\epsilon = \delta/\operatorname{diam}(E)$. As we assume that the stationary Markov process is couplable (Definition 3.49), it is possible to find some n_0 small enough and a probability space $(\overline{\Omega}, \overline{\mathcal{A}}, \overline{\mathbb{P}})$ with a joining $(\mathcal{F}', \mathcal{F}'')$ of $\mathcal{F}$ independent up to n_0 such that $\overline{\mathbb{P}}\left[\overline{T} < +\infty\right] > 1 - \epsilon$ where $\overline{T}$ is defined by

$$\overline{T} = \begin{cases} \inf\{n \mid n_0 \leqslant n \leqslant 0, \rho(X'_n, X''_n) < \delta\} & \text{if this infimum exists;} \\ +\infty & \text{otherwise.} \end{cases}$$

By replacing $(\overline{\Omega}, \overline{\mathcal{A}}, \overline{\mathbb{P}})$ with its independent product with a sufficiently rich probability space, we can assume that we have a sequence $(\widehat{U}_n)_{n \leq 0}$ of independent random variables $\widehat{U}_n$ each uniformly distributed on $[0, 1]$ and that is independent of $\mathcal{F}'_0 \vee \mathcal{F}''_0$. We denote by $\overline{\mathcal{H}}$ the supremum of the filtration $\mathcal{F}' \vee \mathcal{F}''$ with the filtration generated by $(\widehat{U}_n)_{n \leq 0}$.

For the sake of convenience, we assume that γ in the decomposition (1) given by the stochastic self-contractivity assumption (2), is the Lebesgue measure on $[0, 1]$. Moreover we write $f(x, u)$ instead of $\nabla_u(x)$. Let $\Lambda_{x,y}$ be given by the stochastic self-contractivity assumption, and $\Delta_{x,y}$ by Lemma 3.42 applied with $\Lambda_{x,y}$. We define the processes $(\widehat{X}'_n)_{n \leq 0}$, $(\widehat{X}''_n)_{n \leq 0}$, $(\widetilde{U}'_n)_{n \leq 0}$ and $(\widetilde{U}''_n)_{n \leq 0}$ by letting, for $n \leq T$,

$$\begin{cases} (\widetilde{U}'_n, \widetilde{U}''_n) = \Delta_{X'_{n-1}, X''_{n-1}}(\widehat{U}_n), \\ \widehat{X}'_n = X'_n, \\ \widehat{X}''_n = X''_n, \end{cases}$$

and, for n from T to -1,

$$\begin{cases} (\widetilde{U}'_{n+1}, \widetilde{U}''_{n+1}) = \Delta_{\widehat{X}'_n, \widehat{X}''_n}(\widehat{U}_{n+1}), \\ \widehat{X}'_{n+1} = f(\widehat{X}'_n, \widetilde{U}'_{n+1}), \\ \widehat{X}''_{n+1} = f(\widehat{X}''_n, \widetilde{U}''_{n+1}). \end{cases}$$

Clearly, each of $\widetilde{U}'_{n+1}$ and $\widetilde{U}''_{n+1}$ is independent of $\overline{\mathcal{H}}_n$, hence each of the filtrations generated by $(\widetilde{U}'_n)_{n \leq 0}$ and $(\widetilde{U}''_n)_{n \leq 0}$ is immersed in $\overline{\mathcal{H}}$ (Lemma 1.6). Therefore, according to Lemma 3.54, the processes $(\widehat{X}'_n)_{n \leq 0}$ and $(\widehat{X}''_n)_{n \leq 0}$ are two copies of $(X_n)_{n \leq 0}$ and generate jointly immersed isomorphic filtrations, and we know that they are independent up to n_0.

By construction, due to the stochastic self-contractivity (2), we have

$$\overline{\mathbb{E}}\big[\rho(\widehat{X}'_0, \widehat{X}''_0) \mid \mathcal{F}'_n \vee \mathcal{F}''_n \big] \mathbb{1}_{\overline{T}=n} \leq \rho(X'_n, X''_n) \mathbb{1}_{\overline{T}=n}.$$

Hence we obtain

$$\overline{\mathbb{E}}\big[\rho(\widehat{X}'_0, \widehat{X}''_0) \mid (\mathcal{F}' \vee \mathcal{F}'')_T \big] \mathbb{1}_{\overline{T} \neq +\infty} \leq \rho(X'_T, X''_T) \mathbb{1}_{\overline{T} \neq +\infty} \leq \delta,$$

and consequently we have $\overline{\mathbb{E}}\big[\rho(\widehat{X}'_0, \widehat{X}''_0) \mathbb{1}_{\overline{T} \neq +\infty}\big] \leq \delta$. As we have in addition $\overline{\mathbb{E}}\big[\rho(\widehat{X}'_0, \widehat{X}''_0) \mathbb{1}_{\overline{T}=+\infty}\big] \leq \delta$, we finally obtain $\overline{\mathbb{E}}\big[\rho(\widehat{X}'_0, \widehat{X}''_0)\big] \leq 2\delta$; so X_0 is I-cosy. $\square$

Remark on Example 3.51. Let $(X_n)_{n \leq 0}$ be the stationary random walk on the circle with an irrational step. We have seen that Theorem 3.53 applies and thus we know that the filtration $\mathcal{F}$ generated by $(X_n)_{n \leq 0}$ is I-cosy. Note that this filtration is of local product type: for each $n \leq 0$, the random variable $\varepsilon_n := \mathbb{1}_{\{X_n = X_{n-1} + \alpha\}}$ is an independent complement of $\mathcal{F}_{n-1}$ in $\mathcal{F}_n$. As ε_n takes two possible values with equal probability, $\mathcal{F}$ is a *dyadic* filtration, a particular case of homogeneous filtrations (defined in the introduction and in Definition 4.1). Therefore, according to Theorem A stated in the introduction (or to Corollary 4.5), $\mathcal{F}$ is actually a filtration of product type. A generating innovation for this filtration is constructed in [22].

3.6 Example Continued: Split-Word Processes

We discuss about Result 2.43 and we show how to deduce part (a) of Theorem 2.39 assuming this result. In fact, in references [20] and [7], Result 2.43 is deduced from the following result:

Result 2.43'. *If A is finite, then $\mathcal{F}$ is not I-cosy under (Δ), unless μ is degenerate.*

Then Result 2.43 follows from the fact that any filtration of product type is I-cosy (Proposition 3.31). Now we shall prove part (a) of Theorem 2.39 assuming result 2.43'. Consider an alphabet $(A, \mathfrak{A}, \mu)$ containing at least two letters a and b. Let $f : A \to \{a, b\}$ be any measurable function such that $f(\mu)$ assigns positive measure to each of a and b. For a given splitting sequence, consider the split-word process $(X_n, \varepsilon_n)_{n \leq 0}$ on A and define the process $(f(X_n), \varepsilon_n)_{n \leq 0}$. This latter is the split-word process on the alphabet $\{a, b\}$ with the same splitting sequence, and we know that its generated filtration is not I-cosy under condition (Δ) on this splitting sequence. By Lemma 1.6, we can see that the filtration generated by $(f(X_n), \varepsilon_n)_{n \leq 0}$ is immersed in the one generated by $(X_n, \varepsilon_n)_{n \leq 0}$, and thus $(X_n, \varepsilon_n)_{n \leq 0}$ itself does not generate an I-cosy filtration under (Δ), due to inheritance of I-cosiness by immersion (Lemma 3.30).

4 Theorems

In this section, we restate and prove the theorems stated in the introduction, and we return to the example of split-word processes in order to finish the proof of Theorem 2.39. Let us first recall the notion of homogeneous filtrations given in the introduction.

Definition 4.1. A filtration $\mathcal{F} = (\mathcal{F}_n)_{n \leq 0}$ is *homogeneous* if there exists an innovation $(V_n)_{n \leq 0}$ of $\mathcal{F}$ such that for each $n \leq 0$, V_n either has a diffuse law or is uniformly distributed on some finite set.

Thus, any homogeneous filtration is of local product type (Definition 2.3), and a conditionally non-atomic filtration (Definition 3.1) is a particular homogeneous filtration. But note that no homogeneity in time is required: some V_n may be diffuse, others may take two values, others three values, etc.

Theorem 4.4 states the equivalence, separately for each random variable, between I-cosiness and Vershik's first level criterion for homogeneous filtrations. As an intermediate step we use Vershik's self-joining criterion, which has been shown to be equivalent to Vershik's first level criterion in Theorem 2.38. Under the context of essentially separable filtrations, we have seen that Vershik's first level criterion is equivalent to productness (Theorem 2.25), thus Theorem A stated in the introduction follows as a consequence of our Theorem 4.4. Our generalization to locally separable filtrations has no practical interest; however, even in the essentially separable case, Theorem 4.4 is more precise than Theorem A: it asserts that, for a homogeneous filtration, I-cosiness and Vershik's first level criterion are equivalent for a random variable, not only for the whole filtration. For a locally separable filtration, Corollary 4.8 states that I-cosiness for a random variable is equivalent to Vershik's first level criterion in a conditionally non-atomic extension of the filtration; this result is still interesting when restricted to the context of essentially separable filtrations. Theorem 4.9 states the equivalence between I-cosiness and standardness or weak standardness, according as we consider essentially separable filtrations or locally separable filtrations.

4.1 Theorems

The key theorem is Theorem 4.4. All other theorems stated in the introduction will easily derive therefrom. The key step in the proof is to consider Vershik's self-joining criterion (Definition 2.37 and Theorem 2.38) as an intermediate step between I-cosiness and Vershik's first level criterion.

The next two lemmas involve joinings $(\mathcal{F}', \mathcal{F}'')$ of $\mathcal{F}$ *permutational after an integer* (Definition 2.35), as those appearing in Vershik's self-joining criterion. We recall the picture to be kept in mind for such joinings:

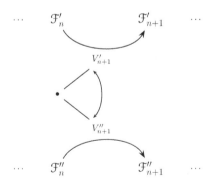

Lemma 4.2. *Let* $\mathcal{F} = (\mathcal{F}_n)_{n \leq 0}$ *be a filtration of local product type and* $(V_n)_{n \leq 0}$ *an innovation of* $\mathcal{F}$*. We assume that for some integer* $n_0 \leq 0$*, each innovation* V_n

is uniformly distributed on a finite set for all $n \in \{n_0 + 1, \ldots, 0\}$. Let (E, ρ) be a Polish metric space and $X \in L^1(\mathcal{F}_0; E)$, and let $(\mathcal{F}', \mathcal{F}'')$ be a joining of $\mathcal{F}$ on a probability space $(\overline{\Omega}, \overline{\mathcal{A}}, \overline{\mathbb{P}})$.

Then there exists a filtration $\mathcal{F}'''$ on $(\overline{\Omega}, \overline{\mathcal{A}}, \overline{\mathbb{P}})$ such that $(\mathcal{F}', \mathcal{F}''')$ is a joining of $\mathcal{F}$ permutational after n_0, and such that we have $\overline{\mathbb{E}}[\rho(X', X''')] \leq \overline{\mathbb{E}}[\rho(X', X'')]$ where X', X'' and X''' are the respective copies of X in $\mathcal{F}'$, $\mathcal{F}''$ and $\mathcal{F}'''$.

Proof. By induction on n, it suffices to show the lemma when assuming that $(\mathcal{F}', \mathcal{F}'')$ is a joining of $\mathcal{F}$ permutational after $n_0 + 1$ for an integer $n_0 < 0$. Let $(V_n')_{n \leq 0}$ and $(V_n'')_{n \leq 0}$ be the respective copies of $(V_n)_{n \leq 0}$ in $\mathcal{F}'$ and $\mathcal{F}''$. Thus we are assuming that $(V_{n_0+2}', \ldots, V_0')$ and $(V_{n_0+2}'', \ldots, V_0'')$ are such that $V_n'' = T_n(V_n')$ for $n \in \{n_0 + 2, \ldots, 0\}$ where the T_n are the $(\mathcal{F}' \vee \mathcal{F}'')_{n-1}$-measurable random transformations given in Definition 2.35.

We shall firstly write the random transformations T_n and the random variables X' and X'' in a convenient form in order to express the conditional expectation $\overline{\mathbb{E}}[\rho(X', X'') \mid \mathcal{F}_{n_0}' \vee \mathcal{F}_{n_0}'']$ as a linear function of the conditional law $\mathcal{L}(V_{n_0+1}', V_{n_0+1}'' \mid \mathcal{F}_{n_0}' \vee \mathcal{F}_{n_0}'')$. By Lemma 2.36, there are two $\mathcal{F}_{n_0+1}$-measurable random variables C_{n_0+1} and D_{n_0+1} such that

$$T_n(\cdot) = \psi_0^n(C_{n_0+1}', D_{n_0+1}'', V_{n_0+2}', \ldots, V_{n-1}', \cdot)$$

for every $n \in \{n_0 + 2, \ldots, 0\}$, where the ψ_0^n are measurable functions. As we have

$$(\mathcal{F}' \vee \mathcal{F}'')_{n_0+1} = (\mathcal{F}' \vee \mathcal{F}'')_{n_0} \vee \sigma(V_{n_0+1}', V_{n_0+1}''),$$

we can take (Lemma 1.1) an $(\mathcal{F}_{n_0}' \vee \mathcal{F}_{n_0}'')$-measurable random variable $\bar{H}_{n_0}$ such that $\sigma(C_{n_0+1}', D_{n_0+1}'') \subset \sigma(\bar{H}_{n_0}, V_{n_0+1}', V_{n_0+1}'')$. On the other hand, we have $\sigma(X) \subset \sigma(X_{n_0}, V_{n_0+1}, \ldots, V_0)$ for some $\mathcal{F}_{n_0}$-measurable random variable X_{n_0} (Lemma 1.1). Finally we take a random variable $\bar{Z}_{n_0}$ such that $\sigma(\bar{Z}_{n_0}) = \sigma(\bar{H}_{n_0}, X_{n_0}', X_{n_0}'')$. Thus, we can write

$$T_n(\cdot) = \psi^n(\bar{Z}_{n_0}, V_{n_0+1}', V_{n_0+1}'', V_{n_0+2}', \ldots, V_{n-1}', \cdot) \tag{3}$$

for each $n \in \{n_0 + 2, \ldots, 0\}$, where ψ^n is measurable, and we can write

$$X' = f(\bar{Z}_{n_0}, V_{n_0+1}', \ldots, V_0')$$

and

$$\begin{aligned} X'' &= f(\bar{Z}_{n_0}, V_{n_0+1}'', \ldots, V_0'') \\ &= f\left(\bar{Z}_{n_0}, V_{n_0+1}'', T_{n_0+2}(V_{n_0+2}'), \ldots, T_0(V_0')\right), \end{aligned}$$

for some Borelian function f. We denote by h the function such that

$$X'' = h(\bar{Z}_{n_0}, V_{n_0+1}', V_{n_0+1}'', V_{n_0+2}', \ldots, V_0') \tag{4}$$

which is obtained by combining the later equality with (3).

Hence, we have

$$\overline{\mathbb{E}}\big[\rho(X',X'') \mid \mathcal{F}'_{n_0} \vee \mathcal{F}''_{n_0}\big] = L_{\tilde{Z}_{n_0}}\big(\mathcal{L}(V'_{n_0+1}, V''_{n_0+1} \mid \mathcal{F}'_{n_0} \vee \mathcal{F}''_{n_0})\big)$$

where, letting F be the state space of V_{n_0+1} and ν a probability on $F \times F$,

$$L_z(\nu) = \int \overline{\mathbb{E}}\big[g_z(v', v'', V'_{n_0+2}, \dots, V'_0)\big]\, d\nu(v', v''),$$

where

$$\begin{aligned}
g_z&(v', v'', V'_{n_0+2}, \dots, V'_0) \\
&= \rho\big(f(z, v', V'_{n_0+2}, \dots, V'_0), h(z, v', v'', V'_{n_0+2}, \dots, V'_0)\big).
\end{aligned}$$

The set of all probability measures on $F \times F$ having both margins equal to the uniform probability measure on the finite set F is convex; by Birkhoff–Von Neumann's theorem (see [1]), its extreme points are the probability measures supported on graphs of permutations of F. The map $\nu \mapsto L_z(\nu)$ is linear on this convex set, thus it reaches its minimum at such an extremal probability measure. For each z, we measurably select a permutation ϕ_z such that L_z attains its minimum at the measure supported by the graph of ϕ_z.

Then we define $V'''_{n_0+1} = \phi_{\tilde{Z}_{n_0}}(V'_{n_0+1})$, and $V'''_n = T'_n(V'_n)$ for $n \in \{n_0 + 2, \dots, 0\}$ where T'_n is obtained by replacing V''_{n_0+1} with V'''_{n_0+1} in the expression (3) of T_n; that is, we put

$$V'''_n = \psi^n_{\tilde{Z}_{n_0}, V'_{n_0+1}, V'''_{n_0+1}, V'_{n_0+2}, \dots, V'_{n-1}}(V'_n).$$

By Lemma 2.4, $(V'''_{n_0+1}, \dots, V'''_0)$ is an innovation of $\mathcal{F}' \vee \mathcal{F}''$ from n_0 to 0. Finally we define the filtration $\mathcal{F}'''$ as the filtration equaling $\mathcal{F}''$ up to time n_0 and for which $(V'''_{n_0+1}, \dots, V'''_0)$ is an innovation from n_0 to 0; that is, we put $\mathcal{F}'''_n = \mathcal{F}''_n$ for $n \leq n_0$ and $\mathcal{F}'''_n = \mathcal{F}''_{n_0} \vee \sigma(V'''_{n_0+1}, \dots, V'''_n)$ for $n \in \{n_0 + 1, \dots, 0\}$.

Thus $(\mathcal{F}', \mathcal{F}''')$ is a joining of $\mathcal{F}$ permutational after n_0. The copy X''' of X in $\mathcal{F}'''$ is obtained by replacing V''_{n_0+1} by V'''_{n_0+1} in (4), and consequently $\overline{\mathbb{E}}[\rho(X', X''') \mid \mathcal{F}'_{n_0} \vee \mathcal{F}''_{n_0}]$ is the minimum value of $L_{\tilde{Z}_{n_0}}$. So we have

$$\overline{\mathbb{E}}\big[\rho(X', X''') \mid \mathcal{F}'_{n_0} \vee \mathcal{F}''_{n_0}\big] \leq \overline{\mathbb{E}}\big[\rho(X', X'') \mid \mathcal{F}'_{n_0} \vee \mathcal{F}''_{n_0}\big],$$

and then $\overline{\mathbb{E}}[\rho(X', X''')] \leq \overline{\mathbb{E}}[\rho(X', X'')]$. □

Lemma 4.3. *Let $\mathcal{F} = (\mathcal{F}_n)_{n \leq 0}$ be a homogeneous filtration. Let (E, ρ) be a Polish metric space and $X \in L^1(\mathcal{F}_0; E)$. Let $(\mathcal{F}', \mathcal{F}'')$ be a joining of $\mathcal{F}$ on $(\overline{\Omega}, \overline{\mathcal{A}}, \overline{\mathbb{P}})$ and $n_0 \leq 0$ an integer.*

Then for any $\epsilon > 0$, there exists a filtration $\mathcal{F}'''$ on $(\overline{\Omega}, \overline{\mathcal{A}}, \overline{\mathbb{P}})$ such that $(\mathcal{F}', \mathcal{F}''')$ is a joining of $\mathcal{F}$ permutational after n_0 such that $\overline{\mathbb{E}}[\rho(X', X''')] \leq \overline{\mathbb{E}}[\rho(X', X'')] + \epsilon$ where X', X'' and X''' are the respective copies of X in $\mathcal{F}'$, $\mathcal{F}''$ and $\mathcal{F}'''$.

Proof. Let F be the subset of $\{n_0+1, \ldots, 0\}$ consisting of those integers n for which V_n is non-atomic. Without loss of generality, we assume that V_n has the uniform law $[0, 1]$ for $n \in F$. Let $\epsilon > 0$. Let k sufficiently large so that, by putting

$$\eta'_n = \sum_{i=0}^{2^k-1} \frac{i}{2^k} \mathbb{1}_{\left\{\frac{i}{2^k} < V'_n \leq \frac{i+1}{2^k}\right\}} \quad \text{for } n \in F$$

and $\eta'_n = V'_n$ for $n \in \{n_0 + 1, \ldots, 0\} \setminus F$, then there exists a random variable R' measurable with respect to $\mathcal{F}'_{n_0} \vee \sigma(\eta'_{n_0+1}, \ldots, \eta'_0)$ and such that $\overline{\mathbb{E}}[\rho(X', R')] \leq \epsilon/4$. Define a filtration $\mathcal{E}' = (\mathcal{E}'_n)_{n \leq 0}$ by letting $\mathcal{E}'_n = \mathcal{F}'_n$ for $n \leq n_0$ and $\mathcal{E}'_n = \mathcal{F}'_{n_0} \vee \sigma(\eta'_{n_0+1}, \ldots, \eta'_n)$ for $n \in \{n_0 + 1, \ldots, 0\}$. By applying the preceding lemma to $\mathcal{E}'$, we obtain some random permutations T_n for $n \in \{n_0 + 1, \ldots, 0\}$ such that by putting $\eta'''_n = T_n(\eta'_n)$ for those n, we have $\overline{\mathbb{E}}[\rho(R', R''')] \leq \overline{\mathbb{E}}[\rho(R', R'')]$ where R''' is the copy of R' in the filtration $\mathcal{E}'''$ defined by $\mathcal{E}'''_n = \mathcal{F}''_n$ for $n \leq n_0$ and $\mathcal{E}'''_n = \mathcal{F}''_{n_0} \vee \sigma(\eta'''_{n_0+1}, \ldots, \eta'''_n)$ for $n \in \{n_0 + 1, \ldots, 0\}$. The T_n are naturally extended to the interval $[0, 1]$ so that we can define $V'''_n = T_n(V'_n)$ for every $n \in \{n_0 + 1, \ldots, 0\}$. Finally one has $\overline{\mathbb{E}}[\rho(X', X''')] \leq \overline{\mathbb{E}}[\rho(X', X'')] + \epsilon$ where X''' is the copy of X' in the filtration $\mathcal{F}'''$ defined by $\mathcal{F}'''_n = \mathcal{F}''_n$ for $n \leq n_0$ and $\mathcal{F}'''_n = \mathcal{F}''_{n_0} \vee \sigma(V'''_{n_0+1}, \ldots, V'''_n)$ for $n \in \{n_0 + 1, \ldots, 0\}$. □

Theorem 4.4. *Let $\mathcal{F} = (\mathcal{F}_n)_{n \leq 0}$ be a homogeneous filtration, E a Polish space and $X \in L^1(\mathcal{F}_0; E)$. Then X is I-cosy if and only if X satisfies Vershik's first level criterion. Consequently, a σ-field $\mathcal{E}_0 \subset \mathcal{F}_0$ is I-cosy if and only if it satisfies Vershik's first level criterion, and the filtration $\mathcal{F}$ is I-cosy if and only if it satisfies Vershik's first level criterion.*

Proof. The last sentence is obvious from definitions. We have seen in Proposition 3.31 that the 'if' part holds more generally true for any filtration of local product type. The preceding lemma shows that I-cosiness of X implies that X satisfies Vershik's self-joining criterion (Definition 2.37), hence the 'only if' part follows from Theorem 2.38. □

Corollary 4.5. *An essentially separable homogeneous filtration is I-cosy if and only if it is of product type.*

Proof. This results from Theorem 4.4 and Vershik's first level criterion (Theorem 2.25). □

The following two corollaries justify some terminology introduced in Sect. 3.

Corollary 4.6. *A filtration is weakly standard conditionally non-atomic according to Definition 3.19 if and only if it is both weakly standard and conditionally non-atomic.*

Proof. Obviously, a weakly standard conditionally non-atomic filtration is weakly standard and conditionally non-atomic. Conversely, let $\mathcal{F}$ be a weakly standard filtration which is conditionally non-atomic. Then $\mathcal{F}$ is I-cosy by Corollary 3.32, and therefore $\mathcal{F}$ satisfies Vershik's first level criterion by Theorem 4.4, thus $\mathcal{F}$ is weakly standard conditionally non-atomic according to Definition 3.19.   □

Corollary 4.7. *A filtration is standard conditionally non-atomic according to Definition 3.1 if and only if it is both standard and conditionally non-atomic.*

Proof. The proof is similar to the one of Corollary 4.6 by using Corollary 4.5 instead of Theorem 4.4.   □

The following corollary of Theorem 4.4 gives a characterization of I-cosiness of a random variable with respect to a locally separable filtration. In assertions (ii) and (iii), we commit the same slight abuse of language as in Corollary 3.47.

Corollary 4.8. *Let $\mathcal{F}$ be a locally separable filtration and $\mathcal{E}_0 \subset \mathcal{F}_0$ a σ-field. Then the following facts are equivalent:*

  (i) *$\mathcal{E}_0$ satisfies the I-cosiness criterion with respect to $\mathcal{F}$.*
 (ii) *$\mathcal{E}_0$ satisfies Vershik's first level criterion with respect to the independent product of $\mathcal{F}$ with a standard non-atomic filtration.*
(iii) *$\mathcal{E}_0$ satisfies Vershik's first level criterion with respect to a conditionally non-atomic extension of $\mathcal{F}$.*

Letting E be a Polish space and $X \in L^1(\mathcal{F}_0; E)$, these three statements with $\mathcal{E}_0$ replaced by X are still equivalent.

Proof. This follows from Theorem 4.4 and Corollary 3.47.   □

Theorem 4.9. *Let $\mathcal{F}$ be a locally separable filtration. The following assertions are equivalent:*

  (a) *$\mathcal{F}$ is I-cosy (Definition 3.29).*
  (b) *The independent product of $\mathcal{F}$ with a standard conditionally non-atomic filtration is weakly standard conditionally non-atomic (Definition 3.19).*
  (c) *$\mathcal{F}$ is weakly standard (Definition 3.21).*

If in addition $\mathcal{F}$ is essentially separable, then these assertions also are equivalent to:

(b)' *The independent product of $\mathcal{F}$ with a standard conditionally non-atomic filtration is itself standard conditionally non-atomic (Definition 3.1);*

(c)' *$\mathcal{F}$ is standard (Definition 3.6)*

Proof. The independent product of a filtration $\mathcal{F}$ with a standard conditionally non-atomic filtration is an extension of $\mathcal{F}$ (Lemma 3.5); that shows that (b) $\Longrightarrow$ (c) and (b)' $\Longrightarrow$ (c)'. Corollary 3.32 shows that (c) $\Longrightarrow$ (a) and (c)' $\Longrightarrow$ (a). It remains to show that (a) $\Longrightarrow$ (b) and (a) $\Longrightarrow$ (b)'. A standard conditionally non-atomic filtration is of product type, hence is I-cosy by Proposition 3.31. Therefore, if $\mathcal{F}$ is I-cosy, then so is its independent product with a standard conditionally non-atomic filtration in view of Proposition 3.46. By Lemma 3.22, this product filtration is conditionally non-atomic. Hence, Theorem 4.4 shows that (a) $\Longrightarrow$ (b) and Corollary 4.5 and Lemma 3.2 show that (a) $\Longrightarrow$ (b)'.   □

Corollary 4.10. *A filtration is standard if and only if it is weakly standard and essentially separable.*

Proof. We have already noticed that a standard filtration obviously is weakly standard and essentially separable. The converse follows from Theorem 4.9.   □

4.2 *Example Continued: Split-Word Processes*

We show how to deduce part (b) of Theorem 2.39 assuming Result 2.45. Consider a split word process $(X_n, \varepsilon_n)_{n \leq 0}$ on the alphabet $[0, 1]$ equipped with the Lebesgue measure. Let μ be a probability measure on $\mathbb{R}$ and f be the right-continuous inverse of the cumulative distribution function of μ. Then the process $(f(X_n), \varepsilon_n)_{n \leq 0}$ is the split-word process on the alphabet $(\mathbb{R}, \mathfrak{B}_{\mathbb{R}}, \mu)$ with the same splitting sequence as $(X_n, \varepsilon_n)_{n \leq 0}$. By Lemma 1.6, the filtration $\mathcal{F}$ generated by $(f(X_n), \varepsilon_n)_{n \leq 0}$ is immersed in the filtration $\mathcal{G}$ generated by $(X_n, \varepsilon_n)_{n \leq 0}$. Consequently, if $\mathcal{G}$ is I-cosy then so is $\mathcal{F}$ (Lemma 3.30), and $\mathcal{F}$ is of product type by Corollary 4.5. If (Δ) does not hold, we know from Result 2.45 that $\mathcal{G}$ is of product type, and consequently is I-cosy (Proposition 3.31). Finally $\mathcal{F}$ is of product type if (Δ) does not hold. As a consequence, this is also true if $(A, \mathfrak{A}, \mu)$ is Polish because every Polish probability space is Lebesgue isomorphic to a probability space on $\mathbb{R}$.

5 Annex

5.1 *Isomorphisms*

This annex gives definitions and elementary lemmas about isomorphisms between probability spaces. The classical definition of an embedding between from a probability space $(\Omega, \mathcal{B}, \mathbb{P})$ into a probability space $(\Omega', \mathcal{A}', \mathbb{P}')$ is given in terms of a Boolean morphism from $\mathcal{B}/\mathbb{P}$ into $\mathcal{A}'/\mathbb{P}$ which preserves probabilities; such an embedding extends uniquely to random variables (see [2]) and then the definition is equivalently rephrased as follows.

Definition 5.1. Let $(\Omega, \mathcal{B}, \mathbb{P})$ and $(\Omega', \mathcal{A}', \mathbb{P}')$ be two probability spaces. We say that an application $\Psi: L^0(\Omega, \mathcal{B}, \mathbb{P}) \rightarrow L^0(\Omega', \mathcal{A}', \mathbb{P}')$, is an *embedding* from $(\Omega, \mathcal{B}, \mathbb{P})$ into $(\Omega', \mathcal{A}', \mathbb{P}')$ if the following two conditions hold:

(i) For all integer $n \geq 1$, for all random variables $X_1, \ldots, X_n$ on $(\Omega, \mathcal{B}, \mathbb{P})$, and all Borelian applications $f: \mathbb{R}^n \rightarrow \mathbb{R}$, one has $\Psi(f(X_1, \ldots, X_n)) = f(\Psi(X_1), \ldots, \Psi(X_n))$.

(ii) Each random variable X on $(\Omega, \mathcal{B}, \mathbb{P})$ has the same law as $\Psi(X)$.

The random variable $\Psi(X)$ is also called the *copy* of the random variable X by the embedding Ψ. We shortly say that Ψ is an embedding from $(\Omega, \mathcal{B}, \mathbb{P})$ into $(\Omega', \mathcal{A}', \mathbb{P}')$, and we shortly write $\Psi: \mathcal{B} \rightarrow \mathcal{A}'$. We say that an embedding

$\Psi: \mathcal{B} \to \mathcal{A}'$ is an *isomorphism* from $\mathcal{B}$ onto $\mathcal{A}'$ if it is surjective. It is trivial that an embedding is linear, injective, and continuous for the topology of the convergence in probability. If $X = \mathbb{1}_B$ is the indicator function of an event $B \in \mathcal{B}$, one can verify that $\Psi(X)$ is the indicator function of an event $B' \in \mathcal{A}'$, which we denote by $\Psi(B)$. One easily verifies that the set $\Psi(\mathcal{B}) := \{\Psi(B) \mid B \in \mathcal{B}\}$ is a σ-field and that Ψ defines an isomorphism from $(\Omega, \mathcal{B}, \mathbb{P})$ into $(\Omega', \Psi(\mathcal{B}), \mathbb{P}')$; thus any $\Psi(\mathcal{B})$-measurable random variable X' has form $X' = \Psi(X)$ for some $\mathcal{B}$-measurable random variable X. If $\mathcal{B} = \sigma(Y)$ for some random variable Y, it is also easy to see that $\Psi(\mathcal{B}) = \sigma(\Psi(Y))$.

Example 5.2. Let $(\Omega, \mathcal{B}, \mathbb{P})$ and $(\Omega^*, \mathcal{C}^*, \mathbb{P}^*)$ be two probability spaces, and let $(\widehat{\Omega}, \widehat{\mathcal{A}}, \widehat{\mathbb{P}}) = (\Omega, \mathcal{B}, \mathbb{P}) \otimes (\Omega^*, \mathcal{C}^*, \mathbb{P}^*)$. The identification with the first factor is the canonical embedding $\iota: \mathcal{B} \to \widehat{\mathcal{A}}$ defined by $\iota(X): (\omega, \omega^*) \mapsto X(\omega)$.

The proof of the following lemma is left to the reader.

Lemma 5.3. *Let $(\Omega, \mathcal{B}, \mathbb{P})$ and $(\Omega', \mathcal{A}', \mathbb{P}')$ be two probability spaces and $\Psi: \mathcal{B} \to \mathcal{A}'$ be an embedding. Let $X \in L^1(\mathcal{B})$ and $\mathcal{C} \subset \mathcal{B}$ be a σ-field. Then $\Psi\left(\mathbb{E}[X \mid \mathcal{C}]\right) = \mathbb{E}'[\Psi(X) \mid \Psi(\mathcal{C})]$.*

As shown below, an embedding also defines uniquely a copy of a random variable taking its values in a Polish space (a topological space is said to be *Polish* if it is separable and admits a complete metrization).

Definition 5.4. Let E be a separable metric space, $(\Omega, \mathcal{B}, \mathbb{P})$ and $(\Omega', \mathcal{A}', \mathbb{P}')$ two probability spaces, $X \in L^0(\mathcal{B}; E)$, and $\Psi: \sigma(X) \to \mathcal{A}'$ an embedding. A random variable $X' \in L^0(\mathcal{A}'; E)$ is denoted by $\Psi(X)$ if one has $\Psi\left(f(X)\right) = f(X')$ for every Borelian function $f: E \to \mathbb{R}$.

It is straightforward to verify that, given another separable metric space and a Borelian function $g: E \to F$, one has $\Psi\left(g(X)\right) = g\left(\Psi(X)\right)$ provided that $\Psi\left(g(X)\right)$ and $\Psi(X)$ exist.

Lemma 5.5. *With the same notations as the preceding definition, when E is Polish, there exists a unique random variable $X' = \Psi(X)$.*

Proof. Any Polish probability space is Lebesgue isomorphic to a probability space on $\mathbb{R}$ (see [6, 25, 29]). Hence, there exist a bimeasurable bijection T from a set $E_0 \subset E$ of full $\mathbb{P}_X$-measure, where $\mathbb{P}_X$ is the law of X, into a set $F_0 \subset \mathbb{R}$ of full μ-measure, where μ is a probability distribution on $\mathbb{R}$, and T satisfies $T(\mathbb{P}_X) = \mu$. If X' is a random variable such that $\Psi\left(f(X)\right) = f(X')$ for all Borelian functions f, then its law is the same as the one of X. Thus the random variables $T(X)$ and $T(X')$ are well-defined and we have $\Psi\left(T(X)\right) = T(X')$. It makes sense to take the image under T^{-1} of this random variable and this yields $X' = T^{-1}\left(\Psi(T(X))\right)$. Thus there is at most one random variable X' satisfying the desiderata. Finally, putting $X' = T^{-1}\left(\Psi(T(X))\right)$ it is easy to verify that the equality $\Psi\left(f(X)\right) = f(X')$ is indeed satisfied for each Borelian function $f: E \to \mathbb{R}$. $\square$

Example 5.6. Let $(\Omega, \mathcal{B}, \mathbb{P})$ and $(\Omega', \mathcal{A}', \mathbb{P}')$ be two probability spaces and $\Psi: \mathcal{B} \to \mathcal{A}'$ an embedding. Let $X_1, \ldots, X_n$ be random variables on $(\Omega, \mathcal{B}, \mathbb{P})$ taking their values in possibly different Polish spaces, and consider $(X_1, \ldots, X_n)$ as a random variable in the product Polish space. Then $\Psi(X_1, \ldots, X_n) = (\Psi(X_1), \ldots, \Psi(X_n))$. Indeed each σ-field $\sigma(X_i)$ is essentially separable so it is possible to write $X_i = g_i(Z_i)$ where Z_i takes its values in $\mathbb{R}$. Thus, $\Psi(f(X_1, \ldots, X_n)) = f(g_1(\Psi(Z_1)), \ldots, g_n(\Psi(Z_n))) = f(\Psi(X_1), \ldots, \Psi(X_n))$.

Lemma 5.7. *Let E be a separable metric space, $(\Omega, \mathcal{B}, \mathbb{P})$ and $(\Omega', \mathcal{A}', \mathbb{P}')$ be two probability spaces, $X \in L^0(\mathcal{B}; E)$ and $X' \in L^0(\mathcal{A}'; E)$. If X and X' have the same law, then $\sigma(X)$ and $\sigma(X')$ are isomorphic and there exists a unique isomorphism $\Psi: \sigma(X) \to \sigma(X')$ such that $\Psi(X) = X'$ (in the sense of Definition 5.4).*

Proof. By Doob's functional representation theorem, any real-valued random variable measurable with respect to $\sigma(X)$ is of the form $f(X)$ for a Borelian function $f: E \to \mathbb{R}$. One easily verifies that we define an isomorphism $\Psi: \sigma(X) \to \sigma(X')$ by putting $\Psi(f(X)) = f(X')$. □

5.2 Copies of Conditional Laws

Let E be a Polish space. On $(\Omega, \mathcal{A}, \mathbb{P})$, if X is a random variable taking its values in E, and if $\mathcal{C} \subset \mathcal{A}$ is a σ-field, then the conditional law $\mathcal{L}[X \mid \mathcal{C}]$ is a $\mathcal{C}$-measurable random variable taking its values in the set $\mathcal{P}(E)$ of probability measures on E, which is Polish in the *weak topology* (see [4]), generated by the maps $\pi_f: \mu \mapsto \mu(f)$ for bounded continuous functions $f: E \to \mathbb{R}$. The associated Borel σ-field on $\mathcal{P}(E)$ is itself generated by the maps π_f for bounded continuous functions $f: E \to \mathbb{R}$. Therefore, the σ-field $\sigma(\mathcal{L}[X \mid \mathcal{C}])$ is generated by the conditional expectations $\mathbb{E}[f(X) \mid \mathcal{C}] = \pi_f(\mathcal{L}[X \mid \mathcal{C}])$ for all bounded continuous functions $f: E \to \mathbb{R}$, and $\mathbb{E}[f(X) \mid \mathcal{C}]$ is $\sigma(\mathcal{L}[X \mid \mathcal{C}])$-measurable for all suitable functions $f: E \to \mathbb{R}$.

Lemma 5.8. *Let $(\Omega, \mathcal{B}, \mathbb{P})$ and $(\Omega', \mathcal{A}', \mathbb{P}')$ be two probability spaces and $\Psi: \mathcal{B} \to \mathcal{A}'$ an embedding. Let E be a Polish space and $\mu: \Omega \times \mathcal{B}_E \to [0, 1]$ a probability kernel from $(\Omega, \mathcal{B})$ to $\mathcal{B}_E$. Thus μ defines a random variable taking its values in the Polish space of probability measures on E. Then one has $\mu' = \Psi(\mu)$ according to Definition 5.4 if and only if $\mu'(f) = \Psi(\mu(f))$ for all bounded continuous functions $f: E \to \mathbb{R}$.*

Proof. If $\mu' = \Psi(\mu)$ then from Definition 5.4 we know that $\mu'(f) = \Psi(\mu(f))$ for all suitable functions f. Conversely, it is well-known that any measure m on a Polish space E is uniquely determined by the values of $m(f)$ for bounded continuous functions $f: E \to \mathbb{R}$. □

The proof of the following lemma is easily derived from Lemma 5.8; we leave it to the reader.

Lemma 5.9. *On a probability space* $(\Omega, \mathcal{C}, \mathbb{P})$, *let* μ *be a random probability on a Polish space* E. *We define the probability* $\widehat{\mathbb{P}} := \mathbb{P} \otimes \mu$ *on the measurable space* $(\widehat{\Omega}, \widehat{\mathcal{B}}) := (\Omega \times E, \mathcal{C} \otimes \mathcal{B}_E)$ *by*

$$\widehat{\mathbb{P}}[\widehat{B}] = \mathbb{E}\Big[\int \mathbb{1}_{\widehat{B}}(\cdot, t)\, d\mu(t)\Big].$$

Then the identification with the first factor $\iota: \mathcal{C} \to \widehat{\mathcal{B}}$ *is an embedding from* $(\Omega, \mathcal{C}, \mathbb{P})$ *into* $(\widehat{\Omega}, \widehat{\mathcal{B}}, \widehat{\mathbb{P}})$, *and one has* $\iota(\mu) = \mathcal{L}[\widehat{V} \mid \widehat{\mathcal{C}}]$ *where* $\widehat{\mathcal{C}} = \iota(\mathcal{C})$ *and* $\widehat{V}$ *is the random variable defined by* $\widehat{V}(\omega, t) = t$.

Lemma 5.10. *Let* $(\Omega, \mathcal{B}, \mathbb{P})$ *and* $(\Omega', \mathcal{A}', \mathbb{P}')$ *be two probability spaces and* $\Psi: \mathcal{B} \to \mathcal{A}'$ *an embedding. Let* E *be a Polish space,* $X \in L^0(\mathcal{B}; E)$ *and* $\mathcal{C} \subset \mathcal{B}$ *be a* σ-*field. Then* $\Psi(\mathcal{L}[X \mid \mathcal{C}]) = \mathcal{L}[\Psi(X) \mid \Psi(\mathcal{C})]$.

Proof. By Lemma 5.8, it suffices to check that the equalities $\Psi\left(\mathbb{E}\left[f(X) \mid \mathcal{C}\right]\right) = \mathbb{E}'\left[f\left(\Psi(X)\right) \mid \Psi(\mathcal{C})\right]$ hold for all bounded continuous functions $f: E \to \mathbb{R}$. This stems from Lemma 5.3. $\square$

Proposition 5.11. *Let* $(\Omega, \mathcal{B}, \mathbb{P})$ *be a probability space and let* $\mathcal{C}_1$, $\mathcal{C}_2$ *be two sub-*σ-*fields of* $\mathcal{B}$. *Let* $(\Omega', \mathcal{A}', \mathbb{P}')$ *be a probability space, and* $\Psi_1: \mathcal{C}_1 \to \mathcal{A}'$, $\Psi_2: \mathcal{C}_2 \to \mathcal{A}'$ *two embeddings. There exists an isomorphism* $\Psi: \mathcal{C}_1 \vee \mathcal{C}_2 \to \Psi_1(\mathcal{C}_1) \vee \Psi_2(\mathcal{C}_2)$ *which simultaneously extends* Ψ_1 *and* Ψ_2 *if and only if one has* $\Psi_1(\mathcal{L}[\mathcal{C}_2 \mid \mathcal{C}_1]) = \mathcal{L}[\Psi_2(\mathcal{C}_2) \mid \Psi_1(\mathcal{C}_1)]$ *for every* $\mathcal{C}_2$-*measurable random variable* C_2.

Proof. The 'only if' part follows from Lemma 5.10. We show the 'if' part. Let X be a $\mathcal{C}_1 \vee \mathcal{C}_2$-measurable random variable. Then, by Lemma 1.1, there exist a $\mathcal{C}_1$-measurable random variable C_1, a $\mathcal{C}_2$-measurable random variable C_2, and a Borelian function f such that $X = f(C_1, C_2)$. If Ψ exists, one must have $\Psi(X) = f(\Psi_1(C_1), \Psi_2(C_2))$. The condition $\Psi_1(\mathcal{L}[C_2 \mid \mathcal{C}_1]) = \mathcal{L}[\Psi_2(C_2) \mid \Psi_1(\mathcal{C}_1)]$ shows that the pair (C_1, C_2) has the same distribution as $(\Psi_1(C_1), \Psi_2(C_2))$, so X has the same law as $f(\Psi_1(C_1), \Psi_2(C_2))$. To show that Ψ is defined without ambiguity, consider that $X = g(D_1, D_2)$ where D_1 is a $\mathcal{C}_1$-measurable random variable, D_2 is a $\mathcal{C}_2$-measurable random variable, and g a Borelian function. The assumption implies $\Psi_1(\mathcal{L}[C_2, D_2 \mid \mathcal{C}_1]) = \mathcal{L}[\Psi_2(C_2, D_2) \mid \Psi_1(\mathcal{C}_1)]$, which implies that the four-tuple (C_1, D_1, C_2, D_2) has the same distribution as $(\Psi_1(C_1), \Psi_1(D_1), \Psi_2(C_2), \Psi_2(D_2))$; so if $f(C_1, C_2) = g(D_1, D_2)$ almost surely then $f(\Psi_1(C_1), \Psi_2(C_2)) = g(\Psi_1(D_1), \Psi_2(D_2))$ almost surely. Checking condition (i) in Definition 5.1 is left to the reader. $\square$

Corollary 5.12. *Let* $(\Omega, \mathcal{B}, \mathbb{P})$ *be a probability space,* $\mathcal{C} \subset \mathcal{B}$ *a* σ-*field, and* V *a* $\mathcal{B}$-*measurable random variable taking values in some Polish space* E. *Let* $(\Omega', \mathcal{A}', \mathbb{P}')$ *be a probability space,* $\Psi_0: \mathcal{C} \to \mathcal{A}'$ *an embedding and* V' *an* $\mathcal{A}'$-*measurable random variable taking values in* E. *Then there exists an isomorphism* $\Psi: \mathcal{C} \vee \sigma(V) \to \Psi_0(\mathcal{C}) \vee \sigma(V')$ *extending* Ψ_0 *and sending* V *to* V' *if and only if one has* $\Psi_0(\mathcal{L}[V \mid \mathcal{C}]) = \mathcal{L}[V' \mid \Psi_0(\mathcal{C})]$.

Proof. The 'only if' part follows from Lemma 5.10. The 'if' part follows from Lemma 5.7, Lemma 5.8, and Proposition 5.11.   □

Corollary 5.13. *On* $(\Omega, \mathcal{A}, \mathbb{P})$, *let* $\mathcal{C} \subset \mathcal{A}$ *be a* σ-*field and let* V *be a random variable. We put* $\mu = \mathcal{L}[V \mid \mathcal{C}]$. *Then, with the notations of Lemma 5.9, there exists an unique isomorphism* $\Psi: \mathcal{C} \vee \sigma(V) \to \hat{\mathcal{C}} \vee \sigma(\widehat{V})$ *such that the restriction of* Ψ *to* $\mathcal{C}$ *equals the canonical embedding* ι *and* $\Psi(V) = \widehat{V}$.

Acknowledgements Financial support from the IAP research network (grant nr. P6/03 of the Belgian government, Belgian Science Policy) is gratefully acknowledged. I am also indebted to M. Émery for helpful and encouraging comments and suggestions on earlier drafts of this paper.

References

1. Bapat, R.B., Raghavan, T.E.S.: Nonnegative matrices and applications. Cambridge University Press, Cambridge (1997)
2. Barlow, M., Émery, M., Knight, F., Song, S., Yor, M.: Autour d'un théorème de Tsirelson sur des filtrations browniennes et non-browniennes. In: Séminaire de Probabilités XXXII. Lecture Notes in Mathematics, vol. 1686, pp. 264–305. Springer, Berlin (1998)
3. Beghdadi-Sakrani, S., Émery, M.: On certain probabilities equivalent to coin-tossing, d'après Schachermayer. Séminaire de Probabilités XXXIII. Lecture Notes in Mathematics, vol. 1709, pp. 240–256. Springer, Berlin (1999)
4. Billingsley, P.: Convergence of Probability Measures. Wiley, New York (1968)
5. Blum, J., Hanson, D.: Further results on the representation problem for stationary stochastic processes with Trivial Tail Field. J. Math. Mech. **12**(6), 935–943 (1963)
6. Bogachev, V.I.: Measure Theory, vol. II. Springer, Berlin (2007)
7. Ceillier, G.: The Filtration of the Split-Word Processes. Preprint (2009)
8. Dellacherie, C., Meyer, P.-A.: Probabilités et potentiel, Chapitres I à IV. Hermann, Paris (1975)
9. Doob, J.L.: Classical Potential Theory and its Probabilistic Counterpart. Springer, New York, (1984)
10. Dubins, L.E., Feldman, J., Smorodinsky, M., Tsirelson, B.: Decreasing sequences of σ-fields and a measure change for Brownian motion. Ann. Probab. **24**, 882–904 (1996)
11. Dudley, R.M.: Real Analysis and Probability. Wadsworth and Brooks/Cole Math Series, Pacific Grove (1989)
12. Émery, M.: Old and new tools in the theory of filtrations. In: Maass, A., Martinez, S., San Martin, J. (eds.) Dynamics and Randomness, pp. 125–146. Kluwer Academic Publishers, Massachusetts (2002)
13. Émery, M.: On certain almost Brownian filtrations. Annales de l'I.H.P. Probabilités et statistiques **41**(3), 285–305 (2005)
14. Émery, M., Schachermayer, W.: On Vershik's standardness criterion and Tsirelson's notion of cosiness. Séminaire de Probabilités XXXV. Lecture Notes in Mathematics, vol. 1755, pp. 265–305. Springer, Berlin (2001)
15. Feldman, J., Smorodinsky, M.: Decreasing sequences of measurable partitions: product type, standard and prestandard. Ergod. Theor. Dyn. Syst. **20**(4), 1079–1090 (2000)
16. Feldman, J., Smorodinsky, M.: Addendum to our paper 'Decreasing sequences of sigma fields: product type, standard, and substandard'. Ergod. Theor. Dyn. Syst. **22**(4), 1329–1330 (2002)
17. Hanson, D.L.: On the representation problem for stationary stochastic processes with Trivial Tail Field. J. Appl. Math. Mech. **12**(2), 294–301 (1963)
18. Heicklen, D.: Bernoullis are standard when entropy is not an obstruction. Isr. J. Math. **107**(1), 141–155 (1998)

19. Kallenberg, O.: Foundations of Modern Probability. Springer, Berlin, New York (1997)
20. Laurent, S.: Filtrations à temps discret négatif. PhD Thesis, Université de Strasbourg, Strasbourg (2004)
21. Laurent, S.: On Vershikian and I-cosy random variables and filtrations. Teoriya Veroyatnostei i ee Primeneniya **55**, 104–132 (2010)
22. Leuridan, C.: Filtration d'une marche aléatoire stationnaire sur le cercle. Séminaire de Probabilités XXXVI. Lecture Notes in Mathematics, vol. 1801, pp. 335–347. Springer, Berlin (2002)
23. Major, P.: On the invariance principle for sums of independent identically distributed random variables. J. Multivariate Anal. **8**, 487–517 (1978)
24. Parry, W.: Decoding with two independent processes. In: Mauldin, R.D., Shortt, R.M., Silva, C.E. (eds.) Measure and Measurable Dynamics, Contemporary Mathematics, vol. 94, pp. 207–209. American Mathematical Society, Providence (1989)
25. Rokhlin, V.A.: On the fundamental ideas of measure theory. Am. Math. Soc. Transl. **71**, 1–53 (1952)
26. Rosenblatt, M.: Stationary processes as shifts of functions of independent random variables. J. Math. Mech. **8**(5), 665–682 (1959)
27. Rosenblatt, M.: Stationary markov chains and independent random variables. J. Math. Mech. **9**(6), 945–949 (1960)
28. Rosenblatt, M.: The representation of a class of two state stationary processes in terms of independent random variables. J. Math. Mech. **12**(5), 721–730 (1963)
29. de la Rue, T.: Espaces de Lebesgue. Séminaire de Probabilités XXVII. Lecture Notes in Mathematics, vol. 1557, pp. 15–21. Springer, Berlin (1993)
30. Schachermayer, W.: On certain probabilities equivalent to wiener measure d'après Dubins, Feldman, Smorodinsky and Tsirelson. In: Séminaire de Probabilités XXXIII. Lecture Notes in Mathematics, vol. 1709, pp. 221–239. Springer, Berlin (1999)
31. Schachermayer, W.: Addendum to the paper 'On Certain Probabilities Equivalent to Wiener Measure d'après Dubins, Feldman, Smorodinsky and Tsirelson'. In: Séminaire de Probabilités XXXVI. Lecture Notes in Mathematics, vol. 1801, pp. 493–497. Springer, Berlin (2002)
32. Smorodinsky, M.: Processes with no standard extension. Isr. J. Math. **107**, 327–331 (1998)
33. Thorrisson, H.: Coupling, Stationarity, and Regeneration. Springer, New York (2000)
34. Tsirelson, B.: Triple points: from non-Brownian filtrations to harmonic measures. Geomet. Funct. Anal. (GAFA) **7**, 1096–1142 (1997)
35. Vershik, A.M.: Theorem on lacunary isomorphisms of monotonic sequences of partitions. Funktsional'nyi Analiz i Ego Prilozheniya. **2**(3), 17–21 (1968) English translation: Functional analysis and its applications. **2**:3, 200–203 (1968)
36. Vershik, A.M.: Decreasing sequences of measurable partitions, and their applications. Sov Math – Dokl, **11**, 1007–1011 (1970)
37. Vershik, A.M.: Continuum of pairwise nonisomorphic diadic sequences. Funktsional'nyi Analiz i Ego Prilozheniya, **5**(3), 16–18 (1971) English translation: Functional analysis and its applications, **5**(3), 182–184 (1971)
38. Vershik, A.M.: Approximation in measure theory (in Russian). PhD Thesis, Leningrad University, Leningrad (1973)
39. Vershik, A.M.: Four definitions of the scale of an automorphism. Funktsional'nyi Analiz i Ego Prilozheniya, **7**(3), 1–17 (1973) English translation: Functional analysis and its applications, **7**(3), 169–181 (1973)
40. Vershik, A.M.: The theory of decreasing sequences of measurable partitions (in Russian). Algebra i Analiz, **6**(4), 1–68 (1994) English translation: St. Petersburg Mathematical Journal, **6**(4), 705–761 (1995)

On Isomorphic Probability Spaces

Claude Dellacherie

Abstract In the appendix to his contribution (Laurent, On standardness and I-cosiness, this volume) to this volume, Stéphane Laurent recalls that if a probability space $(\Omega, \mathcal{A}, \mathbb{P})$ is embedded in another probability space $(\Omega', \mathcal{A}', \mathbb{P}')$, to every r.v. X on Ω the embedding associates a r.v. X' on Ω'. More precisely, his Lemma 5.5 states this property when X is valued in a Polish space E. Michel Émery has asked me the following question: is completeness of E really needed, or does the property more generally hold for separable, non complete metric spaces? By means of a counter-example, this short note shows that completeness cannot be dispensed of.

Keywords Isomorphic probability space · Counterexample

If $(\Omega, \mathcal{A}, \mathbb{P})$ is a probability space, denote by $\mathcal{A}/\mathbb{P}$ the quotient σ-field obtained from $\mathcal{A}$ by identifying any two events A and A' such that $\mathbb{P}(A \triangle A') = 0$. Observe that $\mathcal{A}/\mathbb{P}$ is an abstract σ-field endowed with a probability; it need not be a σ-field of subsets of some set.

Define two probability spaces $(\Omega, \mathcal{A}, \mathbb{P})$ and $(\Omega', \mathcal{A}', \mathbb{P}')$ to be *isomorphic* whenever $\mathcal{A}/\mathbb{P}$ and $\mathcal{A}'/\mathbb{P}'$ are isomorphic, that is, when there exists a bijection from $\mathcal{A}/\mathbb{P}$ to $\mathcal{A}'/\mathbb{P}'$ which preserves Boolean operations, monotone limits, and the probability measures. For instance, if $(\Omega, \bar{\mathcal{A}}, \mathbb{P})$ is obtained from $(\Omega, \mathcal{A}, \mathbb{P})$ by $\mathbb{P}$-completion, then $(\Omega, \bar{\mathcal{A}}, \mathbb{P})$ is isomorphic to $(\Omega, \mathcal{A}, \mathbb{P})$.

It is not difficult to see that if $(\Omega, \mathcal{A}, \mathbb{P})$ and $(\Omega', \mathcal{A}', \mathbb{P}')$ are isomorphic, not only are their events (modulo $\mathbb{P}$-nullity) in one-to-one correspondence, but so are also their random variables (defined up to $\mathbb{P}$-a.s. equality); and this easily extends to r.v. with values in any Polish space. This is a particular instance of Lemma A.5 from the Appendix of [1]. (Two r.v. $X : \Omega \to \mathbb{R}$ and $X' : \Omega' \to \mathbb{R}$ are in correspondence if so are the events $\{X \in B\}$ and $\{X' \in B\}$ for each Borel set $B \subset \mathbb{R}$.)

C. Dellacherie (✉)
Laboratoire Raphaël Salem, C.N.R.S. et Université de Rouen Avenue de l'Université,
BP.12, 76801 Saint-Étienne-du-Rouvray, France
e-mail: claude.dellacherie@aliceadsl.fr

The aim of this short note is to show that this does not extend to random variables with values in a separable, non complete, metric space. We shall exhibit two isomorphic probability spaces $(\Omega, \mathcal{A}, \mathbb{P})$ and $(\Omega', \mathcal{A}', \mathbb{P}')$, a separable (but not complete) metric space E, and a r.v. $X' : \Omega' \to E$ such that no r.v. $X : \Omega \to E$ corresponds to X' by the isomorphism.

Take $(\Omega, \mathcal{A}, \mathbb{P})$ to be the interval $[0, 1]$ endowed with the Lebesgue σ-field and the Lebesgue measure (the Lebesgue σ-field $\mathcal{A}$ is the $\mathbb{P}$-completion of the Borel σ-field $\mathcal{B}$ on $[0, 1]$). Let Ω' be any non-measurable subset of Ω with outer measure $\mathbb{P}^*(\Omega') = 1$; so its inner measure verifies $\mathbb{P}_*(\Omega') < 1$.

The existence of such an Ω' needs the axiom of choice; for instance, Ω' can be constructed as the complementary of a set having exactly one point in each equivalence class modulo $\mathbb{Q}$.

Endow Ω' with the σ-field $\mathcal{A}'$ and the probability $\mathbb{P}'$ inherited from $(\Omega, \mathcal{A}, \mathbb{P})$ as follows: a set $A' \subset \Omega'$ belongs to $\mathcal{A}'$ whenever there exists $A \in \mathcal{A}$ such that $A' = A \cap \Omega'$; then, the probability of A' is defined by $\mathbb{P}'(A') = \mathbb{P}(A)$. This probability does not depend upon the choice of A since, for A_1 and A_2 in $\mathcal{A}$ such that $A_1 \cap \Omega' = A_2 \cap \Omega'$, the symmetric difference $A_1 \triangle A_2$ is negligible, because it does not meet Ω' which has outer measure 1.

The map $A \mapsto A \cap \Omega'$ from $\mathcal{A}$ to $\mathcal{A}'$ can be quotiented by a.s. equality, and realises an isomorphism between $(\Omega, \mathcal{A}, \mathbb{P})$ and $(\Omega', \mathcal{A}', \mathbb{P}')$. Indeed, since $\mathcal{A}$ is $\mathbb{P}$-complete, a subset $A' \subset \Omega'$ is $\mathbb{P}'$-negligible if and only if it belongs to $\mathcal{A}$ and is $\mathbb{P}$-negligible; consequently, $A'_1 \in \mathcal{A}'$ and $A'_2 \in \mathcal{A}'$ are $\mathbb{P}'$-a.s. equal if and only if $A'_1 \triangle A'_2 \in \mathcal{A}$ and $\mathbb{P}(A'_1 \triangle A'_2) = 0$.

As a subset of $[0, 1]$, the space $E = \Omega'$ endowed with the usual distance is a separable metric space; its Borel σ-field is $\mathcal{B}' = \{B \cap \Omega', B \in \mathcal{B}\}$. The identity map $X' : \Omega' \to E$ is measurable since $\mathcal{B}' \subset \mathcal{A}'$; so X' is a random variable on Ω' with values in E. We shall see that *no r.v. $X : \Omega \to E$ corresponds to X' by the isomorphism.* It suffices to establish that *no r.v. $X : \Omega \to E$ can have the same law as X'.*

So let $X : (\Omega, \mathcal{A}) \to (E, \mathcal{B}')$ be any r.v. For each rational interval $I \subset [0,1]$, the inverse image $X^{-1}(I)$ belongs to $\mathcal{A}$; hence $X^{-1}(I) \triangle N_I \in \mathcal{B}$ for some negligible $N_I \in \mathcal{A}$. The (countable) union of all the N_I is negligible, so it is included in some negligible Borel set $N \subset [0, 1]$. For each I, the set $X^{-1}(I) \cap N^c = \left(X^{-1}(I) \triangle N_I\right) \cap N^c$ is in $\mathcal{B}$. This shows that the restriction of X to the Borel set N^c is a Borel map from N^c to $[0, 1]$. Fix some $e \in E$ and put

$$
Y(\omega) = \begin{cases} X(\omega) & \text{if } \omega \in N^c, \\ e & \text{if } \omega \in N; \end{cases}
$$

Y is a Borel map from Ω to $[0, 1]$, with values in E, and with the same law as X. It now suffices to prove that Y *and* X' *cannot have the same law.*

As Y is a Borel map, its image $Y(\Omega)$ is an analytic subset of $[0, 1]$, and a fortiori, by a classical consequence of Choquet's capacitability theorem, $Y(\Omega) \in \mathcal{A}$. From $Y(\Omega) \subset E = \Omega'$, one draws $\mathbb{P}(Y(\Omega)) \leq \mathbb{P}_*(\Omega') < 1$, and so, for some Borel set $B \in \mathcal{B}$, one has $Y(\Omega) \subset B \subset [0, 1]$ and $\mathbb{P}(B) < 1$. The set $B' = B \cap E$ is a Borel subset of the metric space E, and one can write $\mathbb{P}'(X' \in B') = \mathbb{P}'(B \cap \Omega') = \mathbb{P}(B) < 1 = \mathbb{P}(\Omega) = \mathbb{P}(Y \in B')$; this shows that X' and Y do not have the same law.

Reference

1. Séminaire de Probabilités XLIII, Lecture Notes in Math., vol. 2006, Springer, New York (2011)

Cylindrical Wiener Processes

Markus Riedle

Abstract This work is an expository article on cylindrical Wiener processes in Banach spaces. We expose the definition of a cylindrical Wiener process as a specific example of a cylindrical process. For that purpose, we gather results on cylindrical Gaussian measures, γ-radonifying operators and cylindrical processes from different sources and relate them to each other. We continue with introducing a stochastic integral with respect to cylindrical Wiener processes but such that the stochastic integral is only a cylindrical random variable. We need not put any geometric constraints on the Banach space under consideration. To this expository work we add a few novel conclusions on the question when a cylindrical Wiener process is a Wiener process in the original sense and on the relation between different stochastic integrals existing in the literature.

Keywords Cylindrical Wiener process · Cylindrical process · Cylindrical measure · Stochastic integral · Stochastic differential equation · Radonifying operator · Reproducing kernel Hilbert space

1 Introduction

Cylindrical Wiener processes appear in a huge variety of models in infinite dimensional spaces as a source of random noise or random perturbation. But there are various different definitions of cylindrical Wiener processes in the literature. Most of these definitions suffer from the fact that they do not generalise comprehensibly the real-valued definition to the infinite dimensional situation. However, this drawback might be avoided by introducing a cylindrical Wiener process as a specific example of a cylindrical process. A cylindrical process is a generalised stochastic process and it is closely related to cylindrical measures and radonifying operators, see for example the work by Laurent Schwartz in [8] and [9].

M. Riedle (✉)
School of Mathematics, The University of Manchester, Manchester M13 9PL, UK
e-mail: markus.riedle@manchester.ac.uk

C. Donati-Martin et al. (eds.), *Séminaire de Probabilités XLIII*, Lecture Notes in Mathematics 191
2006, DOI 10.1007/978-3-642-15217-7__7, © Springer-Verlag Berlin Heidelberg 2011

Although this approach by cylindrical processes is well known, see for example Kallianpur and Xiong [4] or Métivier and Pellaumail [5], there seems to be no coherent work which relates the fundamentals on cylindrical processes and cylindrical measures with cylindrical Wiener processes in Banach spaces and its stochastic integral. The main objectives of this work are to provide an introduction to the fundamentals and to expose the cylindrical approach for the definition of a cylindrical Wiener process and its stochastic integral. To this expository work we add a few novel conclusions on the question when a cylindrical Wiener process is a Wiener process in the original sense and on the relation between different stochastic integrals existing in the literature.

More in detail, we begin with introducing cylindrical measures which are finitely additive measures on Banach spaces that have σ-additive projections to Euclidean spaces of all dimensions. We continue with considering Gaussian cylindrical measures. This part is based on the monograph Vakhaniya et al. [11]. The next section reviews reproducing kernel Hilbert spaces where we follow the monograph [2] by Bogachev. In the following section γ-radonifying operators are introduced as it can be found in van Neerven [6].

After we have established these fundamentals we introduce cylindrical processes whose probability distributions are naturally described by cylindrical measures. We define a weakly cylindrical Wiener process as a *cylindrical process which is Wiener*. This definition of a weakly cylindrical Wiener process is a straightforward extension of the real-valued situation but it is immediately seen to be too general in order to be analytically tractable. An obvious request is that the covariance operator of the associated Gaussian cylindrical measures exists and has the analogue properties as in the case of ordinary Gaussian measures on infinite-dimensional spaces. This leads to a second definition of a *strongly* cylindrical Wiener process.

For strongly cylindrical Wiener processes we give a representation by a series with independent real-valued Wiener processes. On the other hand, we see, that by such a series a strongly cylindrical Wiener process can be constructed.

The obvious question when a cylindrical Wiener process is actually a Wiener process in the ordinary sense is the objective of the following section and can be answered easily thanks to our approach by the self-suggesting answer: if and only if the underlying cylindrical measure extends to an infinite countably additive set function, i.e. a probability measure.

For modelling random perturbations in models by a cylindrical Wiener process an appropriate definition of a stochastic integral with respect to a cylindrical Wiener process is required. In Hilbert spaces one can define a stochastic integral as a genuine random variable in the Hilbert space, see Da Prato and Zabczyk [3]. However, in Banach spaces there is no general theory of stochastic integration known. We continue our expository work in the final section with introducing a stochastic integral but such that the integral is only a cylindrical random variable. This approach is known and can be found for example in Berman and Root [1]. By requiring that the stochastic integral is only a cylindrical random variable we need not put any geometric constraints on the Banach space under consideration. We finish with two corollaries giving conditions such that the cylindrical probability distribution of the stochastic integral extends to a probability measure. These results relate our cylindrical integral to other well known integrals in the literature.

2 Preliminaries

Throughout this notes let U be a separable Banach space with dual U^*. The dual pairing is denoted by $\langle u, u^* \rangle$ for $u \in U$ and $u^* \in U^*$. If V is another Banach space then $L(U, V)$ is the space of all linear, bounded operators from U to V equipped with the operator norm $\|\cdot\|_{U \to V}$.

The Borel σ-algebra is denoted by $\mathcal{B}(U)$. Let Γ be a subset of U^*. Sets of the form

$$\mathcal{Z}(u_1^*, \ldots, u_n^*, B) := \{u \in U : (\langle u, u_1^* \rangle, \cdots, \langle u, u_n^* \rangle) \in B\},$$

where $u_1^*, \ldots, u_n^* \in \Gamma$ and $B \in \mathcal{B}(\mathbb{R}^n)$ are called *cylindrical sets* or *cylinder with respect to* (U, Γ). The set of all cylindrical sets is denoted by $\mathcal{Z}(U, \Gamma)$, which turns out to be an algebra. The generated σ-algebra is denoted by $\mathcal{C}(U, \Gamma)$ and it is called *cylindrical σ-algebra with respect to* (U, Γ). If $\Gamma = U^*$ we write $\mathcal{Z}(U) := \mathcal{Z}(U, \Gamma)$ and $\mathcal{C}(U) := \mathcal{C}(U, \Gamma)$. If U is separable then both the Borel $\mathcal{B}(U)$ and the cylindrical σ-algebra $\mathcal{C}(U)$ coincide.

A function $\mu : \mathcal{Z}(U) \to [0, 1]$ is called a *cylindrical measure on* $\mathcal{Z}(U)$, if for each finite subset $\Gamma \subseteq U^*$ the restriction of μ on the σ-algebra $\mathcal{C}(U, \Gamma)$ is a probability measure.

For every function $f : U \to \mathbb{C}$ which is measurable with respect to $\mathcal{C}(U, \Gamma)$ for a finite subset $\Gamma \subseteq U^*$ the integral $\int f(u) \, \mu(du)$ is well defined as a Lebesgue integral if it exists. In particular, the characteristic function $\varphi_\mu : U^* \to \mathbb{C}$ of a finite cylindrical measure μ is defined by

$$\varphi_\mu(u^*) := \int e^{i \langle u, u^* \rangle} \, \mu(du) \qquad \text{for all } u^* \in U^*.$$

In contrast to measures on infinite dimensional spaces there is an analogue of Bochner's Theorem for cylindrical measures:

Theorem 1. *A function* $\varphi : U^* \to \mathbb{C}$ *is a characteristic function of a cylindrical measure on* U *if and only if*

(a) $\varphi(0) = 0$.
(b) φ *is positive definite.*
(c) *The restriction of* φ *to every finite dimensional subset* $\Gamma \subseteq U^*$ *is continuous with respect to the norm topology.*

For a finite set $\{u_1^*, \ldots, u_n^*\} \subseteq U^*$ a cylindrical measure μ defines by

$$\mu_{u_1^*, \ldots, u_n^*} : \mathcal{B}(\mathbb{R}^n) \to [0, \infty],$$

$$\mu_{u_1^*, \ldots, u_n^*}(B) := \mu\big(\{u \in U : (\langle u, u_1^* \rangle, \ldots, \langle u, u_n^* \rangle) \in B\}\big)$$

a measure on $\mathcal{B}(\mathbb{R}^n)$. We call $\mu_{u_1^*, \ldots, u_n^*}$ the image of the cylindrical measure μ under the mapping $u \mapsto (\langle u, u_1^* \rangle, \ldots, \langle u, u_n^* \rangle)$. Consequently, we have for the characteristic function $\varphi_{\mu_{u_1^*, \ldots, u_n^*}}$ of $\mu_{u_1^*, \ldots, u_n^*}$ that

$$\varphi_{\mu_{u_1^*, \ldots, u_n^*}}(\beta_1, \ldots, \beta_n) = \varphi_\mu(\beta_1 u_1^* + \cdots + \beta_n u_n^*)$$

for all $\beta_1, \ldots, \beta_n \in \mathbb{R}$.

Cylindrical measures are described uniquely by their characteristic functions and therefore by their one-dimensional distributions μ_{u^*} for $u^* \in U^*$.

3 Gaussian Cylindrical Measures

A measure μ on $\mathcal{B}(\mathbb{R})$ is called Gaussian with mean $m \in \mathbb{R}$ and variance $\sigma^2 \geqslant 0$ if either $\mu = \delta_m$ and $\sigma^2 = 0$ or it has the density

$$f : \mathbb{R} \to \mathbb{R}_+, \qquad f(s) = \frac{1}{\sqrt{2\pi\sigma^2}} \exp\left(-\frac{1}{2\sigma^2}(s - m)^2\right).$$

In case of a multidimensional or an infinite dimensional space U a measure μ on $\mathcal{B}(U)$ is called Gaussian if the image measures μ_{u^*} are Gaussian for all $u^* \in U^*$. Gaussian cylindrical measures are defined analogously but due to some reasons explained below we have to distinguish between two cases: weakly and strongly Gaussian.

Definition 1. A cylindrical measure μ on $\mathcal{Z}(U)$ is called *weakly Gaussian* if μ_{u^*} is Gaussian on $\mathcal{B}(\mathbb{R})$ for every $u^* \in U^*$.

Because of well known properties of Gaussian measures in finite dimensional Euclidean spaces a cylindrical measure μ is weakly Gaussian if and only if $\mu_{u_1^*, \ldots, u_n^*}$ is a Gaussian measure on $\mathcal{B}(\mathbb{R}^n)$ for all $u_1^*, \ldots, u_n^* \in U^*$ and all $n \in \mathbb{N}$.

Theorem 2. *Let μ be a weakly Gaussian cylindrical measure on $\mathcal{Z}(U)$. Then its characteristic function φ_μ is of the form*

$$\varphi_\mu : U^* \to \mathbb{C}, \qquad \varphi_\mu(u^*) = \exp\left(im(u^*) - \tfrac{1}{2}s(u^*)\right), \qquad (1)$$

where the functions $m : U^* \to \mathbb{R}$ and $s : U^* \to \mathbb{R}_+$ are given by

$$m(u^*) = \int_U \langle u, u^* \rangle \, \mu(du), \qquad s(u^*) = \int_U \langle u, u^* \rangle^2 \mu(du) - (m(u^*))^2.$$

Conversely, if μ is a cylindrical measure with characteristic function of the form

$$\varphi_\mu : U^* \to \mathbb{C}, \qquad \varphi_\mu(u^*) = \exp\left(im(u^*) - \tfrac{1}{2}s(u^*)\right),$$

for a linear functional $m : U^* \to \mathbb{R}$ and a quadratic form $s : U^* \to \mathbb{R}_+$, then μ is a weakly Gaussian cylindrical measure.

Proof. Follows from [11, Proposition IV.2.7], see also [11, p. 393]. □

Example 1. Let H be a separable Hilbert space. Then the function

$$\varphi : H \to \mathbb{C}, \qquad \varphi(u) = \exp(-\tfrac{1}{2}\|u\|_H^2)$$

satisfies the condition of Theorem 2 and therefore there exists a weakly Gaussian cylindrical measure γ with characteristic function φ. We call this cylindrical measure *standard Gaussian cylindrical measure on H*. If H is infinite dimensional the cylindrical measure γ is not a measure, see [2, Corollary 2.3.2].

Note, that this example might be not applicable for a Banach space U because then $x \mapsto \|x\|_U^2$ need not be a quadratic form.

For a weakly Gaussian cylindrical measure μ one defines for $u^*, v^* \in U^*$:

$$r(u^*, v^*) := \int_U \langle u, u^* \rangle \langle u, v^* \rangle \, \mu(du) - \int_U \langle u, u^* \rangle \, \mu(du) \int_U \langle u, v^* \rangle \, \mu(du).$$

These integrals exist as μ is a Gaussian measure on the cylindrical σ-algebra generated by u^* and v^*. One defines the *covariance operator Q of μ* by

$$Q : U^* \to (U^*)', \qquad (Qu^*)v^* := r(u^*, v^*) \qquad \text{for all } v^* \in U^*,$$

where $(U^*)'$ denotes the algebraic dual of U^*, i.e. all linear but not necessarily continuous functionals on U^*. Hence, the characteristic function φ_μ of μ can be written as

$$\varphi_\mu : U^* \to \mathbb{C}, \qquad \varphi_\mu(u^*) = \exp\left(im(u^*) - \tfrac{1}{2}(Qu^*)u^*\right).$$

The cylindrical measure μ is called *centred* if $m(u^*) = 0$ for all $u^* \in U^*$.

If μ is a Gaussian measure or more general, a measure of weak order 2, i.e.

$$\int_U |\langle u, u^* \rangle|^2 \, \mu(du) < \infty \qquad \text{for all } u^* \in U^*,$$

then the covariance operator Q is defined in the same way as above. However, in this case it turns out that Qu^* is not only continuous and thus in U^{**} but even in U considered as a subspace of U^{**}, see [11, Theorem III.2.1]. This is basically due to properties of the Pettis integral in Banach spaces. For cylindrical measures we have to distinguish this property and define:

Definition 2. A centred weakly Gaussian cylindrical measure μ on $\mathcal{Z}(U)$ is called *strongly Gaussian* if the covariance operator $Q : U^* \to (U^*)'$ is U-valued.

Below Example 2 gives an example of a weakly Gaussian cylindrical measure which is not strongly. This example can be constructed in every infinite dimensional space in particular in every Hilbert space.

Strongly Gaussian cylindrical measures exhibit an other very important property:

Theorem 3. *For a cylindrical measure μ on $\mathcal{Z}(U)$ the following are equivalent:*

(a) *μ is a continuous linear image of the standard Gaussian cylindrical measure on a Hilbert space.*

(b) *There exists a symmetric positive operator $Q : U^* \to U$ such that*

$$\varphi_\mu(u^*) = \exp\left(-\tfrac{1}{2}\langle Qu^*, u^*\rangle\right) \qquad \text{for all } u^* \in U^*.$$

Proof. See [11, Proposition VI.3.3].

Theorem 3 provides an example of a weakly Gaussian cylindrical measure which is not strongly Gaussian:

Example 2. For a discontinuous linear functional $f : U^* \to \mathbb{R}$ define

$$\varphi : U^* \to \mathbb{C}, \qquad \varphi(u^*) = \exp\left(-\frac{1}{2}(f(u^*))^2\right).$$

Then φ is the characteristic function of a weakly Gaussian cylindrical measure due to Theorem 2 but this measure can not be strongly Gaussian by Theorem 3 because every symmetric positive operator $Q : U^* \to U$ is continuous.

4 Reproducing Kernel Hilbert Space

According to Theorem 3 a centred strongly Gaussian cylindrical measure is the image of the standard Gaussian cylindrical measure on a Hilbert space H under an operator $F \in L(H, U)$. In this section we introduce a possible construction of this Hilbert space H and the operator F.

For this purpose we start with a bounded linear operator $Q : U^* \to U$, which is positive,

$$\langle Qu^*, u^*\rangle \geq 0 \qquad \text{for all } u^* \in U^*,$$

and symmetric,

$$\langle Qu^*, v^*\rangle = \langle Qv^*, u^*\rangle \qquad \text{for all } u^*, v^* \in U^*.$$

On the range of Q we define a bilinear form by

$$[Qu^*, Qv^*]_{H_Q} := \langle Qu^*, v^* \rangle.$$

It can easily be seen that this defines an inner product $[\cdot, \cdot]_{H_Q}$. Thus, the range of Q is a pre-Hilbert space and we denote by H_Q the real Hilbert space obtained by its completion with respect to $[\cdot, \cdot]_{H_Q}$. This space will be called the *reproducing kernel Hilbert space associated with* Q.

In the following we collect some properties of the reproducing kernel Hilbert space and its embedding:

(a) The inclusion mapping from the range of Q into U is continuous with respect to the inner product $[\cdot, \cdot]_{H_Q}$. For, we have

$$\| Qu^* \|_{H_Q}^2 = |\langle Qu^*, u^* \rangle| \leq \| Q \|_{U^* \to U} \| u^* \|^2,$$

which allows us to conclude

$$|\langle Qu^*, v^* \rangle| = \left|[Qu^*, Qv^*]_{H_Q}\right| \leq \| Qu^* \|_{H_Q} \| Qv^* \|_{H_Q}$$
$$\leq \| Qu^* \|_{H_Q} \| Q \|_{U^* \to H_Q} \| v^* \|.$$

Therefore, we end up with

$$\| Qu^* \|_U = \sup_{\|v^*\| \leq 1} |\langle Qu^*, v^* \rangle| \leq \| Q \|_{U^* \to H_Q} \| Qu^* \|_{H_Q}.$$

Thus, the inclusion mapping is continuous on the range of Q and it extends to a bounded linear operator i_Q from H_Q into U.

(b) The operator Q enjoys the decomposition

$$Q = i_Q i_Q^*.$$

For the proof we define $h_{u^*} := Qu^*$ for all $u^* \in U^*$. Then we have $i_Q(h_{u^*}) = Qu^*$ and

$$[h_{u^*}, h_{v^*}]_{H_Q} = \langle Qu^*, v^* \rangle = \langle i_Q(h_{u^*}), v^* \rangle = [h_{u^*}, i_Q^* v^*]_{H_Q}.$$

Because the range of Q is dense in H_Q we arrive at

$$h_{v^*} = i_Q^* v^* \qquad \text{for all } v^* \in U^* \tag{2}$$

which finally leads to

$$Qv^* = i_Q(h_{v^*}) = i_Q(i_Q^* v^*) \qquad \text{for all } v^* \in U^*.$$

(c) By (2) it follows immediately that the range of i_Q^* is dense in H_Q.
(d) the inclusion mapping i_Q is injective. For, if $i_Q h = 0$ for some $h \in H_Q$ it follows that

$$[h, i_Q^* u^*]_{H_Q} = \langle i_Q h, u^* \rangle = 0 \qquad \text{for all } u^* \in U^*,$$

which results in $h = 0$ because of (c).
(e) If U is separable then H_Q is also separable.

Remark 1. Let μ be a centred strongly Gaussian cylindrical measure with covariance operator $Q : U^* \to U$. Because Q is positive and symmetric we can associate with Q the reproducing kernel Hilbert space H_Q with the inclusion mapping i_Q as constructed above. For the image $\gamma \circ i_Q^{-1}$ of the standard Gaussian cylindrical measure γ on H_Q we calculate

$$\begin{aligned}
\varphi_{\gamma \circ i_Q^{-1}}(u^*) &= \int_U e^{i \langle u, u^* \rangle} (\gamma \circ i_Q^{-1})(du) \\
&= \int_{H_Q} e^{i \langle h, i_Q^* u^* \rangle} \gamma(dh) \\
&= \exp\left(-\tfrac{1}{2} \left\| i_Q^* u^* \right\|_{H_Q}^2\right) \\
&= \exp\left(-\tfrac{1}{2} \langle Q u^*, u^* \rangle\right).
\end{aligned}$$

Thus, $\mu = \gamma \circ i_Q^{-1}$ and we have found one possible Hilbert space and operator satisfying the condition in Theorem 3.

But note, that there might exist other Hilbert spaces exhibiting this feature. But the reproducing kernel Hilbert space is characterised among them by a certain "minimal property", see [2].

5 γ-Radonifying Operators

This section follows the notes [6].

Let $(\Omega, \mathcal{A}, P)$ be a probability space with a filtration $\{\mathcal{F}_t\}_{t \geq 0}$ and U be a separable Banach space. The space of all random variables $X : \Omega \to U$ is denoted by $L^0(\Omega; U)$ and the space of all random variables $X : \Omega \to U$ with $E \|X\|_U^p < \infty$ is denoted by $L^p(\Omega; U)$. If $U = \mathbb{R}$ we write $L^0(\Omega)$ and $L^p(\Omega)$.

Let $Q : U^* \to U$ be a positive symmetric operator and H the reproducing kernel Hilbert space with the inclusion mapping $i_Q : H \to U$. If U is a Hilbert space then it is a well known result by Mourier ([11, Theorem IV.2.4]) that Q is the covariance operator of a Gaussian measure on U if and only if Q is nuclear or equivalently if i_Q is Hilbert–Schmidt. By Remark 1 it follows that the cylindrical measure $\gamma \circ i_Q^{-1}$ extends to a Gaussian measure on $\mathcal{B}(U)$ and Q is the covariance operator of this Gaussian measure.

The following definition generalises this property of $i_Q : H \to U$ to define by $Q := i_Q i_Q^*$ a covariance operator to the case when U is a Banach space:

Definition 3. Let γ be the standard Gaussian cylindrical measure on a separable Hilbert space H. A linear bounded operator $F : H \to U$ is called γ-*radonifying* if the cylindrical measure $\gamma \circ F^{-1}$ extends to a Gaussian measure on $\mathcal{B}(U)$.

Theorem 4. *Let γ be the standard Gaussian cylindrical measure on a separable Hilbert space H with orthonormal basis $(e_n)_{n \in \mathbb{N}}$ and let $(G_n)_{n \in \mathbb{N}}$ be a sequence of independent standard real normal random variables. For $F \in L(H, U)$ the following are equivalent:*

(a) *F is γ-radonifying.*
(b) *The operator $FF^* : U^* \to U$ is the covariance operator of a Gaussian measure μ on $\mathcal{B}(U)$.*
(c) *The series $\displaystyle\sum_{k=1}^{\infty} G_k F e_k$ converges a.s. in U.*
(d) *The series $\displaystyle\sum_{k=1}^{\infty} G_k F e_k$ converges in $L^p(\Omega; U)$ for some $p \in [1, \infty)$.*
(e) *The series $\displaystyle\sum_{k=1}^{\infty} G_k F e_k$ converges in $L^p(\Omega; U)$ for all $p \in [1, \infty)$.*

In this situation we have for every $p \in [1, \infty)$:

$$\int_U \|u\|^p \, \mu(du) = E \left\| \sum_{k=1}^{\infty} G_k F e_k \right\|^p .$$

Proof. As in Remark 1 we obtain for the characteristic function of $\nu := \gamma \circ F^{-1}$:

$$\varphi_\nu(u^*) = \exp\left(-\tfrac{1}{2}\langle FF^* u^*, u^*\rangle\right) \qquad \text{for all } u^* \in U^*.$$

This establishes the first equivalence between (a) and (b). The proofs of the remaining part can be found in [6, Proposition 4.2].

To show that γ-radonifying operators generalise Hilbert–Schmidt operators to Banach spaces we prove the result by Mourier mentioned already above. Other proofs only relying on Hilbert space theory can be found in the literature.

Corollary 1. *If H and U are separable Hilbert spaces then the following are equivalent for $F \in L(H, U)$:*

(a) *F is γ-radonifying.*
(b) *F is Hilbert–Schmidt.*

Proof. Let $(e_k)_{k \in \mathbb{N}}$ be an orthonormal basis of H. The equivalence follows immediately from

$$E \left\| \sum_{k=m}^{n} G_k F e_k \right\|^2 = \sum_{k=m}^{n} \| F e_k \|^2$$

for every family $(G_k)_{k \in \mathbb{N}}$ of independent standard normal random variables.

In general, the property of being γ-radonifying is not so easily accessible as Hilbert–Schmidt operators in case of Hilbert spaces. However, for some specific Banach spaces, as L^p or l^p spaces, the set of all covariance operators of Gaussian measures can be also described more precisely, see [11, Theorems V.5.5 and V.5.6].

It turns out that the set of all γ-radonifying operators can be equipped with a norm such that it is a Banach space, see [6, Theorem 4.14].

6 Cylindrical Processes

Similarly to the correspondence between measures and random variables there is an analogue random object associated to cylindrical measures:

Definition 4. A *cylindrical random variable X in U* is a linear map

$$X : U^* \to L^0(\Omega).$$

A cylindrical process X in U is a family $(X(t) : t \geqslant 0)$ of cylindrical random variables in U.

The characteristic function of a cylindrical random variable X is defined by

$$\varphi_X : U^* \to \mathbb{C}, \qquad \varphi_X(u^*) = E[\exp(i X u^*)].$$

The concepts of cylindrical measures and cylindrical random variables match perfectly. Because the characteristic function of a cylindrical random variable is positive-definite and continuous on finite subspaces there exists a cylindrical measure μ with the same characteristic function. We call μ the *cylindrical distribution of X*. Vice versa, for every cylindrical measure μ on $\mathcal{Z}(U)$ there exists a probability space $(\Omega, \mathcal{A}, P)$ and a cylindrical random variable $X : U^* \to L^0(\Omega)$ such that μ is the cylindrical distribution of X, see [11, Sect 3.2 in Chap. VI].

Example 3. A cylindrical random variable $X : U^* \to L^0(\Omega)$ is called weakly Gaussian, if $X u^*$ is Gaussian for all $u^* \in U^*$. Thus, X defines a weakly Gaussian cylindrical measure μ on $\mathcal{Z}(U)$. The characteristic function of X coincides with the one of μ and is of the form

$$\varphi_X(u^*) = \exp(im(u^*) - \tfrac{1}{2}s(u^*))$$

with $m : U^* \to \mathbb{R}$ linear and $s : U^* \to \mathbb{R}_+$ a quadratic form. If X is strongly Gaussian there exists a covariance operator $Q : U^* \to U$ such that

$$\varphi_X(u^*) = \exp(im(u^*) - \tfrac{1}{2}\langle Qu^*, u^* \rangle).$$

Because $\varphi_X(u^*) = \varphi_{Xu^*}(1)$ it follows

$$E[Xu^*] = m(u^*) \qquad \text{and} \qquad \text{Var}[Xu^*] = \langle Qu^*, u^* \rangle.$$

In the same way by comparing the characteristic function

$$\begin{aligned}
\varphi_{Xu^*, Xv^*}(\beta_1, \beta_2) &= E\left[\exp\left(i\left(\beta_1 Xu^* + \beta_2 Xv^*\right)\right)\right] \\
&= E\left[\exp\left(i\left(X\left(\beta_1 u^* + \beta_2 v^*\right)\right)\right)\right]
\end{aligned}$$

for $\beta_1, \beta_2 \in \mathbb{R}$ with the characteristic function of X we may conclude

$$\text{Cov}[Xu^*, Xv^*] = \langle Qu^*, v^* \rangle.$$

Let H_Q denote the reproducing kernel Hilbert space of the covariance operator Q. Then we obtain

$$E\,|Xu^* - m(u^*)|^2 = \text{Var}[Xu^*] = \langle Qu^*, u^* \rangle = \left\| i_Q^* u^* \right\|_{H_Q}^2.$$

The cylindrical process $X = (X(t) : t \geq 0)$ is called *adapted to a given filtration* $\{\mathcal{F}_t\}_{t \geq 0}$, if $X(t)u^*$ is $\mathcal{F}_t$-measurable for all $t \geq 0$ and all $u^* \in U^*$. The cylindrical process X has *weakly independent increments* if for all $0 \leq t_0 < t_1 < \cdots < t_n$ and all $u_1^*, \ldots, u_n^* \in U^*$ the random variables

$$(X(t_1) - X(t_0))u_1^*, \ldots, (X(t_n) - X(t_{n-1}))u_n^*$$

are independent.

Remark 2. Our definition of cylindrical processes is based on the definitions in [1] and [11]. In [5] and [10] cylindrical random variables are considered which have values in $L^p(\Omega)$ for $p > 0$. They assume in addition that a cylindrical random variable is continuous. The continuity of a cylindrical variable is reflected by continuity properties of its characteristic function, see [11, Proposition IV. 3.4]. The notion of weakly independent increments origins from [1].

Example 4. Let $Y = (Y(t) : t \geq 0)$ be a stochastic process with values in a separable Banach space U. Then $\hat{Y}(t)u^* := \langle Y(t), u^* \rangle$ for $u^* \in U^*$ defines a cylindrical process $\hat{Y} = (\hat{Y}(t) : t \geq 0)$. The cylindrical process $\hat{Y}$ is adapted if and only if Y is also adapted and $\hat{Y}$ has weakly independent increments if and only if Y has also independent increments. Both statements are due to the fact that the Borel and the cylindrical σ-algebras coincide for separable Banach spaces due to Pettis' measurability theorem.

An $\mathbb{R}^n$-*valued Wiener process* $B = (B(t) : t \geq 0)$ is an adapted stochastic process with independent, stationary increments $B(t) - B(s)$ which are normally distributed with expectation $E[B(t) - B(s)] = 0$ and covariance $\mathrm{Cov}[B(t) - B(s), B(t) - B(s)] = |t - s| C$ for a non-negative definite symmetric matrix C. If $C = \mathrm{Id}$ we call B a *standard* Wiener process.

Definition 5. An adapted cylindrical process $W = (W(t) : t \geqslant 0)$ in U is a *weakly cylindrical Wiener process*, if

(a) for all $u_1^*, \ldots, u_n^* \in U^*$ and $n \in \mathbb{N}$ the $\mathbb{R}^n$-valued stochastic process

$$\big((W(t)u_1^*, \ldots, W(t)u_n^*) : t \geqslant 0\big)$$

is a Wiener process.

Our definition of a weakly cylindrical Wiener process is an obvious extension of the definition of a finite-dimensional Wiener process and is exactly in the spirit of cylindrical processes. The multidimensional formulation in Definition 5 would be already necessary to define a finite-dimensional Wiener process by this approach and it allows to conclude that a weakly cylindrical Wiener process has weakly independent increments. The latter property is exactly what is needed in addition to an one-dimensional formulation:

Lemma 1. *For an adapted cylindrical process $W = (W(t) : t \geqslant 0)$ the following are equivalent:*

(a) *W is a weakly cylindrical Wiener process*
(b) *W satisfies*

    (i) *W has weakly independent increments*
    (ii) *$(W(t)u^* : t \geqslant 0)$ is a Wiener process for all $u^* \in U^*$*

Proof. We have only to show that (b) implies (a) for which we fix some $u_1^*, \ldots, u_n^* \in U^*$. By linearity we have

$$\beta_1(W(t) - W(s))u_1^* + \cdots + \beta_n(W(t) - W(s))u_n^*$$
$$= (W(t) - W(s)) \left(\sum_{i=1}^{n} \beta_i u_i^* \right),$$

for all $\beta_i \in \mathbb{R}$ which shows that the increments of $((W(t)u_1^*, \ldots, W(t)u_n^*) : t \geqslant 0)$ are normally distributed and stationary. The independence of the increments follows by (i).

Because $W(1)$ is a centred weakly Gaussian cylindrical random variable there exists a weakly Gaussian cylindrical measure μ such that

$$\varphi_{W(1)}(u^*) = E[\exp(iW(1)u^*)] = \varphi_\mu(u^*) = \exp\left(-\tfrac{1}{2}s(u^*)\right)$$

for a quadratic form $s : U^* \to \mathbb{R}_+$. Therefore, one obtains

$$\varphi_{W(t)}(u^*) = E[\exp(iW(t)u^*)] = E[\exp\big(iW(1)(tu^*)\big)] = \exp\left(-\tfrac{1}{2}t^2 s(u^*)\right)$$

for all $t \geqslant 0$. Thus, the cylindrical distributions of $W(t)$ for all $t \geqslant 0$ are only determined by the cylindrical distribution of $W(1)$.

Definition 6. A weakly cylindrical Wiener process $(W(t) : t \geqslant 0)$ is called *strongly cylindrical Wiener process*, if

(b) the cylindrical distribution μ of $W(1)$ is strongly Gaussian.

The additional condition on a weakly cylindrical Wiener process to be strongly requests the existence of an U-valued covariance operator for the Gaussian cylindrical measure. To our knowledge weakly cylindrical Wiener processes are not defined in the literature and (strongly) cylindrical Wiener processes are defined by means of other conditions. Often, these definitions are formulated by assuming the existence of the reproducing kernel Hilbert space. But this implies the existence of the covariance operator. Another popular way for defining cylindrical Wiener processes is by means of a series. We will see in the next chapter that this is also equivalent to our definition.

Later, we will compare a strongly cylindrical Wiener process with an U-valued Wiener process. Also the latter is defined as a direct generalisation of a real-valued Wiener process:

Definition 7. An adapted U-valued stochastic process $(W(t) : t \geqslant 0)$ is called a *Wiener process* if

(a) $W(0) = 0$ P-a.s.
(b) W has independent, stationary increments.
(c) There exists a Gaussian covariance operator $Q : U^* \rightarrow U$ such that

$$W(t) - W(s) \overset{d}{=} N(0, (t-s)Q) \qquad \text{for all } 0 \leqslant s \leqslant t.$$

If U is finite dimensional then Q can be any symmetric, positive semi-definite matrix. In case that U is a Hilbert space we know already that Q has to be nuclear. For the general case of a Banach space U we can describe the possible Gaussian covariance operator by Theorem 4.

It is obvious that every U-valued Wiener process W defines a strongly cylindrical Wiener process $(\hat{W}(t) : t \geqslant 0)$ in U by $\hat{W}(t)u^* := \langle W(t), u^* \rangle$. For the converse question, if a cylindrical Wiener process can be represented in such a way by an U-valued Wiener process we will derive later necessary and sufficient conditions.

7 Representations of Cylindrical Wiener Processes

In this section we derive representations of cylindrical Wiener processes and U-valued Wiener processes in terms of some series. In addition, these representations can also serve as a construction of these processes, see Remark 5.

Theorem 5. *For an adapted cylindrical process $W := (W(t) : t \geq 0)$ the following are equivalent:*

(a) *W is a strongly cylindrical Wiener process.*
(b) *There exist a Hilbert space H with an orthonormal basis $(e_n)_{n \in \mathbb{N}}$, $F \in L(H, U)$ and independent real-valued standard Wiener processes $(B_n)_{n \in \mathbb{N}}$ such that*

$$W(t)u^* = \sum_{k=1}^{\infty} \langle Fe_k, u^* \rangle B_k(t) \qquad in \ L^2(\Omega) \ for \ all \ u^* \in U^*.$$

Proof. (b) $\Rightarrow$ (a) By Doob's inequality we obtain for any $m, n \in \mathbb{N}$

$$E \left[\sup_{t \in [0,T]} \left| \sum_{k=n}^{n+m} \langle Fe_k, u^* \rangle B_k(t) \right|^2 \right] \leq 4E \left| \sum_{k=n}^{n+m} \langle Fe_k, u^* \rangle B_k(T) \right|^2$$

$$= 4T \sum_{k=n}^{n+m} \langle e_k, F^*u^* \rangle^2$$

$$\to 0 \qquad for \ m, n \to \infty.$$

Thus, for every $u^* \in U^*$ the random variables $W(t)u^*$ are well defined and form a cylindrical process $(W(t) : t \geq 0)$. For any $0 = t_0 < t_1 < \cdots < t_m$ and $\beta_k \in \mathbb{R}$ we calculate

$$E \left[\exp \left(i \sum_{k=0}^{m-1} \beta_k (W(t_{k+1})u^* - W(t_k)u^*) \right) \right]$$

$$= \lim_{n \to \infty} E \left[\exp \left(i \sum_{k=0}^{m-1} \beta_k \sum_{l=1}^{n} \langle Fe_l, u^* \rangle (B_l(t_{k+1}) - B_l(t_k)) \right) \right]$$

$$= \lim_{n \to \infty} \prod_{k=0}^{m-1} \prod_{l=1}^{n} E \left[\exp \left(i\beta_k \langle Fe_l, u^* \rangle (B_l(t_{k+1}) - B_l(t_k)) \right) \right]$$

$$= \lim_{n \to \infty} \prod_{k=0}^{m-1} \prod_{l=1}^{n} \exp \left(-\tfrac{1}{2} \beta_k^2 \langle Fe_l, u^* \rangle^2 (t_{k+1} - t_k) \right)$$

$$= \prod_{k=0}^{m-1} \exp \left(-\tfrac{1}{2} \beta_k^2 \| F^*u^* \|_H^2 (t_{k+1} - t_k) \right),$$

which shows that $(W(t)u^* : t \geq 0)$ has independent, stationary Gaussian increments and is therefore established as a real-valued Wiener process. Similarly, one establishes that W has weakly independent increments.

The calculation above of the characteristic function yields

$$E\left[\exp(iW(1)u^*)\right] = \exp\left(-\tfrac{1}{2}\,\|F^*u^*\|_H^2\right) = \exp\left(-\tfrac{1}{2}\langle FF^*u^*, u^*\rangle^2\right).$$

Hence, the process W is a strongly cylindrical Wiener process with covariance operator $Q := FF^*$.

(a) $\Rightarrow$ (b): Let $Q : U^* \to U$ be the covariance operator of $W(1)$ and H its reproducing kernel Hilbert space with the inclusion mapping $i_Q : H \to U$. Because the range of i_Q^* is dense in H and H is separable there exists an orthonormal basis $(e_n)_{n\in\mathbb{N}} \subseteq \text{range}(i_Q^*)$ of H. We choose $u_n^* \in U^*$ such that $i_Q^* u_n^* = e_n$ for all $n \in \mathbb{N}$ and define $B_n(t) := W(t)u_n^*$. Then we obtain

$$E\left|\sum_{k=1}^n \langle i_Q e_k, u^*\rangle B_k(t) - W(t)u^*\right|^2 = E\left[W(t)\left(\sum_{k=1}^n \langle i_Q e_k, u^*\rangle u_k^* - u^*\right)\right]^2$$

$$= t\left\|i_Q^*\left(\sum_{k=1}^n \langle i_Q e_k, u^*\rangle u_k^* - u^*\right)\right\|_H^2$$

$$= t\left\|\sum_{k=1}^n [e_k, i_Q^* u^*]_H e_k - i_Q^* u^*\right\|_H^2$$

$$\to 0 \qquad \text{for } n \to \infty.$$

Thus, W has the required representation and it remains to establish that the Wiener processes $B_n := (B_n(t) : t \geqslant 0)$ are independent. Because of the Gaussian distribution it is sufficient to establish that $B_n(s)$ and $B_m(t)$ for any $s \leqslant t$ and $m, n \in \mathbb{N}$ are independent:

$$E[B_n(s)B_m(t)] = E[W(s)u_n^* W(t)u_m^*]$$
$$= E[W(s)u_n^*(W(t)u_m^* - W(s)u_m^*)] + E[W(s)u_n^* W(s)u_m^*].$$

The first term is zero by Lemma 1 and for the second term we obtain

$$E[W(s)u_n^* W(s)u_m^*] = s\langle Qu_n^*, u_m^*\rangle = s[i_Q^* u_n^*, i_Q^* u_m^*]_{H_Q} = s[e_n, e_m]_{H_Q}.$$

Hence, $B_n(s)$ and $B_m(t)$ are uncorrelated and therefore independent.

Remark 3. The proof has shown that the Hilbert space H in part (b) can be chosen as the reproducing kernel Hilbert space associated to the Gaussian cylindrical distribution of $W(1)$. In this case the function $F : H \to U$ is the inclusion mapping i_Q.

Remark 4. Let H be a separable Hilbert space with orthonormal basis $(e_k)_{k\in\mathbb{N}}$ and $(B_k(t) : t \geqslant 0)$ be independent real-valued Wiener processes. By setting $U = H$

and $F = \mathrm{Id}$ Theorem 5 yields that a strongly cylindrical Wiener process $(W_H(t) : t \geq 0)$ is defined by

$$W_H(t)h = \sum_{k=1}^{\infty} \langle e_k, h \rangle B_k(t) \qquad \text{for all } h \in H.$$

The covariance operator of W_H is $\mathrm{Id} : H \to H$. This is the approach how a cylindrical Wiener process is defined for example in [2] and [7].

If in addition V is a separable Banach space and $F \in L(H, V)$ we obtain by defining

$$W(t)v^* := W_H(t)(F^* v^*) \qquad \text{for all } v^* \in V^*,$$

a strongly cylindrical Wiener process $(W(t) : t \geq 0)$ with covariance operator $Q := FF^*$ according to our Definition 6.

Theorem 6. *For an adapted U-valued process $W := (W(t) : t \geq 0)$ the following are equivalent:*

(a) *W is an U-valued Wiener process.*
(b) *There exist a Hilbert space H with an orthonormal basis $(e_n)_{n \in \mathbb{N}}$, a γ-radonifying operator $F \in L(H, U)$ and independent real-valued standard Wiener processes $(B_n)_{n \in \mathbb{N}}$ such that*

$$W(t) = \sum_{k=1}^{\infty} F e_k B_k(t) \qquad \text{in } L^2(\Omega; U).$$

Proof. (b) $\Rightarrow$ (a): As in the proof of Theorem 5 we obtain by Doob's Theorem (but here for infinite-dimensional spaces) for any $m, n \in \mathbb{N}$

$$E \left[\sup_{t \in [0,T]} \left\| \sum_{k=n}^{n+m} F e_k B_k(t) \right\|^2 \right] \leq 4E \left\| \sum_{k=n}^{n+m} F e_k B_k(T) \right\|^2$$
$$\to 0 \qquad \text{for } m, n \to \infty,$$

where the convergence follows by Theorem 4 because F is γ-radonifying. Thus, the random variables $W(t)$ are well defined and form an U-valued stochastic process $W := (W(t) : t \geq 0)$. As in the proof of Theorem 5 we can proceed to establish that W is an U-valued Wiener process.

(a) $\Rightarrow$ (b): By Theorem 5 there exist a Hilbert space H with an orthonormal basis $(e_n)_{n \in \mathbb{N}}$, $F \in L(H, U)$ and independent real-valued standard Wiener processes $(B_n)_{n \in \mathbb{N}}$ such that

$$\langle W(t), u^* \rangle = \sum_{k=1}^{\infty} \langle F e_k, u^* \rangle B_k(t) \qquad \text{in } L^2(\Omega) \text{ for all } u^* \in U^*.$$

The Itô-Nisio Theorem [11, Theorem 2.4 in Chapter V] implies

$$W(t) = \sum_{k=1}^{\infty} Fe_k B_k(t) \qquad P\text{-a.s.}$$

and a result by Hoffmann-Jorgensen [11, Corollary 2 in Chapter V, Sect. 3.3] yields the convergence in $L^2(\Omega; U)$. Theorem 4 verifies F as γ-radonifying.

Remark 5. In the proofs of the implication from (b) to (a) we established in both Theorems 5 and 6 even more than required: we established the convergence of the series in the specified sense without assuming the existence of the limit process, respectively. This means, that we can read these results also as a construction principle of cylindrical or U-valued Wiener processes without assuming the existence of the considered process a priori.

The construction of these random objects differs significantly in the required conditions on the involved operator F. For a cylindrical Wiener process no conditions are required, however, for an U-valued Wiener process we have to guarantee $Q = FF^*$ to be a covariance operator of a Gaussian measure by assuming F to be γ-radonifying.

8 When is a Cylindrical Wiener Process U-Valued?

In this section we give equivalent conditions for a strongly cylindrical Wiener process to be an U-valued Wiener process. To be more precise a cylindrical random variable $X : U^* \to L^0(\Omega)$ is called *induced by a random variable* $Z : \Omega \to U$, if P-a.s.

$$Xu^* = \langle Z, u^* \rangle \qquad \text{for all } u^* \in U^*.$$

This definition generalises in an obvious way to cylindrical processes.

Because of the correspondence to cylindrical measures the question whether a cylindrical random variable is induced by an U-valued random variable is reduced to the question whether the cylindrical measure extends to a Radon measure ([11, Theorem 3.1 in Chapter VI]). There is a classical answer by Prokhorov ([11, Theorem 3.2 in Chapter VI]) to this question in terms of tightness. A cylindrical measure μ on $\mathcal{Z}(U)$ is called *tight* if for each $\varepsilon > 0$ there exists a compact subset $K = K(\varepsilon) \subseteq U$ such that

$$\mu_{u_1^*,\dots,u_n^*}\left(\{(\beta_1,\dots,\beta_n) \in \{(\langle u, u_1^* \rangle,\dots,\langle u, u_n^* \rangle) : u \in K\}\}\right) \geq 1 - \varepsilon$$

for all $u_1^*,\dots,u_n^* \in U^*$ and all $n \in \mathbb{N}$.

Theorem 7. *For a strongly cylindrical Wiener process* $W := (W(t) : t \geq 0)$ *with covariance operator* $Q = i_Q i_Q^*$ *the following are equivalent:*

(a) *W is induced by an U-valued Wiener process.*

(b) i_Q is γ-radonifying.
(c) The cylindrical distribution of $W(1)$ is tight.
(d) The cylindrical distribution of $W(1)$ extends to a measure.

Proof. (a) $\Rightarrow$ (b) If there exists an U-valued Wiener process $(\tilde{W}(t) : t \geq 0)$ with $W(t)u^* = \langle \tilde{W}(t), u^* \rangle$ for all $u^* \in U^*$, then $\tilde{W}(1)$ has a Gaussian distribution with covariance operator Q. Thus, i_Q is γ-radonifying by Theorem 4.

(b) $\Leftrightarrow$ (c) $\Leftrightarrow$ (d) This is Prokhorov's Theorem on cylindrical measures.

(b) $\Rightarrow$ (a) Due to Theorem 5 there exist an orthonormal basis $(e_n)_{n \in \mathbb{N}}$ of the reproducing kernel Hilbert space of Q and independent standard real-valued Wiener process $(B_k(t) : t \geq 0)$ such that

$$W(t)u^* = \sum_{k=1}^{\infty} \langle i_Q e_k, u^* \rangle B_k(t) \qquad \text{for all } u^* \in U^*.$$

On the other hand, because i_Q is γ-radonifying Theorem 6 yields that

$$\tilde{W}(t) = \sum_{k=1}^{\infty} i_Q e_k B_k(t)$$

defines an U-valued Wiener process $(\tilde{W}(t) : t \geq 0)$. Obviously, we have $W(t)u^* = \langle \tilde{W}(t), u^* \rangle$ for all u^*.

If U is a separable Hilbert space we can replace the condition (b) by
(b') i_Q is Hilbert–Schmidt

because of Corollary 1.

9 Integration

In this section we introduce an integral with respect to a strongly cylindrical Wiener process $W = (W(t) : t \geq 0)$ in U. The integrand is a stochastic process with values in $L(U, V)$, the set of bounded linear operators from U to V, where V denotes a separable Banach space. For that purpose we assume for W the representation according to Theorem 5:

$$W(t)u^* = \sum_{k=1}^{\infty} \langle i_Q e_k, u^* \rangle B_k(t) \qquad \text{in } L^2(\Omega) \text{ for all } u^* \in U^*,$$

where H is the reproducing kernel Hilbert space of the covariance operator Q with the inclusion mapping $i_Q : H \to U$ and an orthonormal basis $(e_n)_{n \in \mathbb{N}}$ of H. The real-valued standard Wiener processes $(B_k(t) : t \geq 0)$ are defined by $B_k(t) = W(t)u_k^*$ for some $u_k^* \in U^*$ with $i_Q^* u_k^* = e_k$.

Definition 8. The set $M_T(U, V)$ contains all random variables $\Phi : [0, T] \times \Omega \to L(U, V)$ such that:

(a) $(t, \omega) \mapsto \Phi^*(t, \omega)v^*$ is $\mathcal{B}[0, T] \otimes \mathcal{A}$ measurable for all $v^* \in V^*$.
(b) $\omega \mapsto \Phi^*(t, \omega)v^*$ is $\mathcal{F}_t$-measurable for all $v^* \in V^*$ and $t \in [0, T]$.
(c) $\int_0^T E \|\Phi^*(s, \cdot)v^*\|_{U^*}^2 \, ds < \infty$ for all $v^* \in V^*$.

As usual we neglect the dependence of $\Phi \in M_T(U, V)$ on ω and write $\Phi(s)$ for $\Phi(s, \cdot)$ as well as for the dual operator $\Phi^*(s) := \Phi^*(s, \cdot)$ where $\Phi^*(s, \omega)$ denotes the dual operator of $\Phi(s, \omega) \in L(U, V)$.

We define the candidate for a stochastic integral:

Definition 9. For $\Phi \in M_T(U, V)$ we define

$$I_t(\Phi)v^* := \sum_{k=1}^{\infty} \int_0^t \langle \Phi(s)i_Q e_k, v^* \rangle \, dB_k(s) \qquad \text{in } L^2(\Omega)$$

for all $v^* \in V^*$ and $t \in [0, T]$.

The stochastic integrals appearing in Definition 9 are the known real-valued Itô integrals and they are well defined thanks to our assumption on Φ. In the next Lemma we establish that the asserted limit exists:

Lemma 2. $I_t(\Phi) : V^* \to L^2(\Omega)$ is a well defined cylindrical random variable in V which is independent of the representation of W, i.e. of $(e_n)_{n \in \mathbb{N}}$ and $(u_n^*)_{n \in \mathbb{N}}$.

Proof. We begin to establish the convergence in $L^2(\Omega)$. For that, let $m, n \in \mathbb{N}$ and we define for simplicity $h(s) := i_Q^* \Phi^*(s)v^*$. Doob's theorem implies

$$E \left| \sup_{0 \leq t \leq T} \sum_{k=m+1}^{n} \int_0^t \langle \Phi(s)i_Q e_k, v^* \rangle \, dB_k(s) \right|^2$$

$$\leq 4 \sum_{k=m+1}^{n} \int_0^T E \, [e_k, h(s)]_H^2 \, ds$$

$$\leq 4 \sum_{k=m+1}^{\infty} \int_0^T E \, [[e_k, h(s)]_H \, e_k, h(s)]_H \, ds$$

$$= 4 \sum_{k=m+1}^{\infty} \sum_{l=m+1}^{\infty} \int_0^T E \, [[e_k, h(s)]_H \, e_k, [e_l, h(s)]_H \, e_l]_H \, ds$$

$$= 4 \int_0^T E \, \|(\mathrm{Id} - \pi_m)h(s)\|_H^2 \, ds,$$

where $\pi_m : H \to H$ denotes the projection onto the span of $\{e_1, \ldots, e_m\}$. Because $\|(\mathrm{Id} - \pi_m) h(s)\|_H^2 \to 0$ for $m \to \infty$ and

$$\int_0^T E \|(\mathrm{Id} - \pi_m) h(s)\|_H^2 \, ds \leqslant \|i_Q^*\|_{U^* \to H}^2 \int_0^T E \|\Phi^*(s, \cdot) v^*\|_{U^*}^2 \, ds < \infty$$

we obtain by Lebesgue's theorem the convergence in $L^2(\Omega)$.

To prove the independence on the chosen representation of W let $(f_l)_{l \in \mathbb{N}}$ be an other orthonormal basis of H and $w_l^* \in U^*$ such that $i_Q^* w_l^* = f_l$ and $(C_l(t) : t \geqslant 0)$ independent Wiener processes defined by $C_l(t) = W(t) w_l^*$. As before we define in $L^2(\Omega)$:

$$\tilde{I}_t(\Phi) v^* := \sum_{l=1}^{\infty} \int_0^t \langle \Phi(s) i_Q f_l, v^* \rangle \, dC_l(s) \qquad \text{for all } v^* \in V^*.$$

The relation $\mathrm{Cov}(B_k(t), C_l(t)) = t \left[i_Q^* u_k^*, i_Q^* w_l^* \right]_H = t [e_k, f_l]_H$ enables us to calculate

$$E \left| I_t(\Phi) v^* - \tilde{I}_t(\Phi) v^* \right|^2$$

$$= E \left| I_t(\Phi) v^* \right|^2 + E \left| \tilde{I}_t(\Phi) v^* \right|^2 - 2E \left[\left(I_t(\Phi) v^* \right) \left(\tilde{I}_t(\Phi) v^* \right) \right]$$

$$= \sum_{k=1}^{\infty} \int_0^t E \langle \Phi(s) i_Q e_k, v^* \rangle^2 \, ds + \sum_{l=1}^{\infty} \int_0^t E \langle \Phi(s) i_Q f_l, v^* \rangle^2 \, ds$$

$$- 2 \sum_{k=1}^{\infty} \sum_{l=1}^{\infty} \int_0^t E \left[\langle \Phi(s) i_Q e_k, v^* \rangle \langle \Phi(s) i_Q f_l, v^* \rangle \left[i_Q^* u_k^*, i_Q^* w_l^* \right]_H \right] ds$$

$$= 2 \int_0^t E \| i_Q^* \Phi^*(s) v^* \|_H^2 \, ds - 2 \int_0^t E \| i_Q^* \Phi^*(s) v^* \|_H^2 \, ds$$

$$= 0,$$

which proves the independence of $I_t(\Phi)$ on $(e_k)_{k \in \mathbb{N}}$ and $(u_k^*)_{k \in \mathbb{N}}$.

The linearity of $I_t(\Phi)$ is obvious and hence the proof is complete.

Our next definition is not very surprising:

Definition 10. For $\Phi \in M_T(U, V)$ we call the cylindrical random variable

$$\int_0^t \Phi(s) \, dW(s) := I_t(\Phi)$$

cylindrical stochastic integral with respect to W.

Because the cylindrical stochastic integral is strongly based on the well known real-valued Itô integral many features can be derived easily. We collect the martingale property and Itô's isometry in the following theorem.

Theorem 8. *Let Φ be in $M_T(U, V)$. Then we have*

(a) *For every $v^* \in V^*$ the family*

$$\left(\left(\int_0^t \Phi(s) \, dW(s) \right) v^* : t \in [0, T] \right)$$

forms a continuous square-integrable martingale.

(b) *The Itô's isometry:*

$$E \left| \left(\int_0^t \Phi(s) \, dW(s) \right) v^* \right|^2 = \int_0^t E \left\| i_Q^* \Phi^*(s) v^* \right\|_H^2 \, ds.$$

Proof. (a) In Lemma 2 we have identified $I_t(\Phi)v^*$ as the limit of

$$M_n(t) := \sum_{k=1}^n \int_0^t \langle \Phi(s) i_Q e_k, v^* \rangle \, dB_k(s),$$

where the convergence takes place in $L^2(\Omega)$ uniformly on the interval $[0, T]$. As $(M_n(t) : t \in [0, T])$ are continuous martingales the assertion follows.

(b) Using Itô's isometry for real-valued stochastic integrals we obtain

$$E \left| \left(\int_0^t \Phi(s) \, dW(s) \right) v^* \right|^2 = \sum_{k=1}^\infty E \left[\int_0^T \langle \Phi(s) i_Q e_k, v^* \rangle \, dB_k(s) \right]^2$$

$$= \sum_{k=1}^\infty \int_0^T E \left[e_k, i_Q^* \Phi^*(s) v^* \right]_H^2 \, ds$$

$$= \int_0^T E \left\| i_Q^* \Phi^*(s) v^* \right\|_H^2 \, ds.$$

An obvious question is under which conditions the cylindrical integral is induced by a V-valued random variable. The answer to this question will also allow us to relate the cylindrical integral with other known definitions of stochastic integrals in infinite dimensional spaces.

From our point of view the following corollary is an obvious consequence. We call a stochastic process $\Phi \in M_T(U, V)$ non-random if it does not depend on $\omega \in \Omega$.

Corollary 2. *For non-random $\Phi \in M_T(U, V)$ the following are equivalent:*

(a) $\int_0^T \Phi(s) \, dW(s)$ *is induced by a V-valued random variable.*
(b) *There exists a Gaussian measure μ on V with covariance operator R such that:*

$$\int_0^T \left\| i_Q^* \Phi^*(s) v^* \right\|_H^2 \, ds = \langle Rv^*, v^* \rangle \qquad \text{for all } v^* \in V^*.$$

Proof. (a) ⇒ (b): If the integral $I_T(\Phi)$ is induced by a V-valued random variable then the random variable is centred Gaussian, say with a covariance operator R. Then Itô's isometry yields

$$\langle Rv^*, v^* \rangle = E\,|I_T(\Phi)v^*|^2 = \int_0^T \left\| i_Q^* \Phi^*(s)v^* \right\|_H^2 \, ds.$$

(b) ⇒ (a): Again Itô's isometry shows that the weakly Gaussian cylindrical distribution of $I_T(\Phi)$ has covariance operator R and thus, extends to a Gaussian measure on V.

The condition (b) of Corollary 2 is derived in van Neerven and Weis [7] as a sufficient and necessary condition for the existence of the stochastic Pettis integral introduced in this work. Consequently, it is easy to see that under the equivalent conditions (a) or (b) the cylindrical integral is induced by the stochastic Pettis integral which is a genuine random variable in the underlying Banach space. Further relation of condition (b) to γ-radonifying properties of the integrand Φ can also be found in [7].

Our next result relates the cylindrical integral to the stochastic integral in Hilbert spaces as introduced in Da Prato and Zabczyk [3]. For that purpose, we assume that U and V are separable Hilbert spaces. Let W be a strongly cylindrical Wiener process in U and let the inclusion mapping $i_Q : H_Q \to U$ be Hilbert–Schmidt. Then there exist an orthonormal basis $(f_k)_{k \in \mathbb{N}}$ in U and real numbers $\lambda_k \geq 0$ such that $Qf_k = \lambda_k f_k$ for all $k \in \mathbb{N}$. For the following we can assume that $\lambda_k \neq 0$ for all $k \in \mathbb{N}$. By defining $e_k := \sqrt{\lambda_k}\, f_k$ for all $k \in \mathbb{N}$ we obtain an orthonormal basis of H_Q and W can be represented as usual as a sum with respect to this orthonormal basis.

Our assumption on i_Q to be Hilbert–Schmidt is not a restriction because in general the integral with respect to a strongly cylindrical Wiener process is defined in [3] by extending U such that i_Q becomes Hilbert–Schmidt.

Corollary 3. *Let W be a strongly cylindrical Wiener process in a separable Hilbert space U with $i_Q : H_Q \to U$ Hilbert–Schmidt. If V is a separable Hilbert space and $\Phi \in M_T(U, V)$ is such that*

$$\sum_{k=1}^{\infty} \lambda_k \int_0^T E\, \|\Phi(s) i_Q e_k\|_V^2 \, ds < \infty,$$

then the cylindrical integral

$$\int_0^T \Phi(s) \, dW(s)$$

is induced by a V-valued random variable. This random variable is the standard stochastic integral in Hilbert spaces of Φ with respect to W.

Proof. By Theorem 7 the cylindrical Wiener process W is induced by an U-valued Wiener process Y. We define U-valued Wiener processes $(Y_N(t) : t \in [0, T])$ by

$$Y_N(t) = \sum_{k=1}^{N} i_Q e_k B_k(t).$$

Theorem 6 implies that $Y_N(t)$ converges to Y in $L^2(\Omega; U)$. By our assumption on Φ the stochastic integrals $\Phi \circ Y_N(T)$ in the sense of Da Prato and Zabczyk [3] exist and converge to the stochastic integral $\Phi \circ Y(T)$ in $L^2(\Omega; V)$, see [3, Ch. 4.3.2].

On the other hand, by first considering simple functions Φ and then extending to the general case we obtain

$$\langle \Phi \circ Y_N(T), v^* \rangle = \sum_{k=1}^{N} \int_0^t \langle \Phi(s) i_Q e_k, v^* \rangle \, dB_k(s)$$

for all $v^* \in V^*$. By Definition 9 the right hand side converges in $L^2(\Omega)$ to

$$\left(\int_0^T \Phi(s) \, dW(s) \right) v^*,$$

whereas at least a subsequence of $(\langle \Phi \circ Y_N(T), v^* \rangle)_{N \in \mathbb{N}}$ converges to $\langle \Phi \circ Y(T), v^* \rangle$ P-a.s..

Based on the cylindrical integral one can consider *linear cylindrical stochastic differential equations*. Of course, a solution will be in general a cylindrical process but there is no need to put geometric constrains on the state space under consideration. If one is interested in classical stochastic processes as solutions for some reasons one can tackle this problem as in our two last results by deriving sufficient conditions guaranteeing that the cylindrical solution is induced by a V-valued random process.

Acknowledgements I thank David Applebaum for his careful review of the original manuscript and many fruitful discussions.

References

1. Berman, N., Root, W.L.: A weak stochastic integral in Banach space with application to a linear stochastic differential equation. Appl. Math. Optimization **10**, 97–125 (1983)
2. Bogachev, V.I.: Gaussian Measures. AMS, Providence, RI (1998)
3. Da Prato, G., Zabczyk, J.: Stochastic Equations in Infinite Dimensions. Cambridge University Press, Cambridge (1992)
4. Kallianpur, G., Xiong, J.: Stochastic Differential Equations in Infinite Dimensional Spaces. IMS, Inst. of Math. Statistics, Hayward, CA (1996)

5. Métivier, M., Pellaumail, J.: Stochastic Integration. Academic, New York (1980)
6. van Neerven, J.: Gaussian sums and γ-radonifying operators. Lecture Notes. fa.its.tudelft.nl/seminar/seminar2002_2003/seminar.pdf (2003)
7. van Neerven, J., Weis, L.: Stochastic integration of functions with values in a Banach space. Stud. Math. **166**(2), 131–170 (2005)
8. Schwartz, L.: Radon Measures on Arbitrary Topological Spaces and Cylindrical Measures. Oxford University Press, London (1973)
9. Schwartz, L.: Seminar Schwartz. Notes on Pure Mathematics. 7. Australian National University, Department of Mathematics, Canberra (1973)
10. Schwartz, L.: Le théorème des trois opérateurs. Ann. Math. Blaise Pascal **3**(1), 143–164 (1996)
11. Vakhaniya, N., Tarieladze, V., Chobanyan, S.: Probability Distributions on Banach Spaces. D. Reidel Publishing Company, Dordrecht (1987)

A Remark on the $1/H$-Variation of the Fractional Brownian Motion

Maurizio Pratelli

Abstract We give an elementary proof of the following property of H-fractional Brownian motion: almost all sample paths have infinite 1/H-variation on every interval.

Keywords Fractional Brownian motion · p-Variation · Ergodic theorem

1 Introduction and Statement of the Result

Let $(B_t)_{t \geq 0}$ be the *Fractional Brownian Motion* with Hurst (or self-similarity) parameter H, $0 < H < 1$ (we refer for instance to [2] or [5] or [8] p. 273 for the definitions): fix $t > 0$ and let $t_k^n = \frac{kt}{n}$ for n integer and $k = 0, \dots, n$. It is well known (see e.g. [5] or [9]) that

$$\lim_{n \to \infty} \sum_{k=0}^{n-1} \left| B_{t_{k+1}^n} - B_{t_k^n} \right|^p \overset{L^1(\Omega)}{=} \begin{cases} +\infty & p < 1/H \\ t\, E\big[|B_1|^{1/H} \big] & p = 1/H \\ 0 & p > 1/H \end{cases}.$$

However, if we define the random variable

$$\mathcal{V}(\omega) = \mathcal{V}_{[0,t]}^{1/H}(\omega) = \sup_{n,\, 0 \leq t_1 < t_2 < \dots < t_n \leq t} \sum_{i=1}^{n-1} \left| B_{t_{i+1}}(\omega) - B_{t_i}(\omega) \right|^{1/H}$$

then $\mathcal{V}(\omega) = +\infty$ a.s. (note that $\mathcal{V}$ is measurable since the paths of the fractional Brownian motion are continuous and therefore the "sup" can be taken over rationals t_i).

M. Pratelli (✉)
Dipartimento di Matematica, Largo B. Pontecorvo 5, 56127 Pisa, Italy
e-mail: pratelli@dm.unipi.it

C. Donati-Martin et al. (eds.), *Séminaire de Probabilités XLIII*, Lecture Notes in Mathematics 215
2006, DOI 10.1007/978-3-642-15217-7_8, © Springer-Verlag Berlin Heidelberg 2011

This property is well known in the case of standard *Brownian Motion* (i.e. $H = 1/2$): it was stated (but without a rigorous proof) by P. Lévy in [6] p. 190, and also the excellent book "Revuz-Yor" quotes the result without a proof (see [8] p. 28). A sketch of the proof (always in the case $H = 1/2$) was given by D. Freedman in [4] p. 48.

For the standard Brownian Motion, there is a more precise (and more technical) result due to Taylor (see [11]): given an increasing function $\psi : [0, +\infty) \to [0, +\infty)$, we can define the ψ–variation of the function f on the interval $[a, b]$ as

$$\sup_{n, a=t_1 < t_2 < ... < t_n = b} \psi\big(|f(t_{i+1}) - f(t_i)|\big)$$

(when $\psi(t) = t^p$ it is called the p–variation).

Taylor showed that the *correct* function for the variation of the paths of the BM is the function $\psi_1(s) = s^2/2\log^* \log^* s$ (where $\log^* s = \max(1, |\log s|)$) in the sense that

$$\mathcal{V}_{\psi^*, [a,b]}(\omega) = \sup_{n, a=t_1 < t_2 < ... < t_n = b} \psi\big(|B_{t_{i+1}}(\omega) - B_{t_i}(\omega)|\big)$$

is a finite r.v. but is infinite if ψ_1 is replaced by any function ψ such that $\psi(s)/\psi_1(s) \to +\infty$ as $s \to 0+$.

The impact of the p-variation of the paths for (stochastic) integration is well highlighted by L. Coutin (see [2]) for the case of FBM and by Dudley and Norvaisa (see [3]) for more general stochastic processes.

In the general case of the Fractional Brownian Motion, it is well known that $p = 1/H$ is a *limit* case, and that sample paths are γ-hölder continuous for any $\gamma < H$. The result of Theorem 1 is known since it is a consequence of Theorem IV.5.1 of [1]: their proof, however, is based on a complex technology (the theory of Besov spaces).

The aim of this short note is to give an elementary complete proof suggested by the argument presented in [4]; I want to thank Sara Biagini and Giorgio Letta for a discussion on the subject.

Let us fix H with $0 < H < 1$, let $(B_t)_{t \geq 0}$ be a FBM with parameter H and continuous sample paths: define

$$\mathcal{V}_{[a,b]}(\omega) = \sup_{n, a=t_1 < t_2 < ... < t_n = b} \big|B_{t_{i+1}}(\omega) - B_{t_i}(\omega)\big|^{1/H}$$

The statement of the result is as follows

Theorem 1. *There exists a null-set $N \subseteq \Omega$ such that, if $\omega \notin N$, then for every $a < b$, $\mathcal{V}_{[a,b]}(\omega) = +\infty$.*

2 The Proof

In the sequel, λ is the Lebesgue measure on $\mathbb{R}^+$ and p^* is a shorthand for $1/H$. Let U be a finite union of disjoint open intervals $]s_i, t_i[$ of $\mathbb{R}^+$ with $s_i, t_i \in \mathbb{Q}$ and let $\mathcal{U}$ be the collection of such subsets of $\mathbb{R}^+$.

For $U = \cup_{i=1}^n]s_i, t_i[$ $(0 \leq s_1 < t_1 \leq s_2 < \dots < t_n)$, let $q_U(\omega) = \sum_{i=1}^n |B_{t_i}(\omega) - B_{s_i}(\omega)|^{p^*}$; note that

$$\mathcal{V}_{[a,b]}(\omega) = \sup_{U \in \mathcal{U}, U \subseteq [a,b]} q_U(\omega)$$

Lemma 1. *Fix* $m > 0$ *and let* $p_m = P(|B_1|^{p^*} \geq m)$. *Let* $Z_n = I_{\{|B_n - B_{n-1}|^{p^*} \geq m\}}$: *then* $M_n = \frac{Z_1 + \dots + Z_n}{n}$ *converges a.s. to* $E[Z_1] = p_m$.

Proof. The sequence of one step increments of B, $X_n = B_n - B_{n-1}$, is stationary, centered Gaussian and with covariance function $R(n) = E[X_1 X_{n+1}]$ which tends to 0 when n goes to infinity (see e.g. [7] p. 274): therefore $(X)_{n \geq 1}$ is ergodic (see [10] p. 413). Now Z_n can be written in the form $Z_n = g(X_n)$ with a borel function g and therefore also $(Z_n)_{n \geq 1}$ is ergodic.

As a consequence of the ergodic theorem (Theorem 3.3 p. 413 of [10]), the sequence of the empirical means $M = (M_n)_{n \geq 1}$ converges a.s. (and in L^1) to $E[Z_1]$.

The key of the proof is the following result:

Lemma 2. *Let* $I =]s, t[$ *be an open interval with* $s, t \in \mathbb{Q}^+$ *and fix* $m > 0$: *let* $p_m = P\{|B_1|^{p^*} \geq m\}$ *and* $r < p_m$. *Then there exists a measurable* $A_m \subseteq \Omega$ *with* $P(A_m) = 1$ *such that for all* $\omega \in A_m$ *there exists* $U_\omega \in \mathcal{U}$ *with the properties:*

1. $U_\omega \subset I$
2. $\lambda(U_\omega) > r \lambda(I)$
3. $q_{U_\omega}(\omega) \geq m \lambda(U_\omega)$

Proof. For $n > 1$ and $i = 0, \dots, n$ let $t_i^n = s + \frac{i}{n}(t - s)$ and $J_i^n =]t_{i-1}^n, t_i^n[$. Set

$$S_n = \sum_{i=1}^n I_{\left\{|B_{t_{i+1}^n} - B_{t_i^n}|^{p^*} \geq m \frac{(t-s)}{n}\right\}}$$

$S_n(\omega)$ counts the number of subintervals J_i^n on which $q_{J_i^n}(\omega) > m \lambda(J_i^n)$. Thanks to the self-similarity property of fBm (see e.g. [7] p. 275), S_n is distributed as $Z_n = \sum_{i=1}^n I_{\{|B_{i+1} - B_i|^{p^*} \geq m\}}$.

By the Lemma 1, $\frac{Z_n}{n} \to p_m$ almost surely, whence $\frac{S_n}{n}$ tends to p_m in probability. Modulo a subsequence,

$$\lim_n \frac{S_n}{n} = p_m \text{ a.s.}$$

Call A_m the set on which the above sequence converges: if $\omega \in A_m$ then there exists n_ω such that $\frac{S_n}{n}(\omega) > r$ for $n \geq n_\omega$.

Select $n \geq n_\omega$ and among the subintervals J_i^n exactly those such that $q_{J_i^n}(\omega) \geq m \frac{(t-s)}{n}$ and let U_ω be their union. Then

$$\lambda(U_\omega) = S_n(\omega) \frac{t-s}{n} > r \, \lambda(I)$$

and

$$q_{U_\omega}(\omega) \geq S_n(\omega) \, m \frac{(t-s)}{n} = m \, \lambda(U_\omega).$$

The same result holds evidently for an element $U \in \mathcal{U}$. Since $\mathcal{U}$ in countable, we have immediately the following result

Corollary 1. *Fix $m > 0$ and set $r = p_m/2$: there exists a measurable set $C_m \subseteq \Omega$ with $\mathbf{P}(C_m) = 1$ such that, if $\omega \in C_m$ and $V \in \mathcal{U}$, there exists $U_\omega \in \mathcal{U}$ and $U_\omega \subset V$ such that $\lambda(U_\omega) > r \, \lambda(V)$ and $q_{U_\omega}(\omega) \geq m \, \lambda(U_\omega)$.*

Lemma 3. *Fix $0 \leq a < b$; $a, b \in \mathbb{Q}$: then $\mathcal{V}_{[a,b]}(\omega) = +\infty$ a.s.*

Proof. Choose $m > 0$ and apply Lemma 2 to $I =]a, b[$: then for every $\omega \in C_m$ there exists $U_\omega^1 \subset I$ such that $q_{U_\omega^1}(\omega) \geq m \, \lambda(U_\omega^1)$ and $\lambda(U_\omega^1) \geq r \, \lambda(I)$.

Now iterate the procedure, that is apply Corollary 1 to $\left(I \setminus \overline{U}_\omega^1\right)$ ($\overline{U}_\omega^1$ is the closure of U_ω^1): thus there exists $U_\omega^2 \subseteq \left(I \setminus \overline{U}_\omega^1\right)$ with $q_{U_\omega^2}(\omega) \geq m \, \lambda(U_\omega^2)$ and $\lambda(U_\omega^2) > r \, \lambda\left(I \setminus \overline{U}_\omega^1\right)$.

At the $(k+1)$-th step, we have a subset U_ω^{k+1} of $\left(I \setminus (\overline{U}_\omega^1 \cup \ldots \cup \overline{U}_\omega^k)\right)$ such that $\lambda(U_\omega^{k+1}) > r \, \lambda\left(I \setminus (\overline{U}_\omega^1 \cup \ldots \cup \overline{U}_\omega^k)\right)$ and $q_{U_\omega^{k+1}}(\omega) \geq m \, \lambda(U_\omega^{k+1})$.

Call $V_\omega^k = U_\omega^1 \cup \ldots \cup U_\omega^k$, then $V_\omega^k \in \mathcal{U}$ and

$$q_{V_\omega^k}(\omega) \geq m \, \lambda(V_\omega^k)$$

Moreover, by induction $\lambda(I \setminus \overline{V}_\omega^k) \leq (1-r)^k (b-a)$ and therefore

$$\sup_k \, q_{V_\omega^k}(\omega) \geq m \, \lim_k \lambda(V_\omega^k) = m \, (b-a)$$

Now, the intersection $C = \bigcap_{m \in \mathbb{N}} C_m$ has probability one and any $\omega \in C$ satisfies

$$\mathcal{V}_{[a,b]}(\omega) \geq m \, (b-a) \quad \forall m \in \mathbb{N}$$

whence the thesis.

If $N_{[a,b]}$ is the null-set $\{\omega \in \Omega \,|\, \mathcal{V}_{[a,b]}(\omega) < +\infty\}$, then the countable union of all $N_{[a,b]}$ with $a < b$; $a, b \in \mathbb{Q}$, satisfies the hypothesis of Theorem 1.

References

1. Ciesielski, Z., Kerkyacharian, G., Roynette, B.: Quelques espaces fonctionnelles associés a des processus Gaussiens. Studia Math. **107**(2), 171–204 (1993)
2. Coutin, L.: An introduction to (stochastic) calculus with respect to fractional Brownian motion. In: Séminaire de Probabilités XL, Lecture Notes in Mathematics, vol. 1899, pp. 3–65. Springer (2007)
3. Dudley, N.R.: An introduction to P-variation and Young integrals. Tech. rep. 1., Maphysto, Center for Mathematical Physics and Stochastics, University of Aarhus (1998)
4. Freedman, D.: Brownian Motion and Diffusion. Holden-Day (1971)
5. Guerra, J., Nualart, D.: The $1/H$-variation of the divergence integral with respect to the fractional Brownian motion for $H > 1/2$ and fractional Bessel processes. Stoch. Process. Appl. **115**(1), 91–115 (2005)
6. Lévy, P.: Processus stochastiques et Mouvement Brownien. Gauthier-Villars (1965)
7. Nualart, D.: The Malliavin Calculus and Related Topics. Springer, New York (2006)
8. Revuz, D., Yor, M.: Continuous Martingales and Brownian Motion, 3rd edn. Springer, Berlin (1999)
9. Rogers, L.C.G.: Arbitrage with fractional Brownian motion. Math. Finance **7**, 95–105 (1997)
10. Shiryaev, A.N.: Probability, 2nd edn. Springer, Berlin (1996)
11. Taylor, S.J.: Exact asymptotic estimates of Brownian path variation. Duke Math. J. **39**, 219–241 (1972)

Simulation of a Local Time Fractional
Stable Motion

Matthieu Marouby

Abstract The aim of this paper is to simulate sample paths of a class of symmetric α-stable processes. This will be achieved by using the series expansion of the processes seen as shot noise series. In our case, as the general term of the series expansion has to be approximated, a first result is needed in shot noise theory. Then, this will lead to a convergence rate of the approximation towards the Local Time Fractional Stable Motion.

Keywords Stable process · Self similar process · Shot noise series · Local time · Fractional Brownian motion · Simulation

AMS 2000 Subject Classification: Primary 60G18, Secondary 60F25, 60E07, 60G52

1 Introduction

Fractional fields have often been used to model irregular phenomena. The simplest one is the fractional Brownian motion introduced in [12] then developed in [17]. More recently, many fractional processes have been studied, usually obtained by a stochastic integration of a deterministic kernel against a random measure (cf. among others [3, 11, 13, 16] and [4]). Many different simulation methods have been discussed in the literature, but shot noise series seem to perfectly fit that kind of problem. Generalized shot noise series were introduced for simulation in [20], further developments were done in [21] and [22] and a general framework was developed in [5]. Moreover, a computer study of the convergence rate of LePage series to α-stable random variables has been done in a particular case in [10].

A shot noise series can be seen as:

$$\sum_{n=1}^{\infty} \Gamma_n^{-1/\alpha} V_n$$

M. Marouby (✉)

Institut de Mathématiques de Toulouse, Université Paul Sabatier, 31062 Toulouse, France

e-mail: marouby@math.univ-toulouse.fr

C. Donati-Martin et al. (eds.), *Séminaire de Probabilités XLIII*, Lecture Notes in Mathematics 221
2006, DOI 10.1007/978-3-642-15217-7_9, © Springer-Verlag Berlin Heidelberg 2011

where (V_n) are i.i.d. random variables and (Γ_n) are the arrival times of a Poisson process. Usually, there is no question about the simulation of V_n. In this paper, we will consider the convergence rate when V_n can not be simulated but only approximated. The quality of the approximation will be allowed to change depending on its impact on the overall process. Indeed, in a shot noise series representation of an α-stable process, the first terms are the most significant so one has to minimize the error made in approximating V_n for small n. Moreover, for small n, $\Gamma_n^{-1/\alpha}$ has infinite q-moments, which means the corresponding terms will need a particular treatment. For large n, $\Gamma_n^{-1/\alpha}$ is smaller, so it is not as useful to approximate V_n with the same precision. The convergence rate towards the limiting process will be shown depending on the approximation of each term of the series.

Subsequently, this result will be applied to study a particular class of processes which is the main interest of the paper. In network traffic modeling, properties like self-similarity, heavy tails and long-range dependance are often needed; see for example [18]. Moreover, empirical studies like [7] have shown the importance of self-similarity and long-range dependance in that area.

In [6], the authors introduced "fractional Brownian motion local time fractional stable motion" as a stochastic integration of a non-deterministic kernel against a random measure, which will be our main interest in the second part. Here we will call it Local Time Fractional Stable Motion (LTFSM). This process has been defined as:

$$\int_{\Omega'} \int_{\mathbb{R}} l(x,t)(\omega') M(d\omega', dx), \text{ for } t \geq 0.$$

In this expression, l is the local time of a fractional Brownian motion of Hurst parameter H defined on $(\Omega', \mathcal{F}', \mathbf{P}')$. M is a symmetric alpha stable random measure (see [23] for more details) with control measure $\mathbf{P}' \times Leb$ (Leb being the Lebesgue measure on $\mathbb{R}$). LTFSM is α-stable but also self-similar and its increments are long-range dependent.

The first step towards understanding LTFSM is naturally to observe its sample paths. Unfortunately, the above expression does not directly provide a way to obtain the sample paths. In the case of Brownian motion local time, this process can be seen as the limit of a discrete random walk with random rewards model. It is not completely satisfying for a few reasons: first, it only works for $H = 1/2$, then, there is no control of the convergence speed rate towards the limit. This is where the tool that we have developed in the first part will be used. In this paper, we will study how we can simulate this process by using the expression given in (5.3) in [6], which can be seen as a shot noise expansion. In fact, up to a multiplicative constant, LTSFM has the same distribution as:

$$\sum_{n=1}^{\infty} \Gamma_n^{-1/\alpha} V_n,$$

where V_n depends on independent copies of the fractional Brownian motion local time. Two kind of approximations will be involved: one from the truncation of the series, the other one from the approximation of the local time which will be dealt with in more details later.

The next section is devoted to the shot noise theory results. In the following part we will see how LTFSM fits in the general frame we have just developed, how the local time is approximated and we will be able to obtain a convergence rate in the case of our process. The last section will be devoted to a quick study of our simulations with a comparison to the random walk with random rewards model that was introduced in [6] in the case $H = 1/2$.

2 Shot Noise Series

In this section, some results on shot noise series will be shown, mainly thanks to Theorem 2.4 in [21].

Assumption 1. *Let E_K be the space of continuous functions defined on a compact subset $K \subset \mathbb{R}$, equipped with the uniform norm denoted by $\|\cdot\|_K$. For $p \geq 1$, we will denote by $\|\cdot\|_{K,p}$ the $L^p(K)$ norm. For $\alpha \in (0,2)$, let us consider the map $h : \mathbb{R}_+ \times E_K \to E_K$ with*

$$h(r, v) = r^{-1/\alpha} v. \tag{1}$$

h is a Borel measurable map. Let $(\Gamma_n)_{n \geq 1}$ be the arrival times of a Poisson process of rate 1 in $\mathbb{R}_+$ and $(V_n)_{n \geq 1}$ be a sequence of i.i.d. symmetric random variables of distribution λ with value in E_K. Let us assume that $(\Gamma_n)_{n \in \mathbb{N}}$ and $(V_n)_{n \in \mathbb{N}}$ are independent. Moreover we will suppose that for all $q \geq \alpha$, there exists M_q such that for all n, $\mathbb{E}[\|V_n\|_K^q] \leq M_q < \infty$.

Proposition 1. *Under Assumption 1, the series $\sum_{n=1}^{\infty} h(\Gamma_n, V_n)$ converges in E_K almost surely.*

Proof. The proof of this proposition simply consists in verifying that the series satisfies the assumptions of Theorem 2.4 in [21]. ∎

Now that the series $\sum_{n=1}^{\infty} h(\Gamma_n, V_n)$ has been proved to be convergent in E_K, it can be considered as a stochastic process defined on the compact set K. Its characteristic function Φ can be computed. By considering the process (Γ_n, V_n) as a marked Poisson process, Poisson process related techniques can be used to prove the next proposition. Details of the proof will be skipped.

Proposition 2. *Under Assumption 1, E_K' being the dual space of E_K, for all $y' \in E_K'$, if Φ is defined by*

$$\Phi(y') := \mathbb{E}\left[e^{i\langle y', \sum_{n=1}^{\infty} h(\Gamma_n, V_n)\rangle}\right],$$

then

$$\Phi(y') = \exp \int \left(e^{i<y',h(r,v)>} - 1 - i < y', h(r, v) > \mathbf{1}_{\|h(r,v)\|_K \leq 1}\right) dr\lambda(dv)$$

where λ is the distribution of V_n.

The following proposition, regarding the L^q distance between the sum of the series and the truncated sum, is proved with inspiration from [14].

Proposition 3. *Under Assumption 1, let us consider the shot noise series*

$$Y = \sum_{n=1}^{\infty} h(\Gamma_n, V_n).$$

Denote by $Y_N = \sum_{n=1}^{N} h(\Gamma_n, V_n)$. Take $q \geq 2$ as defined in Assumption 1. Then, for $N > q/\alpha - 1$, we have

$$\mathbb{E}[|Y(t) - Y_N(t)|^q] \leq \frac{A_q}{N^{q(\frac{2-\alpha}{2\alpha})}},$$

where A_q depends only on q through M_q.

Proof. For $N < P$, let

$$R_{N,P}(t) = \sum_{n=N+1}^{P} h(\Gamma_n, V_n)(t).$$

According to Proposition 1, $R_{N,P}$ converges in E_K uniformly almost surely when P goes to ∞.

As $h(\Gamma_n, V_n)(t) = \Gamma_n^{-1/\alpha} V_n(t)$ and because $V_n(t)$ is symmetric and independent from Γ_n, $h(\Gamma_n, V_n)(t)$ is also symmetric. Therefore, Proposition 2.3 in [15] can be applied to obtain

$$\mathbb{E}\left[\max_{N+1 \leq n \leq P} |R_{N,n}(t)|^q\right] \leq 2\mathbb{E}[|R_{N,P}(t)|^q].$$

Khintchine inequality can now be applied with a little subtlety. Let ε_n be a sequence of i.i.d. Rademacher random variables, independent from everything else. Thus $\varepsilon_n h(\Gamma_n, V_n)(t)$ has the same distribution as $h(\Gamma_n, V_n)(t)$ since it is symmetric. Then, Khintchine inequality claims

$$\mathbb{E}\left[\left|\sum_{n=N+1}^{P} \varepsilon_n h(\Gamma_n, V_n)(t)\right|^q \left|(h(\Gamma_n, V_n)(t))_{n \in \mathbb{N}}\right|\right]^{1/q}$$

$$\leq B_q \left(\sum_{n=N+1}^{P} |h(\Gamma_n, V_n)(t)|^2\right)^{1/2},$$

where $B_q = \sqrt{2}\left(\frac{\Gamma((q+1)/2)}{\sqrt{\pi}}\right)^{1/q}$ for $q \geq 2$. By taking the expected value on both sides of the inequality, then using Minkowski's inequality, we obtain

$$\mathbb{E}\left[\left|\sum_{n=N+1}^{P} h(\Gamma_n, V_n)(t)\right|^q\right] \le B_q^q \left(\sum_{n=N+1}^{P} \mathbb{E}\left[|h(\Gamma_n, V_n)(t)|^q\right]^{2/q}\right)^{q/2}. \quad (2)$$

There is only $\mathbb{E}\left[|h(\Gamma_n, V_n)(t)|^q\right]$ left to compute. As V_n and Γ_n are independent, as we can compute $\mathbb{E}[\Gamma_n^{-q/\alpha}]$ and as $\mathbb{E}[\|V_n\|_K^q] \le M_q < \infty$, we have

$$\mathbb{E}\left[|h(\Gamma_n, V_n)(t)|^q\right] \le M_q \frac{\Gamma\left(n - \frac{q}{\alpha}\right)}{\Gamma(n)}.$$

These results can be used in (2), leading to

$$\mathbb{E}[|R_{N,P}(t)|^q] \le B_q^q M_q \left(\sum_{n=N+1}^{P} \left(\frac{\Gamma\left(n - \frac{q}{\alpha}\right)}{\Gamma(n)}\right)^{2/q}\right)^{q/2}. \quad (3)$$

Using Stirling formulae

$$\sum_{n=N+1}^{+\infty} \left(\frac{\Gamma\left(n - \frac{q}{\alpha}\right)}{\Gamma(n)}\right)^{2/q} \sim \sum_{n=N+1}^{+\infty} \frac{1}{n^{2/\alpha}} \sim \frac{\alpha}{(2-\alpha)N^{2/\alpha-1}},$$

which shows us that the series converges. Let us denote

$$H_{n,q} = \frac{\Gamma\left(n - \frac{q}{\alpha}\right) n^{q/\alpha}}{\Gamma(n)}. \quad (4)$$

It can be easily proved that $\sup_{n \ge N+1} H_{n,q} = H_{N+1,q}$. So $H_{n,q}$ can be bounded uniformly in n by H_q. So, we obtain

$$\mathbb{E}\left[\sup_{M \ge N+1} |R_{N,M}(t)|^q\right] \le \frac{A_q}{N^{q/2(2/\alpha-1)}},$$

with

$$A_q = 2 B_q^q H_q M_q \left(\frac{\alpha}{2-\alpha}\right)^{q/2}.$$

Letting M go to infinity yields the conclusion.   □

Proposition 3 can be extended on E_K equipped with $\|\cdot\|_{K,p}$.

Proposition 4. *Under the same assumptions as Proposition 3 and for $p \ge 1$ and $(N+1)\alpha > q > \max(p, 2)$, there exists A_q such that*

$$\mathbb{E}\left[\|Y - Y_N\|_{K,p}^q\right] \le Vol(K)^{q/p} \frac{A_q}{N^{q\left(\frac{2-\alpha}{2\alpha}\right)}}.$$

where Y and Y_N are defined in Proposition 3.

Proof. According to Hölder's inequality,

$$\|Y - Y_N\|_{K,p} \leq \left(Vol(K)^{1-p/q} \left(\int_K |Y(t) - Y_N(t)|^q \, dt\right)^{p/q}\right)^{1/p},$$

where $Vol(K) = \int_{\mathbb{R}} \mathbf{1}_K$. Thanks to Proposition 3 we have

$$\mathbb{E}\left[\|Y - Y_N\|_{K,p}^q\right] \leq Vol(K)^{q/p-1} \int_K \frac{A_q}{N^{q/2(2/\alpha-1)}} dt.$$

$\square$

The error coming from the truncation is now explicit thanks to the previous propositions. Unfortunately, the distribution of V_n is not always easy to simulate or may even be unknown. From now on, we will consider a sequence of random variables $(W_{n,k})_{n\geq 1}$ such that $\lim_{k\to\infty} W_{n,k} = V_n$ in a sense we will define later. Here, k is the parameter which will control the closeness between V_n and $W_{n,k}$.

Next, we will evaluate the distance between the quantities $\sum_{n=N+1}^{P} h(\Gamma_n, V_n)(t)$ and $\sum_{n=N+1}^{P} h(\Gamma_n, W_{n,k})(t)$ in L^q for $q > 2$. Due to the fact that $\Gamma_n^{-1/\alpha}$ has a finite q-moment if and only if $n > q/\alpha$, we will not always be able to compute the distance between the sums starting at $n = 1$.

The next two propositions allow us to evaluate the error when approximating $V_n(t)$ by $W_{n,k}(t)$, knowing their L^q-distance. The main interest in these propositions is that the distance between V_n and $W_{n,k}$ is allowed to grow with n. Indeed, as $\Gamma_n^{-1/\alpha}$ is decreasing with n, the larger n is, the less significant is the distance between V_n and $W_{n,k}$ in the overall distance between the two sums.

Proposition 5. *Let us have the same assumption as Proposition 3 and take $P > N$ and $\alpha(N + 1) > q \geq 2$. If $(W_{n,k})_{n\geq 1}$ is a sequence of symmetric random variables with values in the space of bounded functions on K such that there exists a constant $M_{q,k}$ with*

$$\mathbb{E}\left[\|V_n - W_{n,k}\|_K^q\right] \leq M_{q,k} n^{q\beta} < \infty,$$

for $\beta < 1/\alpha - 1/2$, then

$$\mathbb{E}\left[\left|\sum_{n=N+1}^{P} h(\Gamma_n, V_n)(t) - h(\Gamma_n, W_{n,k})(t)\right|^q\right]$$

$$\leq A_q' M_{q,k} \left(\frac{1}{N^{2/\alpha-\beta-1}} - \frac{1}{P^{2/\alpha-\beta-1}}\right)^{\frac{q}{2}}$$

where A_q' depends only on q.

Proof. Let us denote

$$R_{N,P}(t) = \sum_{n=N+1}^{P} h(\Gamma_n, V_n)(t) - h(\Gamma_n, W_{n,k})(t).$$

In the same way that we have obtained (3) in the proof of Proposition 3, we have

$$\mathbb{E}[|R_{N,P}(t)|^q] \le B_q^q M_{q,k} \left(\sum_{n=N+1}^{P} \left(n^{q\beta} \frac{\Gamma\left(n - \frac{q}{\alpha}\right)}{\Gamma(n)} \right)^{2/q} \right)^{q/2}.$$

Using the same definition of $H_{n,q}$ as we did in (4), and knowing that $\sup_{n \ge N+1} H_{n,q} = H_{N+1,q}$. We can denote by H_q the uniform bound in n of $H_{n,q}$. Using well-known series-integral comparison we obtain

$$\mathbb{E}[|R_{N,P}(t)|^q] \le A_q' M_{q,k} \left(\frac{1}{N^{2/\alpha - \beta - 1}} - \frac{1}{P^{2/\alpha - \beta - 1}} \right)^{q/2}$$

where $A_q' = B_q^q H_q \left(\frac{\alpha}{2 - \alpha\beta - \alpha} \right)^{q/2}$. $\qquad\qquad\qquad\qquad\qquad\qquad\square$

In the same way we had Proposition 4:

Proposition 6. *Under the same assumptions as Proposition 5 we have,*

$$\mathbb{E}\left[\left\| \sum_{n=N+1}^{P} h(\Gamma_n, V_n) - h(\Gamma_n, W_{n,k}) \right\|_{K,p}^q \right] \le Vol(K)^{q/p} A_q'$$

$$M_{q,k} \left(\frac{1}{N^{2/\alpha - \beta - 1}} - \frac{1}{P^{2/\alpha - \beta - 1}} \right)^{\frac{q}{2}}.$$

In the next part, we will use these results to obtain an approximation and a convergence rate towards the LFTSM. Propositions 3 and 6 will be used to balance both types of errors. In this procedure, an appropriately precise approximation of V_n will be needed to minimize the error between V_n and $W_{n,k}$. As stated before, depending on the values of q and α, we will have to consider the first terms in a specific manner.

3 Application to the Local Time Fractional Stable Motion

In this section, we will apply the results from Sect. 2 to the process defined in the introduction. Our precise working definition, coming from (5.3) in [6], is up to a multiplicative constant:

$$\sum_{n \ge 1} \Gamma_n^{-1/\alpha} G_n e^{X_n'^2/2\alpha} l_n(X_n', t), \tag{5}$$

where

- $(G_n)_{n\geq 1}$ is a sequence of i.i.d. standard Gaussian variables
- $(X'_n)_{n\geq 1}$ is another one
- $(l_n)_{n\geq 1}$ are independent copies of a fractional Brownian motion local time, each one defined on some probability space $(\Omega', \mathcal{F}', P')$
- $(\Gamma_n)_{n\geq 1}$ are the arrival times of a Poisson process of rate 1 on $[0, \infty)$

(G_n), (X'_n), (l_n) and (Γ_n) being independent.

But before that, we will have to prove that this process satisfies the required assumptions.

In fact, this work has to be generalized for functions that are not local times, in particular for the approximated local times. The required setup is a family f_n of functions satisfying the following assumptions: $f_n : \mathbb{R} \times \mathbb{R}_+ \to \mathbb{R}$ are the independent copies of a non negative continuous random function on the probability space $(\Omega', \mathcal{F}', P')$, such that for all $K \subset \mathbb{R}$ compact set, denoting the uniform norm $\|\cdot\|_K$, we have for some $p > \alpha$

$$\mathbb{E}\left[\|f_n(x,\cdot)\|_K^p\right]$$

uniformly bounded in x, and $f_n(\cdot, t)$ has its support included in

$$S_{\rho,n} = \left[\inf_{s\leq t} B_s^{H,n} - \rho, \sup_{s\leq t} B_s^{H,n} + \rho\right], \tag{6}$$

where $(B^{H,n})$ are independent fractional Brownian motions with Hurst parameter H defined on the same probability space as f_n and $\rho \geq 0$. In the following, we will consider

$$Y(t) = \sum_{n\geq 1} \Gamma_n^{-1/\alpha} G_n e^{X'^2_n/2\alpha} f_n(X'_n, t). \tag{7}$$

Remark 1. The local time of a fractional Brownian motion obviously satisfies the support condition with $\rho = 0$. It satisfies the other condition because $l_n(x, \cdot)$ is a non decreasing function so that

$$\sup_{t\in K} l_n(x, t) = l_n(x, t_0),$$

t_0 being the upper bound of K (see for example [9] for more details on local times). It simply claims $\mathbb{E}\left[\|l_n(x, t)\|_K^p\right] \leq t_0^p$.

The following lemma is a direct consequence of the support condition so we will skip its proof.

Lemma 1. *Let f_n satisfy the assumptions stated above, $f_n(x, \cdot)$ be a continuous function on K, $\|\cdot\|_K$ being the uniform norm on K. Let X_n be a sequence of i.i.d. real random variables independent from everything else whose distribution density is φ with respect to Lebesgue measure. For $q > 0$, if for all $a > 0$,*

$$\int_{\mathbb{R}} \varphi(x)^{-q/\alpha+1} e^{-ax^2} < \infty,$$

there exists C such that

$$\mathbb{E}\left[\left\|\varphi(X_n)^{-1/\alpha} f_n(X_n, \cdot)\right\|_K^q\right] \leq C \sup_{x \in \mathbb{R}} \left(\mathbb{E}\left[\|f_n(x, \cdot)\|_K^{qp'}\right]\right)^{1/p'},$$

for some $p' > 1$.

This result still holds if $f_n(x, \cdot)$ is not continuous but bounded and all the other assumptions are still satisfied.

We will need to apply Lemma 1 with $q > \alpha$ later, but we can not apply it to process Y because $\varphi(x) = e^{-x^2}$ is not a suitable density to apply Lemma 1. To avoid this problem, we will work with a slightly different process, having the same distribution as Y. Thus, we will see that the processes Y and Y_φ are identically distributed, where

$$Y_\varphi(t) = \sum_{n \geq 1} \Gamma_n^{-1/\alpha} G_n \varphi(X_n)^{-1/\alpha} f_n(X_n, t),$$

where X_n are i.i.d. random variables, and φ is the density of their distribution which satisfies for $q > 0$, for all $a > 0$, $\int_{\mathbb{R}} \varphi(x)^{-q/\alpha+1} e^{-ax^2} < \infty$.

Proposition 7. *If for some $q > 0$ and for all $a > 0$ we have*

$$\int_{\mathbb{R}} \varphi(x)^{-q/\alpha+1} e^{-ax^2} < \infty,$$

then processes Y_φ and Y have the same distribution.

This is a direct consequence of Proposition 2 when replacing the distribution λ by its expression. After a quick calculation, we can see that φ has no influence on the characteristic function as long as Lemma 1 can be applied for some q.

From now on, we will only use

$$Y(\cdot) = \sum_{n \geq 1} \Gamma_n^{-1/\alpha} G_n e^{2|X_n|/\alpha} f_n(X_n, \cdot), \tag{8}$$

where X_n has a Laplace distribution of parameters $(0, 1/2)$, i.e. its density is $e^{-2|x|}$ with respect to the Lebesgue measure on $\mathbb{R}$. The choice of the Laplace distribution is not at all significant but this distribution is easy to simulate and satisfies the assumptions required.

We will now see how we can apply Propositions 1, 4 and 6.

Remark 2. Denote $V_n = G_n e^{2|X_n|/\alpha} f_n(X_n, \cdot)$. For $K \subset \mathbb{R}_+$ compact,

$$M_q := \mathbb{E}[\|V_n\|_K^q] < \infty.$$

Indeed, by independence of the variables,

$$M_q = \mathbb{E}[|G_n|^q]\mathbb{E}\left[e^{\frac{2|X_n|q}{\alpha}} \|f_n(X_n, t)\|_K^q\right],$$

and the last expectation can be bounded using Lemma 1.

Proposition 8. *Let $K \subset \mathbb{R}_+$ denote a compact set. The series defining the process Y in (8) converges uniformly on K.*

Proof. The proof of this proposition simply consists in verifying that Y satisfies the assumptions of Proposition 1. We only have to check that $\mathbb{E}[\|V_n\|_K^\alpha]$ is bounded, which is the point of the above remark.

Using Remark 2 and Proposition 4, the following corollary is obtained.

Corollary 1. *For $K \subset \mathbb{R}_+$ compact, there exists a constant C such that for $p > 0$ and $P(\alpha + 1) > q > \max(p, 2)$*

$$\mathbb{E}\left[\|Y - Y_P\|_{K,p}^q\right] \leq \frac{C}{P^{q\left(\frac{2-\alpha}{2\alpha}\right)}}.$$

Now, let us study the non-truncated terms. Denoting

$$g_{n,k}(x, t) = \int_\mathbb{R} \varphi_k(y - x)l_n(y, t)dy,$$

where $(\varphi_k)_k$ is an approximate identity with support in $[-1/k, 1/k]$. We will use $\varphi(x) = -|x| + 1$ on $[-1, 1]$, $\varphi = 0$ elsewhere. We will denote $\varphi_k(x) = k\varphi(kx)$. We can rewrite $g_{n,k}$ as

$$g_{n,k}(x, t) = \int_0^t \varphi_k(B_s^{H,n} - x)ds,$$

where $B_s^{H,n}$ is the fractional Brownian motion from which $l_n(x, t)$ is defined. $g_{n,k}$ is the theoretical approximation of the fractional Brownian motion local time. The Dirac function in the classical occupation formula density has been replaced by an approximate identity. We will now denote $I_{n,k}$ the discretisation of this integral calculated with the rectangle method using $m_{n,k}$ points uniformly spread on $[0, T]$.

$$I_{n,k}(x, t) = \frac{T}{m_{n,k}} \sum_{i=0}^{[m_{n,k}t/T]} \varphi_k\left(B_{\frac{iT}{m_{n,k}}}^{H,n} - x\right),$$

where $[x]$ is the floor function. $V_n(\cdot) = G_n e^{2|X_n|/\alpha} l_n(X_n, \cdot)$ will be approximated by $W_{n,k}(\cdot) = G_n e^{2|X_n|/\alpha} I_{n,k}(X_n, \cdot)$.

In the following, C will denote a generic constant.

Proposition 9. *Let K denote a compact subset of $\mathbb{R}_+$. For $q > 0$, for $\beta < 1/\alpha - 1/2$, for all $\delta < \frac{1}{2H} - \frac{1}{2}$, if we take $m_{n,k} = max([k^{\frac{\delta+2}{\delta'}} n^{-\frac{\beta}{\delta'}}], 1)$ with $\delta' < H$, there exists C such that:*

$$\mathbb{E}\left[\|V_n - W_{n,k}\|_K^q\right] \leq \frac{Cn^{q\beta}}{k^{q\delta}}.$$

Proof. In this proof $\kappa(\omega)$ will denote a generic random variable with finite moments of all order.

According to Lemma 1,

$$\mathbb{E}\left[\|V_n - W_{n,k}\|_K^q\right] \leq C \sup_{x \in \mathbb{R}} \left(\mathbb{E}\left[\|l_n(x, \cdot) - I_{n,k}(x, \cdot)\|_K^{qp'}\right]\right)^{1/p'}, \qquad (9)$$

Let us write

$$l_n(x, t) - I_{n,k}(x, t) = (l_n(x, t) - g_{n,k}(x, t)) + (g_{n,k}(x, t) - I_{n,k}(x, t)).$$

First, consider $l_n(x, t) - g_{n,k}(x, t) = \int_{\mathbb{R}} \varphi_k(y - x)(l_n(x, t) - l_n(y, t))dy$. As the fractional Brownian motion is locally non-deterministic, we can apply Theorem 4 in [19], in order to have for all $\delta < \frac{1}{2H} - \frac{1}{2}$, there exists $\kappa(\omega) > 0$ such that

$$|l_n(x, t) - g_{n,k}(x, t)| \leq \frac{\kappa(\omega)}{k^\delta}. \qquad (10)$$

The random variable $\kappa(\omega)$ has finite moments of all orders (see for example [25]).

Now, consider

$$g_{n,k}(x, t) - I_{n,k}(x, t)$$
$$= \sum_{i=1}^{[m_{n,k}t/T]} \int_{(i-1)T/m_{n,k}}^{iT/m_{n,k}} \left(\varphi_k(B_s^{H,n} - x) - \varphi_k\left(B_{\frac{(i-1)T}{m_{n,k}}}^{H,n} - x\right)\right) ds$$
$$+ \int_{[\frac{m_{n,k}t}{T}]\frac{T}{m_{n,k}}}^{t} \varphi_k(B_s^{H,n} - x)ds - \frac{T}{m_{n,k}}\varphi_k\left(B_{[\frac{m_{n,k}t}{T}]\frac{T}{m_{n,k}}}^{H,n} - x\right). \qquad (11)$$

Remark that φ_k is k^2-Lipschitz, and that for $\delta' < H$

$$\sup_{s,t \in K} \frac{\left|B_t^{H,n} - B_s^{H,n}\right|}{|t - s|^{\delta'}}$$

has finite moments of all order. Consequently, using that bound in (11) yields

$$\left| g_{n,k}(x,t) - I_{n,k}(x,t) \right| \leq \frac{k^2 \kappa(\omega)}{m_{n,k}^{\delta'}}. \tag{12}$$

Combining (10) and (12) and by taking $m_{n,k} = [k^{\frac{\delta+2}{\delta'}} n^{-\frac{\beta}{\delta'}}]$ with $\beta < 1/\alpha - 1/2$, there exists $\kappa(\omega)$ with finite moments of all order such that

$$\left| l_n(x,t) - I_{n,k}(x,t) \right| \leq \frac{\kappa(\omega) n^\beta}{k^\delta}. \tag{13}$$

Using the bound (13) in (9) concludes. Note that $\kappa(\omega)$ may depend on n but its expectation $\mathbb{E}[\kappa(\omega)]$ is non-increasing with n.

Using Proposition 6, we get

Corollary 2. *For $p > 0$ and $(N + 1)\alpha > q > \max(p, 2)$, for $\beta < 1/\alpha - 1/2$, for all $\delta < \frac{1}{2H} - \frac{1}{2}$, if we take $m_{n,k} = [k^{\frac{\delta+2}{\delta'}} n^{-\frac{\beta}{\delta'}}]$ with $\delta' < H$, there exists C such that*

$$\mathbb{E}\left[\left\| \sum_{n=N+1}^{P} h(\Gamma_n, V_n) - h(\Gamma_n, W_{n,k}) \right\|_{K,p}^q \right] \leq \frac{C}{k^{q\delta}} \frac{1}{N^{q(2/\alpha - \beta - 1)/2}}.$$

For the last part of this section, let us recall some notations. Y_N is defined by

$$Y_N(t) = \sum_{n=1}^{N} \Gamma_n^{-1/\alpha} G_n e^{2|X_n|/\alpha} l_n(X_n, t),$$

and Y_N converges uniformly almost surely on every compact towards $Y(t) = \lim_{N \to \infty} Y_N(t)$. Let us denote

$$Z_{N,k} = \sum_{n=1}^{N} \Gamma_n^{-1/\alpha} G_n e^{2|X_n|/\alpha} I_{n,k}(X_n, t). \tag{14}$$

We will consider $\mathbb{P}\left(\left\| Y - Z_{P,k} \right\|_{K,p} > 3\tau \right)$ for $(P + 1)\alpha > (N + 1)\alpha > q > \max(p, 2) \geq p \geq 1$.

$$\mathbb{P}\left(\left\| Y - Z_{P,k} \right\|_{K,p} > 3\tau \right) < \mathbb{P}\left(\left\| Y - Y_P \right\|_{K,p} > \tau \right)$$
$$+ \mathbb{P}\left(\left\| (Y_P - Y_N) - (Z_{P,k} - Z_{N,k}) \right\|_{K,p} > \tau \right)$$
$$+ \mathbb{P}\left(\left\| Y_N - Z_{N,k} \right\|_{K,p} > \tau \right).$$

Corollaries 1 and 2 combined with Markov's inequality allow us to evaluate without any difficulties both

$$\mathbb{P}\left(\|Y - Y_P\|_{K,p} > \tau\right) \leq \frac{C}{\tau^q P^{q(1/\alpha - 1/2)}} \tag{15}$$

and

$$\mathbb{P}\left(\|(Y_P - Y_N) - (Z_{P,k} - Z_{N,k})\|_{K,p} > \tau\right) < \frac{C}{\tau^q k^{q\delta} N^{q(1/\alpha - \beta/2 - 1/2)}}. \tag{16}$$

Now, we must study the remaining terms $h(\Gamma_n, V_n) - h(\Gamma_n, W_{n,k})$ for $n \leq N$. Denote

$$\xi'_{n,k}(t) = \Gamma_n^{-1/\alpha} G_n e^{2|X_n|/\alpha}(l_n(x,t) - g_{n,k}(x,t)),$$

and

$$\xi''_{n,k}(t) = \Gamma_n^{-1/\alpha} G_n e^{2|X_n|/\alpha}(g_{n,k}(x,t) - I_{n,k}(x,t)).$$

Since $\sum_{n=1}^N \xi'_{n,k} + \xi''_{n,k} = Y_N(t) - Z_{N,k}(t)$, we have

$$\mathbb{P}\left(\|Y_N - Z_{N,k}\|_{K,p} > \tau\right) \leq \sum_{n=1}^N \mathbb{P}\left(\|\xi'_{n,k}\|_{K,p} > \frac{\tau}{2N}\right)$$
$$+ \sum_{n=1}^N \mathbb{P}\left(\|\xi''_{n,k}\|_{K,p} > \frac{\tau}{2N}\right). \tag{17}$$

By conditioning with respect to $(G_n, X_n, (B_t^{H,n})_{t \geq 0})$, we have

$$\mathbb{P}\left(\|\xi'_{n,k}\|_{K,p} > \frac{\tau}{2N}\right)$$
$$= \mathbb{E}\left[P\left(\Gamma_n < \left(\frac{2N}{\tau}|G_n|e^{2|X_n|/\alpha}\|l_n - g_{n,k}\|_{K,p}\right)^\alpha \Big| (B^{H_n}, G_n, X_n)\right)\right].$$

Then, given the density of the distribution of Γ_n and the bound obtained in (10), the following inequality is obtained:

$$\mathbb{P}\left(\|\xi'_{n,k}\|_{K,p} > \frac{\tau}{2N}\right) \leq \frac{N^\alpha C}{\tau^\alpha k^{\alpha\delta}}. \tag{18}$$

The bound obtained in inequality (12) is still true if $m_{n,k}$ is a random variable only depending on Γ_n but the bound also depends on Γ_n. For $n \leq N$ setting $m_{n,k} = [\Gamma_n^{-1/(\delta'\alpha)} k^{\frac{2+\delta}{\delta'}}]$ we have

$$\mathbb{E}\left[\frac{\Gamma_n^{-q/\alpha}}{m_{n,k}^{\delta'q}}\right] \le \frac{C}{k^{q(\delta+2)}},$$

so we can apply Markov's inequality on the last term using inequality (12). It leads to

$$\mathbb{P}\left(\left\|\xi''_{n,k}\right\|_{K,p} > \frac{\tau}{2N}\right) \le \frac{(2N)^q C}{\tau^q k^{q\delta}}. \tag{19}$$

Once these remarks have been made, we can combine these inequalities to obtain a convergence rate. But first, we must tune the parameters to balance the errors and obtain a manageable expression.

Tuning procedure 2. *We want to approximate process Y with parameters (H,α) by a family $(Z_{P_\varepsilon,k_\varepsilon})_{\varepsilon>0}$, $Z_{P_\varepsilon,k_\varepsilon}$ being a truncated series. In this tuning procedure, the parameters will be set equal to their bounds for the sake of simplicity.*

We can adjust two parameters: P_ε is the size of the truncation and k_ε controls the approximation of the fractional Brownian motion local time approximation. We will make two kinds of errors, one coming from the truncation itself, and one from the approximation of the local time.

The error from the truncation can be controlled if we take $P_\varepsilon \sim C\varepsilon^{-\frac{2\alpha}{2-\alpha}}$ in inequality (15).

The approximation error has two sources: our theoretical approximation of the local time, and the discretisation used to compute the approximation.

We can deal with the first one by taking $k_\varepsilon \sim C\varepsilon^{-1/\delta}$ in inequalities (16), (18) and (19) where $\delta = 1/(2H) - 1/2$ comes from the Hölder continuity of the fractional Brownian motion local time.

After P, we have to fix N. According to our analysis, we must choose $N > q/\alpha - 1$, but otherwise, by combining equations (16), (18) and (19), N is not allowed to vary with ε. Then, N will be a constant later set with computer tests.

Let us recall that m_{n,k_ε} is the number of points used in the discretisation. In the series defining Y, the first terms are the most important. We distinguish two cases. The first N terms, where $m_{n,k}$ is random: $m_{n,k} = [\Gamma_n^{-1/(\delta'\alpha)} k_\varepsilon^{\frac{2+\delta}{\delta'}}]$, where $\delta' = 1/H$ comes from the Hölder continuity of the fractional Brownian motion. For the remaining terms, a high level of precision is not as important, so we will need fewer points in our discretisation. Thus, $m_{n,k} = [k^{\frac{\delta+2}{\delta'}} n^{-\frac{\beta}{\delta'}}]$ with $\beta = 1/\alpha - 1/2$.

Remark 3 (Computational Cost). We will compute the expected value of the computational cost. The predominant operation in the algorithm is the computation of the fractional Brownian motions. There is P computations of a fractional Brownian motion, each one needing $m_{n,k} \log m_{n,k}$ operations (using the Davies–Harte algorithm introduced in [8]). Using the parameters given in the Tuning procedure 2, the overall computational cost is

$$\varepsilon^{-\frac{\delta+\delta'-\beta}{\delta\delta'}} = \varepsilon^{-\frac{2H^2(3\alpha H+2\alpha-2)}{2\alpha(1-H)}}.$$

Theorem 3. *With Tuning procedure 2, we get a convergence rate for the family of processes $Z_{P_\varepsilon, k_\varepsilon}$ defined in (14). For $p \geq 1$ and $q \geq \alpha$ there exists C and C' such that*

$$P\left(\left\|Y - Z_{P_\varepsilon, k_\varepsilon}\right\|_{K,p} > \tau\right) \leq C\left(\frac{\varepsilon}{\tau}\right)^\alpha + C'\left(\frac{\varepsilon}{\tau}\right)^q$$

with $P_\varepsilon \sim C_1 \varepsilon^{-\frac{2\alpha}{2-\alpha}}$, $k_\varepsilon \sim C_2 \varepsilon^{-1/\delta}$ and m_{n,k_ε} as defined in the tuning procedure

Using the same tuning procedure, we have

Theorem 4. *Under the assumptions of Theorem 3, $Z_{P_\varepsilon, k_\varepsilon}$ converges almost surely in $L^p(K)$ towards Y when ε tends to 0.*

Proof. In the following, take $q \geq 2$ and $N > q/\alpha - 1$. Using Minkowski's inequality, we obtain

$$\mathbb{E}\left[\left\|(Y(t) - Y_N(t)) - (Z_{P_\varepsilon, k_\varepsilon} - Z_{N,k_\varepsilon})\right\|_{K,p}^q\right]^{1/q}$$

$$\leq \mathbb{E}\left[\left\|(Y - Y_{P_\varepsilon})\right\|_{K,p}^q\right]^{1/q} + \mathbb{E}\left[\left\|(Y_{P_\varepsilon} - Y_N) - (Z_{P_\varepsilon, k_\varepsilon} - Z_{N,k_\varepsilon})\right\|_{K,p}^q\right]^{1/q}.$$

Thanks to Corollaries 1 and 2, using the expression of P_ε, k_ε and $m_{n,k}$ given in Tuning procedure 2, we get each term bounded by $C\varepsilon$. Thus, we can say that

$$\mathbb{E}\left[\left\|(Y - Y_N) - (Z_{P_\varepsilon, k_\varepsilon} - Z_{N,k_\varepsilon})\right\|_{K,p}^q\right]^{1/q} \leq C\varepsilon. \tag{20}$$

Inequality (20) and Borel-Cantelli lemma imply $Z_{P_\varepsilon, k_\varepsilon} - Z_{N,k_\varepsilon}$ converges almost surely towards $Y - Y_N$ in $L^p(K)$. We only have to prove Z_{N,k_ε} converges almost surely towards Y_N in $L^p(K)$. But according to (13), using k_ε and m_{n,k_ε} as defined in the Tuning procedure, $\left\|l_n(x,t) - I_{n,k}(x,t)\right\|_{K,p}$ converges almost surely towards 0 so $Y_N - Z_{N,k_\varepsilon}$ too since only a finite number of terms is considered. Consequently, $Y_N - Z_{N,k_\varepsilon}$ converges almost surely to 0 in $L^p(K)$. $\square$

4 Simulation

In [6], the authors explained how to simulate the Brownian motion local time stable motion with a random walk with random rewards approach. They are many advantages of our approach against the random walk with random rewards approach. The most obvious is that it is valid for all Hurst parameters and not only $H = 1/2$. Moreover, we have already highlighted that we have a convergence rate, which was not the case previously.

Unfortunately, as one can see in our tuning procedure, the parameters we use, namely P_ε and m_{n,k_ε}, depend on α and H. If α is close to 2, the number of terms P in the sum is too high to be accepted and if H is close to 1, the precision needed

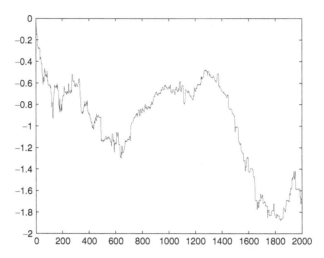

Fig. 1 Trajectory of Y for $H = 0.3$ and $\alpha = 1.2$

in the simulation of the fractional Brownian motion is impossible to achieve. To be more precise, a run of this program can take from seconds (H and α close to 0) to several hours (H or α close to their upper bounds). Even if this method is not perfect for all values of H and α, it is still a major improvement since we can have sample paths for different values of H with a convergence rate. See for example one simulation in Fig. 1.

Those sample paths were obtained by a program written in C. To calibrate the tuning procedure, we must choose all the constants. δ, δ' and β are chosen equals to their bounds. The minimum number of points obtained when simulating a fractional Brownian motion is 1,000, so the same minimal number is used to compute the numerical integration. The choice of N does not change the theoretical convergence rate we had in the previous section. After a testing period, N was fixed arbitrarily at 1,000.

The fractional Brownian motion is simulated using the Davies–Harte algorithm (see [8, 24]). The Gaussian random variables are simulated using the FL algorithm (see [2]). Random variables Γ_n and X_n were simulated using exponential random variables (Γ_n is the sum of n exponential random variables and X_n the difference of two exponential random variables). Exponential random variables were obtained using SA algorithm (see [1]).

According to Theorem 5.1 in [6], the Hölder exponent d of our process is such that $d < 1 - H$. It means that the closer to 1 H is, the less regular our process is. See Fig. 2 for sample paths with different values of H and α constant.

Let us make a more convincing comparison of the two ways we have used to simulate this process. We must take $H = 1/2$, so that both methods can be used. A straightforward computation from the result of Proposition 2 in the case $\alpha = 1$ yields the following characteristic function,

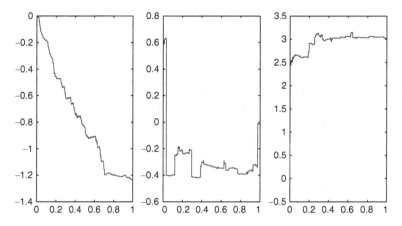

Fig. 2 Trajectories of Y for $H = 0.2$, $H = 0.4$ and $H = 0.6$ with $\alpha = 0.7$

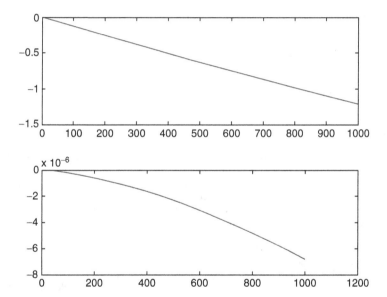

Fig. 3 Plots of $\log(\mathbb{E}\left[e^{iuY(t)}\right])$ with respect to t. Shot noise series method on the first line, and the random walk method on the second

$$\mathbb{E}\left[e^{iuY(t)}\right] = \exp\left(C\,|u|\,t\right).$$

Thus, we are going to check the logarithm of the empirical expectation of this process using the two different methods and see which one is the closest to a straight line. This is done by simulating 10,000 processes using each method, then computing the empirical expectation and taking the logarithm, i.e.:

$$\log \left(\frac{1}{10000} \sum_{k=1}^{10000} e^{i u Y_k(t)} \right).$$

Note that the 10,000 simulations of the process require around forty hours. The result is shown in Fig. 3, the first one being the method developed in this paper, the second the random walk with random rewards method. Notice that the shot noise series method yields a graph closer to a straight line than the random walk method. The scale is very different because in each method, the process is simulated up to a multiplicative constant and the constants from both methods are completely different.

Thus we can conclude that even if the random walk method seems a bit quicker, not only has our method the advantage of being able to consider the $H \neq 1/2$ case, it is also closer to what is theoretically expected.

References

1. Ahrens, J.H., Dieter, U.: Computer methods for sampling from the exponential and normal distributions. Commun. ACM **15**, 873–882 (1972)
2. Ahrens, J.H., Dieter, U.: Extensions of Forsythe's method for random sampling from the normal distribution. Math. Comp. **27**, 927–937 (1973)
3. Benassi, A., Cohen, S., Istas, J.: Identification and properties of real harmonizable fractional Lévy motions. Bernoulli **8**(1), 97–115 (2002)
4. Biermé, H., Meerschaert, M.M., Scheffler, H.-P.: Operator scaling stable random fields. Stoch. Process. Appl. **117**(3), 312–332 (2007)
5. Cohen, S., Lacaux, C., Ledoux, M.: A general framework for simulation of fractional fields. Stoch. Process. Appl. **118**(9), 1489–1517 (2008)
6. Cohen, S., Samorodnistky, G.: Random rewards, fractional brownian local times and stable self-similar processes. Ann. Appl. Probab. **16**, 1432–1461 (2006)
7. Crovella, M.E., Bestavros, A.: Self-similarity in World Wide Web traffic: evidence and possiblecauses. IEEE/ACM Trans. Networking **5**(6), 835–846 (1997)
8. Davies, R.B., Harte, D.S.: Tests for Hurst effect. Biometrika **74**(1), 95–101 (1987)
9. Geman, D., Horowitz, J.: Occupation densities. Ann. Probab. **8**(1), 1–67 (1980)
10. Janicki, A., Kokoszka, P.: Computer investigation of the rate of convergence of Lepage type series to α-stable random variables. Statistics **23**(4), 365–373 (1992)
11. Kaj, I., Taqqu, MS: Convergence to fractional Brownian motion and to the Telecom process: the integral representation approach. In: Vares, M.E., Sidoravicius, V. (eds.) Brazilian Probability School, 10th anniversary volume (2007)
12. Kolmogorov, A.N.: Wienersche Spiralen und einige andere interessante Kurven im Hilbertschen Raum. CR (Doklady) Acad. URSS (NS) **26**(1), 15–1 (1940)
13. Lacaux, C.: Real harmonizable multifractional Lévy motions. Ann. Inst. H. Poincaré Probab. Statist. **40**(3), 259–277 (2004)
14. Lacaux, C.: Series representation and simulation of multifractional Levy motions. Adv. Appl. Prob. **36**(1), 171–197 (2004)
15. Ledoux, M., Talagrand, M.: Probability in Banach Spaces: Isoperimetry and Processes. Springer (1991)
16. Levy, J.B., Taqqu, M.S.: Renewal reward processes with heavy-tailed interrenewal times and heavy-tailed rewards. Bernoulli **6**(1), 23–44 (2000)
17. Mandelbrot, B.B., Ness, J.W.V.: Fractional Brownian Motions, Fractional Noises and Applications. SIAM Rev. **10**(4), 422–437 (1968)

18. Mikosch, T., Resnick, S., Rootzen, H., Stegeman, A.: Is network traffic approximated by stable Levy motion or fractional Brownian motion. Ann. Appl. Probab. **12**(1), 23–68 (2002)
19. Pitt, L.: Local times for gaussian vector fields. Indiana Univ. Math. J. **27**, 309–330 (1978)
20. Rosiński, J.: On path properties of certain infinitely divisible Processes. Stoch. Process. Appl. **33**(1), 73–87 (1987)
21. Rosiński, J.: On series representations of infinitely divisible random vectors. Ann. Probab. **18**(1), 405–430 (1990)
22. Rosiński, J.: Series representations of Lévy processes from the perspective of point processes. In: Lévy Processes, pp. 401–415. Birkhäuser Boston, Boston, MA (2001)
23. Samorodnitsky, G., Taqqu, M.: Stable Non-Gaussian Random Processes. Chapman & Hall (1994)
24. Wood, A.T.A., Chan, G.: Simulation of stationary Gaussian processes in $[0, 1]^d$. J. Comput. Graph. Stat. **3**(4), 409–432 (1994)
25. Xiao, Y.: Strong local nondeterminism and the sample path properties of Gaussian random fields. In: Asymptotic Theory in Probability and Statistics with Applications, pp. 136–176. Higher Education Press, Beijing (2007)

Convergence at First and Second Order of Some Approximations of Stochastic Integrals

Blandine Bérard Bergery and Pierre Vallois

Abstract We consider the convergence of the approximation schemes related to Itô's integral and quadratic variation, which have been developed in Russo and Vallois (Elements of stochastic calculus via regularisation, vol. 1899, pp. 147–185, Springer, Berlin, 2007). First, we prove that the convergence in the a.s. sense exists when the integrand is Hölder continuous and the integrator is a continuous semi-martingale. Second, we investigate the second order convergence in the Brownian motion case.

Keywords Stochastic integration by regularization · Quadratic variation · First and second order convergence · Stochastic Fubini's theorem

2000 MSC: 60F05, 60F17, 60G44, 60H05, 60J65

1 Introduction

We consider a complete probability space $(\Omega, \mathcal{F}, \mathcal{F}_t, P)$, which satisfies the usual hypotheses. The notation (ucp) will stand for the convergence in probability, uniformly on the compact sets in time.

1. Let X be a real continuous $(\mathcal{F}_t)$-semimartingale. In the usual stochastic calculus, the quadratic variation and the stochastic integral with respect to X play a central role. In [11–13], Russo and Vallois extended these notions to continuous processes. Let us briefly recall their main definitions.

Definition 1. Let X be a real-valued continuous process, $(\mathcal{F}_t)$-adapted, and H be a locally integrable process. The forward integral $\int_0^t H d^- X$ is defined as

$$\int_0^t H d^- X = \lim_{\epsilon \to 0} (ucp) \frac{1}{\epsilon} \int_0^t H_u \left(X_{u+\epsilon} - X_u \right) d u,$$

B.B. Bergery (✉) and P. Vallois
Institut de Mathématiques Elie Cartan, Université Henri Poincaré Nancy I, B.P. 239, 54506
Vandoeuvre-lès-Nancy Cedex, France
e-mail: berardb@iecn.u-nancy.fr; Pierre.Vallois@iecn.u-nancy.fr

C. Donati-Martin et al. (eds.), *Séminaire de Probabilités XLIII*, Lecture Notes in Mathematics 241
2006, DOI 10.1007/978-3-642-15217-7_10, © Springer-Verlag Berlin Heidelberg 2011

if the limit exists. The quadratic variation is defined by

$$[X]_t = \lim_{\epsilon \to 0} (ucp) \frac{1}{\epsilon} \int_0^t (X_{u+\epsilon} - X_u)^2 \, du$$

if the limit exists.

In the article, X will stand for a real-valued continuous $(\mathcal{F}_t)$-semimartingale and $(H_t)_{t \geq 0}$ for an $(\mathcal{F}_t)$-progressively measurable process. If H is continuous, then, according to Proposition 1.1 of [13], the limits in (1) exist and coincide with the usual objects. In order to work with adapted processes only, we change $u + \epsilon$ into $(u + \epsilon) \wedge t$ in the above integrals. This change does not affect the limit (cf. (3.3) of [10]). Consequently,

$$\int_0^t H_u dX_u = \lim_{\epsilon \to 0} (ucp) \frac{1}{\epsilon} \int_0^t H_u \left(X_{(u+\epsilon) \wedge t} - X_u \right) du, \tag{1}$$

and

$$\langle X \rangle_t = \lim_{\epsilon \to 0} (ucp) \frac{1}{\epsilon} \int_0^t \left(X_{(u+\epsilon) \wedge t} - X_u \right)^2 du \tag{2}$$

where $\int_0^t H_u dX_u$ is the usual stochastic integral and $\langle X \rangle$ is the usual quadratic variation of X.

2. First, we determine sufficient conditions under which the convergences in (1) and (2) hold in the almost sure sense. Let us mention that some results in this direction have been obtained in [2] and [5].

We say that a process Y is *locally Hölder continuous* if, for all $T > 0$, there exist $\alpha' \in]0, 1]$ and a finite (random) constant C_Y such that

$$|Y_s - Y_u| \leq C_Y |u - s|^{\alpha'} \quad \forall u, \, s \in [0, T], \text{ a.s.} \tag{3}$$

Our first result related to stochastic integral is the following.

Theorem 1. *If $(H_t)_{t \geq 0}$ is adapted and locally Hölder continuous, then*

$$\lim_{\epsilon \to 0} \frac{1}{\epsilon} \int_0^t H_u (X_{(u+\epsilon) \wedge t} - X_u) du = \int_0^t H_u dX_u, \tag{4}$$

in the sense of almost sure convergence, uniformly on the compact sets in time.

Our assumption related to (H_t) is simple but too strong as shows item 1 of Theorem 4 below. In [5], a general result of a.s. convergence of sequences of stochastic integrals has been given. However it cannot be applied to obtain (4) (see Remark 2).

We now consider the convergence of ϵ-integrals to the bracket of X.

Proposition 1. *If X is locally Hölder continuous, then*

$$\lim_{\epsilon \to 0} \frac{1}{\epsilon} \int_0^t (X_{(u+\epsilon)\wedge t} - X_u)^2 du = \langle X \rangle_t, \tag{5}$$

in the sense of almost sure convergence, uniformly on the compact sets in time. Moreover, if K is a continuous process,

$$\lim_{\epsilon \to 0} \frac{1}{\epsilon} \int_0^t K_u (X_{(u+\epsilon)\wedge t} - X_u)^2 du = \int_0^t K_u d \langle X \rangle_u, \tag{6}$$

in the sense of almost sure convergence.

3. Under the assumptions given in Theorem 1, we have an approximation scheme of $\int_0^{\cdot} H_s dX_s$ which converges a.s. According to Remark 1, the (a.s.) rate of convergence of is of order ϵ^{α}, when X has a finite variation and H is α-Hölder continuous. Therefore, it remains to determine the rate of convergence when X is a local martingale. This leads to introduce

$$\Delta_\epsilon(H,t) = \frac{1}{\sqrt{\epsilon}} \left[\frac{1}{\epsilon} \int_0^t H_u (X_{(u+\epsilon)\wedge t} - X_u) du - \int_0^t H_u dX_u \right], \quad t \geq 0 \tag{7}$$

where H is a progressively measurable and locally bounded process.

In order to study the limit in distribution of the family of processes $\big(\Delta_\epsilon(H,t),$ $t \geq 0\big)$ as $\epsilon \to 0$, a two-steps strategy has been adopted. First, we consider the case where $X = H = B$ and B denotes the standard Brownian motion. Second, using a functional theorem of convergence we determine the limit of $\big(\Delta_\epsilon(H,t), t \geq 0\big)$. Note that in [2], some related results have been proven.

(a) Suppose that $X = H = B$. In that case, using stochastic Fubini's theorem (cf. relation (39) with $\Phi = 1$) we have:

$$\Delta_\epsilon(B,t) = -W_\epsilon(t) + R_\epsilon^1(B,t),$$

where

$$W_\epsilon(t) = \int_0^t G_\epsilon(u) dB_u, \qquad G_\epsilon(u) = \frac{1}{\epsilon\sqrt{\epsilon}} \int_{(u-\epsilon)+}^u (B_u - B_s) ds, \tag{8}$$

and

$$R_\epsilon^1(B,t) := \frac{1}{\sqrt{\epsilon}} \int_0^{t\wedge\epsilon} \left(\frac{s}{\epsilon} - 1\right) B_s dB_s.$$

From Lemma 6, the process $R_\epsilon^1(B,\cdot)$ does not contribute to the limit since $R_\epsilon^1(B,\cdot) \xrightarrow{\text{(ucp)}} 0$, as $\epsilon \to 0$. Therefore, the convergence of $\Delta_\epsilon(B,\cdot)$ reduces to the one of W_ϵ. We determine, more generally, in Theorem 2 below the limit of the pair $\big(W_\epsilon, B\big)$.

Theorem 2. $(W_\epsilon(t), B_t)_{t \geqslant 0}$ *converges in distribution to* $(\sigma W_t, B_t)_{t \geqslant 0}$, *as* $\epsilon \to 0$, *where* W *is a standard Brownian motion, independent from* B, *and* $\sigma^2 = \frac{1}{3}$.

(b) We now investigate the convergence of $(\Delta_\epsilon(H, t))_{t \geqslant 0}$. We restrict ourselves to processes H of the type $H_t = H_0 + M_t + V_t$ where:

1. H_0 is $\mathcal{F}_0$-measurable.
2. M_t is a Brownian martingale, i.e. $M_t = \displaystyle\int_0^t \Lambda_s dB_s$, where (Λ_t) is progressively measurable, locally bounded and is right-continuous with left-limits.
3. V is a continuous process, which is Hölder continuous with order $\alpha > 1/2$, vanishing at time 0.

Note that if $V_t = \displaystyle\int_0^t v_s ds$, where $(v_t)_{t \geqslant 0}$ is progressively measurable and locally bounded, then above condition 3 holds with $\alpha = 1$ and in that case, (H_t) is a semi-martingale.

As for X, we assume that it is a Brownian martingale with representation:

$$X_t = \int_0^t \Phi(u) dB_u, \quad t \geq 0 \tag{9}$$

where $(\Phi(u))$ is predictable, locally bounded and right-continuous at 0.
From now on,

(W_t) denote a standard Brownian motion independent from (B_t),

and

$$\sigma := \frac{1}{\sqrt{3}}.$$

Using functional results of convergence (Proposition 3.2 and Theorem 5.1 in [4]) and Theorem 2, we obtain the following result.

Theorem 3. *1. For any* $0 < t_1 < \cdots < t_n$, *the random vector* $(\Delta_\epsilon(H_0, t_1),$ $\ldots, \Delta_\epsilon(H_0, t_n))$ *converges in law to* $\sigma H_0 \Phi(0)(N_0, \cdots, N_0)$, *where* N_0 *is a standard Gaussian r.v, independent from* $\mathcal{F}_0$.
2. If V *is a process which is locally Hölder continuous of order* $\alpha > \frac{1}{2}$, *then* $\Delta_\epsilon(V, t)$ *converges to* 0 *in the ucp sense as* $\epsilon \to 0$.
3. If $M_t = \displaystyle\int_0^t \Lambda_s dB_s$, *then the process* $(\Delta_\epsilon(M, t))_{t \geqslant 0}$ *converges in distribution to* $(\sigma \int_0^t \Lambda_u \Phi(u) dW_u)_{t \geqslant 0}$ *as* $\epsilon \to 0$.
4. If $H_0 = 0$, M *and* V *are as in points* $(2) - (3)$ *above, then* $(\Delta_\epsilon(M + V, t))_{t \geqslant 0}$ *converges in law to* $(\sigma \int_0^t \Lambda_u \Phi(u) dW_u)_{t \geqslant 0}$ *as* $\epsilon \to 0$.

Let us discuss the assumptions of Theorem 3. As for item 2, the conclusion is false if $\alpha \leq 1/2$. Indeed, if we take $V_t = B_t$ then, $t \mapsto V_t$ is α-Hölder with $\alpha < 1/2$, however, as shows Theorem 2, the limit of $(\Delta_\epsilon(V, t))$ equals (σW_t) and is not null. It is likely too strong to suppose that (H_t) is a semimartingale: we can show (see

Proposition 2 below) that $\big(\Delta_\epsilon(H,t),\ t \geq 0\big)$ converges in distribution where $H_t = h(B_t)$ and h is only supposed to be of class C^1. Note that in this case (H_t) is a Dirichlet process. However, if (H_t) is a stepwise and progressively measurable process then, we have the convergence in law of the finite dimensional distributions of $\big(\Delta_\epsilon(H,t),\ t \geq 0\big)$ but this family of processes does not converge in distribution (see Theorem 4 below).

Next, we consider the convergence of $\Delta_\epsilon(h(B),\cdot)$ for a large class of functions h. A function $h : \mathbb{R} \to \mathbb{R}$ is said to subexponential if there exist $C_1, C_2 > 0$ such that

$$|h(x)| \leq C_1 e^{C_2|x|}, \quad x \in \mathbb{R}. \tag{10}$$

Proposition 2. *Suppose that h is a function of class C^1 such that $h(0) = 0$ and h' is subexponential. Then, $\big(\Delta_\epsilon(h(B),t),\ t \geq 0\big)$ converges in distribution as $\epsilon \to 0$ to $\big(\sigma \int_0^t h'(B_s)\Phi(s)dW_s,\ t \geq 0\big)$.*

According to Exercise 3.13, Chap. V in [8] we have:

$$h(B_t) = E\big(h(B_t)\big) + \int_0^t H(t,s)dB_s, \quad t \geq 0$$

where $H(t,s) = \varphi(t,s,B_s)$ and $\varphi(t,s,x) := E\big(h'(x + B_{t-s})\big)$.

Consequently $\big(H(t,s),\ 0 \leq s \leq t\big)$ is progressively measurable but depends on t, therefore item 3 of Theorem 3 cannot be applied.

(c) We now focus on the case where (H_t) is a stepwise and progressively measurable process. We study the a.s. convergence of $\dfrac{1}{\epsilon} \int_0^{\cdot} H_u\big(X_{(u+\epsilon)\wedge t} - X_u\big)du$ towards $\int_0^{\cdot} H_u dX_u$ and the convergence in distribution of $\Delta_\epsilon(H,\cdot)$ as ϵ goes to 0.

Theorem 4. *Let $(a_i)_{i\in\mathbb{N}}$ be an increasing sequence of real numbers which satisfies $a_0 = 0$ and $a_n \to \infty$. Let $h, (h_i)_{i\in\mathbb{N}}$ be r.v.'s such that h_i is $\mathcal{F}_{a_i}$-measurable, h is $\mathcal{F}_0$-measurable. Let H be the progressively measurable and stepwise process:*

$$H_t = h\mathbb{1}_{\{t=0\}} + \sum_{i\geq 0} h_i\,\mathbb{1}_{\{t\in]a_i,a_{i+1}]\}}.$$

1. Suppose that X is continuous, then, $\dfrac{1}{\epsilon} \int_0^t H_s(X_{(s+\epsilon)\wedge t} - X_s)ds$ converges almost surely to $\int_0^t H_s dX_s$, uniformly on the compact sets in time, as $\epsilon \to 0$.

2. Suppose $h = 0$ and X is defined by (9). Associated with a sequence $(N_i)_{i\in\mathbb{N}}$ of i.i.d. r.v's with Gaussian law $\mathcal{N}(0,1)$, independent from B consider the piecewise and left-continuous process:

$$Z_s := \sigma\Big(h_0\Phi(0)N_0\mathbb{1}_{\{0<s\leq a_1\}} + \sum_{i\geq 1}(h_i-h_{i-1})\Phi(a_i)N_i\mathbb{1}_{\{a_i<s\leq a_{i+1}\}}\Big), \quad s>0$$

and $Z_0 = 0$.

Suppose that Φ is right-continuous at any point a_i. Then, for any fixed times $0 \leq s_1 < \cdots < s_n$,

$$\Big((B_s, \; s \geq 0), \big(\Delta_\epsilon(H, s_1), \cdots, \Delta_\epsilon(H, s_n) \big) \Big)$$

converges in law to $\Big((B_s, \; s \geq 0), \big(Z_{s_1}, \cdots, Z_{s_n} \big) \Big)$ as $\epsilon \to 0$.

A weak version of Theorem 4 has been given in Sect. 6.3 of [1].

Note that the family of processes $\big(\Delta_\epsilon(H, t), \; t \geq 0 \big)$ cannot converge in the Skorokhod space to a right continuous process $\big(Z_0(t), \; t \geq 0 \big)$. Indeed, according to Theorem 4, the map $t \in]0, a_1[\mapsto Z_0(t)$ should be constant and not null. This contradicts the fact that $Z_0(0) = 0$.

In [13], convergence in distribution of sequences of stochastic integrals are considered. We discuss in Remark 4 the link between Rootzen's result and ours.

4. Let us finally present our result of convergence in distribution related to the quadratic variation.

Let us consider

$$\Delta_\epsilon^{(2)}(K, t) = \frac{1}{\sqrt{\epsilon}} \left[\frac{1}{\epsilon} \int_0^t K_u (B_{(u+\epsilon)\wedge t} - B_u)^2 du - \int_0^t K_u du \right], \qquad (11)$$

where (K_s) is locally bounded and progressively measurable.

Proposition 3. *Let (K_s) be a predictable, right-continuous with left limits and locally bounded process. Then, $(\Delta_\epsilon^{(2)}(K, t))_{t \geq 0}$ converges in distribution to $(2\sigma \int_0^t K_u dW_u)_{t \geq 0}$, as $\epsilon \to 0$.*

5. Let us briefly detail the organization of the paper. Section 2 contains the proofs of the almost convergence results, i.e. Theorem 1 and Proposition 1. Then, the proof of Theorem 2 (resp. Propositions 2, 3 and Theorems 3, 4) is (resp. are) given in Sect. 3 (resp. Sect. 4).

In the calculations, C will stand for a generic constant (random or not). We will use several times a stochastic version of Fubini's theorem, which can be found in Sect. IV.5 of [8].

2   Proof of Theorem 1 and Proposition 1

We begin with the proof of Theorem 1 in Points **1–4** below. Then, we deduce Proposition 1 from Theorem 1 in Point **5**.

1. Let $T > 0$. We suppose that $(H_t)_{t \geqslant 0}$ is locally Hölder continuous of order α' and we study the almost sure convergence of

$$I_\epsilon(t) := \frac{1}{\epsilon} \int_0^t H_u(X_{(u+\epsilon) \wedge t} - X_u) du \text{ to } I(t) := \int_0^t H_u dX_u,$$

as $\epsilon \to 0$, uniformly on $t \in [0, T]$.

By stopping, we can suppose that $(X_t)_{0 \leqslant t \leqslant T}$ and $\langle X \rangle_T$ are bounded by a constant.

Let $X = X_0 + M + V$ be the canonical decomposition of X, where M is a continuous local martingale and V is an adapted process with finite variation. It is clear that $I_\epsilon(t) - I(t)$ can be decomposed as

$$I_\epsilon(t) - I(t) = \left(\frac{1}{\epsilon} \int_0^t H_u(M_{(u+\epsilon) \wedge t} - M_u) du - \int_0^t H_u dM_u \right)$$

$$+ \left(\frac{1}{\epsilon} \int_0^t H_u(V_{(u+\epsilon) \wedge t} - V_u) du - \int_0^t H_u dV_u \right).$$

Then, Theorem 1 will be proved as soon as $I_\epsilon(t) - I(t)$ converges to 0, in the case where X is either a continuous local martingale or a continuous finite variation process.

We deal with the finite variation case resp. the martingale case in Point **2** resp. Points **3, 4**.

2. Suppose that X has a finite variation, writing $X_{(u+\epsilon) \wedge t} - X_u = \int_u^{(u+\epsilon) \wedge t} dX_s$ and using Fubini's theorem yield to:

$$I_\epsilon(t) - I(t) = \int_0^t \left(\frac{1}{\epsilon} \int_{(s-\epsilon)+}^s H_u du - H_s \right) dX_s,$$

$$= \int_0^t \left(\frac{1}{\epsilon} \int_{(s-\epsilon)+}^s (H_u - H_s) du \right) dX_s - \int_0^{t \wedge \epsilon} \frac{\epsilon - s}{\epsilon} H_s dX_s.$$

Using the Hölder property (3) (in the first integral) and the fact that H is bounded by a constant (in the second integral), we have for all $t \in [0, T]$:

$$|I_\epsilon(t) - I(t)| \leqslant \int_0^T \left(\frac{1}{\epsilon} \int_{(s-\epsilon)+}^s C_H |u - s|^\alpha du \right) d|X|_s + \int_0^\epsilon \frac{\epsilon - s}{\epsilon} C \, d|X|_s$$

$$\leqslant C_H \epsilon^\alpha |X|_T + C(|X|_\epsilon - |X|_0). \tag{12}$$

Consequently, $I_\epsilon(t) - I(t)$ converges almost surely to 0, as $\epsilon \to 0$, uniformly on any compact set in time.

Remark 1. Note that (12) implies that:

$$\sup_{0 \le t \le T} \left| \int_0^t H_s \frac{X_{(s+\epsilon)\wedge t} - X_s}{\epsilon} ds - \int_0^t H_s dX_s \right| \le C\epsilon^\alpha$$

when (H_t) is α-Hölder continuous and X has finite variation.

3. In the two next points, X is a continuous martingale. We prove that there is a sequence $(\epsilon_n)_{n \in \mathbb{N}}$ such that:

$$\lim_{n \to \infty} \sup_{t \in [0,T]} |I_{\epsilon_n}(t) - I(t)| = 0, \quad a.s. \tag{13}$$

We proceed as in step **2** above: observing that $X_{(u+\epsilon)\wedge t} - X_u = \int_u^{(u+\epsilon)\wedge t} dX_s$ and using Fubini's stochastic theorem come to

$$I_\epsilon(t) - I(t) = \int_0^t \left(\frac{1}{\epsilon} \int_{(s-\epsilon)+}^s H_u du - H_s \right) dX_s. \tag{14}$$

Thus, $(I_\epsilon(t) - I(t))_{t \in [0,T]}$ is a continuous local martingale. Moreover, $E(\langle I_\epsilon - I \rangle_t)$ is bounded since H and $\langle X \rangle$ are bounded on $[0, T]$.

Let us introduce $p = \frac{2(1-\alpha)}{\alpha^2} + 1$. This explicit expression of p in terms of α will be used later at the end of Point **4.** Burkhölder–Davis–Gundy inequalities give:

$$E \left(\sup_{t \in [0,T]} |I_\epsilon(t) - I(t)|^p \right) \le c_p E \left[\left(\int_0^T \left(\frac{1}{\epsilon} \int_{(s-\epsilon)+}^s H_u du - H_s \right)^2 d\langle X \rangle_s \right)^{\frac{p}{2}} \right].$$

The Hölder property (3) implies that:

$$\left| \frac{1}{\epsilon} \int_{(s-\epsilon)+}^s H_u du - H_s \right| \le \frac{1}{\epsilon} \int_{s-\epsilon}^s |H_u - H_s| du \le C_H \epsilon^\alpha, \quad \epsilon \le s,$$

$$\left| \frac{1}{\epsilon} \int_{(s-\epsilon)+}^s H_u du - H_s \right| \le \frac{1}{\epsilon} \int_0^s |H_u - H_s| du + \frac{\epsilon - s}{\epsilon} |H_s| \le C\epsilon^\alpha, \quad s < \epsilon.$$

(a) Suppose that in (3), $C_H \le C$ for some C. Consequently,

$$\sup_{0 \le s \le T} \left| \frac{1}{\epsilon} \int_{(s-\epsilon)+}^s H_u du - H_s \right| \le C\epsilon^\alpha \tag{15}$$

and

$$E \left(\sup_{t \in [0,T]} |I_\epsilon(t) - I(t)|^p \right) \le C\epsilon^{\alpha p} E[\langle X \rangle_T]^{\frac{p}{2}} \le C\epsilon^{\alpha p}.$$

Then, for any $\delta > 0$, Markov inequality leads to:

$$P\left(\sup_{t\in[0,T]} |I_\epsilon(t) - I(t)| > \delta\right) \le \frac{C\epsilon^{\alpha p}}{\delta^p}. \tag{16}$$

Let us now define $(\epsilon_n)_{n\in\mathbb{N}^*}$ by $\epsilon_n = n^{-\frac{2}{p\alpha}}$ for all $n > 0$. Replacing ϵ by ϵ_n in (16) comes to:

$$P\left(\sup_{t\in[0,T]} |I_{\epsilon_n}(t) - I(t)| > \delta\right) \le \frac{C}{\delta^p} n^{-2}.$$

Since $\sum_{n=1}^{\infty} n^{-2} < \infty$, the Borel-Cantelli lemma implies (13).

(b) Using localization and Lemma 1 below we can reduce to the case where C_H is bounded by a constant. Then (13) holds.

Lemma 1. *Let (Y_t) be an adapted process and locally Hölder continuous with index α. Then for any $\beta \in]0, \alpha[$ there exists a continuous and adapted process $(Lip(Y, t))$ such that*

$$|Y_u - Y_v| \le Lip(Y, t)|u - v|^\beta, \quad u, v \in [0, t].$$

Proof (Proof of Lemma 1). Set:

$$Lip(Y, t) := \sup_{0 \le u, v \le t} |\widetilde{Y}(u, v)|, \quad t \ge 0$$

where $\widetilde{Y}(u, v) := \dfrac{Y_u - Y_v}{|u - v|^\beta}$ when $u \ne v$ and 0 otherwise.
Lemma 1 follows from the continuity of $\widetilde{Y}$. $\qquad\qquad\square$

4. For all $\epsilon \in]0, 1[$, let $n = n(\epsilon)$ denote the integer such that $\epsilon \in]\epsilon_{n+1}, \epsilon_n]$. Then, we decompose $I_\epsilon(t) - I(t)$ as follows:

$$I_\epsilon(t) - I(t) = (I_\epsilon(t) - I_{\epsilon_n}(t)) + (I_{\epsilon_n}(t) - I(t)).$$

From (13), Theorem 1 is proved if

$$\lim_{\epsilon\to 0} \sup_{t\in[0,T]} |I_{\epsilon_n}(t) - I_\epsilon(t)| = 0, \quad a.s. \quad (n = n(\epsilon)). \tag{17}$$

From the definition of $I_\epsilon(t)$, it is easy to deduce that we have:

$$I_\epsilon(t) - I_{\epsilon_n}(t) = \frac{1}{\epsilon}\left(\int_0^t H_u X_{(u+\epsilon)\wedge t}\,du - \int_0^t H_u X_{(u+\epsilon_n)\wedge t}\,du\right)$$
$$+ \left(\frac{1}{\epsilon} - \frac{1}{\epsilon_n}\right)\left(\int_0^t H_u(X_{(u+\epsilon_n)\wedge t} - X_u)\,du\right).$$

The changes of variable either $v = u + \epsilon$ or $v = u + \epsilon_n$ lead to

$$I_\epsilon(t) - I_{\epsilon_n}(t) = \frac{1}{\epsilon}\int_\epsilon^{t+\epsilon}(H_{v-\epsilon} - H_{v-\epsilon_n})\,X_{v\wedge t}\,dv \tag{18}$$
$$+ \frac{\epsilon_n - \epsilon}{\epsilon\epsilon_n}\left(\int_{\epsilon_n}^t(H_{v-\epsilon_n} - H_v)\,X_v\,dv\right) + R_\epsilon(t),$$

where we gather under the notation $R_\epsilon(t)$ all the remaining terms. Let us observe that $R_\epsilon(t)$ is the sum of terms which are of the form $\frac{1}{\epsilon}\int_a^b \ldots dv$ where $|a - b| \leqslant \epsilon_n - \epsilon$ or $\left(\frac{1}{\epsilon} - \frac{1}{\epsilon_n}\right)\int_a^b \ldots dv$ where $|a - b| \leqslant \epsilon_n$. Since H and X are bounded on $[0, T]$, we have

$$|R_\epsilon(t)| \leqslant C\frac{\epsilon_n - \epsilon}{\epsilon} \quad \forall t \in [0, T]. \tag{19}$$

By Hölder property (3), we get

$$|H_{v-\epsilon} - H_{v-\epsilon_n}| \leqslant C(\epsilon_n - \epsilon)^\alpha, \quad |H_{v-\epsilon_n} - H_v| \leqslant C_H\epsilon_n^\alpha. \tag{20}$$

Since X and H are bounded, we can deduce from (18), (19) and (20) that:

$$|I_\epsilon(t) - I_{\epsilon_n}(t)| \leqslant C\left(\frac{(\epsilon_n - \epsilon)^\alpha}{\epsilon} + \frac{(\epsilon_n - \epsilon)\epsilon_n^\alpha}{\epsilon\epsilon_n} + \frac{\epsilon - \epsilon_n}{\epsilon}\right), \quad \forall t \in [0, T]. \tag{21}$$

Using the definition of ϵ_n, easy calculations lead to:

$$\frac{\epsilon_n - \epsilon}{\epsilon} \leqslant Cn^{-1}, \quad \frac{(\epsilon_n - \epsilon)^\alpha}{\epsilon} \leqslant Cn^{\frac{2(1-\alpha)}{p\alpha} - \alpha}, \quad \frac{(\epsilon_n - \epsilon)\epsilon_n^\alpha}{\epsilon\epsilon_n} \leqslant n^{-\frac{2}{p} - 1 + \frac{2}{p\alpha}} \leqslant n^{\frac{2(1-\alpha)}{p\alpha} - \alpha}.$$

Note that $p = \frac{2(1-\alpha)}{\alpha^2} + 1$ implies that $\frac{2(1-\alpha)}{p\alpha} - \alpha < 0$. As a result, we may deduce that (17) holds. $\square$

Remark 2. Let (H_t) be an progressively measurable process. Suppose for simplicity that (X_t) is a local semimartingale. Let (ϵ_n) denote a sequence of decreasing positive numbers converging to 0 as $n \to \infty$. Applying Theorem 2 in [5] to (14) gives the a.s. convergence of $\sup_{0\leqslant u\leqslant T}\left|I_{\epsilon_n}(u) - I_\epsilon(u)\right|$ to 0 as $n \to \infty$, provided that

$$\sum_{n\geqslant 1}\left(\sup_{0\leqslant u\leqslant T}\left|H_u - \frac{1}{\epsilon_n}\int_{(u-\epsilon_n)_+}^u H_r\,dr\right|\right)^2 < \infty, \quad a.s. \tag{22}$$

Suppose that (H_t) is locally Hölder with index α. According to (15), relation (22) holds if $\sum_{n \geq 1} \epsilon_n^\alpha < \infty$. To simplify the discussion suppose that $\epsilon_n = 1/n^\rho$, with $\rho > 0$. Obviously, the previous sum is finite if and only if $\rho\alpha > 1$.

Note that inequality (21) permit to prove the a.s. of $I_{\epsilon_n}(u)$ as soon as

$$\lim_{n \to \infty} \frac{(\epsilon_n - \epsilon)^\alpha}{\epsilon} = \lim_{n \to \infty} \frac{(\epsilon_n - \epsilon)\epsilon_n^\alpha}{\epsilon \epsilon_n} = \lim_{n \to \infty} \frac{\epsilon - \epsilon_n}{\epsilon} = 0.$$

Since ϵ varies in $[\epsilon_{n+1}, \epsilon_n]$, then

$$\frac{(\epsilon_n - \epsilon)^\alpha}{\epsilon} \leq \frac{(\epsilon_n - \epsilon_{n+1})^\alpha}{\epsilon_{n+1}}.$$

It is easy to prove that

$$\frac{(\epsilon_n - \epsilon_{n+1})^\alpha}{\epsilon_{n+1}} \sim \frac{\rho^\alpha}{n^{(1+\rho)\alpha-\rho}}, \quad n \to \infty.$$

Therefore ρ has to be chosen such that $(1 + \rho)\alpha - \rho > 0$, i.e. $\rho < \dfrac{\alpha}{1 - \alpha}$. Recall that $\rho > 1/\alpha$, then $\dfrac{1}{\alpha} < \dfrac{\alpha}{1 - \alpha}$. This condition is equivalent to $\alpha > \alpha_0 := \dfrac{\sqrt{5} - 1}{2}$. This inequality is not necessarily satisfied since it is only supposed that α belongs to $]0, 1[$. Finally, our Theorem 1 is not a consequence of Theorem 2 of [5].

5. In this item X is supposed to be a locally Hölder continuous semimartingale. Note that replacing X by $X - X_0$ does not change (5). Therefore we may suppose that $X_0 = 0$.

It is clear that $\frac{1}{\epsilon} \int_0^t (X_{(u+\epsilon)\wedge t} - X_u)^2 du$ equals

$$\frac{1}{\epsilon} \left[\int_0^t X_{(u+\epsilon)\wedge t}^2 du - \int_0^t X_u^2 du - 2 \int_0^t X_u(X_{(u+\epsilon)\wedge t} - X_u)du \right].$$

Making the change of variable $v = u + \epsilon$ in the first integral, we easily get:

$$\frac{1}{\epsilon} \int_0^t (X_{(u+\epsilon)\wedge t} - X_u)^2 du = X_t^2 - \frac{1}{\epsilon} \int_0^{t\wedge\epsilon} X_v^2 dv - \frac{2}{\epsilon} \int_0^t X_u(X_{(u+\epsilon)\wedge t} - X_u)du.$$

Since X is continuous, $\frac{1}{\epsilon} \int_0^{t\wedge\epsilon} X_v^2 dv$ tends to 0 a.s, uniformly on $[0, T]$. Therefore, it can be deduced from Theorem 1:

$$\lim_{\epsilon \to 0} \frac{1}{\epsilon} \int_0^t (X_{(u+\epsilon)\wedge t} - X_u)^2 du = X_t^2 - 2 \int_0^t X_u dX_u \quad (a.s.).$$

Itô's formula implies that the right-hand side of the above identity equals to $\langle X \rangle_t$.

Replacing $(u + \epsilon) \wedge t$ by $u + \epsilon$ in either (5) or (6) does not change the limit. Then, identity (5) may be interpreted as follows: the measures $\frac{1}{\epsilon}(X_{u+\epsilon} - X_u)^2 du$ converges a.s. to the measure $d\langle X \rangle_u$. That implies the almost sure convergence of $\frac{1}{\epsilon} \int_0^t K_u(X_{(u+\epsilon)\wedge t} - X_u)^2 du$ to $\int_0^t K_u d\langle X \rangle_u$, for any continuous process K.   □

3 Proof of Theorem 2

Recall that $W_\epsilon(t)$ and $G_\epsilon(t)$ are defined by (8). We study the convergence in distribution of the two dimensional process $(W_\epsilon(t), B_t)$, as $\epsilon \to 0$.

First, we determine the limit in law of $W_\epsilon(t)$. In Point **1** we demonstrate preliminary results. Then, we prove the convergence of the moments of $W_\epsilon(t)$ in Point **2**. By the method of moments, the convergence in law of $W_\epsilon(t)$ for a fixed time is proven in Point **3**. We deduce the finite-dimensional convergence in Point **4**. Finally, Kolmogorov criterion concludes the proof in Point **5**. Then, we briefly sketch in Point **6** the proof of the joint convergence of $(W_\epsilon(t))_{t \geqslant 0}$ and $(B_t)_{t \geqslant 0}$. The approach is close to the one of $(W_\epsilon(t))_{t \geqslant 0}$.

1. We begin by calculating the moments of $W_\epsilon(t)$ and $G_\epsilon(u)$. We denote by $\overset{\mathcal{L}}{=}$ the equality in law.

Lemma 2. $E\left[|G_\epsilon(u)|^2\right] = \frac{(u\wedge\epsilon)^3}{\epsilon^3}\sigma^2$. *Moreover, for all* $k \in \mathbb{N}$, *there exists a constant* m_k *such that* $E\left[|G_\epsilon(u)|^k\right] \leqslant m_k, \forall u \geqslant 0, \epsilon > 0$.

Proof. First, we apply the change of variable $s = u - (u \wedge \epsilon)r$ in (8). Then, using the identity $(B_u - B_{u-v}; 0 \leqslant v \leqslant u) \overset{\mathcal{L}}{=} (B_v; 0 \leqslant v \leqslant u)$ and the scaling property of B, we get

$$G_\epsilon(u) \overset{\mathcal{L}}{=} \frac{(u \wedge \epsilon)\sqrt{u \wedge \epsilon}}{\epsilon\sqrt{\epsilon}} \int_0^1 B_r dr.$$

Since $\int_0^1 B_r dr \overset{\mathcal{L}}{=} \sigma N$, where $\sigma^2 = 1/3$ and N is a standard gaussian r.v, we obtain

$$E\left[|G_\epsilon(u)|^k\right] = \frac{(u \wedge \epsilon)^{\frac{3k}{2}}}{\epsilon^{\frac{3k}{2}}}\sigma^k E\left[|N|^k\right]. \tag{23}$$

Taking $k = 2$ gives $E\left[|G_\epsilon(u)|^2\right] = \frac{(t\wedge\epsilon)^3}{\epsilon^3}\sigma^2$. Using $u \wedge \epsilon \leqslant \epsilon$ and (23), we get $E[|G_\epsilon(u)|^k] \leqslant m_k$ with $m_k = \sigma^k E\left[|N|^k\right]$.   □

Lemma 3. *For all* $k \geqslant 2$, *there exists a constant* $C(k)$ *such that*

$$\forall t \geqslant 0, \quad E\left[|W_\epsilon(t)|^k\right] \leqslant C(k) t^{\frac{k}{2}}.$$

Moreover, for $k = 2$, we have

$$E\left[\left(W_\epsilon(u) - W_\epsilon((u - \epsilon)^+)\right)^2\right] \leqslant \sigma^2\epsilon, \quad \forall u \geqslant 0.$$

Proof. The Burkhölder-Davis-Gundy inequality and (8) give

$$E\left[|W_\epsilon(t)|^k\right] \leqslant c(k)E\left[\left(\int_0^t (G_\epsilon(u))^2\, du\right)^{\frac{k}{2}}\right].$$

Then, Jensen inequality implies:

$$E\left[\left(\int_0^t (G_\epsilon(u))^2\, du\right)^{\frac{k}{2}}\right] \leqslant t^{\frac{k}{2}-1}E\left[\int_0^t |G_\epsilon(u)|^k\, du\right].$$

Finally, applying Lemma 2 comes to

$$E\left[|W_\epsilon(t)|^k\right] \leqslant c(k)m_k t^{\frac{k}{2}}.$$

The case $k = 2$ can be easily treated via (8) and Lemma 2:

$$E\left[\left(W_\epsilon(u) - W_\epsilon((u - \epsilon)^+)\right)^2\right] = \int_{(u-\epsilon)^+}^u E\left[(G_\epsilon(v))^2\right]dv,$$

$$= \int_{(u-\epsilon)^+}^u \sigma^2\frac{(v \wedge \epsilon)^3}{\epsilon^3}dv \leqslant \sigma^2\epsilon.$$

$\square$

2. Let us now study the convergence of the moments of $W_\epsilon(t)$.

Proposition 4.

$$\lim_{\epsilon \to 0} E\left[(W_\epsilon(t))^{2n}\right] = E\left[(\sigma W_t)^{2n}\right], \quad \forall n \in \mathbb{N}, t \geqslant 0. \tag{24}$$

Proof. **(a)** We prove Proposition 4 by induction on $n \geqslant 1$.
For $n = 1$, from Lemma 2, we have:

$$E\left[(W_\epsilon(t))^2\right] = \int_0^t E\left[(G_\epsilon(u))^2\right]du = \int_0^t \sigma^2\frac{(u \wedge \epsilon)^3}{\epsilon^3}du.$$

Then, $E\left[(W_\epsilon(t))^2\right]$ converges to $\sigma^2 t = E[(\sigma W_t)^2]$.
Let us suppose that (24) holds. First, we apply Itô's formula to $(W_\epsilon(t))^{2n+2}$. Second, taking the expectation reduces to 0 the martingale part. Finally, we get

$$E\left[(W_\epsilon(t))^{2n+2}\right] = \frac{(2n + 2)(2n + 1)}{2}\int_0^t E\left[(W_\epsilon(u))^{2n}(G_\epsilon(u))^2\right]du. \tag{25}$$

(b) We admit for a while that

$$E\left[(W_\epsilon(u))^{2n}(G_\epsilon(u))^2\right] \longrightarrow \sigma^2 E\left[(\sigma W_u)^{2n}\right], \quad \forall u \geqslant 0. \qquad (26)$$

Using Cauchy-Schwarz inequality and Lemmas 2, 3 give:

$$E\left[(W_\epsilon(u))^{2n}(G_\epsilon(u))^2\right] \leqslant \sqrt{E\left[(W_\epsilon(u))^{4n}\right]E\left[(G_\epsilon(u))^4\right]}$$

$$\leqslant \sqrt{C(4n)u^{2n}m_4} \leqslant \sqrt{C(4n)m_4u^n}.$$

Consequently, we may apply Lebesgue's theorem to (25), we have

$$\lim_{\epsilon \to 0} E\left[(W_\epsilon(t))^{2n+2}\right] = \frac{(2n+2)(2n+1)}{2}\sigma^2 \int_0^t E\left[(\sigma W_u)^{2n}\right]du,$$

$$= \frac{(2n+2)(2n+1)}{2}\sigma^{2n+2} \int_0^t u^n \frac{(2n)!}{n!\,2^n}du,$$

$$= \frac{(2n+2)!}{(n+1)!\,2^{n+1}}(\sigma\sqrt{t})^{2n+2} = E\left[(\sigma W_t)^{2n+2}\right].$$

(c) We have now to prove (26). If $u = 0$, $E\left[(W_\epsilon(0))^{2n}(G_\epsilon(0))^2\right] = 0 = \sigma^2 E\left[(\sigma W_0)^{2n}\right]$. If $u > 0$, it is clear that:

$$E\left[(W_\epsilon(u))^{2n}(G_\epsilon(u))^2\right] = E\left[(W_\epsilon((u-\epsilon)^+))^{2n}(G_\epsilon(u))^2\right] + \xi_\epsilon(u), \qquad (27)$$

where
$$\xi_\epsilon(u) = E\left[\left\{(W_\epsilon(u))^{2n} - (W_\epsilon((u-\epsilon)^+))^{2n}\right\}(G_\epsilon(u))^2\right].$$

Since $G_\epsilon(u)$ is independent from $\mathcal{F}_{(u-\epsilon)^+}$, we have

$$E\left[(W_\epsilon((u-\epsilon)^+))^{2n}(G_\epsilon(u))^2\right] = E\left[(W_\epsilon((u-\epsilon)^+))^{2n}\right]E\left[(G_\epsilon(u))^2\right].$$

Finally, plugging the identity above in (27) gives:

$$E\left[(W_\epsilon(u))^{2n}(G_\epsilon(u))^2\right] = E\left[(W_\epsilon(u))^{2n}\right]E\left[(G_\epsilon(u))^2\right] + \xi_\epsilon(u) + \tilde{\xi}_\epsilon(u),$$

where
$$\tilde{\xi}_\epsilon(u) = E\left[(W_\epsilon((u-\epsilon)^+))^{2n} - (W_\epsilon(u))^{2n}\right]E\left[(G_\epsilon(u))^2\right].$$

Lemma 2 implies that $E\left[(G_\epsilon(u))^2\right]$ tends to σ^2 as $\epsilon \to 0$. The recurrence hypothesis implies that $E\left[(W_\epsilon(u))^{2n}\right]$ converges to $E\left[(\sigma W_u)^{2n}\right]$ as $\epsilon \to 0$. It remains to prove that $\xi_\epsilon(u)$ and $\tilde{\xi}_\epsilon(u)$ tend to 0 to conclude the proof.

The identity $a^{2n} - b^{2n} = (a - b) \sum_{k=0}^{2n-1} a^k b^{2n-1-k}$ implies that $\xi_\epsilon(u)$ is equal to the sum $\sum_{k=0}^{2n-1} S_k(\epsilon, u)$, where

$$S_k(\epsilon, u) = E\left[\left(W_\epsilon(u) - W_\epsilon((u - \epsilon)^+)\right)(G_\epsilon(u))^2 (W_\epsilon(u))^k\right.$$
$$\left. (W_\epsilon((u - \epsilon)^+))^{2n-1-k}\right].$$

Applying four times the Cauchy–Schwarz inequality yields to:

$$|S_k(\epsilon, u)| \leq \left[E\left(W_\epsilon(u) - W_\epsilon((u - \epsilon)^+)\right)^2\right]^{\frac{1}{2}} \left[E(G_\epsilon(u))^8\right]^{\frac{1}{4}}$$
$$\times \left[E(W_\epsilon(u))^{8k}\right]^{\frac{1}{8}} \left[E(W_\epsilon((u - \epsilon)^+))^{16n-8-8k}\right]^{\frac{1}{8}}.$$

Lemmas 2 and 3 lead to

$$|S_k(\epsilon, u)| \leq C(k)T^{n-\frac{1}{2}}\sqrt{\epsilon}, \quad \forall u \in [0, T].$$

Consequently, $\xi_\epsilon(u)$ tends to 0 as $\epsilon \to 0$. Using the same method, it is easy to prove that $\tilde{\xi}_\epsilon(u)$ tends to 0 as $\epsilon \to 0$. $\square$

3. From Proposition 4, it easy to deduce the convergence in law of $W_\epsilon(t)$ (t being fixed).

Proposition 5. *For any fixed $t \geq 0$, $W_\epsilon(t)$ converges in law to σW_t, as $\epsilon \to 0$.*

Remark 3. Using stochastic Fubini theorem we have

$$W_\epsilon(t) = \frac{1}{\epsilon\sqrt{\epsilon}} \int_0^t \left(\int_0^u (v - (u - \epsilon)+)_+ dB_v\right) dB_u.$$

We keep notation given in [7]. Let us introduce the function f_ϵ:

$$f_\epsilon(u, v) := \frac{1}{\epsilon\sqrt{\epsilon}}(v - (u - \epsilon)+)_+ \mathbb{1}_{\{0 \leq v \leq u \leq t\}}.$$

Consequently $W_\epsilon(t) = J_2^1(f_\epsilon)$.

It is easy to prove that

$$\left(\|f_\epsilon\|_{\Delta_t^2}\right)^2 := \int_0^t \left(\int_0^u f_\epsilon(u, v)^2 dv\right) du = \frac{\epsilon}{12} + \frac{t - \epsilon}{3}, \quad t > \epsilon.$$

Therefore
$$\lim_{\epsilon \to 0} \| f_\epsilon \|_{\Delta_t^2} = \sigma \sqrt{t}.$$

Proposition 3 in [7] ensures that $W_\epsilon(t)$ converges in distribution to σW_t, as $\epsilon \to 0$ if and only if

$$\lim_{\epsilon \to 0} \int_{[0,t]^2} F_\epsilon(s_1, s_2)^2 ds_1 ds_2 = 0 \qquad (28)$$

where

$$F_\epsilon(s_1, s_2) := \int_0^t \big(f_\epsilon(u, s_1) f_\epsilon(u, s_2) + f_\epsilon(s_1, u) f_\epsilon(s_2, u) \big) du.$$

Identity (28) can be shown by tedious calculations. This gives a new proof of Proposition 5.

Let us recall the method of moments.

Proposition 6. *Let* $X, (X_n)_{n \in \mathbb{N}}$ *be r.v's such that* $E(|X|^k) < \infty$, $E(|X_n|^k) < \infty$, $\forall k, n \in \mathbb{N}$ *and*

$$\overline{\lim}_{k \to \infty} \frac{[E(X^{2k})]^{\frac{1}{2k}}}{2k} < \infty. \qquad (29)$$

If for all $k \in \mathbb{N}$, $\lim_{n \to \infty} E(X_n^k) = E(X^k)$, *then* X_n *converges in law to* X *as* $n \to \infty$.

Proof (Proof of Proposition 5). Let $t \geq 0$ be a fixed time. The odd moments of $W_\epsilon(t)$ are null. By Proposition 4, the even moments of $W_\epsilon(t)$ tends to σW_t. Since σW_t is a Gaussian r.v. with variance $\sigma \sqrt{t}$, it is easy to check that (29) holds. As a result, $W_\epsilon(t)$ converges in law to σW_t. $\qquad \square$

4. Next, we prove the finite-dimensionnal convergence.

Proposition 7. *Let* $0 < t_1 < t_2 < \cdots < t_n$. *Then,* $(W_\epsilon(t_1), \ldots, W_\epsilon(t_n))$ *converges in law to* $(\sigma W_{t_1}, \ldots, \sigma W_{t_n})$, *as* $\epsilon \to 0$.

Proof. We take $n = 2$ for simplicity. We consider $0 < t_1 < t_2$ and $\epsilon \in]0, t_1 \wedge (t_2 - t_1)[$. Since $t_1 > \epsilon$, note that $(u - \epsilon)^+ = u - \epsilon$ for $u \in [t_1, t_2]$. We begin with the decomposition:

$$W_\epsilon(t_2) = W_\epsilon(t_1) + \frac{1}{\epsilon \sqrt{\epsilon}} \int_{t_1 + \epsilon}^{t_2} \left(\int_{u-\epsilon}^{u} (B_u - B_s) ds \right) dB_u + R_\epsilon^1(t_1, t_2),$$

where $R_\epsilon^1(t_1, t_2) = \frac{1}{\epsilon \sqrt{\epsilon}} \int_{t_1}^{t_1 + \epsilon} \left(\int_{u-\epsilon}^{u} (B_u - B_s) ds \right) dB_u$. Let us note that $W_\epsilon(t_1)$ is independent from $\frac{1}{\epsilon \sqrt{\epsilon}} \int_{t_1 + \epsilon}^{t_2} \left(\int_{u-\epsilon}^{u} (B_u - B_s) ds \right) dB_u$.

Let us introduce $B'_t = B_{t+t_1} - B_{t_1}, t \geq 0$. B' is a standard Brownian motion. The changes of variables $u = t_1 + v$ and $r = s - t_1$ in $\int_{t_1 + \epsilon}^{t_2} \left(\int_{u-\epsilon}^{u} (B_u - B_s) ds \right) dB_u$ leads to

$$W_\epsilon(t_2) = W_\epsilon(t_1) + \Theta_\epsilon(t_1, t_2) + R_\epsilon^2(t_1, t_2) + R_\epsilon^1(t_1, t_2), \qquad (30)$$

where

$$
\Theta_\epsilon(t_1, t_2) = \frac{1}{\epsilon\sqrt{\epsilon}} \int_0^{t_2-t_1} \left(\int_{(v-\epsilon)+}^v (B_v' - B_r') dr \right) dB_v',
$$

$$
R_\epsilon^2(t_1, t_2) = \frac{1}{\epsilon\sqrt{\epsilon}} \int_0^\epsilon \left(\int_0^v (B_v' - B_r') dr \right) dB_v'.
$$

Straightforward calculation shows that $E\left[\left(R_\epsilon^1(t_1, t_2)\right)^2\right]$ and $E\left[\left(R_\epsilon^2(t_1, t_2)\right)^2\right]$ are bounded by $C\epsilon$. Thus, $R_\epsilon^1(t_1, t_2)$ and $R_\epsilon^1(t_1, t_2)$ converge to 0 in $L^2(\Omega)$. Proposition 5 gives the convergence in law of $\Theta_\epsilon(t_1, t_2)$ to $\sigma(W_{t_2} - W_{t_1})$ and the convergence in law of $W_\epsilon(t_1)$ to σW_{t_1}, as $\epsilon \to 0$.

Since $W_\epsilon(t_1)$ and $\Theta_\epsilon(t_1, t_2)$ are independent, the decomposition (30) implies that $(W_\epsilon(t_1), W_\epsilon(t_2) - W_\epsilon(t_1))$ converges in law to $(\sigma W_{t_1}, \sigma(W_{t_2} - W_{t_1}))$, as $\epsilon \to 0$. Proposition 5 follows immediately. $\qquad \square$

5. We end the proof of the convergence in law of the process $(W_\epsilon(t))_{t\geq 0}$ by showing that the family of the laws of $(W_\epsilon(t))_{t\geq 0}$ is tight as $\epsilon \in]0, 1]$.

Lemma 4. *There exists a constant K such that*

$$
E\left[|W_\epsilon(t) - W_\epsilon(s)|^4\right] \leq K|t - s|^2, \qquad 0 \leq s \leq t, \epsilon > 0.
$$

Proof. Applying Burkhölder–Davis–Gundy inequality, we obtain:

$$
E\left[|W_\epsilon(t) - W_\epsilon(s)|^4\right] \leq cE\left[\left(\int_s^t (G_\epsilon(u))^2 du\right)^2\right]
$$

$$
\leq c(t - s) \int_s^t E\left[(G_\epsilon(u))^4\right] du.
$$

Using Lemma 2, we get $E\left[|W_\epsilon(t) - W_\epsilon(s)|^4\right] \leq cm_4(t - s)^2$ and ends the proof (see Kolmogorov Criterion in Sect. XIII-1 of [8]). $\qquad \square$

6. To prove the joint convergence of $(W_\epsilon(t), B_t)_{t\geq 0}$ to $(\sigma W_t, B_t)_{t\geq 0}$, we mimick the approach developed in Points **1–5** above.

(a) Convergence $(W_\epsilon(t), B_t)$ to $(\sigma W_t, B_t)$, t being fixed. First, we prove that

$$
\lim_{\epsilon \to 0} E(W_\epsilon^p(t) B_t^q) = E((\sigma W_t)^p B_t^q), \qquad p, q \in \mathbb{N}. \tag{31}
$$

Let us note that the limit is null when either p or q is odd.

Using Itô's formula, we get

$$E\left[(W_\epsilon(t))^p\, B_t^q\right] = \frac{p(p-1)}{2}\alpha_1(t,\epsilon) + \frac{q(q-1)}{2}\alpha_2(t,\epsilon) + pq\alpha_3(t,\epsilon),$$

where

$$\alpha_1(t,\epsilon) = \int_0^t E\left[(W_\epsilon(u))^{p-2}\, B_u^q\,(G_\epsilon(u))^2\right] du,$$

$$\alpha_2(t,\epsilon) = \int_0^t E\left[(W_\epsilon(u))^p\, B_u^{q-2}\right] du,$$

$$\alpha_3(t,\epsilon) = \int_0^t E\left[(W_\epsilon(u))^{p-1}\, B_u^{q-1}\, G_\epsilon(u)\right] du.$$

To demonstrate (31), we proceed by induction on q, then by induction on p, q being fixed.

First, we apply (31) with $q-2$ instead of q, then we have directly:

$$\lim_{\epsilon\to 0}\alpha_2(t,\epsilon) = \int_0^t E\left[(\sigma W_u)^p\right] E\left[B_u^{q-2}\right] du.$$

As for $\alpha_1(t,\epsilon)$, we write

$$(W_\epsilon(u))^{p-2} = (W_\epsilon(u))^{p-2} - \left(W_\epsilon((u-\epsilon)^+)\right)^{p-2} + \left(W_\epsilon((u-\epsilon)^+)\right)^{p-2}$$
$$B_u^q = B_u^q - B_{(u-\epsilon)^+}^q + B_{(u-\epsilon)^+}^q.$$

We proceed similarly with $\alpha_3(t,\epsilon)$. Reasoning as in Point **2** and using the two previous identities, we can prove:

$$\lim_{\epsilon\to 0}\alpha_1(t,\epsilon) = \sigma^2 \int_0^t E\left[(\sigma W_u)^{p-2}\right] E\left[B_u^q\right] du \text{ and } \lim_{\epsilon\to 0}\alpha_3(t,\epsilon) = 0.$$

Consequently, when either p or q is odd, then $\lim_{\epsilon\to 0}\alpha_i(t,\epsilon) = 0$, $(i=1,2)$ and therefore:

$$\lim_{\epsilon\to 0} E(W_\epsilon^p(t) B_t^q) = 0 = E((\sigma W_t)^p B_t^q).$$

It remains to determine the limit in the case where p and q are even. Let us denote $p = 2p'$ and $q = 2q'$. Then we have

$$\lim_{\epsilon\to 0}\alpha_1(t,\epsilon) = \int_0^t \sigma^2 \frac{(p-2)!}{2^{p'-1}(p'-1)!} u^{p'-1}\sigma^{p-2}\,\frac{q!}{2^{q'}(q')!} u^{q'} du$$

$$= \frac{(p-2)!\,q!}{2^{p'+q'-1}(p'-1)!(q')!(p'+q')}\sigma^p t^{p'+q'},$$

$$\lim_{\epsilon \to 0} \alpha_2(t, \epsilon) = \int_0^t \frac{p!}{2^{p'} (p')!} \sigma^p u^{p'} \frac{(q-2)!}{2^{q'-1} (q'-1)!} u^{q'-1} du$$

$$= \frac{p! (q-2)!}{2^{p'+q'-1} (p')! (q'-1)! (p'+q')} \sigma^p t^{p'+q'}.$$

Then, it is easy to deduce

$$\lim_{\epsilon \to 0} E\left[(W_\epsilon(t))^p B_t^q\right] = \frac{p!}{2^{p'} (p')!} \sigma^p t^{p'} \frac{q!}{2^{q'} (q')!} t^{q'} = E\left[(\sigma W_t)^p\right] E\left[B_t^q\right].$$

Next, we use a two dimensional version of the method of moments:

Proposition 8. *Let* $X, Y, (Y_n)_{n \in \mathbb{N}} (X_n)_{n \in \mathbb{N}}$ *be r.v. whose moments are finite. Let us suppose that* X *and* Y *satisfy (29) and that* $\forall p, q \in \mathbb{N}$, $\lim_{n \to \infty} E(X_n^p Y_n^q) = E(X^p Y^q)$. *Then,* (X_n, Y_n) *converges in law to* (X, Y) *as* $n \to \infty$.

Since W_t and B_t are Gaussian r.v.'s, they both satisfy (29). Consequently, $(W_\epsilon(t), B_t)$ converges in law to $(\sigma W_t, B_t)$ as $\epsilon \to 0$.

(b) Finite-dimensional convergence. Let $0 < t_1 < t_2$. We prove that the vector $(W_\epsilon(t_1), W_\epsilon(t_2), B_{t_1}, B_{t_2})$ converges in law to $(\sigma W_{t_1}, \sigma W_{t_2}, B_{t_1}, B_{t_2})$. We apply decomposition (30) to $W_\epsilon(t_2)$.

By Point **6(a)**, $(W_\epsilon(t_1), B_{t_1})$ converges in law to $(\sigma W_{t_1}, B_{t_1})$ and $(\Theta_\epsilon(t_1, t_2), B_{t_2} - B_{t_1})$ converges to $(\sigma W_{t_2} - \sigma W_{t_1}, B_{t_2} - B_{t_1})$. Since $(\Theta_\epsilon(t_1, t_2), B_{t_2} - B_{t_1})$ is independent from $(W_\epsilon(t_1), B_{t_1})$, we can conclude that $(W_\epsilon(t_1), W_\epsilon(t_2), B_{t_1}, B_{t_2})$ converges in law to $(\sigma W_{t_1}, \sigma W_{t_2}, B_{t_1}, B_{t_2})$. $\square$

4 Proofs of Theorems 3, 4 and Propositions 2, 3

1. Convergence in distribution of a family of stochastic integrals with respect to W_ϵ.

Denote $C([0, T])$ the set of real valued and continuous functions defined on $[0, T]$. $C([0, T])$ equipped with the uniform norm is a Banach space. Set $\mathcal{B}_c([0, T])$ the Borel σ-field on $C([0, T])$. Let $D([0, T])$ be the space of right-continuous functions with left-limits equipped with the Skorokhod topology.

Consider a predictable, right-continuous with left-limits process (Γ_u) such that:

$$(\Gamma, W_\epsilon) \text{ converges in distribution to } (\Gamma, \sigma W), \quad \epsilon \to 0. \tag{32}$$

In (32), the pair (Γ, W_ϵ) is considered as an element of $D([0, T]) \times C([0, T])$.

Proposition 9. *1. Let* $F : \left(\Omega \times C([0, T]), \ \sigma(B_u, u \geq 0) \otimes \mathcal{B}_c([0, T])\right) \to \mathbb{R}$ *be a bounded an measurable map and such that for any* ω, $F(\omega, \cdot)$ *is continuous. Then:*

$$\lim_{\epsilon \to 0} E\left(F(\cdot, W_\epsilon)\right) = E\left(F(\cdot, \sigma W)\right). \tag{33}$$

2. Under (32), the process $\left(\int_0^t \Gamma_u d W_\epsilon(u) \right)_{t \geq 0}$ converges in distribution to $\left(\sigma \int_0^t \Gamma_u d W_u \right)_{t \geq 0}$ as $\epsilon \to 0$, where (Γ_u) is independent of (W_u).

Proof (Proof of Proposition 9). **(a)** Denote $\mathcal{H}$ the set of $\sigma(B_u, u \geq 0)$-measurable and bounded r.v.'s A such that

$$\lim_{\epsilon \to 0} E\big(A\Theta(W_\epsilon)\big) = E\big(A\Theta(\sigma W)\big) = E(A)E\big(\Theta(\sigma W)\big), \qquad (34)$$

where $\Theta : C\big([0, T]\big) \to \mathbb{R}$ is continuous and bounded.
It is clear that $\mathcal{H}$ is a linear vector space. Let $(A_n, n \geq 1)$ be a sequence of elements in $\mathcal{H}$ which satisfies

(i) $(A_n, n \geq 1)$ converges uniformly to a bounded element A
    either
(ii) $n \mapsto A_n$ is non-decreasing and the limit A is bounded.
    Since

$$
\begin{aligned}
E\big(A\Theta(W_\epsilon)\big) - E\big(A\Theta(\sigma W)\big) = {}& E\big((A - A_n)\Theta(W_\epsilon)\big) \\
& + E\big(A_n\Theta(W_\epsilon)\big) - E\big(A_n\Theta(\sigma W)\big) \\
& E\big((A_n - A)\Theta(\sigma W)\big)
\end{aligned}
$$

we have

$$
\begin{aligned}
\Big| E\big(A\Theta(W_\epsilon)\big) - E\big(A\Theta(\sigma W)\big) \Big| \leq {}& CE\big(|A - A_n|\big) + \Big| E\big(A_n\Theta(W_\epsilon)\big) \\
& - E\big(A_n\Theta(\sigma W)\big) \Big|.
\end{aligned}
$$

Consequently, $A \in \mathcal{H}$.

Consider the set C of r.v.'s of the type $f(B_{t_1}, \cdots, B_{t_n})$ where f is continuous and bounded. Theorem 2 implies that $C \subset \mathcal{H}$. Then, (34) is direct consequence of Theorem T20 p. 28 in [6].

According to Proposition 2.4 in [3], relations (34) and (33) are equivalent.

(b) Denote $F_0 : D\big([0, T]\big) \times C\big([0, T]\big) \to \mathbb{R}$ a bounded and continuous function. Property (32) is a direct consequence of item 1 of Proposition 9 applied with:

$$F(\omega, w) := F_0\big((\Gamma_s(\omega), 0 \leq s \leq T), w\big), \quad w \in C\big([0, T]\big).$$

Recall that W_ϵ is a continuous martingale, which converges in distribution to σW as $\epsilon \to 0$. Then, by Proposition 3.2 of [4], W_ϵ satisfies the condition of uniform tightness. Consequently, from Theorem 5.1 of [4] and (32), we can deduce that for any predictable, right-continuous with left-limits process Γ, $\int_0^{\cdot} \Gamma_u d W_\epsilon(u)$ converges in distribution to $\sigma \int_0^{\cdot} \Gamma_u d W(u)$.

Remark 4. 1. The convergence in item 1 of Proposition 9 corresponds to the stable convergence, cf. [3].

2. According to relation (14), we have

$$\Delta_\epsilon(H,t) = \frac{1}{\sqrt{\epsilon}} \int_0^t \left(\frac{1}{\epsilon} \int_{(s-\epsilon)+}^s H_u du - H_s \right) dB_s.$$

Let us apply the general result obtained in [9]. Let (ϵ_n) be a sequence of positive numbers converging to 0 as $n \to \infty$. For any $t > 0$, suppose:

$$\frac{1}{\epsilon_n} \int_0^t \left(\frac{1}{\epsilon_n} \int_{(s-\epsilon_n)+}^s H_u du - H_s \right)^2 ds \xrightarrow{(P)} \tau(t), \quad n \to \infty \qquad (35)$$

and

$$\sup_{0 \le r \le t} \frac{1}{\sqrt{\epsilon_n}} \left| \int_0^r \left(\frac{1}{\epsilon_n} \int_{(s-\epsilon_n)+}^s H_u du - H_s \right) ds \right| \xrightarrow{(P)} 0, \quad n \to \infty \qquad (36)$$

where $(\tau(t))$ denotes a continuous process and (P) stands for the convergence in probability.

Then, from Theorem 1.2 in [9] we can deduce that

$$\left(\Delta_{\epsilon_n}(H,t), \; t \ge 0 \right) \xrightarrow{(d)} \left(W(\tau(t)) \; t \ge 0 \right), \quad n \to \infty \qquad (37)$$

where (W_t) is a standard Brownian motion independent from $(\tau(t))$.

Suppose that (H_t) is of the type $H_t = H_0 + \int_0^t \Lambda_s dB_s + V_t$, where (Λ_t) and (V_t) satisfy the assumptions given in Theorem 3. Note that

$$\left(\sigma \int_0^t \Lambda_u d W_u, \; t \ge 0 \right) \overset{(d)}{=} \left(W \left(\sigma^2 \int_0^t \Lambda_u^2 du \right), \; t \ge 0 \right).$$

Therefore (37) suggests to prove (35) with

$$\tau(t) := \sigma^2 \int_0^t \Lambda_u^2 du, \quad t \ge 0.$$

We have tried without any success to directly prove (35) and (36). In the particular case $H_t = B_t$, the calculations are tractable. Theorem 1.2 in [9] may be applied: $\left(\Delta_{\epsilon_n}(B,t), t \ge 0 \right)$ converges in distribution to $(\sigma W_t, t \ge 0)$, as $n \to \infty$. However, this result is not sufficient to have the convergence of $\left(\Delta_{\epsilon_n}(H,t), \; t \ge 0 \right)$ since we need the convergence of $\left(\Delta_{\epsilon_n}(B,t), B_t \right)$ and the convergence of the previous pair of processes is not given by Theorem 1.2 in [9].

2. Some preliminary results related to the proof of Theorem 3

Lemma 5. *Let $\big(\xi_\epsilon(t), t \geq 0\big)$ be a family of processes. Suppose there exists a increasing sequence $(T_n)_{n\geq 1}$ of random times such that $T_n \uparrow \infty$ as $n \to \infty$ and for any $n \geq 1$, $\big(\xi_\epsilon(t \wedge T_n), t \geq 0\big)$ converges in the ucp sense to 0, as $\epsilon \to 0$. Then $\big(\xi_\epsilon(t), t \geq 0\big)$ converges in the ucp sense to 0, as $\epsilon \to 0$, i.e. for any $T > 0$,*

$$\sup_{0 \leq s \leq T} |\xi_\epsilon(s)| \to 0 \text{ in probability as } \epsilon \to 0.$$

Lemma 6. *Denote (K_t) an progressively measurable process which is right-continuous at 0, $K_0 = 0$ and locally bounded. Set:*

$$R_\epsilon^1(K, t) := \frac{1}{\sqrt{\epsilon}} \int_0^{t \wedge \epsilon} K_s \Big(\frac{s}{\epsilon} - 1\Big) dB_s, \quad t \geq 0. \tag{38}$$

Then $\big(R_\epsilon^1(K, t), t \geq 0\big)$ converges in the ucp sense to 0.

Proof (Proof of Lemma 6). Since (K_t) is locally bounded there exists a increasing sequence of stopping times $(T_n)_{n\geq 1}$ such that $T_n \uparrow \infty$ as $n \to \infty$ and $|K(t \wedge T_n)| \leq n$, for any $t \geq 0$. Then, according to Lemma 5 it is sufficient to prove that $\big(R_\epsilon^1(K, t), t \geq 0\big)$ converges in the ucp sense to 0 when (K_t) is bounded. In that case, using Doob's inequality we get:

$$E\Big(\sup_{t \in [0,T]} \big(R_\epsilon^1(K, t)\big)^2\Big) \leq \frac{C}{\epsilon} E\Big(\int_0^\epsilon K(s)^2 \Big(\frac{s}{\epsilon} - 1\Big)^2 ds\Big) \leq C \sup_{0 \leq s \leq \epsilon} E\big(K(s)^2\big)$$

where $T > 0$.

Recall that (K_s) is bounded, $s \mapsto K(s)$ is right continuous at 0, $K(0) = 0$, then the dominated convergence theorem implies that $\lim\limits_{\epsilon \to 0} \Big(\sup\limits_{0 \leq s \leq \epsilon} E\big(K(s)^2\big)\Big) = 0$. This proves that $\sup\limits_{t \in [0,T]} \big|R_\epsilon^1(K, t)\big|$ goes to 0 in $L^2(\Omega)$. $\square$

Remark 5. Note that under (9), relation (14) implies that:

$$\Delta_\epsilon(H, t) = \widetilde{\Delta}_\epsilon(H, t) + R_\epsilon^1(H\Phi, t) \tag{39}$$

where $\Delta_\epsilon(H, t)$ has been defined by (7) and

$$\widetilde{\Delta}_\epsilon(H, t) := \frac{1}{\epsilon\sqrt{\epsilon}} \int_0^t \Big(\int_{(u-\epsilon)_+}^u (H_s - H_u) ds\Big) \Phi(u) dB_u. \tag{40}$$

Proof (Proof of Proposition 3). Recall that $\Delta_\epsilon^{(2)}(K, t)$ is defined by (11). Using Itô's formula, we obtain:

$$(B_{(s+\epsilon)\wedge t} - B_s)^2 = 2 \int_s^{(s+\epsilon)\wedge t} (B_u - B_s) dB_u + (s + \epsilon) \wedge t - s.$$

Reporting in $\Delta_\epsilon^{(2)}(K,t)$ and applying stochastic Fubini's theorem lead to

$$\Delta_\epsilon^{(2)}(K,t) = 2 \int_0^t K_u d W_\epsilon(u) + R_\epsilon^1(t) + R_\epsilon^2(t),$$

where

$$R_\epsilon^1(t) := \frac{2}{\epsilon\sqrt{\epsilon}} \int_0^t \left[\int_{(u-\epsilon)+}^u (K_s - K_u)(B_u - B_s)ds \right] dB_u$$

$$R_\epsilon^2(t) := \frac{1}{\epsilon\sqrt{\epsilon}} \int_{(t-\epsilon)+}^t K_s(t - s - \epsilon)ds.$$

Note that Proposition 9 (with $\Gamma = K$) ensures the convergence in distribution of $\int_0^\cdot K_u d W_\epsilon(u)$ to $\sigma \int_0^\cdot K_u d W(u)$.

Since $s \to K_s$ is locally bounded, then $\lim_{\epsilon \to 0} \sup_{t\in[0,T]} |R_\epsilon^2(t)| = 0$ a.s.

To prove that $R_\epsilon^1 \xrightarrow{\text{(ucp)}} 0$, we may assume that K is bounded (cf. Lemma 5). Using the Cauchy–Schwarz and Doob inequalities, we obtain successively:

$$E\left(\sup_{0\le t\le T} (R_\epsilon^1)^2 \right) \le \frac{C}{\epsilon^3} \int_0^T E\left(\left(\int_{(u-\epsilon)+}^u (K_s - K_u)(B_s - B_u)ds \right)^2 \right) du$$

$$\le \frac{C}{\epsilon^2} \int_0^T du \int_{(u-\epsilon)+}^u \sqrt{E\left((K_s - K_u)^4 \right) E\left((B_s - B_u)^4 \right)} ds$$

$$\le C \int_0^T \left(\sup_{s\le u\le(s+\epsilon)\wedge t} E\left((K_s - K_u)^4 \right) \right) ds$$

Since K is bounded and right-continuous, then the term in the right-hand side of the above inequality goes to 0 as $\epsilon \to 0$. $\square$

Proof (Proof of Point (1) *of Theorem 3).* Using (39) we have:

$$\Delta_\epsilon(H_0,t) = H_0\big(\widetilde{\Delta}_\epsilon(1,t) + R_\epsilon^1(\Phi,t)\big) = H_0 R_\epsilon^1(\Phi,t)$$

$$= H_0\Phi(0)N_\epsilon + H_0 R_\epsilon^1\big(\Phi - \Phi(0),t\big)$$

where $\epsilon < t$ and

$$N_\epsilon := \frac{1}{\sqrt{\epsilon}} \int_0^\epsilon \left(\frac{u}{\epsilon} - 1 \right) dB_u, \quad \epsilon < t.$$

The r.v N_ϵ has a centered Gaussian distribution, with variance

$$E(N_\epsilon^2) = \int_0^\epsilon \left(\frac{u}{\epsilon} - 1 \right)^2 \frac{du}{\epsilon} = \frac{1}{3} = \sigma.$$

According to Lemma 6, $R_\epsilon^1\big(\Phi - \Phi(0), \cdot\big) \xrightarrow{\text{(ucp)}} 0$ as $\epsilon \to 0$. $\square$

Proof (Proof of Point (2) of Theorem 3). Since $(H_t) = (V_t)$ is continuous and $V_0 = 0$ then, Lemma 6 applied with $K = \Phi H$ implies that $R_\epsilon^1(\Phi H, \cdot) \xrightarrow{\text{(ucp)}} 0$ as $\epsilon \to 0$.

Let $T > 0$. According to Lemmas 5 and 1, we may suppose that Φ is bounded and:

$$|V_s - V_u| \le C |u - v|^\beta, \quad u, v \in [0, T], \beta > \frac{1}{2}.$$

As a result,

$$\left| \frac{1}{\epsilon \sqrt{\epsilon}} \int_{(u-\epsilon)_+}^u (V_s - V_u) ds \right| \le \frac{1}{\epsilon \sqrt{\epsilon}} \int_{u-\epsilon}^u C |s - u|^\beta \, ds \le C \epsilon^{\beta - \frac{1}{2}}$$

and

$$E\left(\sup_{t \in [0,T]} \left(\widetilde{\Delta}_\epsilon(V, t) \right)^2 \right) \le C \epsilon^{2\beta - 1}.$$

Using (39), item 2 of Theorem 3 follows.   □

Proof (Proof of Point 3 of Theorem 3). **(a)** Recall that $M_t = \int_0^t \Lambda_r d B_r$ and

$$\widetilde{\Delta}_\epsilon(M, t) = -\frac{1}{\epsilon \sqrt{\epsilon}} \int_0^t \left(\int_{(u-\epsilon)_+}^u (M_u - M_s) ds \right) \Phi(u) d B_u.$$

Let $s < u$, we have

$$M_u - M_s = \Lambda_{(u-\epsilon)_+}(B_u - B_s) + \int_s^u \left(\Lambda_r - \Lambda_{(u-\epsilon)_+} \right) d B_r.$$

Using (8) we get:

$$\widetilde{\Delta}_\epsilon(M, t) = -\int_0^t \Lambda_{u_-} \Phi(u) d W_\epsilon(u) + R_\epsilon^2(t) + R_\epsilon^3(t), \tag{41}$$

where

$$R_\epsilon^2(t) := -\int_0^t \left(\Lambda_{(u-\epsilon)_+} - \Lambda_{u_-} \right) \Phi(u) d W_\epsilon(u)$$

and

$$R_\epsilon^3(t) = -\frac{1}{\epsilon \sqrt{\epsilon}} \int_0^t \left(\int_{(u-\epsilon)_+}^u (r - (u - \epsilon)_+)(\Lambda_r - \Lambda_{(u-\epsilon)_+}) d B_r \right) \Phi(u) d B_u.$$

(b) Suppose for a while that R_ϵ^2 and R_ϵ^3 converge in the ucp sense to 0, as $\epsilon \to 0$. Then, Proposition 9 with $\Gamma = \Lambda$ implies that the convergence of $\Delta_\epsilon(H, \cdot)$ to $\sigma \int_0^{\cdot} \Lambda_{u-} \Phi(u) d W_u$. Note that $\int_0^{\cdot} \Lambda_u \Phi(u) d W_u = \int_0^{\cdot} \Lambda_{u-} \Phi(u) d W_u$ a.s.

(c) Let us prove that R_ϵ^3 converge in the ucp sense to 0. The proof related to R_ϵ^2 is similar and easier; it is left to the reader. From Lemma 5, we can suppose that $(\Lambda_u, 0 \le u \le T)$ and $(\Phi(u), 0 \le u \le T)$ are bounded. Then, using Burkhölder-Davis-Gundy and Hölder inequalities we get:

$$E\left(\sup_{0 \le t \le T} R_\epsilon^3(t)^2 \right) \le \frac{C}{\epsilon} E \left(\int_0^T \left\{ \int_{(u-\epsilon)_+}^u \frac{r-(u-\epsilon)_+}{\epsilon} (\Lambda_r - \Lambda_{(u-\epsilon)_+}) d B_r \right\}^2 \right.$$

$$\left. \times \Phi(u)^2 d u \right)$$

$$\le \frac{C}{\epsilon} \int_0^T d u E \left(\left\{ \int_{(u-\epsilon)_+}^u \frac{r-(u-\epsilon)_+}{\epsilon} (\Lambda_r - \Lambda_{(u-\epsilon)_+}) d B_r \right\}^2 \right)$$

$$\le \frac{C}{\epsilon} \int_0^T d u \int_{(u-\epsilon)_+}^u \left(\frac{r-(u-\epsilon)_+}{\epsilon} \right)^2 E\left((\Lambda_r - \Lambda_{(u-\epsilon)_+})^2 \right) d r$$

$$\le C \int_0^T \sup_{(u-\epsilon)_+ \le r < u} \left(E\left((\Lambda_r - \Lambda_{(u-\epsilon)_+})^2 \right) \right) d u.$$

Using the dominated convergence theorem and the fact that $t \mapsto \Lambda_t$ has left-limits we can conclude that the right-hand side in the above inequality goes to 0 as $\epsilon \to 0$. Consequently, $\sup_{0 \le t \le T} |R_\epsilon^3(t)|$ goes to 0 in $L^2(\Omega)$. □

Proof (Proof of Proposition 2). From (39), we have:

$$\Delta_\epsilon(h(B), t) = \widetilde{\Delta}_\epsilon(h(B), t) + R_\epsilon^1(h(B)\Phi, t)$$

where

$$\widetilde{\Delta}_\epsilon(h(B), t) = \frac{1}{\epsilon\sqrt{\epsilon}} \int_0^t \left(\int_{(u-\epsilon)_+}^u \{h(B_s) - h(B_u)\} ds \right) \Phi(u) d B_u.$$

Since:

$$h(B_s) - h(B_u) = (B_s - B_u) \int_0^1 h'(B_u + \theta(B_s - B_u)) d\theta$$

$$= (B_s - B_u) h'(B_u)$$

$$+ (B_s - B_u) \int_0^1 \{h'(B_u + \theta(B_s - B_u)) - h'(B_u)\} d\theta$$

then,

$$\Delta_\epsilon\big(h(B),t\big) = -\int_0^t h'(B_u)\Phi(u)dW_\epsilon(u) + R_\epsilon^1\big(h(B)\Phi,t\big) + R_\epsilon^3(t)$$

where $\big(W_\epsilon(u)\big)$ is the process defined by (8) and

$$R_\epsilon^3(t) := \frac{1}{\sqrt{\epsilon}}\int_0^t \left\{\frac{1}{\epsilon}\int_{(u-\epsilon)_+}^u (B_s - B_u)\left[\int_0^1 \Big\{h'\big(B_u + \theta(B_s - B_u)\big)\right.\right.$$
$$\left.\left. - h'(B_u)\Big\}d\theta\right]ds\right\}\Phi(u)dB_u.$$

Using Proposition 9 (with $\Gamma = h'(B)\Phi$) implies that $\int_0^\cdot h'(B_u)\Phi(u)dW_\epsilon(u)$, con-

verges in distribution to $\sigma \int_0^\cdot h'(B_u)\Phi(u)dW(u)$, as $\epsilon \to 0$. Since $h(0) = 0$,

Lemma 6 may be applied: $R_\epsilon^1\big(h(B)\Phi,\cdot\big) \xrightarrow{\text{(ucp)}} 0$, as $\epsilon \to 0$. We claim that R_ϵ^3
has the same behavior. By localization and Lemma 5 we may suppose that Φ is
bounded. Using Doob's and Hölder inequalities we obtain:

$$E\left(\sup_{0\le t\le T} \big(R_\epsilon^3(t)\big)^2\right) \le \frac{C\delta(h',\epsilon)}{\epsilon^2}\int_0^T \left\{\int_{(u-\epsilon)_+}^u \sqrt{E\big([B_s - B_u]^4\big)}ds\right\} du$$
$$\le C\delta(h',\epsilon)$$

where

$$\delta(\phi,\epsilon) := \sqrt{\sup_{0\le\theta\le 1, 0\le u-\epsilon\le s\le u\le T} E\left(\big\{\phi\big(B_u + \theta(B_s - B_u)\big) - \phi(B_u)\big\}^4\right)}.$$

It can be proved that $\lim_{\epsilon\to 0}\delta(\phi,\epsilon) = 0$ as soon as ϕ is subexponential. As a result,
$\sup_{t\le T}\big|R_\epsilon^3(t)\big|$ goes to 0 in $L^2(\Omega)$ as $\epsilon \to 0$. $\square$

Proof (Proof of Theorem 4). (**a**) The a.s. convergence comes from the continuity of
$t \mapsto X_t$ and the identity

$$\frac{1}{\epsilon}\int_0^t H_s\big(X_{s+\epsilon} - X_s\big)ds = \sum_{j=0}^{i-1} h_j\left(\frac{1}{\epsilon}\int_{a_{j+1}}^{a_{j+1}+\epsilon} X_s ds - \frac{1}{\epsilon}\int_{a_j}^{a_j+\epsilon} X_s ds\right)$$

$$+ h_i\left(\frac{1}{\epsilon}\int_t^{t+\epsilon} X_s ds - \frac{1}{\epsilon}\int_{a_i}^{a_i+\epsilon} X_s ds\right)$$

where $a_i \le t \le a_{i+1}$ and $i \ge 0$.

(b) Let us deal the convergence in distribution. Recall that we supposed that $X = B$. Using the definition of $\Delta_\epsilon(H, t)$, identity (14) and easy calculations we get:

$$\Delta_\epsilon(H, t) = h_0 \left\{ \Phi(0) G_0(\epsilon) + R_\epsilon^1 (\Phi - \Phi(0), \epsilon) \right\}, \quad 0 < t \leq a_1, 0 < \epsilon < t$$

where $R_\epsilon^1 (\Phi - \Phi(0), \epsilon)$ has been defined by (38) and

$$G_0(\epsilon) := \frac{1}{\sqrt{\epsilon}} \int_0^\epsilon \left(\frac{s}{\epsilon} - 1 \right) dB_s.$$

More generally when $t \in]a_i, a_{i+1}]$, $\epsilon < (t - a_i) \wedge (a_i - a_{i-1})$ and $i \geq 1$, we have

$$\Delta_\epsilon(H, t) = \Delta_\epsilon(H, a_i) + (h_i - h_{i-1})\left(\Phi(a_i) G_i(\epsilon) + \widetilde{R}_\epsilon^1 \right)$$

with

$$G_i(\epsilon) := \frac{1}{\sqrt{\epsilon}} \int_{a_i}^{a_i + \epsilon} \left(\frac{s - a_i}{\epsilon} - 1 \right) dB_s$$

$$\widetilde{R}_\epsilon^1 := \frac{1}{\sqrt{\epsilon}} \int_{a_i}^{a_i + \epsilon} \left(\frac{s - a_i}{\epsilon} - 1 \right) \left(\Phi(s) - \Phi(a_i) \right) dB_s$$

As a result for any $t \in]a_i, a_{i+1}]$ we have:

$$\begin{aligned}
\Delta_\epsilon(H, t) = {} & h_0 \Phi(0) G_0(\epsilon) + (h_1 - h_0) \Phi(a_1) G_1(\epsilon) \\
& + \cdots + (h_i - h_{i-1}) \Phi(a_i) G_i(\epsilon) + (h_i - h_{i-1}) \widetilde{R}_\epsilon^1
\end{aligned}$$

where

$$\epsilon < (a_1 - a_0) \wedge \cdots \wedge (a_i - a_{i-1}) \wedge (t - a_i). \tag{42}$$

Recall that Φ has been supposed to be right-continuous at a_i, then Lemma 6 may be applied: $\widetilde{R}_\epsilon^1 \xrightarrow{\text{(ucp)}} 0$, as $\epsilon \to 0$. As a result, the term $\widetilde{R}_\epsilon^1$ gives no contribution to the limit of $\Delta_\epsilon(H, \cdot)$.

Note that $G_i(\epsilon)$ is a Gaussian r.v. with variance $\sigma^2 = 1/3$ and under (42) the r.v.'s $G_0(\epsilon), \cdots, G_i(\epsilon)$ are independent and

$$\lim_{\epsilon \to 0} E\left(B_s G_i(\epsilon) \right) = 0, \quad \forall s \geq 0.$$

Item 2 of Theorem 4 follows. $\qquad\square$

References

1. Bérard Bergery, B.: Approximation du temps local et intégration par régularisation. PhD thesis, Nancy Université (2004–2007)
2. Gradinaru, M., Nourdin, I.: Approximation at first and second order of m-order integrals of the fractional Brownian motion and of certain semimartingales. Electron. J. Probab., **8**(18), 26 pp. (electronic) (2003)

3. Jacod, J., Mémin, J.: Sur un type de convergence intermédiaire entre la convergence en loi et la convergence en probabilité. In Seminar on Probability, XV (Univ. Strasbourg, Strasbourg, 1979/1980), Lecture Notes in Mathematics, vol. 850, pp. 529–546. Springer (1981)

4. Jakubowski, A., Mémin, J., Pagès, G.: Convergence en loi des suites d'intégrales stochastiques sur l'espace $\mathbf{D}^1$ de Skorokhod. Probab. Theor. Relat. Field. **81**(1), 111–137 (1989)

5. Karandikar, R.: On almost sure convergence results in stochastic calculus. In Séminaire de Probabilités, XXXIX, Lecture Notes in Mathematics, vol. 1874, pp. 137–147, Springer, Berlin (2006)

6. Meyer, P.A.: Probabilités et potentiel, Publications de l'Institut de Mathématique de l'Université de Strasbourg, No. XIV. Actualités Scientifiques et Industrielles, No. 1318, Hermann, Paris (1966)

7. Nualart, D., Peccati, G.: Central limit theorems for sequences of multiple stochastic integrals. Ann. Probab. **33**(1), 177–193 (2005)

8. Revuz, D., Yor, M.: Continuous martingales and Brownian motion, volume 293 of Grundlehren der Mathematischen Wissenschaften [Fundamental Principles of Mathematical Sciences]. Third edition. Springer, Berlin (1999)

9. Rootzen, H.: Limit distribution for the error in approximations of stochastic integrals. Ann. Probab. **8**(2), 241–251 (1980)

10. Russo, F., Vallois, P.: Elements of stochastic calculus via regularisation. In Séminaire de Probabilités XL, Lecture Notes in Mathematics, vol. 1899, pp. 147–185, Springer, Berlin (2007)

11. Russo, F., Vallois, P.: Itô formula for C^1-functions of semimartingales. Probab. Theor. Relat. Field. **104**(1), 27–41 (1996)

12. Russo, F., Vallois, P.: Stochastic calculus with respect to continuous finite quadratic variation processes. Stochast. Rep. **70**(1–2), 1–40 (2000)

13. Russo, F., Vallois, P.: The generalized covariation process and Itô formula. Stoch. Process. Appl. **59**(1), 81–104 (1995)

Convergence of Multi-Dimensional Quantized *SDE*'s

Gilles Pagès and Afef Sellami

Abstract We quantize a multidimensional *SDE* (in the Stratonovich sense) by solving the related system of *ODE*'s in which the d-dimensional Brownian motion has been replaced by the components of functional stationary quantizers. We make a connection with rough path theory to show that the solutions of the quantized solutions of the *ODE* converge toward the solution of the *SDE*. On our way to this result we provide convergence rates of optimal quantizations toward the Brownian motion for $\frac{1}{q}$-Hölder distance, $q > 2$, in $L^p(\mathbb{P})$.

Keywords Functional quantization · Stochastic differential equations · Stratonovich stochastic integral · Stationary quantizers · Rough path theory · Itô map · Hölder semi-norm · p-variation

1 Introduction

Quantization is a way to discretize the path space of a random phenomenon: a random vector in finite dimension, a stochastic process in infinite dimension. Optimal Vector Quantization theory (finite-dimensional) random vectors finds its origin in the early 1950s in order to discretize some emitted signal (see [10]). It was further developed by specialists in Signal Processing and later in Information Theory. The infinite dimensional case started to be extensively investigated in the early 2000s by several authors (see e.g. [4, 5, 12, 18–20], etc).

G. Pagès (✉)
Laboratoire de Probabilités et Modèles Aléatoires, Université de Paris 6, Case 188, 4 pl. Jussieu, 75252 Paris Cedex 5, France
e-mail: gilles.pages@upmc.fr

A. Sellami
JP Morgan, London & Laboratoire de Probabilités et Modèles Aléatoires, Université de Paris 6, Case 188, 4 pl. Jussieu, 75252 Paris Cedex 5, France
e-mail: afef.x.sellami@jpmorgan.com

C. Donati-Martin et al. (eds.), *Séminaire de Probabilités XLIII*, Lecture Notes in Mathematics 269
2006, DOI 10.1007/978-3-642-15217-7__11, © Springer-Verlag Berlin Heidelberg 2011

In [20], the functional quantization of a class of Brownian diffusions has been investigated from a constructive point of view. The main feature of this class of diffusions was that the diffusion coefficient was the inverse of the gradient of a diffeomorphism (both coefficients being smooth). This class contains most (non degenerate) scalar diffusions. Starting from a sequence of rate optimal quantizers, some sequences of quantizers of the Brownian diffusion are produced as solutions of (non coupled) *ODE*'s. This approach relied on the Lamperti transform and was closely related to the Doss–Sussman representation formula of the flow of a diffusion as a functional of the Brownian motion. In many situations these quantizers are rate optimal (or almost rate optimal) i.e. that they quantize the diffusion at the same rate $O((\log N)^{-\frac{1}{2}})$ as the Brownian motion itself where N denotes the generic size of the quantizer. In a companion paper (see [27]), some cubature formulas based on some of these quantizers were implemented, namely those obtained from some optimal product quantizers based on the Karhunen–Loève expansion of the Brownian motion, to price some Asian options in a Heston stochastic volatility model. Rather unexpectedly in view of the theoretical rate of convergence, the numerical experiments provided quite good numerical results for some "small" sizes of quantizers. Note however that these numerical implementations included some further speeding up procedures combining the stationarity of the quantizers and the Romberg extrapolation leading to a $O((\log N)^{-\frac{3}{2}})$ rate. Although this result relies on some still pending conjectures about the asymptotics of bilinear functionals of the quantizers, it strongly pleads in favour of the construction of such stationary (rate optimal) quantizers, at least when one has in mind possible numerical applications.

Recently a sharp quantization rate (i.e. including an explicit constant) has been established for a class of not too degenerate one-dimensional Brownian diffusions. However the approach is not constructive (see [4]). On the other hand, the standard rate $O((\log N)^{-\frac{1}{2}})$ has been extended in [22] to general d-dimensional Itô processes, so including d-dimensional Brownian diffusions regardless of their ellipticity properties. This latter approach, based an expansion in the Haar basis, is constructive, but the resulting quantizers are no longer stationary.

Our aim in this paper is to extend the constructive natural approach initiated in [20] to general d-dimensional diffusions in order to produce some rate optimal stationary quantizers of these processes. To this end, we will call upon some seminal results from rough path theory, namely the continuity of the Itô map, to replace the "Doss–Sussman setting". In fact we will show that if one replaces in an *SDE* (written in the Stratonovich sense) the Brownian motion by some elementary quantizers, the solutions of the resulting *ODE*'s make up some rough paths which converge (in p-variation and in the Hölder metric) to the solution of the *SDE*. We use her the rough path theory as a tool and we do not aim at providing new insights on this theory. We can only mention that these rate optimal stationary quantizers can be seen as a new example of rough paths, somewhat less "stuck" to a true path of the underlying process.

This work is devoted to Brownian diffusions which is naturally the prominent example in view of applications, but it seems clear that this could be extended to *SDE* driven e.g. by fractional Brownian motions (however our approach requires to have an explicit form for the Karhunen–Loève basis as far as numerical implementation is concerned).

Now let us be more precise. We consider a diffusion process

$$dX_t = b(t, X_t)\, dt + \sigma(t, X_t) \circ dW_t, \ X_0 = x \in \mathbb{R}^d, \ t \in [0, T],$$

in the Stratonovich sense where $b : [0, T] \times \mathbb{R}^d \to \mathbb{R}^d$ and $\sigma : [0, T] \times \mathbb{R}^d \to \mathcal{M}(d \times d)$ are continuously differentiable with linear growth (uniformly with respect to t) and $W = (W_t)_{t \in [0, T]}$ is a d-dimensional Brownian motion defined on a filtered probability space $(\Omega, \mathcal{A}, \mathbb{P})$. (The fact that the state space and W have the same dimension is in no case a restriction since our result has nothing to do with ellipticity).

Such an *SDE* admits a unique strong solution denoted $X^x = (X_t^x)_{t \in [0, T]}$ (the dependency in ^{x} will be dropped from now to alleviate notations). The $\mathbb{R}^d$-valued process X is pathwise continuous and $\sup_{t \in [0, T]} |X_t| \in L^r(\mathbb{P})$, $r > 0$ (where $|.|$ denotes the canonical Euclidean norm on $\mathbb{R}^d$). In particular X is bi-measurable and can be seen as an $L^r(\mathbb{P})$-Radon random variable taking values in the Banach spaces $(L_{T, \mathbb{R}^d}^p, |.|_{L_T^p})$ where $L_{T, \mathbb{R}^d}^p = L_{\mathbb{R}^d}^p([0, T], dt)$ and $|g|_{L_T^p} = \left(\int_0^T |g(t)|^p dt \right)^{\frac{1}{p}}$ denotes the usual L^p-norm when $p \in [1, \infty)$.

For every integer $N \geq 1$, we can investigate for X the level N $(L^r(\mathbb{P}), L_T^p)$-quantization problem for this process X, namely solving the minimization of the $L^r(\mathbb{P})$-mean $L_{T, \mathbb{R}^d}^p$-quantization error

$$e_{N,r}(X, L^p) := \min \left\{ e_{N,r}(\alpha, X, L^p), \ \alpha \subset L_{T, \mathbb{R}^d}^p, \ \mathrm{card}\, \alpha \leq N \right\} \tag{1}$$

where $e_{N,r}(\alpha, X, L^p)$ denotes the L^r-mean quantization error induced by α, namely

$$e_{N,r}(\alpha, X, L^p) := \left(\mathbb{E} \min_{a \in \alpha} |X - a|_p^r \right)^{\frac{1}{r}} = \left\| \min_{a \in \alpha} |X - a|_{L_{T, \mathbb{R}^d}^p} \right\|_{L^r(\mathbb{P})}.$$

The use of "min" in (1) is justified by the existence of an optimal quantizer solution to that problem as shown in [3, 13] in this infinite dimensional setting. The Voronoi diagram associated to a quantizer α is a Borel partition $(C_a(\alpha))_{a \in \alpha}$ such that

$$C_a(\alpha) \subset \left\{ x \in L_{T, \mathbb{R}^d}^p \mid |x - a|_{L_{T, \mathbb{R}^d}^p} \leq \min_{b \in \alpha} |x - b|_{L_{T, \mathbb{R}^d}^p} \right\}$$

and a functional quantization of X by α is defined by the nearest neighbour projection of X onto α related to the Voronoi diagram

$$\widehat{X}^\alpha := \sum_{a \in \alpha} a \mathbf{1}_{\{X \in C_a(\alpha)\}}.$$

In finite dimension (when considering $\mathbb{R}^d$-valued random vectors instead of $L^p_{T,\mathbb{R}^d}$-valued processes) the answer is provided by the so-called Zador Theorem which says (see [10]) that if $\mathbb{E}|X|^{r+\delta} < +\infty$ for some $\delta > 0$ and if g denotes the absolutely continuous part of its distribution then

$$N^{\frac{1}{d}} e_{N,r}(X, \mathbb{R}^d) \to \widetilde{J}_{r,d} \|g\|^{\frac{1}{r}}_{\frac{d}{d+r}} \qquad \text{as} \qquad N \to \infty \qquad (2)$$

where $\widetilde{J}_{r,d}$ is finite positive real constant obtained as the limit of the normalized quantization error when $X \stackrel{d}{=} U([0,1])$. This constant is unknown except when $d = 1$ or $d = 2$.

A non-asymptotic version of Zador's Theorem can be found e.g. in [22]: for every $r, \delta > 0$ there exists a universal constant $C_{r,\delta} > 0$ and an integer $N_{r,\delta} \geq$ such that, for every random vector $\Omega, \mathcal{A}, \mathbb{P}) \to \mathbb{R}^d$,

$$\forall N \geq N_{r,\delta}, \qquad e_{N,r}(X, \mathbb{R}^d) \leq C_{r,\delta} \|X\|_{r+\delta} N^{-\frac{1}{d}}.$$

The asymptotic behaviour of the $L^s(P)$-quantization error of sequences of L^r-optimal quantizers of a random vector X when $s > r$ has been extensively investigated in [13] and will be one crucial tool to establish our mains results.

In infinite dimension, the case of Gaussian processes was the first to have been extensively investigated, first in the purely quadratic case ($r = p = 2$): sharp rates have been established for a wide family of Gaussian processes including the Brownian motion, the fractional Brownian motions (see [18, 19]). For these two processes sharp rates are also known for $p \in [1, \infty]$ and $r \in (0, \infty)$ (see [4]). More recently, a connection between mean regularity of $t \mapsto X_t$ (from $[0, T]$ into $L^r(\mathbb{P})$) and the quantization rate has been established (see [22]): if the above mapping is μ-Hölder for an index $\mu \in (0, 1]$, then

$$e_{N,r}(X, L^p) = O((\log N)^{-\mu}), \qquad p \in (0, r).$$

Based on this result, some universal quantization rates have been obtained for general Lévy processes with or without Brownian component some of them turning out to be optimal, once compared with the lower bound estimates derived from small deviation theory (see e.g. [11] or [5]). One important feature of interest of the purely quadratic case is that it is possible to construct from the Karhunen–Loève expansion of the process two families of rate optimal (stationary) quantizers, relying on

- Sequences $(\alpha^{(N,prod)})_{N \geq 1}$ of optimal *product quantizers* which are rate optimal i.e. such that $e_{N,r}(\alpha^{(N)}, X, L^2) = O(e_{N,2}(X, L^2))$ (although not with a sharp optimal rate).
- Sequences of *true* optimal quantizers (or at least some good numerical approximations) $(\alpha^{(N,*)})_{N \geq 1}$ i.e. such that $e_{N,r}(\alpha^{(N,*)}, X, L^2) = e_{N,2}(X, L^2)$.

We refer to Sect. 2.1 below for further insight on these objects (both being available on the website www.quantize.math-fi.com).

The main objective of this paper is the following: let $(\alpha^N)_{N \geq 1}$ denote a sequence of rate optimal stationary (see (8) further on) quadratic quantizers of a d'-dimensional standard Brownian motion $W = (W^1, \ldots, W^d)$. Define the sequence $x^N = (x_n^N)_{n=1,\ldots,N}$, $N \geq 1$, of solutions of the ODE's

$$x_n^N(t) = x + \int_0^t b(x_n^N(s))ds + \int_0^t \sigma(x_n^N(s))d\alpha_n^N(s), \quad n = 1, \ldots, N.$$

Then, the finitely valued-process defined by

$$\widetilde{X}^N = \sum_{n=1}^N x_n^N \mathbf{1}_{\{W \in C_n(\alpha^{(N)})\}}$$

converges toward the diffusion X on $[0, T]$ (at least in probability) as $N \to \infty$. This convergence will hold with respect to distance introduced in the rough path theory (see [6, 9, 14, 25, 26]) which always implies convergence with respect to the sup norm. The reason is that our result will appear as an application of (variants of the) the celebrated *Universal Limit Theorem* originally established by T. Lyons in [25]. The distances of interest in rough path theory are related to the $\frac{1}{q}$-Hölder semi-norm or the q-variation semi-norm both when $q > 2$ defined for every $x \in \mathcal{C}([0, T], \mathbb{R}^d)$ by

$$\|x\|_{q,Hol} = T^{\frac{1}{p}} \sup_{0 \leq s < t \leq T} \frac{|x(t) - x(s)|}{|t - s|^{\frac{1}{q}}} \leq +\infty,$$

and

$$\mathrm{Var}_{q,[0,T]}(x) := \sup \left\{ \left(\sum_{0 \leq \ell \leq k-1} |x(t_{\ell+1}) - x(t_\ell)|^q \right)^{\frac{1}{q}}, \right.$$

$$\left. 0 \leq t_0 \leq t_1 \leq \cdots \leq t_k \leq T, k \geq 1 \right\} \leq +\infty$$

respectively. Note that

$$\|x - x(0)\|_{\sup} \leq \mathrm{Var}_{p,[0,T]}(x) \leq \|x\|_{p,Hol}.$$

From a technical viewpoint we aim at applying some continuity results established on the Itô map by several authors (see e.g. [6, 14, 16, 25]) that is the continuity of a solution x of the *ODE* (in a rough path sense)

$$dx_t = f(x_t)dy_t, \quad x_0 = x(0),$$

as a functional of y. However, the above (semi-)norms associated to a function x are not sufficient and the natural space to define such rough ODE is not the "naive" space of paths but a space of enhanced paths, which involves in the case of a multi-dimensional Brownian motion the mutual Lévy areas of its components. Convergence in this space is defined by considering appropriate $\frac{1}{q}$-Hölder and p-variation semi-norms to both the d-dimensional path and the related (pseudo-) Lévy areas (with different values of q and p, see Sect. 3). Our application to quantized SDE's will make extensively use the fact that our functional quantizations of the Brownian motion W will all satisfy a stationary assumption i.e.

$$\widehat{W} = \mathbb{E}(W \mid \sigma(\widehat{W}))$$

so that we will extend the Kolmogorov criterion satisfied by W to its functional quantizers $\widehat{W}$ for free. This approach is rather straightforward and its field of application seems more general than our functional quantization purpose: thus the piecewise affine interpolations of the Brownian motion obviously satisfy such a property (see Appendix).

The paper is organized as follows. In Sect. 2 we provide some short background on functional quantization as well as preliminary elementary results on stochastic integration with respect to a stationary functional quantizer of a d-dimensional standard Brownian motion. In Sect. 3, we define a quantized approximation scheme of an SDE (in the Stratonovich sense) driven by a standard Brownian motion by its functionally quantized counterpart which turns out to be a system of (non-coupled) ODE's. To this end we recall some basic facts on rough path theory, in particular the notion of convergence we need to define on the so-called *multiplicative functionals* involved in the continuity of the Itô map which, when dealing with Brownian motion amounts, to some convergence in Hölder semi-norm of the naive path as well as, roughly speaking, the running (pseudo-)Lévy areas of its components. In Sects. 4 and 5, we establish successively the convergence in the Hölder distance of sequences of optimal stationary quantizations $\widehat{W}$ of the Brownian motion toward W: Sect. 4 is devoted to the convergence of the "regular" paths whereas Sect. 5 deals with the convergence of the running (pseudo-)Lévy areas (and to the global convergence of the couple). In both cases we provide some convergence rate in the $(\log N)^{-a}$, $a \in (0, \frac{1}{2})$ scale which is the natural scale for such convergences since optimal functional quantizations of the Brownian motion are known to converge at a $(\log N)^{-\frac{1}{2}}$-rate for most usual norms (like quadratic pathwise norm on $L^2([0, T], dt)$).

Notations. • For every $d \geq 1$, one denotes $\xi = (\xi^1, \ldots, \xi^d)$ a row vector of $\mathbb{R}^d$. $\mathcal{M}(d \times d)$ will denote the set of square matrices with d lines.
• $|\,.\,|$ denotes the canonical Euclidean norm on $\mathbb{R}^d$.
• We denote $(\mathcal{F}_t^X)_{t \geq 0}$ the augmented natural filtration of a process $X = (X_t)_{t \geq 0}$ (so that it satisfies the usual conditions).

- For a bounded function $f : [0, T] \to \mathbb{R}^d$, $\| f \|_{\sup} := \sup_{t \in [0,T]} |f(t)|$. If f is a Borel function and $p \in [1, +\infty)$, $\| f \|_{L_{T,\mathbb{R}^d}^p} := \left(\int_0^T |f(t)|^p dt \right)^{\frac{1}{p}}$.
- For an $\mathbb{R}^d$-valued bi-measurable process X and $p \in [1, +\infty)$, we denote $\| X \|_p := \| |X|_{L_{T,\mathbb{R}^d}^p} \|_p = \left(\mathbb{E} \int_0^T |X_t|^p dt \right)^{1/p}$.
- We denote $t_k^n = \frac{kT}{2^n}$, $k = 0, \dots, 2^n$, the uniform mesh of the interval $[0, T]$, $T > 0$ and $I_k^n = [t_k^n, t_{k+1}^n]$, $k = 0, \dots, 2^n - 1$.
- $\lfloor x \rfloor$ denotes the lower integral part of $x \in \mathbb{R}$.
- Let $(a_n)_{n \geq 0}$ and $(b_n)_{n \geq 0}$ be two sequences of real numbers: $a_n \sim b_n$ if $a_n = b_n + o(b_n)$ and $a_n \asymp b_n$ if $a_n = O(b_n)$ and $b_n = O(a_n)$.

2 Background and Preliminary Results on Functional Quantization

2.1 Some Background on Functional Quantization

Functional quantization of stochastic processes can be seen as a discretization of the path-space of a process and the approximation (or coding) of a process by finitely many deterministic functions from its path-space. In a Hilbert space setting this reads as follows.

Let $(H, \langle \cdot, \cdot \rangle)$ be a separable Hilbert space with norm $|\cdot|$ and let $X : (\Omega, \mathcal{A}, \mathbb{P}) \to H$ be a random vector taking its values in H with distribution $\mathbb{P}_X$. Assume the integrability condition

$$\mathbb{E} |X|^2 < +\infty. \tag{3}$$

For $N \geq 1$, the L^2-optimal N-quantization problem for X consists in minimizing

$$\left\| \min_{a \in \alpha} |X - a| \right\|_{L^2(\mathbb{P})} = \left(\mathbb{E} \min_{a \in \alpha} |X - a|^2 \right)^{1/2}$$

over all subsets $\alpha \subset H$ with card$(\alpha) \leq N$. Such a set α is called N-codebook or N-quantizer. The minimal quantization error of X at level N is then defined by

$$e_N(X, H) := \inf \left\{ (\mathbb{E} \min_{a \in \alpha} |X - a|^2)^{1/2} : \alpha \subset H, \text{ card}(\alpha) \leq N \right\}. \tag{4}$$

For a given N-quantizer α one defines an associated nearest neighbour projection

$$\pi_\alpha := \sum_{a \in \alpha} a \mathbf{1}_{C_a(\alpha)}$$

and the induced α-(Voronoi)quantization of X by setting

$$\hat{X}^\alpha := \pi_\alpha(X), \tag{5}$$

where $\{C_a(\alpha) : a \in \alpha\}$ is a Voronoi partition induced by α, that is a Borel partition of H satisfying

$$C_a(\alpha) \subset \{x \in H : |x - a| = \min_{b \in \alpha} |x - b|\} \tag{6}$$

for every $a \in \alpha$. Then one easily checks that, for any random vector $X' : \Omega \to \alpha \subset H$,

$$\mathbb{E}|X - X'|^2 \geq \mathbb{E}|X - \hat{X}^\alpha|^2 = \mathbb{E} \min_{a \in \alpha} |X - a|^2$$

so that finally

$$\begin{aligned} e_n(X, H) &= \inf \left\{ \left\| |X - q(X)| \right\|_{L^2(\mathbb{P})}, \, q : H \overset{Borel}{\to} H, \text{card}(q(H)) \leq N \right\} \\ &= \inf \left\{ \left\| |X - Y| \right\|_{L^2(\mathbb{P})}, \, Y : (\Omega, \mathcal{A}) \overset{r.v.}{\to} H, \text{card}(Y(\Omega)) \leq N \right\}. \end{aligned} \tag{7}$$

A typical setting for functional quantization is $H = L_T^2 := L_{\mathbb{R}}^2([0, 1], dt)$ (equipped with $\langle f, g \rangle_2 := \int_0^T fg(t)dt$ and $|f|_{L_T^2} := \sqrt{\langle f, f \rangle_2}$). Thus any (bi-measurable, real-valued) process $X = (X_t)_{t \in [0,T]}$ defined on a probability space $(\Omega, \mathcal{A}, \mathbb{P})$ such that

$$\int_0^T \mathbb{E}(X_t^2)dt < +\infty$$

is a random variable $X : (\Omega, \mathcal{A}, \mathbb{P}) \to L_T^2$. But this Hilbert setting is not the only possible one for functional quantization (see e.g. [5, 12, 21], etc.) since natural Banach spaces like $L_{\mathbb{R}}^p([0, T], dt)$ or $\mathcal{C}([0, T], \mathbb{R})$ are natural path-spaces.

In the purely Hilbert setting the existence of (at least) one *optimal N-quantizer* for every integer $N \geq 1$ is established so that the infimum in (4) holds as a minimum. A typical feature of this quadratic Hilbert framework is the so-called *stationarity* (or self-consistency) property satisfied by such an optimal N-quantizer $\alpha^{(N,*)}$:

$$\hat{X}^{\alpha^{(N,*)}} = \mathbb{E}(X \mid \hat{X}^{\alpha^{(N,*)}}). \tag{8}$$

This property, known as stationarity, will be used extensively throughout the paper.

This existence property holds true in any reflexive Banach space and L^1 path spaces (see [12] for details).

2.2 Constructive Aspects of Functional Quantization of the Brownian Motion

2.2.1 Karhunen–Loève Basis ($d = 1$)

First we consider a scalar Brownian motion $(W_t)_{t \in [0,T]}$ on a probability space $(\Omega, \mathcal{A}, \mathbb{P})$. The two main classes of rate optimal quantizers of the Brownian motion are the product optimal quantizers and the true optimal quantizers. Both are based on the Karhunen–Loève expansion of the Brownian motion given by

$$W_t = \sum_{k \geq 1} \sqrt{\lambda_k}\, \xi_k\, e_k^W(t) \tag{9}$$

where, for every $k \geq 1$,

$$\lambda_k = \left(\frac{T}{\pi(k-1/2)}\right)^2 \quad \text{and} \quad e_k^W(t) = \sqrt{\frac{2}{T}} \sin\left(\frac{t}{\sqrt{\lambda_k}}\right) \tag{10}$$

and

$$\xi_k = \frac{(W \mid e_k^W)_2}{\sqrt{\lambda_k}} = \sqrt{\frac{2}{T}} \int_0^T W_t \, \sin(t/\sqrt{\lambda_k}) \frac{dt}{\sqrt{\lambda_k}}.$$

The sequence $(e_k^W)_{k\geq 1}$ is an orthonormal basis of L_T^2. The system $(\lambda_k, e_k^W)_{k\geq 1}$ can be characterized as the eigensystem of the symmetric positive trace class covariance operator of $f \mapsto (t \mapsto \int_0^T (s \wedge t) \, f(s)ds) \equiv (t \mapsto \mathbb{E}(< f \mid W >_2 W_t)$. In particular this implies that the Gaussian sequence $(\xi_k)_{k\geq 1}$ is pairwise uncorrelated hence i.i.d., $\mathcal{N}(0; 1)$-distributed. The Karhunen–Loève expansion of W plays the role of *PCA* of the process: it is the fastest way to exhaust the variance of W among all expansions on an orthonormal basis.

The convergence of the series in the right hand side of (9) holds in L_T^2 for every $\omega \in \Omega$ and $\mathbb{P}(d\omega)$-*a.s.* for every $t \in [0, T]$. In fact this convergence also holds in $L^2(\mathbb{P})$ and $\mathbb{P}(d\omega)$-*a.s.* for the sup norm over $[0, T]$. The first convergence follows from Theorem 3(*a*) further on applied with $X = W$ and $\mathcal{G}_N = \sigma(\xi_1, \ldots, \xi_N)$ and the second one follows e.g. from [21] $\mathbb{P}(d\omega)$-*a.s.*. In particular the convergence holds in $L^2(d\mathbb{P} \otimes dt)$ or equivalently in $L_{L_T^2}^2(\mathbb{P})$. Note that this basis has already been used in the framework of rough path theory for Gaussian processes, see e.g. [2, 7, 8].

2.2.2 Optimal Product Quantization ($d \geq 1$)

▷ *The one-dimensional case $d = 1$.* The previous expansion of the Brownian motion suggests to define a product quantization of W at level N by

$$\widehat{W}_t^{(N_1, \ldots, N_L)} := \sqrt{\frac{2}{T}} \sum_{k=1}^L \sqrt{\lambda_k}\, \widehat{\xi}_k^{N_k} \sin\left(\frac{t}{\sqrt{\lambda_k}}\right) \tag{11}$$

where $N_1, \ldots, N_L$ are non zero integers satisfying $N_1 \cdots N_L \leq N$ and $\widehat{\xi}_1^{N_1}, \ldots,$ $\widehat{\xi}_L^{N_L}$ are optimal quadratic quantizations of $\xi_1, \ldots, \xi_L$. The resulting (squared) quadratic quantization error reads

$$\| W - \widehat{W}^{(N_1,\ldots,N_L)} \|_2^2 = \sum_{k \geq 1} \frac{\lambda_k}{N_k^2}. \tag{12}$$

An *optimal product N-quantization* $\widehat{W}^{N,prod}$ is obtained as a solution to the following *integral bit allocation optimization* problem for the sequence $(N_k)_{k \geq 1}$:

$$\min \left\{ \| W - \widehat{W}^{(N_1,\ldots,N_L)} \|_2, \, N_1, \ldots, N_L \geq 1, \, N_1 \cdots N_L \leq N, \, L \geq 1 \right\} \tag{13}$$

(see [18] for further details and [27] for the numerical aspects). It is established in [18] (as a special case of a more general result on Gaussian processes) that

$$\frac{1}{T} \| W - \widehat{W}^{N,prod} \|_2 \asymp (\log N)^{-\frac{1}{2}} \tag{14}$$

Furthermore, the critical dimension $L = L_W(N)$ satisfies $L_W(N) \sim \log N$. Numerical experiments carried out in [27] show that

$$\frac{1}{T} \| W - \widehat{W}^{N,prod} \|_2 \approx c_W (\log N)^{-\frac{1}{2}}$$

with $c_W \approx 0.5$ (at least up to $N \leq 10,000$).

It is possible to get a closed form for the underlying optimal product quantizers α^N. First, note that the normal distribution on the real line being log-concave, there is exactly one stationary quadratic quantizer of full size M for every $M \geq 1$ (hence it is the optimal one). So, let $N \geq 1$ and let $(N_k)_{k \geq 1}$ denote its optimal integral bit allocation for the Brownian motion W. For every $N_k \geq 1$, we denote by $\beta^{(N_k)} := \{\beta_{i_k}^{(N_k)}, \, 1 \leq i_k \leq N_k\}$ the unique optimal quantizer of the normal distribution: thus $\alpha(0) = \{0\}$ by symmetry of the normal distribution. Then, the optimal quadratic product N-quantizer $\alpha^{N,prod}$ (of "true size" $N_1 \times \cdots \times N_{L_W(N)} \leq N$) can be described using a multi-indexation as follows:

$$\alpha^{N,prod}_{(n_1,\ldots,n_k,\ldots)}(t) = \sum_{k \geq 1} \beta_{n_k}^{(N_k)} \sqrt{\lambda_k} e_k^W(t), \qquad n_k \in \{1, \ldots, N_k\}, \, k \geq 1.$$

These sums are in fact all finite so that all the functions $\alpha^{N,prod}_{(i_1,\ldots,i_n,\ldots)}$ are $\mathcal{C}^\infty$ with finite variation on every interval of $\mathbb{R}_+$.

Explicit optimal *integral bit allocations* as well as optimal quadratic quantizations (quantizers and their weights) of the scalar normal distribution are available on the website [28]. Note for practical applications that this optimal product quantization is based on one-dimensional quantizations of small size of the scalar normal distribution $\mathcal{N}(0; 1)$. This kind of functional quantization has been applied in [27] to price Asian options in a Heston stochastic volatility model.

▷ *The d-dimensional case.* Assume now $W = (W^1, \ldots, W^d)$ is a d-dimensional Brownian motion. Its optimal product quantization at level $N \geq 1$ will be defined as the optimal product quantization at level $\lfloor N^{\frac{1}{d}} \rfloor$ of each of its d components.

▷ *Additional results on optimal vector quantization of the normal distribution on* $\mathbb{R}^d$. We will extensively make use of the *distortion mismatch* result established in [13] that we recall here only in the d-dimensional Gaussian case. Let Z be an $\mathcal{N}(0; I_d)$ random vector and let α^N be an optimal quadratic quantizer at level N of Z (hence of size N). Then

$$(i) \ \forall p \in (0, 2 + d), \ \forall N \geq 1, \ \|Z - \widehat{Z}^{\alpha^N}\|_p \leq C_{Z,p} N^{-\frac{1}{d}}. \tag{15}$$

$$(ii) \ \forall p \in [2 + d, +\infty), \ \forall \eta \in (0, d + 2), \ \forall N \geq 1,$$

$$\|Z - \widehat{Z}^{\alpha^N}\|_p \leq C_{Z,p,\eta} N^{-\frac{2+d-\eta}{dp}} \tag{16}$$

where $C_{Z,p}$ and $C_{Z,p,\eta}$ are two positive real constants.

2.2.3 Optimal Quantization ($d = 1$)

It is established in [18] (Theorem 3.2) that the quadratic optimal quantization of the one-dimensional Brownian motion reads

$$\widehat{W}_t^{N,opt} = \sqrt{\frac{2}{T}} \sum_{k=1}^{d_W(N)} \sqrt{\lambda_k} \, (\widehat{\zeta}_{d_W(N)}^N)^k \sin\left(\frac{t}{\sqrt{\lambda_k}}\right) \tag{17}$$

where, for every integer $d \geq 1$, $\zeta_d = \mathrm{Proj}_{E_d}^{\perp}(W) \sim \mathcal{N}(0; \mathrm{Diag}(\lambda_1, \ldots, \lambda_d))$ with $E_d := \mathbb{R}\text{-span}\{\sin(./\sqrt{\lambda_1}), \ldots, \sin(./\sqrt{\lambda_d})\}$ and $\widehat{\zeta}_d^N$ is an optimal quadratic quantization of ζ_d at level (or of size) N.

If one considers an optimal quadratic N-quantizer $\beta^N = \{\beta_n^N, \ n = 1, \ldots, N\} \subset \mathbb{R}^{d_W(N)}$ of the distribution $\mathcal{N}(0; \mathrm{Diag}(\lambda_1, \ldots, \lambda_{d_W(N)}))$ (a priori not unique)

$$\alpha_n^{N,opt}(t) = \sum_{k=1}^{d_W(N)} (\beta_n^{(N)})^k \sqrt{\lambda_k} \, e_k^W(t), \qquad n = 1, \ldots, N.$$

Once again this defines a $\mathcal{C}^\infty$ function with finite variation on every interval of $\mathbb{R}_+$. A sharp rate has been obtained in [19] for the resulting optimal quantization error

$$\|W - \widehat{W}^{N,opt}\|_2 \sim T c_W^{opt} (\log N)^{-\frac{1}{2}} \quad \text{as} \quad N \to \infty \tag{18}$$

where $c_W^{opt} = \frac{\sqrt{2}}{\pi} \approx 0.4502$.

The true value of the critical dimension $d_W(N)$ is unknown. A conjecture supported by numerical evidences is that $d_W(N) \sim \log N$. Recently a first step to this conjecture has been established in [23] by showing that

$$\liminf_N \frac{d_W(N)}{\log(N)} \geq \frac{1}{2}.$$

Large scale computations of optimal quadratic quantizers of the Brownian motion have been carried out (up to $N = 10{,}000$ and $d = 10$). They are available on the website [28].

In the d-dimensional setting, several definitions of an *optimal quantization* of the Brownian motion $W = (W^1, \ldots, W^d)$ can be given. For our purpose, it is convenient to adopt the following one:

$$\widehat{W}^{N,opt} := \left(\widehat{W^i \lfloor N^{\frac{1}{d}} \rfloor, opt} \right)_{1 \leq i \leq d}.$$

Its property of interest is that this definition preserves the componentwise independence as well as a stationarity property (see below) since

$$\mathbb{E}\left(W^i \mid \widehat{W}^{N,opt} \right) = \mathbb{E}\left(W^i \mid \widehat{W^i \lfloor N^{\frac{1}{d}} \rfloor, opt} \right) = \widehat{W^i \lfloor N^{\frac{1}{d}} \rfloor, opt}, \quad i = 1, \ldots, d.$$

2.2.4 Wiener Like Integral with Respect to a Stationary Functional Quantization ($d = 1$)

Both types of quantizations defined above share an important property of quantizers: stationarity.

Definition 1. Let $\alpha \subset L^2_T$, $\alpha \neq \emptyset$, be a quantizer. The quantizer α is stationary for the (one-dimensional) Brownian motion W if there is a Voronoi quantization $\widehat{W} := \widehat{W}^\alpha$ induced by α such that

$$\widehat{W} = \mathbb{E}(W \mid \sigma(\widehat{W})) \qquad a.s. \tag{19}$$

where $\mathbb{E}(\,.\mid\mathcal{G})$ denotes the functional conditional expectation given the σ-field $\mathcal{G}$ on $L^2_{L^2_T}(\mathbb{P})$ (see Appendix) and $\sigma(\widehat{W})$ is the σ-field spanned by $\widehat{W}$.

Note that if α is stationary for one Brownian motion, so it is for any Brownian motion since this stationarity property only depends on the Wiener distribution.

In the case of product quantization $\widehat{W}^{N,prod}$, this follows from the stationarity property of the optimal quadratic quantization of the marginals ξ_n (see [18] or [27]). In the case of optimal quadratic quantization $\widehat{W}^{N,opt}$ this follows from the optimality of the quantization of $\zeta_{d_W(N)}$ itself.

We will now define a kind of *Wiener integral with respect* to such *a stationary quantization* $\widehat{W}$ of a *one-dimensional* W. So we assume that $d = 1$ until the end of this section.

First, we must have in mind that if W is an $(\mathcal{F}_t)$-Brownian motion where the filtration $(\mathcal{F}_t)_{t\geq 0}$ satisfies the usual conditions, one can define the Wiener stochastic integral (on $[0, T]$) of any process $\varphi \in L^2([0, T] \times \Omega, \mathcal{B}([0, T]) \otimes \mathcal{F}_0, dt \otimes d\mathbb{P})$ with respect to W. The non-trivial case is when $\mathcal{F}_t^W \neq \mathcal{F}_t$, typically when $\mathcal{F}_t = \mathcal{F}_T^B \vee \mathcal{F}_t^W$, $t \in [0, T]$ where B and W are independent. One can see it as a special case of Itô stochastic integral or as an extended Wiener integral: if $(\varphi(t, \omega))_{(\omega,t)\in\Omega\times[0,T]}$ denotes an elementary process of the form

$$\varphi(t, \omega) := \sum_{k=1}^{n} \varphi_k(\omega) \mathbf{1}_{s_k < t \leq s_{k+1}}, \quad 0 = s_0 < s_1 < \cdots < s_{n-1} < s_n = T$$

where the random variables φ_i are $\mathcal{F}_0$-measurable (hence independent of W). Set

$$I_T(\varphi) := \sum_{k=1}^{n} \varphi_k(W_{s_{k+1}} - W_{s_k}).$$

Then, I_T is an isometry from $L^2_{L^2_T}(\mathbb{P})$ into $L^2(\mathcal{F}_T, \mathbb{P})$. Furthermore, one easily checks that

$$\mathbb{E}\left(\int_0^T \varphi(s, .)dW_s \mid \mathcal{F}_T^W\right) = \int_0^T \mathbb{E}\left(\varphi(s, .) \mid \mathcal{F}_T^W\right) dW_s$$

where $\mathcal{F}_T^W$ denotes the augmented filtration of W at time T. We follow the same lines to define the stochastic integral with respect to a stationary quantizer. Set for the same elementary process φ

$$\widehat{I}_T(\varphi) = \sum_{k=1}^{n} \xi_k(\widehat{W}_{s_{k+1}} - \widehat{W}_{s_k})$$

so that

$$\widehat{I}_T(\varphi) = \sum_{k=1}^{n} \xi_i \, \mathbb{E}(W_{s_{k+1}} - W_{s_k} \mid \widehat{W})$$

$$= \sum_{k=1}^{n} \mathbb{E}\big(\xi_k(W_{s_{k+1}} - W_{s_k}) \mid \mathcal{F}_0 \vee \sigma(\widehat{W})\big)$$

$$= \mathbb{E}\left(\int_0^T \varphi(t, .)dW_t \mid \mathcal{F}_0 \vee \sigma(\widehat{W})\right)$$

where we used that the σ-fields $\sigma(\widehat{W})$ and $\mathcal{F}_0$ are independent since $\widehat{W}$ is a Borel function of W. As a consequence,

$$\|\widehat{I}_T(\varphi)\|_2^2 \leq \|I_T(\varphi)\|_2^2 = \| \, |\varphi|_{L_T^2} \|_2^2.$$

Hence, the linear transformation $\widehat{I}_T$ extends into a linear continuous mapping on the whole set $L_{L_T^2}^2(\mathcal{F}_0, \mathbb{P})$. Furthermore, one checks, first on elementary processes, then on $L_{L_T^2}^2(\mathcal{F}_0, \mathbb{P})$ by continuity of the (functional) conditional expectation, that

$$\mathbb{E}\left(I_T(\varphi) \mid \mathcal{F}_0 \vee \sigma(\widehat{W})\right) = \widehat{I}_T(\varphi).$$

We will denote from now on $\widehat{I}_T(\varphi)(\omega)$ as an integral, namely

$$\widehat{I}_T(\varphi)(\omega) := \int_0^T \varphi(t, \omega) d\widehat{W}_t(\omega).$$

Now set as usual, for every $t \in [0, T]$,

$$\int_0^t \varphi(s, \omega) d\widehat{W}_s(\omega) := \int_0^T \mathbf{1}_{[0,t]}(s)\varphi(s, \omega) d\widehat{W}_s(\omega).$$

One checks using Jensen and Doob Inequality that,

$$\mathbb{E} \sup_{t \in [0,T]} \left| \int_0^t \varphi(s, .) d\widehat{W}_s \right|^2 \leq \mathbb{E} \sup_{t \in [0,T]} \left| \int_0^t \varphi(s, .) dW_s \right|^2$$

$$\leq 4\,\mathbb{E} \int_0^T \varphi^2(s, .)\, ds. \tag{20}$$

Furthermore, as soon as the underlying stationary quantizer α (such that $\widehat{W} = \widehat{W}^\alpha$) is made up with pathwise continuous elements, for every elementary process φ, its integral process

$$\int_0^t \varphi(s, .)\, dW_s = \sum_{k=1}^n \xi_k (\widehat{W}_{s_{k+1} \wedge t} - \widehat{W}_{s_k \wedge t})$$

pathwise continuous as well since $\widehat{W}$ is α-valued. One classically derives, by combining this result with (20) and the everywhere density of elementary processes, that, for every $\varphi \in L_{L_T^2}^2(\mathcal{F}_0, \mathbb{P})$, the process

$$\left(\int_0^t \varphi(s, .) d\widehat{W}_s\right)_{t \in [0,T]} \qquad\qquad \text{admits a continuous modification.}$$

This is always this modification that will be considered from now on. As a matter of fact, if φ_n denotes a sequence of elementary processes in $L^2_{L^2_T}(\mathcal{F}_0, \mathbb{P})$ converging to φ, i.e. satisfying

$$\mathbb{E} \int_0^T (\varphi - \varphi_n)^2 (s, .) ds \longrightarrow 0 \qquad \text{as } n \to \infty.$$

It follows from (20) that the convergence also holds in $L^2_{L^\infty_T}(\mathcal{F}_0, \mathbb{P})$. In particular, there is a subsequence that converges $\mathbb{P}$-*a.s.* for the $\| . \|_{\sup}$ which implies the existence of a continuous modification for $\int_0^t \varphi(s, \omega) d\widehat{W}_s(\omega)$.

Finally, using the characterization of functional conditional expectation (see Appendix), it follows that

$$\mathbb{E}\left(\int_0^{\cdot} \varphi(s, .) d\widehat{W}_s, | \mathcal{F}_0 \vee \sigma(\widehat{W}) \right) = \int_0^{\cdot} \varphi(s, .) d\widehat{W}_s. \tag{21}$$

Proposition 1. *Let W be a (real-valued) $\mathcal{F}_t$-standard Brownian motion. (a) For every $\varphi \in L^2_{L^2_T}(\mathcal{F}_0, \mathbb{P})$*

$$\int_0^t \varphi(s, .) dW_s = \sqrt{\frac{2}{T}} \sum_{k \geq 1} \xi_k \int_0^t \varphi(s, .) \cos(s/\sqrt{\lambda_k}) ds \tag{22}$$

where $\xi_k := (W | e_k^W)_2 / \sqrt{\lambda_k}$ are independent, $\mathcal{N}(0; 1)$-distributed (see (9) and (10)) and independent of φ.

(b) Let $\widehat{W}$ be a stationary quantization of W. For every $\varphi \in L^2_{L^2_T}(\mathcal{F}_0, \mathbb{P})$

$$\int_0^t \varphi(s, .) d\widehat{W}_s = \sqrt{\frac{2}{T}} \sum_{k \geq 1} \frac{(\widehat{W} | e_k^W)_2}{\sqrt{\lambda_k}} \int_0^t \varphi(s, .) \cos(s/\sqrt{\lambda_n}) ds. \tag{23}$$

In particular if $\widehat{W}$ is a product quantization, then

$$\frac{(\widehat{W} | e_k^W)_2}{\sqrt{\lambda_k}} = \frac{\widehat{(W | e_k^W)_2}}{\sqrt{\lambda_k}} = \widehat{\xi}_k.$$

Proof. (a) Set for every $\varphi \in L^2_{L^2_T}(\mathcal{F}_0, \mathbb{P})$,

$$J_T(\varphi) := \sqrt{\frac{2}{T}} \sum_{k \geq 1} \xi_k \sqrt{\lambda_k} \int_0^T \varphi(s, .) d \sin(s/\sqrt{\lambda_k}) \tag{24}$$

$$= \sqrt{\frac{2}{T}} \sum_{k \geq 1} \xi_k \int_0^T \varphi(s,.) \cos(s/\sqrt{\lambda_k}) ds.$$

This defines clearly an isometry from $L^2_{L^2_T}(\mathcal{F}_0, \mathbb{P})$ into the Gaussian space spanned by $(\xi_n)_{n \geq 1}$ since

$$\mathbb{E}(J_T(\varphi)^2) = \frac{2}{T} \sum_{k \geq 1} \mathbb{E}(\xi_k^2) \, \mathbb{E} \left(\int_0^T g(s) \frac{1}{\sqrt{\lambda_k}} \cos(s/\sqrt{\lambda_k}) ds \right)^2 = \mathbb{E} \int_0^T g^2(t) dt.$$

The last equality uses that the sequence $\left(\sqrt{\frac{2}{T}} \cos(\pi(k-\frac{1}{2})t/T) \right)_{k \geq 1}$ is an orthonormal basis of L^2_T. Finally, note that for every $t \in [0, T]$, $J_T(\mathbf{1}_{[0,t]}) = \sqrt{\frac{2}{T}} \sum_{k \geq 1} \sqrt{\lambda_k} \xi_k \sin(t/\sqrt{\lambda_k}) = W_t$. This proves that $J_T = I_T$ i.e. is but the (extended) Wiener integral with respect to W.

(b) This follows by taking the (functional) conditional expectation of (22).   □

2.2.5   Application to Multi-Dimensional Brownian Motions ($d \geq 2$)

Now we apply the above result to a componentwise (stationary) functional quantization of a multi-dimensional standard Brownian motion.

Proposition 2. Let $W =: (W^1, \ldots, W^d)$ denote a d-dimensional standard Brownian motion and let $\widehat{W} := (\widehat{W}^1, \ldots, \widehat{W}^d)$ be a pathwise continuous stationary quantization of W (no optimality is requested here). Then, $\mathbb{P}$-a.s., for every $i \neq j$, $i, j \in \{1, \ldots, d\}$, for every $s, t \in [0, T]$, $0 \leq s \leq t$,

$$\mathbb{E} \left(\int_s^t (W_u^i - W_s^i) dW_u^j \mid \sigma(\widehat{W}) \right) = \int_s^t (\widehat{W}_s^i - \widehat{W}_s^i) d\widehat{W}_u^j.$$

Proof. All the components of $\widehat{W}$ being independent, it is clear one can replace $\sigma(\widehat{W})$ by $\sigma(\widehat{W}^i, \widehat{W}^j)$. Then, the stochastic integral $\int_0^{\cdot} W_s^i dW_s^j$ coincides with the (extended) Wiener integral defined with respect to the filtration $\mathcal{G}_{i,t}^j := \sigma(\mathcal{F}_T^{W^i}, \mathcal{F}_t^{W^j})$ (it is clear that W^j is a $\mathcal{G}_{i,t}^j$-standard Brownian motion still by independence). The result is then a straightforward consequence of (21).   □

Remark. The above result still holds if one considers an additional "0th" component $W_t^0 = t$ to the Brownian motion and to its functional quantization by setting $\widehat{W}_t^0 = t$ as well.

3 Convergence of Quantized *SDE*'s: A Rough Path Approach

3.1 From Itô to Stratonovich

An *SDE*

$$dX_t = b(t, X_t)dt + \sigma(t, X_t)dW_t, \quad X_0 \in L^p_{\mathbb{R}^d}(\mathbb{P})$$

where $b : [0, T] \times \mathbb{R}^d \to \mathbb{R}^d$ and $\sigma : [0, T] \times \mathbb{R}^d \to \mathcal{M}(d \times q)$ are smooth enough functions (e.g. continuously differentiable with bounded differentials) and $W = (W_t)_{t \in [0,T]}$ is a q-dimensional Brownian motion. First note that without loss of generality one may assume that $q = d$ by increasing the dimension of W or adding some identically zero components to X (no ellipticity like assumption is needed here). This *SDE* can be written in the Stratonovich sense as follows

$$dX_t = f(X_t) \circ dW_t, \quad X_0 \in L^p_{\mathbb{R}^d}(\mathbb{P}), \tag{25}$$

where, for notational convenience $W = (W^0, W^1, \ldots, W^q)$ stands for (t, W_t), $X_t = (X^0_t, X^1_t, \ldots, X^d_t)$ stands for (t, X_t) and $f : [0, T] \times \mathbb{R}^d \to \mathcal{M}((d + 1) \times (d + 1))$ (with $f^{0 \cdot}(t, x) = (1, 0, \ldots, 0)$ as 0th line) is a differentiable function with bounded differentials.

Following rough paths theory initiated by T. Lyons ([25]) and developed with many co-authors (see e.g. [9, 14, 16, 26] for an introduction), one can also solve this equation in the sense of rough paths with finite p-variation, $p \geq 2$, since we know (e.g. from the former Kolmogorov criterion) that W *a.s.* does have finite $\frac{1}{q}$-Hölder norm, for any $q > 2$. Namely this means solving an equation formally reading

$$dx_t = f(x_t)d\mathbf{y}_t, \quad x_0 \in \mathbb{R}^d. \tag{26}$$

In this equation $\mathbf{y}$ does not represent the path (null at 0) $y_t = W_t(\omega), t \in [0, T]$ itself but an *enhanced path* embedded in a larger space, also called *geometric multiplicative functional lying on* y with controlled $\frac{1}{q}$-Hölder semi-norm, namely a couple $\mathbf{y} = ((\mathbf{y}^1_{s,t})_{0 \leq s \leq t \leq T}, (\mathbf{y}^2_{s,t})_{0 \leq s \leq t \leq T})$ where $\mathbf{y}^1_{s,t} = y_t - y_s \in \mathbb{R}^{d+1}, 0 \leq s \leq t \leq T$, can be identified with the path (y_t) and $(\mathbf{y}^2_{s,t})_{0 \leq s \leq t \leq T}$ satisfies, $\mathbf{y}^2_{s,t} \in \mathbb{R}^{(d+1)^2}$ for every $0 \leq s \leq u \leq t \leq T$ and the following tensor multiplicative property

$$\mathbf{y}^2_{s,t} = \mathbf{y}^2_{s,u} + \mathbf{y}^2_{u,t} + \mathbf{y}^1_{s,u} \otimes \mathbf{y}^1_{u,t}.$$

Different choices for this functional are possible, leading to different solutions to the above equation (26). The choice that makes coincide *a.s.* the solution of (25) and the pathwise solutions of (26) is given by

$$\mathbf{y}^1_{s,t} = W_t(\omega) - W_s(\omega), \ \mathbf{y}^2_{s,t} := \left(\int_s^t (W^i_u - W^i_s) \circ dW^j_u \right)(\omega) \underset{i,j=0,\ldots,d}{} \tag{27}$$

so that

$$\mathbf{y}_{s,u}^1 \otimes \mathbf{y}_{u,t}^1 = \left(\mathbf{y}_{s,u}^{1,i} \mathbf{y}_{s,u}^{1,j}\right)_{i,j=0,\dots,d}.$$

The term $\mathbf{y}_{s,t}^2$ is but the "running" Lévy areas related to the components of the Brownian motion W. The enhanced path of W will be denoted $\mathbf{W}$ (although we will keep the notation $\mathbf{y}$ in some proofs for notational convenience). One defines, for every $q \geq 1$, the $\frac{1}{q}$-Hölder distance by setting

$$\rho_q(\mathbf{y} - \mathbf{x}) = \|\mathbf{y}^1 - \mathbf{x}^1\|_{q,Hol} + \|\mathbf{y}^2 - \mathbf{x}^2\|_{q/2,Hol}$$

where

$$\|\mathbf{x}^2\|_{q/2,Hol} := T^{\frac{2}{q}} \sup_{0 \leq s < t \leq T} \frac{|\mathbf{x}_{s,t}^2|}{|t-s|^{\frac{2}{q}}}.$$

Remark. Likewise, when $p \in [2,3)$, one defines the *p-variation distance* between two such multiplicative functionals $\mathbf{y}, \mathbf{z}$ is defined by

$$\delta_p(\mathbf{y}, \mathbf{z}) = \text{Var}_{p,[0,T]}(\mathbf{y}^1 - \mathbf{z}^1) + \text{Var}_{p/2,[0,T]}(\mathbf{y}^2 - \mathbf{z}^2)$$

where

$$\text{Var}_{q,[0,T]}(\mathbf{y}^2) := \sup \left\{ \left(\sum_{\ell=0}^{k-1} |\mathbf{y}_{t_\ell,t_{\ell+1}}^2|^q\right)^{\frac{1}{q}}, 0 \leq t_0 \leq t_1 \leq \cdots \leq t_k \leq T, k \geq 1 \right\}.$$

The distance ρ_q has been introduced in [24] although rough path theory was originally developed for the distance δ_p in p-variation. Recently several authors came back to Hölder distances ρ_q (see e.g. [6,9,16]).

The following so-called universal limit theorem (including variants) describes the continuity of the so-called Itô map $\mathbf{y} \mapsto \mathbf{x}$ with respect to both δ_p and ρ_p-distances and will be the key for our main result. It was the starting point of rough path theory initiated by T. Lyons. Several statements (or improvements) can be found e.g. in [9,14,15,25,26]. We state here some versions coming from [14,16].

Theorem 1. *Let $\alpha \in (0,1]$.*
(a) (See [16]) Let $f : [0,T] \times \mathbb{R}^d \to \mathcal{M}((d+1) \times (d+1))$, twice differentiable with a bounded first differential and an α-Hölder second differential. Suppose the multiplicative functional $\mathbf{y}$ satisfies $\|\mathbf{y}^1 - \mathbf{x}^1\|_{q,Hol} + \|\mathbf{y}^2 - \mathbf{x}^2\|_{q/2,Hol} < +\infty$ for $q \in (2, 2+\alpha)$. Then (26) has a unique solution starting at x_0.

When $\mathbf{y} = \mathbf{W}(\omega)$ (i.e. given by (27)), the first component $\mathbf{x}^1 = x$ of the solution $\mathbf{x} = (\mathbf{x}^1, \mathbf{x}^2)$ a.s. coincides with $(X_t(\omega))_{t \in [0,T]}$, solution to the SDE in the Stratonovich sense.

Furthermore, the Itô map $\mathbf{y} \mapsto \mathbf{x}$ is continuous for the Hölder ρ_q distance (and locally Lipschitz in sense described in [16]).

(b) (See [9, 17]) If $f \in \mathcal{C}^2([0, T] \times \mathbb{R}^d, \mathcal{M}((d + 1) \times (d + 1))$ is such that $f.\nabla f$ is bounded with an α-Hölder differential, then the conclusions of claim (a) still hold.

3.2 Quantization of the SDE and Main Result

Let $(\alpha^N)_{N \geq 1}$ denote a sequence of quantizers of the Brownian motion. Each α^N is made up of N functions (or elementary quantizer) $\alpha_n^N : [0, T] \rightarrow \mathbb{R}^d$, $n = 1, \ldots, N$. For convenience a component "0" will be added accordingly to each elementary quantizer α_n^N by setting $\alpha_n^{N,0}(t) = t$ (which exactly quantizes the function $W_t^0 = t$). We assume that every elementary quantizer α_n^N is a continuous function with finite variation over $[0, T]$. The resulting Voronoi quantizer $\widehat{W} = \widehat{W}^{\alpha^N}$ of W reads

$$\widehat{W}_t = \sum_{n=1}^{N} \alpha_n^N(t) \mathbf{1}_{\{W \in C_n(\alpha^N)\}}, \quad t \in [0, T].$$

Our aim is to approximate the diffusion process $(x_t)_{t \in [0,T]}$ solution to the *SDE* (25) by the solution $\widetilde{X}^N$ of the equation

$$d\widetilde{X}_t^N = f(\widetilde{X}_t^N) d\widehat{W}_t, \quad \widetilde{X}_0^N = x_0.$$

as $N \rightarrow \infty$. In fact, a less formal expression is available for the process $\widetilde{X}^N$, namely

$$\widetilde{X}^N = \sum_{n=1}^{N} \widetilde{x}_n^N \mathbf{1}_{\{W \in C_n(\alpha^N)\}}$$

where each x_n^N is solution to the *ODE*

$$d\widetilde{x}_n^N(t) = f(\widetilde{x}_n^N(t)) \, d\alpha_n^N(t), \quad \widetilde{x}_n^N(0) = x_0, \quad n = 1, \ldots, N. \tag{28}$$

Note that X^N is a non-Voronoi quantization of (x_t) (at level N). The starting natural idea was to hope that X^N converges to (x_t) owing to the convergence of $\widehat{W}^N$ toward W...in an appropriate sense. Since we will use the above Theorem 1, we need to prove the convergence of the geometric functional $\widehat{\mathbf{W}}^N$ related to $\widehat{W}$ toward that of W. The quantity $\widehat{\mathbf{W}}^N$ is formally defined by mimicking the definition of $\mathbf{W}$, namely, for every $(s, t) \in [0, T], 0 \leq s < t \leq T$,

$$\widehat{\mathbf{W}}^{1,N}(\omega) := \widehat{W}_t(\omega) - \widehat{W}_s(\omega), \quad \widehat{\mathbf{W}}_{s,t}^{2,N}(\omega) := \left(\int_s^t (\widehat{W}_u^i - \widehat{W}_s^i) d\widehat{W}_u^j \right)_{i,j=0,\ldots,d} (\omega)$$

still with the convention $\widehat{W}_t^{0,N} = t$. The integral must be understood in the usual Stieltjes sense.

Theorem 2. *Let* $(\widehat{W}^N)_{N \geq 1}$ *be a sequence of stationary quadratic functional quantizers of the Brownian motion converging to* W *in* $L^2_{L^2_T}(\mathbb{P})$.

Let f *be like in claims* (a) *or* (b) *in Theorem 1. Consider for every* $N \geq 1$, *the solutions of the quantized ODE*

$$d\widetilde{X}^N_t = f(\widetilde{X}^N_t) \, d\widehat{W}^N_t, \qquad N \geq 1.$$

as defined by (28). *Let* $\mathbf{X}$ *and* $\widetilde{\mathbf{X}}^N$ *denote the enhanced paths of* X, *solution to* (25), *and* $\widetilde{X}^N$ *respectively. Then, for every* $q \in (2, 2 + \alpha)$,

$$\rho_q(\widetilde{\mathbf{X}}^N, \mathbf{X}) \xrightarrow{\mathbb{P}} 0.$$

Furthermore if $r > \frac{2}{3}$ *then*

$$\rho_q(\widetilde{\mathbf{X}}^{\lfloor e^{N^r} \rfloor}, \mathbf{X}) \xrightarrow{a.s.} 0.$$

In view of what precedes this result is, as announced, a straightforward corollary of the continuity of the Itô map established Theorem 1, once the convergence $\rho_q(\widehat{\mathbf{W}}^N, \mathbf{W})$ in probability is established for any $q \in (2, 3)$. A slightly more derailed proof is proposed at the end of Sect. 5.

In fact we will prove a much precise statement concerning the Brownian motion since we will establish for every $q > 2$ the convergence in every $L^p(\mathbb{P})$, $0 < p < \infty$, of $\rho_q(\widehat{\mathbf{W}}^N, \mathbf{W})$ with an explicit $L^p(\mathbb{P})$-rate of convergence in the scale $(\log N)^{-\theta}$, $\theta \in (0, 1)$.

These rates can be transferred to the convergence of the quantized *SDE*, conditionally to some events on which the Itô map is itself Lipschitz continuous for the distances ρ_q. Several results of local Lipschitz continuity have been established recently, especially in [6,9,16,17], although not completely satisfactory from a practical point of view. So we decided not to reproduce (and take advantage of) them here.

The proof is divided into two steps: the convergence for the Hölder semi-norm) of the regular path component is established in Sect. 4 (in which more general processes are considered) and the convergence of approximate Lévy areas in Sect. 5 (entirely devoted to the Brownian case for the sake of simplicity).

Remarks.

- There is a small abuse of notation in the above Theorem since $\widetilde{X}^N$ is not a Voronoi quantizer of X: this quantization of X is defined on the Voronoi partition (for the $L^2_{T,\mathbb{R}^d}$-norm) induced by the quantization of the Brownian motion W.
- The same results holds for the Brownian bridge, the Ornstein-Uhlenbeck process and more generally for continuous Gaussian semi-martingales that satisfy the Kolmogorov criterion.

4 Convergence of the Paths of Processes in Hölder Semi-Norm

4.1 A General Setting Including Stationary Functional Quantization

In this section we investigate the connections between the celebrated Kolmogorov criterion and the tightness of some classes of sequences of processes for the topology of $\frac{1}{q}$-Hölder convergence. In fact this connection is somehow the first step of the rough path theory, but we will look at it in a slightly different way. Whatsoever this naive pathwise convergence is not sufficient to get the continuity of the Itô map in a Brownian framework and we will also have to deal for our purpose with the multiplicative functional (see Sect. 5).

But at this stage we aim at showing that when a sequence $(Y^N)_{N \geq 1}$ satisfies some "stationarity property" with respect to a process Y, several properties of Y can be transferred to the Y^N. Indeed, the same phenomenon will occur for the multiplicative function (see the next section).

If Y satisfies the Kolmogorov criterion and $(\mathcal{G}_N)_{N \geq 1}$ denotes a sequence of sub-σ-fields of $\mathcal{A}$, then a sequence of processes defined by

$$Y^N := \mathbb{E}(Y \mid \mathcal{G}^N), \quad N \geq 1,$$

where the conditional expectation is considered in the functional sense (see Appendix) is (C-tight and) tight for a whole family of topologies induced by convergence in $\frac{1}{q}$-Hölder sense.

Definition 2. Let $p \geq 1$, $\theta > 0$. A process $Y = (Y_t)_{t \in [0,T]}$ satisfies the Kolmogorov criterion $(K_{p,\theta})$ if there is a real constant $C_T^{Kol} > 0$ such that

$$\forall s, t \in [0, T], \qquad \mathbb{E}|Y_t - Y_s|^p \leq C_T^{Kol} |t - s|^{1+\theta} \qquad \text{and} \qquad Y_0 \in L^p(\mathbb{P}).$$

Theorem 3. Let $Y := (Y_t)_{t \in [0,T]}$ be a pathwise continuous process defined on $(\Omega, \mathcal{A}, \mathbb{P})$ satisfying the Kolmogorov criterion $(K_{p,\theta})$. Let $(\mathcal{G}_N)_{N \geq 1}$ be a sequence of sub-σ-fields of $\mathcal{A}$. For every $N \geq 1$ set

$$Y^N := \mathbb{E}(Y \mid \mathcal{G}_N).$$

For every $N \geq 1$, Y^N has a pathwise continuous version satisfying

$$\forall t \in [0, T], \qquad Y_t^N = \mathbb{E}(Y_t \mid \mathcal{G}_N) \quad a.s.$$

Furthermore, if one of the following conditions is satisfied:

(a) $\mathcal{G}_N \subset \mathcal{G}_{N+1}$,
(b) There exists an everywhere dense subset $D \subset [0, T]$ such that

$$\forall t \in [0, T], \quad Y_t^N \xrightarrow{\mathbb{P}} Y_t.$$

(c) $|Y^N - Y|_{L_T^r} \xrightarrow{\mathbb{P}} 0$ *for some* $r \geq 1$,
   then

$$\forall q > \frac{1}{\theta}, \qquad \forall p \in [1, q\theta), \qquad \|Y - Y^N\|_{\sup} + \|Y - Y^N\|_{q,Hol} \xrightarrow{L^p} 0.$$

The proof of the theorem is a variant of the proof of the Kolmogorov criterion for functional tightness of processes. It consists in a string of several lemmas. For the following classical lemma, we refer to [14] (where it is stated and proved for semi-norms in p-variation).

Lemma 1. *Let* x, $y \in C([0, T], \mathbb{R}^d)$ *and let* $q \geq 1$. *Then*

(a) $\|x - x(0)\|_{\sup} \leq \|x\|_{q,Hol}$.
(b) $\|x + y\|_{q,Hol} \leq \|x\|_{p,Hol} + \|y\|_{q,Hol}$ *if* $q \geq 1$,
(c) *For every* $q > q' \geq 1$, $\|x\|_{q,Hol}^q \leq (2\|x\|_{\sup})^{q-q'} \|x\|_{q',Hol}^{q'}$.
(d) *Claims* (a)–(b)–(c) *remain true with the* p-*variation semi-norm* $\mathrm{Var}_{q,[0,T]}$ *instead of the* $\frac{1}{q}$-*Hölder semi-norm.*

Lemma 2. *Let* $p \in [1, \infty)$. *If* Y *satisfies the Kolmogorov criterion* $(K_{p,\theta})$ *then, for every* $N \geq 1$, *the process* Y^N *defined by* $Y_t^N = \mathbb{E}(Y_t \mid \mathcal{G}_N)$ *has a continuous modification which is* $\frac{\theta'}{p}$-*Hölder continuous for every* $\theta' \in (0, \theta)$ *(i.e.* $\|Y^N\|_{\frac{p}{\theta'},Hol} < +\infty$ *a.s.). Furthermore, the sequence* $(Y^N)_{N\geq1}$ *is* C-*tight and for every* $\theta' \in (0, \theta)$, *there exists a random variable* $Z_{\theta'} \in L_{\mathbb{R}}^p(\mathbb{P})$ *such that*

$$\mathbb{P}(d\omega)\text{-}a.s. \ \|Y(\omega)\|_{\frac{p}{\theta'},Hol} \leq Z_{\theta'} \tag{29}$$

and

$$\forall N \geq 1, \ \|Y^N(\omega)\|_{\frac{p}{\theta'},Hol} \leq \mathbb{E}(Z_{\theta'} \mid \mathcal{G}_N)(\omega). \tag{30}$$

In particular, the sequence of Hölder semi-norms $(\|Y^N\|_{\frac{p}{\theta'},Hol})_{N\geq1}$ *is* L^p-*uniformly integrable.*

Remark. As a by-product of the proof we also get that

$$\mathbb{E}(Z_{\theta'}^p) \leq C_{p,T,\theta,\theta'} C_T^{Kol}$$

where $C_{T,p,\theta,\theta'}$ is a finite real constant that only depends upon p, T, θ and θ' (and not on Y or the σ-fields $\mathcal{G}_N$).

Proof. First it follows form the Kolmogorov criterion that for every $N \geq 1$, Y^N admits a continuous modification which is $\frac{\theta'}{p}$-Hölder for every $\theta' \in (0, \theta)$. Moreover the sequence $(Y^N)_{N\geq1}$ is C-tight since every Y^N satisfies the same Kolmogorov criterion $(K_{p,\theta})$ and $Y_0^N = \mathbb{E}(Y_0|\mathcal{G}_N)$ is tight on $\mathbb{R}$ (see [1, 29] p. 26). Now, let s, $t \in [0, T]$, let m, $n \geq 1$ be two fixed integers. First note that

$$\sup_{s,t\in[0,T],\,t\leq s\leq t+\frac{T}{2^n}} |Y_t - Y_s| \leq 2 \sum_{m\geq 0} \max_{0\leq k\leq 2^{n+m}-1} |Y_{t_{k+1}^{n+m}} - Y_{t_k^{n+m}}| \tag{31}$$

and

$$\max_{0\leq k\leq 2^{n+m}-1} |Y_{t_{k+1}^{n+m}} - Y_{t_k^{n+m}}|^p \leq \sum_{k=0}^{2^{n+m}-1} |Y_{t_{k+1}^{n+m}} - Y_{t_k^{n+m}}|^p.$$

For every $\theta' \in (0,\theta)$, set

$$Z_{\theta'} := \frac{2}{T}\left(\sum_{n\geq 0} 2^{n\frac{\theta'}{p}} \sup_{s,t\in[0,T],\,t\leq s\leq t+\frac{T}{2^n}} |Y_t - Y_s|\right). \tag{32}$$

Taking the L^p-norm in (31) yields

$$\|Z_{\theta'}\|_p \leq \left(\frac{2}{T}\right)^{\frac{\theta'}{p}} \sum_{n\geq 0} 2^{n\frac{\theta'}{p}} \left\| \sup_{s,t\in[0,T],\,t\leq s\leq t+\frac{T}{2^n}} |Y_t - Y_s| \right\|_p$$

$$\leq 2 \left(\frac{2}{T}\right)^{\frac{\theta'}{p}} \sum_{n\geq 0} 2^{n\frac{\theta'}{p}} \sum_{\ell\geq 0} \left\| \max_{0\leq k\leq 2^{n+m}-1} |Y_{t_{k+1}^{n+m}} - Y_{t_k^{n+m}}| \right\|_p.$$

On the other hand, owing to the Kolmogorov criterion $(K_{p,\theta})$,

$$\mathbb{E} \max_{0\leq k\leq 2^{n+m}-1} |Y_{t_{k+1}^{n+m}} - Y_{t_k^{n+m}}|^p \leq \sum_{k=0}^{2^{n+m}-1} \mathbb{E}|Y_{t_{k+1}^{n+m}} - Y_{t_k^{n+m}}|^p$$

$$\leq 2^{n+m} C_T^{Kol} 2^{-(n+m)(1+\theta)} T^{-(1+\theta)}$$

$$= C_T^{Kol} T^{-(1+\theta)} 2^{-(n+m)\theta}.$$

Hence

$$\mathbb{E}\, Z_{\theta'}^p \leq C_T^{Kol} C_{p,T,\theta,\theta'} \left(\sum_{n\geq 0}\sum_{m\geq 0} 2^{n\frac{\theta'-\theta}{p}} 2^{-m\frac{\theta}{p}}\right)^p < +\infty$$

where the finite real constant $C_{p,T,\theta,\theta'}$ only depends on p, T, θ and θ'. On the other hand, for every $\delta \in [0,T]$, there exists a integer $n_\delta \geq 1$ such that $2^{-(1+n_\delta)} \leq \delta/T \leq 2^{-n_\delta}$. Hence,

$$\delta^{-\theta'} \sup_{s,t\in[0,T],\,t\leq s\leq t+\delta} |Y_t - Y_s|^p \leq 2^{(1+n_\delta)\theta'} T^{-\theta'}$$

$$\times \sup_{s,t\in[0,T],\,t\leq s\leq t+\frac{T}{2^n}} |Y_t - Y_s|^p \leq Z_{\theta'}^p.$$

Consequently, for every s, $t \in [0, T]$, and every $\omega \in \Omega$,

$$|Y_t(\omega) - Y_s(\omega)| \leq Z_{\theta'}(\omega)|t - s|^{\frac{\theta'}{p}}$$

i.e.

$$\|Y(\omega)\|_{\frac{p}{\theta'},Hol} \leq Z_{\theta'}(\omega).$$

Finally, it follows from Jensen's Inequality that for every $s, t \in \mathbb{Q} \cap [0, T]$,

$$\mathbb{P}(d\omega)\text{-}a.s. \qquad |Y_t^N(\omega) - Y_s^N(\omega)| \leq \mathbb{E}(Z_{\theta'} \,|\, \mathcal{G}_N)(\omega)|t - s|^{\theta'}.$$

In particular this means that, for every $p \geq 1$ and every $\theta' \in (0, \theta)$,

$$\mathbb{P}(d\omega)\text{-}a.s. \qquad \|Y^N(\omega)\|_{\frac{p}{\theta'},Hol} \leq \mathbb{E}(Z_{\theta'} \,|\, \mathcal{G}_N)(\omega) < +\infty$$

and satisfies the L^p-uniform integrability assumption. □

Proof (Proof of Theorem 3). The sequence $(Y^N)_{N \geq 1}$ being C-tight on $(\mathcal{C}([0, T], \mathbb{R}^d), \|.\|_{\sup})$, so is the case of the sequence $(Y^N, Y)_{N \geq 1}$ on $(\mathcal{C}([0, T], \mathbb{R}^{2d}), \|.\|_{\sup})$ since the product topology coincides with the uniform topology. Let $\mathbb{Q} = w\text{-}\lim_N \mathbb{P}_{(Y^{N'}, Y)}$ denote a weak functional limiting value of $(Y^N, Y)_{N \geq 1}$. If $\varXi = (\varXi^1, \varXi^2)$ denotes the canonical process on $(\mathcal{C}([0, T], \mathbb{R}^{2d}), \|.\|_{\sup})$, it is clear that $\mathbb{Q}_{\varXi^2} = \mathbb{P}_Y$.

▷ *Convergence of the sup-norm.* Assume that (c) holds: the functional $y \mapsto |y^1(t) - y^2(t)|_{L_T^r}$ is continuous on $(\mathcal{C}([0, T], \mathbb{R}^{2d}), \|.\|_{\sup})$, consequently, $|\varXi^1 - \varXi^2|_{L_T^r} = 0$ $\mathbb{Q}$-a.s. i.e. $\mathbb{Q} = \mathbb{P}_{(Y,Y)}$ so that $(Y^N, Y) \overset{\mathcal{L}(\|.\|_{\sup})}{\longrightarrow} (Y, Y)$ as $N \to \infty$ which simply means that $\|Y^N - Y\|_{\sup} \overset{\mathbb{P}}{\longrightarrow} 0$. On the other hand, it follows from Lemma 2 that, for every $N \geq 1$,

$$\|Y^N - Y\|_{\sup}^p \leq C_{p,T} \left(\mathbb{E}(Z_{\theta'}^p \,|\, \mathcal{G}_N) + Z_{\theta'}^p\right) \quad a.s.$$

(for a given fixed $\theta' \in (0, \theta)$) which implies that $(\|Y^N - Y\|_{\sup}^p)_{N \geq 1}$ is uniformly integrable. Finally,

$$\mathbb{E}\|Y^N - Y\|_{\sup}^p \longrightarrow 0 \qquad \text{as} \qquad N \to \infty.$$

Assume that (b) holds: it follows that, for every $t_1, \ldots, t_k \in D$, one has $(Y_{t_1}^N, \ldots, Y_{t_k}^N) \overset{\mathbb{P}}{\longrightarrow} (Y_{t_1}, \ldots, Y_{t_k})$, which in turn implies that the convergence $(Y_{t_1}^N, \ldots, Y_{t_k}^N, Y_{t_1}, \ldots, Y_{t_k}) \overset{\mathcal{L}}{\longrightarrow} (Y_{t_1}, \ldots, Y_{t_k}, Y_{t_1}, \ldots, Y_{t_k})$. This means that $\mathbb{Q}$ and $\mathbb{P}_{(Y,Y)}$ have the same finite dimensional marginals i.e. $\mathbb{Q} = \mathbb{P}_{(Y,Y)}$. One concludes like in (c).

If (a) holds, for every $t \in [0, T]$, $Y_t^N \to Y_t$ $\mathbb{P}$-a.s., so that (b) is satisfied.

▷ *Convergence of the Hölder semi-norm.* Let $q \geq 1$. As concerns the convergence of the $\frac{1}{q}$-Hölder semi-norm, one proceeds as follows. Let $q' \in (\frac{p}{\theta}, q)$ and set $\theta' := \frac{p}{q'} \in (0, \theta)$. It follows from Lemma 1(*b*)–(*c*) that

$$\|Y - Y^N\|_{q,Hol} \leq 2^{1-\frac{q'}{q}} \|Y - Y^N\|_{sup}^{1-\frac{q'}{q}} \times \left(\|Y\|_{q',Hol} + \|Y^N\|_{q',Hol}\right)^{\frac{q'}{q}}.$$

Now let $Z := Z_{\theta'}$ be defined by (32). Then,

$$\|Y\|_{q',Hol} + \|Y^N\|_{q',Hol} \leq Z + \left(\mathbb{E}(Z \mid \mathcal{G}_N)\right).$$

Hence, the sequence $(\|Y\|_{q',Hol} + \|Y^N\|_{q',Hol})_{N \geq 1}$, is tight since it is L^p-bounded. On the other hand, $\|Y - Y^N\|_{sup} \xrightarrow{L^p} 0$ so that $\|Y - Y^N\|_{q,Hol} \xrightarrow{\mathbb{P}} 0$ as $N \to \infty$.

Now let $\widetilde{\theta} = \frac{p}{q} \in (0, \theta)$. The same argument as above shows that $\|Y - Y^N\|_{q,Hol} \leq \widetilde{Z} + \mathbb{E}(\widetilde{Z} \mid \mathcal{G}_N)$ where $\widetilde{Z} = Z_{\widetilde{\theta}}$ is still given by (32). As a consequence, $(\|Y - Y^N\|_{q,Hol}^p)_{N \geq 1}$ is uniformly integrable since, for every $N \geq 1$, Jensen's Inequality implies

$$\|Y - Y^N\|_{q,Hol}^p \leq 2^{p-1}\left(\widetilde{Z}^p + \mathbb{E}(\widetilde{Z}^p \mid \mathcal{G}_N)\right)$$

which finally implies that $\|Y - Y^N\|_{q,Hol} \xrightarrow{L^p} 0$. ◇

4.2 Application to Stationary Quantizations of Brownian Motion: Convergence and Rates

Theorem 4. (*a*) *Let* $(\widehat{W}^N)_{N \geq 1}$ *be a sequence of stationary quadratic functional quantizers of a standard d-dimensional Brownian motion W defined by (11) or (17) converging to W in a (purely) quadratic sense, namely* $\| |W - \widehat{W}^N|_{L_T^2} \|_2 \to 0$ *as* $N \to \infty$. *Then, for every* $q > 2$,

$$\forall p \in (0, \infty), \qquad \|W - \widehat{W}^N\|_{q,Hol} \xrightarrow{L^p} 0 \text{ as } N \to \infty.$$

(*b*) *Let* $q > 2$. *If, for every* $N \geq 1$, $\widehat{W}^N$ *is an optimal product quantization at level N. Then, for every* $p \in (0, \infty)$,

$$\left\| \|W - \widehat{W}^N\|_{q,Hol} \right\|_p = o\left((\log N)^{-\frac{3}{2}\min\left(\frac{1}{5}(1-\frac{2}{q}),\frac{1}{p}\right)+\alpha}\right), \qquad \forall \alpha > 0.$$

The proof of this Theorem is a consequence of the above Theorem 3. So we need to get accurate estimates for the increments of the processes $W - \widehat{W}^N$. This is the aim of the following lemma.

Lemma 3. *Let $p \in [2, +\infty)$. Let $\widehat{W}^N$, $N \geq 1$, denote a sequence of optimal product quadratic quantizers. For every $\rho \in (0, \frac{1}{2})$ and every $\varepsilon \in (0, 3)$, for every $s, t \in [0, T]$, $s \leq t$,*

$$\left\| (W_t - W_s) - (\widehat{W}_t^N - \widehat{W}_s^N) \right\|_p \leq C_{\rho, p, T, d, \varepsilon} |t - s|^\rho (\log N)^{-(\frac{1}{2} - \rho) \wedge (\frac{3-\varepsilon}{2p})}. \quad (33)$$

In particular, if $p \in (2, 3)$, then

$$\left\| (W_t - W_s) - (\widehat{W}_t^N - \widehat{W}_s^N) \right\|_p \leq C_{\rho, p, T, d} |t - s|^\rho (\log N)^{-(\frac{1}{2} - \rho)}. \quad (34)$$

Proof. We may assume without loss of generality that we deal with a one-dimensional Brownian motion W, quantized at level $N' = \lfloor N^{\frac{1}{d}} \rfloor$ since everything is done component by component. Set for every $k \geq 1$, $\widetilde{\xi}_k := \xi_k - \widehat{\xi}_k^{N_k}$ where $N_1, \ldots, N_k, \ldots$ denotes the optimal bit allocation of an optimal product quadratic quantization at level N'. Keep in mind that for every $k > L_W(N')$, $N_k = 1$ and that of course $N_1 \cdots N_{L_W(N')} \leq N'$. The random vectors $(\widetilde{\xi}_k)_{k \geq 1}$ are independent and centered.

It follows from the K-L expansion of W and its product quantization that

$$(W_t - W_s) - (\widehat{W}_t^{N'} - \widehat{W}_s^{N'}) = \sum_{k \geq 1} \lambda_k \widetilde{\xi}_k (e_k^W(t) - e_k^W(s)).$$

Then, it follows from the B.D.G. Inequality for discrete time martingales that

$$\left\| (W_t - W_s) - (\widehat{W}_t^{N'} - \widehat{W}_s^{N'}) \right\|_p \leq C_{p, T} \left\| \sum_{k \geq 1} \lambda_k \widetilde{\xi}_k (e_k^W(t) - e_k^W(s))^2 \right\|_{\frac{p}{2}}^{\frac{1}{2}}$$

$$\leq C_{p, T} \left(\sum_{k \geq 1} \lambda_k^{1-\rho} \|\widetilde{\xi}_k\|_p^2 \right)^{\frac{1}{2}} |t - s|^\rho$$

since, for every $k \geq 1$,

$$(e_k^W(t) - e_k^W(s))^2 = \frac{8}{T} \sin^2 \left(\frac{t - s}{\sqrt{\lambda_k}} \right) \cos^2 \left(\frac{t - s}{\sqrt{\lambda_k}} \right) \leq \frac{8}{T} |t - s|^{2\rho} \lambda_k^{-\rho}.$$

The random variables $\widehat{\xi}_k^{N_k}$ being an optimal quadratic quantization of the one-dimensional normal distribution for every $k \in \{1, \ldots, L_W(N')\}$, it follows from (16) that, there exists for every $\varepsilon \in (0, 3)$, a constant $\kappa_{p, \varepsilon}$ such that

$$\forall m \geq 1, \qquad \|\widetilde{\xi}_k\|_p = \|\xi - \widehat{\xi}_k^{N_k}\|_p \leq \kappa_{p, \varepsilon} \frac{1}{N_k^{1 \wedge \frac{3-\varepsilon}{p}}}$$

where $\widehat{\xi}^m$ denotes the (unique) optimal quadratic quantization at level m of a normally distributed scalar random variable ξ. As a consequence,

$$\left\| (W_t - W_s) - (\widehat{W}_t^{N'} - \widehat{W}_s^{N'}) \right\|_p \leq C_{p,T,\varepsilon} |t - s|^\rho \left(\sum_{k \geq 1} \lambda_k^{1-\rho} \frac{1}{N_k^{2(1 \wedge \frac{3-\varepsilon}{\rho})}} \right)^{\frac{1}{2}}.$$

▷ Temporarily assume that $p \in [2, 3)$. One may choose ε so that $1 \wedge \frac{3-\varepsilon}{\rho} = 1$. Now, keeping in mind that $L' := L_W(N') \sim \log N'$ and $\lambda_k \leq c\, k^{-2}$ for a real constant $c > 0$, one gets

$$\sum_k \lambda_n^{1-\rho} \frac{1}{N_k^2} \leq \lambda_{L'}^\rho \sum_{k=1}^{L'} \frac{\lambda_k}{N_k^2} + \sum_{k > L'} \lambda_k^{1-\rho}$$

$$\leq C_\rho \left((\log N')^{2\rho} \sum_{k=1}^{L'} \frac{\lambda_k}{N_k^2} + (\log N)^{2\rho - 1} \right).$$

Now, following e.g. [18], we know that the optimal bit allocation yields

$$\sum_{k=1}^{L'} \frac{\lambda_k}{N_k^2} \leq \frac{C}{T} (\log N')^{-1}$$

so that, finally

$$\left\| (W_t - W_s) - (\widehat{W}_t^{N'} - \widehat{W}_s^{N'}) \right\|_p \leq C_{\rho,p,T} |t - s|^\rho (\log N')^{\rho - \frac{1}{2}}.$$

▷ Assume now that $p \in [3, +\infty)$ and $\varepsilon \in (0, 3)$. Set $\tilde{p} = \frac{p}{3-\varepsilon} > 1$ and $\tilde{q}$ its conjugate exponent. Then, Hölder Inequality implies

$$\sum_{k=1}^{L'} \frac{\lambda_k^{1-\rho}}{N_k^{\frac{2}{p}}} \leq \left(\sum_{k=1}^{L'} \frac{\lambda_k}{N_k^2} \right)^{\frac{1}{\tilde{p}}} \left(\sum_{k=1}^{L'} \lambda_k^{1 - \frac{\rho p}{p-3+\varepsilon}} \right)^{\frac{1}{\tilde{q}}}.$$

We inspect now three possible cases for ρ.

- If $0 < \rho < \frac{1}{2}(1 - \frac{3-\varepsilon}{p})$, then $1 - \frac{\rho p}{p-3+\varepsilon} > \frac{1}{2}$ so that $\sum_{k \geq 1} \lambda_k^{1 - \frac{\rho p}{p-3+\varepsilon}} < +\infty$, which in turn implies that

$$\sum_{k=1}^{L'} \frac{\lambda_k^{1-\rho}}{N_k^{\frac{2}{p}}} \leq C_{\rho,p,T} \left(\log N' \right)^{-\frac{3-\varepsilon}{p}}.$$

Furthermore $1 - \frac{\rho}{2} > \frac{3-\varepsilon}{p}$.

- If $\frac{1}{2}(1 - \frac{3-\varepsilon}{p}) < \rho < \frac{1}{2}$, then, $1 - \frac{\rho}{2} < \frac{3-\varepsilon}{p}$ and $1 - \frac{\rho p}{p-3+\varepsilon} = \frac{1}{2}$ so that

$$\sum_{k\geq 1} \lambda_k^{1-\frac{\rho p}{p-3+\varepsilon}} < +\infty$$

$$\sum_{k=1}^{L'} \frac{\lambda_k^{1-\rho}}{N_k^{\frac{2}{p}}} \leq C_{\rho,p,T} \left(\log N' \right)^{-\frac{3-\varepsilon}{p}} \times \left(L_W(N')^{\frac{2\rho p}{p-3+\varepsilon}-1} \right)^{1-\frac{3-\varepsilon}{p}}$$

$$= C_{\rho,p,T} \left(\log N' \right)^{2\rho-1}.$$

- If $\frac{1}{2}(1 - \frac{3-\varepsilon}{p}) = \rho < \frac{1}{2}$, then $1 - \frac{\rho}{2} = \frac{3-\varepsilon}{p}$ and $1 - \frac{\rho p}{p-3+\varepsilon} = \frac{1}{2}$ so that
$\sum_{k=1}^{L'} \lambda_k^{1-\frac{\rho p}{p-3+\varepsilon}} \leq C_{\rho,p,T} \log\log N'$ (keep in mind $L' = L_W(N') \sim \log N'$).
Hence, for every $\varepsilon' \in (0, \varepsilon)$,

$$\sum_{k=1}^{L'} \frac{\lambda_k^{1-\rho}}{N_k^{\frac{2}{p}}} = o\left((\log N')^{-\frac{3-\varepsilon'}{p}} \right).$$

As conclusion, we get that

$$\left(\sum_k \lambda_k^{1-\rho} \|\widetilde{\xi}_k\|_p^2 \right)^{\frac{1}{2}} \leq \left(\sum_k \lambda_k^{1-\rho} \frac{1}{N_k^{2(1\wedge\frac{3-\varepsilon}{p})}} \right)^{\frac{1}{2}} = O\left((\log N')^{-(\frac{1}{2}-\rho)\wedge(\frac{3-\varepsilon}{2p})} \right)$$

$$(35)$$

which completes the proof since $\log(1 + N') > \frac{1}{d} \log N$ (which implies $\log N' > \frac{1}{d} \log(N/2)$). $\qquad\square$

Proof (Proof of Theorem 4). (*a*) Owing to the monotonicity of the L^p-norms, it is enough to show that, the announced convergence holds for every $q > 2$ and every $p > \frac{2q}{q-2}$ or equivalently for every $p > 2$ and every $q > \frac{2p}{p-2}$. This statement follows for the $\frac{1}{q}$-Hölder (semi-)norm follows from Theorem 3(*c*). Indeed W satisfies the Kolmogorov $K_{p,\theta}$ with $\theta = p/2 - 1$. On the other hand, it follows from [13] that, for any sequence of (Voronoi) quantizations $\widehat{W}^N$ at level N converging in $L^2_{L^2_T}(\mathbb{P})$ toward W, this convergence also holds in the *a.s.* sense. So Criterion(*c*) is fulfilled.

(*b*) Let $q > 2$. The process $W - \widehat{W}^N$ satisfies $K_{p,\rho p-1}$ for every $\rho \in (\frac{1}{p}, \frac{1}{2})$ with "Kolmogorov constants"

$$C_{T,p}^{Kol} = C_{p,T,\rho,d,\varepsilon} (\log N)^{-p[(\frac{1}{2}-\rho)\wedge(\frac{3-\varepsilon}{2p})]}, \quad \varepsilon \in (0,3).$$

We wish to apply Lemma 2 (and the remark that follows).

$\triangleright$ Assume $0 < p < \frac{5q}{q-2}$. Then there exists $\eta > 0$ such that $p < p' = \frac{5q}{q-2+\eta}$. Set $\theta' = \frac{p'}{q}$. One checks that $\frac{1}{p'} + \frac{1}{q} < \frac{1}{2}$ so that there exists $\eta' > 0$ such that

$\rho = \frac{1}{p'} + \frac{1}{q} + \eta' < \frac{1}{2}$. Elementary computations show that $\frac{1}{2} - \rho < \frac{3}{2p}$. Let $\varepsilon \in (0, 3)$ such that $\frac{1}{2} - \rho < \frac{3}{2p} - \varepsilon$. Consequently, Lemma 2 (and the remark that follows) imply that

$$\left\| \|W - \widehat{W}^N\|_{q,Hol} \right\|_{p'} \leq C_{q,\eta,\eta',T,\varepsilon} (\log N)^{-(\frac{1}{2}-\rho)}$$

and for any small enough $\alpha > 0$, one my specify η, η' and ε so that $\frac{1}{2} - \rho = \frac{3}{10}(1 - \frac{2}{q}) - \alpha$. Finally this bounds holds true for $p \in (0, p')$ since the L^p-norm is non-decreasing.

▷ Now, if $p \geq \frac{5q}{q-2}$, one checks that $\frac{3}{2p} \geq \frac{1}{2} - \left(\frac{1}{p} + \frac{1}{q} \right)$. It becomes impossible to specify $\rho \in (0, \frac{1}{2})$ so that $\theta' = \frac{p}{q} < \theta = \rho p - 1$ and $1 - \rho > \frac{3}{2p}$. So the same specifications as above lead to

$$\left\| \|W - \widehat{W}^N\|_{q,Hol} \right\|_{p'} \leq C_{q,\eta,\eta',\varepsilon,T} (\log N)^{-\frac{3-\varepsilon}{2p}}$$

which yields the announced result.   □

5 Convergence of Stationary Quantizations of the Brownian Motion for the ρ_q-Hölder Distance

In view of what will be needed to apply this theorem to the Brownian motion and its functional quantizations, we need to prove a counterpart of Lemmas 2 and 3 for $\mathbf{W}^2_{s,t}$. However, for the sake of simplicity, by contrast with the previous section, we will only deal with the case of the Brownian motion and its stationary quantizations.

The main result of this section is the following Theorem.

Theorem 5. *Let $q > 2$.*

(a) Let $(\widehat{W}^N)_{N\geq 1}$ be a sequence of stationary quadratic functional quantizers of a standard d-dimensional Brownian motion W defined by (11) or (17) converging to W in a (purely) quadratic sense, namely $\| \|W - \widehat{W}^N|_{L^2_T} \|_2 \to 0$ as $N \to \infty$. Then,

$$\forall q > 2, \quad \forall p > 0, \quad \left\| \rho_q(\mathbf{W}, \widehat{\mathbf{W}}^N) \right\|_p \longrightarrow 0 \quad as \ N \to \infty.$$

(b) Let $q > 2$. Assume that, for every $N \geq 1$, $\widehat{W}^N$ is an optimal product quantization at level N of W. Then, for every $q > 2$ and every $p > 0$,

$$\left\| \|\mathbf{W}^2 - \widehat{\mathbf{W}}^{2,N}\|_{\frac{q}{2},Hol} \right\|_p = o\left((\log N)^{-\frac{3}{2} \min \left(\frac{2}{7}(1-\frac{2}{q}), \frac{1}{p} \right) + \alpha} \right), \quad \forall \alpha > 0,$$

so that, finally,

$$\left\| \rho_q(\mathbf{W}, \widehat{\mathbf{W}}^N) \right\|_p = o\left((\log N)^{-\frac{3}{2}\min\left(\frac{1}{q}(1-\frac{2}{q}),\frac{1}{p}\right)+\alpha} \right), \ \forall\, \alpha > 0.$$

(c) If $r > \frac{2}{3}$, then

$$\rho_q(\mathbf{W}, \widehat{\mathbf{W}}^{\lfloor e^{N^r} \rfloor}) = o\left(N^{-(\frac{3}{2}r-1)\frac{q-2}{7q}+\alpha} \right) \ \forall\, \alpha > 0, \ \mathbb{P}\text{-}a.s.$$

Note that the result of interest for our purpose (convergence on multi-dimensional stochastic integrals) corresponds to $q \in (2,3)$. The proposition below appears as the counterpart of Lemma 2 on the way to the proof.

Proposition 3. *Let* $p > 2$.
(a) *Let* $\mathbf{W}^2_{s,t}$ *be defined by* (27). *For every* $\tilde{\theta}' \in (0, p-1)$, *there exists a random variable* $Z^{(2)}_{\tilde{\theta}'} \in L^p$ *such that*

$$\mathbb{P}\text{-}a.s. \quad \forall\, s,\, t \in [0,T], \qquad |\mathbf{W}^2_{s,t}| \le Z^{(2)}_{\tilde{\theta}'} |t-s|^{\frac{\tilde{\theta}'}{p}}.$$

(b) *Let*

$$\widehat{\mathbf{W}}^{2,N}_{s,t}(\omega) = \left(\int_s^t (\widehat{W}^i_u - \widehat{W}^i_s) d\,\widehat{W}^j_u \right)_{i,j=0,\dots,d} (\omega), \qquad s,\, t \in [0,T], \ s \le t,$$

where $\widehat{W} = \widehat{W}^N$ *is a stationary quantization of* W *(the integration holds in the Stieltjes sense). Then, for every* $p > 2$ *and every* $\tilde{\theta}' \in (0, p-1)$,

$$\mathbb{P}\text{-}a.s. \quad \forall\, s,\, t \in [0,T], \qquad |\widehat{\mathbf{W}}^{2,N}_{s,t}| \le \mathbb{E}(Z^{(2)}_{p,\tilde{\theta}'} | \mathcal{G}_N)|t-s|^{\frac{\tilde{\theta}'}{p}}.$$

(c) *Let* $\widetilde{\mathbf{W}}^{2,N}_{s,t} = \mathbf{W}^2_{s,t} - \widehat{\mathbf{W}}^{2,N}_{s,t}$ *where* $\widehat{W} = \widehat{W}^N$ *is now an optimal quadratic product quantization of* W *at level* N. *Then, if* $p > \frac{1}{\rho}$, *for every* $\tilde{\theta}' \in (0, p(\rho + \frac{1}{2}) - 2)$, *for every* $\varepsilon \in (0,3)$ *and every* $\delta > 0$, *there exists a real constant* $C_{\rho,p,T,d,\varepsilon,\delta} > 0$ *such that*

$$\left\| \sup_{s,t\in[0,T]} \frac{|\widetilde{\mathbf{W}}^{2,N}_{s,t}|}{|t-s|^{\frac{\tilde{\theta}'}{p}}} \right\|_p \le C_{\rho,p,T,d,\varepsilon,\delta} (\log N)^{-(\frac{1}{2}-\rho)\wedge\frac{3-\varepsilon}{2(p+\delta)}}.$$

Proof. (a) The random variable $Z^{(2)}_{\tilde{\theta}'}$ of interest is defined by

$$Z^{(2)}_{\tilde{\theta}'} := \frac{2}{T} \sum_{n\ge 0} 2^{n\frac{\tilde{\theta}'}{p}} \sup_{s\le t\le s+\frac{T}{2^n}} |\mathbf{W}^2_{s,t}|.$$

Let $s,\, t \in [0,T],\, s \le t \le s + \frac{T}{2^n}$. We know from the multiplicative tensor property that, for every $u \in [s,t]$,

$$\mathbf{W}^2_{s,t} = \mathbf{W}^2_{s,u} + \mathbf{W}^2_{u,t} + W_{s,u} \otimes W_{u,t}$$

and that, for every $i, j \in \{0, \dots, d\}$,

$$|W_{s,u}^i \otimes W_{u,t}^j| \leq \frac{1}{2}(|W_{s,u}^i|^2 + |W_{u,t}^j|^2).$$

To evaluate $\sup_{t \in [s, s + \frac{T}{2^n}]} |\mathbf{W}_{s,t}^{2;\cdot}|$, we may restrict to dyadic numbers owing to the continuity in (s, t) of $\mathbf{W}_{s,t}^2$. As a consequence, we have, still following the classical scheme of Kolmogorov criterion

$$\sup_{t \in [s, s + \frac{T}{2^n}]} |\mathbf{W}_{s,t}^2| \leq 2 \sum_{m \geq 0} \max_{0 \leq k \leq 2^{n+m}-1} |\mathbf{W}_{t_k^{n+m}, t_{k+1}^{n+m}}^2|$$

$$+ \max_{0 \leq k \leq 2^{n+m}-1} |W_{t_k^{n+m}, t_{k+1}^{n+m}}|^2.$$

Now

$$\mathbb{E} \max_{0 \leq k \leq 2^{n+m}-1} |\mathbf{W}_{t_k^{n+m}, t_{k+1}^{n+m}}^2|^p \leq \sum_{\ell=0}^{2^{m+n}-1} \mathbb{E} |\mathbf{W}_{t_\ell^{n+m}, t_{\ell+1}^{n+m}}^2|^p$$

and

$$\mathbb{E} \max_{0 \leq k \leq 2^{n+m}-1} |W_{t_k^{n+m}, t_{k+1}^{n+m}}|^p \leq \sum_{\ell=0}^{2^{m+n}-1} \mathbb{E} |W_{t_\ell^{n+m}, t_{\ell+1}^{n+m}}|^p$$

where the norms $|.|$ are the canonical Euclidean norms on the spaces $\mathcal{M}((d+1), (d+1))$ and $\mathbb{R}^{d+1}$ respectively.

It is clear that, for every $i \neq j, i, j \geq 1$ and every $t \geq s$,

$$\|\mathbf{W}_{s,t}^{2,ij}\|_p = \left\| \int_s^t (W_u^i - W_s^i) dW_u^j \right\|_p$$

$$\leq \left\| \int_s^t (W_u^i - W_s^i) dW_u^j \right\|_p$$

$$\leq C_p^{BDG} \left\| \int_s^t (W_u^i - W_s^i)^2 du \right\|_{\frac{p}{2}}^{\frac{1}{2}}$$

$$\leq C_p |t' - t|$$

whereas

$$\| |W_{t'} - W_t|^2 \|_p = |t' - t| \| |W_1| \|_p = C_{p,d} |t' - t|.$$

Noting that $W_t^0 = t$ and, if $i = j, 1 \leq i \leq d, \mathbf{W}_{s,t}^{2,ii} = \frac{1}{2}(W_t^i - W_s^i)^2$ shows that the above upper-bound still holds for $i = j$ and i or $j = 0$. Consequently, we also have

$$\|\mathbf{W}_{s,t}^{2,ij}\|_p \leq C_{p,d} |t' - t|.$$

Consequently

$$\mathbb{E} \max_{0 \le k \le 2^{n+m}-1} |\mathbf{W}^2_{t_k^{n+m},t_{k+1}^{n+m}}|^p \le C_{p,d} \sum_{k=0}^{2^{n+m}-1} \left(\frac{T}{2^{n+m}}\right)^p = C_{p,d,T} 2^{(n+m)(1-p)}$$

so that

$$\|Z^{(2)}_{\tilde{\theta}'}\|_p \le C_{p,d,T} \sum_{n \ge 0} 2^{n\frac{\tilde{\theta}'}{p}} \sum_{m \ge 0} 2^{(n+m)(\frac{1}{p}-1)} = C_{p,d,T} \sum_{n \ge 0} 2^{n(\frac{\tilde{\theta}'}{p}-1)} < +\infty$$

since $\tilde{\theta}' < p - 1$.

On the other hand, one has obviously

$$\sup_{s,t \in [0,T], s \neq t} \frac{|\mathbf{W}^2_{s,t}|}{|t-s|^{\frac{\tilde{\theta}'}{p}}} \le Z^{(2)}_{\tilde{\theta}'} < +\infty \qquad a.s.$$

Lemma $2(a)$ applied to W (which satisfies $(K_{p,\frac{p}{2}-1})$) yields for every $\theta' \in (0, \frac{p}{2}-1)$ the existence of $Z^{(1)} \in L^p(\mathbb{P})$ such that

$$\sup_{s,t \in [0,T], s \le t} \frac{|\mathbf{W}^1_{s,t}|}{|t-s|^{\frac{\theta'}{p}}} \le Z^{(1)}_{\theta'} \qquad a.s.$$

As a consequence, combining these two results shows that, for every $q > \frac{2p}{p-2}$,

$$\rho_q(\mathbf{W}, 0) < Z = Z^{(1)}_{\theta'} + Z^{(2)}_{\tilde{\theta}'} \in L^p(\mathbb{P})$$

where $Z^{(1)}$ is related to $\theta' = \frac{p}{q} \in (0, \frac{p}{2} - 1)$ and $Z^{(2)}$ is related to $\tilde{\theta}' = \frac{2p}{q} \in (0, p-2)$.

(b) If $i \neq j$, $0 \le i,j \le d$, it follows from Proposition 2 that $\widehat{\mathbf{W}}^{2,ij,N}_{s,t} = \mathbb{E}(\widehat{\mathbf{W}}^{2,ij,N}_{s,t} \mid \mathcal{G}_N)$ where $\mathcal{G}_N = \sigma(\widehat{W})$ and $\widehat{W}^N = (\widehat{W}^{i,N})_{1 \le i \le d}$ is an optimal product quantization at level N (which means that for each component W^i, $\widehat{W}^{i,N}$ is an optimal product quantization at level $N' = \lfloor N^{\frac{1}{d}} \rfloor$).

When $i = j \ge 1$, $|\widehat{\mathbf{W}}^{2,ii,N}_{s,t}| \le \frac{1}{2}\mathbb{E}\left((W^i_t - W^i_s)^2 \mid \mathcal{G}_N\right)$. One derives that

$$\frac{|\widehat{\mathbf{W}}^{2,ii,N}_{s,t}|}{|t-s|^{\frac{\tilde{\theta}'}{p}}} \le \mathbb{E}\left(\frac{|\mathbf{W}^{2,ii}_{s,t}|}{|t-s|^{\frac{\tilde{\theta}'}{p}}} \mid \mathcal{G}_N\right) \le \mathbb{E}\left((Z^{(2)}_{\tilde{\theta}'})^{\frac{\tilde{\theta}'}{p}} \mid \mathcal{G}_N\right).$$

When $i = j = 0$, $\widehat{\mathbf{W}}^{2,ii,N} = \mathbf{W}^{2,ii} = \frac{1}{2}(t-s)^2$.

(c) In this claim, the random variable $Z^{(2),N}_{\tilde{\theta}'}$ of interest is defined by

$$\widetilde{Z}_{\theta'}^{(2),N} = \frac{2}{T} \sum_{n \geq 0} 2^{n\frac{\bar{\theta}'}{p}} \sup_{s \leq t \leq s + \frac{T}{2^n}} |\widetilde{\mathbf{W}}_{s,t}^{2,N}|$$

and we aim at showing that it lies in $L^p(\mathbb{P})$ with a control on its L^p-norm as a function of N. One first derives for $\widetilde{\mathbf{W}}_{s,t}^{2,N}$ the straightforward identity when $s \leq u \leq t$

$$\widetilde{\mathbf{W}}_{s,t}^{2,N} = \widetilde{\mathbf{W}}_{s,u}^{2,N} + \widetilde{\mathbf{W}}_{u,t}^{2,N} + \widetilde{W}_{s,u,t}^{N}$$

where

$$\widetilde{W}_{s,u,t}^{N} = W_{s,u} \otimes W_{u,t} - \widehat{W}_{s,u}^{N} \otimes \widehat{W}_{u,t}^{N}$$
$$= (W_{s,u} - \widehat{W}_{s,u}^{N}) \otimes W_{u,t} + \widehat{W}_{s,u}^{N} \otimes (W_{u,t} - \widehat{W}_{u,t}^{N}) \quad (36)$$

with $W_{r,s} := W_r - W_s$ if $r \geq s$, etc. One derives from (36) that

$$|\widetilde{\mathbf{W}}_{s,t}^{2,N}| \leq 2 \sum_{m \geq 0} \max_{0 \leq k \leq 2^{n+m}-1} |\widetilde{\mathbf{W}}_{t_k^{n+m}, t_{k+1}^{n+m}}^{2,N}| \quad (37)$$

$$+ 2 \sum_{\substack{m,m' \geq 0}} \max_{\substack{0 \leq k \leq 2^{n+m}-1 \\ 0 \leq k' \leq 2^{n+m'}-1}} |W_{t_k^{n+m}, t_{k+1}^{n+m}}^{2,N} - \widehat{W}_{t_k^{n+m}, t_{k+1}^{n+m}}^{2,N}| |W_{t_{k'}^{n+m}, t_{k'+1}^{n+m}}| \quad (38)$$

$$+ 2 \sum_{\substack{m,m' \geq 0}} \max_{\substack{0 \leq k \leq 2^{n+m}-1 \\ 0 \leq k' \leq 2^{n+m'}-1}} |W_{t_k^{n+m}, t_{k+1}^{n+m}}^{2,N} - \widehat{W}_{t_k^{n+m}, t_{k+1}^{n+m}}^{2,N}| |\widehat{W}_{t_{k'}^{n+m}, t_{k'+1}^{n+m}}|. \quad (39)$$

where we used that $|u \otimes v| \leq |u||v|$.

We will first deal with deal with the first term in (37). We note that

$$\mathbb{E} \max_{0 \leq k \leq 2^{n+m}-1} |\widetilde{\mathbf{W}}_{t_k^{n+m}, t_{k+1}^{n+m}}^{2,N}|^p \leq \sum_{0 \leq k \leq 2^{n+m}-1} \mathbb{E}|\widetilde{\mathbf{W}}_{t_k^{n+m}, t_{k+1}^{n+m}}^{2,N}|^p.$$

Let $s, t \in [0, T]$, $s \leq t$ and $i, j \in \{1, \ldots, d\}$, $i \neq j$. One checks that the following decomposition holds

$$\widetilde{\mathbf{W}}_{s,t}^{2,ij,N} = \underbrace{\int_s^t W_{s,u}^i d(W_u^j - \widehat{W}_u^{j,N})}_{(A)} + \underbrace{\int_s^t \widehat{W}_{u,t}^{j,N} d(W_u^i - \widehat{W}_u^{i,N})}_{(B)}.$$

Let us focus on (A). First not that, owing to Proposition 1 applied with $\mathcal{F}_t = \sigma(W_u^i, u \in [0, T], W_s^j, s \leq t)$,

$$(A) = \sum_{n \geq 1} \bar{\xi}_n^j \int_s^t W_{s,u}^i \cos\left(\frac{u}{\sqrt{\lambda_n}}\right) du.$$

Using that W^i and W^j are independent, one derives that (A) is the terminal value of a martingale with respect to the filtration $\sigma(\xi^j_k, \, k \leq n, \, W^i_u, 0 \leq u \leq T), n \geq 1$ so that combining B.D.G. and Minkowski inequalities yields, with the notations of Lemma 3,

$$\mathbb{E}(|(A)|^p) \leq C_p^{BDG} \mathbb{E}\left(\sum_{n \geq 1}(\widetilde{\xi}^j_n)^2 \left(\int_s^t W^i_{s,u} \cos\left(\frac{u}{\sqrt{\lambda_n}}\right) du\right)^2\right)^{\frac{p}{2}}$$

$$\leq C_p^{BDG} \left(\sum_{n \geq 1}\|\widetilde{\xi}^j_n\|_p^2 \left\|\int_s^t W^i_{s,u} \cos\left(\frac{u}{\sqrt{\lambda_n}}\right) du\right\|_p^2\right)^{\frac{p}{2}}$$

where $\widetilde{\xi}_n = \xi_n - \widehat{\xi}_n^{N_n}$ and $N_1, \ldots, N_n, \ldots$ denote the optimal bit allocation of an optimal quadratic product quantization at level N' (keep in mind that $N_k = 1, k > L_B(N')$ and $N_1 \cdots N_{L_B(N')} \leq N'$ (B scalar Brownian motion). Now an elementary integration by parts yields

$$\int_s^t W^i_{s,u} \cos\left(\frac{u}{\sqrt{\lambda_n}}\right) du = \sqrt{\lambda_n} \int_s^t \left(\sin\left(\frac{t}{\sqrt{\lambda_n}}\right) - \sin\left(\frac{u}{\sqrt{\lambda_n}}\right)\right) dW^i_u$$

so that, for every $\rho \in (0, \frac{1}{2})$, one checks that, owing to the *BDG* Inequality,

$$\left\|\int_s^t W^i_{s,u} \cos\left(\frac{u}{\sqrt{\lambda_n}}\right) du\right\|_p \leq C_p^{BDG} C_{p,\rho} \lambda_n^{\frac{1-\rho}{2}} |t-s|^{\frac{1}{2}+\rho}.$$

Finally, for every $\varepsilon \in (0, 3)$,

$$\|(A)\|_p \leq C_{p,T,\rho,\varepsilon} \left(\sum_{n \geq 1} \lambda_n^{1-\rho} \|\widetilde{\xi}_n\|_p^2\right)^{\frac{1}{2}} |t-s|^{\frac{1}{2}+\rho}.$$

One shows likewise the same inequality for (B) once noted that

$$\int_s^t \widehat{W}^{i,N}_{s,u} \cos\left(\frac{u}{\sqrt{\lambda_n}}\right) du = \mathbb{E}\left(\int_s^t W^i_{s,u} \cos\left(\frac{u}{\sqrt{\lambda_n}}\right) du \mid \mathcal{F}_T^{\widehat{W}^{i,N}}\right)$$

which implies

$$\left\|\int_s^t \widehat{W}^{i,N}_{s,u} \cos\left(\frac{u}{\sqrt{\lambda_n}}\right) du\right\|_p \leq \left\|\int_s^t W^i_{s,u} \cos\left(\frac{u}{\sqrt{\lambda_n}}\right) du\right\|_p.$$

Consequently, for every $\varepsilon \in (0, 3)$,

$$\|\widetilde{\mathbf{W}}^{2,ij,N}_{s,t}\|_p \leq C_{p,\rho,T}\left(\sum_{n\geq 1}\lambda_n^{1-\rho}\|\tilde{\xi}_n\|_p^2\right)^{\frac{1}{2}}|t-s|^{\frac{1}{2}+\rho}$$

$$\leq C_{p,T,\rho,d,\varepsilon}(\log N)^{-(\frac{1}{2}-\rho)\wedge\frac{3-\varepsilon}{2p}}|t-s|^{\frac{1}{2}+\rho}. \qquad (40)$$

If $i = j \geq 1$, then

$$\widetilde{\mathbf{W}}^{2,ii,N}_{s,t} = \frac{1}{2}\left((W^i_{s,t})^- (\widehat{W}^{i,N}_{s,t})^2\right)$$

so that, using again Hölder Inequality,

$$\|\widetilde{\mathbf{W}}^{2,ii,N}_{s,t}\|_p = \frac{1}{2}\|W^i_{s,t} - \widehat{W}^{i,N}_{s,t}\|_{p+\delta}\|W^i_{s,t} - \widehat{W}^{i,N}_{s,t}\|_{p(1+\frac{p}{\delta})}$$

and one gets the same bounds as in the case $i \neq j$.

If i or $j = 0$, one gets similar bounds: we leave the details to the reader. Finally, one gets that, for every $i, j \in \{0, \ldots, d\}$,

$$\|\widetilde{\mathbf{W}}^{2,N}_{s,t}\|_p \leq C_{p,\rho,T,d,\varepsilon,\delta}(\log N)^{-(\frac{1}{2}-\rho)\wedge\frac{3-\varepsilon}{2(p+\delta)}}|t-s|^{\frac{1}{2}+\rho}.$$

By standard computations similar to those detailed in Lemma 2, we get

$$\sum_{m\geq 0}\|\max_{0\leq k\leq 2^{n+m}-1}|\widetilde{\mathbf{W}}^{2,N}_{t_k^{n+m},t_{k+1}^{n+m}}|\|_p \leq C_{p,\rho,T,d,\varepsilon,\delta}(\log N)^{-(\frac{1}{2}-\rho)\wedge\frac{3-\varepsilon}{2(p+\delta)}}2^{-n(\frac{1}{2}+\rho)}.$$

Let us pass now to the two other sums. We will focus on (38) since both behave and can be treated similarly.

$$\max_{\substack{0\leq k\leq 2^{n+m}-1 \\ 0\leq k'\leq 2^{n+m'}-1}}|W^{2,N}_{t_k^{n+m},t_{k+1}^{n+m}} - \widehat{W}^{2,N}_{t_k^{n+m},t_{k+1}^{n+m}}|^p|W_{t_{k'}^{n+m},t_{k'+1}^{n+m}}|^p$$

$$\leq \sum_{\substack{0\leq k\leq 2^{n+m}-1 \\ 0\leq k'\leq 2^{n+m'}-1}}|W^{2,N}_{t_k^{n+m},t_{k+1}^{n+m}} - \widehat{W}^{2,N}_{t_k^{n+m},t_{k+1}^{n+m}}|^p|W_{t_{k'}^{n+m},t_{k'+1}^{n+m}}|^p.$$

Now for every $s, u, t \in [0, T]$, $s \leq u \leq t$, it follows from Hölder Inequality that

$$\|\,|W_{s,u} - \widehat{W}^N_{s,u}|\,|W_{u,t}|\,\|_p \leq \|W_{s,u} - \widehat{W}^N_{s,u}\|_{p+\delta}\|W_{u,t}\|_{p(1+p/\delta)}$$

$$\leq C_{p,\delta}\|W_{s,u} - \widehat{W}^N_{s,u}\|_{p+\delta}|t-u|^{\frac{1}{2}}.$$

Using Inequality (33) from Lemma 3, we get for every $p > 2$, every $\rho \in (0, \frac{1}{2})$, every $\varepsilon \in (0, 3)$, and every s, $t \in [0, T]$, $s \leq t$,

$$\left\| W^i_{s,t} - \widehat{W}^{i,N}_{s,t} \right\|_p \leq C_{\rho,p,T,d,\varepsilon}|t - s|^\rho (\log N)^{-(\frac{1}{2}-\rho)\wedge(\frac{3-\varepsilon}{2p})}.$$

Now,

$$\mathbb{E} \max_{\substack{0\leq k\leq 2^{n+m}-1 \\ 0\leq k'\leq 2^{n+m'}-1}} \left| W^{2,N}_{t^{n+m}_k,t^{n+m}_{k+1}} - \widehat{W}^{2,N}_{t^{n+m}_k,t^{n+m}_{k+1}} \right|^p \left| W_{t^{n+m}_{k'},t^{n+m}_{k'+1}} \right|^p$$

$$\leq \left(C_{\rho,p,T,d,\delta,\varepsilon}(\log N)^{-(\frac{1}{2}-\rho)\wedge(\frac{3-\varepsilon}{2(p+\delta)})}\right)^p 2^{(n+m)(1-\rho p)}2^{(n+m')(1-\frac{p}{2})}$$

and we use that $\rho > \frac{1}{p}$ and $p > 2$ to show that

$$\sum_{m,m'\geq 0} \max_{\substack{0\leq k\leq 2^{n+m}-1 \\ 0\leq k'\leq 2^{n+m'}-1}} \left\| \left| W^{2,N}_{t^{n+m}_k,t^{n+m}_{k+1}} - \widehat{W}^{2,N}_{t^{n+m}_k,t^{n+m}_{k+1}} \right|^p \left| W_{t^{n+m}_{k'},t^{n+m}_{k'+1}} \right| \right\|_p$$

$$\leq C_{\rho,p,T,d,\delta,\varepsilon}(\log N)^{-(\frac{1}{2}-\rho)\wedge\frac{3-\varepsilon}{2(p+\delta)}} 2^{n(\frac{2}{p}-(\frac{1}{2}+\rho))}.$$

Finally, we get

$$\mathbb{E}\left(\widetilde{Z}^{(2),N}_{\tilde{\theta}'}\right)^p \leq C_{\rho,p,T,d,\varepsilon,\delta}\left(\log N\right)^{-p(\frac{1}{2}-\rho)\wedge\frac{3-\varepsilon}{2(p+\delta)}}$$

as soon as $\tilde{\theta}' \in (0, \tilde{\theta})$ with $\tilde{\theta} = p(\rho + \frac{1}{2}) - 2$. Now, it follows by standard arguments that

$$\sup_{s,t\in[0,T]} |\widetilde{\mathbf{W}}^{2,N}_{s,t}| \leq \widetilde{Z}^{(2),N}_{\tilde{\theta}'}|t - s|^{\frac{\tilde{\theta}'}{p}}$$

so that, finally

$$\left\| \sup_{s,t\in[0,T]} \frac{|\widetilde{\mathbf{W}}^{2,N}_{s,t}|}{|t - s|^{\frac{\tilde{\theta}'}{p}}} \right\|_p \leq C_{\rho,p,T,d,\varepsilon,\delta}\left(\log N\right)^{-(\frac{1}{2}-\rho)\wedge\frac{3-\varepsilon}{2(p+\delta)}}.$$

$\square$

Now, we are in position to prove the main result of this section.

Proof of Theorem 5. (a) *Given Theorem 4, this amounts to proving that* $\|\mathbf{W}^2 - \widehat{\mathbf{W}}^{2,N}\|_{\frac{q}{2},Hol}$ *converges to* 0 *in every* $L^p(\mathbb{P})$. *This easily follows from Proposition 3(a)–(b).*

(b) *We inspect successively four cases to maximize* $\min(1 - \rho, \frac{3}{2p})$ *in* ρ *when it is possible.*

▷ $q \in (2, 4)$ and $p < \frac{7q}{2(q-2)}$. Let p' be defined by $\frac{1}{p'} = \frac{2(q-2)}{7q} + \frac{\alpha}{2}$ with $\alpha > 0$ small enough so that $p' > p$ and $\frac{1}{p'} + \frac{1}{q} < \frac{1}{2}$. Then set $\rho' = \frac{2}{q} + \frac{2}{p'} - \frac{1}{2} + \frac{\alpha}{2}$ (note that $\rho' > \frac{1}{p'}$). One checks that $\frac{1}{2} - \rho' = 1 - 2(\frac{1}{p'} + \frac{1}{q}) = \frac{3}{7}(1 - \frac{2}{q}) - \alpha \in (0, \frac{3-\varepsilon}{2(p'+\delta)} \wedge \frac{1}{2})$ at least for any small enough α, $\delta = \delta(\alpha, q) > 0$ and $\varepsilon = \varepsilon(\alpha, q) > 0$. Now, Proposition 3($c$) applied with $\tilde{\theta}' = \frac{2p'}{q} < p'(\rho' + \frac{1}{2}) - 2$ yields the announced asymptotic rate for $\left\| \|\mathbf{W}^2 - \widehat{\mathbf{W}}^{2,N}\|_{\frac{q}{2}, Hol} \right\|_p$, $p < p'$, since $L^p(\mathbb{P})$-norms are non-decreasing in p.

▷ $q \in (2, 4)$ and $p \geq \frac{7q}{2(q-2)}$. One sets the same specifications as above for ρ but with $p' = p$. Then $1/2 - \rho > \frac{3}{2p}$ and choose $\varepsilon = \varepsilon(q, \alpha) > 0$ and $\delta = \delta(q, \alpha) > 0$ small enough so that $\frac{3-\varepsilon}{2(p+\delta)} \leq \frac{3}{2p} + \alpha$.

▷ $q \in [4, 20/3)$. Then $\frac{7q}{2(q-2)} < \frac{2q}{q-4}$ and one checks that the cases $p \in (2, \frac{7q}{2(q-2)})$ and $p \in [\frac{7q}{2(q-2)}, \frac{2q}{q-4})$ can be solved as above. If $p \geq \frac{2q}{q-4}$ (hence ≥ 5), no optimization in ρ is possible i.e. any admissible ρ satisfies $\frac{1}{2} - \rho > \frac{3}{2p}$.

▷ $q \geq 20/3$ i.e. $\frac{7q}{2(q-2)} > \frac{2q}{q-4}$. If $p < \frac{2q}{q-4}$, set p' such that $\frac{1}{p'} = \frac{q-4}{2q} + \alpha'/2$, $\alpha' > 0$ small enough and $\rho' = \frac{2}{q} + \frac{2}{p'} - \frac{1}{2} + \frac{\alpha}{2}$. Doing as above yields $\min(1 - \rho, \frac{3}{2p}) = \frac{2}{q} + \alpha$ for an arbitrary small $\alpha > 0$. Note that this quantity is greater than $\frac{3}{7}(1 - \frac{2}{q}) + \alpha$ (so in that case our exponent is not optimal). If $p \geq \frac{2q}{q-4}$, we proceed to no optimization in ρ.

(c) This is a consequence of Borel-Cantelli's Lemma by considering $p > \frac{7q}{q-2}$. □

Now we conclude by proving Theorem 2.

Proof (Proof of Theorem 2). First we check using Proposition 3 that $\rho_q(\widehat{\mathbf{W}}^N, 0)$ and $\rho_q(\mathbf{W}, 0)$ are *a.s.* finite since they are integrable. Now we may apply Theorem 1 which yields the announced result. □

Acknowledgements We thank A. Lejay for helpful discussions and comments about several versions of this work and F. Delarue and S. Menozzi for initiating our first "meeting" with rough path theory.

References

1. Billingsley, P.: Convergence of probability measure. Wiley Series in Probability and Mathematical Statistics. Wiley, New York, 253 p. (1968)
2. Coutin, L., Victoir, N.: Enhanced Gaussian processes and applications. ESAIM P&S **13**, 247–269 (2009)
3. Cuesta-Albertos, J.A., Matrán, C.: The strong law of large numbers for k-means and best possible nets of Banach valued random variables. Probab. Theory Relat. Field. **78**, 523–534 (1988)
4. Dereich, S.: The coding complexity of diffusion processes under $L^p[0, 1]$-norm distortion, preprint. Stoch. Process. Appl. **118**(6), 938–951 (2008)

5. Dereich S., Fehringer F., Matoussi A., Scheutzow M.: On the link between small ball probabilities and the quantization problem for Gaussian measures on Banach spaces. J. Theor. Probab. **16**, 249–265 (2003)
6. Friz, P.: Continuity of the Itô-map for Hölder rough paths with applications to the support theorem in Hölder norm. Probability and partial differential equations in modern applied mathematics, IMA Vol. Math. Appl. vol. 140, pp. 117–135. Springer, New York (2005)
7. Friz, P., Victoir, N.: Differential equations driven by Gaussian signals. Ann. Inst. Henri Poincaré Probab. Stat. **46**(2), 369–413 (2010)
8. Friz, P., Victoir, N.: Differential equations driven by Gaussian signals II. Available at arxiv:0711.0668 (2007)
9. Friz, P., Victoir, N.: Multidimensional stochastic differential equations as rough paths: theory and applications, Cambridge Studies in Advanced Mathematics, 670 p. (2010)
10. Graf, S., Luschgy, H.: Foundations of quantization for probability distributions. Lecture Notes in Mathematics, vol. 1730. Springer, Berlin (2000)
11. Graf, S., Luschgy, H., Pagès, G.: Functional quantization and small ball probabilities for Gaussian processes. J. Theor. Probab. **16**(4), 1047–1062 (2003)
12. Graf, S., Luschgy, H., Pagès, G.: Optimal quantizers for Radon random vectors in a Banach space. J. Approx. Theor. **144**, 27–53 (2007)
13. Graf, S., Luschgy, H., Pagès, G.: Distortion mismatch in the quantization of probability measures. ESAIM P&S **12**, 127–153 (2008)
14. Lejay, A.: An introduction to rough paths. Séminaire de Probabilités YXXVII. Lecture Notes in Mathematics, vol. 1832, pp. 1–59 (2003)
15. Lejay, A.: Yet another introduction to rough paths. Séminaire de Probabilités XLII. Lecture Notes in Mathematics, vol. 1979, pp. 1–101 (2009)
16. Lejay, A.: On rough differential equations. Electron. J. Probab. **14**(12), 341–364 (2009)
17. Lejay, A.: Global solutions to rough differential equations with unbounded vector fields, prepub. INRIA 00451193 (2010)
18. Luschgy, H., Pagès, G.: Functional quantization of stochastic processes. J. Funct. Anal. **196**, 486–531 (2002)
19. Luschgy, H., Pagès, G.: Sharp asymptotics of the functional quantization problem for Gaussian processes Ann. Probab. **32**, 1574–1599 (2004)
20. Luschgy, H., Pagès, G.: Functional quantization of a class of Brownian diffusions: a constructive approach. Stoch. Process. Appl. **116**, 310–336 (2006)
21. Luschgy, H., Pagès, G.: High resolution product quantization for Gaussian processes under sup-norm distortion. Bernoulli **13**(3), 653–671 (2007)
22. Luschgy, H., Pagès, G.: Functional quantization and mean pathwise regularity with an application to Lévy processes. Ann. Appl. Probab. **18**(2), 427–469 (2008)
23. Luschgy, H., Pagès, G., Wilbertz, B.: Asymptotically optimal quantization schemes for Gaussian processes. ESAIM P&S **12**, 127–153 (2008)
24. Lyons, T.: Interpretation and solutions of ODE's driven by rough signals. Proc. Symp. Pure 1583 (1995)
25. Lyons, T.: Differential Equations driven by rough signals. Rev. Mat. Iberoamericana **14**(2), 215–310 (1998)
26. Lyons, T., Caruana, M.J., Lévy, T.: Differential equations driven by rough paths. Lecture Notes in Mathematics, vol. 1908, 116 p. (Notes from T. Lyons's course at École d'été de Saint-Flour (2004).) (2007)
27. Pagès, G., Printems, J.: Functional quantization for numerics with an application to option pricing. Monte Carlo Methods Appl. **11**(4), 407–446 (2005)
28. Pagès, G., Printems, J.: Website devoted to vector and functional optimal quantization: www.quantize.maths-fi.com (2005)
29. Revuz, D., Yor, M.: Continuous martingales and Brownian motion, 3rd edn. Grundlehren der Mathematischen Wissenschaften [Fundamental Principles of Mathematical Sciences], **293**, 602 p. Springer, Berlin (1999)

Appendix: Functional Conditional Expectation

Let $(Y_t)_{t\in[0,T]}$ be a bi-measurable process defined on a probability space $(\Omega, \mathcal{A}, \mathbb{P})$ such that

$$\int_0^T \mathbb{E}(Y_t^2)dt < +\infty.$$

One can consider Y as a random variable $Y : (\Omega, \mathcal{A}, \mathbb{P}) \to L_T^2 := L^2([0,T], dt)$ and more precisely as an element of the Hilbert space

$$L_{L_T^2}^2(\Omega, \mathcal{A}, \mathbb{P}) := \left\{ Y : (\Omega, \mathcal{A}, \mathbb{P}) \to L_T^2, \ \mathbb{E}|Y|_{L_T^2}^2 < +\infty \right\}$$

where $|f|_{L_T^2}^2 = \int_0^T f^2(t)dt$. For the sake of simplicity, one denotes $\|Y\|_2 := \sqrt{\mathbb{E}|Y|_{L_T^2}^2}$. If $\mathcal{B}$ denotes a sub-σ-field of $\mathcal{A}$ (containing all $\mathbb{P}$-negligible sets of $\mathcal{A}$) then $L_{L_T^2}^2(\Omega, \mathcal{B}, \mathbb{P})$ is a closed sub-space of $L_{L_T^2}^2(\Omega, \mathcal{A}, \mathbb{P})$ and one can define the *functional conditional expectation* of Y by

$$\mathbb{E}(Y \mid \mathcal{B}) := \mathrm{Proj}_{L_{L_T^2}^2(\Omega, \mathcal{B}, \mathbb{P})}^{\perp}(Y).$$

Functional conditional expectation can be extended to bi-measurable processes Y such that $\|Y\|_1 := \mathbb{E}|Y|_{L_T^1} < +\infty$ following the approach used for $\mathbb{R}^d$-valued random vectors. Then, $\mathbb{E}(Y \mid \mathcal{B})$ is characterized by: for every $\mathcal{B}([0,T]) \otimes \mathcal{B}$-bi-measurable process $Z = (Z_t)_{t\in[0,T]}$, bounded by 1,

$$\mathbb{E}\int_0^T Z_t\, Y_t\, dt = \mathbb{E}\int_0^T Z_t\, \mathbb{E}(Y \mid \mathcal{B})_t\, dt.$$

In particular, owing to the Fubini theorem, this implies that as soon as the process $(\mathbb{E}(Y_t \mid \mathcal{B}))_{t\in[0,T]}$ has a $\mathcal{B}([0,T]) \otimes \mathcal{B}$ bi-measurable version, the functional conditional expectation could be defined by setting

$$\mathbb{E}(Y \mid \mathcal{B})_t(\omega) = \mathbb{E}(Y_t \mid \mathcal{B})(\omega), \qquad (\omega, t) \in \Omega \times [0,T].$$

Examples: (*a*) Let $\mathcal{B} := \sigma(\mathcal{N}_{\mathcal{A}}, B_i, \ i \in I)$ where $(B_i)_{i\in I}$ is a finite measurable partition of Ω such that $\mathbb{P}(B_i) > 0, i \in I$.
(*b*) Let $Y := (W_t)_{t\in[0,T]}$ a standard Brownian motion in $\mathbb{R}^d$ and let $\mathcal{B} := \sigma(W_{t_1}, \ldots, W_{t_n})$ where $0 = t_0 < t_1 < \ldots < t_n = T$. Then

$$\forall t \in [t_k, t_{k+1}), \qquad \mathbb{E}(W \mid \mathcal{B})_t = W_{t_k} + \frac{t - t_k}{t_{k+1} - t_k}(W_{t_{k+1}} - W_{t_k}).$$

Asymptotic Cramér's Theorem and Analysis on Wiener Space

Ciprian A. Tudor

Abstract We prove an asymptotic Cramér's theorem, that is, if the sequence $(X_n + Y_n)_{n \geq 1}$ converges in law to the standard normal distribution and for every $n \geq 1$ the random variables X_n and Y_n are independent, then $(X_n)_{n \geq 1}$ *and* $(Y_n)_{n \geq 1}$ converge in law to a normal distribution. Then we compare this result with recent criteria for the central convergence obtained in terms of Malliavin derivatives.

Keywords Multiple stochastic integrals · Limit theorems · Malliavin calculus · Stein's method

2000 AMS Classification Numbers: 60G15, 60H05, 60F05, 60H07

1 Introduction

The sum of two independent random variables with Gaussian distribution is a Gaussian random variable. A famous result by Harald Cramér [1] says that the converse implication is also true. Namely, if the law of $X + Y$ is Gaussian and X and Y are independent random variables, then X and Y are Gaussian. We study in this paper the following problem: given two sequences of centered square integrable random variables $(X_n)_{n \geq 1}$ and $(Y_n)_{n \geq 1}$ such that $\mathbf{E}X_n^2 \to_{n \to \infty} c_1$ and $\mathbf{E}Y_n^2 \to_{n \to \infty} c_2$ with $c_1, c_2 > 0$ and $c_1 + c_2 = 1$ and assuming that for every $n \geq 1$, X_n and Y_n are independent and $X_n + Y_n \to_{n \to \infty} N(0, 1)$ in law, can we get the convergence of X_n to the normal law $N(0, c_1)$ *and* the convergence of Y_n to the normal law $N(0, c_2)$? We will say in this case that the central limit of the sum is decoupled. A partial answer has been given in [9]: here the authors proved that the central limit for the sum implies the central limit for each term when the random variables X_n and Y_n lives in a Wiener chaos of fixed order. In this work we will prove this result for a very general class of random variables.

C.A. Tudor (✉)
Laboratoire Paul Painlevé, Université de Lille 1, 59655 Villeneuve d'Ascq, France
e-mail: tudor@math.univ-lille1.fr

C. Donati-Martin et al. (eds.), *Séminaire de Probabilités XLIII*, Lecture Notes in Mathematics 309
2006, DOI 10.1007/978-3-642-15217-7_12, © Springer-Verlag Berlin Heidelberg 2011

Then we will try to understand this asymptotic Cramér's theorem from the perspective of some recent ideas from [3] and [4] related to the Stein's method on Wiener space and some older ideas from [8, 11, 12] where the independence of random variables is characterized in terms of the Malliavin derivatives. Let $(\Omega, \mathcal{F}, P)$ be a probability space and let $(W_t)_{t \in [0,1]}$ be a Wiener process on this space. Recall that a result in [3] says that a sequence of Malliavin differentiable (with respect to W) random variables X_n (defined on Ω) converges to the normal law $N(0, 1)$ if and only if

$$\mathbf{E}\left(f_z'(X_n)(1 - \langle DX_n, D(-L)^{-1}X_n\rangle)\right) \to_{n \to \infty} 0$$

where we denoted by D the Malliavin derivative with respect to W, by L the Ornstein-Uhlenbeck generator and by f_z the solution of the Stein's equation (for fixed $z \in \mathbb{R}$)

$$1_{(-\infty, z]}(x) - P(Z \le z) = f'(x) - xf(x), \qquad x \in \mathbb{R}. \tag{1}$$

(Throughout this paper we denote by $\langle \cdot, \cdot \rangle$ the scalar product in $L^2([0,1])$.) In particular, if $\mathbf{E}\left(1 - \langle DX_n, D(-L)^{-1}X_n\rangle\right)^2 \to_{n \to \infty} 0$ then X_n converges to $N(0, 1)$ as $n \to \infty$ by using Schwarz's inequality and the fact that f_z' is bounded (actually it suffices to have $\mathbf{E}\left|1 - \langle DX_n, D(-L)^{-1}X_n\rangle\right| \to_{n \to \infty} 0$).

Let us describe the basic idea to treat the convergence of sums of independent random variables to the normal law. Let X_n, Y_n be two sequences as above (that means Malliavin differentiable with $\mathbf{E}X_n^2 \to_{n \to \infty} c_1 > 0$, $\mathbf{E}Y_n^2 \to_{n \to \infty} c_2 > 0$ and $c_1 + c_2 = 1$). The fact that $X_n + Y_n \to_{n \to \infty} N(0, 1)$ (in law) implies that

$$\mathbf{E}\left(f_z'(X_n + Y_n)(1 - \langle D(X_n + Y_n), D(-L)^{-1}(X_n + Y_n)\rangle)\right) \to_{n \to \infty} 0. \tag{2}$$

Suppose now that X_n and Y_n are independent for every n. A result by Üstunel and Zakai ([11], Theorem 3) says that in this case

$$\mathbf{E}(\langle DX_n, D(-L)^{-1}Y_n\rangle|X_n) = 0 \text{ and } \mathbf{E}(\langle DY_n, D(-L)^{-1}X_n\rangle|Y_n) = 0 \text{ a.s.} \tag{3}$$

The relation (3) induces the idea that the summands containing $\langle DX_n, D(-L)^{-1}Y_n\rangle$ and $\langle DY_n, D(-L)^{-1}X_n\rangle$ could be eliminated from (2). We will see that it is not immediate and that actually a stronger condition than the independence of X_n and Y_n is necessary in order to do this. Therefore our first step is to introduce some classes of independent random variables X, Y such that the "mixed" terms $\langle DX, D(-L)^{-1}Y\rangle$ and $\langle DY, D(-L)^{-1}X\rangle$ vanish. A first class that we consider here is the class of so-called *strongly independent* random variables for which every multiple integral in the chaos decomposition of X is independent of every multiple integral in the chaos decomposition of Y. We will see that if X and Y are strongly independent, then $\langle DX, D(-L)^{-1}Y\rangle = \langle DY, D(-L)^{-1}X\rangle = 0$ almost surely. Another class we consider is the class of random variables X, Y that are differentiable in the Malliavin sense and such that X is independent of the couple $(Y, \langle DY, D(-L)^{-1}Y\rangle)$ and Y is independent of the couple $(X, \langle DX, D(-L)^{-1}X\rangle)$. We will say in this case that the couple (X, Y) belongs to the class $\mathcal{A}$. A couple of strongly independent random variables belongs to $\mathcal{A}$ and in this sense this class is an intermediary class between the independent and strongly independent random variables. For couples in $\mathcal{A}$ we will show that $\mathbf{E}(\langle DX, D(-L)^{-1}Y \rangle | X + Y) = \mathbf{E}(\langle DY, D(-L)^{-1}X \rangle | X + Y) = 0$ almost surely and it is then again possible to cancel the mixed terms in (2). We will prove, by an elementary argument coming from the original Cramér's theorem and without using Malliavin calculus, that for independent random variables the asymptotic Cramér's theorem holds. But for random variables in these classes (in the class $\mathcal{A}$ or strongly independent) we can give further results by using the tools of the Malliavin calculus. Concretely, we will treat the following problem: suppose that the sum $X_n + Y_n$ converges to the normal law in a strong sense, that is, the upper bound $\mathbf{E}\left(1 - \langle D(X_n + Y_n), D(-L)^{-1}(X_n + Y_n) \rangle\right)^2$ converges to zero as $n \to \infty$. We can interpret this by saying that the sum $X_n + Y_n$ is "close" to $N(0, 1)$, not in the sense of the rate of convergence but in the sense that $X_n + Y_n$ belongs to a subset of the set of the sequences of random variables converging to $N(0, 1)$. Then can we obtain the strong convergence of X_n and Y_n to the normal law, that is $\mathbf{E}\left(c_1 - G_{X_n}\right)^2 \to_{n\to\infty} 0$ and $\mathbf{E}\left(c_2 - G_{Y_n}\right)^2 \to_{n\to\infty} 0$, where G_{X_n} is given by (12)? We prove that this property is true for strongly independent random variables while for couples in the class $\mathcal{A}$ a supplementary assumption is necessary in order to ensure the strong convergence of X_n and Y_n from the convergence of $X_n + Y_n$.

The organization of the paper is as follows. Section 2 contains preliminaries on the stochastic calculus of variations. In Sect. 3 we prove the asymptotic Cramér's theorem by using an elementary argument while Sect. 4 contains some thoughts on this theorem from the perspective of recent results on central limit theorem obtained via Malliavin calculus.

2 Preliminaries

Let $(W_t)_{t\in[0,1]}$ be a classical Wiener process on a standard Wiener space $(\Omega, \mathcal{F}, \mathbf{P})$. If $f \in L^2([0, 1]^n)$ with $n \geq 1$ integer, we introduce the multiple Wiener-Itô integral of f with respect to W. The basic references are the monographs [2] or [6]. Let $f \in \mathcal{S}_n$ be an elementary function with n variables that can be written as

$$f = \sum_{i_1,\dots,i_n} c_{i_1,\dots,i_n} 1_{A_{i_1} \times \dots \times A_{i_n}}$$

where the coefficients satisfy $c_{i_1,\dots,i_n} = 0$ if two indices i_k and i_l are equal and the sets $A_i \in \mathcal{B}([0, 1])$ are pairwise disjoint. For such a step function f we define

$$I_n(f) = \sum_{i_1,\dots,i_n} c_{i_1,\dots i_n} W(A_{i_1}) \dots W(A_{i_n})$$

where we put $W(A) = \int_0^1 1_A(s)dW_s$. It can be seen that the application I_n constructed above from $\mathcal{S}_n$ to $L^2(\Omega)$ is an isometry on $\mathcal{S}_n$, i.e.

$$\mathbf{E}[I_n(f)I_m(g)] = n!\langle f,g\rangle_{L^2([0,1]^n)} \text{ if } m = n \tag{4}$$

and

$$\mathbf{E}[I_n(f)I_m(g)] = 0 \text{ if } m \neq n.$$

Since the set $\mathcal{S}_n$ is dense in $L^2([0,1]^n)$ for every $n \geq 1$ the mapping I_n can be extended to an isometry from $L^2([0,1]^n)$ to $L^2(\Omega)$ and the above properties hold true for this extension. It also holds that

$$I_n(f) = I_n(\tilde{f}) \tag{5}$$

where $\tilde{f}$ denotes the symmetrization of f defined by $\tilde{f}(x_1,\ldots,x_n) = \frac{1}{n!}\sum_{\sigma\in\mathbf{S}_n} f(x_{\sigma(1)},\ldots,x_{\sigma(n)})$. We will need the general formula for calculating products of Wiener chaos integrals of any orders m,n for any symmetric integrands $f \in L^2([0,1]^{\otimes m})$ and $g \in L^2([0,1]^{\otimes n})$; it is

$$I_m(f)I_n(g) = \sum_{\ell=0}^{p\wedge q} \ell! C_m^\ell C_n^\ell I_{m+n-2\ell}(f \otimes_\ell g) \tag{6}$$

where the contraction $f \otimes_\ell g$ is defined by

$$(f \otimes_\ell g)(s_1,\ldots,s_{m-\ell},t_1,\ldots,t_{n-\ell})$$
$$= \int_{[0,T]^{m+n-2\ell}} f(s_1,\ldots,s_{m-\ell},u_1,\ldots,u_\ell)$$
$$g(t_1,\ldots,t_{n-\ell},u_1,\ldots,u_\ell)du_1\ldots du_\ell. \tag{7}$$

Note that the contraction $(f \otimes_\ell g)$ is an element of $L^2([0,1]^{m+n-2\ell})$ but it is not necessary symmetric. We will denote by $(f\tilde{\otimes}_\ell g)$ its symmetrization.

We recall that any square integrable random variable which is measurable with respect to the σ-algebra generated by W can be expanded into an orthogonal sum of multiple stochastic integrals

$$F = \sum_{n\geq 0} I_n(f_n) \tag{8}$$

where $f_n \in L^2([0,1]^n)$ are (uniquely determined) symmetric functions and $I_0(f_0) = \mathbf{E}[F]$.

We denote by D the Malliavin derivative operator that acts on smooth functionals of the form $F = g(W(\varphi_1),\ldots,W(\varphi_n))$ (here g is a smooth function with compact support and $\varphi_i \in L^2([0,1])$ for $i = 1,..,n$)

$$DF = \sum_{i=1}^n \frac{\partial g}{\partial x_i}(W(\varphi_1),\ldots,W(\varphi_n))\varphi_i.$$

We can define the i th Malliavin derivative $D^{(i)}$ defined iteratively. The operator $D^{(i)}$ can be extended to the closure $\mathbb{D}^{p,2}$ of smooth functionals with respect to the norm

$$\|F\|_{p,2}^2 = \mathbf{E}F^2 + \sum_{i=1}^{p} \mathbf{E}\|D^i F\|_{L^2([0,1]^i)}^2$$

The adjoint of D is denoted by δ and is called the divergence (or Skorohod) integral. Its domain $Dom(\delta)$ coincides with the class of stochastic processes $u \in L^2(\Omega \times [0,1])$ such that

$$|\mathbf{E}\langle DF, u \rangle| \leq c\|F\|_2$$

for all $F \in \mathbb{D}^{1,2}$ and $\delta(u)$ is the element of $L^2(\Omega)$ characterized by the duality relationship

$$\mathbf{E}(F\delta(u)) = \mathbf{E}\langle DF, u \rangle.$$

For adapted integrands, the divergence integral coincides with the classical Itô integral.

Let L be the Ornstein–Uhlenbeck operator defined on $Dom(L) = \mathbb{D}^{2,2}$

$$LF = -\sum_{n \geq 0} n I_n(f_n)$$

if F is given by (8). There exists a connection between δ, D and L in the sense that a random variable F belongs to the domain of L if and only if $F \in \mathbb{D}^{1,2}$ and $DF \in Dom(\delta)$ and then $\delta DF = -LF$. Also we will need in the paper the integration by parts formula

$$F\delta(u) = \delta(Fu) + \langle DF, u \rangle \tag{9}$$

whenever $F \in \mathbb{D}^{1,2}$, $u \in Dom(\delta)$ and $\mathbf{E}F^2 \int_0^1 u_s^2 ds < \infty$.

3 Asymptotic Cramér's Theorem

We start by proving an asymptotic version of the Cramér's theorem [1]. A particular case (when the sequences X_n and Y_n are multiple integrals in a Wiener chaos of fixed order) has been proven in [9], Corollary 1. Our proof is based on the Cramér's theorem (see [1]) and an idea from [7].

Theorem 1. *Suppose that* $(X_n)_{n \geq 1}$ *and* $(Y_n)_{n \geq 1}$ *are two sequences of centered random variables in* $L^2(\Omega)$ *such that* $\mathbf{E}X_n^2 \to_{n \to \infty} c_1 > 0$ *and* $\mathbf{E}Y_n^2 \to_{n \to \infty} c_2 > 0$ *with* $c_1 + c_2 = 1$. *Assume that for every* $n \geq 1$, *the random variables* X_n *and* Y_n *are independent. Then*

$$X_n + Y_n \to N(0, 1) \Leftrightarrow (X_n \to N(0, c_1) \text{ and } Y_n \to N(0, c_2)).$$

Proof. One direction is trivial. Let us assume that $X_n + Y_n \to_{n\to\infty} N(0, 1)$. We will prove that $X_n \to_{n\to\infty} N(0, c_1)$ and $Y_n \to_{n\to\infty} N(0, c_2)$. Since $\mathbf{E}X_n^2 \to_{n\to\infty} c_1$ and $\mathbf{E}Y_n^2 \to_{n\to\infty} c_2$ it follows that the sequence $(X_n, Y_n)_{n\geq 1}$ is bounded in $L^2(\Omega)$. By Prohorov's theorem it suffices to prove that for any subsequence which converges in distribution to some random vector (F, G), then we must have $F \sim N(0, c_1), G \sim N(0, c_2)$ and F, G are independent. Let us consider such an arbitrary sequence (X_{n_k}, Y_{n_k}) which converges in law to (F, G) as $k \to \infty$. Because X_{n_k} and Y_{n_k} are independent for each k, it is clear that F and G are independent. Since $X_n + Y_n \to_{n\to\infty} N(0, 1)$ it follows that $F + G \sim N(0, 1)$.

Cramér's theorem implies that $F \sim N(0, c_1)$ and $G \sim N(0, c_2)$. $\qquad \square$

This result can be extended to finite and even infinite sums of independent random variables.

Proposition 1. *Suppose that for every $n \geq 1$, $X^n = \sum_{k\geq 1} X_k^n$ where for every n the random variables $X_n^k, k \geq 1$ are mutually independent and the series is convergent for every ω. Assume also that X_k^n are centered for every $n, k \geq 1$ and $\mathbf{E}(X_k^n)^2 \to_{n\to\infty} c_k > 0$ for every $k \geq 1$. Suppose that X^n converging in law to $N(0, 1)$ as $n \to \infty$. Then for every $k \geq 1$ the sequence X_k^n converges to the normal law as $n \to \infty$.*

Proof. Since $X^n = X_1^n + \sum_{k\geq 2} X_k^n$ and the two summands are independent, Theorem 1 implies that X_1^n converges to the normal law. Inductively, the conclusion can be obtained. $\qquad \square$

Remark 1. When, for every $n \geq 1$, $X_n = I_{k_1}(f^n)$ and $Y_n = I_{k_2}(g^n)$ are multiple stochastic integrals (possibly of different orders, that can also vary with n) we can go further by proving the following result. If $\mathbf{E}(X_n + Y_n)^2 \to_{n\to\infty} 1$ and $X_n + Y_n$ converges in law to $N(0, 1)$, if $\varliminf_n \mathbf{E}X_n^2 > 0$ and $\varliminf_n \mathbf{E}Y_n^2 > 0$ then

$$d_{Kol}(X_n, N(0, \mathbf{E}X_n^2)) \to_{n\to\infty} 0 \text{ and } d_{Kol}Y_n, N(0, \mathbf{E}Y_n^2) \to_{n\to\infty} 0 \quad (10)$$

Here d_{Kol} means the Kolmogorov distance (recall that the Kolmogorov distance between the law of two random variables U and V is given by $d_{Kol}(U, V) = \sup_{x\in\mathbb{R}} |P(U \leq x) - P(V \leq x)|$). That is, there is an asymptotic Cramér's theorem even if the variances of X_n and Y_n do not converge a priori. Relation (10) can be proved as follows. First, recall the following bound when X lives in a chaos of fixed order (see e.g. [5])

$$d_{Kol}X, N(0, \mathbf{E}X^2) \leq \frac{\left(|\mathbf{E}X^4 - 3(\mathbf{E}X^2)^2|\right)^{\frac{1}{2}}}{\mathbf{E}X^2} =: \frac{(|k_4(X)|)^{\frac{1}{2}}}{\mathbf{E}X^2} \quad (11)$$

where $k_4(X)$ is the fourth cumulant of X. It is immediate, by the definition of the cumulant, that $k_4(X+Y) = k_4(X)+k_4(Y)$ if X and Y are independent. Moreover, it follows from [5], identity (3.31) that $k_4(X) \geq 0$ if X is a multiple integral. Hence, if $\mathbf{E}(X_n + Y_n)^2 \to_{n\to\infty} 1$ and $X_n + Y_n$ converges in law to $N(0, 1)$, then

$$k_4(X_n) + k_4(Y_n) = k_4(X_n + Y_n) = \mathbf{E}(X_n + Y_n)^4 - 3(\mathbf{E}(X_n + Y_n)^2)^2 \to_{n \to \infty} 0$$

and this implies that $k_4(X_n) \to_{n \to \infty} 0$ and $k_4(Y_n) \to_{n \to \infty} 0$. The convergence (10) is obtained by using (11) and the hypothesis $\underline{\lim}_n \mathbf{E}X_n^2 > 0$ and $\underline{\lim}_n \mathbf{E}Y_n^2 > 0$.

4 Decoupling Central Limit Under Strong Independence

Let us regard Theorem 1 from the perspective of the results in [3]. In this part all random variables are centered. We recall some facts related to the convergence of a sequence of random variables to the normal law in terms of the Malliavin calculus. For any random variable $X \in \mathbb{D}^{1,2}$ we denote by

$$G_X := \langle DX, D(-L)^{-1}X \rangle. \tag{12}$$

The following result is a slight extension of Proposition 3.1 in [3]. See also Theorem 3 in [10].

Proposition 2. Let $(X_n)_{n \geq 1}$ be a sequence of square integrable random variables such that $\mathbf{E}X_n^2 \to_{n \to \infty} c > 0$. Then the following are equivalent:

1. The sequence $(X_n)_{n \geq 1}$ converges in law, an $n \to \infty$, to the normal random variable $N(0, c)$, $c > 0$
2. For every $t \in \mathbb{R}$, $\mathbf{E}\left(e^{itX_n}(c - G_{X_n})\right) \to_{n \to \infty} 0$
3. $\mathbf{E}\left((c - G_{X_n})|X_n\right) \to_{n \to \infty} 0$ a.s.
4. For every $z \in \mathbb{R}$, $\mathbf{E}\left(f_z'(X_n)(c - G_{X_n})\right) \to_{n \to \infty} 0$

Proof. We follow the scheme $1. \Rightarrow 2. \Rightarrow 3. \Rightarrow 4. \Rightarrow 1.$ The implications $1. \Rightarrow 2.$ and $3. \Rightarrow 4. \Rightarrow 1$ follow exactly as in [10], Theorem 3. Concerning $2. \Rightarrow 3.$, set $F_n = c - G_{X_n}$ for every $n \geq 1$. The random variable $\mathbf{E}(F_n|X_n)$ is the Radon–Nykodim derivative with respect to P of the measure $Q_n(A) = \mathbf{E}(F_n 1_A)$, $A \in \sigma(X_n)$. Relation 2. means that $\mathbf{E}\left(e^{itX_n}\mathbf{E}(F_n/X_n)\right) = \mathbf{E}_{Q_n}(e^{itX_n}) \to_{n \to \infty} 0$ and hence $\int_{\mathbb{R}} e^{ity}d(Q_n \circ X_n^{-1})(y) \to_{n \to \infty} 0$. This implies that $Q_n(A) = \mathbf{E}(F_n 1_A) \to_{n \to \infty} 0$ for any $A \in \sigma(X_n)$ or $\mathbf{E}(F_n|X_n) \to_{n \to \infty} 0$.

As an immediate consequence we have (see also [3]).

Corollary 1. Suppose that $(X_n)_{n \geq 1}$ is a sequence of random variables such that $\mathbf{E}X_n^2 \to_{n \to \infty} c$. suppose that

$$\mathbf{E}(c - G_{X_n})^2 \to_{n \to \infty} 0. \tag{13}$$

Then $X_n \to_{n \to \infty} N(0, c)$.

Remark 2. In the case when the variables X_n live in a fixed Wiener chaos, $X_n = I_k(f_n)$, then the convergence in distribution of X_n to the normal law is equivalent to (13), see [7].

Assume that $(X_n)_{n \geq 1}$ and $(Y_n)_{n \geq 1}$ are two sequences of random variables such that: (a) for every $n \geq 1$ the random variables X_n and Y_n are independent and (b) $X_n + Y_n \to N(0, 1)$ in distribution as $n \to \infty$. The quantity $G_{X_n + Y_n}$, which plays a central role, can be written as

$$G_{X_n + Y_n} = G_{X_n} + G_{Y_n} + \langle DX_n, D(-L)^{-1}Y_n \rangle + \langle DY_n, D(-L)^{-1}X_n \rangle.$$

The force of the Cramér's theorem can be observed here: the fact that

$$\mathbf{E}\left(c_1 - G_{X_n} + c_2 - G_{X_n}\right.$$
$$\left. -\langle DX_n, D(-L)^{-1}Y_n \rangle - \langle DY_n, D(-L)^{-1}X_n \rangle | X_n + Y_n\right)$$

converges to zero implies that $\mathbf{E}(c_1 - G_{X_n}|X_n)$ and $\mathbf{E}(c_2 - G_{Y_n}|Y_n)$ both converge to zero. It is not obvious to prove this directly. Note also that the independence of X_n and Y_n does not guarantee a priori that the terms $\mathbf{E}(\langle DY_n, D(-L)^{-1}X_n \rangle | X_n + Y_n)$ $\mathbf{E}(\langle DX_n, D(-L)^{-1}Y_n \rangle | X_n + Y_n)$ vanish. But the situation when these two terms vanish is also interesting and we will analyze this case in the sequel. We will see that it requires a slightly stronger assumption than the independence of X_n and Y_n. We introduce the following concept.

Definition 1. Two random variables $X = \sum_{n \geq 0} I_n(f_n)$ and $Y = \sum_{m \geq 0} I_m(g_m)$ are called *strongly independent* if for every $m, n \geq 0$, the random variables $I_n(f_n)$ and $I_m(g_m)$ are independent.

Remark 3. Let us recall the criterion for the independence of two multiple integrals given in [11]: Let $X' = I_p(f)$ and $Y' = I_q(g)$ where $f \in L^2([0, 1]^p)$ and $g \in L^2([0, 1]^q)$ $(p, q \geq 1)$ are symmetric functions. Then X' and Y' are independent if and only if

$$f \otimes_1 g = 0 \text{ almost everywhere on } [0, 1]^{p+q-2}.$$

As a consequence two random variables X and Y as in Definition 1 are strongly independent if and only if for every $m, n \geq 1$, $f_n \otimes_1 g_m = 0$ almost everywhere on $[0, 1]^{m+n-2}$.

Let us also note that the class of strongly independent random variables is strictly included in the class of independent random variables. Indeed, consider

$$X_1 = \sqrt{2} I_1 \left(1_{[\frac{1}{2}, 1]}\right) \text{ and } Y_1 = \sqrt{2} \int_0^{\frac{1}{2}} sign(W_s) dW_s.$$

Then X_1 and Y_1 are independent standard normal random variables. Define

$$X = \frac{1}{\sqrt{2}}(X_1 + Y_1) \text{ and } Y = \frac{1}{\sqrt{2}}(X_1 - Y_1).$$

Then X, Y are also independent standard normal but they are not strongly independent because for example the chaoses of order one of X and Y are not independent (note that the random variable $\int_0^{\frac{1}{2}} sign(W_s)dW_s$ has only even order chaos components).

Lemma 1. *Assume that $X, Y \in \mathbb{D}^{1,2}$ and X, Y are strongly independent. Then*

$$\langle DX, D(-L)^{-1}Y \rangle = 0 \quad a.s. .$$

Proof. Suppose first that $X = I_n(f)$ and $Y = I_m(g)$. Then, since $D_\alpha X = nI_{n-1}(f(\cdot, \alpha))$ and $D_\alpha(-L)^{-1}Y = I_{m-1}(g(\cdot, \alpha))$, using (6)

$$\langle DX, D(-L)^{-1}Y \rangle = n \int_0^1 d\alpha I_{n-1}(f(\cdot, \alpha)) I_{m-1}(g(\cdot, \alpha)) = m \sum_{k=0}^{(m-1) \wedge (n-1)}$$

$$k! C_{m-1}^k C_{n-1}^k \int_0^1 d\alpha I_{m+n-2-2k}(f(\cdot, \alpha) \otimes_k g(\cdot, \alpha))$$

$$= \sum_{k=0}^{(m-1) \wedge (n-1)} k! C_{m-1}^k C_{n-1}^k I_{m+n-2-2k}(f \otimes_{k+1} g)$$

and this is equal to zero from the characterization of the independence of two multiple integrals (see Remark 3). The extension to the general case is immediate since, if $X = \sum_n I_n(f_n)$ and $Y = \sum_m I_m(g_m)$,

$$\langle DX, D(-L)^{-1}Y \rangle = \sum_{m,n} \langle DI_n(f_n), D(-L)^{-1}I_m(g_m) \rangle.$$

$\square$

In view of Lemma 1, the Proposition 2 can be formulated for strongly independent random variables as follows: Suppose that $(X_n)_{n \geq 1}$ and $(Y_n)_{n \geq 1}$ are two sequences of centered strongly independent random variables such that $\mathbf{E}X_n^2 \to_{n \to \infty} c_1$ and $\mathbf{E}Y_n^2 \to_{n \to \infty} c_2$ where $c_1, c_2 > 0$ are such that $c_1 + c_2 = 1$. Then the following affirmations are equivalent:

1. The sequence $(X_n + Y_n)_{n \geq 1}$ converges in law to a standard normal random variable as $n \to \infty$
2. For every $t \in \mathbb{R}$, $\mathbf{E}\left(e^{it(X_n+Y_n)}(c_1 - G_{X_n} + c_2 - G_{X_n})\right) \to_{n \to \infty} 0$
3. $\mathbf{E}(c_1 - G_{X_n} + c_2 - G_{Y_n} | X_n + Y_n) \to_{n \to \infty} 0$
4. For every $z \in \mathbb{R}$, $\mathbf{E}\left(f_z'(X_n + Y_n)(c_1 - G_{X_n} + c_2 - G_{Y_n})\right) \to_{n \to \infty} 0$

Let us assume now that the two sequences of Theorem 1 are strongly independent. We will also assume that the convergence of the sum $X_n + Y_n$ to $N(0, 1)$ is strong in the sense that $\mathbf{E}(1 - G_{X_n+Y_n})^2$ converges to zero as $n \to \infty$. We can say, somehow, that in this case the sum $X_n + Y_n$ is rather close to the normal law since the upper bound of the distance between it and $N(0, 1)$ goes to zero. We will prove that this implies that the convergence of X_n and Y_n to the normal law is also strong.

Remark 4. The case of multiple stochastic integrals can be easily understood. Suppose that $X_n = I_k(f^n)$ and $Y_n = I_l(g^n)$ where for every $n \geq 1$ the kernels f^n, g^n are in $L^2([0, 1]^k)$ and $L^2([0, 1]^l)$ respectively. Assume that $EX_n^2 \to_{n\to\infty} c_1 > 0$ and $EY_n^2 \to_{n\to\infty} c_2 > 0$ such that $c_1 + c_2 = 1$. Then if $X_n + Y_n \to_{n\to\infty} N(0, 1)$ and X_n, Y_n are independent (thus strongly independent) it follows that $X_n \to N(0, c_1)$ and $Y_n \to N(0, c_2)$ and by Remark 2, $E(c_1 - G_{X_n})^2 \to_{n\to\infty} 0$ and $E(c_2 - G_{Y_n})^2 \to_{n\to\infty} 0$, so the convergence of X_n and Y_n to the normal distribution is strong.

We will also need the following lemma.

Lemma 2. *Assume that $X, Y \in \mathbb{D}^{1,2}$ and X, Y are strongly independent. Then the random variables G_X and G_Y are strongly independent.*

Proof. Let us assume once again that $X = I_n(f)$ and $Y = I_m(g)$. The result can easily be extended to the general case. We have

$$G_X = n \sum_{k=0}^{n-1} \left(C_{n-1}^k\right)^2 I_{2n-2-2k}(f \otimes_{k+1} f)$$

and

$$G_Y = m \sum_{l=0}^{m-1} l! \left(C_{m-1}^l\right)^2 I_{2m-2-2l}(g \otimes_{l+1} g).$$

It suffices to show that for every $k = 1, .., n - 1$ and $l = 1, .., m - 1$ the random variables $I_{2n-2k}(f \otimes_k f)$ and $I_{2m-2l}(g \otimes_l g)$ are independent or equivalently

$$(f \tilde{\otimes}_k f) \otimes_1 (g \tilde{\otimes}_l g) = 0 \text{ a.e. on } [0, 1]^{2m-2k+2m-2l-2}.$$

But since

$$(f \tilde{\otimes}_k f)(x_1, .., x_{2n-2k})$$
$$= \sum_{\sigma \in S_{2n-2k}} \int_{[0,1]^k} f(u_1, .., u_k, x_{\sigma(1)}, .., x_{\sigma(n-k)})$$
$$f(u_1, .., u_k, x_{\sigma(n-k+1)}, .., x_{\sigma(2n-2k)}) du_1 .. du_k$$

and

$$(g \tilde{\otimes}_l g)(y_1, .., y_{2m-2l})$$
$$= \sum_{\rho \in S_{2m-2l}} \int_{[0,1]^l} g(v_1, .., v_l, y_{\rho(1)}, .., y_{\rho(m-l)})$$
$$g(v_1, .., v_l, y_{\rho(m-l+1)}, .., y_{\rho(2m-2l)}) dv_1 .. dv_l$$

then $(f \tilde{\otimes}_k f) \otimes_1 (g \tilde{\otimes}_l g) = 0$ almost everywhere on $[0, 1]^{2m-2k+2m-2l-2}$ by using Fubini and the fact that

$$\int_0^1 d\alpha f(u_1, .., u_k, x_1, .., x_{n-k-1}, \alpha) g(v_1, .., v_l, y_1, .., y_{m-l+1}, \alpha) = 0$$

for almost all u_1, v_i, x_i, y_i. The general case demands to prove that $(f_n \tilde{\otimes}_k f_{n'}) \otimes_1 (g_m \tilde{\otimes}_l g_{m'}) = 0$ almost everywhere for every $n, n', m, m' \geq 1$ and for every $k = 1, .., n \wedge n'$ and $l = 1, .., m \wedge m'$ and this can be done similarly as above (note that the fact that $k, l \geq 1$ and the value zero is excluded is essential for the proof). $\square$

Proposition 3. *Suppose that $(X_n)_{n\geq 1}$ and $(Y_n)_{n\geq 1}$ are two sequences of centered strongly independent random variables such that $\mathbf{E} X_n^2 \to_{n\to\infty} c_1$ and $\mathbf{E} Y_n^2 \to_{n\to\infty} c_2$ where $c_1, c_2 > 0$ are such that $c_1 + c_2 = 1$. Then $\mathbf{E}(1 - G_{X_n+Y_n})^2 \to_{n\to\infty} 0$ if and only if*

$$\mathbf{E}(c_1 - G_{X_n})^2 \to_{n\to\infty} 0 \text{ and } \mathbf{E}(c_2 - G_{Y_n})^2 \to_{n\to\infty} 0.$$

Proof. By using Lemmas 1 and 2 we have

$$\mathbf{E}(1 - G_{X_n+Y_n})^2 = \mathbf{E}(c_1 - G_{X_n})^2 + \mathbf{E}(c_2 - G_{Y_n})^2$$

and the conclusion is immediate. $\square$

We will study next if the result in Proposition 3 can be obtained by relaxing the hypothesis on the strong independence of X_n and Y_n for every n. As we have seen, the strong independence of two variables X and Y implies that $\langle DX, D(-L)^{-1} Y \rangle = \langle DY, D(-L)^{-1} X \rangle = 0$ a.s. But in order to eliminate the "mixed" terms we only need $\mathbf{E}(\langle DX, D(-L)^{-1} Y \rangle | X + Y) = \mathbf{E}(\langle DY, D(-L)^{-1} X \rangle | X + Y) = 0$ a.s. We therefore introduce an intermediary class between the class of independent random variables and the class of strongly independent random variables for which this property holds.

Definition 2. We will say that a couple (X, Y) of two random variables in the space $\mathbb{D}^{1,2}$ belongs to the class $\mathcal{A}$ if the vector X is independent of the vector (Y, G_Y) and Y is independent of the vector (X, G_X).

We will give now examples of random variables in $\mathcal{A}$. First we recall the following result from [11].

Lemma 3. *Let $X \in \mathbb{D}^{1,2}$ and $Y, Z \in L^2(\Omega)$. Then X is independent of the pair (Y, Z) if and only if for every $\alpha, \beta \in \mathbb{R}$*

$$\mathbf{E}\left(\langle DX, D(-L)^{-1} e^{i(\alpha Z + \beta Y)}\rangle | X\right) = 0 \text{ a.s.}.$$

We show that a couple of strongly independent random variables is in the set $\mathcal{A}$. We consider first the case of multiple integrals.

Lemma 4. *Suppose that $X = I_p(f)$ and $Y = I_q(g)$ where $f \in L^2([0, 1]^p)$ and $g \in L^2([0, 1]^q)$ $(p, q \geq 1)$ are symmetric functions. Assume that X and Y are independent. Then (X, Y) belongs to the class $\mathcal{A}$.*

Proof. We will prove that X is independent of the couple (Y, G_Y). Similarly it will follow that Y is independent of (X, G_X). We prove that

$$\langle DX, D(-L)^{-1} e^{i(\alpha Y + \beta G_Y)} \rangle = 0 \text{ a.s.}$$

or, since $D(-L)^{-1} = (I + L)^{-1} D$,

$$\langle DX, (I + L)^{-1} D e^{i(\alpha Y + \beta G_Y)} \rangle = 0 \text{ a.s.}$$

Note that $D e^{i(\alpha Y + \beta G_Y)} = e^{i(\alpha Y + \beta G_Y)} (i\alpha DY + i\beta D G_Y)$. First we will show that

$$\langle DX, e^{i(\alpha Y + \beta G_Y)} DY \rangle = 0 \text{ a.s.}$$

Assume that the random variable $e^{i(\alpha Y + \beta G_Y)}$ admits the chaos expansion $e^{i(\alpha Y + \beta G_Y)} = \sum_{N \geq 0} I_N(h_N)$ (in the sense that its real part and its imaginary part admit such a decomposition). Then

$$e^{i(\alpha Y + \beta G_Y)} D_\alpha Y = q \sum_{N \geq 0} I_N(h_N) q I_{q-1}(g(\cdot, \alpha))$$

$$= q \sum_{N \geq 0} \sum_{r=0}^{N \wedge (q-1)} r! C_{q-1}^r C_N^r I_{N+q-1-2r}(h_N \otimes_r g(\cdot, \alpha))$$

and

$$(I + L)^{-1} e^{i(\alpha Y + \beta G_Y)} D_\alpha Y = q \sum_{N \geq 0} \sum_{r=0}^{N \wedge (q-1)}$$
$$r! C_{q-1}^r C_N^r (1 + N + (q-1) - 2r)^{-1} I_{N+q-1-2r}(h_N \otimes_r g(\cdot, \alpha)).$$

Therefore

$$\langle DX, (I + L)^{-1} D e^{i(\alpha Y + \beta G_Y)} \rangle$$
$$= pq \sum_{N \geq 0} \sum_{r=0}^{N \wedge (q-1)} r! C_{q-1}^r C_N^r (1 + N + (q-1) - 2r)^{-1}$$
$$\times \sum_{a=0}^{(N+q-1-2r) \wedge (p-1)} I_{N+q-1-2r+p-2a} \int_0^1 \left((h_N \tilde{\otimes}_r g(\cdot, \alpha)) \otimes_a f(\cdot, \alpha) \right) d\alpha.$$

Above, $(h_N \tilde{\otimes}_r g(\cdot, \alpha))$ means the symmetrization of the function $(t_1, \ldots, t_{N+q-1-2r}) \to (h_N \otimes_r g(t_1, \ldots, t_{N+q-1-2r}, \alpha)$ for fixed α. In other words the above symmetrization does not affect the variable α. By interchanging the order of integration to integrate first with respect to α we will obtain that the last quantity is zero. Similarly it will follow that $\langle DX, e^{i(\alpha Y + \beta G_Y)} D G_Y \rangle$ is almost surely zero. $\square$

We can extend the previous result to the case of infinite chaos expansion.

Lemma 5. *Assume that X, Y are two strongly independent random variables in $\mathbb{D}^{1,2}$. Then (X, Y) belongs to $\mathcal{A}$.*

Proof. The proof follows the lines of the proof of Lemma 4. In order to check that

$$\langle DX, (I + L)^{-1} D e^{i(\alpha Y + \beta G_Y)}\rangle = 0 \text{ a.s.}$$

we write

$$\langle DX, (I + L)^{-1} D e^{i(\alpha Y + \beta G_Y)}\rangle$$

$$= \sum_{p,q \geq 1} pq \sum_{N \geq 0} \sum_{r=0}^{N \wedge (q-1)} r! C_{q-1}^r C_N^r (1 + N + (q-1) - 2r)^{-1}$$

$$\times \sum_{a=0}^{(N+q-1-2r) \wedge (p-1)} I_{N+q-1-2r+p-2a} \int_0^1 \left((h_N \tilde{\otimes}_r g(\cdot, \alpha)) \otimes_a f(\cdot, \alpha) \right) d\alpha$$

and we can finish as in the proof of the previous lemma. □

An interesting property of the couples in $\mathcal{A}$ is that the conditional expectation given $X + Y$ of the mixed scalar products $\langle DX, D(-L)^{-1}Y\rangle$ and $\langle DY, D(-L)^{-1}X\rangle$ vanish.

Lemma 6. *Assume that (X, Y) belongs to the class $\mathcal{A}$. Then*

$$\mathbf{E}\left(e^{it(X+Y)} \langle DX, D(-L)^{-1}Y\rangle \right) = 0 \text{ a.s.}$$

Proof. We have

$$\mathbf{E}\left(e^{it(X+Y)} \langle DX, D(-L)^{-1}Y\rangle \right) = \mathbf{E}\frac{1}{it} \langle D e^{itX}, e^{itY} D(-L)^{-1}Y\rangle$$

$$= \frac{1}{it}\mathbf{E}\left(e^{itX} \delta(e^{itY} D(-L)^{-1}Y) \right)$$

$$= \frac{1}{it}\mathbf{E}\left(e^{itX} \left(e^{itY} \delta D(-L)^{-1}Y \right.\right.$$
$$\left.\left. -it e^{itY} \langle DY, D(-L)^{-1}Y\rangle \right) \right)$$

$$= \frac{1}{it}\mathbf{E}\left(e^{itX} \left(e^{itY} Y - it e^{itY} G_Y \right) \right)$$

where we used the fact that since $e^{itY} \in \mathbb{D}^{1,2}$ and $D(-L)^{-1}Y \in Dom(\delta)$, then $e^{itY} D(-L^{-1})Y \in Dom(\delta)$ and by (9) $\delta(e^{itY}(D(-L)^{-1}Y)) = e^{itY} \delta(D(-L)^{-1}Y)$ $-it e^{itY} \langle DY, D(-L)^{-1}Y\rangle$. By using the fact that (X, Y) belongs to the class $\mathcal{A}$ we obtain

$$\mathbf{E}\left(e^{it(X+Y)} \langle DX, D(-L)^{-1}Y\rangle \right) = \frac{1}{it}\mathbf{E}(e^{itX}) \left(\mathbf{E}(e^{itY}Y) - it\mathbf{E}(e^{itY}G_Y) \right).$$

Now, going in the converse direction

$$\mathbf{E}(e^{itY}Y) - it\mathbf{E}(e^{itY}G_Y) = \mathbf{E}\delta(e^{itY}(D(-L)^{-1}Y)) = 0.$$

$\square$

We are now answering the following question: let $(X_n)_{n \geq 1}$ and $(Y_n)_{n \geq 1}$ be two sequences of random variables such that for every $n \geq 1$ the couple (X_n, Y_n) is in the class $\mathcal{A}$. Suppose that the sum $X_n + Y_n$ converges to the normal law and is such that the upper bound from Stein's method is attained, in the sense that $\mathbf{E}(1 - G_{X_n+Y_n})^2$ converges to zero. Could we then conclude that both X_n and Y_n converge in a strong sense to the normal laws $N(0, c_1)$ and $N(0, c_2)$ respectively? We will see that this is true in some particular case under a supplementary hypothesis on the sequences X_n and Y_n.

4.1 Wiener Chaos Stable Random Variables

Let us denote another class of families of random variables where the central limit of the sum implies central limit for each component. The idea is to assume a property on the filtration generated by $X_n + Y_n$. Let us denote by J_n the orthogonal projection of $L^2(\Omega)$ on the n-th Wiener chaos. We recall the following definition (see [8, 12]).

Definition 3. We will say that a sigma-algebra $\tau \subset \mathcal{F}$ is Wiener chaos stable if for every n, $J_n(L^2(\tau)) \subset L^2(\tau)$. In other words, if a random variable $F \in L^2(\tau)$ admits the chaos decomposition $F = \sum_{n \geq 0} I_n(f_n)$ then for every $n \geq 0$ the random variable $I_n(f_n)$ is τ-measurable.

Remark 5. The Wiener chaos stable property for sigma-algebras is equivalent to the L^{-1}-stable property. Recall that a sigma-algebra τ is L^{-1} stable if $L^{-1}(L_0^2(\tau)) \subset L_0^2(\tau)$ where $L_0^2(\tau)$ is the set of τ-measurable square integrable random variables with zero expectation. As a matter of fact, the sigma-algebra generated by $I_p(f), \langle DI_p(f), h_1 \rangle, \langle DI_p(f), h_2 \rangle, \ldots, \langle D^{p-1}I_p(f), h_{i_1} \otimes .. \otimes h_{i_{p-1}} \rangle$, where $h_i, i \geq 1$ is a complete orthogonal sequence in $L^2[0, 1]$, is Wiener stable (see [8, 12]).

Theorem 2. *Suppose that for every* $X_n = \sum_{n \geq 1} I_k(f_k^n)$ *and* $Y_n = \sum_{l \geq 1} I_l(g_l^n)$ *are such that* $\mathbf{E}X_n^2 \to_{n \to \infty} c_1$ *and* $\mathbf{E}Y_n^2 \to_{n \to \infty} c_2$ *(such that* $c_1, c_2 > 0$ *and* $c_1 + c_2 = 1$*). Assume that*

i. *For every* $n \geq 1$ *the couple* (X_n, Y_n) *belongs to the class* $\mathcal{A}$
ii. *For every* $n \geq 1$ *the sigma-algebras* $\sigma(X_n)$ *and* $\sigma(Y_n)$ *are Wiener chaos stable*

Assume also that $\mathbf{E}(1 - G_{X_n+Y_n})^2 \to_{n \to \infty} 0$. *Then*

$$\mathbf{E}(c_1 - G_{X_n})^2 \to_{n \to \infty} 0 \text{ and } \mathbf{E}(c_2 - G_{Y_n})^2 \to_{n \to \infty} 0.$$

Proof. We will show that under assumption ii, the random variable G_{X_n} belongs to $\sigma(X_n)$ for every $n \geq 1$. Since X_n is $\sigma(X_n)$ measurable and $\sigma(X_n)$ is Wiener chaos stable, we get that $I_k(f_k^n)$ is $\sigma(X_n)$ measurable for every n, k. Consequently, $I_k^n(f_k^n)I_l^n(f_l^n)$ is $\sigma(X_n)$ measurable for every n, k, l and by using the product formula we will have that

$$I_{k+l-2r}\left(f_k^n \otimes_r f_l^n\right)$$

is $\sigma(X_n)$ measurable for every $n, k, l \geq 1$ and $r = 0, .., k \wedge l$. As a consequence we can easily obtain that G_{X_n} is measurable with respect to $\sigma(X_n)$ and similarly G_{Y_n} is measurable with respect to $\sigma(Y_n)$. Assume now that $\mathbf{E}(1 - G_{X_n+Y_n})^2 \to_{n\to\infty} 0$. The asymptotic Cramér's Theorem 1 together with Proposition 2 imply that $\mathbf{E}(c_1 - G_{X_n}|X_n) \to 0$ and $\mathbf{E}(c_2 - G_{Y_n}|Y_n) \to 0$ a.s. and by the measurability of G_{X_n} and G_{Y_n} we obtain the conclusion. □

4.2 Vectorial Convergence of $X_n + Y_n$ and $G_{X_n} + G_{Y_n}$

A second class of sequences of random variables for which the central limit can be broken in order to ensure the strong convergence of each summand is inspired by [4].

Theorem 3. *Assume that* $\mathbf{E}X_n^2 \to c_1$ *and* $\mathbf{E}Y_n^2 \to c_2$ *(such that* $c_1, c_2 > 0$ *and* $c_1 + c_2 = 1$*). Assume that for every* $n \geq 1$ *the couple of random variables* (X_n, Y_n) *belongs to* $\mathcal{A}$*. Suppose moreover that the vector*

$$\left(X_n + Y_n, \frac{c_1 - G_{X_n} + c_2 - G_{Y_n}}{\mathbf{E}\left((c_1 - G_{X_n})^2 + (c_2 - G_{Y_n})^2\right)^{\frac{1}{2}}}\right) \tag{14}$$

converges as $n \to \infty$ *to the vector* (N_1, N_2) *where* N_1, N_2 *are standard normal random variables with correlation* ρ*. Then* $X_n + Y_n \to_{n\to\infty} N(0, 1)$ *implies that*

$$\mathbf{E}(c_1 - G_{X_n})^2 \to_{n\to\infty} 0 \text{ and } \mathbf{E}(c_2 - G_{Y_n})^2 \to_{n\to\infty} 0$$

Proof. On one hand, we have that

$$\mathbf{E}\left(f_z'(X_n + Y_n)(c_1 - G_{X_n} + c_2 - G_{Y_n})\right) \to_{n\to\infty} 0. \tag{15}$$

On the other hand, from the convergence of the vector (14) we get

$$\mathbf{E}\left(f_z'(X_n + Y_n)(c_1 - G_{X_n} + c_2 - G_{Y_n})a_n^{-1}\right) \to f_z'(N_1)N_2$$

where $a_n = \mathbf{E}\left((c_1 - G_{X_n})^2 + (c_2 - G_{Y_n})^2\right)^{\frac{1}{2}}$. It follows from the proof of Theorem 3.1 in [4] that we can find a constant $c \in (0, 1)$ such that

$$\left| \mathbf{E} \left(f'_z(X_n + Y_n)(c_1 - G_{X_n} + c_2 - G_{Y_n}) \right) \right|^2 \geq c \mathbf{E} \left((c_1 - G_{X_n})^2 + (c_2 - G_{Y_n})^2 \right)^2.$$
$$(16)$$

By combining the relations (15) and (16), we obtain that $\mathbf{E} \left((c_1 - G_{X_n})^2 + (c_2 - G_{Y_n})^2 \right)^2 \to_{n \to \infty} 0$ and this gives the convergence of X_n and Y_n to $N(0, c_1)$ and $N(0, c_2)$ respectively.   □

4.3 Random Variables with Independent Chaos Components

In this part we prove that in the case when the chaotic components appearing in the decomposition of X_n are mutually independent (and the same is true for Y_n) then the central limit of the sum implies the central limit of the summands (in a strong sense) under simple independence.

Proposition 4. *Assume that for every* $n \geq 1$, $X_n = \sum_{k \geq 1} I_k(f_k^n)$ *and* $Y_n = \sum_{l \geq 1} I_l(g_l^n)$ *and*

$$\mathbf{E}X_n^2 \to_{n \to \infty} c_1, \qquad \mathbf{E}Y_n^2 \to_{n \to \infty} c_2$$

with $c_1, c_2 > 0$ *and* $c_1 + c_2 = 1$. *Suppose that the following conditions are fulfilled*

i. For every $n \geq 1$ *the random variables* X_n *and* Y_n *are independent*
ii. For every $n \geq 1$, *the random variables* $(I_k(f_k^n))_{k \geq 1}$ *are pairwise independent the same holds for* $(I_k(g_l^n))_{l \geq 1}$

Then $X_n + Y_n \to N(0, 1)$ *implies*

$$\mathbf{E}(c_1 - G_{X_n})^2 \to_{n \to \infty} 0 \text{ and } \mathbf{E}(c_2 - G_{Y_n})^2 \to_{n \to \infty} 0.$$

Proof. The Theorem 1 implies that $X_n \to N(0, c_1)$ and $Y_n \to N(0, 1)$ in law. Corollary 1 and Assumption ii. gives that for every k the sequence $I_k(f_l^n)$ converges to a normal law as $k \to \infty$. Finally, we use Remark 2 and Lemmas 1, 2.   □

Acknowledgments We wish to thank the anonymous referee for valuable comments on our manuscript.

References

1. Cramér, H.: Über eine Eigenschaft der normalen Verteilungsfunction. Math. Z. **41**(2), 405–414 (1936)
2. Malliavin, P.: Stochastic Analysis. Springer, Berlin (2002)
3. Nourdin, I., Peccati, G.: Stein's method on Wiener chaos. Probab. Theor. Relat. Field. **145**(1), 75–118 (2009)
4. Nourdin, I., Peccati, G.: Stein's method and exact Berry-Esséen asymptotics for functionals of Gaussian fields. Ann. Probab. **37**(6), 2200–2230 (2009)

5. Nourdin, I., Peccati, G.: Stein's method meets Malliavin calculus: a short survey with new estimates in Recent Advances in Stochastic Dynamics and Stochastic Analysis. Interdisciplinary Mathematical Sciences - Vol. 8 edited by J. Duan, S.Luo and C.Wang, World Scientific (2008)
6. Nualart, D.: Malliavin Calculus and Related Topics, 2nd edn. Springer, Berlin (2006)
7. Nualart, D., Ortiz-Latorre, S.: Central limit theorems for multiple stochastic integrals and Malliavin calculus. Stoch. Proc. Appl. **118**, 614–628 (2008)
8. Nualart, D., Üstunel, A.S., Zakai, M.: Some relations among classes of σ fields on Wiener space. Probab. Theor. Relat. Field. **84**, 119–129 (1990)
9. Peccati, G., Tudor, C.A.: Gaussian limits for vector-valued multiple stochastic integrals. Séminaire de Probabilités, **XXXIV**, 247–262 (2004)
10. Tudor, C.A.: On the Structure of Gaussian Random Variables. (preprint) (2009) http://arxiv.org/PS_cache/arxiv/pdf/0907/0907.2501v2.pdf
11. Ustunel, A.S., Zakai, M.: On independence and conditioning on Wiener space. Ann. Probab. **17**(4), 1441–1453 (1989)
12. Ustunel, A.S., Zakai, M.: On the structure of independence on Wiener space. J. Funct. Anal. **90**, 113–137 (1990)

Moments of the Gaussian Chaos

Joseph Lehec

Abstract This paper deals with Latała's estimation of the moments of Gaussian chaoses. It is shown that his argument can be simplified significantly using Talagrand's generic chaining.

Keywords Wiener chaos · Metric entropy · Chaining

1 Introduction

In the article [3], Latała obtains an upper bound on the moments of the Gaussian chaos

$$Y = \sum a_{n_1,\dots,n_d} g_{n_1} \cdots g_{n_d},$$

where $g_1, g_2, \dots$ is a sequence of independent standard Gaussian random variables and the $a_{n_1,\dots,n_d}$ are real numbers. His bound his sharp up to constants depending only on the order d of the chaos. The purpose of the present paper is to give another proof of Latała's result.

Observe that the case $d = 1$ is easy, since

$$\big(\mathrm{E}|\sum a_i g_i|^p\big)^{1/p} = \big(\sum a_i^2\big)^{1/2}\big(\mathrm{E}|g_1|^p\big)^{1/p} \sim \sqrt{p}\big(\sum a_i^2\big)^{1/2}.$$

When $d = 2$, Latała recovers a result by Hanson and Wright [2] which involves the operator and the Hilbert–Schmidt norms of the matrix $a = (a_{ij})$

$$\big(\mathrm{E}|\sum a_{ij} g_i g_j|^p\big)^{1/p} \sim \sqrt{p}\|a\|_{\mathrm{HS}} + p\|a\|_{\mathrm{op}}.$$

It is known (see [5]) that the moments of the *decoupled* chaos

J. Lehec (✉)
CEREMADE (UMR CNRS 7534), Université Paris-Dauphine, Place du maréchal de Lattre de Tassigny, 75016 Paris, France
e-mail: lehec@ceremade.dauphine.fr

C. Donati-Martin et al. (eds.), *Séminaire de Probabilités XLIII*, Lecture Notes in Mathematics 327
2006, DOI 10.1007/978-3-642-15217-7_13, © Springer-Verlag Berlin Heidelberg 2011

$$\tilde{Y} = \sum a_{n_1,\dots,n_d} g_{n_1,1} \cdots g_{n_d,d}$$

where $(g_{i,j})$ is a family of standard independent Gaussian variables, are comparable to those of Y wih constants depending only on d. Using this fact and reasoning by induction on the order d of the chaos, Latała shows that the problem boils down to the estimation of the supremum of a complicated Gaussian process. Given a set T and a Gaussian process $(X_t)_{t \in T}$, estimating $\mathrm{E} \sup_T X_t$ amounts to studying the metric space (T, d) where d is given by the formula

$$\mathrm{d}(s,t) = \big(\mathrm{E}(X_s - X_t)^2\big)^{1/2}.$$

Dudley's estimate for instance, asserts that if the process is centered (meaning that $\mathrm{E} X_t = 0$ for all $t \in T$) then there exists a universal constant C such that

$$\mathrm{E} \sup T_t \leq C \int_0^\infty \sqrt{\log N(T, \mathrm{d}, \epsilon)} \, \mathrm{d}\epsilon,$$

where the entropy number $N(T, \mathrm{d}, \epsilon)$ is the smallest number of balls (for the distance d) of radius ϵ needed to cover T. Let us refer to Fernique [1] for a proof of this inequality and several applications. However, Dudley's inequality is not sharp: there exist Gaussian processes for which the integral is much larger than the expectation of the sup. Unfortunately, the phenomenon occurs here. Latała is able to give precise bounds for the entropy numbers, but Dudley's integral does not give the correct order of magnitude. Something finer is needed.

The precise estimate of the supremum of a Gaussian process in terms of metric entropy was found by Talagrand. This was the famous *Majorizing Measure Theorem* [6], which is now called *Generic chaining*, see the book [7]. Latała did not manage to use Talagrand's theory, and his proof contains a lot of tricky entropy estimates to beat the Dudley bound. We find this part of his paper very hard to read, and our purpose is to short-circuit it using Talagrand's generic chaining.

Lastly, let us mention that we disagree with P. Major who released an article on arXiv[1] in which he claims that Latała's proof is incorrect. The present paper is all about understanding Latała's work, not correcting it.

2 Notations, Statement of Latała's Result

2.1 Tensor Products, Mixed Injective and L_2 Norms

To avoid heavy multi-indices notations, it is convenient to use tensor products. If X and Y are finite dimensional normed spaces, the notation $X \otimes^\epsilon Y$ stands for the injective tensor product of X and Y, so that $X \otimes^\epsilon Y$ is isometric to $\mathcal{L}(X^*, Y)$ equipped with the operator norm. If X and Y are Euclidean spaces, we denote by

---

[1] http://arxiv.org/abs/0803.1453

$X \otimes^2 Y$ their Euclidean tensor product. Moreover, in this case we identify X and X^*, so that $X \otimes^2 Y$ is isometric to $\mathcal{L}(X, Y)$ equipped with the Hilbert–Schmidt norm.

Throughout the article $[d]$ denotes the set $\{1, \ldots, d\}$. Let $E_1, \ldots, E_d$ be Euclidean spaces. Given a non-empty subset $I = \{i_1, \ldots, i_p\}$ of $[d]$, we let

$$E_I = E_{i_1} \otimes^2 \cdots \otimes^2 E_{i_p}.$$

Also, by convention $E_\emptyset = \mathbb{R}$. The notation $\|\cdot\|_I$ stands for the norm of E_I and

$$B_I = \{x \in E_I; \ \|x\|_I \leq 1\}$$

for its unit ball. Let $A \in E_{[d]}$ and $\mathcal{P} = \{I_1, \ldots, I_k\}$ be a partition of $[d]$, we let $\|A\|_\mathcal{P}$ be the norm of A as an element of the space

$$E_{I_1} \otimes^\epsilon \cdots \otimes^\epsilon E_{I_k}.$$

When $d = 2$ for instance, the tensor A can be seen as a linear map from E_1 to E_2, then $\|A\|_{\{1\}\{2\}}$ and $\|A\|_{\{1,2\}}$ are the operator and Hilbert–Schmidt norms of A, respectively. Let us give another example: assume that $d = 3$ and that $E_1 = E_2 = E_3 = L_2(\mu)$ for some measure μ. Then for any $f \in E_1 \otimes E_2 \otimes E_3$ which we identify $L_2(\mu^{\otimes 3})$, we have

$$\|f\|_{\{1\}\{2,3\}} = \sup\left(\int f(x, y, z) u(x) v(y, z) \ d\mu(x) d\mu(y) d\mu(z)\right),$$

where the sup is taken over all u, v having L_2 norms at most 1. Going back to the general setting, let us define for a non-empty subset I of $[d]$ and an element $x \in E_I$ the contraction $\langle A, x \rangle$ to be the image of x by A, when A is seen as an element of $\mathcal{L}(E_I, E_{[d]\setminus I})$. Then for every partition $\mathcal{P} = \{I_1, \ldots, I_k\}$ we have

$$\|A\|_\mathcal{P} = \sup\{\langle A, x_1 \otimes \cdots \otimes x_k \rangle; \ x_j \in B_{I_j}\}.$$

If $\mathcal{Q} = \{J_1, \ldots, J_l\}$ is a finer partition than $\mathcal{P}$ (this means that any element of $\mathcal{Q}$ is contained in an element of $\mathcal{P}$) then

$$\{x_1 \otimes \cdots \otimes x_l, \ x_j \in B_{J_j}\} \subset \{y_1 \otimes \cdots \otimes y_k, \ y_j \in B_{I_j}\},$$

hence $\|A\|_\mathcal{Q} \leq \|A\|_\mathcal{P}$. In particular,

$$\|A\|_{\{1\}\cdots\{d\}} \leq \|A\|_\mathcal{P} \leq \|A\|_{[d]}.$$

2.2 Moments of the Gaussian Chaos

If $\mathcal{P}$ is a partition of $[d]$, its cardinality card $\mathcal{P}$ is the number of subsets of $[d]$ in $\mathcal{P}$. Let $E_1, \ldots, E_d$ be Euclidean spaces and $A \in E_{[d]}$. Let $X_1, \ldots, X_d$ be independent random vectors such that for all i, the vector X_i is a standard Gaussian vector of E_i. The (real) random variable

$$Z = \langle A, X_1 \otimes \cdots \otimes X_d \rangle$$

is called *decoupled* Gaussian chaos of order d. Here is the main result of Latała.

Theorem 1. *There exists a constant α_d depending only on d such that for all $p \geq 1$*

$$\left(E |Z|^p \right)^{1/p} \leq \alpha_d \sum_{\mathcal{P}} p^{\frac{\mathrm{card}\,\mathcal{P}}{2}} \|A\|_{\mathcal{P}},$$

the sum running over all partitions $\mathcal{P}$ of $[d]$.

The following theorem and corollary are intermediate results from which the previous theorem shall follow; however we believe they are of independent interest.

Theorem 2. *Let $F_1, \ldots, F_{k+1}$ be Euclidean spaces, let $A \in F_{[k+1]}$ and X be a standard Gaussian vector on F_{k+1}, recall that $\langle A, X \rangle \in F_1 \otimes \cdots \otimes F_k$. Then for all $\tau \in (0, 1)$:*

$$E \|\langle A, X \rangle\|_{\{1\}\cdots\{k\}} \leq \beta_k \sum_{\mathcal{P}} \tau^{k - \mathrm{card}\,\mathcal{P}} \|A\|_{\mathcal{P}},$$

where the sum runs over all partitions $\mathcal{P}$ of $[k + 1]$ and the constant β_k depends only on k.

Corollary 3. *Under the same hypothesis, we have for all $p \geq 1$*

$$\left(E \|\langle A, X \rangle\|_{\{1\}\cdots\{k\}}^p \right)^{1/p} \leq \delta_k \sum_{\mathcal{P}} p^{\frac{\mathrm{card}\,\mathcal{P} - k}{2}} \|A\|_{\mathcal{P}}.$$

Proof. Let $f : x \in F_{k+1} \mapsto \|\langle A, x \rangle\|_{\{1\}\cdots\{k\}}$. Let us use the concentration property of the Gaussian measure, which asserts that Lipschitz functions are close to their means with high probability. More precisely, letting $m = E f(X)$, we have for all $p \geq 1$

$$\left(E |f(X) - m|^p \right)^{1/p} \leq \delta' \sqrt{p} \|f\|_{\mathrm{lip}},$$

where $\|f\|_{\mathrm{lip}}$ is the Lipschitz constant of f and δ' is a universal constant. We refer to [4] for more details on this inequality. Noting that

$$\|f\|_{\mathrm{lip}} = \sup_{x \in B_{k+1}} \|\langle A, x \rangle\|_{\{1\}\cdots\{k\}} = \|A\|_{\{1\}\cdots\{k+1\}}.$$

and using the triangle inequality, we get

$$\left(E |f(X)|^p \right)^{1/p} \leq E f(X) + \delta' \sqrt{p} \|A\|_{\{1\}\cdots\{k+1\}}.$$

The result then follows from the upper bound on $\mathrm{E}\, f(X)$ given by Theorem 2 with $\tau = p^{-1/2}$. $\square$

Let us prove Theorem 1. We proceed by induction on d. When $d = 1$, the random variable $\langle A, X_1 \rangle$ is, in law, equal to the Gaussian variable of variance $\|A\|^2_{\{1\}}$. The p-th moment of the standard Gaussian variable being of order $\sqrt{p}$, we get

$$\left(\mathrm{E}|\langle A, X_1 \rangle|^p\right)^{1/p} \leq \alpha \sqrt{p} \|A\|_{\{1\}}$$

for some universal α, hence the theorem for $d = 1$.

Assume that the result holds for chaoses of order $d - 1$. From now on, if $I = \{i_1, \ldots, i_r\}$ is a subset of $[d]$ we denote the tensor $X_{i_1} \otimes \cdots \otimes X_{i_r}$ by X_I. Notice that

$$\langle A, X_{[d]} \rangle = \langle \langle A, X_d \rangle, X_{[d-1]} \rangle$$

and apply the induction assumption to the matrix $B = \langle A, X_d \rangle$. This yields

$$\mathrm{E}\left(|\langle B, X_{[d-1]} \rangle|^p \big| X_d\right) \leq \alpha_{d-1}^p \left(\sum_{\mathcal{P}} p^{\frac{\operatorname{card} \mathcal{P}}{2}} \|B\|_{\mathcal{P}}\right)^p,$$

where the sum runs over all partitions $\mathcal{P}$ of $[d - 1]$. Taking expectation and the p-th root, we obtain

$$
\begin{aligned}
\left(\mathrm{E}|\langle A, X_{[d]} \rangle|^p\right)^{1/p} &\leq \alpha_{d-1} \left(\mathrm{E}\left(\sum_{\mathcal{P}} p^{\frac{\operatorname{card} \mathcal{P}}{2}} \|\langle A, X_d \rangle\|_{\mathcal{P}}\right)^p\right)^{1/p} \\
&\leq \alpha_{d-1} \sum_{\mathcal{P}} p^{\frac{\operatorname{card} \mathcal{P}}{2}} \left(\mathrm{E}\|\langle A, X_d \rangle\|_{\mathcal{P}}^p\right)^{1/p},
\end{aligned}
\tag{1}
$$

by the triangle inequality. Let $\mathcal{P} = \{I_1, \ldots, I_k\}$ be a partition of $[d - 1]$. Let $F_i = E_{I_i}$ for $i \in [k]$ and $F_{k+1} = E_d$. The tensor A can be seen as an element of $F_{[k+1]}$, let us rename it A' when we do so. Corollary 3 gives

$$\left(\mathrm{E}\|\langle A', X_d \rangle\|_{\{1\}\cdots\{k\}}^p\right)^{1/p} \leq \delta_k p^{-\frac{k}{2}} \sum_{\mathcal{Q}} p^{\frac{\operatorname{card} \mathcal{Q}}{2}} \|A'\|_{\mathcal{Q}},$$

where the sum is taken over all partitions $\mathcal{Q}$ of $[k]$. Going back to the space $E_{[d]}$, this inequality translates as

$$\left(\mathrm{E}\|\langle A, X_d \rangle\|_{\mathcal{P}}^p\right)^{1/p} \leq \delta_k p^{-\frac{k}{2}} \sum_{\mathcal{Q}} p^{\frac{\operatorname{card} \mathcal{Q}}{2}} \|A\|_{\mathcal{Q}},
\tag{2}$$

and this time the sum runs over partitions $\mathcal{Q}$ of $[d]$ such that the partition

$$\{I_1, \ldots, I_k, \{d\}\}$$

is finer than $\mathcal{Q}$. However, the inequality still holds if we take the sum over all partitions of $[d]$ instead. We plug (2) into (1), the numbers $p^{\frac{\text{card}\,\mathcal{P}}{2}}$ cancel out and we get the desired inequality with constant

$$\alpha_d = \alpha_{d-1} \sum_{\mathcal{P}} \delta_{\text{card}\,\mathcal{P}},$$

where the sum is taken over all partitions $\mathcal{P}$ of $[d-1]$.

So it is enough to prove Theorem 2, this is the purpose of the rest of the article.

3 The Generic Chaining

Let $F_1, \dots, F_{k+1}$ be Euclidean spaces, let $A \in F_{[k+1]}$ and X be a standard Gaussian vector of F_{k+1}. For $i \in [k]$ let B_i be the unit ball of F_i, let $T = B_1 \times \dots \times B_k$. Recall that for $x = (x_1, \dots, x_k) \in T$, the notation $x_{[k]}$ stands for the tensor $x_1 \otimes \dots \otimes x_k$. Note that

$$\mathrm{E}\|\langle A, X\rangle\|_{\{1\}\cdots\{k\}} = \mathrm{E}\sup_{x\in T}\langle A, x_{[k]} \otimes X\rangle = \mathrm{E}\sup_{x\in T}\langle\langle A, x_{[k]}\rangle, X\rangle. \qquad (3)$$

Notice also that $(P_x)_{x\in T} = \big(\langle\langle A, x_{[k]}\rangle, X\rangle\big)_{x\in T}$ is a Gaussian process. To estimate $\mathrm{E}\sup_T P_x$, we shall study the metric space (T, d), where

$$\mathrm{d}(x, y) = \big(\mathrm{E}(P_x - P_y)^2\big)^{1/2}.$$

This distance can be computed explicitly. Indeed

$$\mathrm{d}(x, y)^2 = \mathrm{E}\langle\langle A, x_{[k]} - y_{[k]}\rangle, X\rangle^2 = \|\langle A, x_{[k]} - y_{[k]}\rangle\|^2_{\{k+1\}}. \qquad (4)$$

The *generic chaining*, introduced by Talagrand, will be our main tool. We sketch briefly the main ideas of the theory and refer to Talagrand's book [7] for details.

Let (T, d) be a metric space. If S is a subset of T we let $\delta_{\mathrm{d}}(S)$ be the diameter of S

$$\delta_{\mathrm{d}}(S) = \sup_{s,t\in S} \mathrm{d}(s, t).$$

Given a sequence $(\mathcal{A}_n)_{n\in\mathbb{N}}$ of partitions of T and an element $t \in T$, we let $A_n(t)$ be the unique element of $\mathcal{A}_n$ containing t.

Definition 4. Let

$$\gamma_{\mathrm{d}}(T) = \inf\Big(\sup_{t\in T} \sum_{n=0}^{\infty} \delta_{\mathrm{d}}\big(A_n(t)\big)2^{n/2}\Big),$$

where the infimum is over all sequences of partitions $(A_n)_{n \in \mathbb{N}}$ of T satisfying the cardinality condition

$$A_0 = \{T\} \quad \text{and} \quad \forall n \geq 1, \text{ card } A_n \leq 2^{2^n}. \tag{5}$$

Notice that $\gamma_{\mathrm{d}}(T) \geq \delta_{\mathrm{d}}(T)$. In particular, if the metric is not trivial then $\gamma_{\mathrm{d}}(T)$ is non-zero. Thus there exists a sequence of partitions $(A_n)_{n \in \mathbb{N}}$ satisfying the cardinality condition and

$$\sup_{t \in T} \sum_{n=0}^{\infty} \delta_{\mathrm{d}}\big(A_n(t)\big) 2^{n/2} \leq 2\gamma_{\mathrm{d}}(T).$$

We recall the all important

Theorem 5 (Majorizing Measure). *There exists a universal constant κ such that for any Gaussian process $(X_t)_{t \in T}$ that is centered (meaning $\mathrm{E}\, X_t = 0$ for all $t \in T$) we have*

$$\tfrac{1}{\kappa} \gamma_{\mathrm{d}}(T) \leq \mathrm{E} \sup_{t \in T} X_t \leq \kappa \gamma_{\mathrm{d}}(T),$$

where the metric d *is defined by* $\mathrm{d}(s,t) = \big(\mathrm{E}(X_s - X_t)^2\big)^{1/2}$.

Here are two simple lemmas.

Lemma 6. *Let (T, d) be a metric space. Let $a, b \geq 1$, and $(A_n)_{n \in \mathbb{N}}$ be a sequence of partitions of T satisfying*

$$\forall n \in \mathbb{N}, \text{ card } A_n \leq 2^{a+b2^n}.$$

Letting $\gamma = \sup_{t \in T} \sum_{n=0}^{\infty} \delta_{\mathrm{d}}\big(A_n(t)\big) 2^{n/2}$, we have

$$\gamma_{\mathrm{d}}(T) \leq \rho\big(\sqrt{ab}\, \delta_{\mathrm{d}}(T) + \sqrt{b}\, \gamma\big),$$

for some universal ρ.

Proof. Let p, q be the smallest integers satisfying $a \leq 2^p$ and $b \leq 2^q$. Let

$$\mathcal{B}_n = \begin{cases} \{T\} & \text{if } n \leq p+q \\ A_{n-q-1} & \text{if } n \geq p+q+1. \end{cases}$$

If $n \geq p+q+1$ then $p \leq n-1$ so

$$\text{card } \mathcal{B}_n \leq 2^{2^p + 2^{n-1}} \leq 2^{2^n}.$$

Thus the sequence $(\mathcal{B}_n)_{n \in \mathbb{N}}$ satisfies (5). On the other hand, for all $t \in T$

$$\sum_{n=0}^{\infty} \delta_{\mathrm{d}}\big(B_n(t)\big)2^{n/2} = \sum_{n=0}^{p+q} \delta_d(T)2^{n/2} + \sum_{n=p}^{\infty} \delta_{\mathrm{d}}\big(A_n(t)\big)2^{\frac{n+q+1}{2}}$$

$$\leq \frac{2^{\frac{p+q+1}{2}}-1}{\sqrt{2}-1}\delta_{\mathrm{d}}(T) + 2^{\frac{q+1}{2}}\gamma.$$

Moreover $2^p \leq 2a$ and $2^q \leq 2b$, hence the result. $\qquad\square$

Lemma 7. *Let* $\mathrm{d}_1,\ldots,\mathrm{d}_N$ *be distances defined on* T *and let* $\mathrm{d} = \sum \mathrm{d}_i$. *Then*

$$\gamma_{\mathrm{d}}(T) \leq \rho'\sqrt{N}\sum_{i=1}^{N}\gamma_{\mathrm{d}_i}(T),$$

where ρ' *is a universal constant.*

Proof. For all $i \in [N]$, there exists a sequence $(\mathcal{A}_n^i)_{n\in\mathbb{N}}$ of partitions of T satisfying the cardinality condition (5) and

$$\sup_{t\in T}\sum_{n=0}^{\infty}\delta_{\mathrm{d}_i}\big(A_n^i(t)\big)2^{n/2} \leq 2\gamma_{\mathrm{d}_i}(T).$$

Then let

$$\mathcal{A}_n = \{A^1 \cap \cdots \cap A^N, \; A^i \in \mathcal{A}_n^i\}.$$

This clearly defines a sequence of partitions of T, and for all n we have

$$\mathrm{card}\,\mathcal{A}_n \leq 2^{N2^n}. \tag{6}$$

On the other hand, for all $t \in T$ and $i \in [N]$ we have $A_n(t) \subset A_n^i(t)$, so

$$\delta_{\mathrm{d}}\big(A_n(t)\big) \leq \sum_{i=1}^{N}\delta_{\mathrm{d}_i}\big(A_n(t)\big) \leq \sum_{i=1}^{N}\delta_{\mathrm{d}_i}\big(A_n^i(t)\big).$$

Consequently

$$\sup_{t\in T}\sum_{n=0}^{\infty}\delta_{\mathrm{d}}\big(A_n(t)\big)2^{n/2} \leq 2\sum_{i=1}^{N}\gamma_{\mathrm{d}_i}(T). \tag{7}$$

By the previous lemma, (6) and (7) yield the result. $\qquad\square$

4 Proof of Theorem 2

The proof is by induction on k. When $k = 1$ the theorem is a consequence of the following: let $A \in F_1 \otimes F_2$ and X be a standard Gaussian vector on F_2, then

$$\mathrm{E}\|\langle A, X\rangle\|_{\{1\}} \leq \big(\mathrm{E}\|\langle A, X\rangle\|_{\{1\}}^2\big)^{1/2} = \|A\|_{\{1,2\}}.$$

Assume that $k \geq 2$ and that the theorem holds for any $k' < k$. Let $A \in F_{[k+1]}$. Recall that for $i \in [k]$ the unit ball of F_i is denoted by B_i and the product $B_1 \times \cdots \times B_k$ by T. Let I be a *non-empty* subset of $[k]$ and d_I be the pseudo-metric on T defined by

$$d_I(x, y) = \|\langle A, x_I - y_I \rangle\|_{[k+1] \setminus I}. \tag{8}$$

By the majorizing measure theorem and the (3) and (4), Theorem 2 is equivalent to

Theorem 8. *For all $\tau \in (0, 1)$*

$$\gamma_{d_{[k]}}(T) \leq \beta'_k \sum_{\mathcal{P}} \tau^{k - \text{card}\,\mathcal{P}} \|A\|_{\mathcal{P}},$$

with a sum running over all partitions $\mathcal{P}$ of $[k + 1]$.

Our purpose is to prove Theorem 8 by induction on k. Let τ be a fixed positive real number and let d^τ be the following metric:

$$d^\tau = \sum_{\emptyset \subsetneq I \subsetneq [k]} \tau^{k - \text{card}\,I} d_I. \tag{9}$$

Let us sketch the argument. First we use an entropy estimate and the generic chaining to compare $\gamma_{d_{[k]}}(T)$ and $\gamma_{d^\tau}(T)$, then we use the induction assumption to estimate the latter.

Here is the crucial entropy estimate of Latała [3, Corollary 2].

Lemma 9. *Let $S \subset T$, let $\tau \in (0, 1)$ and $\epsilon = \delta_{d^\tau}(S) + \tau^k \|A\|_{[k+1]}$. Then*

$$N(S, d_{[k]}, \epsilon) \leq 2^{c_k \tau^{-2}},$$

for some constant c_k depending only on k.

Let us postpone the proof to the last section.

Let $(\mathcal{B}_n)_{n \in \mathbb{N}}$ be a sequence of partitions of T satisfying the cardinality condition (5) and

$$\sup_{t \in T} \sum_{n=0}^\infty \delta_{d^\tau}(B_n(t)) 2^{n/2} \leq 2\gamma_{d^\tau}(T). \tag{10}$$

Let $n \in \mathbb{N}$ and $B \in \mathcal{B}_n$, set $\tau_n = \min(\tau, 2^{-n/2})$ and $\epsilon_n = \delta_{d^{\tau_n}}(B) + \tau_n^k \|A\|_{[k+1]}$. Observe that $\tau_n^{-2} \leq \tau^{-2} + 2^n$ and apply Lemma 9 to B and τ_n:

$$N(B, d_{[k]}, \epsilon_n) \leq 2^{c_k \tau_n^{-2}} \leq 2^{c_k \tau^{-2} + c_k 2^n}.$$

Therefore we can find a partition $\mathcal{A}_B$ of B whose cardinality is controlled by the number above and such that any $R \in \mathcal{A}_B$ satisfies

$$\delta_{d_{[k]}}(R) \leq 2\epsilon_n \leq 2\delta_{d^\tau}(B) + 2\tau_n^k \|A\|_{[k+1]}.$$

Indeed $\tau_n \leq \tau$ implies that $d^{\tau_n} \leq d^\tau$. Then we let $\mathcal{A}_n = \cup\{A_B; \ B \in \mathcal{B}_n\}$. This clearly defines a sequence of partitions of T which satisfies

$$\text{card } \mathcal{A}_n \leq 2^{c_k \tau^{-2} + c_k 2^n} \ \text{card } \mathcal{B}_n \leq 2^{c_k \tau^{-2} + (c_k+1)2^n}, \tag{11}$$

$$\delta_{d_{[k]}}(A_n(t)) \leq 2\delta_{d^\tau}(B_n(t)) + 2\tau_n^k \|A\|_{[k+1]}, \tag{12}$$

for all $n \in \mathbb{N}$ and $t \in T$. Recall that $\tau_n = \min(\tau, 2^{-n/2})$, an easy computation shows that

$$\sum_{n=0}^{\infty} \tau_n^k 2^{n/2} \leq C\tau^{k-1}$$

for some universal C. Therefore, for all $t \in T$, we have

$$\sum_{n=0}^{\infty} \delta_{d_{[k]}}(A_n(t)) 2^{n/2} \leq 2 \sum_{n=0}^{\infty} (\delta_{d^\tau}(B_n(t)) + \tau_n^k \|A\|_{[k+1]}) 2^{n/2},$$

$$\leq 4\gamma_{d^\tau}(T) + 2C\tau^{k-1}\|A\|_{[k+1]}.$$

By (11) and applying Lemma 6, we get for some constant C_k depending only on k

$$\gamma_{d_{[k]}}(T) \leq C_k (\gamma_{d^\tau}(T) + \tau^{k-1}\|A\|_{[k+1]} + \tau^{-1}\delta_{d_{[k]}}(T)),$$

$$\leq 2C_k (\gamma_{d^\tau}(T) + \tau^{k-1}\|A\|_{[k+1]} + \tau^{-1}\|A\|_{\{1\}\cdots\{k+1\}}). \tag{13}$$

Indeed

$$\delta_{d_{[k]}}(T) = 2 \sup_{x \in T} \|\langle A, x \rangle\|_{\{k+1\}} = 2\|A\|_{\{1\}\cdots\{k+1\}}.$$

We have not used the induction assumption yet. Let $I = \{i_1, \ldots, i_p\}$ be a subset of $[k]$, different from $\emptyset$ and $[k]$. For $j \in [p]$ let $F'_j = F_{i_j}$ and let $F'_{p+1} = F_{[k+1]\backslash I}$. Since $p < k$ we can apply inductively Theorem 8 to the tensor A seen as an element of $F'_{[p+1]}$. For all $\tau \in (0,1)$

$$\gamma_{d_I}(T) \leq \beta'_p \sum_{\mathcal{Q}} \tau^{p-\text{card }\mathcal{Q}} \|A\|_{\mathcal{Q}}, \tag{14}$$

where the sum runs over all partitions $\mathcal{Q}$ of $[k+1]$ such that the partition $\{i_1\}, \ldots, \{i_p\}, [k+1]\backslash I$ is finer than $\mathcal{Q}$. Again, the inequality is still true if we take the sum over all partitions of $[k+1]$ instead. According to Lemma 7 and since γ is clearly homogeneous, we have

$$\gamma_{d^\tau}(T) \leq \rho'\sqrt{N} \sum_{\emptyset \subsetneq I \subsetneq [k]} \tau^{k-\text{card }I} \gamma_{d_I}(T)$$

where N is the number of subsets of $[k]$ which are different from $\emptyset$ and $[k]$, namely $2^k - 2$. By (14) we get

$$\gamma_{d\tau}(T) \leq D_k \sum_{\mathcal{P}} \tau^{k-\operatorname{card}\mathcal{P}} \|A\|_{\mathcal{P}},$$

for some D_k depending only on k. This, together with (13), concludes the proof of Theorem 8.

In the last section we prove Lemma 9, this is essentially Latała's proof.

5 Proof of the Entropy Estimate

Let $x = (x_1, \dots, x_k) \in F_1 \times \cdots \times F_k$, let $|x_i|$ be the norm of x_i in F_i. Let $X_1, \dots, X_k$ be independent standard Gaussian vectors on $F_1, \dots, F_k$, respectively.

Lemma 10. *For all semi-norm $\|\cdot\|$ on $F_{[k]}$, we have*

$$P\left(\|X_{[k]} - x_{[k]}\| \leq E \sum_{\emptyset \subsetneq I \subset [k]} 4^{\operatorname{card} I} \|X_I \otimes x_{[k]\setminus I}\|\right) \geq 2^{-k} e^{-\frac{1}{2}\sum_{i=1}^{k}|x_i|^2}.$$

Proof. Let us start with an elementary remark. Let $x \in \mathbb{R}^n$, let K be a symmetric subset of $\mathbb{R}^n$, and γ_n be the standard Gaussian measure on $\mathbb{R}^n$. Then

$$\gamma_n(x + K) \geq \gamma_n(K)e^{-\frac{1}{2}|x|^2}. \tag{15}$$

Indeed, the symmetry of K and the convexity of the exponential function imply that

$$\int_{x+K} e^{-\frac{1}{2}|z|^2}\, dz = \int_K \frac{1}{2}(e^{-\frac{1}{2}|x+y|^2} + e^{-\frac{1}{2}|x-y|^2})\, dy$$
$$\geq \int_K e^{-\frac{1}{2}(|x|^2 + |y|^2)}\, dy$$

which proves (15).

Let us prove the lemma by induction on k.

If $k = 1$, applying (15) to $K = \{y \in F_1,\ \|y\| \leq 4\,E\|X_1\|\}$ and $x = x_1$, we get

$$P(\|X_1 - x_1\| \leq 4\,E\|X_1\|) \geq e^{-\frac{1}{2}|x_1|^2}\, P(\|X_1\| \leq 4\,E\|X_1\|).$$

Besides, by Markov we have $P(\|X_1\| \geq 4\,E\|X_1\|) \leq \frac{1}{4} \leq \frac{1}{2}$, hence the result for $k = 1$.

Let $k \geq 2$ and assume that the result holds for $k - 1$. Let

$$S = \sum_{\emptyset \subsetneq I \subset [k-1]} 4^{\operatorname{card} I} \|X_I \otimes x_{[k-1]\setminus I} \otimes X_k\|$$

$$T = \sum_{\emptyset \subsetneq I \subset [k-1]} 4^{\operatorname{card} I} \|X_I \otimes x_{[k-1]\setminus I} \otimes x_k\|$$

and let A, B and C be the events

$$A = \{\|x_{[k-1]} \otimes (X_k - x_k)\| \leq 4\,\mathrm{E}\|x_{[k-1]} \otimes X_k\|\}$$
$$B = \{\|(X_{[k-1]} - x_{[k-1]}) \otimes X_k\| \leq \mathrm{E}(S \mid X_k)\}$$
$$C = \{\mathrm{E}(S \mid X_k) \leq 4\,\mathrm{E}\,S + \mathrm{E}\,T\}.$$

By the following triangle inequality

$$\|X_{[k]} - x_{[k]}\| \leq \|x_{[k-1]} \otimes (X_k - x_k)\| + \|(X_{[k-1]} - x_{[k-1]}) \otimes X_k\|,$$

when A, B and C occur we have

$$\|X_{[k]} - x_{[k]}\| \leq 4\,\mathrm{E}\|x_{[k-1]} \otimes X_k\| + 4\,\mathrm{E}\,S + \mathrm{E}\,T$$
$$= \mathrm{E} \sum_{\emptyset \subsetneq I \subset [k]} 4^{\mathrm{card}\,I} \|X_I \otimes x_{[k]\setminus I}\|.$$

Assume that X_k is deterministic, and apply the induction assumption to the spaces $F_1, \ldots, F_{k-1}$ and to the semi-norm $\|y\|_1 = \|y \otimes X_k\|$ for all $y \in F_{[k-1]}$, then

$$P(B \mid X_k) \geq 2^{-k+1} e^{-\frac{1}{2}\sum_{i=1}^{k-1}|x_i|^2}.$$

Since A and C depend only on X_k, this implies that

$$P(A \cap B \cap C) \geq P(A \cap C)2^{-k+1}e^{-\frac{1}{2}\sum_{i=1}^{k-1}|x_i|^2}.$$

So it is enough to prove that $P(A \cap C) \geq 2^{-1}e^{-\frac{1}{2}|x_k|^2}$. For all $y \in F_k$ we let

$$\|y\|_2 = \|x_{[k-1]} \otimes y\|,$$
$$\|y\|_3 = \mathrm{E} \sum_{\emptyset \subsetneq I \subset [k-1]} 4^{\mathrm{card}\,I} \|X_I \otimes x_{[k-1]\setminus I} \otimes y\|.$$

So that

$$A = \{\|X_k - x_k\|_2 \leq 4\,\mathrm{E}\|X_k\|_2\},$$
$$C = \{\|X_k\|_3 \leq 4\,\mathrm{E}\|X_k\|_3 + \|x_k\|_3\}.$$

Let

$$K = \{y \in F_k,\ \|y\|_2 \leq 4\,\mathrm{E}\|X_k\|_2\} \cap \{y \in F_k,\ \|y\|_3 \leq 4\,\mathrm{E}\|X_k\|_3\},$$

then, by the triangle inequality, the event $X_k \in x_k + K$ is included in $A \cap C$. Using (15), we get

$$P(A \cap C) \geq P(X_k \in x_k + K) \geq e^{-\frac{1}{2}|x_k|^2} P(X_k \in K).$$

Therefore, it is enough to prove that $P(X_k \in K) \geq \frac{1}{2}$, and this is a simple application of Markov again.   □

Let F_{k+1} be another Euclidean space and let $A \in F_{[k+1]}$. Recall that for $I = \{i_1, \ldots, i_p\} \subset [k+1]$, we let

$$F_I = F_{i_1} \otimes^2 \cdots \otimes^2 F_{i_p}$$

and $\|\cdot\|_I$ be the corresponding (Euclidean) norm. Our purpose is to apply the previous lemma to the semi-norm defined by $\|y\| = \|\langle A, y\rangle\|_{\{k+1\}}$, for all $y \in F_{[k]}$. Notice that for all $x \in F_1 \times \cdots \times F_k$ and for all $\emptyset \subsetneq I \subsetneq [k]$

$$\mathbb{E}\|X_I \otimes x_{[k]\setminus I}\| \leq \left(\mathbb{E}\|X_I \otimes x_{[k]\setminus I}\|^2\right)^{1/2}$$
$$= \|\langle A, x_{[k]\setminus I}\rangle\|_{I \cup \{k+1\}},$$

which, according to the definition (8), is equal to $d_{[k]\setminus I}(0, x)$. In the same way, when $I = [k]$

$$\mathbb{E}\|\langle A, X_{[k]}\rangle\|_{\{k+1\}} \leq \|A\|_{[k+1]}.$$

We let the reader check that Lemma 10 then implies the following: for all $\tau \in (0, 1)$ and $x \in T$, letting $\epsilon_x = d^\tau(x, 0) + \tau^k\|A\|_{[k+1]}$, we have

$$P\big(d_{[k]}(x, \tau X) \leq \epsilon_x/2\big) \geq 2^{-c_k \tau^{-2}} \tag{16}$$

for some constant c_k depending only on k.

Lemma 9 follows easily from this observation. Indeed let $S \subset T$, since S and its translates have the same entropy numbers, we can assume that $0 \in S$. Then $\epsilon_x \leq \epsilon := \delta_{d^\tau}(S) + \tau^k\|A\|_{[k+1]}$ for all $x \in S$. Let S' be a subset of S satisfying:

(i) For all $x, y \in S'$, $d_{[k]}(x, y) \geq \epsilon$.
(ii) The set S' is maximal (for the inclusion) with this property.

By maximality S' is an ϵ-net of S, so $N(S, d_{[k]}, \epsilon) \leq \text{card } S'$. On the other hand, by (i) the balls (for $d_{[k]}$) of radius $\epsilon/2$ centered at different points of S' do not intersect. This, together with (16), implies that

$$2^{-c_k \tau^{-2}} \text{ card } S' \leq \sum_{x \in S'} P\big(d_{[k]}(x, \tau X) \leq \epsilon/2\big) \leq 1,$$

hence the result.

References

1. Fernique, X.: Fonctions aléatoires gaussiennes, vecteurs aléatoires gaussiens. Université de Montréal, Centre de Recherches Mathématiques, Montreal, QC (1997)
2. Hanson, D.L., Wright, F.T. : A bound on tail probabilities for quadratic forms of independant random variables. Ann. Math. Stat. **42**, 1079–1083 (1971)

3. Latała, R.: Estimates of moments and tails of Gaussian chaoses. Ann. Probab. **34**(6), 2315–2331 (2006)
4. Ledoux, M.: The Concentration of Measure Phenomenon. American Mathematical Society, Providence, RI (2001)
5. de la Peña V.M., Montgomery-Smith, S.: Bounds for the tail probabilities of U-statistics and quadratic forms. Bull. Am. Math. Soc. **31**, 223–227 (1994)
6. Talagrand, M.: Regularity of gaussian process. Acta Math. **159**, 99–149 (1987)
7. Talagrand, M.: The generic chaining. Upper and lower bounds of stochastic processes. Springer Monographs in Mathematics. Springer-Verlag, Berlin (2005)

The Lent Particle Method for Marked Point Processes

Nicolas Bouleau

Abstract Although introduced in the case of Poisson random measures (cf. Bouleau and Denis [2, 3]), the lent particle method applies as well in other situations. We study here the case of marked point processes. In this case the Malliavin calculus (here in the sense of Dirichlet forms) operates on the marks and the point process does not need to be Poisson. The proof of the method is even much simpler than in the case of Poisson random measures. We give applications to isotropic processes and to processes whose jumps are modified by independent diffusions.

Keywords Poisson random measure · Lent particle method · Marked point process · Isotropic process · Dirichlet form · Energy Image Density property · Lévy process · Wiener space · Ornstein-Uhlenbeck form · Malliavin calculus

1 Construction of the Upper Dirichlet Structure

(a) *Marked point processes.* Let $(X, \mathcal{X})$ and $(Y, \mathcal{Y})$ be two measurable spaces such that $\{x\} \in \mathcal{X} \ \forall x \in X$ and $\{y\} \in \mathcal{Y} \ \forall y \in Y$.

Let $\mathfrak{C}(X)$ be the configuration space of X i.e. the space of countable sum m of Dirac masses such that $m\{x\} \in \{0, 1\} \ \forall x \in X$, so that m may be indentified with its support. $\mathfrak{C}(X)$ is equipped with the smallest σ-field $\mathfrak{F}_X$ s.t. the maps $\omega \mapsto card(\omega \cap A)$ be measurable for any $A \in \mathcal{X}$.

Similarly we consider $\mathfrak{C}(X \times Y)$ equipped with $\mathfrak{F}_{X \times Y}$.

Let μ be a probability measure on $(Y, \mathcal{Y})$ and $\mathbb{Q}$ a probability measure on $(\mathfrak{C}(X), \mathfrak{F}_X)$. Let us denote by M the random measure on X with law $\mathbb{Q}$.

For F a function $\mathfrak{F}_{X \times Y}$-measurable and bounded, we may define a linear operator S by putting

$$S(F) = \int F((x_1, y_1), \dots, (x_n, y_n), \dots) \, \mu(dy_1) \cdots \mu(dy_n) \cdots$$

N. Bouleau (✉)
Ecole des Ponts ParisTech, Paris, France
e-mail: bouleau@enpc.fr

C. Donati-Martin et al. (eds.), *Séminaire de Probabilités XLIII*, Lecture Notes in Mathematics 341
2006, DOI 10.1007/978-3-642-15217-7_14, © Springer-Verlag Berlin Heidelberg 2011

the integral does not depend on the order of the numbering. $S(F)$ is $\mathfrak{F}_X$-measurable. Thus by

$$\mathbb{P}(F) = \int S(F)\,d\mathbb{Q}$$

we define a probability measure on $(\mathcal{C}(X \times Y), \mathfrak{F}_{X \times Y})$. We will say that $\mathbb{P}$ is the law of the random measure M marked by μ. It will be convenient to denote $N = M \odot \mu$ this random measure of law $\mathbb{P}$.

(b) *Dirichlet structure on a marked point process.* We suppose that the measure μ is such that there exists a local Dirichlet structure with carré du champ $(Y, \mathcal{Y}, \mu, \mathbf{d}, \gamma)$. Although not necessary, we assume for simplicity that constants belong to $\mathbf{d}_{loc}$ (see Bouleau–Hirsch [5], Chap. I, Definition 7.1.3.)

$$1 \in \mathbf{d}_{loc} \text{ which implies } \gamma[1] = 0.$$

By the same argument as the theorem on products of Dirichlet structures ([5], Chap. V, Sect. 2.2), the domain

$$\mathbb{D} = \{F \in L^2(\mathbb{P}), \text{ for } \mathbb{Q}\text{-a.e. } m = \sum \varepsilon_{x_i}, \forall i, \text{ for } \mu\text{-a.e.} u_1, \ldots, \mu\text{-a.e.} u_{i-1},$$

$$\mu\text{-a.e.} u_{i+1}, \ldots$$

$$F((x_1, u_1), \ldots, (x_{i-1}, u_{i-1}), (x_i, \cdot), (x_{i+1}, u_{i+1}) \ldots) \in \mathbf{d}$$

$$\text{and } \mathbb{E}_{\mathbb{P}}\left[\sum_i (\gamma[F])(u_i)\right] < +\infty\}$$

and the operator $\Gamma[F] = \sum_i (\gamma[F])(u_i)$ define a local Dirichlet structure

$$(\mathcal{C}(X \times Y), \mathfrak{F}_{X \times Y}, \mathbb{P}, \mathbb{D}, \Gamma).$$

(c) Let us recall the *Energy Image Density* property. For a σ-finite measure ν on some measurable space, a Dirichlet form on $L^2(\nu)$ with carré du champ γ is said to satisfy (EID) if for any d and for any $\mathbb{R}^d$-valued function U whose components are in the domain of the form

$$U_*[(\det\gamma[U, U^t]) \cdot \nu] \ll \lambda^d$$

where U_* denotes taking the image measure by U, det denotes the determinant, and λ^d the Lebesgue measure on $\mathbb{R}^d$.

For a local Dirichlet structure with carré du champ, the above property is always true for real-valued functions in the domain of the form (Bouleau [1], Bouleau–Hirsch [5], Chap. I, Sect. 7). It has been conjectured in 1986 (Bouleau–Hirsch [4], p. 251) that (EID) were true for any $\mathbb{R}^d$-valued function whose components are in the domain of the form for any local Dirichlet structure with carré du champ. This has been shown for the Wiener space equipped with the Ornstein-Uhlenbeck form and for some other structures by Bouleau–Hirsch (cf. [5], Chap. II, Sect. 5 and Chap. V, Example 2.2.4) and also for the Poisson space by A. Coquio [6] when the intensity measure is the Lebesgue measure on an open set, and in more general cases in [2] thanks to a result of Song [8]. But this conjecture being at present neither refuted nor proved in full generality, the property has to be established in every particular setting.

Lemma 1. *If the structure* $(Y, \mathcal{Y}, \mu, \mathbf{d}, \gamma)$ *is such that any finite product* $(Y, \mathcal{Y}, \mu,$ $\mathbf{d}, \gamma)^n$, $n \in \mathbb{N}$, *satisfies then the structure* $(\mathfrak{C}(X \times Y), \mathfrak{F}_{X \times Y}, \mathbb{P}, \mathbb{D}, \Gamma)$ *satisfies* (EID).

Proof. This is an application of Proposition 2.2.3 and Theorem 2.2.1 of Chap. V of [5].

(d) *The lent particle method.* Let us denote ϖ the current point of the space $\mathfrak{C}(X \times Y)$, and let us introduce the operators

$$\varepsilon^+_{(x,u)} \varpi = \varpi \cup \{(x, u)\} \qquad \varepsilon^-_{(x,u)} \varpi = \varpi \cap \{(x, u)\}^c$$

then we have the lent particle formula

$$\forall F \in \mathbb{D} \qquad \Gamma[F] = \int \varepsilon^- \gamma \varepsilon^+ F \, dN \qquad (1)$$

Proof. For $F \in \mathbb{D}$ we have

$$\varepsilon^+ F = F((x, u), (x_1, u_1), \ldots, (x_i, u_i), \ldots)$$
$$\gamma \varepsilon^+ F = \gamma[F((x, .), (x_1, u_1), \ldots, (x_i, u_i), \ldots)](u)$$

and $\int \varepsilon^- \gamma \varepsilon^+ F \, dN$ is the sum, when (x, u) varies among the points $(x_i, u_i) \in \varpi$ of the preceding result. This makes

$$\sum_i \gamma_i[F],$$

exactly what we obtained by the product construction. This shows also, by the definition of $\mathbb{D}$, that the integral $\int \varepsilon^- \gamma \varepsilon^+ F \, dN$ exists and belongs to $L^1(\mathbb{P})$. $\square$

(e) *Gradient.* Let us explain how could be done the construction of a gradient for the structure $(\mathfrak{C}(X \times Y), \mathfrak{F}_{X \times Y}, \mathbb{P}, \mathbb{D}, \Gamma)$ starting from a gradient for the structure $(Y, \mathcal{Y}, \mu, \mathbf{d}, \gamma)$.

Let us suppose that the structure $(Y, \mathcal{Y}, \mu, \mathbf{d}, \gamma)$ is such that the Hilbert space $\mathbf{d}$ be separable. Then by a result of Mokobodzki (see Bouleau–Hirsch [5], Exercise 5.9, p. 242) this Dirichlet structure admits a gradient operator in the sense that there exists a separable Hilbert space H and a continuous linear map D from $\mathbf{d}$ into $L^2(Y, \mu; H)$ such that

- $\forall u \in \mathbf{d}, \|D[u]\|^2_H = \gamma[u]$.
- If $F : \mathbb{R} \to \mathbb{R}$ is Lipschitz then

$$\forall u \in \mathbf{d}, \ D[F \circ u] = (F' \circ u) Du.$$

- If F is $\mathcal{C}^1$ (continuously differentiable) and Lipschitz from $\mathbb{R}^d$ into $\mathbb{R}$ (with $d \in \mathbb{N}$) then

$$\forall u = (u_1, \cdots, u_d) \in \mathbf{d}^d, \ D[F \circ u] = \sum_{i=1}^{d} (F_i' \circ u) D[u_i].$$

As only the Hilbertian structure of H plays a role, we can choose for H a space $L^2(R, \mathcal{R}, \rho)$ where $(R, \mathcal{R}, \rho)$ is a probability space such that the dimension of the vector space $L^2(R, \mathcal{R}, \rho)$ is infinite. As usual, we identify $L^2(\mu; H)$ and $L^2(Y \times R, \mathcal{Y} \otimes \mathcal{R}, \mu \times \rho)$ and we denote the gradient D by $\flat$:

$$\forall u \in \mathbf{d}, \ Du = u^\flat \in L^2(Y \times R, \mathcal{Y} \otimes \mathcal{R}, \mu \times \rho).$$

Without loss of generality, we assume moreover that operator $\flat$ takes its values in the orthogonal space of 1 in $L^2(R, \mathcal{R}, \rho)$, in other words we take for H the orthogonal of 1. So that we have

$$\forall u \in \mathbf{d}, \ \int u^\flat d\rho = 0 \ \mu\text{-a.e.}$$

Finally, by the hypothesis on γ we have

$$1 \in \mathbf{d}_{loc} \text{ which implies } \gamma[1] = 0 \text{ and } 1^\flat = 0.$$

With these tools and hypotheses we obtain easily a gradient for the structure $(\mathfrak{C}(X \times Y), \mathfrak{F}_{X \times Y}, \mathbb{P}, \mathbb{D}, \Gamma)$. We have to follow the same construction as above replacing the measure $\mathbb{Q} \times \mu^N$ by the measure $\mathbb{Q} \times \mu^N \times \rho^N$. This yields a random measure $N \odot \rho = M \odot \mu \times \rho$ defined under the probability measure $\mathbb{P} \times \rho^N$.

Now it is straightforward to show that the formula

$$F^\sharp = \int \varepsilon^- (\varepsilon^+ F)^\flat \, dN \odot \rho$$

for $F \in \mathbb{D}$ defines a gradient for the structure $(\mathfrak{C}(X \times Y), \mathfrak{F}_{X \times Y}, \mathbb{P}, \mathbb{D}, \Gamma)$ with values in $L^2(\mathbb{P} \times \rho^N)$. The existence of the integral $\int \varepsilon^- (\varepsilon^+ F)^\flat \, dN \odot \rho$ comes from the fact that it is controlled by that of $\int \varepsilon^- \gamma \varepsilon^+ F \, dN$ thanks to

$$\rho^N \left\{ \left(\int \varepsilon^- (\varepsilon^+ F)^\flat \, dN \odot \rho \right)^2 \right\} = \int \int (\varepsilon^- (\varepsilon^+ F)^\flat)^2 d\rho dN = \int \varepsilon^- \gamma [\varepsilon^+ F] dN$$

(similar formula as in Corollary 12 of [2]).

Example. If $F = e^{-N(f)}$, then

$$\varepsilon^+_{(x,u)} F = e^{-N(f)} e^{-f(x,u)}$$

$$\gamma \varepsilon^+_{(x,u)} F = e^{-2N(f)} e^{-2f(x,u)} \gamma[f]$$

$$\int \varepsilon^- \gamma \varepsilon^+ F \, dN = e^{-2N(f)} N(\gamma[f]) \qquad (= e^{-2N(f)} \Gamma[N(f)])$$

$(\Gamma[N(f)] = N(\gamma[f]))$ even in the non Poissonian case).

Let us summarize this construction which gives a result, similar to Theorem 17 of [2], obtained much more easily here for marked point processes than for random Poisson measures.

Theorem 1. *The carré du champ operator of the upper Dirichlet structure* $(\mathfrak{C}(X \times Y), \mathfrak{F}_{X \times Y}, \mathbb{P}, \mathbb{D}, \Gamma)$ *satisfies* $\forall F \in \mathbb{D}$

$$\Gamma[F] = \int \varepsilon^- \gamma[\varepsilon^+ F] dN$$

and this structure satisfies as soon as every finite product $(Y, \mathcal{Y}, \mu, \mathbf{d}, \gamma)^n$ *satisfies* (EID).

2 Application to Isotropic Processes

Let us consider a Lévy process $Z = (Z^1, Z^2)$ with values in $\mathbb{R}^2$ and Lévy measure $\sigma(dx, dy) = \nu(dr)\tau(d\theta)$ where τ is the uniform probability on the circle. Let us suppose that Z is centered without Gaussian part and that σ integrates $r^2 = x^2 + y^2$. Let N be the Poisson measure such that for any h_1 and h_2 in $L^2(ds)$

$$\int_0^t h_1(s)dZ_s^1 + h_2(s)dZ_s^2 = \int 1_{[0,t]}(s)(h_1(s)x + h_2(s)y) \, \tilde{N}(dsdxdy).$$

Let us construct the upper Dirichlet structure starting from the classical structure on the unit circle with domain H^1. And let us consider as illustration the very simple functional $F = Z_t = (r_t \cos\theta_t, r_t \sin\theta_t)$

$$\varepsilon^+_{(t_0, r_0, \theta_0)} F = (Z_t^1 + 1_{t \geq t_0} r_0 \cos\theta_0, Z_t^2 + 1_{t \geq t_0} r_0 \sin\theta_0)$$

$$\gamma \varepsilon^+ F = 1_{t \geq t_0} \begin{pmatrix} \sin^2\theta_0 & \cos\theta_0 \sin\theta_0 \\ \cos\theta_0 \sin\theta_0 & \cos^2\theta_0 \end{pmatrix} r_0^2$$

$$\Gamma[F] = \int \varepsilon^- \gamma \varepsilon^+ F \, dN = \int_0^t r^2 \begin{pmatrix} \sin^2\theta & \cos\theta \sin\theta \\ \cos\theta \sin\theta & \cos^2\theta \end{pmatrix} N(dsdrd\theta).$$

As soon as ν has an infinite mass, $\forall t > 0, \exists r_1 \neq 0, r_2 \neq 0$ et $\theta_1 \neq \theta_2$ s.t.

$$\Gamma[F] \geq r_1^2 \wedge r_2^2 \begin{pmatrix} \sin^2 \theta_1 + \sin^2 \theta_2 & \cos \theta_1 \sin \theta_1 + \cos \theta_2 \sin \theta_2 \\ \cos \theta_1 \sin \theta_1 + \cos \theta_2 \sin \theta_2 & \cos^2 \theta_1 + \cos^2 \theta_2 \end{pmatrix}$$

in the sense of positive symmetric matrices. Hence it follows that

$$\det \Gamma[F] \geq (r_1^2 \wedge r_2^2)^2 \sin^2(\theta_1 - \theta_2) > 0.$$

So that Z_t possesses a density on $\mathbb{R}^2$, as soon as $\nu(\mathbb{R}_+^*) = +\infty$. This result is probably known although not contained in the criterion of Sato [7] which supposes ν absolutely continuous. (Here ν may be possibly a weighted sum of Dirac masses because it does not carry any Dirichlet form).

The measure on the circle need not to be uniform provided that it carries a Dirichlet form such that its n-th powers satisfy (EID). The idea generalizes obviously replacing the circle by a d-dimensional sphere.

Actually, the process Z does not need to be Lévy. The method applies as well for instance to a real process purely discontinuous if we modify its jumps by i.i.d. transformations.

3 Insight on Transform of Lévy Processes by Diffusions

Since the Wiener measure is a probability measure we may take for $(Y, \mathcal{Y}, \mu)$ the Wiener space equipped with the Ornstein–Uhlenbeck structure. We know that is fulfilled as asked in Theorem 2.

Let us consider the SDE

$$X_t^x = x + \sum_{j=1}^d \int_0^t A_j(X_\tau^x, x) dB_\tau^j + \int_0^t B(X_\tau^x, x) d\tau \tag{2}$$

where $x \in \mathbb{R}^m$. The coefficients are $C^1 \cap Lip$ with respect to the first argument.

Let us take for $(X, \mathcal{X})$ the Euclidean space $(\mathbb{R}_+ \times \mathbb{R}^m, \mathcal{B}(\mathbb{R}_+ \times \mathbb{R}^m))$. Let M be a random Poisson measure on $\mathbb{R}_+ \times \mathbb{R}^m$ with intensity $ds \times \nu$ and law $\mathbb{Q}$ associated with a Lévy process Z. We denote $\varpi = \sum_\alpha \varepsilon_{(s_\alpha, x_\alpha)}$ the current point of $\mathfrak{C}(X)$.

Equation (2) is not that of a homogeneous Markov process because of the second argument in the coefficients. We can nevertheless define $\Pi_{t,x}(d\xi)$ to be the law of X_t^x and $\nu\Pi_t = \int \nu(dx)\Pi_{t,x}$ to be the law of X_t starting with the measure ν.

Lemma 2. *If the coefficients A_j, B are Lipschitz with respect to the first argument with constant independent of x and vanish at zero, the transition Π_t preserves Lévy measures and measures integrating $x \mapsto |x| \wedge 1$.*

Proof. By Gronwall lemma for $p = 1$ or $p = 2$, $\mathbb{E}|X_t^x|^p \leq k|x|^p e^{kt}$, this means that $\nu\Pi_t$ is a Lévy measure for any Lévy measure ν and the lemma is proved. $\square$

The transformed Lévy process $(T_t(Z))_s$ whose jumps are modified independently by the diffusion (2), which is a Lévy process with Lévy measure $\nu \Pi_t$, is a functional F of the marked point process. Let us suppose for simplicity that the jumps of Z are summable, i.e. that ν integrates $x \mapsto |x| \wedge 1$, then F may be written

$$F = \int_{[0,s] \times \mathbb{R}^m \times Y} X_t^x(y) N(dsdxdy)$$

with as above $N = M \odot \mu$. The lent particle formula gives

$$F^\sharp = \int_{[0,s] \times \mathbb{R}^m \times Y \times R} (X_t^x)^\flat \, d(N \odot \rho)$$

and

$$\Gamma[F] = \int_{[0,s] \times \mathbb{R}^m \times Y} \gamma[X_t^x] \, dN.$$

Now $(X_t^x)^\flat$ and $\gamma[X_t^x]$ are known by the usual Malliavin calculus : $(.)^\flat$ is a gradient on the Wiener space associated with the O-U structure, for which we can choose (cf. [5]) the operator defined by

$$\left(\int h(s) dB_s^j \right)^\flat = \int h(s) d\hat{B}_s^j \quad h \in L^2(\mathbb{R}_+)$$

where $\hat{B}^j$ are independent copies of B^j.

$$(X_t^x)^\flat = K_t \int_0^t K_v^{-1} \sigma(X_v^x, x) \cdot d\hat{B}_v$$

$$\gamma[X_t^x] = K_t \left[\int_0^t K_v^{-1} \sigma(X_v^x, x) \sigma^*(X_v^x, x)(K_v^{-1})^* dv \right] K_t^*$$

where σ is the matrix whose columns are the A_j $j = 1, \ldots, d$ and K the continuous invertible matrix valued process solution of

$$K_t^x = I + \sum_{j=1}^d \int_0^t \partial A^j(X_v^x, x) K_v^x dB_v^j + \int_0^t \partial B(X_v^x, x) K_v^x dv.$$

where ∂A^j and ∂B are the Jacobian matrices with respect to the first argument.
We can write

$$\Gamma[F] = \int_{[0,s] \times \mathbb{R}^m \times Y} \left(K_t^x \left[\int_0^t (K_v^x)^{-1} \sigma(X_v^x, x) \sigma^*(X_v^x, x)(K_v^x)^{-1*} dv \right] (K_t^x)^* \right)$$
$$(y) M \odot \mu(dudxdy)$$

By the (EID) property, for F to possess a density it suffises that the vector space $\mathcal{V}$ spanned by the column vectors of the matrices

$$\left(K_t^x(K_v^x)^{-1}\sigma(X_v^x,x)\right)(y) \qquad 0 \le v \le t, \quad x \in \mathbb{R}^m, \quad y \in Y,$$

be m-dimensional a.s.

If we restrict the study to the case where the diffusion coefficients do not depend on the first argument $A_j(X_u^x,x) = A_j(x)$, i.e. for the SDE

$$X_t^x = x + \sum_{j=1}^d A_j(x)B_t^j + \int_0^t B(X_v^x,x)dv$$

then, taking v close to t, the space $\mathcal{V}$ contains the vectors

$$A_j(\Delta Z_u) \qquad j = 1,\ldots,d \qquad u \in JT(Z)$$

where $JT(Z)$ denotes the jump times of Z before s and we have

Proposition 1. *Let us suppose the Lévy measure v infinite. If the vectors $A_j(x)$ are such that for any infinite sequence $x_n \in \mathbb{R}^m$, $x_n \ne 0$, tending to 0, the vector space spanned by the vectors*

$$A_j(x_n), \quad j = 1,\ldots,d, \quad n \in \mathbb{N}$$

is m-dimensional then the Lévy process $(T_t(Z))_s$ has a density on $\mathbb{R}^m$.

Proof. The result comes from the above condition by the fact that Z has infinitely many jumps of size near zero.   □

As in part 2, the fact that Z be a Lévy process does not really matter. The method applies to the transform of the jumps of any process as soon as the perturbations are i.i.d and carry a Dirichlet form yielding (EID).

References

1. Bouleau, N.: Décomposition de l'énergie par niveau de potentiel. In: Lecture Notes in Mathematics, vol. 1096, Springer, Berlin (1984)
2. Bouleau, N., Denis, L.: Energy image density property and local gradient for Poisson random measures. J. Funct. Anal. **257**(4), 1144–1174 (2009)
3. Bouleau, N., Denis, L.: Application of the lent particle method to Poisson driven SDE's. Probab. Theor. Relat. Field (2010) online first DOI 10.1007/s00440-010-0303-x
4. Bouleau, N., Hirsch, F.: Formes de Dirichlet générales et densité des variables aléatoires réelles sur l'espace de Wiener. J. Funct. Anal. **69**(2), 229–259 (1986)
5. Bouleau, N., Hirsch, F.: Dirichlet Forms and Analysis on Wiener Space. De Gruyter, Berlin (1991)

6. Coquio, A.: Formes de Dirichlet sur l'espace canonique de Poisson et application aux équations différentielles stochastiques. Ann. Inst. Henri. Poincaré **19**(1), 1–36 (1993)
7. Sato, K.-I.: Absolute continuity of multivariate distributions of class L. J. Multivariate Anal. **12**(1), 89–94 (1982)
8. Song, S.: Admissible vectors and their associated Dirichlet forms. Potential Anal. **1**(4), 319–336 (1992)

Ewens Measures on Compact Groups and Hypergeometric Kernels

Paul Bourgade, Ashkan Nikeghbali, and Alain Rouault

Abstract On unitary compact groups the decomposition of a generic element into product of reflections induces a decomposition of the characteristic polynomial into a product of factors. When the group is equipped with the Haar probability measure, these factors become independent random variables with explicit distributions. Beyond the known results on the orthogonal and unitary groups ($O(n)$ and $U(n)$), we treat the symplectic case. In $U(n)$, this induces a family of probability changes analogous to the biassing in the Ewens sampling formula known for the symmetric group. Then we study the spectral properties of these measures, connected to the pure Fisher-Hartvig symbol on the unit circle. The associated orthogonal polynomials give rise, as n tends to infinity to a limit kernel at the singularity.

Keywords Decomposition of Haar measure · Random matrices · Characteristic polynomials · Ewens sampling formula · Correlation kernel

1 Introduction

In this paper, $U(n, K)$ is the unitary group over $K = \mathbb{R}, \mathbb{C}$ or $\mathbb{H}$ (the set of real quaternions).

P. Bourgade
Institut Telecom & Université Paris 6, 46 rue Barrault, 75634 Paris Cedex 13, France
e-mail: paulbourgade@gmail.com

A. Nikeghbali
Institut für Mathematik, Universität Zürich, Winterthurerstrasse 190, CH-8057 Zürich, Switzerland
e-mail: ashkan.nikeghbali@math.uzh.ch

A. Rouault (✉)
Université Versailles-Saint Quentin, LMV, Bâtiment Fermat, 45 avenue des Etats-Unis, 78035 Versailles Cedex, France
e-mail: alain.rouault@math.uvsq.fr

C. Donati-Martin et al. (eds.), *Séminaire de Probabilités XLIII*, Lecture Notes in Mathematics 351
2006, DOI 10.1007/978-3-642-15217-7_15, © Springer-Verlag Berlin Heidelberg 2011

Let U be distributed with the Haar measure on $U(n, \mathbb{C})$. The random variable $\det(\mathrm{Id}_n - U)$ has played a crucial role in recent years in the study of some connections between random matrix theory and analytic number theory (see [21] for more details). In [10], the authors show that $\det(\mathrm{Id}_n - U)$ can be decomposed as a product of n independent random variables:

$$\det(\mathrm{Id}_n - U) \overset{\text{law}}{=} \prod_{k=1}^{n} \left(1 - e^{i\omega_k} \sqrt{B_{1,k-1}} \right), \tag{1}$$

where $\omega_1, \ldots, \omega_n, B_{1,0}, \ldots, B_{1,n-1}$ are independent, the $\omega_k's$ being uniformly distributed on $(-\pi, \pi)$ and the $B_{1,j}$'s $(0 \leq j \leq n - 1)$ being beta distributed with parameters 1 and j (with the convention that $B_{1,0} = 1$). In particular, from such a decomposition, fundamental quantities such as the Mellin-Fourier transform of $\det(\mathrm{Id}_n - U)$ follow at once. The main ingredient to obtain the decomposition (1) is a recursive construction of the Haar measure using complex reflections. In particular, every $U \in U(n, \mathbb{C})$ can be decomposed as a product of n independent reflections. More precisely, it is proved in [10] that if $s_1, \ldots, s_n$ are n independent random variables such that for every $k \leq n$, s_k is uniformly distributed on the kth dimensional unit sphere $\mathscr{S}^k$ in $\mathbb{C}^k$ and if $R^{(k)}$ is the reflection of $\mathbb{C}^k$ mapping s_k onto the first vector of the canonical basis, then

$$R^{(n)} \begin{pmatrix} \mathrm{Id}_1 & 0 \\ 0 & R^{(n-1)} \end{pmatrix} \cdots \begin{pmatrix} \mathrm{Id}_{n-2} & 0 \\ 0 & R^{(2)} \end{pmatrix} \begin{pmatrix} \mathrm{Id}_{n-1} & 0 \\ 0 & R^{(1)} \end{pmatrix} \sim \mu_{U(n,\mathbb{C})},$$

where $\mu_{U(n,\mathbb{C})}$ stands for the Haar measure on $U(n, \mathbb{C})$. At this stage two remarks are in order. First, a similar method works to generate the Haar measure on the orthogonal group $O(n, \mathbb{R})$ (see [10]) and this was already noticed by Mezzadri in [24] using Householder reflections. But as already noticed in [10], Householder reflections would not work for $U(n, \mathbb{C})$ (see next section for more details). Moreover in [10], a decomposition such as (1) could not be obtained for the symplectic group $USp(2n, \mathbb{C})$, which also plays an important role in the connections between random matrix theory and the study of families of L functions (see [19, 20]). Indeed, there does not seem to be a natural way to generate recursively the Haar measure on this group.

Question 1. Is there any decomposition of $\det(\mathrm{Id}_n - U)$ as a product of independent variables of the type (1), when U is drawn from $USp(2n, \mathbb{C})$, according to the Haar measure?

In this paper we shall prove that, in a sense to be made precise, if a subgroup $\mathcal{G}$ of $U(n, K)$ contains enough reflections, then one can recursively generate the Haar measure and obtain a decomposition of the type (1) for $\det(\mathrm{Id}_n - U)$, $U \in \mathcal{G}$. In particular this will apply to $U(n, \mathbb{H})$ which can be identified with the symplectic group, hence answering question 1 above. Our recursive decomposition of the

Haar measure also applies to the symmetric group. This leads us to our second remark concerning the generation of the Haar measure obtained in [10] and explained above. Indeed, this way of generating an element of $U(n, \mathbb{C})$ which is Haar distributed by choosing a vector $(s_1, \ldots, s_n)$ of independent variables from $\mathscr{S}^1 \times \ldots \times \mathscr{S}^n$, each s_i being uniformly distributed, is reminiscent of the generation of a random permutation according to the so-called Chinese restaurant process which we briefly describe (see [29] for a complete treatment). Let $[n]$ denote the set $\{1, \ldots, n\}$ and $\mathcal{S}_n$ the symmetric group of order n. It is known that for $n \geq 2$, every permutation $\sigma \in \mathcal{S}_n$ can be decomposed in the following way:

$$\sigma = \tau_n \circ \cdots \circ \tau_2 \tag{2}$$

where for $k = 2, \ldots, n$, either τ_k is the identity or τ_k is the transposition (k, m_k) for some $m_k \in [k - 1]$. In the first case we will say by extension that it is the transposition (k, m_k) with $m_k = k$. This decomposition is unique, see Tsilevich [32], the lemma p. 4075. It corresponds to the Chinese restaurant generation of a permutation. Let us consider cycles as "tables". Integer 1 goes to the first table. If $\tau_2 \neq \text{Id}$, then integer 2 goes to the first table, at the left of 1. If $\tau_2 = \text{Id}$, it goes to a new table. When integers $1, \ldots, k$ are placed, then $k + 1$ goes to a new table if $\tau_{k+1} = \text{Id}$, and goes to the left of $\tau_{k+1}(k + 1) = m_{k+1}$ if not. We get a bijection between $[1] \times [2] \times \cdots \times [n] \to \mathcal{S}_n$. It is projective (or consistent) in the sense that if σ is in $\mathcal{S}_{n+1}$ the restriction of σ to $[n]$ is in $\mathcal{S}_n$.

In this setting, the number of cycles k_σ of a permutation σ is the number of tables, i.e. the number of Id in (2) i.e.[1]

$$k_\sigma = \sum_1^n \xi_r, \tag{3}$$

where $\xi_r = 1(\tau_r = \text{Id})$. For a matricial rewriting, we make a change of basis. Let $e'_j = e_{n-j+1}$ and let $R^{(k)}$ be the restriction of τ_k to $[k]$. Then the product in (2) is represented by

$$R^{(n)} \begin{pmatrix} \text{Id}_1 & 0 \\ 0 & R^{(n-1)} \end{pmatrix} \cdots \begin{pmatrix} \text{Id}_{n-2} & 0 \\ 0 & R^{(2)} \end{pmatrix}.$$

If at each stage, the integer m_k is chosen uniformly in $[k]$, then the induced measure on $\mathcal{S}_n$ is the uniform distribution denoted by $\mu_{\mathcal{S}_n}$.

Actually, one can more generally generate in this way the Ewens measure on $\mathcal{S}_n$ (see Tsilevich [32] and Pitman [29]). The Ewens measure $\mu^{(\theta)}$, $\theta > 0$, is a deformation of $\mu_{\mathcal{S}_n}$ obtained by performing a change of probability measure or a sampling in the following way:

$$\mu_n^\theta(\sigma) = \frac{\theta^{k_\sigma}}{(\theta)_n} \cdot \mu_{\mathcal{S}_n}(\sigma). \tag{4}$$

---

[1] The other construction of a random permutation named Feller's coupling ([3]) uses the variables in the reverse order $\xi_n, \cdots, \xi_1$, but this construction is not projective.

To generate μ_n^θ, one has to pick n integers $m_1, m_2, \ldots, m_n$, independently, from $[1] \times \cdots \times [n]$ according to the probability distribution

$$\mathbb{P}(m_k = k) = \frac{\theta}{\theta + k - 1}, \quad \mathbb{P}(m_k = j) = \frac{1}{\theta + k - 1} \quad j = 1, \cdots, k - 1.$$

Question 2. Is there an analogue of the Ewens measure on the unitary group $U(n, \mathbb{C})$?

We shall see in this paper that there indeed exists an analogue of the Ewens measure on $U(n, \mathbb{C})$: more precisely we generalize (4) to unitary groups and a particular class of their subgroups. The analogue of transpositions are reflections and the weight of the sampling is now $\det(\mathrm{Id} - U)^{\overline{\delta}} \det(\mathrm{Id} - \overline{U})^{\delta}, \delta \in \mathbb{C}, \mathfrak{Re}(\delta) > -1/2$, so that the measure $\mu_{U(n)}^{(\delta)}$ on $U(n)$, which is defined by

$$\mathbb{E}_{\mu_{U(n)}^{(\delta)}} (f(U)) = \frac{\mathbb{E}_{\mu_{U(n)}} \left(f(U) \det(\mathrm{Id} - U)^{\overline{\delta}} \det(\mathrm{Id} - \overline{U})^{\delta} \right)}{\mathbb{E}_{\mu_{U(n)}} \left(\det(\mathrm{Id} - U)^{\overline{\delta}} \det(\mathrm{Id} - \overline{U})^{\delta} \right)}$$

for any test function f, is the analogue of the Ewens measure. Such samplings with $\delta \in \mathbb{R}$ have already been studied on the finite-dimensional unitary group by Hua [18], and results about the infinite dimensional case (on complex Grassmannians) were given by Pickrell ([27, 28]). More recently, Neretin [26] also considered this measure, introducing the possibility $\delta \in \mathbb{C}$. Borodin and Olshanski [7] have used the analogue of this measure in the framework of the infinite dimensional unitary group and proved ergodic properties. Forrester and Witte in [34] referred to this measure as the cJUE distribution. We also studied this ensemble in [12] in relation with the theory of orthogonal polynomials on the unit circle. Following [12,34] we shall call the ensemble of unitary matrices endowed with this sampled measure the *circular Jacobi ensemble*.

It is natural to ask whether the circular Jacobi ensemble has some interesting properties: indeed, the case $\delta = 0$ corresponds to the Haar measure and it is well known this ensemble enjoys many remarkable spectral properties. For instance, the point process associated to the eigenvalues is determinantal and the associated rescaled kernel converges to the sine kernel. The projection of the measures $\mu_{U(n)}^{(\delta)}$ on the spectrum has the density

$$\frac{1}{\mathcal{Z}_n} \prod_{j=1}^{n} w^{\mathbb{T}}(e^{i\theta_j}) \prod_{1 \le i < j \le n} |e^{i\theta_i} - e^{i\theta_j}|^2$$

where the weight $w^{\mathbb{T}}$ on $\mathbb{T} = \{e^{i\theta}, \theta \in [-\pi, \pi]\}$ is defined by

$$w^{\mathbb{T}}(e^{i\theta}) = (1 - e^{i\theta})^{\overline{\delta}} (1 - e^{-i\theta})^{\delta} = (2 - 2\cos\theta)^a e^{-b(\pi \operatorname{sgn}\theta - \theta)},$$

($\delta = a + ib$) and $\mathcal{Z}_n$ is a normalization constant. Note that when $b \ne 0$, an asymmetric singularity at 1 occurs. The statistical properties of the θ_k's depend on the successive orthonormal polynomials (φ_k) with respect to the normalized version $\widetilde{w}^{\mathrm{T}}$ of w^{T} and the normalized reproducing kernel

$$\widetilde{K}_n^{\mathrm{T}}(e^{i\theta}, e^{i\tau}) = \sqrt{\widetilde{w}^{\mathrm{T}}(e^{i\theta})\widetilde{w}^{\mathrm{T}}(e^{i\tau})} \sum_{\ell=0}^{n-1} \overline{\varphi_\ell(e^{i\theta})}\varphi_\ell(e^{i\tau}).$$

In [7] the authors consider the image of $\mu_{U(n)}^{(\delta)}$ by the Cayley transform on the set of Hermitian matrices and make a thorough study of the spectral properties of this random matrix ensemble. In particular they prove that the eigenvalues form a determinantal process and show that the associated rescaled kernel converges to some hypergeometric kernel. As expected, we shall see that the eigenvalues process of the circular Jacobi ensemble is also determinantal and for every n, we identify the hypergeometric kernel $K_n^{(\delta)}$ associated with it.

Question 3. Is there an appropriate rescaling of the kernels $K_n^{(\delta)}$ such that the rescaled kernels converge to some kernel $K_\infty^{(\delta)}$?

We shall see that the answer to Question 3 is positive and that the kernel $K_\infty^{(\delta)}$ is a *confluent hypergeometric kernel*, with a natural connection to that obtained by Borodin and Olshanski in [7] on the set of Hermitian matrices. The case $\delta = 0$ corresponds to the sine kernel.

The weight w^{T} is a generic example leading to a singularity

$$c^{(+)}|\theta|^{2a} \mathbb{1}_{\theta>0} + c^{(-)}|\theta|^{2a} \mathbb{1}_{\theta<0}$$

at $\theta = 0$, with distinct positive constants $c^{(+)}$ and $c^{(-)}$. The confluent hypergeometric kernel, depending on the two parameters a and $b = \frac{1}{2\pi}\log(c^{(-)}/c^{(+)})$, is actually universal for the measures presenting the above singularity, as proved in a forthcoming paper, following the method initiated by Lubinsky ([22, 23]). For a universality result when δ is real see [30].

The layout of the paper is as follows. In Sect. 2 we present the generation by reflections and deduce a splitting formula for the characteristic polynomial (Theorem 2). As an application, we define the generalized Ewens measure depending on the complex parameter δ (Theorem 3). Section 3 is devoted to a study of the kernel which governs the correlations of eigenvalues when the unitary group is equipped with this measure and its asymptotics (Theorem 5). The main properties of the families of hypergeometric functions $_2F_1$ and $_1F_1$ are recalled in the Appendix.

2 Generating the Haar Measure and the Generalized Ewens Measure

2.1 Complex Reflections

Reflections play a central role in the generation of the Haar measure for the classical compact groups. In the case of $O(n)$ the decomposition into a product of reflections is well known, see [15] and other references as explained in [24]. Householder reflections are generally used in the case of $O(n)$, but they are not suitable for $U(n, \mathbb{C})$. Indeed, recall that Householder reflections are of the form $H_v = \mathrm{Id} - 2v \langle v | \cdot \rangle$. For every unit y, it is possible to choose v such that $H_v y = \alpha e_1$ with $\alpha = \pm \frac{y_1}{|y_1|}$, where e_1 is the first element of the canonical basis. So when the ground field is $\mathbb{C}$, then $\alpha \neq 1$ in general and there does not exist a Householder reflection which maps y onto e_1, whereas this can always be achieved when the ground field is $\mathbb{R}$. That is why it is not possible to directly extend the arguments in [24] to $U(n, \mathbb{C})$. In [10] and [12] it is proposed to use complex (resp. quaternionic) proper reflections, that is norm preserving automorphisms of $\mathbb{C}^n$ (resp. $\mathbb{H}^n$) that leave exactly one hyperplane pointwise fixed. So a reflection will be either the identity or a unitary transformation U such that $I - U$ is of rank one. It may be written as

$$s_{a,\lambda}(y) = y - a \frac{(1 - \lambda)\langle a, y \rangle}{|a|^2}$$

where $a \in \mathbb{H}^n$ and $\lambda \in \mathbb{H}$ with $|\lambda| = 1$ (λ is the second eigenvalue). If $x \neq e_1$, there exists a reflection mapping e_1 onto x. It is enough to take $a = e_1 - x$ and $\lambda = -(1 - x_1)(1 - \bar{x}_1)^{-1}$ where $x_1 = \langle e_1, x \rangle$.

2.2 Generating the Haar Measure on $U(n, K)$ and on some of its Subgroups

We first give conditions under which an element of a subgroup of $U(n, K)$ (under the Haar measure) can be generated as a product of independent reflections. This will lead to some remarkable identities for the characteristic polynomial.

Let $(e_1, \ldots, e_n)$ be an orthonormal basis of $\mathbb{K}^n$. Let $\mathcal{G}$ be a subgroup of $U(n, K)$ and for all $1 \leq k \leq n - 1$, let

$$\mathcal{H}_k = \{G \in \mathcal{G} \mid G(e_j) = e_j, \ 1 \leq j \leq k\},$$

the subgroup of $\mathcal{G}$ which stabilizes $e_1, \ldots, e_k$. We set $\mathcal{H}_0 = \mathcal{G}$. For a generic compact group $\mathcal{A}$, we write $\mu_{\mathcal{A}}$ for the unique Haar probability measure on $\mathcal{A}$. Finally for all $1 \leq k \leq n$ let p_k be the map $U \mapsto U(e_k)$.

Proposition 1. *Let $G \in \mathcal{G}$ and $H \in \mathcal{H}_1$ be independent random matrices, and assume that $H \sim \mu_{\mathcal{H}_1}$. Then $GH \sim \mu_{\mathcal{G}}$ if and only if $G(e_1) \sim p_1(\mu_{\mathcal{G}})$.*

Proof. The proof is exactly the same as in [10] Proposition 2.1, changing $U(n + 1)$ into $\mathcal{G}$ and $U(n)$ into $\mathcal{H}$.

Definition 1. A sequence $(v_0, \ldots, v_{n-1})$ of probability measures on $\mathcal{G}$ is said to be coherent with $\mu_{\mathcal{G}}$ if for all $0 \leq k \leq n - 1$,

$$v_k(\mathcal{H}_k) = 1 \quad \text{and} \quad p_{k+1}(v_k) = p_{k+1}(\mu_{\mathcal{H}_k}).$$

In the following, $\nu_0 \star \nu_1 \star \cdots \star \nu_{n-1}$ stands for the law of a random variable $H_0 H_1 \ldots H_{n-1}$ where all H_i's are independent and $H_i \sim \nu_i$. Now we can provide a general method to generate an element of $\mathcal{G}$ endowed with its Haar measure.

Theorem 1. *If $\mathcal{G}$ is a subgroup of $U(n, K)$ and $(\nu_0, \ldots, \nu_{n-1})$ is a sequence of coherent measures with $\mu_\mathcal{G}$, then we have:*

$$\mu_\mathcal{G} = \nu_0 \star \nu_1 \star \cdots \star \nu_{n-1}.$$

Proof. It is sufficient to prove by induction on $1 \leq k \leq n$ that

$$\nu_{n-k} \star \nu_{n-k+1} \star \cdots \star \nu_{n-1} = \mu_{\mathcal{H}_{n-k}},$$

which gives the desired result for $k = n$. If $k = 1$ this is obvious. If the result is true at rank k, it remains true at rank $k + 1$ by a direct application of Proposition 1 to the groups $\mathcal{H}_{n-k-1}$ and its subgroup $\mathcal{H}_{n-k}$.

As an example, take the orthogonal group $O(n)$. Let $\mathscr{S}_{\mathbb{R}}^{(k)}$ be the unit sphere $\{x \in \mathbb{R}^k \mid |x| = 1\}$ and, for $s_k \in \mathscr{S}_{\mathbb{R}}^{(k)}$, let $R^{(k)}$ be the matrix of the reflection which transforms s_k into e_1. If s_k is uniformly distributed on $\mathscr{S}_{\mathbb{R}}^{(k)}$ and if all the s_k are independent, then by Theorem 1, the matrix

$$R^{(n)} \begin{pmatrix} 1 & 0 \\ 0 & R^{(n-1)} \end{pmatrix} \cdots \begin{pmatrix} \mathrm{Id}_{n-2} & 0 \\ 0 & R^{(2)} \end{pmatrix} \begin{pmatrix} \mathrm{Id}_{n-1} & 0 \\ 0 & R^{(1)} \end{pmatrix}.$$

is $\mu_{O(n)}$ distributed.

2.3 Splitting of the Characteristic Polynomial

In view to phrase a general version of formula (1) which is proved in [10], we need the following definition:

Definition 2. Note $\mathcal{R}_k$ the set of elements in $\mathcal{H}_k$ which are reflections. If for all $0 \leq k \leq n - 1$

$$\{R(e_{k+1}) \mid R \in \mathcal{R}_k\} = \{H(e_{k+1}) \mid H \in \mathcal{H}_k\},$$

the group $\mathcal{G}$ will be said to satisfy condition (R) (R standing for reflection).

Remark 1. It is easy to see that $U(n, K)$ and $\mathcal{S}_n$ satisfy condition (R). In the next subsection we shall see more examples.

Lemma 1. *Let $\mathcal{G}$ be a subgroup of $U(n, K)$ which satisfies condition (R). Let $G \in \mathcal{G}$. Then there exist reflections $R_k \in \mathcal{R}_k$, $0 \leq k \leq n - 1$, such that*

$$G = R_0 R_1 \ldots R_{n-1}. \tag{5}$$

Proof. This result has been established in [12] when $\mathcal{G} = U(n, \mathbb{C})$. The proof in this more general case goes exactly along the same line.

The following deterministic lemma is a key result to obtain a decomposition of $\det(\mathrm{Id}_n - U)$ as a product of independent random variables:

Lemma 2. *If for $k = 1, \ldots, n - 1$, $R_k \in \mathcal{R}_k$, then*

$$\det(\mathrm{Id}_n - R_0 \cdots R_{n-1}) = \prod_{k=0}^{n-1} (1 - \langle e_{k+1}, R_k(e_{k+1}) \rangle) . \tag{6}$$

Proof. We start with $\det(\mathrm{Id}_n - RH) = (\det H) \det(H^* - R)$. Since H (hence H^*), stabilizes e_1, we have

i) $(H^* - R)(e_1) = e_1 - R(e_1) =: a$ (say),
ii) for $w \perp e_1$, $H^*(w) \perp e_1$ and since R is a reflection, $R(w) - w$ is a scalar multiple of a.

By the multilinearity of the determinant, we get

$$\det(H^* - R) = \langle e_1, e_1 - R(e_1) \rangle \det(\pi(H^*) - \mathrm{Id}_{n-1})$$

which yields

$$\det(\mathrm{Id}_n - RH) = (1 - \langle e_1, R(e_1) \rangle) \det(\mathrm{Id}_{n-1} - \pi(H)) .$$

Iterating, we can conclude.   □

The following result now follows immediately from Theorem 1 and Lemmas 1 and 2.

Theorem 2. *Let $\mathcal{G}$ be a subgroup of $U(n, K)$ satisfying condition (R), and let $(v_0, \ldots, v_{n-1})$ be coherent with $\mu_{\mathcal{G}}$. If $G \sim \mu_{\mathcal{G}}$, then*

$$\det(\mathrm{Id} - G) \stackrel{\mathrm{law}}{=} \prod_{k=0}^{n-1} (1 - \langle e_{k+1}, H_k(e_{k+1}), \rangle) .$$

where $H_k \sim v_k$, $0 \le k \le n - 1$, are independent.

2.4 Applications

2.4.1 The Symmetric Group

Consider now $\mathcal{S}_n$ the group of permutations of size n. An element $\sigma \in \mathcal{S}_n$ can be identified with the matrix $(\delta^j_{\sigma(i)})_{1 \le i, j \le n}$ (δ is Kronecker's symbol). It is clear that 1 is eigenvalue of this matrix, with eigenvector $e_1 + \cdots + e_n$. Ben Hambly et al. [17] considered the characteristic polynomial at $s \neq 1$. To make relevant our problem of determinant splitting, we introduce wreath products, following the definition of Wieand [33].

Let F be a subgroup of $\mathbb{T} = \{x \in \mathbb{C} \mid |x|^2 = 1\}$, endowed with the Haar probability measure μ_F. Then the wreath product $F \wr \mathcal{S}_n$ provides another example of determinant-splitting. An element of F^n can be thought of as a function from the set $[n]$ to F. The group $\mathcal{S}_n$ acts on F^n in the following way: if $f = (f(1), \ldots, f(n)) \in F^n$ and $\sigma \in \mathcal{S}_n$, define $f_\sigma \in F^n$ to be the function $f_\sigma = f \circ \sigma^{-1}$. Finally take the product on F^n to be $(f(1), \ldots, f(n)) \cdot (g(1), \ldots, g(n)) = (fg(1), \ldots, fg(n))$. The wreath product of F by $\mathcal{S}_n$, denoted $F \wr \mathcal{S}_n$, is the group of elements $\{(f; s) : f \in F^n, \sigma \in \mathcal{S}_n\}$ with multiplication

$$(f; \sigma) \cdot (h; \sigma') = (f h_\sigma; \sigma \sigma').$$

If we represent $(f; \sigma)$ by the matrix $(f(i) \delta_{\sigma(j)}^i)_{1 \leq i, j \leq n}$, then the product in $F \wr \mathcal{S}_n$ corresponds to the usual matricial product which makes $F \wr \mathcal{S}_n$ a subgroup of $U(n, \mathbb{C})$. The usual examples are $F = \{1\}$, $F = \mathbb{Z}_2$ and $F = \mathbb{T}$.

Corollary 1. *Let $G \in \mathcal{G}(= F \wr \mathcal{S}_n)$ be $\mu_{\mathcal{G}}$ distributed. Then*

$$\det(\mathrm{Id}_n - G) \overset{\text{law}}{=} \prod_{j=1}^n \left(1 - \varepsilon_j X_j\right),$$

with $\varepsilon_1, \ldots, \varepsilon_n, X_1, \ldots, X_n$ independent random variables, the ε_j's μ_F distributed, $\mathbb{P}(X_j = 1) = 1/j$, $\mathbb{P}(X_j = 0) = 1 - 1/j$.

Proof. We apply Theorem 2. As reflections correspond now to transpositions, condition (R) holds. Moreover $R_k(e_{k+1})$ is uniformly distributed on the set $F e_{k+1} \cup \cdots \cup F e_n$, so that $\langle e_{k+1}, R_k(e_{k+1}) \rangle$ is 0 with probability $1 - \frac{1}{n-k}$ and otherwise, it is uniform on F. □

Remark 2. Notice that if $G = (f; \sigma)$ with $\sigma = \tau_n \circ \cdots \circ \tau_2$ (cf. (2)), then X_j is the indicator function of $\tau_j = \mathrm{Id}$.

2.4.2 Unitary and Orthogonal Groups

Take $\mathcal{G} = U(n, \mathbb{C})$. Then $\mu_{\mathcal{H}_k} = f_k(\mu_{U(n-k, \mathbb{C})})$ where $f_k : A \in U(n - k, \mathbb{C}) \mapsto \mathrm{Id}_k \oplus A$. As all reflections with respect to a hyperplane of $\mathbb{C}^{n-k}$ are elements of $U(n - k, \mathbb{C})$, one can apply Theorem 1 and Lemma 2. The Hermitian products $\langle e_k, h_k(e_k) \rangle$ are distributed as the first coordinate of the first vector of an element of $U(n - k, \mathbb{C})$, that is to say the first coordinate of the $(n - k)$-dimensional unit complex sphere with uniform measure:

$$\langle e_{k+1}, H_k(e_{k+1}) \rangle \overset{\text{law}}{=} e^{i\omega_n} \sqrt{B_{1, n-k-1}}$$

with ω_n uniform on $(-\pi, \pi)$ and independent of $B_{1,n-k-1}$, a beta variable with parameters 1 and $n - k - 1$.

Therefore, as a consequence of Theorem 2, we obtain the following decomposition formula derived in [10]. For $g \in U(n, \mathbb{C})$ which is $\mu_{U(n,\mathbb{C})}$ distributed, one has

$$\det(\mathrm{Id}_n - G) \stackrel{\mathrm{law}}{=} \prod_{k=1}^{n} \left(1 - e^{i\omega_k} \sqrt{B_{1,k-1}}\right),$$

with $\omega_1, \ldots, \omega_n, B_{1,0}, \ldots, B_{1,n-1}$ independent random variables, the ω_k's uniformly distributed on $(-\pi, \pi)$ and the $B_{1,j}$'s $(0 \le j \le n-1)$ being beta distributed with parameters 1 and j (by convention, $B_{1,0} = 1$).

A similar reasoning may be applied to $SO(2n)$ (with the complex unit spheres replaced by the real ones) to yield the following: let $G \in SO(2n)$ be $\mu_{SO(2n)}$ distributed, then (Corollary 6.2 in [10])

$$\det(\mathrm{Id}_{2n} - G) \stackrel{\mathrm{law}}{=} 2 \prod_{k=2}^{2n} \left(1 - \epsilon_k \sqrt{B_{\frac{1}{2}, \frac{k-1}{2}}}\right).$$

2.4.3 The Quaternionic Group

Our goal with this example is to solve Question 1 which was raised in the Introduction. To this end we establish an analogous to Lemma 2 and use the fact that $U(n, \mathbb{H}) \cong \mathrm{USp}(2n)$ which is also denoted $Sp(n)$, see for instance [24] Theorem 2. Then we apply Theorem 1. Let us give details. Recall that the symplectic group $\mathrm{USp}(2n, \mathbb{C})$ is defined as $\mathrm{USp}(2n, \mathbb{C}) = \{U \in U(2n, \mathbb{C}) \mid U J_n\,{}^t U = J_n\}$, with

$$J_n = \begin{pmatrix} 0 & \mathrm{Id}_n \\ -\mathrm{Id}_n & 0 \end{pmatrix}. \tag{7}$$

Let

$$\phi : \begin{cases} \mathbb{H} & \to & M(2, \mathbb{C}) \\ a + ib + jc + kd & \mapsto & \begin{pmatrix} a + ib & c + id \\ -c + id & a - ib \end{pmatrix} \end{cases},$$

be the usual representation of quaternions. It is a continuous injective ring morphism such that $\phi(\bar{x}) = \phi(x)^*$. It induces the ring morphism

$$\Phi : \begin{cases} M(n, \mathbb{H}) & \to & M(2n, \mathbb{C}) \\ (a_{ij})_{1 \le i,j \le n} & \mapsto & (\phi(a_{ij}))_{1 \le i,j \le n} \end{cases}.$$

In particular

$$\Phi(U(n, \mathbb{H})) = \{G \in U(2n, \mathbb{C}) : G \tilde{Z}_n\,{}^t G = \tilde{Z}_n\}$$

where $\tilde{Z}_n = J_1 \oplus \cdots \oplus J_1$ and $J_1 = \begin{pmatrix} 0 & 1 \\ -1 & 0 \end{pmatrix}$. Since $\tilde{Z}_n$ is conjugate to J_n, defined by (7), the set $\Phi(U(n, \mathbb{H}))$ is therefore conjugate to $\mathrm{USp}(2n, \mathbb{C})$. We can therefore consider $\det(I - \Phi(G))$

Lemma 3. *If for $k = 1, \ldots, n-1$, $R_k \in \mathcal{R}_k$, then*

$$\det(\mathrm{Id}_{2n} - \Phi(R_0 \cdots R_{n-1})) = \prod_{k=0}^{n-1} \det(\mathrm{Id}_2 - \phi(\langle e_{k+1}, R_k(e_{k+1})\rangle)). \tag{8}$$

Proof. Let us first remark that the canonical basis $e_1, \ldots, e_n$ of $\mathbb{H}^n$ is mapped by Φ into the canonical basis $\varepsilon_1, \ldots, \varepsilon_{2n}$ of $\mathbb{C}^{2n}$, where the $2n \times 2$ matrix $[\varepsilon_{2k-1}, \varepsilon_{2k}]$ is exactly $\Phi(e_k)$. Moreover, if R is a proper reflection (leaving invariant an hyperplane), $\Phi(R)$ is a bireflection of $\mathbb{C}^{2n}$ i.e. a unitary transformation leaving invariant a vector space of codimension 2.

We start with

$$\det\left((\mathrm{Id}_{2n} - \Phi(RH)\right) = \det\left(\mathrm{Id}_{2n} - \Phi(R)\Phi(H)\right) = \det \Phi(H) \det\left(\Phi(H^*) - \Phi(R)\right)$$

Since H (hence H^*) stabilizes e_1, then $\Phi(H)$ (and $\Phi(H)^*$) stabilizes ε_1 and ε_2, so we have:

i) $(H^* - R)(e_1) = e_1 - R(e_1) =: a = [a_1, a_2]$ (say), hence, for $i = 1, 2$,

$$(\Phi(H^*) - \Phi(R))(\varepsilon_i) = \varepsilon_i - \Phi(R)(\varepsilon_i) =: a_i .$$

ii) Assume that $\langle e_1, w \rangle = 0$. Trivially, $\langle e_1, H^*(w) \rangle = 0$ hence $\Phi(H^*)(w)$ is a matrix whose column vectors are orthogonal to ε_1 and ε_2. Moreover, since R is a quaternionic reflection, $R(w) - w$ is a (right) scalar multiple of a (see [14] Proposition 1.6), so $\Phi(R(w) - w)$ is a $2n \times 2$ matrix whose columns are in Span (a_1, a_2).

By the multilinearity of the determinant, we get

$$\det\left(\Phi(H)^* - \Phi(R)\right) = \det\left(\langle \epsilon_i, a_j \rangle_{1 \le i,j \le 2}\right) \det(\pi(H^*) - \mathrm{Id}_{2n-2})$$

which yields

$$\det(\mathrm{Id}_n - \Phi(RH)) = \det(\mathrm{Id}_2 - \phi(\langle e_1, R(e_1)\rangle)) \det(\mathrm{Id}_{2n-2} - \pi(H)) .$$

Iterating, we can conclude.   □

Corollary 2 (Symplectic group). *Let $G \in \mathrm{USp}(2n, \mathbb{C})$ be $\mu_{\mathrm{USp}(2n,\mathbb{C})}$ distributed. Then*

$$\det(\mathrm{Id}_{2n} - G) \overset{\text{law}}{=} \prod_{k=1}^{n} \left((a_k - 1)^2 + b_k^2 + c_k^2 + d_k^2\right),$$

where the vectors (a_k, b_k, c_k, d_k), $1 \leq k \leq n$ are independent and (a_k, b_k, c_k, d_k) are 4 coordinates of the 4k-dimensional real unit sphere endowed with the uniform measure.

Remark 3. We have $(a_k, b_k, c_k, d_k) \overset{\text{law}}{=} \dfrac{1}{\sqrt{\mathcal{N}_1^2 + \cdots + \mathcal{N}_{4k}^2}} (\mathcal{N}_1, \mathcal{N}_2, \mathcal{N}_3, \mathcal{N}_4)$, with the

$\mathcal{N}_i'$s i.i.d. $\mathcal{N}(0, 1)$. Now, since for $p < q$

$$\frac{\mathcal{N}_1^2 + \cdots + \mathcal{N}_p^2}{\mathcal{N}_1^2 + \cdots + \mathcal{N}_q^2} \overset{\text{law}}{=} B_{\frac{p}{2}, \frac{q-p}{2}} ,$$

we get the somehow more tractable identity in law

$$\det(\text{Id}_{2n} - G) \overset{\text{law}}{=} \prod_{k=1}^{n} \left(\left(1 + \epsilon_k \sqrt{B_{\frac{1}{2}, 2k-\frac{1}{2}}} \right)^2 + \left(1 - B_{\frac{1}{2}, 2k-\frac{1}{2}} \right) B'_{\frac{3}{2}, 2k-2} \right),$$

with all variables independent, $\mathbb{P}(\epsilon_k = 1) = \mathbb{P}(\epsilon_k = -1) = 1/2$.

This method can be applied to other interesting groups such as $\text{USp}(2n, \mathbb{R}) = \{u \in U(2n, \mathbb{R}) \mid uz\,{}^t u = z\}$ thanks to the morphism

$$\phi : \begin{cases} \mathbb{C} & \to M(2, \mathbb{R}) \\ a + ib \mapsto \begin{pmatrix} a & -b \\ b & a \end{pmatrix} \end{cases}.$$

The traditional representation of the quaternions in $M(4, \mathbb{R})$

$$\phi : \begin{cases} \mathbb{C} & \to & M(4, \mathbb{R}) \\ a + ib + jc + kd \mapsto & \begin{pmatrix} a & -b & -c & -d \\ b & a & -d & -c \\ c & d & a & -b \\ d & -c & b & a \end{pmatrix} \end{cases}$$

gives another identity in law for a compact subgroup of $U(4n, \mathbb{R})$.

2.5 The Generalized Ewens Measure

In this section we wish to define a generalization of the Ewens measure on $U(n, K)$ and some of its subgroups which will agree with the classical definition on the symmetric group. We first recall the definition of the Ewens measure on the symmetric group and how it can be generated.

2.5.1 The Ewens Measure on $\mathcal{S}_n$

Recall (see (2) Section 1) that every permutation $\sigma \in \mathcal{S}_n$ can be decomposed in the following way:

$$\sigma = \tau_n \circ \cdots \circ \tau_2 \tag{9}$$

where for $k = 2, \ldots, n$, τ_k is either the identity or the transposition (k, m_k) for some $m_k \in [k-1]$. In the first case we will say by extension that it is the transposition (k, m_k) with $m_k = k$. The number of cycles in the decomposition of σ is denoted k_σ. The system of Ewens measures of parameter $\theta > 0$ consists in choosing the $m_k, k = 1, \ldots, n$ independently, with distribution

$$\mathbb{P}(m_k = k) = \frac{\theta}{\theta + k - 1} \ ; \ \mathbb{P}(m_k = j) = \frac{1}{\theta + k - 1}, j = 1, \ldots, k - 1.$$

It is known that the induced probability on $\mathcal{S}_n$ is

$$\mu_n^\theta(\sigma) = \frac{\theta^{k_\sigma}}{(\theta)_n}. \tag{10}$$

2.5.2 The Generalized Ewens Measure

In the following, $\mathcal{G}$ is any subgroup of $U(n, K)$. Take $\delta \in \mathbb{C}$ such that

$$0 < \mathbb{E}_{\mu_\mathcal{G}} \left(\det(\mathrm{Id}_n - G)^{\overline{\delta}} \det(\mathrm{Id}_n - \overline{G})^\delta \right) < \infty. \tag{11}$$

For $0 \le k \le n - 1$ we note

$$\exp_\delta^{(k)} : \begin{cases} \mathcal{G} \to \mathbb{R}^+ \\ G \mapsto (1 - \langle e_{k+1}, G(e_{k+1}) \rangle)^{\overline{\delta}} \overline{(1 - \langle e_{k+1}, G(e_{k+1}) \rangle)}^\delta \end{cases}.$$

Moreover, define $\det_\delta$ as the function

$$\det_\delta : \begin{cases} \mathcal{G} \to \mathbb{R}^+ \\ G \mapsto \det(\mathrm{Id}_n - G)^{\overline{\delta}} \det(\mathrm{Id}_n - \overline{G})^\delta \end{cases}.$$

Then the following generalization of Theorem 1 (which corresponds to the case $\delta = 0$) holds. However, note that, contrary to Theorem 1, in the following result we need that the coherent measures be supported by the set of reflections.

Theorem 3 (Generalized Ewens sampling formula). *Let $\mathcal{G}$ be a subgroup of $U(n, K)$ checking condition (R) and (11). Let $(\nu_0, \ldots, \nu_{n-1})$ be a sequence of measures coherent with $\mu_\mathcal{G}$, with $\nu_k(\mathcal{R}_k) = 1$. We note $\mu_\mathcal{G}^{(\delta)}$ the $\det_\delta$-sampling of $\mu_\mathcal{G}$ and $\nu_k^{(\delta)}$ the $\exp_\delta^{(k)}$-sampling of ν_k. Then*

$$v_0^{(\delta)} \star v_1^{(\delta)} \star \cdots \star v_{n-1}^{(\delta)} = \mu_{\mathcal{G}}^{(\delta)},$$

i.e., for all test functions f on $\mathcal{G}$,

$$\mathbb{E}_{v_0^{(\delta)} \star \cdots \star v_{n-1}^{(\delta)}} (f(R_0 R_1 \ldots R_{n-1})) = \frac{\mathbb{E}_{\mu_{\mathcal{G}}} \left(f(G) \det(\mathrm{Id}_n - G)^{\overline{\delta}} \det(\mathrm{Id}_n - \overline{G})^{\delta} \right)}{\mathbb{E}_{\mu_{\mathcal{G}}} \left(\det(\mathrm{Id}_n - G)^{\overline{\delta}} \det(\mathrm{Id}_n - \overline{G})^{\delta} \right)}.$$

Proof. From Theorem 1, $G \overset{\text{law}}{=} R_0 \ldots R_{n-1}$, hence

$$\mathbb{E}_{\mu_{\mathcal{G}}} \left(f(G) \det(\mathrm{Id}_n - G)^{\overline{\delta}} \det(\mathrm{Id}_n - \overline{G})^{\delta} \right)$$

$$= \mathbb{E}_{v_0 \star \cdots \star v_{n-1}} \left(f(R_0 \ldots R_{n-1}) \det(\mathrm{Id}_n - R_0 \ldots R_{n-1})^{\overline{\delta}} \det(\mathrm{Id}_n - \overline{R_0 \ldots R_{n-1}})^{\delta} \right).$$

From Lemma 2, $\det(\mathrm{Id}_n - R_0 \ldots R_{n-1}) = \prod_{k=0}^{n-1}(1 - \langle e_{k+1}, R_k(e_{k+1}) \rangle)$, hence

$$\mathbb{E}_{v_0 \star \cdots \star v_{n-1}} \left(f(R_0 \ldots R_{n-1}) \det(\mathrm{Id}_n - R_0 \ldots R_{n-1})^{\overline{\delta}} \det(\mathrm{Id}_n - \overline{R_0 \ldots R_{n-1}})^{\delta} \right)$$

$$= \mathbb{E}_{v_0 \star \cdots \star v_{n-1}} \left(f(R_0 \ldots R_{n-1}) \prod_{k=0}^{n-1} \exp_{\delta}^{(k)}(R_k) \right).$$

By the definition of the measures $v_k^{(\delta)}$, this is the desired result. □

Before exploring properties of this measure, let us give two examples of δ-samplings.

First we check that we can recover the classical Ewens measure on the symmetric group. Consider $\mathcal{G} = \mathbb{Z}_2 \wr \mathcal{S}_n$. For $\delta > 0$, the δ-sampling in $\mathbb{Z}_2 \wr \mathcal{S}_n$ induces a $\theta = 2^{2\delta-1}$ sampling on $\mathcal{S}_n$.

Proposition 2. *For $\delta > 0$, the pushforward of $\mu_{\mathbb{Z}_2 \wr \mathcal{S}_n}^{(\delta)}$ by the projection $(f, \sigma) \mapsto \sigma$ is μ_n^{θ} with $\theta = 2^{2\delta-1}$.*

Similarly, if we associate with each transposition of the decomposition (9) a Rademacher variable, we get easily a sequence of reflections, and if v_k denotes the kth corresponding measure, then the system $(v_0, \cdots, v_{n-1})$ is coherent with $\mu_{\mathbb{Z}_2 \wr \mathcal{S}_n}$. The pushforward of $v_k^{(\delta)}$ under the projection is a transposition biased by θ, so we recover the Ewens sampling formula.

Proof. Recall that the generic element of $\mathbb{Z}_2 \wr \mathcal{S}_n$ is denoted (f, σ). Let $\mathcal{C}(\sigma)$ the set of cycles of σ. If $c = (d_1, \ldots, d_j)$ is such a cycle, let $\ell(c) = j$ and $w(f; c) = \prod_1^j f(d_j)$. Then it is clear that

$$\det(x \, \mathrm{Id}_n - (f; \sigma)) = \prod_{c \in \mathcal{C}(\sigma)} \left(x^{\ell(c)} - w(f; c) \right),$$

and in particular,

$$\det\left(\mathrm{Id}_n - (f;\sigma)\right) = \begin{cases} 0 & \text{if } \exists c \in \mathcal{C}(\sigma) : w(f;c) = 1 \\ 2^{k_\sigma} & \text{if } \forall c \in \mathcal{C}(\sigma) : w(f;c) = -1. \end{cases} \tag{12}$$

Let $\mathbb{P}$ stand for $\mu_{\mathbb{Z}_2 \wr S_n}$ i.e. the uniform distribution on $\mathbb{Z}_2 \wr S_n$. For any test function F

$$\mathbb{E}\left(F(\sigma) |\det(\mathrm{Id}_n - (f,\sigma))|^{2\delta}\right) = \mathbb{E}\left[F(\sigma)\mathbb{E}\left(|\det(\mathrm{Id}_n - (f,\sigma))|^{2\delta}|\sigma\right)\right].$$

Now, conditionally on σ, the weights of the cycles are independent Rademacher variables (i.e. ± 1 with probability $1/2$). So,

$$\mathbb{P}\left(\cap_{c \in \mathcal{C}(\sigma)} \{w(f,\sigma) = -1\}|\sigma\right) = 2^{-k_\sigma}$$

and, due to (12)

$$\mathbb{E}\left(|\det(\mathrm{Id}_n - (f,\sigma))|^{2\delta}|\sigma\right) = 2^{(2\delta-1)k_\sigma},$$

which easily yields

$$\mathbb{E}_{\mu_{\mathbb{Z}_2 \wr S_n}^{(\delta)}} F(\sigma) = \int_{S_n} F(\sigma) d\mu_n^\theta(\sigma).$$

$\square$

The fundamental example remains $U(n, \mathbb{C})$. In the following section, we will study the determinantal sructure of this model for $\mathfrak{Re}\,\delta > -1/2$. In [12] a precise analysis of the reflections involved in the decomposition is given. The case $\delta = 1$ has a specific interest. If $(\theta_1, \ldots, \theta_n)$ are the eigenangles of a unitary matrix, we have

$$|\det(\mathrm{Id} - U)|^2 = \prod_{j=1}^n |1 - e^{i\theta_j}|^2,$$

which, thanks to the density of the eigenangles, yields

$$\mathbb{E}_{\mu_{U(n)}^{(1)}} (f(\theta_1, \ldots, \theta_n))$$

$$= \mathrm{cst} \int_{(-\pi,\pi)^n} f(\theta_1, \ldots, \theta_n) \prod_{j<k} |e^{i\theta_j} - e^{i\theta_k}|^2 \prod_{l=1}^n |1 - e^{i\theta_l}|^2 d\theta_1 \ldots d\theta_n.$$

This means that the distribution of the eigenangles $(\theta_1, \ldots, \theta_n)$ of a random matrix drawn according to $\mu_{U(n)}^{(1)}$ is the same as the distribution of the n first eigenangles $(\theta_1, \cdots, \theta_n)$ of a random matrix drawn according to $\mu_{U(n+1,\mathbb{C})}$, conditionally on $\theta_{n+1} = 0$, or, as seen in [34], as the distribution of $(\theta_1 - \theta_{n+1}, \cdots, \theta_n - \theta_{n+1})$.

More generally, in [11], Bourgade gives a geometrical characterisation of this kind of measures for $\delta/2 \in \mathbb{N}$, defining the notion of conditional Haar measure.

Remark 4. A generalized Ewens sampling formula could also be stated for $\Phi(\mathcal{G})$, with $\mathcal{G}$ checking condition (R) and Φ the ring morphism previously defined.

3 A Hypergeometric Kernel

In this section, we study the correlations of the point process of eigenvalues under the measure $\mu_{U(n,\mathbb{C})}^{(\delta)}$ and answer Question 3 (see Introduction) asked by Borodin-Olshanski in [7] Sect. 8. Let us recall some basic facts on determinantal processes and correlations, referring to the books [1] 4.2 or [6] or [16] Chap. 4.

Let $\Lambda = \mathbb{R}$ or $\mathbb{T} = \{z \in \mathbb{C} : |z| = 1\} = \{e^{i\theta}; \theta \in [-\pi, \pi]\}$ and let us fix an integer n. The collection of eigenvalues $(\lambda_1, \ldots, \lambda_n)$ of a random $n \times n$ Hermitian (resp. unitary) matrix can be viewed as a point process on Λ, i.e. a random counting measure $\nu_n = \delta_{\lambda_1} + \cdots + \delta_{\lambda_n}$. Let us consider a simple point process ν on Λ. If there exists a sequence of locally integrable functions ρ_k such that for any mutually disjoint family of subsets $D_1, \ldots, D_k$ of Λ

$$\mathbb{E}\left[\prod_{i=1}^{k} \nu(D_i)\right] = \int_{\prod_{i=1}^{k} D_i} \rho_k(x_1, \ldots, x_k) \mathrm{d}x_1 \ldots \mathrm{d}x_k$$

then the functions ρ_k are called the correlation functions, or joint intensities of the point process. In this case, the process is said to be determinantal with kernel K if its correlation functions ρ_k are given by

$$\rho_k(x_1, \ldots, x_k) = \det_{i,j=1}^{k} K(x_i, x_j).$$

For $\nu = \nu_n$ we denote the correlations by $\rho_{k,n}$ for $k \leq n$. When the joint density of the eigenvalues is proportional to

$$\prod_{k=1}^{n} w(x_k) \prod_{1 \leq j < k \leq n} |x_k - x_j|^2$$

for some weight w, the orthogonal poynomial method shows that the point process of eigenvalues is determinantal. The use of Cayley transform allows to connect Hermitian matrices and unitary matrices. We give a detailed description of the consequence of this connection for the corresponding eigenvalue processes in Sect. 3.1, and its impact on the circular Jacobi ensemble in Sect. 3.2. Finally, we study the asymptotic behavior in Sect. 3.3.

3.1 Determinantal Processes and Cayley Transform

We follow the approach of Forrester ([16] 2.5 and 4.1.4). We start with a weight (positive integrable function) $w^{\mathbb{T}}$ on $\mathbb{T}$. The pushforward of the measure

$$\prod_{j=1} w^{\mathbb{T}}(e^{i\theta_j}) \prod_{1 \leq j < k \leq n} |e^{i\theta_k} - e^{i\theta_j}|^2 d\theta_1 \cdots d\theta_n$$

by the stereographic projection (Cayley transform)

$$\lambda = i\frac{1 - e^{i\theta}}{1 + e^{i\theta}} = \tan\frac{\theta}{2} \; ; \; e^{i\theta} = \frac{1 + i\lambda}{1 - i\lambda}$$

gives the measure

$$2^{n^2} \prod_{j=1}^{n} w^{\mathbb{T}}\left(\frac{1 + i\lambda_j}{1 - i\lambda_j}\right)(1 + \lambda_j^2)^{-n} \prod_{1 \leq j < k \leq n} |\lambda_k - \lambda_j|^2 d\lambda_1 \cdots d\lambda_n.$$

We define the weight $w^{\mathbb{R}}$ on $\mathbb{R}$ as

$$w^{\mathbb{R}}(x) = (1 + x^2)^{-n} w^{\mathbb{T}}\left(\frac{1 + ix}{1 - ix}\right).$$

Conversely

$$w^{\mathbb{T}}(e^{i\theta}) = \left(\cos\frac{\theta}{2}\right)^{2n} w^{\mathbb{R}}\left(\tan\frac{\theta}{2}\right).$$

If the monomials $1, x, \ldots, x^n$ are in $L^2(w^{\mathbb{R}}(x)dx)$, then the orthogonal polynomial method gives

$$\frac{1}{Z_n^{\mathbb{R}}} \prod_{j=1}^{n} w^{\mathbb{R}}(\lambda_j) \prod_{1 \leq j < k \leq n} |\lambda_k - \lambda_j|^2 = \det\left(\widetilde{K}_n^{\mathbb{R}}(\lambda_j, \lambda_k)\right)_{1 \leq j,k \leq n}$$

where $Z_n^{\mathbb{R}}$ is a normalization constant and where

$$\widetilde{K}_n^{\mathbb{R}}(x, y) = \sqrt{w^{\mathbb{R}}(x)w^{\mathbb{R}}(y)} \, K_n^{\mathbb{R}}(x, y)$$

$$K_n^{\mathbb{R}}(x, y) = \sum_{\ell=0}^{n-1} p_\ell^{\mathbb{R}}(x) p_\ell^{\mathbb{R}}(y)$$

and the $p_\ell^{\mathbb{R}}$ are orthonormal with respect to the measure $w^{\mathbb{R}}(x)dx$. The Christoffel–Darboux formula gives another expression for the kernel

$$K_n^{\mathbb{R}}(x, y) = \frac{\kappa_{n-1}}{\kappa_n} \frac{p_n^{\mathbb{R}}(x) p_{n-1}^{\mathbb{R}}(y) - p_{n-1}^{\mathbb{R}}(x) p_n^{\mathbb{R}}(y)}{x - y}$$

where κ_j is the coefficient of x^j in $p_j^{\mathbb{R}}(x)$. In terms of the monic orthogonal polynomials $P_0, \cdots, P_n$, this yields

$$K_n^{\mathbb{R}}(x, y) = \sum_{\ell=0}^{n-1} \frac{P_\ell(x) P_\ell(y)}{\|P_\ell\|^2} \tag{13}$$

$$= \frac{P_n^{\mathbb{R}}(x) P_{n-1}^{\mathbb{R}}(y) - P_{n-1}^{\mathbb{R}}(x) P_n^{\mathbb{R}}(y)}{\|P_{n-1}\|^2 (x - y)}. \tag{14}$$

Besides, on the unit circle, we consider the polynomials φ_ℓ (resp. Φ_ℓ) orthonormal (resp. monic orthogonal) with respect to the measure $w^{\mathrm{T}}(e^{i\theta}) d\theta$, and their reciprocal defined by

$$\Phi_\ell^\star(z) = z^\ell \overline{\Phi_\ell(1/\bar{z})}, \quad \varphi_\ell^\star(z) = z^\ell \overline{\varphi_\ell(1/\bar{z})}.$$

We have then

$$\frac{1}{\mathcal{Z}_n^{\mathrm{T}}} \prod_{j=1}^n w^{\mathrm{T}}(e^{i\theta_j}) \prod_{1 \le j < k \le n} |e^{i\theta_k} - e^{i\theta_j}|^2 = \det\left(\widetilde{K}_n^{\mathrm{T}}(e^{i\theta_j}, e^{i\theta_k})\right)_{1 \le j, k \le n}$$

with

$$\widetilde{K}_n^{\mathrm{T}}(z, \zeta) = \sqrt{w^{\mathrm{T}}(z) w^{\mathrm{T}}(\zeta)} \, K_n^{\mathrm{T}}(z, \zeta)$$

and

$$K_n^{\mathrm{T}}(z, \zeta) = \sum_{\ell=0}^{n-1} \overline{\varphi_\ell(z)} \varphi_\ell(\zeta).$$

The Christoffel–Darboux formula is now

$$K_n^{\mathrm{T}}(z, \zeta) = \frac{\overline{\varphi_n^*(z)} \varphi_n^*(\zeta) - \overline{\varphi_n(z)} \varphi_n(\zeta)}{1 - \bar{z}\zeta} \tag{15}$$

(see [31] 1.12 and 3.2), or

$$K_n^{\mathrm{T}}(z, \zeta) = \frac{\overline{\Phi_n^*(z)} \Phi_n^*(\zeta) - \overline{\Phi_n(z)} \Phi_n(\zeta)}{\|\Phi_n\|^2 (1 - \bar{z}\zeta)}. \tag{16}$$

The kernel $\widetilde{K}_n^{\mathbb{R}}$ (resp. $\widetilde{K}_n^{\mathrm{T}}$) rules the correlation function $\rho_{n,m}^{\mathbb{R}}(\lambda_1, \ldots, \lambda_m)$ (resp. $\rho_{n,m}^{\mathbb{C}}(e^{i\theta_1}, \ldots, e^{i\theta_m})$) for $m = 1, \ldots, n$.

3.2 Our Weights and their Characteristics

For the sake of simplicity we use the polygamma symbol

$$\Gamma \begin{bmatrix} a, b, \cdots \\ c, d, \cdots \end{bmatrix} := \frac{\Gamma(a)\Gamma(b)\cdots}{\Gamma(c)\Gamma(d)\cdots}.$$

For $\delta = a + ib \in \mathbb{C}$ with $a > -1/2$, we will consider two weights on $(-\pi, \pi)$

$$w_1^{\mathrm{T}}(e^{i\theta}) = (1 - e^{i\theta})^{\bar{\delta}}(1 - e^{-i\theta})^{\delta} = (2 - 2\cos\theta)^a e^{-b(\pi \operatorname{sgn}\theta - \theta)} \tag{17}$$

$$w_2^{\mathrm{T}}(e^{i\theta}) = (1 + e^{i\theta})^{\bar{\delta}}(1 + e^{-i\theta})^{\delta} = (2 + 2\cos\theta)^a e^{-b\theta} \tag{18}$$

These are "pure" Fisher-Hartwig functions. We can go from w_1^{T} to w_2^{T} by the transform

$$\theta \mapsto \tau := -\theta + \pi(\operatorname{sgn}\theta) \tag{19}$$

which carries the discontinuity in $\theta = 0$ to the edges $\pm\pi$, so that

$$e^{i\theta} = -e^{-i\tau} \text{ and } w_1^{\mathrm{T}}(e^{i\theta}) = w_2^{\mathrm{T}}(e^{-i\tau}). \tag{20}$$

For $a > -1/2$, the Fourier coefficients of w_1 are known ([9] Lemma 2.1)

$$\frac{1}{2\pi} \int_{-\pi}^{\pi} w_1^{\mathrm{T}}(e^{i\theta}) e^{-in\theta} d\theta = (-1)^n \Gamma \begin{bmatrix} 1 + \delta + \bar{\delta} \\ \bar{\delta} - n + 1, \delta + n + 1 \end{bmatrix}.$$

With

$$c(\delta) = \frac{1}{2\pi} \Gamma \begin{bmatrix} 1 + \delta, 1 + \bar{\delta} \\ 1 + \delta + \bar{\delta} \end{bmatrix},$$

the function $\widetilde{w}_1^{\mathrm{T}}(e^{i\theta}) = c(\delta) w_1^{\mathrm{T}}(e^{i\theta})$ is a probability density on $(-\pi, \pi)$. For w_2, we note that

$$\int_{-\pi}^{\pi} w_1^{\mathrm{T}}(e^{i\theta}) e^{-in\theta} d\theta = (-1)^n \int_{-\pi}^{\pi} w_2^{\mathrm{T}}(e^{i\tau}) e^{in\tau} d\tau.$$

Moreover we go from one system of polynomials to the other by the mapping $z \mapsto -z$.

It is known from [4] p. 304 and [5] p. 31–34 that for $n \geq 0$ the nth orthonormal polynomial with respect to $\widetilde{w}_1^{\mathrm{T}}(e^{i\theta}) d\theta$ is

$$\Phi_n(z) = \Gamma \begin{bmatrix} \bar{\delta} + n, \bar{\delta} + 1 \\ \bar{\delta} + n + 1, \delta \end{bmatrix} {}_2F_1\left(\begin{matrix} -n, \bar{\delta} + 1 \\ 1 - n - \delta \end{matrix}; z \right) \tag{21}$$

with

$$\|\Phi_n\|^2 = \Gamma \begin{bmatrix} \delta + \bar{\delta} + n + 1, n + 1, \bar{\delta} + 1, \delta + 1 \\ \bar{\delta} + n + 1, \delta + n + 1, \delta + \bar{\delta} + 1 \end{bmatrix}, \tag{22}$$

(see also [16] Proposition 4.8 in the case δ real). With the complement formula (48) we get the other form

$$\Phi_n(z) = \Gamma \begin{bmatrix} \delta + \bar{\delta} + 1 + n, \bar{\delta} + 1 \\ \bar{\delta} + n + 1, \delta + \bar{\delta} + 1 \end{bmatrix} {}_2F_1 \left(\begin{matrix} -n, \bar{\delta} + 1 \\ \delta + \bar{\delta} + 1 \end{matrix} ; 1 - z \right). \tag{23}$$

In view of (47) and (21) we identify Φ_n^* as

$$\Phi_n^*(z) = {}_2F_1 \left(\begin{matrix} -n, \bar{\delta} \\ -n - \delta \end{matrix} ; z \right), \tag{24}$$

or, using (48) again

$$\Phi_n^*(z) = \Gamma \begin{bmatrix} \delta + \bar{\delta} + 1 + n, \delta + 1 \\ \delta + n + 1, \delta + \bar{\delta} + 1 \end{bmatrix} {}_2F_1 \left(\begin{matrix} -n, \bar{\delta} \\ \delta + \bar{\delta} + 1 \end{matrix} ; 1 - z \right). \tag{25}$$

Borodin and Olshanski considered the following weight on $\mathbb{R}$:

$$2^{-\delta - \bar{\delta}} w_2^{\mathbb{R}}(x) = (1 + \mathrm{i}x)^{-\delta - n}(1 - \mathrm{i}x)^{-\bar{\delta} - n}. \tag{26}$$

Since this weight depends on n, the reference measure has only a finite set of moments so that there is only a finite set of orthogonal polynomials (these are the pseudo-Jacobi polynomials)

$$p_m(x) = (x - \mathrm{i})^m {}_2F_1 \left(\begin{matrix} -m, \delta + n - m \\ \delta + \bar{\delta} + 2n - 2m \end{matrix} ; \frac{2}{1 + \mathrm{i}x} \right) \tag{27}$$

$m < a + n - \frac{1}{2}$. Let us call $\widetilde{K}_{2,n}^{\mathbb{R}}$ the corresponding kernel.

3.3 Asymptotic Behavior

For the weight $w_2^{\mathbb{R}}$, Borodin and Olshanski considered the (thermodynamic) scaling limit $\lambda \mapsto n\lambda$ and proved ([7] Theorem 2.1):

Theorem 4 (Borodin-Olshanski). *Let $\mathfrak{Re}\,\delta > -1/2$.*

1. We have

$$\lim_n (\mathrm{sgn}\, x \, \mathrm{sgn}\, y)^n n \widetilde{K}_{2,n}^{\mathbb{R}}(nx, ny) = \widetilde{K}_\infty^{\mathbb{R}}(x, y) \tag{28}$$

uniformly for x, y in compact sets of $\mathbb{R}^\star \times \mathbb{R}^\star$, where (for $x \neq y$)

$$\widetilde{K}^{\mathbb{R}}_{\infty}(x, y) := \frac{1}{2\pi} \Gamma \begin{bmatrix} \delta + 1, \bar{\delta} + 1 \\ \delta + \bar{\delta} + 1, \delta + \bar{\delta} + 2 \end{bmatrix} \frac{\widetilde{P}(x)Q(y) - Q(x)\widetilde{P}(y)}{x - y} \quad (29)$$

$$\widetilde{P}(x) = \left| \frac{2}{x} \right|^{\frac{\delta + \bar{\delta}}{2}} e^{-\frac{i}{x} + \pi \frac{(\delta - \bar{\delta}) \operatorname{sgn} x}{4}} {}_1F_1 \left(\frac{\delta}{\delta + \bar{\delta} + 1}; \frac{2i}{x} \right) \quad (30)$$

$$Q(x) = \frac{2}{x} \left| \frac{2}{x} \right|^{\frac{\delta + \bar{\delta}}{2}} e^{-\frac{i}{x} + \pi \frac{(\delta - \bar{\delta}) \operatorname{sgn} x}{4}} {}_1F_1 \left(\frac{\delta + 1}{\delta + \bar{\delta} + 2}; \frac{2i}{x} \right). \quad (31)$$

2. *The limiting correlation is given by*

$$\lim_n n^m \rho^{\mathbb{R}}_{n,m}(n\lambda_1, \cdots, n\lambda_m) = \det \left(\widetilde{K}^{\mathbb{R}}_{\infty}(\lambda_i, \lambda_j) \right)_{1 \le i, j \le m}. \quad (32)$$

The kernel $\widetilde{K}^{\mathbb{R}}_{\infty}(1/x, 1/y)$ is called the *confluent hypergeometric kernel* in [8].

For the circular model, we choose the set-up w_1 for the sake of consistency with the above sections. The singularity is in $z = 1$ i.e. $\theta = 0$. To study the asymptotic behavior of the point process on $\mathbb{T}$ at the singularity (edge) we have two ways: either take the thermodynamic scaling $\theta \mapsto \theta/n$, or use the result on $\mathbb{R}$.

Theorem 5. *Let $\mathfrak{Re} \delta > -1/2$.*

1. *With the weight w_1,*

$$\lim_n n^{-1} \widetilde{K}^{\mathbb{T},1}_n (e^{i\theta/n}, e^{i\tau/n}) = \widetilde{K}^{\mathbb{T}}_{\infty}(\theta, \tau) \quad (33)$$

with, for $\theta \ne \tau$

$$\widetilde{K}^{\mathbb{T}}_{\infty}(\theta, \tau) = \frac{1}{2i\pi} \Gamma \begin{bmatrix} 1 + \delta, 1 + \bar{\delta} \\ 1 + \delta + \bar{\delta}, 1 + \delta + \bar{\delta} \end{bmatrix} \frac{P^{\mathbb{T}}(\theta) \overline{P^{\mathbb{T}}(\tau)} - \overline{P^{\mathbb{T}}(\theta)} P^{\mathbb{T}}(\tau)}{\theta - \tau}$$

$$(34)$$

where

$$P^{\mathbb{T}}(\theta) := |\theta|^{\frac{\delta + \bar{\delta}}{2}} e^{i\frac{\theta}{2} - \frac{\pi}{4}(\delta - \bar{\delta}) \operatorname{sgn} \theta} {}_1F_1 \left(\frac{\delta}{\delta + \bar{\delta} + 1}; -i\theta \right) = \widetilde{P}(-2\theta^{-1}),$$

$$(35)$$

and

$$\widetilde{K}^{\mathbb{T}}_{\infty}(\theta, \theta) = \frac{|\theta|^{\delta + \bar{\delta}}}{2\pi} \Gamma \begin{bmatrix} 1 + \delta, 1 + \bar{\delta} \\ 1 + \delta + \bar{\delta}, 1 + \delta + \bar{\delta} \end{bmatrix} \mathfrak{Re}$$

$$\left[{}_1F_1 \left(\frac{\delta}{\delta + \bar{\delta} + 1}; -i\theta \right) \left[{}_1F_1 \left(\frac{\bar{\delta}}{\delta + \bar{\delta} + 1}; i\theta \right) - 2 \, {}_1F_1 \left(\frac{\bar{\delta} + 1}{\delta + \bar{\delta} + 2}; i\theta \right) \right] \right]$$

$$(36)$$

2. *The limiting correlation is given by*

$$\lim_n n^m \rho_{n,m}^{T,1}(e^{i\theta_1/n}, \cdots, e^{i\theta_m/n}) = \det\left(\widetilde{K}_\infty^T(\theta_i, \theta_j)\right)_{1 \le i,j \le m}. \quad (37)$$

Proof. We begin with a direct proof of (33) when $\theta \ne \tau$, and then proceed with the proof of (33) when $\theta = \tau$, which directly yields (37) and we end with an alternate proof of (37) using (32) and the Cayley transform.

1) The following lemma describes the asymptotical behavior of the quantities entering in the kernel.

Lemma 4. *When $n \to \infty$*

$$\lim_n \|\Phi_n\|^2 = \Gamma\begin{bmatrix} \bar{\delta} + 1, \delta + 1 \\ \delta + \bar{\delta} + 1 \end{bmatrix} \quad (38)$$

Moreover if $n\theta_n \to \theta$, then (uniformly for θ in a compact set)

$$\lim_n n^{-\delta} \Phi_n(e^{i\theta_n}) = \Gamma\begin{bmatrix} \bar{\delta} + 1 \\ \delta + \bar{\delta} + 1 \end{bmatrix} {}_1F_1\left(\begin{matrix} \bar{\delta} + 1 \\ \delta + \bar{\delta} + 1 \end{matrix}; i\theta\right), \quad (39)$$

$$\lim_n n^{-\bar{\delta}} \Phi_n^\star(e^{i\theta_n}) = \Gamma\begin{bmatrix} \delta + 1 \\ \delta + \bar{\delta} + 1 \end{bmatrix} {}_1F_1\left(\begin{matrix} \bar{\delta} \\ \delta + \bar{\delta} + 1 \end{matrix}; i\theta\right), \quad (40)$$

$$\lim_n n^{-\bar{\delta}+1} (\Phi_n^\star)'(e^{i\theta_n}) = \bar{\delta}\Gamma\begin{bmatrix} \delta + 1 \\ \delta + \bar{\delta} + 2 \end{bmatrix} {}_1F_1\left(\begin{matrix} \bar{\delta} + 1 \\ \delta + \bar{\delta} + 2 \end{matrix}; i\theta\right). \quad (41)$$

Proof. Let us first recall that, as $n \to \infty$,

$$\frac{\Gamma(c + n)}{\Gamma(n)} \sim n^c, \quad (42)$$

which gives immediately (38). The limits in (39) and (40) are then consequences of (23), (25) and the limiting relation (50). Besides, in view of (49) and (25),

$$(\Phi_n^\star)'(z) = \frac{n\bar{\delta}}{\delta + \bar{\delta} + 1}\Gamma\begin{bmatrix} \delta + \bar{\delta} + n, \delta + 1 \\ \delta + \bar{\delta} + 1, \delta + n + 1 \end{bmatrix} {}_2F_1\left(\begin{matrix} -n + 1, \bar{\delta} + 1 \\ \delta + \bar{\delta} + 2 \end{matrix}; 1 - z\right).$$

It remains to apply (50).   □

A) For $\theta \ne \tau$, we have, by the Christoffel–Darboux formula (15):

$$\lim_n i(\theta - \tau)\Gamma(\delta + \bar{\delta} + 1)n^{-(\delta + \bar{\delta} + 1)} K_n^{T,1}(e^{i\theta/n}, e^{i\tau/n})$$

$$= {}_1F_1\left(\begin{matrix} \delta \\ \delta + \bar{\delta} + 1 \end{matrix}; -i\theta\right) {}_1F_1\left(\begin{matrix} \bar{\delta} \\ \delta + \bar{\delta} + 1 \end{matrix}; i\tau\right)$$

$$-{}_1F_1\left(\begin{array}{c}\delta+1\\\delta+\bar{\delta}+1\end{array};-i\theta\right){}_1F_1\left(\begin{array}{c}\bar{\delta}+1\\\delta+\bar{\delta}+1\end{array};i\tau\right)$$

Now, applying the Kummer's formula (52)

$$_1F_1\left(\begin{array}{c}\delta+1\\\delta+\bar{\delta}+1\end{array};-i\theta\right)=e^{-i\theta}\,{}_1F_1\left(\begin{array}{c}\bar{\delta}\\\delta+\bar{\delta}+1\end{array};i\theta\right)$$

$$_1F_1\left(\begin{array}{c}\bar{\delta}+1\\\delta+\bar{\delta}+1\end{array};i\tau\right)=e^{i\tau}\,{}_1F_1\left(\begin{array}{c}\delta\\\delta+\bar{\delta}+1\end{array};-i\tau\right)$$

Besides we have (recall that we used $\widetilde{w}_1$)

$$\frac{\widetilde{K}_n^{\mathrm{T},1}(e^{i\theta/n},e^{i\tau/n})}{K_n^{\mathrm{T},1}(e^{i\theta/n},e^{i\tau/n})}=c(\delta)\sqrt{w_1(e^{i\theta/n})w_1(e^{i\tau/n})}$$

and from the very definition of w_1

$$\lim n^{2(\delta+\bar{\delta})}w_1(e^{i\theta/n})w_1(e^{i\tau/n})=|\theta\tau|^{2\mathfrak{Re}\delta}e^{-\mathfrak{Im}\delta\pi(\operatorname{sgn}\theta+\operatorname{sgn}\tau)}$$

We conclude that (33) holds true.

<u>B) On the diagonal</u> In the following z and ζ are elements of $\mathbb{T}$. If F and G are differentiable functions on $\mathbb{T}$, the de l'Hospital rule gives

$$\lim_{\zeta\to z}\frac{F(z)G(\zeta)-F(\zeta)G(z)}{z-\zeta}=F'(z)G(z)-F(z)G'(z).$$

Taking

$$F(z)=z^{-n}\Phi_n(z)\,,\ G(z)=\overline{\Phi_n(z)}\,,$$

so that

$$F'(z)=-nz^{-n-1}\Phi_n(z)+z^{-n}\Phi_n'(z)\,,\ G'(z)=-z^{-2}\overline{\Phi_n'(z)}$$

we get the value of the kernel on the diagonal:

$$\lim_{\zeta\to z}\frac{\overline{\Phi_n^*(z)}\Phi_n^*(\zeta)-\overline{\Phi_n(z)}\Phi_n(\zeta)}{1-\bar{z}\zeta}=-n|\Phi_n(z)|^2+2\mathfrak{Re}[\overline{\Phi_n(z)}z\Phi_n'(z)]$$

$$=n|\Phi_n^*(z)|^2-2\mathfrak{Re}[\overline{\Phi_n^*(z)}z(\Phi_n^*)'(z)].$$

$$(43)$$

It remains to apply the lemma.

Notice that

$$\lim_{n}n^{-(1+\delta+\bar{\delta})}K_n^{\mathrm{T},1}(1,1)=\frac{1}{\Gamma(\delta+\bar{\delta}+2)}.$$

2) *Alternate proof of (37)*

The pushforward of the measure

$$\rho_n^{\mathbb{R},2}(x_1,\cdots,x_n)dx_1\ldots dx_n$$

by the Cayley transform is

$$2^{-n}\rho_n^{\mathbb{R},2}\left(\tan\frac{\theta_1}{2},\cdots,\tan\frac{\theta_n}{2}\right)\prod_{k=1}^{n}\cos^{-2}\frac{\theta_k}{2}\,d\theta_1\ldots d\theta_n$$

which, at the level of kernels gives

$$\rho_{n,m}^{\mathbb{T},2}(e^{i\theta_1},\cdots,e^{i\theta_m})=\det\left[\widetilde{K}_n^{\mathbb{R},2}\left(\tan\frac{\theta_i}{2},\tan\frac{\theta_j}{2}\right)\frac{1}{2\cos\theta_i\cos\theta_j}\right]_{1\leq i,j\leq m}.$$

Coming back to the superscript 1 with the help of (19) we obtain

$$\rho_{n,m}^{\mathbb{T},1}(e^{i\theta_1},\cdots,e^{i\theta_m}))=\det\left[H_n(\theta_i),H_n(\theta_j)\right]_{1\leq i,j\leq m}$$

with

$$H_n(\theta,\theta')=\widetilde{K}_n^{\mathbb{R},2}\left(-\cot\frac{\theta}{2},-\cot\frac{\theta'}{2}\right)\frac{1}{2|\sin\frac{\theta}{2}\sin\frac{\theta'}{2}|}.$$

Let us rescale the angles. Since $\lim_n n\tan\frac{\theta}{n}=\theta$, $\lim_n n\tan\frac{\theta'}{n}=\theta'$ and since the limit in (28) is uniform on compact subsets, we get

$$\lim\frac{1}{n}H_n\left(\frac{\theta}{n},\frac{\theta'}{n}\right)=\frac{2}{|\theta\theta'|}\widetilde{K}_\infty^{\mathbb{R}}\left(-\frac{2}{\theta},-\frac{2}{\theta'}\right).$$

We remark that $P^{\mathbb{T}}(\theta)=\widetilde{P}(x)$ with $x\theta=-2$. Moreover, from (53), we have

$$\frac{i}{\bar{\delta}+\delta+1}Q(x)=\overline{P^{\mathbb{T}}(\theta)}-P^{\mathbb{T}}(\theta)$$

so that, if $\tau=-2/y$

$$\frac{i}{\bar{\delta}+\delta+1}\left[\widetilde{P}(x)Q(y)-\widetilde{P}(y)Q(x)\right]=P^{\mathbb{T}}(\theta)\overline{P^{\mathbb{T}}(\tau)}-P^{\mathbb{T}}(\tau)\overline{P^{\mathbb{T}}(\theta)}$$

and consequently

$$\frac{\theta\tau}{2}\widetilde{K}_\infty^{\mathbb{T}}(\theta,\tau)=\widetilde{K}_\infty^{\mathbb{R}}(x,y). \tag{44}$$

$\square$

Remark 5. 1. To have a graphical point of view of this kernel, we refer to [13] p. 56–60.

2. In [25], the behavior of the limiting kernel on $\mathbb{R}$ is used to study asymptotics of the maximal eigenvalue of the generalized Cauchy ensemble.
3. An easy computation shows that for δ real, $\delta > -1/2$, we recover the Bessel kernel

$$K_\infty^T = \frac{\pi}{2}\sqrt{\theta\tau}\frac{J_{\delta+\frac{1}{2}}(\frac{\pi\theta}{2})J_{\delta-\frac{1}{2}}(\frac{\pi\tau}{2}) - J_{\delta-\frac{1}{2}}(\frac{\pi\theta}{2})J_{\delta+\frac{1}{2}}(\frac{\pi\tau}{2})}{2(\theta-\tau)},$$

and for $\delta = 0$ the sine kernel

$$K_\infty^T = \frac{\sin(\frac{\theta-\tau}{2})}{\pi(\theta-\tau)}.$$

4 Appendix: Hypergeometric Functions

For a classical reference on hypergeometric functions, see [2].

The Gauss hypergeometric function is defined as

$$_2F_1\left(\begin{matrix}a,b\\c\end{matrix};z\right) = \sum_{k=0}^{\infty}\frac{(a)_k(b)_k}{(c)_k}\frac{z^k}{k!} \tag{45}$$

where $(x)_n$ stands for the Pochhammer symbol $(x)_k = x(x+1)\dots(x+k-1)$, with the convention $(x)_0 = 1$. When $a = -n \in -\mathbb{N}_0$, it is a polynomial

$$_2F_1\left(\begin{matrix}-n,b\\c\end{matrix};z\right) = \sum_{k=0}^{n}(-1)^k\binom{n}{k}\frac{(b)_k}{(c)_k}z^k. \tag{46}$$

The following relations are useful:

$$z^n\, _2F_1\left(\begin{matrix}-n,b\\c\end{matrix};z^{-1}\right) = (-1)^n\frac{(b)_n}{(c)_n}\, _2F_1\left(\begin{matrix}-n,-n-c+1\\-n-b+1\end{matrix};z\right) \tag{47}$$

$$_2F_1\left(\begin{matrix}-n,b\\c\end{matrix};1-z\right) = \frac{(c-b)_n}{(c)_n}\, _2F_1\left(\begin{matrix}-n,b\\-n+b+1-c\end{matrix};z\right) \tag{48}$$

$$\frac{d}{dz}\, _2F_1\left(\begin{matrix}a,b\\c\end{matrix};z\right) = \frac{ab}{c}\, _2F_1\left(\begin{matrix}a+1,b+1\\c+1\end{matrix};z\right). \tag{49}$$

It is known that, uniformly for z in a compact set, for b,c fixed

$$\lim_{N}\, _2F_1\left(\begin{matrix}-N,b\\c\end{matrix};-\frac{z}{N}\right) = {_1F_1}\left(\begin{matrix}b\\c\end{matrix};z\right) \tag{50}$$

where

$$_1F_1\begin{pmatrix} b \\ c \end{pmatrix}; z = \sum_{k=0}^{\infty} \frac{(b)_k}{(c)_k} \frac{z^k}{k!} \tag{51}$$

is the confluent hypergeometric function.

It satisfies Kummer's formula:

$$e^z \, _1F_1\begin{pmatrix} a \\ c \end{pmatrix}; -z = \, _1F_1\begin{pmatrix} c-a \\ c \end{pmatrix}; z, \tag{52}$$

the recursion formula

$$_1F_1\begin{pmatrix} a \\ c \end{pmatrix}; z = \, _1F_1\begin{pmatrix} a-1 \\ c \end{pmatrix}; z + \frac{z}{c} \, _1F_1\begin{pmatrix} a \\ c+1 \end{pmatrix}; z, \tag{53}$$

and the derivative formula

$$\frac{d}{dz} \, _1F_1\begin{pmatrix} a \\ c \end{pmatrix}; z = \frac{a}{c} \, _1F_1\begin{pmatrix} a+1 \\ c+1 \end{pmatrix}; z. \tag{54}$$

Acknowledgement A.N.'s work is supported by the Swiss National Science Foundation (SNF) grant 200021_119970/1.

A.R's work is partly supported by the ANR project Grandes Matrices Aléatoires ANR-08-BLAN-0311-01.

References

1. Anderson, G.W., Guionnet, A., Zeitouni, O.: An Introduction to Random Matrices. Cambridge University Press, Cambridge (2010)
2. Andrews, G.E., Askey, R.A., Roy, R.: Special functions. Encyclopedia of Mathematics and its Applications, vol. 71. Cambridge University Press, Cambridge (1999)
3. Arratia, R., Barbour, A.D., Tavaré, S.: Logarithmic combinatorial structures: a probabilistic approach. EMS Monographs in Mathematics, vol. 1. European Mathematical Society Publishing House, Zürich (2003)
4. Askey, R.A. (ed.): Gabor Szegö: Collected papers, vol. I. Birkhäuser, Basel (1982)
5. Basor, E.L., Chen, Y.: Toeplitz determinants from compatibility conditions. Ramanujan J. **16**, 25–40 (2008)
6. Blower, G. Random matrices: high dimensional phenomena. London Mathematical Society Lecture Note Series, vol. 367. Cambridge University Press (2009)
7. Borodin, A., Olshanski, G.: Infinite random matrices and Ergodic measures. Commun. Math. Phys. **203**, 87–123 (2001)
8. Borodin, A., Deift, P.: Fredholm determinants, Jimbo-Miwa-Ueno-functions, and representation theory. Commun. Pure Appl. Math. **55**, 1160–1230 (2005)
9. Böttcher, A., Silbermann, B.: Toeplitz matrices and determinants with Fisher-Hartwig symbols. J. Funct. Anal. **63**(2), 178–214 (1985)
10. Bourgade, P., Hughes, C.P., Nikeghbali, A., Yor, M.: The characteristic polynomial of a random unitary matrix: a probabilistic approach. Duke Math. J. **145**(1), 45–69 (2008)

11. Bourgade, P.: Conditional Haar measures on classical compact groups. Ann. Probab. **37**(4), 1566–1586 (2009)
12. Bourgade, P., Nikeghbali, A., Rouault, A.: Circular Jacobi ensembles and deformed Verblunski coefficients. Int. Math. Res. Not. **2009**(23), 4357–4394 (2009)
13. Bourgade, P.: A propos des matrices aléatoires et des fonctions L. Thesis, ENST Paris (2009) available online at http://tel.archives-ouvertes.fr/tel-00373735/fr/
14. Cohen, A.M.: Finite quaternionic reflection groups. J. Algebra **64**(2), 293–324 (1980)
15. Diaconis, P., Shahshahani, M.: The subgroup algorithm for generating uniform random variables. Probab. Eng. Inform. Sci. **1**, 15–32 (1987)
16. Forrester, P.J.: Log-Gases and Random Matrices, Book available online at http://www.ms. unimelb.edu.au/~matpjf/matpjf.htmlhttp://www.ms.unimelb.edu.au/~matpjf/matpjf.html
17. Hambly, B.M., Keevash, P., O'Connell, N., Stark, D.: The characteristic polynomial of a random permutation matrix. Stoch. Process. Appl. **90**, 335–346 (2000)
18. Hua, L.K.: Harmonic Analysis of Functions of Several Complex Variables in the Classical Domains. Science Press, Peking (1958) Transl. Math. Monographs 6, Am. Math. Soc., 1963.
19. Katz, N.M., Sarnak, P.: Random Matrices, Frobenius Eigenvalues and Monodromy American Mathematical Society, vol. 45. Colloquium Publications (1999)
20. Katz, N.M., Sarnak, P.: Zeros of zeta functions and symmetry. Bull. Am. Soc. **36**, 1–26 (1999)
21. Keating, J.P., Snaith, N.C.: Random matrix theory and $\zeta(1/2 + it)$. Commun. Math. Phys. **214**, 57–89 (2000)
22. Levin, E., Lubinsky, D.: Universality limits involving orthogonal polynomials on the unit circle, Comput. Meth. Funct. Theory **7**, 543–561 (2007)
23. Lubinsky, D.: Mutually Regular Measures have Similar Universality Limits. In: Neamtu, M., Schumaker, L.(eds.) Proceedings of 12th Texas Conference on Approximation Theory, pp. 256–269. Nashboro Press, Nashville (2008)
24. Mezzadri, F.: How to generate random matrices from the classical compact groups. Notices Am. Math. Soc. **54**(5), 592–604 (2007)
25. Najnudel, J., Nikeghbali, A., Rubin, F.: Scaled limit and rate of convergence for the largest Eigenvalue from the generalized Cauchy random matrix ensemble. J. Stat. Phys. **137** (2009)
26. Neretin, Yu.A.: Hua type integrals over unitary groups and over projective limits of unitary groups. Duke Math. J. **114**, 239–266 (2002)
27. Pickrell, D.: Measures on infinite-dimensional Grassmann manifolds. J. Funct. Anal. **70**(2), 323–356 (1987)
28. Pickrell, D.: Mackey analysis of infinite classical motion groups. Pacific J. Math. **150**, 139–166 (1991)
29. Pitman, J.: Combinatorial stochastic processes. Ecole d'Eté de Probabilités (Saint-Flour, 2002), Lecture Notes in Mathematics, vol. 1875. Springer, (2006)
30. Rambour, Ph., Seghier, A.: Comportement asymptotique des polynômes orthogonaux associes à un poids ayant un zéro d'ordre fractionnaire sur le cercle. Applications aux valeurs propres d'une classe de matrices aléatoires unitaires, arXiv:math.FA/0904/0904.0777v2 (2009)
31. Simon, B.: The Christoffel–Darboux kernel, In: "Perspectives in PDE, Harmonic Analysis and Applications," a volume in honor of V.G. Maz'ya's 70th birthday, Proceedings of Symposia in Pure Mathematics, vol. 79, pp. 295–335 (2008)
32. Tsilevich, N.V.: Distribution of cycle lengths of infinite permutations. Zap. Nauchn. Sem. (POMI), **223**, 148–161, 339 (1995). Translation in J. Math. Sci. **87**(6), 4072–4081 (1997)
33. Wieand, K.: Permutation matrices, wreath products, and the distribution of eigenvalues. J. Theor. Probab. **16**, 599–623 (2003)
34. Witte, N.S., Forrester, P.J.: Gap probabilities in the finite and scaled Cauchy random matrix ensembles. Nonlinearity, **13**, 1965–1986 (2000)

Discrete Approximation of the Free Fock Space

Stéphane Attal and Ion Nechita

Abstract We prove that the free Fock space $\mathcal{F}\left(L^2(\mathbb{R}^+;\mathbb{C})\right)$, which is very commonly used in Free Probability Theory, is the continuous free product of copies of the space $\mathbb{C}^2$. We describe an explicit embedding and approximation of this continuous free product structure by means of a discrete-time approximation: the free toy Fock space, a countable free product of copies of $\mathbb{C}^2$. We show that the basic creation, annihilation and gauge operators of the free Fock space are also limits of elementary operators on the free toy Fock space. When applying these constructions and results to the probabilistic interpretations of these spaces, we recover some discrete approximations of the semi-circular Brownian motion and of the free Poisson process. All these results are also extended to the higher multiplicity case, that is, $\mathcal{F}\left(L^2(\mathbb{R}^+;\mathbb{C}^N)\right)$ is the continuous free product of copies of the space $\mathbb{C}^{N+1}$.

Key words and Phrases: Free probability · Free Fock space · Toy Fock space · Limit theorems

2000 Mathematics Subject Classification: Primary 46L54. Secondary 46L09, 60F05

1 Introduction

In [1] it is shown that the symmetric Fock space $\Gamma_s(L^2(\mathbb{R}^+;\mathbb{C}))$ is actually the continuous tensor product $\otimes_{t\in\mathbb{R}^+}\mathbb{C}^2$. This result is obtained by means of an explicit embedding and approximation of the space $\Gamma_s(L^2(\mathbb{R}^+;\mathbb{C}))$ by countable tensor products $\otimes_{n\in h\mathbb{N}}\mathbb{C}^2$, when h tends to 0. The result contains explicit approximation of the basic creation, annihilation and second quantization operators by means of elementary tensor products of 2 by 2 matrices.

When applied to probabilistic interpretations of the corresponding spaces (e.g. Brownian motion, Poisson processes), one recovers well-known approximations of these processes by random walks. This means that these different probabilistic

S. Attal (✉) and I. Nechita
Université de Lyon, Université de Lyon 1, Institut Camille Jordan, CNRS UMR 5208, 43, Boulevard du 11 novembre 1918, 69622 Villeurbanne Cedex, France
e-mail: attal@math.univ-lyon1.fr; nechita@math.univ-lyon1.fr

C. Donati-Martin et al. (eds.), *Séminaire de Probabilités XLIII*, Lecture Notes in Mathematics 379
2006, DOI 10.1007/978-3-642-15217-7_16, © Springer-Verlag Berlin Heidelberg 2011

situations and approximations are all encoded by the approximation of the three basic quantum noises: creation, annihilation and second quantization operators.

These results have found many interesting applications and developments in quantum statistical mechanics, for they furnished a way to obtain quantum Langevin equations describing the dissipation of open quantum systems as a continuous-time limit of basic Hamiltonian interactions of the system with the environment: repeated quantum interactions (cf. [4, 7, 8] for example).

When considering the fermionic Fock space, even if it has not been written anywhere, it is easy to show that a similar structure holds, after a Jordan–Wigner transform on the spin-chain representation.

It is thus natural to wonder if, in the case of the free Fock space, a similar structure, a similar approximation and similar probabilistic interpretations exist. Whereas the continuous tensor product structure of the bosonic Fock space exhibits natural tensor-independence structure, it is natural to think that the free Fock space will exhibit a similar property with respect to free independence, as defined in Free Probability Theory [12, 15].

A key element of our construction is the free product of Hilbert spaces. We needed to make explicit the constructions of countable free products, as a first step. We introduce the free toy Fock space, the smallest non-commutative probability space supporting a sequence of free Bernoulli random variables. Then, by an approximation method, we define the structure of continuous free products of Hilbert spaces. This structure appears to be exactly the natural one which describes the free Fock space and its basic operators.

The paper is structured as follows. Section 2 contains a brief review of the Fock space construction in Free Probability Theory; Sects. 3 and 4 deal with free products of Hilbert spaces and the construction of the discrete free toy Fock space. Sections 5 and 6 contain the main techniques and results of the paper: the embedding of the toy Fock space in the free Fock space and the main approximation result, Theorem 1. In Sect. 7, we develop applications of our results to free probability and finally, Sect. 8 contains a generalization of the setup to the higher multiplicity case $\mathcal{F}\big(L^2(\mathbb{R}^+; \mathbb{C}^N)\big)$.

2   Free Probability and the Free Fock Space

Let us start by recalling the general framework of non commutative probability theory. A non commutative probability space is a couple $(\mathcal{A}, \varphi)$, where $\mathcal{A}$ is a complex $*$−algebra (in general non commutative) and φ is a faithful positive linear form such that $\varphi(1) = 1$. We shall call the elements of $\mathcal{A}$ non commutative random variables. The distribution of a family $(x_i)_{i \in I}$ of self-adjoint random variables of $\mathcal{A}$ is the function which maps any non-commutative polynomial $P \in \mathbb{C}\langle X_i | i \in I \rangle$ to its moment $\varphi(P((x_i)_{i \in I}))$. Thus, the map φ should be considered as the analogue of the expectation from classical probability theory. From this abstract framework, one can easily recover the setting of classical probability theory by considering a commutative algebra $\mathcal{A}$ (see [10, 12, 14]).

In order to have an interesting theory, one needs a notion of independence for non commutative probability spaces. However, classical (or tensor) independence is not adapted in this more general setting. *Free independence* was introduced by Voiculescu in the 1980s in order to tackle some problems in operator algebras, and has found many applications since, mainly in random matrix theory. Freeness provides rules for computing mixed moments of random variables when only the marginal distributions are known. More precisely, unital sub-algebras $(\mathcal{A}_i)_{i \in I}$ of $\mathcal{A}$ are called *free* (or *freely independent*) if $\varphi(a_1 \cdots a_n) = 0$ for all $n \in \mathbb{N}$ and $a_i \in \mathcal{A}_{j(i)}$ whenever $\varphi(a_i) = 0$ for all i and neighboring a_i do not come from the same sub-algebra: $j(1) \neq j(2) \neq \cdots \neq j(n)$. This definition allows one to compute mixed moments of elements coming from different algebras $\mathcal{A}_i$, when only the distributions inside each algebra $\mathcal{A}_i$ are known. Note that freeness is a highly non commutative property: two free random variables commute if and only if one of them is constant.

A remarkable setting in which freeness appears naturally is provided by creation and annihilation operators on the full Fock space. Let us now briefly describe this construction. Consider a complex Hilbert space $\mathcal{H}$ and define

$$\mathcal{F}(\mathcal{H}) = \bigoplus_{n=0}^{\infty} \mathcal{H}^{\otimes n},$$

where $\mathcal{H}^{\otimes 0}$ is a one-dimensional Hilbert space we shall denote by $\mathbb{C}\,\Omega$. $\Omega \in \mathcal{F}(\mathcal{H})$ is a distinguished norm one vector which is called the *vacuum vector* and it will play an important role in what follows. For each $f \in \mathcal{H}$, we define the left *creation* operator $\ell(f)$ and the left *annihilation* operator $\ell^*(f)$ by

$$l(f)\Omega = f, \quad l(f)e_1 \otimes \cdots \otimes e_n = f \otimes e_1 \otimes \cdots \otimes e_n;$$
$$l^*(f)\Omega = 0, \quad l^*(f)e_1 \otimes \cdots \otimes e_n = \langle f, e_1 \rangle e_2 \otimes \cdots \otimes e_n.$$

For every $T \in \mathcal{B}(\mathcal{H})$, the *gauge (or second quantization) operator* $\Lambda(T) \in \mathcal{B}(\mathcal{F}(\mathcal{H}))$ is defined by

$$\Lambda(T)\Omega = 0, \quad \Lambda(T)e_1 \otimes \cdots \otimes e_n = T(e_1) \otimes e_2 \otimes \cdots \otimes e_n.$$

All these operators are bounded, with $\|l(f)\| = \|l^*(f)\| = \|f\|$ and $\|\Lambda(T)\| = \|T\|$. On the space $\mathcal{B}(\mathcal{F}(\mathcal{H}))$ of bounded operators on the full Fock space we consider the vector state given by the vacuum vector

$$\tau(X) = \langle \Omega, X\Omega \rangle, \quad X \in \mathcal{B}(\mathcal{F}(\mathcal{H})).$$

The usefulness of the preceding construction when dealing with freeness comes from the following result ([12], Ex. 7.26, pp. 110).

Proposition 1. *Let $\mathcal{H}$ be a complex Hilbert space and consider the non commutative probability space $(\mathcal{B}(\mathcal{F}(\mathcal{H})), \tau)$. Let $\mathcal{H}_1, \ldots, \mathcal{H}_n$ be a family of orthogonal* *subspaces of* $\mathcal{H}$, *and, for each i, let* $\mathcal{A}_i$ *be the unital* ∗*-algebra generated by the set of operators*

$$\{l(f)|f \in \mathcal{H}_i\} \cup \{\Lambda(T)|T \in \mathcal{B}(\mathcal{H}), T(\mathcal{H}_i) \subset \mathcal{H}_i \text{ and } T \text{ vanishes on } \mathcal{H}_i^\perp\}.$$

Then the algebras $\mathcal{A}_1, \ldots, \mathcal{A}_n$ *are free in* $(\mathcal{B}(\mathcal{F}(\mathcal{H})), \tau)$.

In the present note, we shall be concerned mostly with the case of $\mathcal{H} = L^2(\mathbb{R}_+; \mathbb{C})$, the complex Hilbert space of square integrable complex valued functions; in Sect. 8 we shall consider the more general case of $L^2(\mathbb{R}_+; \mathbb{C}^N)$. Until then, we put $\Phi = \mathcal{F}(L^2(\mathbb{R}_+; \mathbb{C}))$, and we call this space the *free (or full) Fock space*. An element $f \in \Phi$ admits a decomposition $f = f_0 \Omega + \sum_{n \geq 1} f_n$, where $f_0 \in \mathbb{C}$ and $f_n \in L^2(\mathbb{R}_+^n)$. In this particular case we shall denote the creation (resp. annihilation) operators by a^+ (resp. a^-):

$$a^+(h)\Omega = h, \quad a^+(h)f_n = [(x_1, x_2, \ldots, x_n, x_{n+1}) \mapsto h(x_1)f_n(x_2, \ldots, x_{n+1})],$$

$$a^-(h)\Omega = 0, \quad a^-(h)f_n = [(x_2, \ldots, x_n) \mapsto \int h(x)f_n(x, x_2 \ldots, x_n)dx],$$

where h is an arbitrary function of $L^2(\mathbb{R}_+)$. For a bounded function $b \in L^\infty(\mathbb{R}_+)$, let $a^\circ(b)$ be the gauge operator associated to the operator of multiplication by b:

$$a^\circ(b)\Omega = 0, \quad a^\circ(b)f_n = [(x_1, x_2, \ldots, x_n) \mapsto b(x_1)f_n(x_1, \ldots, x_n)],$$

and $a^\times(b)$ the scalar multiplication by $\int b$:

$$a^\times(b)\Omega = \int b(x)dx \cdot \Omega,$$

$$a^\times(b)f_n = [(x_1, x_2, \ldots, x_n) \mapsto \left(\int b(x)dx\right) \cdot f_n(x_1, \ldots, x_n)].$$

Finally, we note $\mathbf{1}_t = \mathbf{1}_{[0,t)}$ the indicator function of the interval $[0, t)$ and, for all $t \in \mathbb{R}_+$ and $\varepsilon \in \{+, -, \circ, \times\}$, we put $a_t^\varepsilon = a^\varepsilon(\mathbf{1}_{[0,t)})$. Obviously, $a_t^\times = t \cdot \text{Id}$.

3 The Free Product of Hilbert Spaces

In the previous section we have seen that the algebras generated by creation, annihilation and gauge operators acting on orthogonal subspaces of a Hilbert space $\mathcal{H}$ are free in the algebra of bounded operators acting on the full Fock space $\mathcal{F}(\mathcal{H})$. However, one would like, given a family of non commutative probability spaces, to construct a larger algebra which contains the initial algebras as sub-algebras which are freely independent. In classical probability (usual) independence is achieved by taking the tensor products of the original probability spaces. This is the reason why classical independence is sometimes called tensor independence. In the free probability theory, there is a corresponding construction called the *free product*. Let us recall briefly this construction (see [14, 15] for further details).

Consider a family $(\mathcal{H}_i, \Omega_i)_{i \in I}$ where the $\mathcal{H}_i$ are complex Hilbert spaces and Ω_i is a distinguished norm one vector of $\mathcal{H}_i$. Let $\mathcal{K}_i$ be the orthocomplement of Ω_i in $\mathcal{H}_i$ and define the free product

$$(\mathcal{H}, \Omega) = \underset{i \in I}{\bigstar}(\mathcal{H}_i, \Omega_i) := \mathbb{C}\Omega \oplus \bigoplus_{n \geq 1} \bigoplus_{i_1 \neq i_2 \neq \cdots \neq i_n} \mathcal{K}_{i_1} \otimes \cdots \otimes \mathcal{K}_{i_n}, \qquad (1)$$

where the direct sums are orthogonal and, as usual, $\|\Omega\| = 1$. As in [14] (Sects. 1.9 and 1.10), we proceed with the identification of the algebras of bounded operators $\mathcal{B}(\mathcal{H}_i)$ inside $\mathcal{B}(\mathcal{H})$. To this end, we shall identify an operator $T_i \in \mathcal{B}(\mathcal{H}_i)$, with the operator $\tilde{T}_i \in \mathcal{B}(\mathcal{H})$ which acts in the following way:

$$\tilde{T}_i(\Omega) = T_i(\Omega_i) \qquad (2)$$

$$\tilde{T}_i(k_i \otimes k_{j_1} \otimes \cdots \otimes k_{j_n}) = T_i(k_i) \otimes k_{j_1} \otimes \cdots \otimes k_{j_n} \qquad (3)$$

$$\tilde{T}_i(k_{j_1} \otimes \cdots \otimes k_{j_n}) = T_i(\Omega_i) \otimes k_{j_1} \otimes \cdots \otimes k_{j_n} \qquad (4)$$

where $j_1 \neq i$ and we identify an element of $\mathcal{H}_i$ with the corresponding element of $\mathcal{H}$. The main interest of this construction is the following straightforward result.

Proposition 2. *The algebras* $\{\mathcal{B}(\mathcal{H}_i)\}_{i \in I}$ *are free in* $(\mathcal{B}(\mathcal{H}), \tau)$.

Proof. Consider a sequence $T_{i(1)}, \ldots, T_{i(n)}$ of elements of $\mathcal{B}(\mathcal{H}_{i(1)}), \ldots, \mathcal{B}(\mathcal{H}_{i(n)})$ respectively such that $i(1) \neq i(2) \neq \cdots \neq i(n)$ and $\langle \Omega_{i(k)}, T_{i(k)}\Omega_{i(k)} \rangle = 0$ for all k. By the definition of freeness, it suffices to show that $\langle \Omega, \tilde{T}_{i(1)} \cdots \tilde{T}_{i(n)}\Omega \rangle = 0$. Using the previously described embedding, we get $\tilde{T}_{i(n)}\Omega = T_{i(n)}\Omega_{i(n)}$.

Since $i(n-1) \neq i(n)$ and $\tilde{T}_{i(n)}\Omega \notin \mathbb{C}\Omega$, it follows that $\tilde{T}_{i(n-1)}\tilde{T}_{i(n)}\Omega = [T_{i(n-1)}\Omega_{i(n-1)}] \otimes [T_{i(n)}\Omega_{i(n)}]$. Continuing this way, it is easy to see that $\tilde{T}_{i(1)} \cdots \tilde{T}_{i(n)}\Omega = [T_{i(1)}\Omega_{i(1)}] \otimes \cdots \otimes [T_{i(n)}\Omega_{i(n)}]$, and the conclusion follows. $\square$

We look now at the free Fock space of a direct sum of Hilbert spaces. In the symmetric case (see [1]), it is known that one has to take the tensor product of the symmetric Fock spaces in order to obtain the Fock space of the sum. The free setting admits an analogous *exponential property*, where instead of the tensor product one has to use the free product introduced earlier.

Lemma 1. *Consider a family of orthogonal Hilbert spaces* $(\mathcal{H}_i)_{i \in I}$. *Then*

$$\mathcal{F}(\oplus_{i \in I} \mathcal{H}_i) = \bigstar_{i \in I} \mathcal{F}(\mathcal{H}_i). \qquad (5)$$

Proof. Fix for each $\mathcal{H}_i$ an orthonormal basis $(X^j(i))_{j \in B(i)}$. Then, an orthonormal basis of $\mathcal{F}(\oplus \mathcal{H}_i)$ is given by $\{\Omega\} \cup \{X^{j_1}(i_1) \otimes \cdots \otimes X^{j_n}(i_n)\}$, where $n \geq 1, i_k \in I$ and $j_k \in B(i_k)$ for all $1 \leq k \leq n$. One obtains a Hilbert space basis of $\bigstar \mathcal{F}(\mathcal{H}_i)$ by grouping adjacent elements of $X^{j_1}(i_1) \otimes \cdots \otimes X^{j_n}(i_n)$ with the same i-index (i.e. belonging to the same $\mathcal{H}_i$). Details are left to the reader. $\square$

4 The Free Toy Fock Space

In this section we introduce the *free toy Fock space*, the main object of interest in our paper. From a probabilistic point of view, it is the "smallest" non commutative probability space supporting a free identically distributed countable family of Bernoulli random variables (see Sect. 7).

The free toy Fock space is a countable free product of two-dimensional complex Hilbert spaces: in (1), take $\mathcal{H}_i = \mathbb{C}^2$ for all i. In order to keep track of which copy of $\mathbb{C}^2$ we are referring to, we shall label the ith copy with $\mathbb{C}^2_{(i)}$. Each copy is endowed with the canonical basis $\{\Omega_i = (1,0)^\top, X_i = (0,1)^\top\}$. Since the orthogonal complement of the space $\mathbb{C}\,\Omega_i$ is simply $\mathbb{C}\,X_i$, we obtain the following simple definition of the free toy Fock space $T\Phi$:

$$(T\Phi, \Omega) := \bigstar_{i \in \mathbb{N}} (\mathbb{C}^2_{(i)}, \Omega_i) = \mathbb{C}\,\Omega \oplus \bigoplus_{n \geq 1} \bigoplus_{i_1 \neq \cdots \neq i_n} \mathbb{C}\,X_{i_1} \otimes \cdots \otimes \mathbb{C}\,X_{i_n},$$

where, as usual, Ω is the identification of the vacuum reference vectors Ω_i ($\|\Omega\| = 1$). Note that the orthonormal basis of $T\Phi$ given by this construction is indexed by the set of all finite (eventually empty) words with letters from $\mathbb{N}$ with the property that neighboring letters are distinct. More formally, a word $\sigma = [i_1, i_2, \ldots, i_n] \in \mathbb{N}^n$ is called *adapted* if $i_1 \neq i_2 \neq \cdots \neq i_n$. By convention, the empty word $\emptyset$ is adapted. We shall denote by $\mathcal{W}_n$ (resp. $\mathcal{W}_n^*$) the set of all words (resp. adapted words) of size n and by $\mathcal{W}$ (resp. $\mathcal{W}^*$) the set of all words (resp. adapted words) of finite size (including the empty word). For a word $\sigma = [i_1, i_2, \ldots, i_n]$, let X_σ be the tensor $X_{i_1} \otimes X_{i_2} \otimes \cdots \otimes X_{i_n}$ and put $X_\emptyset = \Omega$. With this notation, an orthonormal basis of $T\Phi$ is given by $\{X_\sigma\}_{\sigma \in \mathcal{W}^*}$.

We now turn to operators on $\mathbb{C}^2_{(i)}$ and their embedding into $\mathcal{B}(T\Phi)$. We are interested in the following four operators acting on $\mathbb{C}^2$:

$$a^+ = \begin{bmatrix} 0 & 0 \\ 1 & 0 \end{bmatrix}, \quad a^- = \begin{bmatrix} 0 & 1 \\ 0 & 0 \end{bmatrix}, \quad a^\circ = \begin{bmatrix} 0 & 0 \\ 0 & 1 \end{bmatrix}, \quad a^\times = \begin{bmatrix} 1 & 0 \\ 0 & 0 \end{bmatrix}.$$

For $\varepsilon \in \{+, -, \circ, \times\}$, we shall denote by a_i^ε the image of a^ε acting on the ith copy of $\mathbb{C}^2$, viewed (by the identification described earlier in (2)–(4)) as an operator on $T\Phi$. The action of these operators on the orthonormal basis of $T\Phi$ is rather straightforward to compute ($\sigma = [\sigma_1, \ldots, \sigma_n]$ is an arbitrary non-empty adapted word and $\mathbf{1}$ is the indicator function):

$$a_i^+ \Omega = X_i, \quad a_i^+ X_\sigma = \mathbf{1}_{\sigma_1 \neq i} X_{[i,\sigma]}; \tag{6}$$

$$a_i^- \Omega = 0, \quad a_i^- X_\sigma = \mathbf{1}_{\sigma_1 = i} X_{[\sigma_2, \ldots, \sigma_n]}; \tag{7}$$

$$a_i^\circ \Omega = 0, \quad a_i^\circ X_\sigma = \mathbf{1}_{\sigma_1 = i} X_\sigma; \tag{8}$$

$$a_i^\times \Omega = \Omega, \quad a_i^\times X_\sigma = \mathbf{1}_{\sigma_1 \neq i} X_\sigma. \tag{9}$$

5 Embedding of the Toy Fock Space into the Full Fock Space

Our aim is now to show that the free toy Fock space can be realized as a closed subspace of the full (or free) Fock space $\Phi = \mathcal{F}(L^2(\mathbb{R}_+; \mathbb{C}))$ of square integrable functions. What is more, to each partition of $\mathbb{R}_+$ we shall associate such an embedding, and, as we shall see in the next section, when the diameter of the partition becomes small, one can approximate the full Fock space with the (much simpler) toy Fock space. These results follow closely similar constructions in the bosonic case (see [1] or [11], Chap. II).

Let $S = \{0 = t_0 < t_1 < \cdots < t_n < \cdots \}$ be a partition of $\mathbb{R}_+$ of diameter $\delta(S) = \sup_i |t_{i+1} - t_i|$. The main idea of [1] was to decompose the symmetric Fock space of $L^2(\mathbb{R}_+)$ along the partition S. In our free setting we have an analogous exponential property (see (5)):

$$\Phi = \underset{i \in \mathbb{N}}{\bigstar} \Phi_i,$$

where $\Phi_i = \mathcal{F}(L^2[t_i, t_{i+1}))$, the countable free product being defined with respect to the vacuum functions. Inside each Fock space Φ_i, we consider two distinguished functions: the vacuum function Ω_i and the normalized indicator function of the interval $[t_i, t_{i+1})$:

$$X_i = \frac{\mathbf{1}_{[t_i, t_{i+1})}}{\sqrt{t_{i+1} - t_i}} = \frac{\mathbf{1}_{t_{i+1}} - \mathbf{1}_{t_i}}{\sqrt{t_{i+1} - t_i}}.$$

These elements span a two-dimensional vector space $\mathbb{C}\,\Omega_i \oplus \mathbb{C}\,X_i$ inside each Φ_i. The toy Fock space associated to the partition S is the free product of these two-dimensional vector spaces:

$$T\Phi(S) = \underset{i \in \mathbb{N}}{\bigstar} (\mathbb{C}\,\Omega_i \oplus \mathbb{C}\,X_i).$$

$T\Phi(S)$ is a closed subspace of the full Fock space Φ and it is naturally isomorphic (as a countable free product of two-dimensional spaces) to the abstract free toy Fock space $T\Phi$ defined in the previous section. It is spanned by the orthonormal family $\{X_\sigma\}_{\sigma \in \mathcal{W}^*}$, where $X_\sigma = X_\sigma(S)$ is defined by

$$X_\sigma = X_{\sigma_1} \otimes X_{\sigma_2} \otimes \cdots \otimes X_{\sigma_n} = \left[(x_1, \ldots, x_n) \mapsto \frac{\prod_{j=1}^n \mathbf{1}_{[t_{\sigma_j}, t_{\sigma_j+1})}(x_j)}{\prod_{j=1}^n \sqrt{t_{\sigma_j+1} - t_{\sigma_j}}} \right],$$

with $\sigma = [\sigma_1, \ldots, \sigma_n]$. We shall denote by $P_S \in \mathcal{B}(\Phi)$ the orthogonal projector on $T\Phi(S)$. For a function $f \in \Phi$, which admits a decomposition $f = f_0 \Omega + \sum_{n \geqslant 1} f_n$ with $f_0 \in \mathbb{C}$ and $f_n \in L^2(\mathbb{R}_+^n)$, the action of P_S is straightforward to compute:

$$P_S f = f_0 \Omega + \sum_{n \geqslant 1} \sum_{\sigma \in \mathcal{W}_n^*} \langle X_\sigma, f_n \rangle X_\sigma, \tag{10}$$

where the scalar products are taken in the corresponding L^2 spaces.

We ask now how the basic operators a_t^ε, $\varepsilon \in \{+, -, \circ, \times\}$, $t \in \mathbb{R}^+$ of the free Fock space relate to their discrete counterparts a_i^ε. In order to do this, we consider the following rescaled restrictions of a_t^+, a_t^- and a_t° on the toy Fock space $T\Phi(S)$:

$$a_i^+(S) = P_S \frac{a_{t_{i+1}}^+ - a_{t_i}^+}{\sqrt{t_{i+1} - t_i}} P_S = P_S a^+ \left(\frac{\mathbf{1}_{[t_i, t_{i+1})}}{\sqrt{t_{i+1} - t_i}} \right) P_S; \tag{11}$$

$$a_i^-(S) = P_S \frac{a_{t_{i+1}}^- - a_{t_i}^-}{\sqrt{t_{i+1} - t_i}} P_S = P_S a^- \left(\frac{\mathbf{1}_{[t_i, t_{i+1})}}{\sqrt{t_{i+1} - t_i}} \right) P_S; \tag{12}$$

$$a_i^\circ(S) = P_S (a_{t_{i+1}}^\circ - a_{t_i}^\circ) P_S = P_S a^\circ \left(\mathbf{1}_{[t_i, t_{i+1})} \right) P_S. \tag{13}$$

The operators $a_i^\varepsilon(S) \in \mathcal{B}(\Phi)$ are such that $a_i^\varepsilon(S)(T\Phi(S)) \subset T\Phi(S)$ and they vanish on $T\Phi(S)^\perp$, so one can also see them as operators on $T\Phi(S)$. For $\varepsilon = \times$, one can not define $a_i^\times(S)$ from $a_t^\times$ as it was done in (11)–(13). Instead, we define it as the linear extension of $a_i^\times$ (via the isomorphism $T\Phi \simeq T\Phi(S)$) which vanishes on $T\Phi(S)^\perp$. Hence, $a_i^\times(S) = P_S(\mathrm{Id} - a_i^\circ(S)) P_S$.

Proposition 3. *For $\varepsilon \in \{+, -, \circ, \times\}$, the operators $a_i^\varepsilon(S)$, acting on the toy Fock space $T\Phi(S)$, behave in the same way as their discrete counterparts a_i^ε, i.e. they satisfy (6)–(9).*

Proof. For each $\sigma = [\sigma_1, \sigma_2, \ldots, \sigma_n] \in \mathcal{W}^*$, consider the corresponding basis function of $T\Phi(S)$:

$$X_\sigma(S) = \frac{\mathbf{1}_\sigma(S)}{\prod_{j=1}^n \sqrt{t_{\sigma_j + 1} - t_{\sigma_j}}},$$

where $\mathbf{1}_\sigma(S)$ is the indicator function of the rectangle $\times_{j=1}^n [t_{\sigma_j}, t_{\sigma_j + 1})$. We have:

$$a_i^+(S) X_\sigma(S) = P_S \frac{a^+ (\mathbf{1}_{[t_i, t_{i+1})})}{\sqrt{t_{i+1} - t_i}} X_\sigma(S) = P_S X_{[i,\sigma]}(S) = \mathbf{1}_{\sigma_1 \neq i} X_{[i,\sigma]}(S),$$

$$a_i^-(S) X_\sigma(S) = P_S \frac{a^- (\mathbf{1}_{[t_i, t_{i+1})})}{\sqrt{t_{i+1} - t_i}} X_\sigma(S) = P_S \mathbf{1}_{\sigma_1 = i} X_{[\sigma_2, \ldots, \sigma_n]}(S)$$

$$= \mathbf{1}_{\sigma_1 = i} X_{[\sigma_2, \ldots, \sigma_n]}(S),$$

$$a_i^\circ(S) X_\sigma(S) = P_S a^\circ (\mathbf{1}_{[t_i, t_{i+1})}) X_\sigma(S) = P_S \mathbf{1}_{\sigma_1 = i} X_\sigma(S) = \mathbf{1}_{\sigma_1 = i} X_\sigma(S).$$

These relations are identical to the action of the corresponding operators a_i^ε on the abstract toy Fock space $T\Phi \simeq T\Phi(S)$ (compare to (6)–(8)). For $a_i^\times(S)$, the conclusion is immediate from the last equation above and its definition:

$$a_i^\times(S) X_\sigma(S) = P_S[\mathrm{Id} - a_i^\circ(S)] X_\sigma(S) = X_\sigma(S) - \mathbf{1}_{\sigma_1 = i} X_\sigma(S)$$

$$= \mathbf{1}_{\sigma_1 \neq i} X_\sigma(S).$$

$\square$

6 Approximation Results

This section contains the main result of this work, Theorem 1. We show that the toy Fock space $T\Phi(S)$ together with its operators a_i^ε approach the full Fock space Φ and its operators a_t^ε when the diameter of the partition S approaches 0.

Let us consider a sequence of partitions $S_n = \{0 = t_0^{(n)} < t_1^{(n)} < \cdots < t_k^{(n)} < \cdots\}$ such that $\delta(S_n) \to 0$. In order to lighten the notation, we put $T\Phi(n) = T\Phi(S_n)$, $P_n = P_{S_n}$ and $a_i^\varepsilon(n) = a_i^\varepsilon(S_n)$.

Theorem 1. *For a sequence of partitions S_n of $\mathbb{R}_+$ such that $\delta(S_n) \to 0$, one has the following approximation results:*

1. *For every $f \in \Phi$, $P_n f \to f$*
2. *For all $t \in \mathbb{R}_+$, the operators*

$$a_t^\pm(n) = \sum_{i:t_i^{(n)} \leq t} \sqrt{t_{i+1}^{(n)} - t_i^{(n)}}\, a_i^\pm(n),$$

$$a_t^\circ(n) = \sum_{i:t_i^{(n)} \leq t} a_i^\circ(n),$$

$$a_t^\times(n) = \sum_{i:t_i^{(n)} \leq t} \left(t_{i+1}^{(n)} - t_i^{(n)}\right) a_i^\times(n)$$

converge strongly, when $n \to \infty$, to $a_t^\pm$, a_t° and $a_t^\times$ respectively

Proof. For the fist part, consider a (not necessarily adapted) word $\sigma = [\sigma_1, \ldots, \sigma_k]$ and denote by $1_\sigma^{(n)}$ the indicator function of the rectangle $\times_{j=1}^k [t_{\sigma_j}^{(n)}, t_{\sigma_j+1}^{(n)})$ of $\mathbb{R}_+^k$. It is a classical result in integration theory that, since $\delta(S_n) \to 0$, the simple functions $\{1_\sigma^{(n)}\}_{\sigma \in W_k, n \geq 1}$ are dense in $L^2(\mathbb{R}_+^k)$ for all k. It is obvious that the result still holds when replacing W_k with the smaller set of adapted words W_k^*. Since $P_n f$ is an element of the vector space generated by the set $\{1_\sigma^{(n)}\}_{\sigma \in W_k^*}$, the result of the first part follows.

As for the second statement of the theorem, let us start by treating the case of a_t^+. For fixed n and t, let $t^{(n)} = t_{i+1}^{(n)}$, where i is the last index appearing in the definition of $a_t^+(n)$, i.e. $t_i^{(n)} \leq t < t_{i+1}^{(n)}$. With this notation, we have $a_t^+(n) = \sum_{i:t_i^{(n)} \leq t} \sqrt{t_{i+1}^{(n)} - t_i^{(n)}}\, a_i^+(n) = P_n a_{t^{(n)}}^+ P_n$. Hence, for any function $f \in \mathcal{F}$, we obtain:

$$\|a_t^+(n)f - a_t^+ f\|$$
$$= \|P_n a_{t^{(n)}}^+ P_n f - a_t^+ f\|$$
$$\leq \|P_n a_{t^{(n)}}^+ P_n f - P_n a_{t^{(n)}}^+ f\| + \|P_n a_{t^{(n)}}^+ f - P_n a_t^+ f\| + \|P_n a_t^+ f - a_t^+ f\|$$
$$\leq \|P_n a_{t^{(n)}}^+\| \|(P_n - I)f\| + \|P_n a^+ 1_{[t,t^{(n)})}\| \|f\| + \|(P_n - I)(a_t^+ f)\|.$$

By the first point, $P_n \to I$ strongly, hence the first and the third terms above converge to 0. The norm of the operator appearing in the second term is bounded by the L^2 norm of $\mathbf{1}_{[t,t^{(n)})}$ which is infinitely small when $n \to \infty$. Hence, the entire quantity converges to 0 and we obtained the announced strong convergence. The proof adapts easily to the cases of a_t^- and a_t°.

Finally, recall that $a_i^\times(n) = P_n(\mathrm{Id} - a_i^\circ(n)) P_n$. Hence, with the same notation as above,

$$\sum_{i:t_i^{(n)} \leqslant t} \left(t_{i+1}^{(n)} - t_i^{(n)} \right) a_i^\times(n) = t^{(n)} P_n - \sum_{i:t_i^{(n)} \leqslant t} \left(t_{i+1}^{(n)} - t_i^{(n)} \right) a_i^\circ(n).$$

The second term above converges to zero in the strong operator topology thanks to the factor $t_{i+1}^{(n)} - t_i^{(n)}$ which is less than $\delta(S_n)$, and thus we are left only with $t^{(n)} P_n$ which converges, by the first point, to $t \cdot \mathrm{Id}$.   □

7 Applications to Free Probability Theory

This section is more probabilistic in nature. We use the previous approximation result to show that the free Brownian motion and the free Poisson operators can be approached, in the strong operator topology, by sums of free Bernoulli-distributed operators living on the free toy Fock space. We obtain, as corollaries, already known free Donsker-like convergence results.

Let us start by recalling some basic facts about free noises and their realization on the free Fock space Φ. The free Brownian motion W_t and the free Poisson process N_t were constructed in [13] as free analogues of the classical Brownian motion (or Wiener process) and, respectively, classical Poisson jump processes. Recall that a process with stationary and freely independent increments is a collection of non commutative self-adjoint random variables $(X_t)_t$ with the following properties:

1. For all $s < t$, $X_t - X_s$ is free from the algebra generated by $\{X_u, u \leqslant s\}$
2. The distribution of $X_t - X_s$ depends only on $t - s$

A free Brownian motion is a process with stationary and freely independent increments $(W_t)_t$ such that the distribution of $W_t - W_s$ is a *semi-circular* random variable of mean 0 and variance $t - s$. Recall that a standard (i.e. mean zero and variance one) semicircular random variable has distribution

$$d\mu(x) = \frac{1}{2\pi} \sqrt{4 - x^2} \mathbf{1}_{[-2,2]}(x) dx.$$

If X is a standard semicircular random variable, then $\sqrt{t-s}X$ is semicircular of variance $(t - s)$. In an analogous manner, a free Poisson process is a process with stationary and freely independent increments $(N_t)_t$ such that the distribution of

$N_t - N_s$ is a *free Poisson* random variable of parameter $\lambda = t - s$. In general, the density of a free Poisson random variable is given by

$$dv_\lambda(x) = \begin{cases} \dfrac{\sqrt{4\lambda-(x-1-\lambda)^2}}{2\pi x}\chi(x)dx & \text{if } \lambda \geqslant 1, \\[2ex] (1-\lambda)\delta_0 + \dfrac{\sqrt{4\lambda-(x-1-\lambda)^2}}{2\pi x}\chi(x)dx & \text{if } 0 < \lambda < 1, \end{cases}$$

where χ is the indicator function of the interval $[(1-\sqrt{\lambda})^2, (1+\sqrt{\lambda})^2]$.

The free Brownian motion and the free Poisson process can be realized on the full Fock space Φ as $W_t = a_t^+ + a_t^-$ and, respectively, $N_t = a_t^+ + a_t^- + a_t^\circ + t \cdot \text{Id}$. Generalization of these processes and stochastic calculus were considered in [5,6,9].

For the sake of simplicity, throughout this section we shall consider the sequence of partitions $S_n = \{k/n; k \in \mathbb{N}\}$; obviously $\delta(S_n) = \frac{1}{n} \to 0$. The following result is an easy consequence of Theorem 1.

Proposition 4. *Consider the operators* $X_i^{(n)} = a_i^+(n) + a_i^-(n)$, $i \in \mathbb{N}$, *acting on* Φ. *Then*

1. *For all $n \geqslant 1$, the family $\{X_i^{(n)}\}_{i\in\mathbb{N}}$ is a free family of Bernoulli random variables of distribution $\frac{1}{2}\delta_{-1} + \frac{1}{2}\delta_1$.*
2. *For all $t \in \mathbb{R}_+$, the operator*

$$W_t^{(n)} = \frac{1}{\sqrt{n}}\sum_{i=0}^{\lfloor nt \rfloor} X_i^{(n)}$$

converges in the strong operator topology, when $n \to \infty$, to the operator of free Brownian motion $W_t = a_t^+ + a_t^-$.

Let us show now that the strong operator convergence implies the convergence in distribution of the corresponding processes. Let $t_1, \ldots, t_s \in \mathbb{R}_+$ and $k_1, \ldots, k_s \in \mathbb{N}$. Since, by the previous result, $W_t^{(n)} \to W_t$ strongly, and multiplication is jointly strongly continuous on bounded subsets, we get that $(W_{t_1}^{(n)})^{k_1} \cdots (W_{t_s}^{(n)})^{k_s} \to W_{t_1}^{k_1} \cdots W_{t_s}^{k_s}$ strongly. Strong convergence implies convergence of the inner products $\langle \Omega, \cdot \Omega \rangle$ and thus the following corollary (which is a direct consequence of the Free Central Limit Theorem, see [12], Lecture 8 or [15]) holds.

Corollary 1. *The distribution of the family $\{W_t^{(n)}\}_{t\in\mathbb{R}_+}$ converges, as n goes to infinity, to the distribution of a free Brownian motion $\{W_t\}_{t\in\mathbb{R}_+}$.*

We move on to the free Poisson process N_t and we state the analogue of Proposition 4.

Proposition 5. *Consider the operators* $Y_i^{(n)} = a_i^+(n) + a_i^-(n) + \sqrt{n}a_i^\circ(n) + \frac{1}{\sqrt{n}}a_i^\times(n)$, *acting on* Φ. *Then*

1. *For all $n \geq 1$, the family $\{Y_i^{(n)}\}_{i \in \mathbb{N}}$ is a free family of Bernoulli random variables of distribution $\frac{1}{n+1}\delta_{\frac{n+1}{\sqrt{n}}} + \frac{n}{n+1}\delta_0$.*

2. *For all $t \in \mathbb{R}_+$, the operator*

$$N_t^{(n)} = \frac{1}{\sqrt{n}} \sum_{i=0}^{\lfloor nt \rfloor} Y_i^{(n)}$$

converges strongly, when $n \to \infty$, to the operator of the free Poisson process $N_t = a_t^+ + a_t^- + a_t^\circ + a_t^\times$.

Proof. As an operator on $\mathbb{C}^2$, $Y_i^{(n)}$ has the form

$$Y_i^{(n)} = \begin{bmatrix} \frac{1}{\sqrt{n}} & 1 \\ 1 & \sqrt{n} \end{bmatrix}.$$

For all $k \geq 1$, the kth moment of $Y_i^{(n)}$ is easily seen to be given by the formula

$$\langle \Omega, (Y_i^{(n)})^k \Omega \rangle = \frac{1}{n+1}\left(\frac{n+1}{\sqrt{n}}\right)^k,$$

which is the same as the kth moment of the probability distribution $\frac{1}{n+1}\delta_{\frac{n+1}{\sqrt{n}}} + \frac{n}{n+1}\delta_0$, and the first part follows. For the second part, we have

$$N_t^{(n)} = \frac{1}{\sqrt{n}} \sum_{i=0}^{\lfloor nt \rfloor} Y_i^{(n)} = \sum_{i\,;\,t_i^{(n)} \leq t} \left[\frac{1}{\sqrt{n}}a_i^+(n) + \frac{1}{\sqrt{n}}a_i^-(n) + a_i^\circ(n) + \frac{1}{n}a_i^\times(n)\right]$$

$$= \sum_{i\,;\,t_i^{(n)} \leq t} \sqrt{t_{i+1}^{(n)} - t_i^{(n)}} \left(a_i^+(n) + a_i^-(n)\right) + \sum_{i\,;\,t_i^{(n)} \leq t} a_i^\circ(n)$$

$$+ \sum_{i\,;\,t_i^{(n)} \leq t} \left(t_{i+1}^{(n)} - t_i^{(n)}\right) a_i^\times(n).$$

Using Theorem 1, one obtains $N_t^{(n)} \to N_t$ in the strong operator topology.   □

Again, we obtain as a corollary the convergence in distribution of the process $(N_t^{(n)})_t$ to the free Poisson process, which is in fact a reformulation of the Free Poisson limit theorem ([12], pp. 203).

Corollary 2. *The distribution of the family $\{N_t^{(n)}\}_{t \in \mathbb{R}_+}$ converges, as n goes to infinity, to the distribution of a free Poisson process $\{N_t\}_{t \in \mathbb{R}_+}$.*

8 Higher Multiplicities

We generalize now the previous construction of the free toy Fock space by replacing $\mathbb{C}^2$ with the $N + 1$-dimensional complex Hilbert space $\mathbb{C}^{N+1}$. Much of what was done in $\mathbb{C}^2$ extends easily to the generalized case, so we only sketch the construction, leaving the details to the reader (for an analogous setup in the symmetric Fock space, see [3]). In what follows, $N \geqslant 1$ is a fixed integer, called the *multiplicity* of the Fock space.

Start with a countable family of copies of $\mathbb{C}^{N+1}$, each endowed with a fixed basis $(\Omega, X^1, \ldots, X^N)$. We shall sometimes note $X^0 = \Omega$. We introduce the free toy Fock space of multiplicity N (see Sect. 4):

$$T\Phi = \underset{i \in \mathbb{N}}{\bigstar}\, \mathbb{C}^{N+1}(i),$$

where the countable tensor product is defined with respect to the stabilizing sequence of vectors $\Omega(i) \in \mathbb{C}^{N+1}(i)$. An orthonormal basis of this space is indexed by the set $\mathcal{W}^{N*}$ of generalized adapted words $\sigma = [(i_1, j_1), (i_2, j_2), \ldots, (i_n, j_n)]$, where $n \in \mathbb{N}$, $i_1 \neq i_2 \neq \cdots \neq i_n$ and $j_1, \ldots, j_n \in \{1, \ldots, N\}$, the corresponding basis element being $X_\sigma = X^{j_1}(i_1) \otimes X^{j_2}(i_2) \otimes \cdots \otimes X^{j_n}(i_n)$.

On each copy of $\mathbb{C}^{N+1}$ we introduce the matrix units a_j^i defined by

$$a_j^i X^k = \delta_{ik} X^j, \quad i, j, k = 0, 1, \ldots, N.$$

We shall now show how the discrete structure of the free toy Fock space of multiplicity N approximates the free Fock space $\Phi = \mathcal{F}(L^2(\mathbb{R}_+; \mathbb{C}^N))$. To this end, consider a partition $\mathcal{S} = \{0 = t_0 < t_1 < \cdots < t_n < \cdots\}$ of $\mathbb{R}_+$ and recall the decomposition of the free Fock space of multiplicity N as a free product of "smaller" Fock spaces:

$$\mathcal{F}(L^2(\mathbb{R}_+; \mathbb{C}^N)) = \underset{i \in \mathbb{N}}{\bigstar}\, \mathcal{F}(L^2([t_i, t_{i+1}); \mathbb{C}^N)).$$

In each factor of the free product we consider $N + 1$ distinguished functions: the constant function Ω_i (sometimes denoted by $X^0(i)$) and the normalized indicator functions

$$X^j(i) = \frac{\mathbf{1}^j_{[t_i, t_{i+1})}}{\sqrt{t_{i+1} - t_i}} = \frac{\mathbf{1}^j_{t_{i+1}} - \mathbf{1}^j_{t_i}}{\sqrt{t_{i+1} - t_i}}, \quad 1 \leqslant j \leqslant N,$$

where $\mathbf{1}^j_A(x) = (0, \ldots, 0, 1, 0, \ldots, 0)^\top$ with the 1 in the jth position if $x \in A$ and 0 otherwise. For a generalized word $\sigma = [(i_1, j_1), (i_2, j_2), \ldots, (i_n, j_n)]$, define the element $X_\sigma(\mathcal{S}) \in \Phi$ by

$$X_\sigma(\mathcal{S}) = X^{j_1}(i_1) \otimes \cdots \otimes X^{j_n}(i_n) = [(x_1, \ldots, x_n) \mapsto \frac{\prod_{k=1}^n \mathbf{1}^{j_k}_{[t_{i_k}, t_{i_k}+1)}(x_k)}{\prod_{k=1}^n \sqrt{t_{i_k}+1 - t_{i_k}}}],$$

with $\sigma = [(i_1, j_1), (i_2, j_2), \ldots, (i_n, j_n)]$. The toy Fock space associated to $\mathcal{S}$ (denoted by $T\Phi(\mathcal{S})$) is the span of $X_\sigma(\mathcal{S})$ for all generalized adapted words $\sigma \in \mathcal{W}^{N*}$. $T\Phi(\mathcal{S})$ is a closed subspace of the full Fock space Φ and it is naturally isomorphic to the abstract toy Fock space of multiplicity N, $T\Phi$. For a given sequence of refining partitions $\mathcal{S}_n$ whose diameters converge to zero, the toy Fock spaces and the operators a^i_j approximate the Fock space Φ and its corresponding operators (compare with Theorem 1):

Theorem 2. *Let Φ be the free Fock space of multiplicity N and $\mathcal{S}_n$ a sequence of refining partitions of $\mathbb{R}_+$ such that $\delta(\mathcal{S}_n) \to 0$. Then one has the following approximation results:*

1. For every $f \in \Phi$, $P_n f \to f$
2. For $i, j \in \{0, 1, \ldots, N\}$, define $\varepsilon_{ij} = \frac{1}{2}(\delta_{0i} + \delta_{0j})$. Then, for all $t \in \mathbb{R}_+$, the operators

$$\sum_{k : t_k^{(n)} \le t} (t_{k+1}^{(n)} - t_k^{(n)})^{\varepsilon_{ij}} a^i_j(k)$$

converge strongly, when $n \to \infty$, to $a^i_j(t)$

8.1 An Example for N = 2

Let us end this section by constructing an approximation of a two-dimensional free Brownian motion constructed on a free Fock space of multiplicity $N = 2$. To this end, define the free Fock space $\Phi = \mathcal{F}(L^2(\mathbb{R}_+; \mathbb{C}^2))$ and its discrete approximation, the free toy Fock space $T\Phi = \star_{k \in \mathbb{N}} \mathbb{C}^3_{(k)}$. The simplest realization of two freely independent free Brownian motions on Φ is the pair of operator processes $W_1(\cdot), W_2(\cdot) \in \mathcal{B}(\Phi)$ defined by:

$$W_1(t) = a^0_1(t) + a^1_0(t) \text{ and } W_2(t) = a^0_2(t) + a^2_0(t).$$

First of all, it is obvious that both $W_1(\cdot)$ and $W_2(\cdot)$ are free Brownian motions (see Sect. 7). Moreover, the families $(W_1(t))_t$ and $(W_2(t))_t$ are freely independent since the functions $\mathbf{1}^1_s$ and $\mathbf{1}^2_t$ are orthogonal in $\mathcal{F}(L^2(\mathbb{R}_+; \mathbb{C}^2))$ (see Proposition 1). We consider, as we did in Sect. 7, the sequence of refining partitions $\mathcal{S}_n = \{k/n; k \in \mathbb{N}\}$. We introduce the following two families of operators:

$$Y_1(k) = a^0_1(k) + a^1_0(k),$$
$$Y_2(k) = a^0_2(k) + a^2_0(k),$$

and respectively

$$Z_1(k) = a^0_1(k) + a^1_0(k) - a^2_2(k),$$

$$Z_2(k) = a_2^0(k) + a_0^2(k) - [a_2^1(k) + a_1^2(k) + a_2^2(k)],$$

for $k \in \mathbb{N}$. It follows from Theorem 2 that for all $t \in \mathbb{R}_+$, both families are approximations of a two-dimensional Brownian motion:

$$\frac{1}{\sqrt{n}} \left(\sum_{i=0}^{\lfloor nt \rfloor} Y_1(n), \sum_{i=0}^{\lfloor nt \rfloor} Y_2(n) \right) \xrightarrow[n \to \infty]{} (W_1(t), W_2(t))$$

and

$$\frac{1}{\sqrt{n}} \left(\sum_{i=0}^{\lfloor nt \rfloor} Z_1(n), \sum_{i=0}^{\lfloor nt \rfloor} Z_2(n) \right) \xrightarrow[n \to \infty]{} (W_1(t), W_2(t)),$$

where the limits hold in the strong operator topology. However, the building blocks of these approximating processes have completely different behaviors at fixed k. To start, note that the self-adjoint operators $Y_1(k)$ and $Y_2(k)$, represented, in the basis (Ω, X^1, X^2), by the hermitian matrices

$$Y_1 = \begin{bmatrix} 0 & 1 & 0 \\ 1 & 0 & 0 \\ 0 & 0 & 0 \end{bmatrix} \text{ and } Y_2 = \begin{bmatrix} 0 & 0 & 1 \\ 0 & 0 & 0 \\ 1 & 0 & 0 \end{bmatrix}$$

do not commute. Hence, they do not admit a classical joint distribution, i.e. it does not exist a probability measure μ on $\mathbb{R}^2$ such that

$$\int_{\mathbb{R}^2} y_1^m y_2^n d\mu(y_1, y_2) = \langle \Omega, Y_1^m Y_2^n \Omega \rangle. \tag{14}$$

On the contrary, for each k, the operators $Z_1(k)$ and $Z_2(k)$, which act on $\mathbb{C}^3$ as the matrices

$$Z_1 = \begin{bmatrix} 0 & 1 & 0 \\ 1 & 0 & 0 \\ 0 & 0 & -1 \end{bmatrix} \text{ and } Z_2 = \begin{bmatrix} 0 & 0 & 1 \\ 0 & 0 & -1 \\ 1 & -1 & -1 \end{bmatrix},$$

commute and they admit the following classical joint distribution (in the sense of (14)):

$$\mu = \frac{1}{2} \delta_{(1,0)} + \frac{1}{3} \delta_{(-1,1)} + \frac{1}{6} \delta_{(-1,-2)}.$$

More details on high multiplicity Fock spaces and the analogue construction in the commutative case can be found in [2, 3].

Acknowledgements We thank the referee for several helpful remarks that improved the presentation of the paper.

References

1. Attal, S.: Approximating the Fock space with the toy Fock space. Séminaire de Probabilités, XXXVI. Lecture Notes in Mathematics, vol. 1801, pp. 477–491. Springer, Berlin (2003)
2. Attal, S., Émery, M.: Équations de structure pour des martingales vectorielles. Séminaire de Probabilités, XXVIII. Lecture Notes in Mathematics, vol. 1583, pp. 256–278. Springer, Berlin (1994)
3. Attal, S., Pautrat, Y.: From $(n + 1)$-level atom chains to n-dimensional noises. Ann. Inst. H. Poincaré Probab. Statist. **41**(3), 391–407 (2005)
4. Attal, S., Pautrat, Y.: From repeated to continuous quantum interactions. Ann. Henri Poincaré **7**(1), 59–104 (2006)
5. Biane, P., Speicher, R.: Stochastic calculus with respect to free Brownian motion and analysis on Wigner space. Probab. Theor. Relat. Field. **112**(3), 373–409 (1998)
6. Bożejko, M., Speicher, R.: An example of a generalized Brownian motion. Comm. Math. Phys. **137**(3), 519–531 (1991)
7. Bruneau, L., Joye, A., Merkli, M.: Asymptotics of repeated interaction quantum systems. J. Funct. Anal. **239**(1), 310–344 (2006)
8. Bruneau, L., Pillet, C.-A.: Thermal relaxation of a QED cavity, preprint available at http://hal. archives-ouvertes.fr/hal-00325206/en/
9. Glockner, P., Schürmann, M., Speicher, R.: Realization of free white noises. Arch. Math. (Basel) **58**(4), 407–416 (1992)
10. Hiai, F., Petz, D.: The semicircle law, free random variables and entropy. Mathematical Surveys and Monographs, vol. 77, 376 pp. American Mathematical Society, Providence, RI (2000)
11. Meyer, P.-A.: Quantum probability for probabilists, 2nd edn. Lecture Notes in Mathematics, vol. 1538, 312 pp. Springer (1995)
12. Nica, A., Speicher, R.: Lectures on the combinatorics of free probability. Cambridge University Press, Cambridge (2006)
13. Speicher, R.: A new example of "independence" and "white noise", Probab. Theor. Relat. Field. **84**(2), 141–159 (1990)
14. Voiculescu, D.V.: Lecture notes on free probability. Lecture Notes in Mathematics, vol. 1738, pp. 279–349, Springer, Berlin (2000)
15. Voiculescu, D.V., Dykema, K., Nica, A.: Free random variables. CRM Monographs Series No. 1. American Mathematical Society, Providence, RI (1992)

Convergence in the Semimartingale Topology and Constrained Portfolios

Christoph Czichowsky, Nicholas Westray, and Harry Zheng

Abstract Consider an $\mathbb{R}^d$-valued semimartingale S and a sequence of $\mathbb{R}^d$-valued S-integrable predictable processes H^n valued in some closed convex set $\mathcal{K} \subset \mathbb{R}^d$, containing the origin. Suppose that the real-valued sequence $H^n \cdot S$ converges to X in the semimartingale topology. We would like to know whether we may write $X = H^0 \cdot S$ for some $\mathbb{R}^d$-valued, S-integrable process H^0 valued in $\mathcal{K}$? This question is of crucial importance when looking at superreplication under constraints. The paper considers a generalization of the above problem to $\mathcal{K} = \mathcal{K}(\omega, t)$ possibly time dependent and random.

Key words: Stochastic integrals · Semimartingale topology · Integrands in polyhedral and continuous convex constraints · Optimization under constraints · Mathematical finance

MSC 2000 Classification: 60G07, 60H05, 28B20, 91B28

1 Introduction

The Émery distance between two real-valued semimartingales $X = (X_t)_{t \geq 0}$ and $Y = (Y_t)_{t \geq 0}$ is defined by

C. Czichowsky
Department of Mathematics, ETH Zurich, Rämistrasse 101, CH-8092 Zurich, Switzerland
e-mail: christoph.czichowsky@math.ethz.ch

N. Westray
Department of Mathematics, Humboldt Universität Berlin, Unter den Linden 6,
10099 Berlin, Germany
e-mail: westray@math.hu-berlin.de

H. Zheng (✉)
Department of Mathematics, Imperial College, London SW7 2AZ, UK
e-mail: h.zheng@imperial.ac.uk

C. Donati-Martin et al. (eds.), *Séminaire de Probabilités XLIII*, Lecture Notes in Mathematics 395
2006, DOI 10.1007/978-3-642-15217-7__17, © Springer-Verlag Berlin Heidelberg 2011

$$d(X, Y) = \sup_{|J| \le 1} \left(\sum_{n \ge 1} 2^{-n} \mathbb{E} \left[1 \wedge |(J \cdot (X - Y))_n| \right] \right)$$

where $(J \cdot X)_t := \int_0^t J_u dX_u$, and the supremum is taken over all predictable processes J bounded by 1. Émery [1] shows that with respect to this metric the space of semimartingales is complete. For a given $\mathbb{R}^d$-valued semimartingale S we write $\mathcal{L}(S)$ for the space of $\mathbb{R}^d$-valued, S-integrable, predictable processes $\mathcal{H}$ and $\mathcal{L}^a_{loc}(S)$ for those processes in $\mathcal{L}(S)$ for which $\mathcal{H} \cdot S$ is locally bounded from below. By construction $\mathcal{H} \cdot S$ is a real-valued semimartingale being the vector stochastic integral of an $\mathbb{R}^d$-valued process $\mathcal{H}$ with respect to the $\mathbb{R}^d$-valued semimartingale S; see Jacod and Shiryaev [5] Sect. III.6. We write $L(S)$ for the space of all equivalence classes in $\mathcal{L}(S)$ with respect to the quasi-norm (in the sense of Yosida [18] Definition I.2.2),

$$d_S(\mathcal{H}^1, \mathcal{H}^2) = d(\mathcal{H}^1 \cdot S, \mathcal{H}^2 \cdot S).$$

We define $L^a_{loc}(S)$ analogously. Hence in $L(S)$ we identify all processes $\mathcal{H} \in \mathcal{L}(S)$ that yield the same stochastic integral $\mathcal{H} \cdot S$. Mémin [10] Theorem V.4 shows that $L(S)$ is a complete topological vector space with respect to d_S. Equivalently, the space of stochastic integrals

$$\{H \cdot S \mid H \in L(S)\}$$

is closed in the semimartingale topology.

A natural question to ask is the following. Given a sequence $H^n \in L(S)$ and a process $H^0 \in L(S)$ with $d_S(H^n, H^0)$ converging to 0 and a closed convex set $\mathcal{K}$, where $\mathcal{K} \subset \mathbb{R}^d$ contains the origin, suppose that $H^n_t \in \mathcal{K}$ a.s. for all t and $n \in \mathbb{N}$, can we deduce that $H^0_t \in \mathcal{K}$ a.s. for all t? More precisely, we assume that each class $H^n \in L(S)$ has a representative $\mathcal{H}^n \in \mathcal{L}(S)$ with $\mathcal{H}^n_t \in \mathcal{K}$ a.s. for all t, and we ask whether the class H^0 admits a representative $\mathcal{H}^0$ such that $\mathcal{H}^0_t \in \mathcal{K}$ a.s. for all t. For brevity we write $H^0 \in L(S)$ and $H^0_t \in \mathcal{K}$ a.s. for all t, etc.

This question is closely related to finding a constrained optional decomposition for a given process. Form the set of $\mathcal{K}$-valued integrands

$$\mathcal{J} := \left\{ H \in L^a_{loc}(S) \mid H_t \in \mathcal{K} \text{ a.s. for all } t \right\}$$

(where we have slightly abused notation as pointed out above). This defines a family of semimartingales via

$$\mathcal{S} := \{H \cdot S \mid H \in \mathcal{J}\}. \tag{1.1}$$

Föllmer and Kramkov [2] characterize those locally bounded below processes Z which may be written as

$$Z = Z_0 + H_Z \cdot S - C_Z \tag{1.2}$$

for some increasing nonnegative optional process C_Z and some $H_Z \in \mathcal{J}$. In mathematical finance S is the discounted asset price process and Z is typically related to some contingent claim. The existence of a decomposition (1.2) means that Z can be superreplicated by a $\mathcal{K}$-valued portfolio H_Z with initial endowment Z_0. Karatzas and Žitković [7], in the context of utility maximization with consumption and random endowment, and Pham [11, 12], in the setting of utility maximization and shortfall risk minimization, apply [2] Theorem 4.1 to deduce the existence of a constrained optimal solution. A crucial condition on the set $\mathcal{S}$ needed for [2] Theorem 4.1 to hold is the following:

Assumption 1.1 ([2, Assumption 3.1]). If $H^n \cdot S$ is a sequence in $\mathcal{S}$, uniformly bounded from below, which converges in the semimartingale topology to X then $X \in \mathcal{S}$.

Consider the set $\mathcal{S}$ defined by (1.1). Let us discuss whether it satisfies the above assumption. Suppose $H^n \cdot S$ is a uniformly bounded below sequence converging in the semimartingale topology to X. Since $\{H \cdot S \mid H \in L(S)\}$ is closed in the semimartingale topology and convergence in that topology implies uniform convergence on compacts in probability it follows that there exists $H^0 \in L^a_{loc}(S)$ with $X = H^0 \cdot S$. Thus we see that it is sufficient to check whether $\mathcal{S}$ verifies the following:

Assumption 1.2. If $H^n \cdot S$ is a sequence in $\mathcal{S}$ which converges in the semimartingale topology to $H^0 \cdot S$ then $H^0 \cdot S \in \mathcal{S}$.

When $\mathcal{K} \neq \mathbb{R}^d$ we are led to investigate whether we can find a representative $\mathcal{H}^0$ of the class H^0 which is $\mathcal{K}$-valued. This is precisely the problem considered in the main result of the present paper, Theorem 3.5. Note that we only require one, not every, representative of the limit class in $L(S)$ to be $\mathcal{K}$-valued.

In [11, 12] it is shown that pointwise properties are preserved under the additional conditions that S is continuous and satisfies $d[S, S]_t = \sigma_t dt$ for a matrix-valued process σ_t, assumed positive definite a.s. for all t. In this case positive definiteness implies that all components of the integrands converge pointwise a.s. for each t and therefore the closedness of $\mathcal{K}$ gives that the limit is again in $\mathcal{K}$. In incomplete markets σ_t is generally only positive semi-definite and one cannot argue in this way.

In [7] it is implicitly assumed that Assumption 1.2 is valid when $\mathcal{K}$ is an arbitrary (fixed) closed convex cone $\mathcal{K}$ and S is any $\mathbb{R}^d$-valued semimartingale satisfying an absence of arbitrage assumption. In Sect. 2 we give a counterexample to show that this is false in general. We show that (without imposing extra conditions on S) to obtain a positive answer to the question of whether the limit class of integrands admits a representative which is $\mathcal{K}$-valued, one must restrict $\mathcal{K}$. In fact it is sufficient that $\mathcal{K}$ be either a continuous or a polyhedral set.

The main contribution of this paper is to show that for these two choices of $\mathcal{K}$ the sets $\mathcal{S}$ defined by (1.1) satisfy Assumption 1.2 and [2] Theorem 4.1 may be applied. This covers many examples currently in the literature. In particular, as shown in Sect. 4, that of no-short-selling constraints and upper and lower bounds on the number of shares of each asset held, listed as examples in [2] without proof.

The layout of the paper is as follows: Sect. 2 contains two insightful examples, Sect. 3 provides the main result (Theorem 3.5), Sect. 4 gives some applications and Sects. 5 and 6 contain the proof.

2   Motivating Examples

To illustrate some of the problems that can arise we give the following examples. They show that, without further conditions on S, for arbitrary closed convex sets it may not be possible to find an appropriate representative of the limiting class H^0. Note that in this section and throughout, all vectors are column vectors.

2.1   Example 1

Let W denote a standard 1-dimensional Brownian motion and τ be the stopping time defined by

$$\tau := \inf\{t \geq 0 \mid |W_t| = 1\}.$$

We let $S := (W^\tau, 0)^\top$ so that we have

$$d\langle S, S\rangle_t = \begin{pmatrix} 1 & 0 \\ 0 & 0 \end{pmatrix} \mathbb{I}_{\{t \leq \tau\}} dt + \begin{pmatrix} 0 & 0 \\ 0 & 0 \end{pmatrix} \mathbb{I}_{\{t > \tau\}} dt$$

$$:= C\mathbb{I}_{\{t \leq \tau\}} dt + \mathbf{0}\mathbb{I}_{\{t > \tau\}} dt.$$

The matrix C is not positive definite and the kernel of C is given by $\mathrm{Ker}(C) = \{\mu(0,1)^\top \mid \mu \in \mathbb{R}\}$. Set

$$\mathcal{K} := \left\{ (x, y)^\top \in (-1, \infty) \times \mathbb{R} \;\middle|\; y \geq \frac{1}{x+1} - 1 \right\}.$$

This is a closed convex set containing the origin. We define, for $n \in \mathbb{N}$, a sequence of (constant) processes valued on the boundary of $\mathcal{K}$,

$$\mathcal{H}^n := (-1 + 1/n, n - 1)^\top.$$

The $\mathcal{H}^n$ do not converge pointwise, their norms are unbounded. Observe that projecting $\mathcal{H}^n$ onto the orthogonal complement of $\mathrm{Ker}(C)$ gives another representative of H^n which we call $\hat{\mathcal{H}}^n$. Thus we have

$$\hat{\mathcal{H}}^n = (-1 + 1/n, 0)^\top,$$
$$\hat{\mathcal{H}}^n \cdot S = \mathcal{H}^n \cdot S, \, n \in \mathbb{N}.$$

If we define the (constant) process $\mathcal{H}^0 = (-1, 0)^\top$ then

$$\mathcal{H}^n \cdot S = (1 - 1/n)(\mathcal{H}^0 \cdot S).$$

It then follows that $\mathcal{H}^n \cdot S$ converges in the semimartingale topology to $\mathcal{H}^0 \cdot S$.

We seek a representative of the class H^0 which is $\mathcal{K}$-valued. By construction, the stochastic integral of each $\mathbb{R}^2$-valued predictable process, valued in $\mathrm{Ker}(C)$ on $\{t \leq \tau\}$ $d\mathbb{P} \otimes dt$-a.e., is zero. This implies that the equivalence class of the process $0 \in \mathcal{L}(S)$ consists up to $d\mathbb{P} \otimes dt$-a.e. equality of the processes $\mathcal{G}^1(0, 1)^\top + \mathcal{G}^2(1, 0)^\top \mathbb{I}_{\{t > \tau\}}$, where $\mathcal{G}^1$ and $\mathcal{G}^2$ are some real-valued predictable processes. Since adding a representative of 0 to some element of $\mathcal{L}(S)$ does not change its equivalence class, we obtain that the equivalence class H of any given $\mathcal{H} \in \mathcal{L}(S)$ is given up to $d\mathbb{P} \otimes dt$-a.e. equality by

$$H = \{\mathcal{H} + \mathcal{G}^1(0, 1)^\top + \mathcal{G}^2(1, 0)^\top \mathbb{I}_{\{t > \tau\}} \mid \mathcal{G}^1, \mathcal{G}^2$$
$$\text{real-valued predictable processes}\}.$$

However, due to the vertical asymptote of $\mathcal{K}$ at $x = -1$ we have

$$\{(-1, 0)^\top + \mu(0, 1)^\top\} \cap \mathcal{K} = \emptyset \text{ for all } \mu \in \mathbb{R}.$$

In particular, adding vectors valued in the kernel of C to $\mathcal{H}^0$ will never give a $\mathcal{K}$-valued integrand and therefore one cannot find an appropriate representative $\tilde{\mathcal{H}}^0$ of H^0. This illustrates that, without making further assumptions on S, one cannot allow arbitrary closed convex sets. The crucial point here is that $\mathcal{K}$ is not a continuous set. One may hope that by imposing some other restriction, for instance that $\mathcal{K}$ is a cone, the closure property may still be proved. The following example shows that this is not the case.

2.2 Example 2

Let $W = (W^1, W^2, W^3)^\top$ be a 3-dimensional Brownian motion and set $Y = \sigma \cdot W$ where

$$\sigma = \begin{pmatrix} 1 & 0 & 0 \\ 0 & 1 & -1 \\ 0 & -1 & 1 \end{pmatrix}.$$

The matrices $\sigma^\top$ and $C := \sigma\sigma^\top$ have nontrivial kernel spanned by $w = \frac{1}{\sqrt{2}}(0, 1, 1)^\top$ i.e.,

$$\mathrm{Ker}(C) = \mathrm{Ker}(\sigma\sigma^\top) = \mathbb{R}w = \left\{\frac{1}{\sqrt{2}}(0, \mu, \mu)^\top \mid \mu \in \mathbb{R}\right\}.$$

As in the previous example we obtain that the stochastic integral of each $\mathbb{R}^3$-valued predictable process, valued in $\text{Ker}(C)$ $d\mathbb{P} \otimes dt$-a.e., is zero and therefore that the equivalence class H of any given $\mathcal{H} \in \mathcal{L}(Y)$ is given up to $d\mathbb{P} \otimes dt$-a.e. equality by

$$H = \{\mathcal{H} + \mathcal{G}w \mid \mathcal{G} \text{ a real-valued predictable process}\}.$$

Let $\mathcal{K}$ be the closed convex cone

$$\{(x, y, z)^\top \in \mathbb{R}^3 \mid x^2 + y^2 \leq z^2, \, z \geq 0\}.$$

Choose a sequence $(z_n)_{n \in \mathbb{N}}$ in $[1, \infty)$ tending to infinity and define the sequence of (constant) processes $(\mathcal{H}^n)_{n \in \mathbb{N}}$ by $\mathcal{H}^n = (1, \sqrt{z_n^2 - 1}, z_n)^\top$ for each $n \in \mathbb{N}$. Then each $\mathcal{H}^n$ is $\mathcal{K}$-valued and we obtain

$$\mathbb{E}\left[((\mathcal{H}^n \cdot Y)_t - W_t^1)^2\right] = \mathbb{E}\left[\langle \mathcal{H}^n \cdot Y - W^1, \mathcal{H}^n \cdot Y - W^1\rangle_t\right]$$

$$= \int_0^t (\mathcal{H}_s^n - (1, 0, 0)^\top)^\top C_s(\mathcal{H}_s^n - (1, 0, 0)^\top) ds$$

$$= \int_0^t 2\left[2z_n^2\left(1 - \sqrt{1 - \frac{1}{z_n^2}}\right) - 1\right] ds$$

$$= \int_0^t 2\left[2z_n^2\left(1 - \left\{1 - \frac{1}{2z_n^2} + O\left(\frac{1}{z_n^4}\right)\right\}\right) - 1\right] ds$$

$$= tO\left(\frac{1}{z_n^2}\right). \tag{2.1}$$

Hence $\mathcal{H}^n \cdot Y$ converges to W^1 locally in $\mathcal{M}^{2,1}(\mathbb{P})$ (the space of $\mathbb{P}$-square-integrable 1-dimensional martingales) and thus by [10] Theorem IV.5 also in the semimartingale topology. However the (constant) process $(1, 0, 0)^\top$ having stochastic integral $(1, 0, 0)^\top \cdot Y = W^1$ is not $\mathcal{K}$-valued.

Recall that we identify processes in $\mathcal{L}(Y)$ yielding the same stochastic integral and therefore it would be sufficient to find one predictable process equivalent to $(1, 0, 0)^\top$ and valued in $\mathcal{K}$ a.s. for each t. As discussed above the equivalence class of $(1, 0, 0)^\top$ is, up to $d\mathbb{P} \otimes dt$-a.e. equality,

$$\{(1, 0, 0)^\top + \mathcal{G}w \mid \mathcal{G} \text{ a real-valued predictable process}\}.$$

For every fixed t and ω these have the form

$$\left(1, \frac{\mathcal{G}_t(\omega)}{\sqrt{2}}, \frac{\mathcal{G}_t(\omega)}{\sqrt{2}}\right)^\top.$$

It follows from the definition of $\mathcal{K}$ that there is no predictable process equivalent to $(1, 0, 0)^\top$ which is $\mathcal{K}$-valued. Defining the stopping time

$$\tau := \inf\left\{t \geq 0 \mid \|W_t\|_{\mathbb{R}^3} = 1\right\}.$$

and setting $S = Y^\tau$ and defining $\mathcal{S}$ by (1.1) gives an example which does not satisfy Assumption 1.1. (For reference observe that we write $\|\cdot\|_{\mathbb{R}^d}$ for the Euclidean norm on $\mathbb{R}^d$). Thus we have a counterexample to the implicit claim in [7] as S is a bounded martingale and therefore satisfies the no arbitrage assumption therein.

Exactly as in the previous example convergence of the stochastic integrals $\mathcal{H}^n \cdot S$ does not necessarily imply that the representatives $\mathcal{H}^n$ satisfying the constraints converge pointwise. Therefore one cannot argue using the pointwise closedness of $\mathcal{K}$ to obtain that the limit is again valued in $\mathcal{K}$. Indeed, in this case $\|\mathcal{H}^n\|_{\mathbb{R}^3} = \sqrt{2}z_n$ and the sequence of representatives actually diverges. Thus, for a general closed convex cone $\mathcal{K}$, one cannot show that $\mathcal{S}$ is closed in the semimartingale topology.

However although the $\mathcal{H}^n$ need not converge pointwise, we can always find a related sequence of representatives $\hat{\mathcal{H}}^n$ that do. The issue then is that these need not be $\mathcal{K}$-valued anymore. To obtain the $\hat{\mathcal{H}}^n$, as in the previous example, we project onto the orthogonal complement of $\mathrm{Ker}(C) = \mathrm{Ker}(\sigma\sigma^\top)$. The eigenvalue decomposition of C is given by $C = P^\top D P$ with

$$P = \begin{pmatrix} 1 & 0 & 0 \\ 0 & \frac{1}{\sqrt{2}} & -\frac{1}{\sqrt{2}} \\ 0 & \frac{1}{\sqrt{2}} & \frac{1}{\sqrt{2}} \end{pmatrix}, \quad D = \begin{pmatrix} 1 & 0 & 0 \\ 0 & 4 & 0 \\ 0 & 0 & 0 \end{pmatrix}.$$

Define $\mathcal{G}^n := P\mathcal{H}^n$ and recall from (2.1) that

$$\left(\mathcal{H}^n - (1,0,0)^\top\right)^\top C \left(\mathcal{H}^n - (1,0,0)^\top\right) \tag{2.2}$$

converges to 0 in $d\mathbb{P} \otimes dt$-measure, hence $d\mathbb{P} \otimes dt$-a.e. along a subsequence, also indexed by n. Using the decomposition of C and writing $\mathcal{G}^{n,i}$ for the ith component of $\mathcal{G}^n$, (2.2) is equivalent to

$$(\mathcal{G}^{n,1} - 1)^2 + 4(\mathcal{G}^{n,2})^2 \longrightarrow 0, \quad d\mathbb{P} \otimes dt\text{-a.e.}$$

Therefore, $d\mathbb{P} \otimes dt$-a.e., $\mathcal{G}^{n,1}$ converges to 1 and $\mathcal{G}^{n,2}$ to 0. The vectors $v_1 = (1,0,0)^\top$, $v_2 = \frac{1}{\sqrt{2}}(0,1,-1)^\top$ together with $w = \frac{1}{\sqrt{2}}(0,1,1)^\top$ form an orthonormal basis of $\mathbb{R}^3$. The decomposition of $\mathcal{H}^n$ with respect to v_1, v_2 and w is given by

$$\mathcal{H}^n = \left((\mathcal{H}^n)^\top v_1\right)v_1 + \left((\mathcal{H}^n)^\top v_2\right)v_2 + \left((\mathcal{H}^n)^\top w\right)w$$
$$= \mathcal{G}^{n,1}v_1 + \mathcal{G}^{n,2}v_2 + \mathcal{G}^{n,3}w.$$

If we now define $\hat{\mathcal{H}}^n = \mathcal{G}^{n,1}v^1 + \mathcal{G}^{n,2}v^2$ then we obtain that, for each $n \in \mathbb{N}$, $\mathcal{H}^n - \hat{\mathcal{H}}^n$ is valued in $\mathrm{Ker}(C)$ $d\mathbb{P} \otimes dt$-a.e. and therefore $\mathcal{H}^n$ and $\hat{\mathcal{H}}^n$ are in the same equivalence class. Moreover, $\hat{\mathcal{H}}^n$ converges to $(1,0,0)^\top$ $d\mathbb{P} \otimes dt$-a.e., and hence for all ω and t if we set $\hat{\mathcal{H}}^n$ equal to $(1,0,0)^\top$ on the null set where this convergence does not hold.

Motivated by these examples we now study those convex sets for which we can use such pointwise convergence to deduce the existence of a $\mathcal{K}$-valued representative of the limit class.

3 Main Results

We work on the filtered probability space $(\Omega, \mathcal{F}, (\mathcal{F}_t), \mathbb{P})$, which is assumed to satisfy the usual conditions. This space supports an $\mathbb{R}^d$-valued semimartingale S, which one may think of as an asset price process. For notational simplicity the index t is valued in $[0, 1]$, the extension to the case when $t \in \mathbb{R}_+$ is straightforward. The predictable σ-field on $\Omega \times [0, 1]$ generated by all left-continuous adapted processes is denoted by $\mathcal{P}$. For an $\mathbb{R}^d$-valued S-integrable predictable process $H \in L(S)$ we write $H \cdot S$ for the stochastic integral of H with respect to S and refer to [5] Sect. III.6 for the theory of stochastic integration of vector valued processes.

We consider possibly random and time dependent $\mathcal{K} = \mathcal{K}(\omega, t)$ and for this we need the notion of a measurable multivalued mapping, taken from Rockafellar [13, 15] and Wagner [16]. Recently the idea to formulate constraints via measurable multivalued mappings has been used in Karatzas and Kardaras [6] to study the numéraire portfolio under convex constraints. Let $\mathcal{T}$ be a set together with $\mathcal{A}_\mathcal{T}$, a σ-field of subsets of $\mathcal{T}$. We write $2^{\mathbb{R}^d}$ for the power set of $\mathbb{R}^d$.

Definition 3.1. A multivalued mapping $F : \mathcal{T} \to 2^{\mathbb{R}^d}$ is called *measurable* if, for all closed subsets Q of $\mathbb{R}^d$,

$$F^{-1}(Q) := \{t \in \mathcal{T} \mid F(t) \cap Q \neq \emptyset\} \in \mathcal{A}_\mathcal{T}.$$

When $F(t)$ is a closed (convex) set for all $t \in \mathcal{T}$ it is said to be *closed (convex)*.

We say F is *predictably measurable* when $\mathcal{T} = \Omega \times [0, 1]$ and $\mathcal{A}_\mathcal{T} = \mathcal{P}$. Motivated by the examples in Sect. 2 we place more restrictions on $\mathcal{K}$ in order to obtain a positive answer to the question posed in the introduction. The following definition is from Gale and Klee [3].

Definition 3.2. A convex set $Q \subset \mathbb{R}^d$ is called *continuous* if it is closed and its support function

$$\rho(u) = \sup_{q \in Q} \left(q^\top u \right)$$

is continuous for all vectors $u \in \mathbb{R}^d$ with $\|u\|_{\mathbb{R}^d} = 1$.

The set Q may be unbounded and hence we allow ρ to take the value $+\infty$, with continuity at u for which $\rho(u) = +\infty$ defined in the usual way.

We can treat another type of $\mathcal{K}$, for this we use the definition of a polyhedral convex set, taken from Rockafellar [14]. By Theorem 19.1 therein this coincides with that of a finitely generated convex set.

Definition 3.3. A closed convex set $Q \subset \mathbb{R}^d$ is called *polyhedral* if there exists $m \in \mathbb{N}$, real numbers $r_1, \ldots, r_m$ and vectors $p_1, \ldots, p_m$ such that

$$Q = \left\{ q \in \mathbb{R}^d \mid p_i^\top q \leq r_i \text{ for } 1 \leq i \leq m \right\}.$$

We make the following assumption throughout the rest of this paper.

Assumption 3.4. The multivalued mapping $\mathcal{K}$ is closed convex and predictably measurable with $0 \in \mathcal{K}(\cdot, t)$ a.s. for all t. For each t, $\mathcal{K}(\cdot, t)$ is a.s. either continuous or a polyhedral set.

With all the necessary preliminaries introduced we can now state our main result.

Theorem 3.5. *Let $\mathcal{K}$ satisfy Assumption 3.4 and $H^n \in L(S)$ be a sequence of predictable processes with*

(i) *$H_t^n \in \mathcal{K}(\cdot, t)$ a.s. for all t and $n \in \mathbb{N}$, i.e., there exists a representative $\mathcal{H}^n$ of each equivalence class H^n such that $\mathcal{H}_t^n \in \mathcal{K}(\cdot, t)$ a.s. for all t.*
(ii) *$H^n \cdot S$ converges in the semimartingale topology to some semimartingale X.*

Then there exists H^0 in $L(S)$ such that $X = H^0 \cdot S$ and $H_t^0 \in \mathcal{K}(\cdot, t)$ a.s. for all t. More precisely, there exists a representative $\mathcal{H}^0$ of the equivalence class H^0 such that $\mathcal{H}_t^0 \in \mathcal{K}(\cdot, t)$ a.s. for all t.

Before we proceed to the proof we give some important situations where one can apply Theorem 3.5.

4 Applications

4.1 Optional Decomposition Under Constraints

We suppose there exists $m \in \mathbb{N}$ and that for $1 \leq i \leq m$ there are predictable processes I_i and G_i, valued in $\mathbb{R}^d$ and $\mathbb{R}_+$ respectively. We define a set of integrands $\mathcal{J}_1$ via

$$\mathcal{J}_1 := \left\{ H \in L_{loc}^a(S) \,\middle|\, I_i^\top H^i \leq G_i \text{ for } 1 \leq i \leq m \right\}. \tag{4.1}$$

In the above all the inequalities are to be understood in the sense of Assumption 3.4. For example, in the above, we have that for $1 \leq i \leq m$,

$$I_i(\cdot, t)^\top H^i(\cdot, t) \leq G_i(\cdot, t) \quad \mathbb{P} - \text{a.s for all } t.$$

By comparison with Definition 3.3 this is equivalent to saying that the integrand H is valued in the closed convex polyhedral set

$$\mathcal{K}_1(\omega, t) := \left\{ k \in \mathbb{R}^d \;\middle|\; \left(I_i(\omega, t) \right)^{\top} k \leq G_i(\omega, t) \text{ for } 1 \leq i \leq m \right\}.$$

Contained within this framework are a very large class of examples. In particular no short selling as well as upper and lower bounds on the number of shares of each asset held; both given in [2] as examples to which their Theorem 4.1 applies. Define

$$\mathcal{S}_1 := \{ H \cdot S \mid H \in \mathcal{J}_1 \}.$$

As discussed in the introduction we need only show that $\mathcal{S}_1$ satisfies Assumption 1.2, more precisely that if $H^n \cdot S$ converges to $H^0 \cdot S$ then there exists an appropriate representative of the limiting class. This will follow from Theorem 3.5 once we show that the set $\mathcal{K}_1(\omega, t)$ is predictably measurable.

The mapping $\pi_i((\omega, t), x) := I_i(\omega, t)^{\top} x$ is continuous in x and predictably measurable. Moreover the mapping

$$\mathcal{K}'_i(\omega, t) := \{ k \in \mathbb{R} \mid k \leq G_i(\omega, t) \}$$

is closed and predictably measurable. By definition we have

$$\mathcal{K}_1(\omega, t) = \bigcap_{i=1}^{m} \left\{ k \in \mathbb{R}^d \;\middle|\; \pi_i((\omega, t), k) \in \mathcal{K}'_i(\omega, t) \right\}$$

and the result now follows from Definition 5.4 and Lemmas 5.3 and 5.5 (see Sect. 5.3). We have shown that sets defined by (4.1) are valid examples to which the optional decomposition theorem under constraints applies.

4.2   Utility Maximization

In [7], within the framework of a utility maximization problem the authors propose the set

$$\mathcal{J}_2 := \left\{ H \in L^a_{loc}(S) \mid x + H \cdot S \geq 0 \text{ and } H_t \in \mathcal{K}_2 \text{ a.s. for all } t \in [0, 1] \right\},$$

where $\mathcal{K}_2$ is a closed convex cone in $\mathbb{R}^d$ containing the origin. The set $\mathcal{S}_2 := \{ x + H \cdot S \mid H \in \mathcal{J}_2 \}$ is the family of nonnegative wealth processes with initial capital x and cone constraints on the investment strategy.

The existence of a solution to the utility maximization problem posed in [7] depends crucially on Proposition 2.13 therein, a dual characterization of superreplicable consumption processes. It is established by an application of [2, Theorem 4.1] to $\mathcal{S}_2$. It is not immediately clear from [7] that $\mathcal{S}_2$ satisfies Assumption 1.2, more specifically that a representative $\mathcal{H}^0$ of the limit class H^0 may be chosen to be

$\mathcal{K}_2$-valued. As illustrated by Example 2 in Sect. 2, this is not true for general closed convex cones.

However when $\mathcal{K}_2$ is additionally assumed polyhedral, we can show the existence of a suitable representation. Indeed one has that $\mathcal{K}_2$ is now a closed convex polyhedral cone containing 0 which is independent of (ω, t) and hence predictably measurable. Applying Theorem 3.5 now shows that Assumption 1.2 holds for $\mathcal{S}_2$.

One may also use Theorem 3.5 in utility maximization on the whole real line under cone constraints and we refer the reader to Westray and Zheng [17] for more details.

5 Measurable Selection

We review some results on stochastic processes, separation of convex sets and measurable selection. We then prove Lemma 5.9 which is crucial in establishing our main result, Theorem 3.5.

5.1 Stochastic Processes

We define all processes up to indistinguishability. We use the phrase "for all t", implicitly meaning "for all $t \in [0, 1]$". Throughout we write X for the process $(X_t)_{t \in [0,1]}$. We reserve n for sequences and i for components of vectors in the sense that $X^{n,i}$ denotes the process formed from the ith-component of X^n.

The $\mathbb{R}^d$-valued semimartingale S may be decomposed as $S = \tilde{M} + \tilde{A}$ where $\tilde{M}$ is a $\mathbb{P}$-local martingale and $\tilde{A}$ a process of finite variation. We write $\mathcal{M}^{2,d}(\mathbb{P})$ for the space of d-dimensional square-integrable martingales and $\mathcal{A}^{1,d}(\mathbb{P})$ for the space of d-dimensional predictable processes of integrable variation on the space $(\Omega, \mathcal{F}, (\mathcal{F}_t), \mathbb{P})$. These are turned into Banach spaces by equipping them with the norms

$$\|\tilde{M}\|_{\mathcal{M}^{2,d}(\mathbb{P})} = \left(\mathbb{E}\left[\sum_{i=1}^{d} \langle \tilde{M}^i, \tilde{M}^i \rangle_1 \right] \right)^{\frac{1}{2}}, \quad \|\tilde{A}\|_{\mathcal{A}^{1,d}(\mathbb{P})} = \mathbb{E}\left[\sum_{i=1}^{d} \int_0^1 |d\tilde{A}_t^i| \right].$$

Here $\tilde{M}^i$ is a $\mathbb{P}$-square-integrable martingale for $1 \leq i \leq d$ and in the above we have written $\langle \tilde{M}^i, \tilde{M}^i \rangle$ for the predictable compensator of $[\tilde{M}^i, \tilde{M}^i]$ satisfying,

$$\mathbb{E}[(\tilde{M}_t^i)^2] = \mathbb{E}[[\tilde{M}^i, \tilde{M}^i]_t] = \mathbb{E}[\langle \tilde{M}^i, \tilde{M}^i \rangle_t] \text{ for all } t \text{ and } 1 \leq i \leq d.$$

In addition, for each i, $|d\tilde{A}^i|$ denotes the differential of the total variation process $|\tilde{A}^i|$.

By [5, Propositions II.2.9 and II.2.29] there exist an increasing $\mathbb{Q}$-integrable, predictable process V, an $\mathbb{R}^d$-valued predictable process B together with a predictable process C, taking values in the set of symmetric positive semi-definite $d \times d$ matrices, such that for $1 \leq i, j \leq d$

$$A^i = B^i \cdot V \text{ and } \langle M^i, M^j \rangle = C^{ij} \cdot V. \tag{5.1}$$

By adding t to V and applying the Radon-Nikodym theorem, exactly as in the proof of [5, Proposition II.2.29], we may, without loss of generality, assume that V is strictly increasing. There exist many processes V, B and C satisfying (5.1), but our results do not depend on the specific choice we make.

We are only interested in the representation of a real-valued semimartingale X, which is the limit of a sequence of stochastic integrals $H^n \cdot S$ converging in the semimartingale topology. Hence we can, as in the proof of [10, Theorem V.4], switch to an equivalent probability measure $\mathbb{Q}$ and find a subsequence, also indexed by n, such that $S = M + A \in \mathcal{M}^{2,d}(\mathbb{Q}) \oplus \mathcal{A}^{1,d}(\mathbb{Q})$, $H^n \cdot S \in \mathcal{M}^{2,1}(\mathbb{Q}) \oplus \mathcal{A}^{1,1}(\mathbb{Q})$ and $H^n \cdot S$ converges to X in $\mathcal{M}^{2,1}(\mathbb{Q}) \oplus \mathcal{A}^{1,1}(\mathbb{Q})$. It then follows that X is given by $H^0 \cdot S$ for some $H^0 \in L^2(M, \mathbb{Q}) \cap L^1(A, \mathbb{Q})$.

Let us explain the previous notation. $\mathcal{L}^2(M, \mathbb{Q})$ (compare $\mathcal{L}(S)$) is the set of M-integrable predictable processes $\mathcal{H}$ for which

$$\mathbb{E}_{\mathbb{Q}}\big[\langle \mathcal{H} \cdot M, \mathcal{H} \cdot M \rangle_1\big] = \mathbb{E}_{\mathbb{Q}}\left[\sum_{i,j=1}^d \big((\mathcal{H}^i \mathcal{H}^j) \cdot \langle M^i, M^j \rangle\big)_1\right] < \infty.$$

The set of equivalence classes in $\mathcal{L}^2(M, \mathbb{Q})$ with respect to the relation

$$\mathcal{H}^1 \sim \mathcal{H}^2 \text{ iff } \mathbb{E}_{\mathbb{Q}}\big[\langle (\mathcal{H}^1 - \mathcal{H}^2) \cdot M, (\mathcal{H}^1 - \mathcal{H}^2) \cdot M \rangle_1\big] = 0$$

is then denoted $L^2(M, \mathbb{Q})$. For the set of A-integrable predictable processes $\mathcal{H}$ such that

$$\mathbb{E}_{\mathbb{Q}}\left[\int_0^1 |d(\mathcal{H} \cdot A)_t|\right] < \infty$$

we write $\mathcal{L}^1(A, \mathbb{Q})$. As in the martingale case $L^1(A, \mathbb{Q})$ is then the set of equivalence classes in $\mathcal{L}^1(A, \mathbb{Q})$ with respect to the relation

$$\mathcal{H}^1 \sim \mathcal{H}^2 \text{ iff } \mathbb{E}_{\mathbb{Q}}\left[\int_0^1 |d((\mathcal{H}^1 - \mathcal{H}^2) \cdot A)_t|\right] = 0.$$

It follows that a predictable process $\mathcal{H}$ in $L^2(M, \mathbb{Q}) \cap L^1(A, \mathbb{Q})$ has stochastic integral $\mathcal{H} \cdot S = 0$ if and only if

$$\mathbb{E}_{\mathbb{Q}}\big[\langle \mathcal{H} \cdot M, \mathcal{H} \cdot M \rangle_1\big] + \mathbb{E}_{\mathbb{Q}}\left[\int_0^1 |d(\mathcal{H} \cdot A)_t|\right] = 0.$$

Using (5.1) this may be equivalently written as

$$\mathbb{E}_{\mathbb{Q}}\left[\int_0^1 \mathcal{H}_t^\top C_t \mathcal{H}_t \, dV_t\right] + \mathbb{E}_{\mathbb{Q}}\left[\int_0^1 |\mathcal{H}_t^\top B_t| \, dV_t\right] = 0.$$

Therefore the equivalence class of 0 in $L^2(M, \mathbb{Q}) \cap L^1(A, \mathbb{Q})$ consists of those predictable processes which are valued in $\mathrm{Ker}(B) \cap \mathrm{Ker}(C) \, d\mathbb{Q} \otimes dV$-a.e. If we add processes which are equivalent to 0 we do not change the equivalence class. Hence there exist different predictable processes in $\mathcal{L}^2(M, \mathbb{Q}) \cap \mathcal{L}^1(A, \mathbb{Q})$, not $d\mathbb{Q} \otimes dV$-a.e. equal, which have the same stochastic integral whenever $\mathrm{Ker}(B) \cap \mathrm{Ker}(C)$ is nontrivial.

Recall that the stochastic integral is invariant under a change to an equivalent probability measure so that any representation we derive under $\mathbb{Q}$ also holds under $\mathbb{P}$.

5.2   Separation Theorems

Here we collect separation theorems which are used in the proof of Theorem 3.5 and motivate Assumption 3.4. First we consider continuous sets, as in Definition 3.2, and refer to [3] for further discussion.

Given two compact convex subsets it is known that their sum and the convex hull of their union are again compact and convex. If in addition they are disjoint then there exists a linear functional which strongly separates them.

In [3] it is shown that these three properties hold for a wider class of sets. This is the class of continuous sets and contains convex compact sets as a proper subclass. An example of a continuous set which is not compact is the area enclosed by a parabola, e.g. $\{(x, y)^\top \in \mathbb{R}^2 \mid y \geq x^2\}$. The key theorem on continuous sets we use is the following.

Theorem 5.1 ([8, Theorem 2]). *Let Q_1, Q_2 be disjoint nonempty convex subsets of $\mathbb{R}^d$. If Q_1 is continuous and Q_2 is closed then they can be strongly separated. That is to say there exist $\phi \in \mathbb{R}^d$, $a \in \mathbb{R}$ and $\delta > 0$ such that*

$$\phi^\top q_1 \geq a + \delta > a \geq \phi^\top q_2, \text{ for all } q_1 \in Q_1, q_2 \in Q_2.$$

There are however many sets which are not continuous. In fact any cone with a ray in its boundary is not continuous. In Sect. 4 it is shown that the case when $\mathcal{K}$ is a cone is of interest, thus we want to find a restriction on the type of convex sets which allows us to prove our result and includes some interesting examples. The class we consider is polyhedral sets, see Definition 3.3, and we refer to [4] and [14] for further details and properties. We are particularly interested in separation theorems for these sets and the important result is the following.

Theorem 5.2 ([14, Corollary 19.3.3]). *Let Q_1, Q_2 be disjoint nonempty polyhedral convex subsets of $\mathbb{R}^d$. Then they can be strongly separated; there exist $\phi \in \mathbb{R}^d$, $a \in \mathbb{R}$ and $\delta > 0$ such that*

$$\phi^\top q_1 \geq a + \delta > a \geq \phi^\top q_2, \text{ for all } q_1 \in Q_1, q_2 \in Q_2.$$

5.3 Measurable Selection

We recall some results from measurable selection and refer the reader to [13, 15, 16] for more details. The setup is as in Definition 3.1, we have a set $\mathcal{T}$ together with $\mathcal{A}_{\mathcal{T}}$, a σ-field of subsets of $\mathcal{T}$. We begin with a result on the intersection of measurable mappings.

Lemma 5.3 ([13, Corollary 1.3]). *Let* $\{F_i \mid i \in I\}$ *be a countable collection of closed measurable multivalued mappings from* $\mathcal{T}$ *to* $2^{\mathbb{R}^d}$. *Then the multivalued mapping*

$$F : t \to \bigcap_{i \in I} F_i(t)$$

is measurable.

The following is taken from Rockafellar [15].

Definition 5.4. A mapping $\pi(t, x) : \mathcal{T} \times \mathbb{R}^d \to \mathbb{R}^m$ is called *Carathéodory* if $\pi(t, x)$ is measurable with respect to t and continuous with respect to x.

The next lemma shows that Carathéodory mappings preserve measurability.

Lemma 5.5 ([15, Corollary 1Q]). *Let* $F' : \mathcal{T} \to 2^{\mathbb{R}^m}$ *be a closed measurable multivalued mapping and* $\pi : \mathcal{T} \times \mathbb{R}^d \to \mathbb{R}^m$ *be a Carathéodory mapping. Then* $F : \mathcal{T} \to 2^{\mathbb{R}^d}$ *given by*

$$F(t) = \{x \in \mathbb{R}^d \mid \pi(t, x) \in F'(t)\}$$

is a closed measurable multivalued mapping.

Definition 5.6. A *measurable selector* for F is a measurable function $f : \mathcal{T} \to \mathbb{R}^d$ such that $f(t) \in F(t)$ for all $t \in \mathcal{T}$.

In Sect. 6 we are interested in finding such a measurable selector. The major theorem in this area is the following.

Theorem 5.7 ([9]). *Let* $F : \mathcal{T} \to 2^{\mathbb{R}^d}$ *be a measurable multivalued mapping such that* $F(t)$ *is a nonempty closed set for all* $t \in \mathcal{T}$. *Then there exists a measurable selector for* F.

Remark 5.8. As in [13], if $F : \mathcal{T} \to 2^{\mathbb{R}^d}$ is a closed measurable multivalued mapping, then the set

$$\mathcal{T}_0 = \{t \in \mathcal{T} \mid F(t) \neq \emptyset\}$$

is measurable. The restriction of F to $\mathcal{T}_0$ is then a measurable multivalued mapping $F_0 : \mathcal{T}_0 \to 2^{\mathbb{R}^d}$ of the type to which Theorem 5.7 applies.

The setting in which we apply these results is with $\mathcal{T} = \Omega \times [0, 1]$ and $\mathcal{A}_{\mathcal{T}} = \mathcal{P}$, the predictable σ-field on $\Omega \times [0, 1]$. As described after Definition 3.1, any multivalued mapping F measurable with respect to $\mathcal{P}$ is called *predictably measurable*. This is to emphasize the fact that any measurable selector of F is a predictable process. For a predictable process $\mathcal{H}$ define the multivalued mapping

$$F_{\mathcal{H}}(\omega, t) = \{\mathcal{H}_t(\omega) + \mathrm{Ker}(B_t(\omega)) \cap \mathrm{Ker}(C_t(\omega))\} \cap \mathcal{K}(\omega, t),$$

where B and C are from (5.1). Note that in the above for a vector $q \in \mathbb{R}^d$ we set

$$\mathrm{Ker}(q) := \{p \in \mathbb{R}^d \mid q^\top p = 0\}.$$

The following result is the principal one of this section.

Lemma 5.9. *The multivalued mapping* $F_{\mathcal{H}}$ *is closed and predictably measurable.*

Proof. $F_{\mathcal{H}}$ is closed as the intersection of two closed sets. The multivalued mapping $\mathcal{K}(\omega, t)$ is predictably measurable by Assumption 3.4. Since we have

$$F_{\mathcal{H}}(\omega, t) = \{\mathcal{H}_t(\omega) + \mathrm{Ker}(B_t(\omega)) \cap \mathrm{Ker}(C_t(\omega))\} \cap \mathcal{K}(\omega, t)$$
$$= \{\mathcal{H}_t(\omega) + \mathrm{Ker}(B_t(\omega))\} \cap \{\mathcal{H}_t(\omega) + \mathrm{Ker}(C_t(\omega))\} \cap \mathcal{K}(\omega, t),$$

by Lemma 5.3 we only need to show the measurability of

$$F^1(\omega, t) := \{\mathcal{H}_t(\omega) + \mathrm{Ker}(B_t(\omega))\} \text{ and } F^2(\omega, t) := \{\mathcal{H}_t(\omega) + \mathrm{Ker}(C_t(\omega))\}.$$

Define the Carathéodory mappings

$$\pi^1\big((\omega, t), x\big) = \big(B_t(\omega)\big)^\top \big(x - \mathcal{H}_t(\omega)\big),$$
$$\pi^2\big((\omega, t), x\big) = C_t(\omega)\big(x - \mathcal{H}_t(\omega)\big).$$

A calculation shows that

$$F^1(\omega, t) = \Big\{x \in \mathbb{R}^d \mid \pi^1\big((\omega, t), x\big) \in \{0\}\Big\},$$
$$F^2(\omega, t) = \Big\{x \in \mathbb{R}^d \mid \pi^2\big((\omega, t), x\big) \in \{0\}\Big\}.$$

Thus F^1 and F^2 are the preimages of the closed predictably measurable multivalued mapping $F' \equiv \{0\}$ (0 taken in $\mathbb{R}$ and $\mathbb{R}^d$ respectively). Thus by Lemma 5.5 they are predictably measurable and the proof is complete.

6 Proof of Theorem 3.5

The main difficulty (as in Sect. 2) comes from the fact that the pointwise constraints need only be satisfied for one representative $\mathcal{H}^n$ within the equivalence class H^n.

However, we only assume convergence of the equivalence classes and this does not necessarily imply pointwise convergence of the representatives $\mathcal{H}^n$ which satisfy the constraints.

Proof (Proof of Theorem 3.5). Let H_t^n be in $\mathcal{K}(\cdot, t)$ a.s. for all t and $n \in \mathbb{N}$. Suppose that the sequence $H^n \cdot S$ converges in the semimartingale topology to X. As discussed in Sect. 5.1 we can find a measure $\mathbb{Q}$ equivalent to $\mathbb{P}$ and a subsequence, also indexed by n, such that $S = M + A \in \mathcal{M}^{2,d}(\mathbb{Q}) \oplus \mathcal{A}^{1,d}(\mathbb{Q})$ and

$$(H^n - H^0) \cdot S \longrightarrow 0 \text{ in } \mathcal{M}^{2,1}(\mathbb{Q}) \oplus \mathcal{A}^{1,1}(\mathbb{Q}).$$

From the proof of [10] Theorem V.4 we may pass to a subsequence, also indexed by n and find representatives $\hat{\mathcal{H}}^n$ and $\hat{\mathcal{H}}^0$ of the corresponding equivalence classes H^n and H^0 such that $\hat{\mathcal{H}}_t^n(\omega)$ converges to $\hat{\mathcal{H}}_t^0(\omega)$ for all (ω, t). For each $n \in \mathbb{N}$ the stochastic integrals of $\mathcal{H}^n \cdot S$ and $\hat{\mathcal{H}}^n \cdot S$ coincide and thus their difference $(\mathcal{H}^n - \hat{\mathcal{H}}^n)$ is valued in $\text{Ker}(B) \cap \text{Ker}(C) \, d\mathbb{Q} \otimes dV$-a.e. Consider the predictable and $d\mathbb{Q} \otimes dV$-null set

$$\Lambda := \bigcup_{n=1}^{\infty} \{(\omega, t) \in \Omega \times [0, 1] \mid (\mathcal{H}^n(\omega) - \hat{\mathcal{H}}_t^n(\omega)) \notin \text{Ker}(B_t(\omega)) \cap \text{Ker}(C_t(\omega))\}.$$

We set $\mathcal{H}^n$ and $\hat{\mathcal{H}}^n$, for $n \in \mathbb{N}$, as well as $\hat{\mathcal{H}}^0$, to be zero on Λ. This does not change the stochastic integrals with respect to S and now, in addition, $(\mathcal{H}^n - \hat{\mathcal{H}}^n)$ is valued in $\text{Ker}(B) \cap \text{Ker}(C)$ for all (ω, t) and $n \in \mathbb{N}$. Since $0 \in \mathcal{K}(\cdot, t)$ a.s. for all t the $\mathcal{H}^n$ remain in $\mathcal{K}(\cdot, t)$ a.s. for all t and for all $n \in \mathbb{N}$. Observe also that $\hat{\mathcal{H}}^n$ now converges pointwise to $\hat{\mathcal{H}}^0$ for all (ω, t). Define the multivalued mapping

$$F_{\hat{\mathcal{H}}^0}(\omega, t) = \left\{ \hat{\mathcal{H}}_t^0(\omega) + \text{Ker}(B_t(\omega)) \cap \text{Ker}(C_t(\omega)) \right\} \cap \mathcal{K}(\omega, t), \qquad (6.1)$$

which is closed and predictably measurable by Lemma 5.9. We want to make $\hat{\mathcal{H}}_t^0$ valued in $\mathcal{K}(\cdot, t)$ a.s. without altering the stochastic integral. We must therefore add a predictable process valued in $\text{Ker}(B_t(\omega)) \cap \text{Ker}(C_t(\omega))$ to get back into $\mathcal{K}(\omega, t)$, which motivates the choice of $F_{\hat{\mathcal{H}}^0}$.

We now want to apply Theorem 5.7 to find a measurable selector for $F_{\hat{\mathcal{H}}^0}$. In particular we must check that the mapping $F_{\mathcal{H}^0}$ defined by (6.1) is nonempty.

Fix (ω, t) such that for all $n \in \mathbb{N}$, $\mathcal{H}_t^n(\omega)$ is in $\mathcal{K}(\omega, t)$ and $\mathcal{K}(\omega, t)$ is either a continuous or a polyhedral set. Suppose for a contradiction that $F_{\hat{\mathcal{H}}^0}(\omega, t) = \emptyset$. By Theorem 5.1 or Theorem 5.2 the sets $\mathcal{K}(\omega, t)$ and

$$\left\{ \hat{\mathcal{H}}_t^0(\omega) + \text{Ker}(B_t(\omega)) \cap \text{Ker}(C_t(\omega)) \right\}$$

may be strongly separated. In particular there exist $\phi \in \mathbb{R}^d$, $a \in \mathbb{R}$ and $\delta > 0$ such that for all $k \in \mathcal{K}(\omega, t)$ and $q \in \text{Ker}(B_t(\omega)) \cap \text{Ker}(C_t(\omega))$

$$\phi^\top k \geq a + \delta > a \geq \phi^\top (\hat{\mathcal{H}}_t^0(\omega) + q). \tag{6.2}$$

Since, for each $n \in \mathbb{N}$, $\mathcal{H}_t^n(\omega) \in \mathcal{K}(\omega, t)$ and $\mathcal{H}_t^n(\omega) - \hat{\mathcal{H}}_t^n(\omega)$ is an element of $\mathrm{Ker}(B_t(\omega)) \cap \mathrm{Ker}(C_t(\omega))$, (6.2) implies that for all $n \in \mathbb{N}$,

$$\phi^\top \mathcal{H}_t^n(\omega) \geq a + \delta > a \geq \phi^\top \big(\hat{\mathcal{H}}_t^0(\omega) + \mathcal{H}_t^n(\omega) - \hat{\mathcal{H}}_t^n(\omega)\big).$$

It then follows that, for all $n \in \mathbb{N}$,

$$\phi^\top \big(\hat{\mathcal{H}}_t^n(\omega) - \hat{\mathcal{H}}_t^0(\omega)\big) \geq \delta.$$

However we now have a contradiction as $\hat{\mathcal{H}}_t^n(\omega)$ converges to $\hat{\mathcal{H}}_t^0(\omega)$. Recall the definition of $F_{\hat{\mathcal{H}}^0}$ from (6.1) and define the set

$$\Gamma := \big\{(\omega, t) \in \Omega \times [0, 1] \,\big|\, F_{\hat{\mathcal{H}}^0}(\omega, t) \neq \emptyset \big\}.$$

By the above reasoning we have that $F_{\hat{\mathcal{H}}^0}(\omega, t) \neq \emptyset$ for those $(\omega, t) \in \Omega \times [0, 1]$ such that $\mathcal{H}_t^n(\omega)$ is in $\mathcal{K}(\omega, t)$ for all $n \in \mathbb{N}$ and $\mathcal{K}(\omega, t)$ is either a continuous or a polyhedral set. Since these conditions hold $\mathbb{P}$-a.s. for all t, and hence $\mathbb{Q}$-a.s. for all t, we can find, for each t, a $\mathbb{Q}$-null set Λ_t^1 such that

$$\left\{ (\omega, t) \in \Omega \times [0, 1] \,\middle|\, \omega \in \big(\Lambda_t^1\big)^c \right\} \subset \Gamma.$$

Exactly as in Remark 5.8 the restriction of $F_{\hat{\mathcal{H}}^0}$ to Γ is a closed, nonempty, predictably measurable, multivalued mapping to which Theorem 5.7 applies. Thus we get a measurable selector $\mathcal{I}$, a predictable process defined on Γ, with $\mathcal{I}_t(\omega) \in \mathcal{K}(\cdot, t)$. We now construct a representative $\mathcal{H}^0$ of H^0 which is in $\mathcal{K}(\cdot, t)$ $\mathbb{Q}$-a.s. for all t by setting

$$\mathcal{H}^0 := \hat{\mathcal{H}}^0 \mathbb{I}_{\Gamma^c} + \mathcal{I}\mathbb{I}_\Gamma.$$

The stochastic integral is invariant under a change to an equivalent probability measure, i.e., $H^0 \cdot S$ is the same under $\mathbb{P}$ as under $\mathbb{Q}$. In particular H^0 is valued in $\mathcal{K}(\cdot, t)$ $\mathbb{P}$-a.s. for all t and this completes the proof. $\qquad \square$

Acknowledgements We thank Freddy Delbaen, Dmitry Kramkov, Aleksandar Mijatović and Martin Schweizer for helpful advice and valuable discussions on the content of this paper.

Nicholas Westray was supported via an EPSRC grant and Christoph Czichowsky acknowledges financial support from the National Centre of Competence in Research "Financial Valuation and Risk Management" (NCCR FINRISK), Project D1 (Mathematical Methods in Financial Risk Management).

References

1. Émery, M.: Une topologie sur l'espace des semimartingales. In: Séminaire de Probabilités, XIII (Univ. Strasbourg, Strasbourg, 1977/1978). Lecture Notes in Mathematics, vol. 721, pp. 260–280. Springer, Berlin (1979)
2. Föllmer, H., Kramkov, D.: Optional decompositions under constraints. Probab. Theory Relat. Field. **109**(1), 1–25 (1997)
3. Gale, D., Klee, V.: Continuous convex sets. Math. Scand. **7**, 379–391 (1959)
4. Goldman, A.J., Tucker, A.W.: Polyhedral convex cones. In: Linear Equalities and Related Systems. Annals of Mathematics Studies, vol. 38, pp. 19–40. Princeton University Press, Princeton, NJ (1956)
5. Jacod, J., Shiryaev A.N.: Limit theorems for stochastic processes. Grundlehren der Mathematischen Wissenschaften [Fundamental Principles of Mathematical Sciences], vol. 288, 2nd edn. Springer, Berlin (2003)
6. Karatzas, I., Kardaras, C.: The numéraire portfolio in semimartingale financial models. Finance Stochast. **11**(4), 447–493 (2007)
7. Karatzas, I., Žitković, G.: Optimal consumption from investment and random endowment in incomplete semimartingale markets. Ann. Probab. **31**(4), 1821–1858 (2003)
8. Klee, V.: Maximal separation theorems for convex sets. Trans. Am. Math. Soc. **134**, 133–147 (1968)
9. Kuratowski, K., Ryll-Nardzewski, C.: A general theorem on selectors. Bull. Acad. Polon. Sci. Sér. Sci. Math. Astronom. Phys. **13**, 397–403 (1965)
10. Mémin, J.: Espaces de semi martingales et changement de probabilité. Z. Wahrsch. Verw. Gebiete. **52**(1), 9–39 (1980)
11. Mnif, M., Pham, H.: Stochastic optimization under constraints. Stochast. Process. Appl. **93**(1), 149–180 (2001)
12. Pham, H.: Minimizing shortfall risk and applications to finance and insurance problems. Ann. Appl. Probab. **12**(1), 143–172 (2002)
13. Rockafellar, R.T.: Measurable dependence of convex sets and functions on parameters. J. Math. Anal. Appl. **28**, 4–25 (1969)
14. Rockafellar, R.T.: Convex Analysis. Princeton Mathematical Series, vol. 28. Princeton University Press, Princeton, NJ (1970)
15. Rockafellar, R.T.: Integral functionals, normal integrands and measurable selections. In: Nonlinear Operators and the Calculus of Variations (Summer School. Univ. Libre Bruxelles, Brussels, 1975). Lecture Notes in Mathematics, vol. 543, pp. 157–207. Springer, Berlin (1976)
16. Wagner, D.H.: Survey of measurable selection theorems. SIAM J. Contr. Optim. **15**(5), 859–903 (1977)
17. Westray, N., Zheng, H.: Constrained non-smooth utility maximization without quadratic inf convolution. Stochast. Process. Appl. **119**(5), 1561–1579 (2009)
18. Yosida, K.: Functional Analysis. Grundlehren der Mathematischen Wissenschaften, vol. 123, 5th edn. Springer, Berlin (1978)

Closedness in the Semimartingale Topology for Spaces of Stochastic Integrals with Constrained Integrands

Christoph Czichowsky and Martin Schweizer

Abstract Let S be an $\mathbb{R}^d$-valued semimartingale and (ψ^n) a sequence of C-valued integrands, i.e. predictable, S-integrable processes taking values in some given closed set $C(\omega, t) \subseteq \mathbb{R}^d$ which may depend on the state ω and time t in a predictable way. Suppose that the stochastic integrals $(\psi^n \cdot S)$ converge to X in the semimartingale topology. When can X be represented as a stochastic integral with respect to S of some C-valued integrand? We answer this with a necessary and sufficient condition (on S and C), and explain the relation to the sufficient conditions introduced earlier in (Czichowsky, Westray, Zheng, Convergence in the semimartingale topology and constrained portfolios, 2010; Mnif and Pham, Stochastic Process Appl 93:149–180, 2001; Pham, Ann Appl Probab 12:143–172, 2002). The existence of such representations is equivalent to the closedness (in the semimartingale topology) of the space of all stochastic integrals of C-valued integrands, which is crucial in mathematical finance for the existence of solutions to most optimisation problems under trading constraints. Moreover, we show that a predictably convex space of stochastic integrals is closed in the semimartingale topology if and only if it is a space of stochastic integrals of C-valued integrands, where each $C(\omega, t)$ is convex.

Key words: Stochastic integrals · Constrained strategies · Semimartingale topology · Closedness · Predictably convex · Projection on predictable range · Predictable correspondence · Optimisation under constraints · Mathematical finance

MSC 2000 Classification Numbers: 60G07, 60H05, 28B20, 91B28.

C. Czichowsky and M. Schweizer (✉)
Department of Mathematics, ETH Zurich, Rämistrasse 101, CH-8092 Zurich, Switzerland
e-mail: christoph.czichowsky@math.ethz.ch; martin.schweizer@math.ethz.ch

C. Donati-Martin et al. (eds.), *Séminaire de Probabilités XLIII*, Lecture Notes in Mathematics 413
2006, DOI 10.1007/978-3-642-15217-7__18, © Springer-Verlag Berlin Heidelberg 2011

1 Introduction

In mathematical finance, proving the existence of a solution to optimisation problems like superreplication, utility maximisation or quadratic hedging usually boils down to the same abstract problem: One must show that a subsequence of (predictably) convex combinations of an optimising sequence of wealth processes, i.e. stochastic integrals with respect to the underlying price process S, converges and that the limit is again a wealth process, i.e. can be represented as a stochastic integral with respect to S. As the space of *all* stochastic integrals is closed in the semimartingale topology, this is the suitable topology to work with.

For applications, it is natural to include trading constraints by requiring the strategy (integrand) to lie pointwise in some set C; this set is usually convex to keep the above procedure applicable, and one would like it to depend on the state and time as well. Examples of interest include no shortselling, no borrowing or nonnegative wealth constraints; see e.g. [4, 16]. As pointed out by Delbaen [8] and Karatzas and Kardaras [16], a natural and convenient formulation of constraints is in terms of *correspondences*, i.e. set-valued functions. This is the approach we also advocate and use here.

For motivation, consider a sequence of (predictably convex combinations of) strategies and suppose (as usually happens by the convexification trick) that this converges pointwise. Each strategy is predictable, so constraints should also be "predictable" in some sense. To have the limit still satisfy the same restrictions as the sequence, the constraints should moreover be of the form "closure of a sequence $(\psi^n(\omega, t))$ of random points," since this is where the limit will lie. But if each $\psi^n(\omega, t)$ is a predictable process, the above closure is then a predictable correspondence by the Castaing representation (see Proposition 2.3). This explains why correspondences come up naturally.

In our constrained optimisation problem, assuming that we have predictable, convex, closed constraints, the same procedure as in the unconstrained case yields a sequence of wealth processes (integrals) converging to some limit which is a candidate for the solution of our problem. (We have cheated a little in the motivation – the integrals usually converge, not the integrands.) This limit process is again a stochastic integral, but it still remains to check that the corresponding trading strategy also satisfies the constraints. In abstract terms, one asks whether the limit of a sequence of stochastic integrals of constrained integrands can again be represented as a stochastic integral of some constrained integrand or, equivalently, if the space of stochastic integrals of constrained integrands is closed in the semimartingale topology. We illustrate by a *counterexample* that this is not true in general, since it might happen that some assets become redundant, i.e. can be replicated on some predictable set by trading in the remaining ones. This phenomenon occurs when there is linear dependence between the components of S.

As in [3,4,19,21], one could resolve this issue by simply assuming that there are no redundant assets; then the closedness result is true for all constraints formulated via closed (and convex) sets. Especially in Itô process models with a Brownian filtration, such a non-redundancy condition is useful (e.g. when working with artificial market completions), but it can be restrictive. Alternatively, as in [6, 15, 25], one can study only constraints given by polyhedral or continuous convex sets. While most constraints of practical interest are indeed polyhedral, this is conceptually unsatisfactory as one does not recover all results from the case when there are no redundant assets. A good formulation should thus account for the interplay between the constraints C and redundancies in the assets S.

To realise this idea, we use the *projection on the predictable range* of S. This is a predictable process taking values in the orthogonal projections in $\mathbb{R}^d$; it has been introduced in [8, 9, 24], and allows us to uniquely decompose each integrand into one part containing all relevant information for its stochastic integral and another part having stochastic integral zero. This reduces our problem to the question whether or not the projection of the constraints on the predictable range is closed. Convexity is not relevant for that aspect. Since that approach turns out to give a necessary and sufficient condition, we recover all previous results in [4, 6, 15, 19, 21] as special cases; and in addition, we obtain for constant constraints $C(\omega, t) \equiv C$ that closedness of the space of C-constrained integrands holds for *all* semimartingales if and only if all projections of C in $\mathbb{R}^d$ are closed. The well-known characterisation of polyhedral cones thus implies in particular that the closedness result for *constant* convex cone constraints is true for arbitrary semimartingales if and only if the constraints are polyhedral.

For a general constraint set $C(\omega, t)$ which is closed and convex, the set of stochastic integrals of C-constrained integrands is the prime example of a predictably convex space of stochastic integrals. By adapting arguments from [8], we show that this is in fact the *only* class of predictably convex spaces of stochastic integrals which are closed in the semimartingale topology. So this paper makes both mathematical contributions to stochastic calculus and financial contributions in the modelling and handling of trading constraints for optimisation problems from mathematical finance.

The remainder of the article is organised as follows. In Sect. 2, we formulate the problem in the terminology of stochastic processes and provide some results on measurable correspondences and measurable selectors. These are needed to introduce and handle the constraints. Section 3 contains a counterexample which illustrates where the difficulties arise and motivates in a simple setting the definition of the projection on the predictable range. The main results discussed above are established in Sect. 4. Section 5 gives the construction of the projection on the predictable range as well as two proofs omitted in Sect. 4. Finally, Sect. 6 briefly discusses some related work.

2 Problem Formulation and Preliminaries

Let $(\Omega, \mathcal{F}, P)$ be a probability space with a filtration $\mathbb{F} = (\mathcal{F}_t)_{0 \le t < \infty}$ satisfying the usual conditions of completeness and right-continuity. For all notation concerning stochastic integration, we refer to the book of Jacod and Shiryaev [14].

Set $\overline{\Omega} := \Omega \times [0, \infty)$. The space of all $\mathbb{R}^d$-valued semimartingales is denoted by $\mathcal{S}^{0,d}(P) := \mathcal{S}^0(P; \mathbb{R}^d)$, or simply $\mathcal{S}(P)$ if the dimension is clear. The *Émery distance* (see [10]) of two semimartingales X and Y is

$$d(X, Y) = \sup_{|\vartheta| \leq 1} \left(\sum_{n \in \mathbb{N}} 2^{-n} E\left[1 \wedge |(\vartheta \cdot (X - Y))_n| \right] \right),$$

where $(\vartheta \cdot X)_t := \int_0^t \vartheta_s dX_s$ stands for the *vector stochastic integral*, which is by construction a real-valued semimartingale, and the supremum is taken over all $\mathbb{R}^d$-valued predictable processes ϑ bounded by 1. With this metric, $\mathcal{S}(P)$ is a complete topological vector space, and the corresponding topology is called the *semimartingale topology*. For brevity, we say "in $\mathcal{S}(P)$" for "in the semimartingale topology". For a given $\mathbb{R}^d$-valued semimartingale S, we write $\mathcal{L}(S)$ for the space of $\mathbb{R}^d$-valued, S-integrable, predictable processes ϑ and $L(S)$ for the space of equivalence classes $[\vartheta] = [\vartheta]^S = \{\varphi \in \mathcal{L}(S) \mid \varphi \cdot S = \vartheta \cdot S\}$ of processes in $\mathcal{L}(S)$ which yield the same stochastic integral with respect to S, identifying processes equal up to P-indistinguishability. By Theorem V.4 in [20], the space of stochastic integrals $\{\vartheta \cdot S \mid \vartheta \in \mathcal{L}(S)\}$ is closed in $\mathcal{S}(P)$. Equivalently, $L(S)$ is a complete topological vector space with respect to $d_S([\vartheta], [\varphi]) = d(\vartheta \cdot S, \varphi \cdot S)$, where ϑ and φ are representatives of the equivalence classes $[\vartheta]$ and $[\varphi]$.

In this paper, we generalise the above closedness result from [20] to integrands restricted to lie in a given closed set, in the following sense. Let $C(\omega, t)$ be a non-empty, closed subset of $\mathbb{R}^d$ which may depend on ω and t in a predictably measurable way. Definition 2.2 below makes this precise: C should be a *predictable correspondence with closed values*. Denote by

$$\mathcal{C} := \mathcal{C}^S := \{\psi \in \mathcal{L}(S) \mid \psi(\omega, t) \in C(\omega, t) \text{ for all } (\omega, t)\} \qquad (2.1)$$

the set of *C-valued* or *C-constrained integrands* for S. If (ψ^n) is a sequence in $\mathcal{C}^S$ such that $(\psi^n \cdot S)$ converges to some X in the semimartingale topology, does there exist a ψ in $\mathcal{C}^S$ such that $X = \psi \cdot S$? In general, the answer is negative, as a simple counterexample in the next section illustrates, and so we ask under which conditions the above is true. By the closedness in $\mathcal{S}(P)$ of the space of all stochastic integrals, the limit X can always be represented as some stochastic integral $\vartheta \cdot S$. Thus it is enough to decide whether or not there exists for the limit class $[\vartheta]$ a representative ψ which is C-valued. Equivalently, one can ask whether $\mathcal{C}^S \cdot S$ is closed in $\mathcal{S}(P)$ or if the corresponding set

$$[\mathcal{C}] := [\mathcal{C}]^S := \{[\vartheta] \in L(S) \mid [\vartheta] \cap \mathcal{C} \neq \emptyset\}$$

of equivalence classes of elements of $\mathcal{C}^S$ is closed in $(L(S), d_S)$.

As already explained, this question arises naturally in mathematical finance for various optimisation problems under trading constraints; see [5, 11, 15, 19, 21, 22]. But not all papers make it equally clear whether the procedure outlined in the introduction can be or is being used. For [15, 19], this is clarified in [5]. Under additional assumptions, the closedness of $\mathcal{C}^S \cdot S$ in the semimartingale topology is sufficient to apply the results of Föllmer and Kramkov [11] on the optional decomposition under constraints, which give a dual characterisation of payoffs that can be superreplicated by constrained trading strategies. This is used in [17,21,22] to prove the existence of solutions to constrained utility maximisation problems. The results in [11] are formulated more generally for sets of (special) semimartingales which are predictably convex.

Definition 2.1. A set $\mathfrak{S}$ of semimartingales is called *predictably convex* if we have $h \cdot X + (1 - h) \cdot Y \in \mathfrak{S}$ for all X and Y in $\mathfrak{S}$ and all $[0, 1]$-valued predictable processes h. Analogously, a set $\mathfrak{C} \subseteq \mathcal{L}(S)$ of integrands is *predictably convex* if $h\vartheta + (1 - h)\varphi \in \mathfrak{C}$ for all ϑ and φ in $\mathfrak{C}$ and all $[0, 1]$-valued predictable processes h.

The prime example of predictably convex sets of integrands is given by C-constrained integrands when C is convex-valued. Theorem 4.11 below shows that *all* predictably convex spaces $\mathfrak{C}$ of integrands must be of this form if $\mathfrak{C} \cdot S$ is in addition closed in $\mathcal{S}(P)$.

To formulate precisely the assumptions on the (random and time-dependent) set C, we adapt the language of measurable correspondences to our framework of predictable measurability and recall for later use some of the results in this context. Note that the general results we exploit do not depend on special properties of the predictable σ-field on $\overline{\Omega}$. However, we do use that the range space $\mathbb{R}^d$ is metric and σ-compact; this ensures by Proposition 1A in [23] or the proof of Lemma 18.2 in [1] that weak measurability and measurability for a closed-valued correspondence coincide in our setting.

Definition 2.2. A mapping $C : \overline{\Omega} \to 2^{\mathbb{R}^d}$ is called an *($\mathbb{R}^d$-valued) correspondence*. Its *domain* is $\mathrm{dom}(C) := \{(\omega, t) \mid C(\omega, t) \neq \emptyset\}$. We call a correspondence C *predictable* if $C^{-1}(F) := \{(\omega, t) \mid C(\omega, t) \cap F \neq \emptyset\}$ is a predictable set for each closed $F \subseteq \mathbb{R}^d$. A correspondence has *predictable graph* if its graph $\mathrm{gr}(C) := \{(\omega, t, x) \in \overline{\Omega} \times \mathbb{R}^d \mid x \in C(\omega, t)\}$ is in $\mathcal{P} \otimes \mathcal{B}(\mathbb{R}^d)$. A *predictable selector* of a predictable correspondence C is a predictable process ψ which satisfies $\psi(\omega, t) \in C(\omega, t)$ for all $(\omega, t) \in \mathrm{dom}(C)$.

The following results ensure the existence of predictable selectors in all situations relevant for us.

Proposition 2.3. (Castaing). *For a correspondence $C : \overline{\Omega} \to 2^{\mathbb{R}^d}$ with closed values, the following are equivalent:*

(1) C is predictable.
(2) $\mathrm{dom}(C)$ is predictable and there exists a Castaing representation of C, i.e. a sequence (ψ^n) of predictable selectors of C such that

$$C(\omega, t) = \overline{\{\psi^1(\omega, t), \psi^2(\omega, t), \ldots\}} \qquad \text{for each } (\omega, t) \in \mathrm{dom}(C).$$

Proof. See Corollary 18.14 in [1] or Theorem 1B in [23]. ∎

Proposition 2.4. (Aumann). *Let $C : \overline{\Omega} \to 2^{\mathbb{R}^d}$ be a correspondence with nonempty values and predictable graph and μ a finite measure on $(\overline{\Omega}, \mathcal{P})$. Then there exists a predictable process ψ with $\psi(\omega, t) \in C(\omega, t)$ μ-a.e.*

Proof. See Corollary 18.27 in [1].   □

The proof of Proposition 2.4 is based on the following result on projections to which we refer later.

Proposition 2.5. *Let* $(R, \mathcal{R}, \mu)$ *be a* σ*-finite measure space,* $\mathcal{R}_\mu$ *the* σ*-field of* μ*-measurable sets and* A *in* $\mathcal{R}_\mu \otimes \mathcal{B}(\mathbb{R}^d)$. *Then the projection* $\pi_R(A)$ *of* A *on* R *belongs to* $\mathcal{R}_\mu$.

Proof. See Theorem 18.25 in [1].   □

Measurability and graph measurability of a correspondence are linked as follows.

Proposition 2.6. *Let* $C : \overline{\Omega} \to 2^{\mathbb{R}^d} \setminus \{\emptyset\}$ *be a correspondence. If* C *is predictable, its closure correspondence* $\overline{C}$ *given by* $\overline{C}(\omega, t) := \overline{C(\omega, t)}$ *has a predictable graph.*

Proof. See Theorem 18.6 in [1].   □

Since we require in (2.1) for our integrands ψ that $\psi(\omega, t) \in C(\omega, t)$ for all (ω, t), we shall assume, as motivated in the introduction, that C *is predictable and has closed values*. Then Proposition 2.3 guarantees the existence of predictable selectors. Moreover, we shall use that predictable measurability of a correspondence is preserved under transformations by Carathéodory functions and is stable under countable unions and intersections. Recall that a function $f : \overline{\Omega} \times \mathbb{R}^n \to \mathbb{R}^m$ is called *Carathéodory* if $f(\omega, t, x)$ is predictable with respect to (ω, t) and continuous in x.

Proposition 2.7. *Let* $C : \overline{\Omega} \to 2^{\mathbb{R}^d}$ *be a predictable correspondence with closed values and* $f : \overline{\Omega} \times \mathbb{R}^m \to \mathbb{R}^d$ *and* $g : \overline{\Omega} \times \mathbb{R}^d \to \mathbb{R}^m$ *Carathéodory functions. Then* C' *and* C'' *given by*

$$C'(\omega, t) = \{y \in \mathbb{R}^m \mid f(\omega, t, y) \in C(\omega, t)\}$$

and

$$C''(\omega, t) = \overline{\{g(\omega, t, x) \mid x \in C(\omega, t)\}}$$

are predictable correspondences with closed values.

Proof. See Corollaries 1P and 1Q in [23].   □

Proposition 2.8. *Let* $C^n : \overline{\Omega} \to 2^{\mathbb{R}^d}$ *for each* $n \in \mathbb{N}$ *be a predictable correspondence with closed values and define the correspondences* C' *and* C'' *by* $C'(\omega, t) = \bigcap_{n \in \mathbb{N}} C^n(\omega, t)$ *and* $C''(\omega, t) = \bigcup_{n \in \mathbb{N}} C^n(\omega, t)$. *Then* C' *and* C'' *are predictable and* C' *is closed-valued.*

Proof. See Theorem 1M in [23] and Lemma 18.4 in [1].   □

To establish a relation between predictably convex spaces of integrands and C-valued integrands, we later use the following result, which is a reformulation of the contents of Theorem 5 in [8]. We view an $\mathbb{R}^d$-valued predictable process on Ω as a $\mathcal{P}$-measurable $\mathbb{R}^d$-valued mapping on $\overline{\Omega}$, take some probability μ on $(\overline{\Omega}, \mathcal{P})$ and denote by $\overline{B(0,r)}^{L^\infty}$ and $\overline{B(0,r)}$ the closures of a ball of radius r in $L^\infty(\overline{\Omega}, \mathcal{P}, \mu; \mathbb{R}^d)$ and in $\mathbb{R}^d$, respectively. Predictable convexity is understood as in the second part of Definition 2.1.

Proposition 2.9. *Let $\mathfrak{K}$ be a predictably convex and μ-weak*-compact subset of $\overline{B(0,r)}^{L^\infty}$ with $0 \in \mathfrak{K}$. Then there exists a predictable correspondence $K : \overline{\Omega} \rightarrow 2^{\overline{B(0,r)}} \setminus \{\emptyset\}$, whose values are convex and compact and contain zero, such that*

$$\mathfrak{K} = \left\{ \vartheta \in L^\infty(\overline{\Omega}, \mathcal{P}, \mu; \mathbb{R}^d) \,\Big|\, \vartheta(\omega, t) \in K(\omega, t) \ \mu\text{-a.e.} \right\}.$$

Proof. In the proof of Theorem 5 in [8], the set $\mathcal{C}^\lambda$ defined there for $\lambda > 0$ contains zero and is by Lemmas 10 and 11 in [8] a predictably convex and weak*-compact subset of $\overline{B(0,\lambda)}^{L^\infty}$. No other properties of $\mathcal{C}^\lambda$ are used. So we can modify the proof of Theorem 5 in [8] by replacing the use of the Radon–Nikodým theorem of Debreu and Schmeidler (Theorem 2 in [7]) with that of Artstein (Theorem 9.1 in [2]). This yields that $K := \Phi^r$ constructed in that proof is predictably measurable and has not only (as argued in [8]) predictable graph. Replacing the correspondence K coming from this construction by $K \cap \overline{B(0,r)}$ then gives that K is valued in $2^{\overline{B(0,r)}}$. $\qquad\square$

3 A Motivating Example

In this section, we give a simple example of a semimartingale Y and a predictable correspondence C with non-empty, closed, convex cones as values such that $C^Y \cdot Y$ is not closed in $\mathcal{S}(P)$. This illustrates where the problems with our basic question arise and suggests a way to overcome them. The example is the same as Example 2.2 in [6], but we use it here for a different purpose and with different emphasis.

Let $W = (W^1, W^2, W^3)^\top$ be a 3-dimensional Brownian motion and $Y = \sigma \cdot W$, where

$$\sigma = \begin{pmatrix} 1 & 0 & 0 \\ 0 & 1 & -1 \\ 0 & -1 & 1 \end{pmatrix}.$$

The matrix σ and hence $\hat{c} = \sigma\sigma^\top$ have a non-trivial kernel spanned by $w = \frac{1}{\sqrt{2}}(0, 1, 1)^\top$, i.e. $\mathrm{Ker}(\hat{c}) = \mathrm{Ker}(\sigma) = \mathbb{R}w = \mathrm{span}\{w\}$. By construction, the stochastic integral of each $\mathbb{R}^3$-valued predictable process valued in $\mathrm{Ker}(\hat{c})$ $dP \otimes dt$-a.e. is zero, and vice versa. Thus the equivalence class $[\vartheta]^Y$ of any given $\vartheta \in \mathcal{L}(Y)$ is given by

$$[\vartheta]^Y = \{\vartheta + hw \mid h \text{ is a real-valued predictable process}\}$$

up to $dP \otimes dt$-a.e. equality, since adding a representative of 0 to some element of $\mathcal{L}(Y)$ does not change its equivalence class. Let K be the closed and convex cone

$$K = \{(x, y, z)^\top \in \mathbb{R}^3 \mid x^2 + y^2 \leq z^2, z \geq 0\}$$

and C the (constant) predictable correspondence with non-empty and closed values given by $C(\omega, t) = K$ for all $(\omega, t) \in \overline{\Omega}$. Define the sequence of (constant) processes (ψ^n) by $\psi^n = (1, \sqrt{n^2 - 1}, n)^\top$ for each $n \in \mathbb{N}$. In geometric terms, K is a circular cone around the z-axis, and (ψ^n) is a sequence of points on its surface going to infinity. (Instead of n, any sequence $z_n \to \infty$ in $[1, \infty)$ would do as well.) Each ψ^n is C-valued, and we compute

$$\psi^n \cdot Y = (\sigma \psi^n) \cdot W = W^1 + (\sqrt{n^2 - 1} - n)(W^2 - W^3).$$

Using this explicit expression yields by a simple calculation that $\psi^n \cdot Y \to W^1$ locally in $\mathcal{M}^2(P)$ and therefore in $\mathcal{S}(P)$; see [6] for details. However, the (constant) process $e_1 := (1, 0, 0)^\top$ leading to the limiting stochastic integral $e_1 \cdot Y = W^1$ does not have values in C, and since its equivalence class is $\{e_1 + hw \mid h$ is a real-valued predictable process$\}$, also no other integrand equivalent to e_1 does. Thus $C^Y \cdot Y$ is not closed in $\mathcal{S}(P)$.

To see why this causes problems, define $\tau := \inf\{t > 0 \mid |W_t| = 1\}$ and set $S := Y^\tau$. The arguments above then imply that the sequence $(\psi^n \cdot Y^\tau)$ is bounded from below (uniformly in n, t, ω) and converges in $\mathcal{S}(P)$ to $(W^1)^\tau$, which cannot be represented as $\psi \cdot S$ for any C-valued integrand ψ. Thus the set $C^S \cdot S$ does not satisfy Assumption 3.1 of the optional decomposition theorem under constraints in [11]. But for instance the proof of Proposition 2.13 in [17] (see p. 1835) explicitly uses that result of [11] in a setting where constrained integrands could be given by C-valued integrands as above. So technically, the argument in [17] is not valid without further assumptions (and Theorem 4.5 and Corollary 4.9 below show ways to fix this).

What can we learn from the counterexample? The key point is that *the convergence of stochastic integrals $\psi^n \cdot Y$ need not imply the pointwise convergence of their integrands*. Without constraints, this causes no problems; by Mémin's theorem, the limit is still *some* stochastic integral of Y, here $e_1 \cdot Y$. But if we insist on having C-valued integrands, the example shows that we ask for too much. Since K is closed, we can deduce above that $(|\psi^n|)$ must diverge (otherwise we should get along a subsequence a limit, which would be C-valued by closedness), and in fact $|\psi^n| = \sqrt{2}n \to \infty$. But at the same time, $(\sigma \psi^n)$ converges to $e_1 = (1, 0, 0)^\top$ – and this observation brings up the key idea of not looking at ψ^n, but at suitable projections of ψ^n linked (via σ) to the integrator Y.

To make this precise, denote the orthogonal projection on $\mathrm{Im}(\sigma \sigma^\top)$ by

$$\Pi^Y = \mathbb{1}_{d \times d} - ww^\top = \begin{pmatrix} 1 & 0 & 0 \\ 0 & \frac{1}{2} & -\frac{1}{2} \\ 0 & -\frac{1}{2} & \frac{1}{2} \end{pmatrix}.$$

Then $\Pi^Y \psi^n = \left(1, \frac{1}{2}(\sqrt{n^2-1}-n), -\frac{1}{2}(\sqrt{n^2-1}-n)\right)^\top$ converges to the limit integrand $(1,0,0)^\top = e_1$. We might worry about the obvious fact that $\Pi^Y \psi^n$ does not take values in C; but for the stochastic integrals, this does not matter because $(\Pi^Y \psi^n) \cdot Y = \psi^n \cdot Y$. Indeed, any $\vartheta \in \mathcal{L}(Y)$ can be written as a sum $\vartheta = \Pi^Y \vartheta + (ww^\top)\vartheta$ of one part with values in $\mathrm{Im}(\sigma\sigma^\top)$ and another part orthogonal to the first one; and since $\sigma^\top w = 0$ implies that $((ww^\top)\vartheta) \cdot Y = (\vartheta^\top ww^\top \sigma)^\top \cdot W = 0$, the claim follows. Going a little further, we even have for any $\vartheta \in \mathcal{L}(Y)$ and any $\mathbb{R}^d$-valued predictable process φ that

$$\varphi \in \mathcal{L}(Y) \text{ with } \varphi \cdot Y = \vartheta \cdot Y \iff \Pi^Y \varphi = \Pi^Y \vartheta \ dP \otimes dt\text{-a.e.,} \qquad (3.1)$$

by using that $\mathrm{Ker}(\sigma\sigma^\top) \cap \mathrm{Im}(\sigma\sigma^\top) = \{0\}$ and that $\sigma^\top(\Pi^Y v) = \sigma^\top v$ for all $v \in \mathbb{R}^d$ to check the Y-integrability of φ. The significance of (3.1) is that the stochastic integral $\vartheta \cdot Y$ is uniquely determined by $\Pi^Y \vartheta$, and so $\Pi^Y \vartheta$ gives a "minimal" choice of a representative of the equivalence class $[\vartheta]^Y$. Moreover, Π^Y gives via (3.1) a simple way to decide whether or not a given $\mathbb{R}^d$-valued predictable process φ belongs to the equivalence class $[\vartheta]^Y$.

Coming back to the set K, we observe that

$$\Pi^Y K = \left\{ \left(x, \frac{1}{2}(y-z), -\frac{1}{2}(y-z)\right)^\top \ \middle| \ x^2 + y^2 \le z^2, \ z \ge 0 \right\}$$

is the projection of the cone K on the plane through the origin and with the normal vector $(0,1,1)^\top$. In geometric terms, the projection of each horizontal slice of the cone transforms the circle above the x-y-plane into an ellipse in the projection plane having the origin as a point of its boundary. As we move up along the z-axis, the circles become larger, and so do the ellipses which in addition flatten out towards the line through the origin and the point $e_1 = (1,0,0)^\top$. But since they never reach that line although they come arbitrarily close, $\Pi^Y K$ is not closed in $\mathbb{R}^d$ – and this is the source of all problems in our counterexample. It explains why the limit $e_1 = \lim_{n\to\infty} \Pi^Y \psi^n$ is not in $\Pi^Y K$, which implies by (3.1) that there cannot exist any C-valued integrand ψ such that $\Pi^Y \psi = e_1$. But the insight about $\Pi^Y K$ also suggests that if we assume for a predictable correspondence C that

$$\Pi^Y C(\omega, t) \text{ is closed } dP \otimes dt\text{-a.e.,} \qquad (3.2)$$

we ought to get that $\mathcal{C}^Y \cdot Y$ is closed in $\mathcal{S}(P)$. This indeed works (see Theorem 4.5), and it turns out that condition (3.2) is not only sufficient, but also necessary.

The above explicit computations rely on the specific structure of Y, but they nevertheless motivate the approach for a general semimartingale S. We are going to define a predictable process Π^S taking values in the orthogonal projections in $\mathbb{R}^d$ and satisfying (3.1) with $dP \otimes dt$ replaced by a suitable measure on $(\overline{\Omega}, \mathcal{P})$ to control the stochastic integrals with respect to S. The process Π^S will be called the *projection on the predictable range* and will allow us to formulate and prove our main results in the next section.

4 Main Results

This section contains the main results (Theorems 4.5 and 4.11) as well as some consequences and auxiliary results. Before we can formulate and prove them, we need some facts and results about the projection on the predictable range of S. For the reader's convenience, the actual construction of Π^S is postponed to Sect. 5.

As in [14], Theorem II.2.34, each semimartingale S has the *canonical representation*

$$S = S^c + \widetilde{A} + [x\mathbb{1}_{\{|x|\leq 1\}}] * (\mu - \nu) + [x\mathbb{1}_{\{|x|>1\}}] * \mu$$

with the jump measure μ of S and its predictable compensator ν. Then the triplet (b, c, F) of predictable characteristics of S consists of a predictable $\mathbb{R}^d$-valued process b, a predictable nonnegative-definite matrix-valued process c and a predictable process F with values in the set of Lévy measures such that

$$\widetilde{A} = b \cdot B, \qquad [S^c, S^c] = c \cdot B \qquad \text{and} \qquad \nu = F \cdot B, \tag{4.1}$$

where $B := \sum_{i=1}^{d} \left([S^c, S^c]^{i,i} + \mathrm{Var}(\widetilde{A}^i)\right) + (|x|^2 \wedge 1) * \nu$.

Note that B is locally bounded since it is predictable and increasing. Therefore $P \otimes B$ is σ-finite on $(\overline{\Omega}, \mathcal{P})$ and there exists a probability measure P_B equivalent to $P \otimes B$. By the construction of the stochastic integral, S-integrable, predictable processes which are P_B-a.e. equal yield the same stochastic integral with respect to S (up to P-indistinguishability). Put differently, $\varphi = \vartheta$ P_B-a.e. implies for the equivalence classes in $L(S)$ that $[\varphi] = [\vartheta]$. But the converse is not true; a sufficient and necessary condition involves the projection Π^S on the predictable range of S, as we shall see below. Because S is now (in contrast to Sect. 3) a general semimartingale, the actual construction of Π^S and the proof of its properties become more technical and are postponed to the next section. We give here merely the definition and two auxiliary results.

Definition 4.1. The *projection on the predictable range of S* is a predictable process $\Pi^S : \overline{\Omega} \to \mathbb{R}^{d \times d}$ which takes values in the orthogonal projections in $\mathbb{R}^d$ and has the following property: If $\vartheta \in L(S)$ and φ is predictable, then φ is in $L(S)$ with $\varphi \cdot S = \vartheta \cdot S$ if and only if $\Pi^S \vartheta = \Pi^S \varphi$ P_B-a.e. We choose and fix one version of Π^S.

Remark 4.2. There are many possible choices for a process B satisfying (4.1). However, the definition of Π^S is independent of the choice of B in the sense that (with obvious notation) $\Pi^{S,B} \vartheta = \Pi^{S,B} \varphi$ P_B-a.e. if and only if $\Pi^{S,B'} \vartheta = \Pi^{S,B'} \varphi$ $P_{B'}$-a.e. This is because stochastic integrals of S do not depend on the choice of B.

As illustrated by the example in Sect. 3, the convergence in $\mathcal{S}(P)$ of stochastic integrals does not imply in general that the integrands converge P_B-a.e. But like in the example, a subsequence of the projections of the integrands on the predictable range does.

Lemma 4.3. *Let (ϑ^n) be a sequence in $L(S)$ such that $\vartheta^n \cdot S \to \vartheta \cdot S$ in $\mathcal{S}(P)$. Then there exists a subsequence (n_k) such that $\Pi^S \vartheta^{n_k} \to \Pi^S \vartheta$ P_B-a.e.*

Lemma 4.4. *Let $C : \overline{\Omega} \to 2^{\mathbb{R}^d} \setminus \{\emptyset\}$ be a predictable correspondence with closed values and such that the projection on the predictable range of S is not closed, i.e.*

$$\widetilde{F} = \{(\omega, t) \in \overline{\Omega} \mid \Pi^S(\omega, t)C(\omega, t) \text{ is not closed}\}$$

has outer P_B-measure > 0. Then there exist $\vartheta \in \mathcal{L}(S)$ and a sequence (ψ^n) of C-valued integrands such that $\psi^n \cdot S \to \vartheta \cdot S$ in $\mathcal{S}(P)$, but there is no C-valued integrand ψ such that $\psi \cdot S = \vartheta \cdot S$. Equivalently, there exists a sequence $([\psi^n])$ in $[C]^S$ such that $[\psi^n] \overset{L(S)}{\to} [\vartheta]$ but $[\vartheta] \notin [C]^S$, i.e. $[C]^S$ is not closed in $L(S)$.

Lemmas 4.3 and 4.4 as well as the existence of Π^S will be shown in Sect. 5. Admitting that, we can now prove our first main result; related work in [16] is discussed in Sect. 6. Recall the definition of $\mathcal{C} := \mathcal{C}^S$ from (2.1).

Theorem 4.5. *Let $C : \overline{\Omega} \to 2^{\mathbb{R}^d} \setminus \{\emptyset\}$ be a predictable correspondence with closed values. Then $\mathcal{C}^S \cdot S$ is closed in $\mathcal{S}(P)$ if and only if the projection of C on the predictable range of S is closed, i.e. $\Pi^S(\omega, t)C(\omega, t)$ is closed P_B-a.e. Equivalently: There exists a C-valued integrand ψ with $X = \psi \cdot S$ for any sequence (ψ^n) of C-valued integrands with $\psi^n \cdot S \to X$ in $\mathcal{S}(P)$ if and only if the projection of C on the predictable range of S is closed.*

Proof. "$\Rightarrow$": This implication follows immediately from Lemma 4.4.

"$\Leftarrow$": Let (ψ^n) be a sequence in $\mathcal{C}$ with $\psi^n \cdot S \to X$ in $\mathcal{S}(P)$. Then there exist by Mémin's theorem $\vartheta \in \mathcal{L}(S)$ with $X = \vartheta \cdot S$ and by Lemma 4.3 a subsequence, again indexed by n, with $\Pi^S \psi^n \to \Pi^S \vartheta$ P_B-a.e. So it remains to show that we can find a C-valued representative ψ of the limit class $[\vartheta] = [\Pi^S \vartheta]$. To that end, we observe that the P_B-a.e. closedness of $\Pi^S(\omega, t)C(\omega, t)$ implies that $\Pi^S \vartheta = \lim_{n \to \infty} \Pi^S \psi^n \in \Pi^S C$ P_B-a.e. By Proposition 2.7, the correspondences given by $\{\Pi^S(\omega, t)\vartheta(\omega, t)\}$, $C'(\omega, t) = \{\Pi^S(\omega, t)\vartheta(\omega, t)\} \cap \Pi^S(\omega, t)C(\omega, t)$ and

$$C''(\omega, t) = \{z \in \mathbb{R}^d \mid \Pi^S(\omega, t)z \in C'(\omega, t)\} \cap C(\omega, t)$$

are predictable and closed-valued. Indeed, $\Pi^S \vartheta$ is a predictable process, and $\{z \in \mathbb{R}^d \mid \Pi^S(\omega, t)z \in C'(\omega, t)\}$ and $\Pi^S C = \overline{\Pi^S C}$ are the pre-image and (the closure of) the image of a closed-valued correspondence under a Carathéodory function, respectively. Thus C' and C'' are the intersections of two predictable and closed-valued correspondences and therefore predictable by Proposition 2.8. So there exists by Proposition 2.3 a predictable selector ψ of C'' on

$$\text{dom}(C'') = \{(\omega, t) \mid \Pi^S(\omega, t)\vartheta(\omega, t) \in \Pi^S(\omega, t)C(\omega, t)\}.$$

This ψ can be extended to a C-valued integrand by using any predictable selector on the P_B-nullset $(\text{dom}(C''))^c$. By construction, ψ is then in $\mathcal{C}$ and satisfies $\Pi^S \psi = \Pi^S \vartheta$ P_B-a.e., so that $\psi \in [\vartheta]$ by the definition of Π^S. This completes the proof. $\qquad\square$

Theorem 4.5 gives as necessary and sufficient condition for the closedness of the space of C-constrained integrals of S that the projection of the constraint set C on the predictable range of S is closed. This uses information from both the semimartingale S and the constraints C, as well as their interplay. We shall see below how this allows to recapture several earlier results as special cases.

Corollary 4.6. *Suppose that $S = S_0 + M + A$ is in $\mathcal{S}^2_{loc}(P)$ and define the process a via $A = a \cdot B$. If*

$$[0]^M = \{ha \mid h \text{ is real-valued and predictable}\} \tag{4.2}$$

up to P_B-a.e. equality, then $\mathcal{C}^S \cdot S$ is closed in $\mathcal{S}(P)$ for all predictable correspondences $C : \overline{\Omega} \to 2^{\mathbb{R}^d} \setminus \{\emptyset\}$ with closed values.

Proof. By Lemma 5.1 below, (4.2) implies $[0]^S = [0]^M \cap [0]^A = \{0\}$ and therefore $\Pi^S = \mathbb{1}_{d \times d}$ by (5.2) below. So the projection of any closed-valued correspondence C on the predictable range of S is closed, which gives the assertion by Theorem 4.5. $\square$

In applications from mathematical finance, S often satisfies the so-called *structure condition (SC)*, i.e. $S = S_0 + M + A$ is in $\mathcal{S}^2_{loc}(P)$ and there exists an $\mathbb{R}^d$-valued predictable process $\lambda \in \mathcal{L}^2_{loc}(M)$ such that $A = \lambda \cdot \langle M, M \rangle$ or, equivalently, $a = \hat{c}\lambda$ P_B-a.e.; this is a weak no-arbitrage type condition. In this situation, Lemma 5.1 below gives $[0]^M \subseteq [0]^A$, and thus condition (4.2) holds if and only if $[0]^M = \{0\}$ (up to P_B-a.e. equality), which means that $\hat{c}$ is P_B-a.e. invertible. This is the case covered in Lemma 3.1 in [21], where one has conditions only on S but not on C. Basically this ensures that there are no redundant assets, i.e. every stochastic integral is realised by exactly one integrand (up to P_B-a.e. equality).

The opposite extreme is to place conditions only on C that ensure closedness of $\mathcal{C}^S \cdot S$ for arbitrary semimartingales S, as in Theorem 3.5 of [6]. We recover this as a special case in the following corollary; note that in a slight extension over [6], the constraints need not be convex. Recall that a closed convex set $K \subseteq \mathbb{R}^d$ is called *continuous* if its support function $\delta(v|K) = \sup_{w \in K} w^\top v$ is continuous for all vectors $v \in \mathbb{R}^d$ with $|v| = 1$; see [13].

Corollary 4.7. *Let $C : \overline{\Omega} \to 2^{\mathbb{R}^d} \setminus \{\emptyset\}$ be a predictable correspondence with closed values. Then $\mathcal{C}^Y \cdot Y$ is closed in $\mathcal{S}(P)$ for all semimartingales Y if with probability 1, for all $t \geq 0$ all projections $\Pi C(\omega, t)$ of $C(\omega, t)$ are closed in $\mathbb{R}^d$. In particular, if with probability 1, every $C(\omega, t)$, $t \geq 0$, is compact, or polyhedral, or a continuous and convex set, then $\mathcal{C}^Y \cdot Y$ is closed in $\mathcal{S}(P)$ for all semimartingales Y.*

Proof. If a set is compact or polyhedral, all its projections have the same property (see Corollary 2.15 in [18]) and are thus closed. For a continuous convex set, every projection is closed by Theorem 1.3 in [13]. Now if with probability 1, for all $t \geq 0$ all projections $\Pi C(\omega, t)$ of $C(\omega, t)$ are closed, the projection $\Pi^Y C$ of C on the predictable range of every semimartingale Y is closed $P \otimes B^Y$-a.e. So $\mathcal{C}^Y \cdot Y$ is closed in $\mathcal{S}(P)$ by Theorem 4.5. $\square$

Combining Theorem 4.5 with the example in Sect. 3, we obtain the following corollary. It is formulated for fixed sets K, but can probably be generalised to predictable correspondences C by using measurable selections.

Corollary 4.8. *Suppose* $(\Omega, \mathcal{F}, P)$ *is sufficiently rich. Fix* $K \subseteq \mathbb{R}^d$ *and define as in* (2.1) $\mathcal{K}^Y = \{\psi \in \mathcal{L}(Y) \mid \psi(\omega, t) \in K \text{ for all } (\omega, t)\}$. *Then* $\mathcal{K}^Y \cdot Y$ *is closed in* $\mathcal{S}(P)$ *for all* $\mathbb{R}^d$-*valued semimartingales* Y *if and only if all projections* ΠK *of* K *in* $\mathbb{R}^d$ *are closed.*

Proof. The "if" part follows immediately from Theorem 4.5. For the converse, assume by way of contradiction that there is a projection Π in $\mathbb{R}^d$ such that ΠK is not closed. Let W be a d-dimensional Brownian motion and set $Y = \Pi^\top \cdot W$. Then Π is the projection on the predictable range of Y, and therefore $\mathcal{K}^Y \cdot Y$ is not closed by Theorem 4.5. □

If the constraints are not only convex, but also cones, a characterisation of convex polyhedra due to Klee [18] gives an even sharper result.

Corollary 4.9. *Let* $K \subseteq \mathbb{R}^d$ *be a closed convex cone. Then* $\mathcal{K}^Y \cdot Y$ *is closed in* $\mathcal{S}(P)$ *for all* $\mathbb{R}^d$-*valued semimartingales* Y *if and only if* K *is polyhedral.*

Proof. By Corollary 4.8, $\mathcal{K}^Y \cdot Y$ is closed in $\mathcal{S}(P)$ if and only if all projections ΠK are closed in $\mathbb{R}^d$. But Theorem 4.11 in [18] says that all projections of a convex cone are closed in $\mathbb{R}^d$ if and only if that cone is polyhedral. □

Remark 4.10. Armed with the last result, we can briefly come back to the proof of Proposition 2.13 in [17]. We have already pointed out in Sect. 3 that the argument in [17] uses the optional decomposition under constraints from [11], without verifying its Assumption 3.1. In view of Corollary 4.9, we can now be more precise: The argument in [17] as it stands (i.e. without assumptions on S) only works for *polyhedral* cone constraints; for others, one could by Corollary 4.9 construct a semimartingale S giving a contradiction.

We now turn to our second main result. Recall again the definition of $\mathcal{C}$ from (2.1) and note that for a correspondence C with convex values, $\mathcal{C}$ is the prime example of a predictably convex space of integrands. The next theorem shows that this is actually the only class of predictably convex integrands if we assume in addition that the resulting space $\mathcal{C} \cdot S$ of stochastic integrals is closed in $\mathcal{S}(P)$. The result and its proof are inspired from Theorems 3 and 4 in [8], but require quite a number of modifications.

Theorem 4.11. *Let* $\mathfrak{C} \subseteq \mathcal{L}(S)$ *be non-empty. Then* $\mathfrak{C} \cdot S$ *is predictably convex and closed in the semimartingale topology if and only if there exists a predictable correspondence* $C : \overline{\Omega} \to 2^{\mathbb{R}^d} \setminus \{\emptyset\}$ *with closed convex values such that the projection of* C *on the predictable range of* S *is closed, i.e.* $\Pi^S(\omega, t)C(\omega, t)$ *is closed* P_B-*a.e., and such that we have* $\mathfrak{C} \cdot S = \mathcal{C}^S \cdot S$, *i.e.*

$$\mathfrak{C} \cdot S = \{\psi \cdot S \mid \psi \in \mathfrak{C}\}$$
$$= \{\psi \cdot S \mid \psi \in \mathcal{L}(S) \text{ and } \psi(\omega, t) \in C(\omega, t) \text{ for all } (\omega, t)\}.$$

Proof. "⇐": The pointwise convexity of C immediately implies that $\mathcal{C}^S \cdot S$ is predictably convex, and closedness follows from Theorem 4.5.

"⇒": Like at the end of Sect. 2, we view predictable processes on Ω as $\mathcal{P}$-measurable random variables on $\overline{\Omega} = \Omega \times [0, \infty)$. Since we are only interested in a non-empty space of stochastic integrals with respect to S, we lose no generality if we replace $\mathfrak{C}$ by $\{\vartheta - \varphi \in \mathcal{L}(S) \mid \vartheta \in [\mathfrak{C}]\}$ for some $\varphi \in \mathfrak{C}$ and identify this with a subspace of $L^0(\overline{\Omega}, \mathcal{P}, P_B; \mathbb{R}^d)$ which contains zero. Indeed, if the assertion is true for $\mathfrak{C} - \varphi$ with a correspondence $\widetilde{C}$, it is also true for $\mathfrak{C}$ with $C = \widetilde{C} + \varphi$, which is a predictable correspondence by Proposition 2.7. In order to apply Proposition 2.9, we truncate $\mathfrak{C}$ to get

$$\mathfrak{C}^q = \{\psi \in \mathfrak{C} \mid \|\psi\|_{L^\infty} \le q\} = \mathfrak{C} \cap \overline{B(0, q)}^{L^\infty} \qquad \text{for } q \in \mathbb{Q}_+.$$

Then $\mathfrak{C}^q$ inherits predictable convexity from $\mathfrak{C}$ and is thus a convex subset of $\overline{B(0, q)}^{L^\infty}$. Moreover, $\mathfrak{C}^q$ is closed with respect to convergence in P_B-measure since its elements are uniformly bounded by q and $\mathfrak{C} \cdot S$ is closed in $\mathcal{S}(P)$; this uses the fact, easily proved via dominated convergence separately for the M- and A-integrals, that for any uniformly bounded sequence of integrands (ψ^n) converging pointwise, the stochastic integrals converge in $\mathcal{S}(P)$. By a well-known application of the Krein–Šmulian and Banach–Alaoglu theorems (see Theorems A.62 and A.63 and Lemma A.64 in [12]), $\mathfrak{C}^q$ is thus weak*-compact, and Proposition 2.9 gives a predictable correspondence $C^q : \overline{\Omega} \to 2^{\overline{B(0,q)}} \setminus \{\emptyset\}$ with convex compact values containing zero such that

$$\mathfrak{C}^q = \{\psi \in L^0(\overline{\Omega}, \mathcal{P}, P_B; \mathbb{R}^d) \mid \psi(\omega, t) \in C^q(\omega, t) \ P_B\text{-a.e.}\}.$$

By the definition of $\mathfrak{C}^q$ we obtain, after possibly modifying the sets on a P_B-nullset, that

$$C^{q_2}(\omega, t) \cap \overline{B(0, q_1)} = C^{q_1}(\omega, t) \qquad \text{for all } (\omega, t) \in \overline{\Omega} \qquad (4.3)$$

for $0 < q_1 \le q_2 < \infty$ by Lemma 12 in [8], since the graph of each C^q is predictable by Proposition 2.6. Using the characterisation of closed sets in metric spaces as limit points of converging sequences implies with (4.3) that the correspondence C given by

$$C(\omega, t) := \bigcup_{q \in \mathbb{Q}_+} C^q(\omega, t)$$

has closed values. Moreover, each $C(\omega, t)$ is convex as the union of an increasing sequence of convex sets, and it only remains to show that $\mathfrak{C} \cdot S = C \cdot S$.

Suppose first that ψ is in $\mathfrak{C}$. By predictable convexity and since $0 \in \mathfrak{C}$, $\psi^n := \mathbb{1}_{\{|\psi| \le n\}} \psi$ is in $\mathfrak{C}^n$ and therefore C^n- and hence C-valued. Since (ψ^n) converges pointwise to ψ, the closedness of C implies that ψ is C-valued, so that $\psi \in C$ and $\mathfrak{C} \cdot S \subseteq C \cdot S$. Conversely, if ψ is in C, then $\psi^n := \mathbb{1}_{\{|\psi| \le n\}} \psi$ is C^n-valued and hence in $\mathfrak{C}^n \subseteq \mathfrak{C}$. But $(\psi^n \cdot S)$ converges to $\psi \cdot S$ in $\mathcal{S}(P)$ and $\mathfrak{C} \cdot S$ is closed in $\mathcal{S}(P)$. So the limit $\psi \cdot S$ is in $\mathfrak{C} \cdot S$ and hence $\psi \in \mathfrak{C}$ and $C \cdot S \subseteq \mathfrak{C} \cdot S$. Finally, $C \cdot S = \mathfrak{C} \cdot S$ is closed in $\mathcal{S}(P)$, and therefore $\Pi^S C$ is closed P_B-a.e. by Theorem 4.5. This completes the proof. □

Remark 4.12. (1) Theorem 4.11 can be used as follows. Start with any convex-valued correspondence C, form the space $C \cdot S$ of corresponding stochastic integrals and take its closure in $S(P)$. Then Theorem 4.11 tells us that we can realise this closure as a space of stochastic integrals from $\widetilde{C}$-constrained integrands, for some predictable correspondence $\widetilde{C}$ with convex and closed values. In other words, $\overline{C \cdot S}^{S(P)} = \widetilde{C} \cdot S$; and one possible choice of $\widetilde{C}$ is $\widetilde{C} = (\Pi^S)^{-1}(\overline{C})$. Another possible choice would be $\widetilde{C} = \overline{C + \mathfrak{N}}$, where $\mathfrak{N}$ denotes the correspondence of null investments for S; see Sect. 6.

(2) If we assume in Theorem 4.11 that $\mathfrak{C} \subseteq \mathcal{L}^p_{loc}(S)$ for $p \in [1, \infty)$, then $\mathfrak{C} \cdot S \subseteq S^p_{loc}(P)$, and $\mathfrak{C} \cdot S$ is closed in $S^p(P)$ if and only if there exists C as in the theorem. This can be useful for applications (e.g., mean–variance hedging under constraints, with $p = 2$).

5 Projection on the Predictable Range

In this section, we construct the projection Π^S on the predictable range of a general semimartingale S in continuous time. The idea to introduce such a projection comes from [9, 24], where it was used to prove the fundamental theorem of asset pricing in discrete time. It was also used for a continuous local martingale in [8] to investigate the structure of m-stable sets and in particular the set of risk-neutral measures.

As already explained before Definition 4.1, a *sufficient* condition for $\varphi \cdot S = \vartheta \cdot S$ (up to P-indistinguishability) or, equivalently, $\varphi = \vartheta$ in $L(S)$ or $[\varphi] = [\vartheta]$, is that $\varphi = \vartheta$ P_B-a.e. If we again view predictable processes on Ω as $\mathcal{P}$-measurable random variables on $\overline{\Omega} = \Omega \times [0, \infty)$, i.e. elements of $\mathcal{L}^0(\overline{\Omega}, \mathcal{P}; \mathbb{R}^d)$, then $\varphi = \vartheta$ P_B-a.e. is the same as saying that $\varphi = \vartheta$ in $L^0(\overline{\Omega}, \mathcal{P}, P_B; \mathbb{R}^d)$. But to get a *necessary and sufficient* condition for $[\vartheta] = [\varphi]$, we need to understand not only what $0 \in \mathcal{L}(S)$ looks like, but rather the precise structure of (the equivalence class) $[0]$. This is achieved by Π^S.

The construction of Π^S basically proceeds by generalising that of Π^Y in the example in Sect. 3 and adapting the steps in [9] to continuous time. The idea is as follows. We start by characterising the equivalence class $[0]$ as a linear subspace of $L^0(\overline{\Omega}, \mathcal{P}, P_B; \mathbb{R}^d)$. Since this subspace satisfies a certain stability property, we can construct predictable processes $e^1, \dots, e^d$ which form an "orthonormal basis" of $[0]$ in the sense that $[0]$ equals up to P_B-a.e. equality their linear combinations with predictable coefficients, i.e.

$$[0] = \left\{ \sum_{j=1}^{d} h^j e^j \ \middle|\ h^1, \dots, h^d \text{ are real-valued predictable} \right\} \tag{5.1}$$

up to P_B-a.e. equality. But these linear combinations contribute 0 to the integral with respect to S; so we filter them out to obtain the part of the integrand which determines the stochastic integral, by defining

$$\Pi^S := \mathbb{1}_{d \times d} - \sum_{j=1}^{d} e^j (e^j)^\top. \tag{5.2}$$

This construction then yields the projection on the predictable range as in Definition 4.1.

To describe $[0] = [0]^S$ as a linear subspace of $L^0(\overline{\Omega}, \mathcal{P}, P_B; \mathbb{R}^d)$, we exploit that although we work with a general semimartingale S, we can by Lemma I.3 in [20] switch to an equivalent probability Q under which S is locally square-integrable. Since the stochastic integral and hence $[0]^S$ are invariant under a change to an equivalent measure, any representation we obtain $Q \otimes B$-a.e. also holds P_B-a.e., as $P_B \sim P \otimes B \sim Q \otimes B$. Let $S = S_0 + M^Q + A^Q$ be the canonical decomposition of S under Q into an $\mathbb{R}^d$-valued square-integrable Q-martingale $M^Q \in \mathcal{M}_0^{2,d}(Q)$ null at 0 and an $\mathbb{R}^d$-valued predictable process $A^Q \in \mathcal{A}^{1,d}(Q)$ of Q-integrable variation $\mathrm{Var}(A^Q)$ also null at 0. By Propositions II.2.9 and II.2.29 in [14], there exist an increasing, locally Q-integrable, predictable process B^Q, an $\mathbb{R}^d$-valued process a^Q and a predictable $\mathbb{R}^{d \times d}$-valued process $\hat{c}^Q$ whose values are positive semidefinite symmetric matrices such that

$$(A^Q)^i = (a^Q)^i \cdot B^Q \quad \text{and} \quad \langle (M^Q)^i, (M^Q)^j \rangle^Q = (\hat{c}^Q)^{ij} \cdot B^Q \tag{5.3}$$

for $i, j = 1, \ldots, d$. By expressing the semimartingale characteristics of S under Q by those under P via Girsanov's theorem, writing A^Q and $\langle M^Q, M^Q \rangle^Q$ in terms of semimartingale characteristics and then passing to differential characteristics with B as predictable increasing process, we obtain that we can and do choose $B^Q = B$ in (5.3); see Theorem III.3.24 and Propositions II.2.29 and II.2.9 in [14]. Using the canonical decomposition of S under Q as auxiliary tool then allows us to give the following characterisation of $[0]^S$.

Lemma 5.1. *Let $Q \sim P$ such that $S = S_0 + M^Q + A^Q \in \mathcal{S}_{loc}^2(Q)$. Then*

(1) $[0]^{M^Q} = \{ \varphi \in \mathcal{L}^0(\overline{\Omega}, \mathcal{P}; \mathbb{R}^d) \mid \hat{c}^Q \varphi = 0 \ P_B\text{-a.e.} \}$.
(2) $[0]^{A^Q} = \{ \varphi \in \mathcal{L}^0(\overline{\Omega}, \mathcal{P}; \mathbb{R}^d) \mid (a^Q)^\top \varphi = 0 \ P_B\text{-a.e.} \}$.
(3) $[0]^S = [0]^{M^Q} \cap [0]^{A^Q}$.

Moreover, $[0]^{M^Q}$, $[0]^{A^Q}$ and $[0]^S$ all do not depend on Q.

Proof. The last assertion is clear since the stochastic integral of a semimartingale (like M^Q, A^Q, S) is invariant under a change to an equivalent measure. Because also $P_B \sim Q \otimes B$, we can argue for the rest of the proof under the measure Q. Then the inclusions "⊇" follow immediately from the definition of the stochastic integral with respect to a square-integrable martingale and a finite variation process, since the conditions on the right-hand side ensure that φ is in $\mathcal{L}^2(M^Q)$ and $\mathcal{L}^1(A^Q)$. For the converse, we start with $\varphi \in [0]^S$ and set $\varphi^n := \mathbb{1}_{\{|\varphi| \le n\}} \varphi$. Then $\varphi^n \cdot S = 0$ implies that $\varphi^n \cdot M^Q = 0$ and $\varphi^n \cdot A^Q = 0$ by the uniqueness of the Q-canonical decomposition of $\varphi^n \cdot S$; this uses that φ^n is bounded. Therefore we can reduce the proof of "$\subseteq$" for (3) to that for (1) and (2). So assume now that φ is in either $[0]^{M^Q}$ or $[0]^{A^Q}$ so that $\varphi^n \cdot M^Q = 0$ or $\varphi^n \cdot A^Q = 0$. But φ^n is bounded, hence in $\mathcal{L}^2(M^Q)$ or $\mathcal{L}^1(A^Q)$, for each n, and by the construction of the stochastic integral, we obtain that $\hat{c}^Q \varphi^n = 0$ or $(a^Q)^\top \varphi^n = 0$ $Q \otimes B$-a.e. and hence P_B-a.e. Since (φ^n) converges pointwise to φ, the inclusions "$\subseteq$" for (1) and (2) follow by passing to the limit. $\qquad\square$

The following technical lemma, which is a modification of Lemma 6.2.1 in [9], gives the announced "orthonormal basis" of $[0]^S$ in the sense of (5.1).

Lemma 5.2. *Let $U \subseteq L^0\left(\overline{\Omega}, \mathcal{P}, P_B; \mathbb{R}^d\right)$ be a linear subspace which is closed with respect to convergence in P_B-measure and satisfies the following stability property:*

$$\varphi^1 \mathbb{1}_F + \varphi^2 \mathbb{1}_{F^c} \in U \qquad \text{for all } \varphi^1 \text{ and } \varphi^2 \text{ in } U \text{ and } F \in \mathcal{P}.$$

Then there exist $e^j \in L^0\left(\overline{\Omega}, \mathcal{P}, P_B; \mathbb{R}^d\right)$ for $j = 1, \ldots, d$ such that

(1) $\{e^{j+1} \neq 0\} \subseteq \{e^j \neq 0\}$ for $j = 1, \ldots, d-1$.
(2) $|e^j(\omega, t)| = 1$ or $|e^j(\omega, t)| = 0$.
(3) $(e^j)^\top e^k = 0$ for $j \neq k$.
(4) $\varphi \in U$ if and only if there are $h^1, \ldots, h^d$ in $L^0\left(\overline{\Omega}, \mathcal{P}, P_B; \mathbb{R}\right)$ such that $\varphi = \sum_{j=1}^d h^j e^j$, i.e.

$$U = \left\{ \sum_{j=1}^d h^j e^j \;\middle|\; h^1, \ldots, h^d \text{ are real-valued predictable} \right\}.$$

Proof. The predictable processes $e^1, \ldots, e^d$ with the properties (1)–(4) are the column vectors of the measurable projection-valued mapping constructed in Lemma 6.2.1 in [9]. Therefore their existence follows immediately from the construction given there. $\qquad\square$

By Lemma I.3 in [20], there always exists a probability measure Q as in Lemma 5.1, and therefore the space $[0]^S$ satisfies the assumptions of Lemma 5.2. So we take a "basis" $e^1, \ldots, e^d$ as in the latter result and define Π^S as in (5.2) by

$$\Pi^S := \mathbb{1}_{d \times d} - \sum_{j=1}^d e^j (e^j)^\top.$$

Then $\Pi^S(\omega, t)$ is the projection on the orthogonal complement of the linear space spanned in $\mathbb{R}^d$ by $e^1(\omega, t), \ldots, e^d(\omega, t)$ so that $\Pi^S(\omega, t)\gamma$ is orthogonal to all $e^i(\omega, t)$ for each $\gamma \in \mathbb{R}^d$; and Lemma 5.2 says that each element of $[0]^S$ is a (random and time-dependent) linear combination of $e^1, \ldots, e^d$, and vice versa. In particular, $\vartheta - \Pi^S \vartheta$ is in $[0]^S$ for every predictable $\mathbb{R}^d$-valued ϑ. The next result shows that Π^S satisfies the properties required in Definition 4.1. Note that Π^S is only defined up to P_B-nullsets since the e^j are; so we have to choose one version for Π^S to be specific.

Lemma 5.3. (Projection on the predictable range of S). *For a semimartingale S, the projection Π^S on the predictable range of S exists, i.e. there exists a predictable process $\Pi^S : \overline{\Omega} \to \mathbb{R}^{d \times d}$ which takes values in the orthogonal projections in $\mathbb{R}^d$ and has the following property: If $\vartheta \in \mathcal{L}(S)$ and ψ is an $\mathbb{R}^d$-valued predictable process, then*

$$\psi \in \mathcal{L}(S) \text{ with } \psi \cdot S = \vartheta \cdot S \quad \Longleftrightarrow \quad \Pi^S \psi = \Pi^S \vartheta \quad P_B\text{-a.e.} \qquad (5.4)$$

Proof. If we define Π^S as above, Lemma 5.2 implies that Π^S is predictable and valued in the orthogonal projections in $\mathbb{R}^d$, and it only remains to check (5.4). So take $\vartheta \in \mathcal{L}(S)$ and assume first that $\Pi^S \vartheta = \Pi^S \psi$ P_B-a.e. The definition of Π^S and Lemma 5.1 then yield that $\vartheta - \Pi^S \vartheta$ and $\Pi^S \vartheta - \Pi^S \psi$ are in $[0]^S$, which implies that $\Pi^S \vartheta = \vartheta - (\vartheta - \Pi^S \vartheta)$ and $\Pi^S \psi$ are in $\mathcal{L}(S)$ and also that $\vartheta \cdot S = (\Pi^S \vartheta) \cdot S = (\Pi^S \psi) \cdot S$. Because also $\psi - \Pi^S \psi$ is in $[0]^S \subseteq \mathcal{L}(S)$, we conclude that $\psi \in \mathcal{L}(S)$ with $\vartheta \cdot S = \psi \cdot S$. Conversely, if $\psi \cdot S = \vartheta \cdot S$, then $\psi - \vartheta \in [0]^S$, and we always have $(\psi - \vartheta) - \Pi^S(\psi - \vartheta) \in [0]^S$. Therefore $\Pi^S(\psi - \vartheta) \in [0]^S$ which says by Lemma 5.2 that for P_B-a.e. (ω, t), $\Pi^S(\psi - \vartheta)(\omega, t)$ is a linear combination of the $e^i(\omega, t)$. But the column vectors of Π^S are orthogonal to $e^1, \ldots, e^d$ for each fixed (ω, t), and so we obtain $\Pi^S(\psi - \vartheta) = 0$ P_B-a.e., which completes the proof. $\qquad \square$

With the existence of the projection on the predictable range established, it remains to prove Lemmas 4.3 and 4.4, which we recall for convenience.

Lemma 4.3. *Let (ϑ^n) be a sequence in $\mathcal{L}(S)$ such that $\vartheta^n \cdot S \to \vartheta \cdot S$ in $\mathcal{S}(P)$. Then there exists a subsequence (n_k) such that $\Pi^S \vartheta^{n_k} \to \Pi^S \vartheta$ P_B-a.e.*

Proof. As in the proof of Theorem V.4 in [20], we can switch to a probability measure $Q \sim P$ such that $\frac{dQ}{dP}$ is bounded, $S - S_0 = M^Q + A^Q$ is in $\mathcal{M}^{2,d}(Q) \oplus \mathcal{A}^{1,d}(Q)$ and $\vartheta^n \cdot S \to \vartheta \cdot S$ in $\mathcal{M}^{2,d}(Q) \oplus \mathcal{A}^{1,d}(Q)$ along a subsequence, again indexed by n. Since $\vartheta^n \cdot S \to \vartheta \cdot S$ in $\mathcal{M}^{2,1}(Q) \oplus \mathcal{A}^{1,1}(Q)$, we obtain by using (4.1) with $B^Q = B$ that

$$E_Q \left[\int_0^\infty (\vartheta_s^n - \vartheta_s)^\top \hat{c}_s^Q (\vartheta_s^n - \vartheta_s) dB_s + \int_0^\infty |(\vartheta_s^n - \vartheta_s)^\top a_s^Q| dB_s \right] \longrightarrow 0$$

as $n \to \infty$, which implies that there exists a subsequence, again indexed by n, such that

$$(\vartheta^n - \vartheta)^\top \hat{c}^Q (\vartheta^n - \vartheta) \to 0 \quad \text{and} \quad |(\vartheta^n - \vartheta)^\top a^Q| \to 0 \quad Q \otimes B\text{-a.e.} \qquad (5.5)$$

Since $P_B \sim Q \otimes B$, Lemma 5.1 gives

$$[0]^S = \{ \varphi \in \mathcal{L}^0(\overline{\Omega}, \mathcal{P}; \mathbb{R}^d) \mid \hat{c}^Q \varphi = 0 \text{ and } (a^Q)^\top \varphi = 0 \quad Q \otimes B\text{-a.e.} \}.$$

Let $e^1, \ldots, e^d$ be predictable processes from Lemma 5.2 which satisfy properties (1)–(4) for $[0]^S$ and set

$$U = \left\{ \psi \in \mathcal{L}^0\left(\overline{\Omega}, \mathcal{P}; \mathbb{R}^d\right) \,\Big|\, \psi^\top \varphi = 0 \quad Q \otimes B\text{-a.e. for all } \varphi \in [0]^S \right\},$$

$$V = \left\{ \psi \in \mathcal{L}^0\left(\overline{\Omega}, \mathcal{P}; \mathbb{R}^d\right) \,\Big|\, \psi^\top \varphi = 0 \quad Q \otimes B\text{-a.e. for all } \varphi \in [0]^{M^Q} \right\}$$

so that loosely speaking, $U^\perp = [0]^S$ and $V^\perp = [0]^{M^Q}$. Then $[0]^{M^Q} \cap U$ and $[0]^{A^Q} \cap V$ satisfy the assumptions of Lemma 5.2 and thus there exist predictable processes $u^1, \ldots, u^d$ and $v^1, \ldots, v^d$ with the properties (1)–(4) for $[0]^{M^Q} \cap U$ and $[0]^{A^Q} \cap V$, respectively. By the definition of U and V we also obtain, using $[0]^S = [0]^{M^Q} \cap [0]^{A^Q}$, that

$$(e^j)^\top u^k = (e^j)^\top v^k = (u^j)^\top v^k = 0 \quad Q \otimes B\text{-a.e. for } j,k = 1, \ldots, d$$

and

$$[0]^{M^Q} = \left\{ \sum_{j=1}^d h^j e^j + \sum_{k=1}^d h^{d+k} u^k \,\Big|\, h^1, \ldots, h^{2d} \text{ real-valued predictable} \right\},$$

$$[0]^{A^Q} = \left\{ \sum_{j=1}^d h^j e^j + \sum_{k=1}^d h^{d+k} v^k \,\Big|\, h^1, \ldots, h^{2d} \text{ real-valued predictable} \right\}$$

up to $Q \otimes B$-a.e. equality. Therefore Π^{M^Q} and Π^{A^Q} can be written as

$$\Pi^{M^Q} = \mathbb{1}_{d \times d} - \sum_{j=1}^d e^j (e^j)^\top - \sum_{k=1}^d u^k (u^k)^\top,$$

$$\Pi^{A^Q} = \mathbb{1}_{d \times d} - \sum_{j=1}^d e^j (e^j)^\top - \sum_{k=1}^d v^k (v^k)^\top,$$

and we have

$$\left(\sum_{k=1}^d v^k (v^k)^\top \right) \Pi^{A^Q} \vartheta^n = \left(\sum_{k=1}^d v^k (v^k)^\top \right) \vartheta^n, \tag{5.6}$$

all up to $Q \otimes B$-a.e. equality. Since $\Pi^{M^Q}(\vartheta^n - \vartheta)$ and $\Pi^{A^Q}(\vartheta^n - \vartheta)$ are by Lemma 5.1 $Q \otimes B$-a.e. valued in $\mathrm{Im}(\hat{c}^Q)$ and $\mathrm{Im}\big((a^Q)^\top\big)$, respectively, (5.5) yields $\Pi^{M^Q}\vartheta^n \to \Pi^{M^Q}\vartheta$ and $\Pi^{A^Q}\vartheta^n \to \Pi^{A^Q}\vartheta$ $Q \otimes B$-a.e. From the latter convergence and (5.6), it follows that

$$\left(\sum_{k=1}^d v^k (v^k)^\top \right) \vartheta^n \to \left(\sum_{k=1}^d v^k (v^k)^\top \right) \vartheta \quad Q \otimes B\text{-a.e.},$$

and since $Q \otimes B \sim P_B$ and

$$\Pi^S = \Pi^{M^Q} + \sum_{k=1}^{d} v^k (v^k)^\top \qquad Q \otimes B\text{-a.e.,}$$

we obtain that $\Pi^S \vartheta^n \to \Pi^S \vartheta$ P_B-a.e. by combining everything. □

The only result whose proof is now still open is Lemma 4.4. This provides the general (and fairly abstract) version of the counterexample in Sect. 3, as well as the necessity part for the equivalence in Theorem 4.5.

Lemma 4.4. *Let $C : \overline{\Omega} \to 2^{\mathbb{R}^d} \setminus \{\emptyset\}$ be a predictable correspondence with closed values and such that the projection on the predictable range of S is not closed, i.e.*

$$\widetilde{F} = \{(\omega, t) \in \overline{\Omega} \mid \Pi^S(\omega, t) C(\omega, t) \text{ is not closed}\}$$

has outer P_B-measure > 0. Then there exist $\vartheta \in \mathcal{L}(S)$ and a sequence (ψ^n) of C-valued integrands such that $\psi^n \cdot S \to \vartheta \cdot S$ in $\mathcal{S}(P)$, but there is no C-valued integrand ψ such that $\psi \cdot S = \vartheta \cdot S$. Equivalently, there exists a sequence $([\psi^n])$ in $[C]^S$ such that $[\psi^n] \overset{L(S)}{\to} [\vartheta]$ but $[\vartheta] \notin [C]^S$, i.e. $[C]^S$ is not closed in $L(S)$.

Proof. The basic idea is to construct a $\vartheta \in \mathcal{L}(S)$ which is valued in $\overline{\Pi^S C} \setminus \Pi^S C$ on some $F \in \mathcal{P}$ with $F \subseteq \widetilde{F}$ and $P_B(F) > 0$, and in C on F^c. Then there exists no C-valued integrand $\psi \in [\vartheta]$ by the definition of Π^S since $\Pi^S \vartheta \notin \Pi^S C$ on F; but one can construct a sequence (ψ^n) of C-valued integrands with $\Pi^S \psi^n \to \Pi^S \psi$ pointwise since $\Pi^S \vartheta \in \overline{\Pi^S C}$. However, this is technically a bit more involved for several reasons: While C, $\Pi^S C$ and $\overline{\Pi^S C}$ are all predictable, $(\Pi^S C)^c$ need not be; so $\widetilde{F}$ need not be predictable, and one cannot use Proposition 2.3 to obtain a predictable selector. In addition, $\overline{\Pi^S C} \setminus \Pi^S C$ need not be closed-valued.

We first argue that $\widetilde{F}$ is $\mathcal{P}_{P_B}$-measurable. Let $\overline{B(0,n)}$ be a closed ball of radius n in $\mathbb{R}^d$. Then $\Pi^S(C \cap \overline{B(0,n)})$ is compact-valued as C is closed-valued. Since C is predictable and $\Pi^S(\omega, t)x$ with $x \in \mathbb{R}^d$ is a Carathéodory function, $\overline{\Pi^S C}$ is predictable by Proposition 2.7. By the same argument,

$$\Pi^S \big(C \cap \overline{B(0,n)}\big) = \overline{\Pi^S \big(C \cap \overline{B(0,n)}\big)}$$

is predictable since $C \cap \overline{B(0,n)}$ is, and then so is $\Pi^S C = \bigcup_{n=1}^{\infty} \Pi^S \big(C \cap \overline{B(0,n)}\big)$ as a countable union of predictable correspondences; see Proposition 2.8. Then Proposition 2.6 implies that $\overline{\Pi^S C}$ and $\Pi^S \big(C \cap \overline{B(0,n)}\big)$ have predictable graph; hence so does $\Pi^S C$. Therefore $\mathrm{gr}(\overline{\Pi^S C}) \cap \big(\mathrm{gr}(\Pi^S C)\big)^c$ is $\mathcal{P} \otimes \mathcal{B}(\mathbb{R}^d)$-measurable, and so by Proposition 2.5,

$$\begin{aligned}
\widetilde{F} &= \{(\omega, t) \in \overline{\Omega} \mid \Pi^S(\omega, t) C(\omega, t) \text{ is not closed}\} \\
&= \{(\omega, t) \in \overline{\Omega} \mid \overline{\Pi^S(\omega, t) C(\omega, t)} \setminus \Pi^S(\omega, t) C(\omega, t) \neq \emptyset\} \\
&= \pi_{\overline{\Omega}}\Big(\mathrm{gr}(\overline{\Pi^S C}) \cap \big(\mathrm{gr}(\Pi^S C)\big)^c\Big)
\end{aligned}$$

is indeed $\mathcal{P}_{P_B}$-measurable. Thus there exists a predictable set $F \subseteq \widetilde{F}$ with $P_B(F) > 0$.

Now fix some C-valued integrand $\widetilde{\psi} \in \mathcal{L}(S)$ and define the correspondence C' by

$$
C'(\omega, t) = \begin{cases} \overline{\Pi^S(\omega,t)C(\omega,t)} \setminus \Pi^S(\omega,t)C(\omega,t) & \text{for } (\omega,t) \in F, \\ \widetilde{\psi}(\omega,t) & \text{else.} \end{cases}
$$

Then C' has non-empty values and predictable graph and therefore admits a P_B-a.e. predictable selector ϑ by Proposition 2.4. By possibly subtracting a predictable P_B-nullset from F, we can without loss of generality assume that ϑ takes values in C'. Moreover, the predictable sets $F_n := F \cap \{|\vartheta| \leq n\}$ increase to F and so we can, by shrinking F to some F_n if necessary, assume that ϑ is uniformly bounded in (ω, t) on F. Let $\{\varphi^m \mid m \in \mathbb{N}\}$ be a Castaing representation of C as in Proposition 2.3. Then $\overline{\Pi^S C} = \overline{\{\Pi^S \varphi^m \mid m \in \mathbb{N}\}}$, and because $\vartheta \in \overline{\Pi^S C}$, we can find for each $n \in \mathbb{N}$ a predictable process ψ^n such that $\Pi^S(\omega,t)\psi^n(\omega,t) \in \vartheta(\omega,t) + \overline{B(0,\frac{1}{n})}$ on F and $\psi^n = \widetilde{\psi}$ on F^c. Note that on F, we have $\vartheta \in \overline{\Pi^S C} \subseteq \Pi^S \mathbb{R}^d$ and therefore $\Pi^S \vartheta = \vartheta$; so $\Pi^S \vartheta = \mathbb{1}_F \vartheta + \mathbb{1}_{F^c} \Pi^S \widetilde{\psi}$ and this shows that $\Pi^S \psi^n \to \Pi^S \vartheta$ uniformly in (ω, t) by construction. Since $\Pi^S \vartheta \in \mathcal{L}(S)$ because ϑ is bounded on F, we thus first get $\Pi^S \psi^n \in \mathcal{L}(S)$, hence $\psi^n \in \mathcal{L}(S)$, and then also that $\psi^n \cdot S \to \vartheta \cdot S$ in $\mathcal{S}(P)$ by dominated convergence. But now $\{\Pi^S \vartheta\} \cap \Pi^S C = \emptyset$ on F shows by Lemma 5.3 that there exists no C-valued integrand $\psi \in [\vartheta]$ and therefore $[\vartheta] \notin [C]^S$. This ends the proof. $\square$

6 Related Work

We have already explained how our results generalise most of the existing literature on optimisation problems under constraints. In this section, we discuss the relation to the work of Karatzas and Kardaras [16].

We start by introducing the terminology of [16]. For a given S with triplet (b, c, F), the linear subspace of *null investments* $\mathfrak{N}$ is given by the predictable correspondence

$$
\mathfrak{N}(\omega, t) := \big\{ z \in \mathbb{R}^d \ \big| \ z^\top c(\omega,t) = 0, \ z^\top b(\omega,t) = 0
$$
$$
\text{and } F(\omega,t)(\{x \mid z^\top x \neq 0\}) = 0 \big\}
$$

(see Definition 3.6 in [16]). Note that we use F instead of ν and that our B is slightly different than in [16]. But this does not affect the definition of $\mathfrak{N}$. As in Definition 3.7 in [16], a correspondence $C : \overline{\Omega} \to 2^{\mathbb{R}^d}$ is said to *impose predictable closed convex constraints* if:

(0) $\mathfrak{N}(\omega,t) \subseteq C(\omega,t)$ for all $(\omega,t) \in \overline{\Omega}$

(1) $C(\omega,t)$ is a closed and convex set for all $(\omega,t) \in \overline{\Omega}$

(2) C is predictable

To avoid confusion, we call constraints with (0)–(2) *KK-constraints* in the sequel.

In the comment following their Theorem 4.4 on p. 467 in [16], Karatzas and Kardaras (KK) remark that $C \cdot S$ is closed in $\mathcal{S}(P)$ if C describes KK-constraints. For comparison, our Theorem 4.5 starts with C which is predictable and has closed values, and shows that $C \cdot S$ is then closed in $\mathcal{S}(P)$ if and only if $\Pi^S C$ is closed P_B-a.e. So we do not need convexity of C, and our condition on C and S is not only sufficient, but also necessary.

Before explaining the connections in more detail, we make the simple but important observation that

$$\text{(0) plus (1) imply that } C + \mathfrak{N} = C \text{ (for all } (\omega,t) \in \overline{\Omega}). \tag{6.1}$$

Indeed, each $\mathfrak{N}(\omega,t)$ is a linear subspace, hence contains 0, and so $C \subseteq C + \mathfrak{N}$. Conversely, $\frac{1}{\varepsilon}z \in \mathfrak{N} \subseteq C$ for every $z \in \mathfrak{N}$ and $\varepsilon > 0$ due to (0); so for every $c \in C$, $(1 - \varepsilon)c + z \in C$ by convexity and hence $c + z = \lim_{\varepsilon \searrow 0}(1 - \varepsilon)c + z$ is in C by closedness, giving $C + \mathfrak{N} \subseteq C$.

As a matter of fact, KK say, but do not explicitly prove, that $C \cdot S$ is closed in $\mathcal{S}(P)$. However, the clear hint they give suggests the following reasoning. Let (ϑ^n) be a sequence in C such that $(\vartheta^n \cdot S) \to X$ in $\mathcal{S}(P)$. By the proof of Theorem V.4 in [20], there exist $\widetilde{\vartheta}^n \in [\vartheta^n]$ and $\vartheta \in \mathcal{L}(S)$ such that $\vartheta \cdot S = X$ and $\widetilde{\vartheta}^n \to \vartheta$ P_B-a.e. From the description of $\mathfrak{N}$ in Section 3.3 in [16], $\widetilde{\vartheta}^n \in [\vartheta^n]$ translates into $\widetilde{\vartheta}^n - \vartheta^n \in \mathfrak{N}$ P_B-a.e. or $\widetilde{\vartheta}^n \in \vartheta^n + \mathfrak{N}$ P_B-a.e. Because each ϑ^n has values in C, (6.1) thus shows that each $\widetilde{\vartheta}^n$ can be chosen to be C-valued, and by the closedness of C, the same is then true for the limit ϑ of $(\widetilde{\vartheta}^n)$. Hence we are done.

In order to relate the KK result to our work, we now observe that

$$\text{(0) plus (1) imply that } \Pi^S C \text{ is closed } P_B\text{-a.e.}$$

To see this, we start with the fact that the null investments $\mathfrak{N}$ and $[0]^S$ are linked by

$$[0]^S = \{\varphi \mid \varphi \text{ is } \mathbb{R}^d\text{-valued predictable with } \varphi \in \mathfrak{N} \ P_B\text{-a.e.}\}; \tag{6.2}$$

see Sect. 3.3 in [16]. Recalling that Π^S is the projection on the orthogonal complement of $[0]^S$, we see from (6.2) that the column vectors of Π^S are P_B-a.e. a generating system of $\mathfrak{N}^\perp$ so that the projection of $\vartheta \in \mathcal{L}(S)$ on the predictable range of S can be alternatively defined P_B-a.e. as a predictable selector of the closed-valued predictable correspondence $\{\vartheta + \mathfrak{N}\} \cap \mathfrak{N}^\perp$ or P_B-a.e. as the pointwise projection $\Pi^{\mathfrak{N}(\omega,t)}\vartheta(\omega,t)$ in $\mathbb{R}^d$ of $\vartheta(\omega,t)$ on $\mathfrak{N}(\omega,t)$, which is always a predictable process. This yields $\Pi^S C = \{C + \mathfrak{N}\} \cap \mathfrak{N}^\perp$ P_B-a.e.; but by (6.1), $C + \mathfrak{N} = C$ due to (0) and (1), and so $\Pi^S C$ is P_B-a.e. closed like C and $\mathfrak{N}^\perp$.

In the KK notation, we could reformulate our Theorem 4.5 as saying that for a predictable and closed-valued C, the space $C \cdot S$ is closed in $\mathcal{S}(P)$ if and only if $C + \mathfrak{N}$ is closed P_B-a.e. This is easily seen from the argument above showing that $\Pi^S C = \{C + \mathfrak{N}\} \cap \mathfrak{N}^\perp$ P_B-a.e. If C is also convex-valued, (0) is a simple and intuitive sufficient condition; it seems however more difficult to find an elegant formulation without convexity.

The difference between our constraints and the KK formulation in [16] is as follows. We fix a set C of constraints and demand that the strategies should lie in C pointwise, so that $\vartheta(\omega, t) \in C(\omega, t)$ for all (ω, t). KK in contrast only stipulate that $\vartheta(\omega, t) \in C(\omega, t) + \mathfrak{N}(\omega, t)$ or, equivalently, that $[\vartheta] \in [\mathcal{C}]$. At the level of wealth (which is as usual in mathematical finance modelled by the stochastic integral $\vartheta \cdot S$), this makes no difference since all $\mathfrak{N}$-valued processes have integral zero. But for practical checking and risk management, it is much simpler if one can just look at the strategy ϑ and tick off pointwise whether or not it lies in C. If S has complicated redundancy properties, it may be quite difficult to see whether one can bring ϑ into C by adding something from $\mathfrak{N}$. Of course, when discussing the closedness of the space of integrals $\vartheta \cdot S$, we face the same level of difficulty when we have to check whether $\Pi^S C$ is closed P_B-a.e. But for actually working with given strategies, we believe that our formulation of constraints is more natural and simpler to handle.

Acknowledgements We thank an anonymous referee for careful reading and helpful suggestions. Financial support by the National Centre of Competence in Research "Financial Valuation and Risk Management" (NCCR FINRISK), Project D1 (Mathematical Methods in Financial Risk Management) is gratefully acknowledged. The NCCR FINRISK is a research instrument of the Swiss National Science Foundation.

References

1. Aliprantis, C.D., Border, K.C.: Infinite Dimensional Analysis, 3rd edn. Springer, Berlin (2006)
2. Artstein, Z.: Set-valued measures. Trans. Am. Math. Soc. **165**, 103–125 (1972)
3. Cuoco, D.: Optimal consumption and equilibrium prices with portfolio constraints and stochastic income. J. Econ. Theory **72**, 33–73 (1997)
4. Cvitanić, J., Karatzas, I.: Convex duality in constrained portfolio optimization. Ann. Appl. Probab. **2**, 767–818 (1992)
5. Czichowsky, C., Schweizer, M.: On the Markowitz problem with cone constraints. Working paper, ETH Zurich (2010, in preparation), http://www.math.ethz.ch/~czichowc
6. Czichowsky, C., Westray, N., Zheng, H.: Convergence in the semimartingale topology and constrained portfolios. In: Séminaire de Probabilités XLIII (2010, this volume)
7. Debreu, G., Schmeidler, D.: The Radon–Nikodým derivative of a correspondence. In: Proceedings of the Sixth Berkeley Symposium on Mathematical Statistics and Probability, vol. II: Probability Theory, pp. 41–56. University of California Press, Berkeley, CA (1972)
8. Delbaen, F.: The structure of m-stable sets and in particular of the set of risk neutral measures. In: In memoriam Paul-André Meyer: Séminaire de Probabilités XXXIX. Lecture Notes in Mathematics, vol. 1874, pp. 215–258. Springer, Berlin (2006)
9. Delbaen, F., Schachermayer, W.: The Mathematics of Arbitrage. Springer Finance. Springer, Berlin (2006)
10. Emery, M.: Une topologie sur l'espace des semimartingales. In: Séminaire de Probabilités XIII. Lecture Notes in Mathematics, vol. 721, pp. 260–280. Springer, Berlin (1979)

11. Föllmer, H., Kramkov, D.: Optional decompositions under constraints. Probab. Theory Relat. Field. **109**, 1–25 (1997)
12. Föllmer, H., Schied, A.: Stochastic Finance. An Introduction in Discrete Time. de Gruyter Studies in Mathematics, vol. 27. Walter de Gruyter & Co., Berlin, second revised and extended edition (2004)
13. Gale, D., Klee, V.: Continuous convex sets. Math. Scand. **7**, 379–391 (1959)
14. Jacod, J., Shiryaev, A.N.: Limit Theorems for Stochastic Processes. Grundlehren der Mathematischen Wissenschaften, vol. 288, 2nd edn. Springer, Berlin (2003)
15. Jin, H., Zhou, X.Y.: Continuous-time Markowitz's problems in an incomplete market, with no-shorting portfolios. In: Benth, F.E., et al. (eds.) Stochastic Analysis and Applications. Proceedings of the Second Abel Symposium, Oslo, 2005, pp. 435–459. Springer, Berlin (2007)
16. Karatzas, I., Kardaras, C.: The numéraire portfolio in semimartingale financial models. Finance Stochast. **11**, 447–493 (2007)
17. Karatzas, I., Žitković, G.: Optimal consumption from investment and random endowment in incomplete semimartingale markets. Ann. Probab. **31**, 1821–1858 (2003)
18. Klee, V.: Some characterizations of convex polyhedra. Acta Math. **102**, 79–107 (1959)
19. Labbé, C., Heunis, A.J.: Convex duality in constrained mean-variance portfolio optimization. Adv. Appl. Probab. **39**, 77–104 (2007)
20. Mémin, J.: Espaces de semi martingales et changement de probabilité. Z. Wahrsch. verw. Gebiete **52**, 9–39 (1980)
21. Mnif, M., Pham, H.: Stochastic optimization under constraints. Stochastic Process. Appl. **93**, 149–180 (2001)
22. Pham, H.: Minimizing shortfall risk and applications to finance and insurance problems. Ann. Appl. Probab. **12**, 143–172 (2002)
23. Rockafellar, R.T.: Integral functionals, normal integrands and measurable selections. In: Nonlinear Operators and the Calculus of Variations. Lecture Notes in Mathematics, vol. 543, pp. 157–207. Springer, Berlin (1976)
24. Schachermayer, W.: A Hilbert space proof of the fundamental theorem of asset pricing in finite discrete time. Insur. Math. Econ. **11**, 249–257 (1992)
25. Westray, N., Zheng, H.: Constrained nonsmooth utility maximization without quadratic inf convolution. Stochastic Process. Appl. **119**, 1561–1579 (2009)

On Martingales with Given Marginals
and the Scaling Property

David Baker and Marc Yor

Abstract In this short note, we draw the reader's attention to a number of constructions of martingales with given marginals.

Key words: Martingales · Brownian scaling · Marginal distributions

(a) In recent years, a number of papers have been devoted to the study of the following *martingale-marginal problems* (MMP): let μ denote a probability measure on $\mathbb{R}$ which satisfies:

$$\int_{\infty}^{\infty} |x| d\mu(x) < \infty, \text{ and } \int_{-\infty}^{\infty} x \, d\mu(x) = 0. \tag{1}$$

The MMP is to find a martingale $(M_t, t \geq 0)$ such that:

$$M_1 \sim \mu \tag{2}$$

and which enjoys the Brownian scaling property:

$$\forall c \in \mathbb{R}, (M_{c^2 t}, t \geq 0) \overset{(law)}{=} (c M_t, t \geq 0). \tag{3}$$

(b) Of course, this MMP is a particular case of the more general problem of finding a martingale (M_t) with given marginals $(\mu_t)_{t \geq 0}$. This problem has been solved, at least in theoretical terms, by Kellerer [5] who proved (following previous important studies by Strassen and Doob) that a necessary and sufficient condition on the family $(\mu_t)_{t \geq 0}$ for the existence of M is that $(\mu_t)_{t \geq 0}$ be increasing in the convex order.

D. Baker (✉)
Laboratoire de Probabilités et Modèles Aléatoires, Université Pierre et Marie Curie, Université Paris 06 and CNRS, UMR 7599, 4, Place Jussieu, 75252 Paris Cedex 05, France
e-mail: david.baker@etu.upmc.fr

M. Yor
Institut Universitaire de France and Laboratoire de Probabilités et Modèles Aléatoires, Université Pierre et Marie Curie, Université Paris 06 and CNRS, UMR 7599, 4, Place Jussieu, 75252 Paris Cedex 05, France

C. Donati-Martin et al. (eds.), *Séminaire de Probabilités XLIII*, Lecture Notes in Mathematics 437
2006, DOI 10.1007/978-3-642-15217-7_19, © Springer-Verlag Berlin Heidelberg 2011

Recently, Lowther [6] has offered a uniqueness result for the Kellerer framework. He achieves uniqueness by adding the condition that the marginals (μ_t) be weakly continuous and by restricting the martingales used for fitting to ones which are almost continuous diffusions.

The monograph *Peacocks and associated martingales* [4] is entirely devoted to explicit constructions of (M_t) for given (μ_t).

(c) However, let us come back to the particular MMP (2)–(3), which although simpler than the general problem, is still of interest, and has the merit of having only one parameter μ, instead of the family (μ_t).

For a given μ, we denote by $\mathcal{M}_{\mu,s}$, the set consisting of (the laws of) martingales which satisfy (2)–(3).

In part B of [3], when μ admits a second moment, a Skorokhod embedding construction provides a purely discontinuous element of $\mathcal{M}_{\mu,s}$. A different Skorokhod embedding based on the Azéma–Yor algorithm is provided in Madan–Yor [7] and studied in detail in part A of [3], for measures μ such that:

$$D_\mu(a) \equiv \frac{a}{\psi_\mu(a)}, a \geq 0, \text{ is increasing}$$

where ψ_μ denotes the Hardy Littlewood function associated to μ:

$$\psi_\mu(a) = \frac{1}{\mu([a,\infty))} \int_{[a,\infty[} x d\mu(x) \equiv \mathbb{E}[M_1 | M_1 \geq a].$$

(d) Given that the constructions made in [3] only generate purely discontinuous martingales, it is natural to study $\mathcal{M}_{\mu,s}^{(c)}$ the set of continuous martingales, with the Brownian scaling property, such that: $M_1 \sim \mu$. A natural question is: for which probability laws μ which satisfy (1), is it true that $\mathcal{M}_{\mu,s}^c$ is non-empty?

When this is the case, we say that μ is continuous(ly) admissible and we write: μ (or X) is CA (or: belongs to CA).

Here are some remarks, and our best knowledge so far about the probabilities μ which belong to CA:

(i) if $\mu \sim X$ is CA and Y is integrable and is independent from X, then XY is CA (Proof: If (M_t) belongs to $\mathcal{M}_{X,s}^c$, then $(YM_t, t \geq 0)$, with Y independent from (M_t), also belongs to $\mathcal{M}_{XY,s}^c$).
Thus, if μ is CA, then all its "multiples" that is, the laws v of variables of the form: YX, with Y independent from X, and $X \sim \mu$, are CA.

(ii) Albin [1] has shown that if g is the standard gaussian, then $\mathcal{M}_{g,s}^{(c)}$ contains other elements than Brownian motion. Baker, Donati-Martin, Yor [2] develop Albin's method further.

(iii) Many elements of $\mathcal{M}_{\mu,s}^{(c)}$ are constructed in Madan–Yor [7] and Chapter 6 of the Monograph. In particular the "classical" symmetric laws i.e.: symmetrized beta (on $[-1, 1]$), symmetrized gamma, symmetrized powers of gamma, are CA!

References

1. Albin, J.M.P.: A continuous non-Brownian motion martingale with Brownian motion marginal distributions. Stat. Probab. Lett. **78**(6), 682–686 (2008)
2. Baker, D., Donati-Martin, C., Yor, M.: Extending Albin's construction of continuous martingales with Brownian marginals and scaling. Séminaire de Probabilités XLIII, in this volume, Lecture Notes in Math. Vol. 2006, Springer Verlag (2011)
3. Hirsch, F., Profeta, C., Roynette, B., Yor, M.: Constructing self-similar martingales via two Skorohod embeddings. Séminaire de Probabilités XLIII, in this volume, Lecture Notes in Math, Vol., 2006, Springer Verlag, (2011)
4. Hirsch, F., Profeta, C., Roynette, B., Yor, M.: Peacocks and associated martingales. Monograph in preparation (2010)
5. Kellerer, H.G.: Markov-Komposition und eine Anwendung auf Martingale. Math. Ann. **198**(3), 99–122 (1972)
6. Lowther, G.: Fitting martingales to given marginals. Preprint arXiv:0808.2319 (2008)
7. Madan, D.B., Yor, M.: Making Markov martingales meet marginals: with explicit constructions. Bernoulli **8**(4), 509–536 (2002)

A Sequence of Albin Type Continuous Martingales with Brownian Marginals and Scaling

David Baker, Catherine Donati-Martin, and Marc Yor

Dedicated to Lester Dubins (1921–2010) to whom the third author owes a lot.

Abstract Closely inspired by Albin's method which relies ultimately on the duplication formula for the Gamma function, we exploit Gauss' multiplication formula to construct a sequence of continuous martingales with Brownian marginals and scaling.

Key words: Martingales · Brownian marginals

1 Motivation and Main Results

(1.1) Knowing the law of a "real world" random phenomena, i.e. random process, $(X_t, t \geq 0)$ is often extremely difficult and in most instances, one avails only of the knowledge of the 1-dimensional marginals of $(X_t, t \geq 0)$. However, there may be many different processes with the same given 1-dimensional marginals.

In the present paper, we make explicit a sequence of continuous martingales $(M_m(t), t \geq 0)$ indexed by $m \in \mathbb{N}$ such that for each m:

(i) $(M_m(t), t \geq 0)$ enjoys the Brownian scaling property: for any $c > 0$,

$$(M_m(c^2 t), t \geq 0) \overset{(law)}{=} (c M_m(t), t \geq 0)$$

(ii) $M_m(1)$ is standard Gaussian.

D. Baker (✉), C. Donati-Martin
Laboratoire de Probabilités et Modèles Aléatoires, Université Pierre et Marie Curie
Université Paris 06 and CNRS, UMR 7599, 4, Place Jussieu, 75252 Paris Cedex 05, France
e-mail: david.baker@etu.upmc.fr; catherine.donati@upmc.fr

M. Yor
Laboratoire de Probabilités et Modèles Aléatoires, Université Pierre et Marie Curie
Université Paris 06 and CNRS, UMR 7599, 4, Place Jussieu, 75252 Paris Cedex 05, France

Institut Universitaire de France

C. Donati-Martin et al. (eds.), *Séminaire de Probabilités XLIII*, Lecture Notes in Mathematics 441
2006, DOI 10.1007/978-3-642-15217-7_20, © Springer-Verlag Berlin Heidelberg 2011

Note that, combining (i) and (ii), we get, for any $t > 0$

$$M_m(t) \overset{(law)}{=} B_t,$$

where $(B_t, t \geq 0)$ is a Brownian motion, i.e. M_m admits the same 1-dimensional marginals as Brownian motion.

(1.2) Our main result is the following extension of Albin's construction [1] from $m = 1$ to any integer m.

Theorem 1. *Let $m \in \mathbb{N}$. Then, there exists a continuous martingale $(M_m(t), t \geq 0)$ which enjoys (i) and (ii) and is defined as follows:*

$$M_m(t) = X_t^{(1)} \ldots X_t^{(m+1)} Z_m \tag{1}$$

where $(X_t^{(i)}, t \geq 0)$, for $i = 1, \ldots, m+1$, are independent copies of the solution of the SDE

$$dX_t = \frac{1}{m+1} \frac{dB_t}{X_t^m}; \quad X_0 = 0 \tag{2}$$

and, furthermore, Z_m is independent from $(X^{(1)}, \ldots, X^{(m+1)})$ and

$$Z_m \overset{(law)}{=} (m+1)^{1/2} \left(\prod_{j=0}^{m-1} \beta \left(\frac{1+2j}{2(m+1)}, \frac{m-j}{m+1} \right) \right)^{\frac{1}{2(m+1)}} \tag{3}$$

where $\beta(a, b)$ denotes a beta variable with parameter (a, b) with density

$$\frac{\Gamma(a+b)}{\Gamma(a)\Gamma(b)} x^{a-1} (1-x)^{b-1} 1_{[0,1]}(x)$$

and the beta variables on the right-hand side of (3) are independent.

Remark. For $m = 1$, $Z_1 = \sqrt{2} \left(\beta(\frac{1}{4}, \frac{1}{2}) \right)^{1/4}$ and we recover the distribution of $Y := Z_1$ given by (2) in [1].

(1.3) For the convenience of the reader, we also recall that, if one drops the continuity assumption when searching for martingales $(M(t); t \geq 0)$ satisfying (i) and (ii), then, the Madan-Yor construction [5] based on the "Azéma-Yor under scaling" method provides such a martingale.

Precisely, starting from a Brownian motion $(B_u, u \geq 0)$ and denoting $S_u = \sup_{s \leq u} B_s$, introduce the family of stopping times

$$\tau_t = \inf\{u, S_u \geq \psi_t(B_u)\}$$

where ψ_t denotes the Hardy-Littlewood function associated with the centered Gaussian distribution μ_t with variance t, i.e.

$$\psi_t(x) = \frac{1}{\mu_t([x,\infty[)} \int_x^\infty y \exp\left(-\frac{y^2}{2t}\right) \frac{dy}{\sqrt{2\pi t}}$$

$$= \sqrt{t} \exp\left(-\frac{x^2}{2t}\right) / \mathcal{N}(x/\sqrt{t})$$

where $\mathcal{N}(a) = \int_a^\infty \exp(-\frac{y^2}{2})dy$. Then, $M_t = B_{\tau_t}$ is a martingale with Brownian marginals.

Another solution has been given by Hamza and Klebaner [4].

(1.4) In Sect. 3, we prove that Theorem 1 is actually the best we can do in our generalisation of Albin's construction: we cannot generalize (1) by allowing the $X^{(i)}$'s to be solution of (2) associated to different m_i's.

Finally, we study the asymptotic behavior of $X_t^{(1)} \ldots X_t^{(m+1)}$ as $m \longrightarrow \infty$.

2 Proof of Theorem 1

Step 1: For $m \in \mathbb{R}$ and $c \in \mathbb{R}$, we consider the stochastic equation:

$$dX_t = c\frac{dB_t}{X_t^m}, \quad X_0 = 0.$$

This equation has a unique weak solution which can be defined as a time-changed Brownian motion

$$(X_t) \overset{(law)}{=} W(\alpha^{(-1)}(t))$$

where W is a Brownian motion starting from 0 and $\alpha^{(-1)}$ is the (continuous) inverse of the increasing process

$$\alpha(t) = \frac{1}{c^2} \int_0^t W_u^{2m} du.$$

We look for $k \in \mathbb{N}$ and c such that $(X_t^{2k}, t \geq 0)$ is a squared Bessel process of some dimension d. It turns out, by application of Itô's formula, that we need to take $k = m+1$ and $c = \frac{1}{m+1}$. Thus, we find that $(X_t^{2(m+1)}, t \geq 0)$ is a squared Bessel process with dimension $d = k(2k-1)c^2 = \frac{2m+1}{m+1}$.

Note that the law of a BESQ(d) process at time 1 is well known to be that of $2\gamma_{d/2}$, where γ_a denotes a gamma variable with parameter a. Thus, we have:

$$|X_1| \overset{(law)}{=} \left(2\gamma_{\frac{2m+1}{2(m+1)}}\right)^{\frac{1}{2(m+1)}} \tag{4}$$

Step 2: We now discuss the scaling property of the solution of (2). From the scaling property of Brownian motion, it is easily shown that, for any $\lambda > 0$, we get:

$$(X_{\lambda t}, t \geq 0) \overset{(law)}{=} (\lambda^\alpha X_t, t \geq 0)$$

with $\alpha = \frac{1}{2(m+1)}$, that is, the process $(X_t, t \geq 0)$ enjoys the scaling property of order $\frac{1}{2(m+1)}$.

Step 3: Consequently, if we multiply $m + 1$ independent copies of the process $(X_t, t \geq 0)$ solution of (2), we get a process

$$Y_t = X_t^{(1)} \dots X_t^{(m+1)}$$

which is a martingale and has the scaling property of order $\frac{1}{2}$.

Step 4: Finally, it suffices to find a random variable Z_m independent of the processes $X_t^{(1)}, \dots, X_t^{(m+1)}$ and which satisfies:

$$N \overset{(law)}{=} X_1^{(1)} \dots X_1^{(m+1)} Z_m \tag{5}$$

where N denotes a standard Gaussian variable. Note that the distribution of any of the $X_1^{(i)}$'s is symmetric. We shall take $Z_m \geq 0$; thus, the distribution of Z_m shall be determined by its Mellin transform $\mathcal{M}(s) = \mathbb{E}(Z_m^s)$. From (5), $\mathcal{M}(s)$ satisfies:

$$\mathbb{E}[(2\gamma_{1/2})^{s/2}] = \left(\mathbb{E}\left[(2\gamma_{d/2})^{s/2(m+1)} \right] \right)^{m+1} \mathcal{M}(s)$$

with $d = \frac{2m+1}{m+1}$, that is:

$$2^{s/2} \frac{\Gamma(\frac{1+s}{2})}{\Gamma(\frac{1}{2})} = 2^{s/2} \left(\frac{\Gamma\left(\frac{d}{2} + \frac{s}{2(m+1)} \right)}{\Gamma(\frac{d}{2})} \right)^{m+1} \mathcal{M}(s)$$

that is precisely:

$$\frac{\Gamma(\frac{1+s}{2})}{\Gamma(\frac{1}{2})} = \left(\frac{\Gamma\left(\frac{2m+1+s}{2(m+1)} \right)}{\Gamma(\frac{2m+1}{2(m+1)})} \right)^{m+1} \mathcal{M}(s). \tag{6}$$

Now, we recall Gauss multiplication formula ([2], see also [3])

$$\Gamma(kz) = \frac{k^{kz-1/2}}{(2\pi)^{\frac{k-1}{2}}} \prod_{j=0}^{k-1} \Gamma\left(z + \frac{j}{k} \right) \tag{7}$$

which we apply with $k = m + 1$ and $z = \frac{1+s}{2(m+1)}$. We then obtain, from (7)

$$\frac{\Gamma(\frac{1+s}{2})}{\sqrt{\pi}} = \frac{(m+1)^{s/2}}{(2\pi)^{m/2}} \frac{1}{\sqrt{\pi}} \prod_{j=0}^{m} \Gamma\left(\frac{1+s+2j}{2(m+1)}\right) \tag{8}$$

$$= (m+1)^{s/2} \prod_{j=0}^{m} \left(\frac{\Gamma\left(\frac{1+s+2j}{2(m+1)}\right)}{\Gamma\left(\frac{1+2j}{2(m+1)}\right)}\right) \tag{9}$$

since the two sides of (8) are equal to 1 for $s = 0$. We now plug (9) into (6) and obtain

$$(m+1)^{s/2} \prod_{j=0}^{m} \left(\frac{\Gamma\left(\frac{1+s+2j}{2(m+1)}\right)}{\Gamma\left(\frac{1+2j}{2(m+1)}\right)}\right) = \left(\frac{\Gamma\left(\frac{2m+1+s}{2(m+1)}\right)}{\Gamma\left(\frac{2m+1}{2(m+1)}\right)}\right)^{m+1} \mathcal{M}(s) \tag{10}$$

We note that for $j = m$, the same term appears on both sides of (10), thus (10) may be written as:

$$(m+1)^{s/2} \prod_{j=0}^{m-1} \left(\frac{\Gamma\left(\frac{1+s+2j}{2(m+1)}\right)}{\Gamma\left(\frac{1+2j}{2(m+1)}\right)}\right) = \left(\frac{\Gamma\left(\frac{2m+1+s}{2(m+1)}\right)}{\Gamma\left(\frac{2m+1}{2(m+1)}\right)}\right)^{m} \mathcal{M}(s) \tag{11}$$

In terms of independent gamma variables, the left-hand side of (11) equals:

$$(m+1)^{s/2} \mathbb{E}\left[\left(\prod_{j=0}^{m-1} \gamma_{\frac{1+2j}{2(m+1)}}^{(j)}\right)^{\frac{s}{2(m+1)}}\right] \tag{12}$$

whereas the right-hand side of (11) equals:

$$\mathbb{E}\left[\left(\prod_{j=0}^{m-1} \gamma_{\frac{1+2m}{2(m+1)}}^{(j)}\right)^{\frac{s}{2(m+1)}}\right] \mathcal{M}(s) \tag{13}$$

where the $\gamma_{a_j}^{(j)}$ denote independent gamma variables with respective parameters a_j. Now, from the beta-gamma algebra, we deduce, for any $j \leq m - 1$:

$$\gamma_{\frac{1+2j}{2(m+1)}}^{(j)} \stackrel{(law)}{=} \gamma_{\frac{1+2m}{2(m+1)}}^{(j)} \beta\left(\frac{1+2j}{2(m+1)}, \frac{m-j}{m+1}\right).$$

Thus, we obtain, again by comparing (12) and (13):

$$\mathcal{M}(s) = (m+1)^{s/2} \, \mathbb{E}\left[\left(\prod_{j=0}^{m-1} \beta\left(\frac{1+2j}{2(m+1)}, \frac{m-j}{m+1}\right)\right)^{\frac{s}{2(m+1)}}\right]$$

which entails:

$$\mathbb{E}[Z_m^s] = (m+1)^{s/2} \, \mathbb{E}\left[\left(\prod_{j=0}^{m-1} \beta\left(\frac{1+2j}{2(m+1)}, \frac{m-j}{m+1}\right)\right)^{\frac{s}{2(m+1)}}\right]$$

that is, equivalently,

$$Z_m \stackrel{(law)}{=} (m+1)^{1/2} \left(\prod_{j=0}^{m-1} \beta\left(\frac{1+2j}{2(m+1)}, \frac{m-j}{m+1}\right)\right)^{\frac{1}{2(m+1)}}$$

3 Some Remarks About Theorem 1

3.1 A Further Extension

We tried to extend Theorem 1 by taking a product of independent martingales $X^{(i)}$, solution of (2) with different m_i's. Here are the details of our attempt. We are looking for the existence of a variable Z such that the martingale

$$M(t) = \left(\prod_{j=0}^{p-1} X_t^{(m_j)}\right) Z$$

satisfies the properties i) and ii). Here $p, (m_j)_{0 \le j \le p-1}$ are integers and $X^{(m_j)}$ is the solution of the EDS (2) associated to m_j, the martingales being independent for j varying. In order that M enjoys the Brownian scaling property, we need the following relation

$$\sum_{j=0}^{p-1} \frac{1}{m_j + 1} = 1. \tag{14}$$

Following the previous computations, see (6), the Mellin transform $\mathcal{M}(s)$ of Z should satisfy

$$\frac{\Gamma\left(\frac{1+s}{2}\right)}{\Gamma\left(\frac{1}{2}\right)} = \left(\prod_{j=0}^{p-1} \frac{\Gamma\left(\frac{2m_j+1+s}{2(m_j+1)}\right)}{\Gamma\left(\frac{2m_j+1}{2(m_j+1)}\right)}\right) \mathcal{M}(s). \tag{15}$$

We recall (see (9)) the Gauss multiplication formula

$$\frac{\Gamma\left(\frac{1+s}{2}\right)}{\sqrt{\pi}} = p^{s/2} \prod_{j=0}^{p-1} \left(\frac{\Gamma\left(\frac{1+s+2j}{2p}\right)}{\Gamma\left(\frac{1+2j}{2p}\right)} \right) \tag{16}$$

To find $\mathcal{M}(s)$ from (15), (16), we give some probabilistic interpretation:

$$\frac{\Gamma\left(\frac{1+s+2j}{2p}\right)}{\Gamma\left(\frac{1+2j}{2p}\right)} = \mathbb{E}\left[\gamma_{(1+2j)/2p}^{s/2p}\right]$$

whereas

$$\frac{\Gamma\left(\frac{2m_j+1+s}{2(m_j+1)}\right)}{\Gamma\left(\frac{2m_j+1}{2(m_j+1)}\right)} = \mathbb{E}\left[\gamma_{(1+2m_j)/2(m_j+1)}^{s/2(m_j+1)}\right].$$

Thus, we would like to factorize

$$\gamma_{(1+2j)/2p}^{1/2p} \stackrel{(law)}{=} \gamma_{(1+2m_j)/2(m_j+1)}^{1/2(m_j+1)} z_{m_j,p}^{(j)} \tag{17}$$

for some variable $z_{m_j,p}^{(j)}$ to conclude that

$$Z = p^{1/2} \prod_{j=0}^{p-1} z_{m_j,p}^{(j)}.$$

It remains to find under which condition the identity (17) may be fulfilled. We write

$$\gamma_{(1+2j)/2p} \stackrel{(law)}{=} \gamma_{(1+2m_j)/2(m_j+1)}^{p/(m_j+1)} (z_{m_j,p}^{(j)})^{2p}. \tag{18}$$

Now, if $\frac{1+2j}{2p} < \frac{1+2m_j}{2(m_j+1)}$, we may apply the beta-gamma algebra to obtain

$$\gamma_{(1+2j)/2p} \stackrel{(law)}{=} \gamma_{(1+2m_j)/2(m_j+1)} \beta\left(\frac{1+2j}{2p}, \frac{1+2m_j}{2(m_j+1)} - \frac{1+2j}{2p}\right)$$

but in (18), we need to have on the right-hand side $\gamma_{(1+2m_j)/2(m_j+1)}^{p/(m_j+1)}$ instead of $\gamma_{(1+2m_j)/2(m_j+1)}$.

However, it is known that

$$\gamma_a \stackrel{(law)}{=} \gamma_a^c \gamma_{a,c}$$

for some variable $\gamma_{a,c}$ independent of γ_a for any $c \in (0, 1]$. This follows from the self-decomposable character of $\ln(\gamma_a)$. Thus, we seem to need $\frac{p}{m_j+1} \leq 1$. But, this condition is not compatible with (14) unless $m_j = m = p - 1$.

3.2 Asymptotic Study

We study the behavior of the product $X_1^{(1)} \ldots X_1^{(m+1)}$, resp. Z_m, appearing in the right-hand side of the equality in law (5), when $m \longrightarrow \infty$. Recall from (4) that

$$|X_1| \overset{(law)}{=} \left(2\gamma_{\frac{2m+1}{2(m+1)}} \right)^{\frac{1}{2(m+1)}} .$$

We are thus led to consider the product

$$\Theta_{a,b,c}^{(p)} = \left(\prod_{i=1}^{p} \gamma_{a-b/p}^{(i)} \right)^{c/p}$$

where in our set up of Theorem 1, $p = m + 1, a = 1, b = c = 1/2$.

$$
\begin{aligned}
\mathbb{E}\left[\left(\Theta_{a,b,c}^{(p)} \right)^s \right] &= \prod_{i=1}^{p} \mathbb{E}\left[\left(\gamma_{a-b/p}^{(i)} \right)^{cs/p} \right] \\
&= \left(\frac{\Gamma\left(a - \frac{b}{p} + \frac{cs}{p} \right)}{\Gamma\left(a - \frac{b}{p} \right)} \right)^p \\
&= \exp\left[p \left(\ln\left(\Gamma\left(a + \frac{cs-b}{p} \right) \right) - \ln\left(\Gamma\left(a - \frac{b}{p} \right) \right) \right) \right] \\
&\xrightarrow[p \to \infty]{} \exp\left(\frac{\Gamma'(a)}{\Gamma(a)} cs \right) .
\end{aligned}
$$

Thus, it follows that

$$\Theta_{a,b,c}^{(p)} \xrightarrow[p \to \infty]{\mathbb{P}} \exp\left(\frac{\Gamma'(a)}{\Gamma(a)} c \right) ,$$

implying that

$$|X_1^{(1)} \ldots X_1^{(m+1)}| \xrightarrow[m \to \infty]{\mathbb{P}} \exp(-\gamma/2) \tag{19}$$

and

$$\exp(-\gamma/2) Z_m \xrightarrow[m \to \infty]{(law)} |N|. \tag{20}$$

where $\gamma = -\Gamma'(1)$ is the Euler constant.

We now look for a central limit theorem for $\Theta_{a,b,c}^{(p)}$. We consider the limiting distribution of

$$\sqrt{p} \left\{ \frac{c}{p} \sum_{i=1}^{p} \ln\left(\gamma_{a-b/p}^{(i)} \right) - c \frac{\Gamma'(a)}{\Gamma(a)} \right\} .$$

$$\mathbb{E}\left(\exp\left[cs\sqrt{p}\left\{\frac{1}{p}\sum_{i=1}^{p}\ln\left(\gamma_{a-b/p}^{(i)}\right)-\frac{\Gamma'(a)}{\Gamma(a)}\right\}\right]\right)$$

$$=\mathbb{E}\left[\prod_{i=1}^{p}\left(\gamma_{a-b/p}^{(i)}\right)^{cs/\sqrt{p}}\right]\exp\left(-cs\sqrt{p}\frac{\Gamma'(a)}{\Gamma(a)}\right)$$

$$=\mathbb{E}\left[\left(\gamma_{a-b/p}^{(i)}\right)^{cs/\sqrt{p}}\right]^{p}\exp\left(-cs\sqrt{p}\frac{\Gamma'(a)}{\Gamma(a)}\right)$$

$$=\left(\frac{\Gamma\left(a-\frac{b}{p}+\frac{cs}{\sqrt{p}}\right)}{\Gamma\left(a-\frac{b}{p}\right)}\right)^{p}\exp\left(-cs\sqrt{p}\frac{\Gamma'(a)}{\Gamma(a)}\right)$$

$$=\exp\left[p\left(\ln\left(\Gamma\left(a-\frac{b}{p}+\frac{cs}{\sqrt{p}}\right)\right)-\ln\left(\Gamma\left(a-\frac{b}{p}\right)\right)\right)-cs\sqrt{p}\frac{\Gamma'(a)}{\Gamma(a)}\right]$$

$$=\exp\left(\frac{c^2s^2}{2}(\ln(\Gamma))''(a)+O(m^{-1/2})\right)$$

We thus obtain that

$$\sqrt{p}\left\{\frac{c}{m}\sum_{i=1}^{m}\ln(\gamma_{a-b/m}^{(i)})-c\frac{\Gamma'(a)}{\Gamma(a)}\right\}\overset{(law)}{\longrightarrow}N(0,\sigma^2)\tag{21}$$

where $N(0,\sigma^2)$ denotes a centered Gaussian variable with variance:

$$\sigma^2=c^2(\ln(\Gamma))''(a)=c^2\left[\frac{\Gamma''(a)}{\Gamma(a)}-\left(\frac{\Gamma'(a)}{\Gamma(a)}\right)^2\right].$$

or, equivalently

$$\left(\Theta_{a,b,c}^{(p)}\exp\left(\frac{\Gamma'(a)}{\Gamma(a)}c\right)\right)^{\sqrt{p}}\overset{(law)}{\underset{p\to\infty}{\longrightarrow}}\exp(N(0,c^2(\ln(\Gamma))''(a))).\tag{22}$$

References

1. Albin, J.M.P.: A continuous non-Brownian motion martingale with Brownian motion marginal distributions. Stat. Probab. Lett. **78**(6), 682–686 (2008)
2. Andrews, G., Askey, R., Roy, R.: Special functions. Encyclopedia of Mathematics and its Applications, vol. 71. Cambridge University Press, Cambridge (1999)
3. Chaumont, L., Yor, M.: Exercises in probability. A guided tour from measure theory to random processes, via conditioning. Cambridge Series in Statistical and Probabilistic Mathematics, vol. 13. Cambridge University Press, Cambridge (2003)
4. Hamza, K., Klebaner, F.: A family of non-Gaussian martingales with Gaussian marginals. J. Appl. Math. Stoch. Anal. ID 92723, 19 pp. (2007)
5. Madan, D., Yor, M.: Making Markov martingales meet marginals: with explicit constructions. Bernoulli **8**(4), 509–536 (2002)

Constructing Self-Similar Martingales via Two Skorokhod Embeddings

Francis Hirsch, Christophe Profeta, Bernard Roynette, and Marc Yor

Abstract With the help of two Skorokhod embeddings, we construct martingales which enjoy the Brownian scaling property and the (inhomogeneous) Markov property. The second method necessitates randomization, but allows to reach any law with finite moment of order 1, centered, as the distribution of such a martingale at unit time. The first method does not necessitate randomization, but an additional restriction on the distribution at unit time is needed.

Key words: Convex order · Hardy-Littlewood functions · Karamata's representation theorem · Schauder fixed point theorem · Self-similar martingales · Skorokhod embeddings

2000 MSC: Primary: 60EXX, 60G18, 60G40, 60G44, 60J25, 60J65; Secondary: 26A12, 47H10

F. Hirsch (✉)
Laboratoire d'Analyse et Probabilités, Université d'Evry - Val d'Essonne, Boulevard F. Mitterrand, 91025 Evry Cedex, France
e-mail: francis.hirsch@univ-evry.fr

C. Profeta and B. Roynette
Institut Elie Cartan, Université Henri Poincaré, B.P. 239, 54506 Vandœuvre-lès-Nancy Cedex, France
e-mail: bernard.roynette@iecn.u-nancy.fr

M. Yor
Laboratoire de Probabilités et Modèles Aléatoires, Université Paris 06 & CNRS UMR 7599, 4, Place Jussieu, 75252 Paris Cedex 05, France
e-mail: deaproba@proba.jussieu.fr

C. Donati-Martin et al. (eds.), *Séminaire de Probabilités XLIII*, Lecture Notes in Mathematics 451
2006, DOI 10.1007/978-3-642-15217-7_21, © Springer-Verlag Berlin Heidelberg 2011

1　General Introduction

1.1　Our General Program

This work consists of two parts, Sects. 2 and 3 which both have the same purpose, i.e.: to construct a large class of martingales $(M_t, t \geq 0)$ which satisfy the two additional properties:

(a) $(M_t, t \geq 0)$ enjoys the Brownian scaling property:

$$\forall c > 0, \quad (M_{c^2 t}, t \geq 0) \overset{\text{(law)}}{=} (c M_t, t \geq 0)$$

(b) $(M_t, t \geq 0)$ is (inhomogeneous) Markovian.

The paper by Madan and Yor [MY02] developed three quite different methods to achieve this aim. In the following Sects. 2 and 3, we further develop two different Skorokhod embedding methods for the same purpose. In the end, the family of laws $\mu \sim M_1$ which are reached in Sect. 2 is notably bigger than in [MY02], while the method in Sect. 3 allows to reach any centered probability measure μ (with finite moment of order 1).

1.2　General Facts about Skorokhod Embeddings

For ease of the reader, we recall briefly the following facts:

• Consider a real valued, integrable and centered random variable X. Realizing a *Skorokhod embedding* of X into the Brownian motion B, consists in constructing a stopping time τ such that:

$(Sk1)$　$B_\tau \overset{\text{(law)}}{=} X$

$(Sk2)$　$(B_{u \wedge \tau}, u \geq 0)$　is a uniformly integrable martingale.

There are many ways to realize such a Skorokhod embedding. J. Obłój [Obl04] numbered twenty one methods scattered in the literature. These methods separate (at least) in two kinds:

– The time τ is a stopping time relative to the natural filtration of the Brownian motion B;
– The time τ is a stopping time relative to an enlargement of the natural filtration of the Brownian motion, by addition of extra random variables, independent of B.

In the second case, the stopping time τ is called a *randomized* stopping time. We call the corresponding embedding a *randomized Skorokhod embedding*.

- Suppose that, for every $t \geq 0$, there exists a stopping time τ_t satisfying (Sk1) and (Sk2) with $\sqrt{t}\,X$ replacing X. If *the family of stopping times* $(\tau_t, t \geq 0)$ *is a.s. increasing*, then the process $(B_{\tau_t}, t \geq 0)$ is a martingale and, for every fixed $t \geq 0$ and for every $c > 0$,

$$B_{\tau_{c^2 t}} \overset{\text{(law)}}{=} c\,\sqrt{t}\,X \overset{\text{(law)}}{=} c\,B_{\tau_t}\;,$$

which, a priori, is a weaker property than the scaling property (a). Nevertheless, the process $(B_{\tau_t}, t \geq 0)$ appears to be a good candidate to satisfy (a), (b) and $B_{\tau_1} \overset{\text{(law)}}{=} X$.

- Section 2 consists in using the Azéma-Yor algorithm, which yields a Skorokhod embedding of the first kind, whereas Sect. 3 hinges on a Skorokhod embedding of the second kind, both in order to obtain martingales $(B_{\tau_t}, t \geq 0)$ which satisfy (a) and (b).

Of course at the beginning of each section, we shall give more details, pertaining to the corresponding embedding, so that Sects. 2 and 3 may be read independently.

1.3 Examples of Such Martingales

The most famous examples of martingales satisfying (a) and (b) are, without any contest, Brownian motion $(B_t, t \geq 0)$ and the Azéma martingale $(\xi_t := \text{sgn}(B_t) \sqrt{t - g_t},\ t \geq 0)$ where $g_t := \sup\{s \leq t; B_s = 0\}$.

The study of the latter martingale $(\xi_t, t \geq 0)$, originally discovered by Azéma [Azé85], was then developed by Emery [Éme89, Éme90], Azéma-Yor [AY89], Meyer [Mey89a]. In particular, M. Emery established that Azéma martingale enjoys the Chaotic Representation Property (CRP). This discovery and subsequent studies were quite spectacular because, until then, it was commonly believed that the only two martingales which enjoy the CRP were Brownian motion and the compensated Poisson process. In fact, it turns out that a number of other martingales enjoying the CRP, together with (a) and (b), could be constructed, and were the subject of studies by P.A. Meyer [Mey89b], M. Emery [Éme96], M. Yor [Yor97, Chap. 15]. The structure equation concept played quite an important role there. However, we shall not go further into this topic, which lies outside the scope of the present paper.

1.4 Relations with Peacocks

Since X is an integrable and centered r.v., the process $(\sqrt{t}X, t \geq 0)$ is increasing in the convex order (see [HPRY10]). We call it a peacock. It is known from Kellerer [Kel72] that to any peacock $(\Pi_t, t \geq 0)$, one can associate a (Markovian) martingale $(M_t, t \geq 0)$ such that, for any fixed $t \geq 0$, $M_t \overset{\text{(law)}}{=} \Pi_t$, i.e.: $(M_t, t \geq 0)$ and $(\Pi_t, t \geq 0)$ have the same one-dimensional marginals. Given a peacock, it is generally difficult to exhibit an associated martingale. However, in the particular case $\Pi_t = \sqrt{t} X$ which we consider here, the process $(B_{\tau_t}, t \geq 0)$ presented above provides us with an associated martingale.

1.5 A Warning

It may be tempting to think that the whole distribution of a martingale $(M_t, t \geq 0)$ which satisfies (a) and (b) is determined by the law of M_1. This is quite far from being the case, as a number of recent papers shows; the interested reader may look at Albin [Alb08] (see also in this volume Baker-Donati-Martin-Yor [BDMY10] who develop Albin's construction further), Oleszkiewicz [Ole08], Hamza-Klebaner [HK07]. We thank David Baker (personal communication, 2009) and David Hobson (personal communication, 2009) for pointing out, independently, these papers to us.

2 Construction via the Skorokhod Embedding of Azéma-Yor

2.1 Introduction

2.1.1 Program

The methodology developed in this section is AYUS (=Azéma-Yor Under Scaling), following the terminology in [MY02]. Precisely, given a r.v. X with probability law μ, we shall use the Azéma-Yor embedding algorithm simultaneously for all distributions μ_t indexed by $t \geq 0$ where:

$$\forall t \geq 0, \quad \mu_t \sim \sqrt{t} X. \tag{1}$$

More precisely, if $(B_t, t \geq 0)$ denotes a Brownian motion and $(S_t := \sup_{u \leq t} B_u,$ $t \geq 0)$, we seek probability measures μ such that the family of stopping times:

$$T_{\mu_t} := \inf\{u \geq 0; S_u \geq \psi_{\mu_t}(B_u)\}$$

where

$$\psi_{\mu_t}(x) = \frac{1}{\mu_t([x, +\infty[)} \int_{[x,+\infty[} y \mu_t(dy)$$

increases, or equivalently, the family of functions $\left(\psi_{\mu_t}(x) = \sqrt{t}\psi_\mu\left(\frac{x}{\sqrt{t}}\right)\right)_{t \geq 0}$ increases (pointwise in x). (Since $\mu = \mu_1$, we write ψ_μ for ψ_{μ_1}). This program was already started in Madan-Yor, who came up with the (easy to prove) necessary and sufficient condition on μ:

$$a \longmapsto D_\mu(a) := \frac{a}{\psi_\mu(a)} \text{ is increasing on } \mathbb{R}_+. \tag{$M \cdot Y$}$$

Our main contribution in this Sect. 2 is to look for nice, easy to verify, sufficient conditions on μ which ensure that $(M \cdot Y)$ is satisfied. Such a condition has been given in [MY02] (Theorems 4 and 5). In the following Sect. 2:

- We discuss further this result of Theorem 4 by giving equivalent conditions for it; this study has a strong likeness with (but differs from) Karamata's representation theorem for slowly varying functions (see, e.g. Bingham–Goldie–Teugels [BGT89, Chap. 1, Theorems 1.3.1 and 1.4.1])
- Moreover, we also find different sufficient conditions for $(M \cdot Y)$ to be satisfied.

   With the help of either of these conditions, it turns out that many subprobabilities μ on $\mathbb{R}_+$ satisfy $(M \cdot Y)$; in particular, all beta and gamma laws satisfy $(M \cdot Y)$.

2.1.2 A Forefather

A forefather of the present paper is Meziane–Yen–Yor [MYY09], where a martingale $(M_t, t \geq 0)$ which enjoys (a) and (b) and is distributed at time 1 as $\varepsilon\sqrt{g}$ with ε a Bernoulli r.v. and g an independent arcsine r.v. was constructed with the same method. Thus, the martingale $(M_t, t \geq 0)$ has the same one-dimensional marginals as Azéma's martingale $(\xi_t, t \geq 0)$ presented in Sect. 1.3 although the laws of M and ξ differ. Likewise in [MY02], Madan and Yor construct a purely discontinuous martingale $(N_t, t \geq 0)$ which enjoys (a) and (b) and has the same one-dimensional marginals as a Brownian motion $(B_t, t \geq 0)$.

2.1.3 Plan

The remainder of this part is organized as follows: Sects. 2.2–2.4 deal with the case of measures μ with support in $] - \infty, 1]$, 1 belonging to the support of μ, while Sect. 2.5 deals with a generic measure μ whose support is $\mathbb{R}$. More precisely:

- First, Sect. 2.2 consists in recalling the Azéma-Yor algorithm and the Madan-Yor condition $(M \cdot Y)$, and then presenting a number of important quantities associated with μ, whether or not $(M \cdot Y)$ is satisfied. Elementary relations between these quantities are established, which will ease up our discussion later on.
- Section 2.3: when $(M \cdot Y)$ is satisfied, it is clear that there exists a subprobability ν_μ on $]0, 1[$ such that:

$$D_\mu(a) = \nu_\mu(]0, a[), \quad a \in [0, 1]. \tag{2}$$

We obtain relations between quantities relative to μ and ν_μ.
In particular:

- In Sect. 2.3.4, we establish a one-to-one correspondence between two sets of probabilities μ and ν.
- Section 2.3.5 consists in the study in the particular case of $(M \cdot Y)$ when:

$$\frac{a}{\psi_\mu(a)} = \frac{1}{Z} \int_0^{+\infty} (1 - e^{-ax}) \rho(dx)$$

for certain positive measures ρ, where $Z = \int_0^{+\infty} (1 - e^{-x}) \rho(dx)$ is the normalizing constant which makes: $D_\mu(a) = \dfrac{a}{\psi_\mu(a)}$ a distribution function on $[0, 1]$.

- Section 2.3.6 gives another formulation of this correspondence.

- Section 2.4 consists in the presentation of a number of conditions $(S_0) - (S_5)$ and subconditions (S_i') which suffice for the validity of $(M \cdot Y)$.
- Section 2.5 tackles the case of a measure μ whose support is $\mathbb{R}$, and gives a sufficient condition for the existence of a probability ν_μ which satisfies (2).
- Finally, in Sect. 2.6, many particular laws μ are illustrated in the form of graphs. We also give an example where $(M \cdot Y)$ is not satisfied, which, given the preceding studies, seems to be rather the exception than the rule.

2.2 General Overview of this Method

2.2.1 The Azéma-Yor Algorithm for Skorokhod Embedding

We start by briefly recalling the Azéma-Yor algorithm for Skorokhod embedding. Let μ be a probability on $\mathbb{R}$ such that:

$$\int_{-\infty}^{+\infty} |x| \mu(dx) < \infty \quad \text{and} \quad \int_{-\infty}^{+\infty} x\mu(dx) = 0. \tag{3}$$

We define its Hardy-Littlewood function ψ_μ by:

$$\psi_\mu(x) = \frac{1}{\mu([x, +\infty[)} \int_{[x, +\infty[} y\mu(dy).$$

In the case where there exists $x \geq 0$ such that $\mu([x, +\infty[) = 0$, we set $\alpha = \inf\{x \geq 0; \mu([x, +\infty[) = 0\}$ and $\psi_\mu(x) = \alpha$ for $x \geq \alpha$. Let $(B_t, t \geq 0)$ be a standard Brownian motion. Azéma-Yor [AY79] introduced the stopping time:

$$T_\mu := \inf\{t \geq 0; S_t \geq \psi_\mu(B_t)\}$$

where $S_t := \sup_{s \leq t} B_s$ and showed:

Theorem 1 ([AY79]).
(1) $(B_{t \wedge T_\mu}, t \geq 0)$ is a uniformly integrable martingale.

(2) The law of B_{T_μ} is μ: $B_{T_\mu} \sim \mu$.

To prove Theorem 1, Azéma-Yor make use of the martingales:

$$\left(\varphi(S_t)(S_t - B_t) + \int_{S_t}^{+\infty} dx\varphi(x), \ t \geq 0 \right)$$

for any $\varphi \in L^1(\mathbb{R}_+, dx)$. Rogers [Rog81] shows how to derive Theorem 1 from excursion theory, while Jeulin-Yor [JY81] develop a number of results about the laws of $\int_0^{T_\mu} h(B_s)ds$ for a generic function h.

2.2.2 A Result of Madan-Yor

Madan-Yor [MY02] have exploited this construction to find martingales $(X_t, t \geq 0)$ which satisfy (a) and (b). More precisely:

Proposition 1. Let X be an integrable and centered r.v. with law μ. Let, for every $t \geq 0$, $\widetilde{X}_t := \sqrt{t}X$. Denote by μ_t the law of $\widetilde{X}_t$ and by ψ_t $(= \psi_{\mu_t})$ the Hardy-Littlewood function associated to μ_t:

$$\psi_t(x) = \frac{1}{\mu_t([x, +\infty[)} \int_{[x,+\infty[} y\mu_t(dy) = \sqrt{t}\psi_1\left(\frac{x}{\sqrt{t}}\right)$$

and by $T_t^{(\mu)}$ the Azéma-Yor stopping time (which we shall also denote T_{ψ_t}):

$$T_t^{(\mu)} := \inf\{u \geq 0; S_u \geq \psi_t(B_u)\} \tag{4}$$

(Theorem 1 asserts that $B_{T_t^{(\mu)}} \sim \mu_t$). We assume furthermore:

$$t \longmapsto T_t^{(\mu)} \text{ is a.s. increasing} \tag{I}$$

Then:

(1) *The process* $\left(X_t^{(\mu)} := B_{T_t^{(\mu)}}, t \ge 0\right)$ *is a martingale, and an (inhomogeneous) Markov process.*

(2) *The process* $\left(X_t^{(\mu)}, t \ge 0\right)$ *enjoys the Brownian scaling property, i.e for every* $c > 0$:

$$\left(X_{c^2 t}^{(\mu)}, t \ge 0\right) \overset{(law)}{=} \left(cX_t^{(\mu)}, t \ge 0\right)$$

In particular, $\left(X_t^{(\mu)} := B_{T_t^{(\mu)}}, t \ge 0\right)$ *is a martingale associated to the peacock* $(\sqrt{t}X, t \ge 0)$ *(see Introduction).*

Proof (Proof of Proposition 1). Point (1) is clear. See in particular [MY02] where the infinitesimal generator of $(X_t^{(\mu)}, t \ge 0)$ is computed. It is therefore sufficient to prove Point (2). Let $c > 0$ be fixed.

(*i*) From the scaling property of Brownian motion:

$$(S_{c^2 t}, B_{c^2 t}, \, t \ge 0) \overset{(law)}{=} (cS_t, cB_t, \, t \ge 0),$$

and the definition (4) of T_{ψ_t}, we deduce that:

$$\left(B_{T_{\psi_t}}, t \ge 0\right) \overset{(law)}{=} \left(cB_{T_{\psi_t^{(c)}}}, t \ge 0\right) \tag{5}$$

with $\psi_t^{(c)}(x) := \dfrac{1}{c}\psi_t(cx).$

(*ii*) An elementary computation yields:

$$\psi_t(x) = \sqrt{t}\psi\left(\frac{x}{\sqrt{t}}\right) \tag{6}$$

with $\psi := \psi_1 = \psi_\mu$. We obtain from (6) that:

$$\psi_{c^2 t}^{(c)}(x) = \frac{1}{c}\psi_{c^2 t}(cx) = \sqrt{t}\psi\left(\frac{x}{\sqrt{t}}\right) = \psi_t(x). \tag{7}$$

Finally, gathering (*i*) and (*ii*), it holds:

$$\left(X_{c^2 t}^{(\mu)} := B_{T_{\psi_{c^2 t}}}, t \ge 0\right) \overset{(law)}{=} \left(cB_{T_{\psi_{c^2 t}^{(c)}}}, t \ge 0\right) \quad \text{(from (5))}$$

$$\overset{(law)}{=} \left(cB_{T_{\psi_t}} = cX_t^{(\mu)}, t \ge 0\right) \quad \text{(from (7))}.$$

$\square$

Remark 1 (Due to P. Vallois). It is easy to prove (see [AY79, Proposition 3.6]) that, for every $x \geq 0$:

$$\mathbb{P}\left(S_{T_\mu} \geq x\right) = \exp\left(-\int_0^x \frac{dy}{y - \psi_\mu^{-1}(y)}\right)$$

where ψ_μ^{-1} is the right-continuous inverse of ψ_μ. Replacing μ by μ_t in this formula, and using $\psi_{\mu_t}^{-1}(y) = \sqrt{t}\,\psi_\mu^{-1}\left(\frac{y}{\sqrt{t}}\right)$, we obtain:

$$\mathbb{P}\left(S_{T_{\mu_t}} \geq x\right) = \exp\left(-\int_0^x \frac{dy}{y - \psi_{\mu_t}^{-1}(y)}\right)$$

$$= \exp\left(-\int_0^x \frac{dy}{y - \sqrt{t}\,\psi_\mu^{-1}\left(\frac{y}{\sqrt{t}}\right)}\right)$$

$$= \exp\left(-\int_0^{\frac{x}{\sqrt{t}}} \frac{du}{u - \psi_\mu^{-1}(u)}\right).$$

Thus, for every $x \geq 0$, the function $t \longmapsto \mathbb{P}\left(S_{T_{\mu_t}} \geq x\right)$ is increasing and, consequently, even without the $(M \cdot Y)$ hypothesis (see Lemma 1 below), the process $(S_{T_t^{(\mu)}}, t \geq 0)$ is stochastically increasing. Note that, under $(M \cdot Y)$, this process is a.s. increasing, which of course implies that it is stochastically increasing.

2.2.3 Examples

In the paper [MYY09] which is a forefather of the present paper, the following examples were studied in details:

(i) The dam-drawdown example:

$$\mu_t(dx) = \frac{1}{\sqrt{t}} \exp\left(-\frac{1}{\sqrt{t}}(x + \sqrt{t})\right) 1_{[-\sqrt{t}, +\infty[}(x)dx$$

which yields to the stopping time $T_t := \inf\{u \geq 0; S_u - B_u = \sqrt{t}\}$. Recall that from Lévy's theorem, $(S_u - B_u, u \geq 0)$ is a reflected Brownian motion.

(ii) The "BES(3)-Pitman" example:

$$\mu_t(dx) = \frac{1}{2t} 1_{[-\sqrt{t}, \sqrt{t}]}(x)dx$$

which corresponds to the stopping time $T_t := \inf\{u \geq 0; 2S_u - B_u = \sqrt{t}\}$. Recall that from Pitman's Theorem, $(2S_u - B_u, u \geq 0)$ is distributed as a Bessel process of dimension 3 started from 0.

(*iii*) The Azéma-Yor "fan", which is a generalization of the two previous examples:

$$\mu^{(\alpha)}(dx) = \frac{\alpha}{\sqrt{t}}\left(\alpha - \frac{(1-\alpha)x}{\sqrt{t}}\right)^{\frac{2\alpha-1}{1-\alpha}} 1_{\left[-\sqrt{t},\frac{\alpha\sqrt{t}}{1-\alpha}\right]}(x)dx, \qquad (0 < \alpha < 1)$$

which yields to the stopping time $T_t^{(\alpha)} := \inf\{u \geq 0; S_u = \alpha(B_u + \sqrt{t})\}$. Example (*i*) is obtained by letting $\alpha \to 1^-$.

2.2.4 The $(M \cdot Y)$ Condition

Proposition 1 highlights the importance of condition (I) (see Proposition 1 above) for our search of martingales satisfying conditions (a) and (b). We now wish to be able to read "directly" from the measure μ whether (I) is satisfied or not. The answer to this question is presented in the following Lemma:

Lemma 1 ([MY02], Lemma 3). *Let $X \sim \mu$ satisfy (3). We define:*

$$D_\mu(x) := \frac{x\overline{\mu}(x)}{\int_{[x,+\infty[} y\mu(dy)} \quad \text{with } \overline{\mu}(x) := \mathbb{P}(X \geq x) = \int_{[x,+\infty[} \mu(dy).$$

Then (I) is satisfied if and only if :

$$x \longmapsto D_\mu(x) \text{ is increasing on } \mathbb{R}_+. \qquad (M \cdot Y)$$

Proof (Proof of Lemma 1). Condition (I) is equivalent to the increase, for any given $x \in \mathbb{R}$, of the function $t \longmapsto \psi_t(x)$. From (6), $\psi_t(x) = \sqrt{t}\psi\left(\frac{x}{\sqrt{t}}\right)$, hence, if $x \leq 0$, since ψ is a positive and increasing function, $t \longmapsto \psi_t(x)$ is increasing. For $x > 0$, we set $a_t = \frac{x}{\sqrt{t}}$; thus:

$$\psi_t(x) = \sqrt{t}\psi\left(\frac{x}{\sqrt{t}}\right) = x\frac{\psi(a_t)}{a_t},$$

and, $t \longmapsto a_t$ being a decreasing function of t, condition (I) is equivalent to the increase of the function $a \longmapsto \frac{a}{\psi(a)} = D_\mu(a)$. $\square$

A remarkable feature of this result is that (I) only depends on the restriction of μ to $\mathbb{R}_+$. This is inherited from the asymmetric character of the Azéma-Yor construction in which $\mathbb{R}_+$, via $(S_u, u \geq 0)$ plays a special role.

Now, let $\widetilde{\mu}$ be a probability on $\mathbb{R}$ satisfying (3). Since our aim is to obtain conditions equivalent to $(M \cdot Y)$, i.e.:

$$x \longmapsto D_{\widetilde{\mu}}(x) := \frac{x \int_{[x,+\infty[} \widetilde{\mu}(dy)}{\int_{[x,+\infty[} y \widetilde{\mu}(dy)} \quad \text{increases on } \mathbb{R}_+, \tag{8}$$

it suffices to study $D_{\widetilde{\mu}}$ on $\mathbb{R}_+$. Clearly, this function (on $\mathbb{R}_+$) depends only on the restriction of $\widetilde{\mu}$ to $\mathbb{R}_+$, which we denote by μ. Observe that $(M \cdot Y)$ remains unchanged if we replace μ by $\lambda \mu$ where λ is a positive constant.

Besides, we shall restrict our study to the case where $\widetilde{\mu}$ is carried by $]-\infty, k]$, i.e. where $\mu = \widetilde{\mu}_{|\mathbb{R}_+}$ is a subprobability on $[0, k]$. To simplify further, but without loss of generality, we shall take $k = 1$ and assume that 1 belongs to the support of μ. In Sect. 2.5, we shall study briefly the case where μ is a measure whose support is $\mathbb{R}_+$.

2.2.5 Notation

In this section, we present some notation which shall be in force throughout the remainder of the paper. Let μ be a positive measure on $[0, 1]$, with finite total mass, and whose support contains 1. We denote by $\overline{\mu}$ and $\overline{\overline{\mu}}$, respectively its tail and its double tail functions:

$$\overline{\mu}(x) = \int_{[x,1]} \mu(dy) = \mu([x, 1]) \quad \text{and} \quad \overline{\overline{\mu}}(x) = \int_x^1 \overline{\mu}(y)dy.$$

Note that $\overline{\mu}$ is left-continuous, $\overline{\overline{\mu}}$ is continuous, and $\overline{\mu}$ and $\overline{\overline{\mu}}$ are both decreasing functions. Furthermore, it is not difficult to see that a function $\Lambda : [0, 1] \longrightarrow \mathbb{R}_+$ is the double tail function of a positive finite measure on $[0, 1]$ if and only if Λ is a convex function on $[0, 1]$, left-differentiable at 1, right-differentiable at 0, and satisfying $\Lambda(1) = 0$.

We also define the tails ratio u_μ associated to μ:

$$u_\mu(x) = \overline{\mu}(x)/\overline{\overline{\mu}}(x), \qquad x \in [0, 1[.$$

Here is now a lemma of general interest which bears upon positive measures:

Lemma 2 ([Pie80] or [RY99], Chap. VI Lemma 5.1).

(1) *For every $x \in [0, 1[$:*

$$\overline{\mu}(x) = \overline{\overline{\mu}}(0)u_\mu(x) \exp\left(-\int_0^x u_\mu(y)dy\right) \tag{9}$$

and u_μ is left-continuous.

(2) *Let $v : [0, 1[\longrightarrow \mathbb{R}_+$ be a left-continuous function such that, for all $x \in [0, 1[$:*

$$\overline{\mu}(x) = \overline{\overline{\mu}}(0)v(x) \exp\left(-\int_0^x v(y)dy\right)$$

Then, $v = u_\mu$.

Proof (Proof of Lemma 2).

(1) We first prove Point (1). For $x \in [0, 1[$, we have:

$$-\int_0^x \frac{\overline{\mu}(y)}{\overline{\overline{\mu}}(y)}dy = \int_0^x \frac{d\overline{\overline{\mu}}(y)}{\overline{\overline{\mu}}(y)} = \left[\log \overline{\overline{\mu}}(y)\right]_0^x = \log \overline{\overline{\mu}}(x) - \log \overline{\overline{\mu}}(0),$$

hence,

$$\overline{\overline{\mu}}(0)u_\mu(x) \exp\left(-\int_0^x u_\mu(y)dy\right) = \overline{\overline{\mu}}(0)\frac{\overline{\mu}(x)}{\overline{\overline{\mu}}(x)} \exp\left(\log \frac{\overline{\overline{\mu}}(x)}{\overline{\overline{\mu}}(0)}\right) = \overline{\mu}(x).$$

(2) We now prove Point (2). Let $U_\mu(x) := \int_0^x u_\mu(y)dy$ and $V(x) := \int_0^x v(y)dy$. Relation (9) implies:

$$u_\mu(x) \exp\left(-U_\mu(x)\right) = v(x) \exp\left(-V(x)\right), \quad \text{i.e.}$$
$$\left(\exp\left(-U_\mu(x)\right)\right)' = \left(\exp\left(-V(x)\right)\right)', \quad \text{hence}$$
$$\exp\left(-U_\mu(x)\right) = \exp\left(-V(x)\right) + c.$$

(The above derivatives actually denote left-derivatives). Now, since, $U_\mu(0) = V(0) = 0$, we obtain $c = 0$ and $U_\mu = V$. Then, differentiating, and using the fact that u_μ and v are left-continuous, we obtain: $u_\mu = v$.   □

Remark 2. Since $\overline{\mu}$ is a decreasing function, we see, by differentiating (9), that the function u_μ satisfies:

– If $\overline{\mu}$ is differentiable, then so is u_μ and $u'_\mu \le u_\mu^2$,
– More generally, the distribution on $]0, 1[$: $u_\mu^2 - u'_\mu$, is a positive measure.

Note that if $\mu(dx) = h(x)dx$, then:

$$u_\mu^2 - u'_\mu = \left(\overline{\mu}/\overline{\overline{\mu}}\right)^2 - \left(\frac{\overline{\mu}^2 - h\overline{\overline{\mu}}}{\overline{\overline{\mu}}^2}\right) = h/\overline{\overline{\mu}} \ge 0.$$

By (9), we have, for any $x \in [0, 1[$,

$$\overline{\overline{\mu}}(x) = \overline{\overline{\mu}}(0) \exp\left(-\int_0^x u_\mu(y)dy\right).$$

Since $\overline{\overline{\mu}}(0) = \int_{[0,1]} y\mu(dy) > 0$ and $\overline{\overline{\mu}}(1) = 0$, we obtain:

$$\forall x < 1, \quad \int_0^x u_\mu(y)dy < \infty \quad \text{and} \quad \int^{1^-} u_\mu(y)dy = +\infty \tag{10}$$

As in Sect. 2.2.1, we now define the Hardy-Littlewood function ψ_μ associated to μ:

$$\begin{cases} \psi_\mu(a) = \dfrac{1}{\mu([a,1])} \displaystyle\int_{[a,1]} y\mu(dy), & a \in [0, 1[\\ \psi_\mu(1) = 1 \end{cases}$$

and the Madan-Yor function associated to μ:

$$D_\mu(a) = \frac{a}{\psi_\mu(a)}, \qquad a \in [0, 1].$$

In particular, $D_\mu(1) = 1$ and $D_\mu(0) = 0$. Note that, integrating by parts:

$$\overline{\overline{\mu}}(a) = \int_a^1 (y - a)\mu(dy) = \overline{\mu}(a)\left(\psi_\mu(a) - a\right),$$

hence, $u_\mu(a) = \dfrac{1}{\psi_\mu(a) - a}$ and, consequently:

$$D_\mu(a) = \frac{a}{\psi_\mu(a) - a + a} = \frac{a}{(1/u_\mu(a)) + a} = \frac{au_\mu(a)}{au_\mu(a) + 1}. \tag{11}$$

We sum up all the previous notation in a table, for future references:

$\mu(dx)$	A finite positive measure on $[0, 1]$ whose support contains 1.
$\overline{\mu}(a) = \mu([a, 1])$	Tail function associated to μ
$\overline{\overline{\mu}}(a) = \int_a^1 \overline{\mu}(x)dx$	Double tail function associated to μ
$u_\mu(a) = \overline{\mu}(a)/\overline{\overline{\mu}}(a) = \dfrac{1}{\psi_\mu(a) - a}$	Tails ratio function associated to μ
$\psi_\mu(a) = \dfrac{1}{\overline{\mu}(a)} \displaystyle\int_{[a,1]} x\mu(dx)$	Hardy-Littlewood function associated to μ
$D_\mu(a) = \dfrac{a}{\psi_\mu(a)} = \dfrac{au_\mu(a)}{au_\mu(a) + 1}$	Madan-Yor function associated to μ

2.3   Some Conditions Which Are Equivalent to $(M \cdot Y)$

2.3.1   A Condition Which Is Equivalent to $(M \cdot Y)$

Let μ denote a positive measure on $[0, 1]$, with finite total mass, and whose support contains 1. We now study the condition $(M \cdot Y)$ in more details.

2.3.2   Elementary Properties of D_μ

(i) From the obvious inequalities, for $x \in [0, 1]$:

$$x \overline{\mu}(x) = x \int_{[x,1]} \mu(dy) \leq \int_{[x,1]} y \mu(dy) \leq \int_{[x,1]} \mu(dy) = \overline{\mu}(x)$$

we deduce that ψ_μ and D_μ are left-continuous on $]0, 1]$, and for every $x \in [0, 1]$,

$$x \leq \psi_\mu(x) \leq 1 \quad \text{and} \quad x \leq D_\mu(x) \leq 1. \tag{12}$$

(ii) We now assume that μ admits a density h; then:

  – If h is continuous at 0, then: $D'_\mu(0^+) = \overline{\mu}(0)/\overline{\overline{\mu}}(0)$,

  – If h is continuous at 1, and $h(1) > 0$, then: $D'_\mu(1^-) = \dfrac{1}{2}$,

  – If h admits, in a neighborhood of 1, the equivalent:

$$h(1 - x) \underset{x \to 0}{=} C x^\alpha + o(x^\alpha), \text{ with } C, \alpha > 0 \text{ then: } D'_\mu(1^-) = \frac{1}{2 + \alpha}.$$

These three properties are consequences of the following formula, which holds at every point where h is continuous:

$$\frac{D'_\mu(x)}{D_\mu(x)} = \frac{1}{x} - h(x) \frac{1 - D_\mu(x)}{\overline{\mu}(x)}.$$

2.3.3   A Condition Which Is Equivalent to $(M \cdot Y)$

Theorem 2. *Let μ be a finite positive measure on $[0, 1]$ whose support contains 1, and u_μ its tails ratio. The following assertions are equivalent:*

  (i) *D_μ is increasing on $[0, 1]$, i.e. $(M \cdot Y)$ holds.*
  (ii) *There exists a probability measure v_μ on $]0, 1[$ such that:*

$$\forall a \in [0, 1], \quad D_\mu(a) = v_\mu(]0, a[). \tag{13}$$

  (iii) *$a \longrightarrow a u_\mu(a)$ is an increasing function on $[0, 1[$.*

Proof (Proof of Theorem 2). Of course, the equivalence between (*i*) and (*ii*) holds, since $D_\mu(0) = 0$ and $D_\mu(1) = 1$. As for the equivalence between (*i*) and (*iii*), it follows from (11):

$$D_\mu(a) = \frac{au_\mu(a)}{au_\mu(a) + 1}.$$

□

Remark 3.

(1) The probability measure v_μ defined via (13) enjoys some particular properties. Indeed, from (13), it satisfies

$$\frac{v_\mu(]0, a[)}{a} = \frac{D_\mu(a)}{a} = \frac{1}{\psi_\mu(a)}.$$

Thus, since the function ψ_μ is increasing on $[0, 1]$, the function $a \longmapsto \frac{v_\mu(]0, a[)}{a}$ is decreasing on $[0, 1]$, and $\lim_{a \to 0+} \frac{v_\mu(]0, a[)}{a} = \frac{1}{\psi_\mu(0)}$.

(2) From (11), we have $v_\mu(]0, a[) = D_\mu(a) = \frac{au_\mu(a)}{au_\mu(a) + 1}$, hence, for every $a \in]0, 1[$:

$$u_\mu(a) = \frac{v_\mu(]0, a[)}{av_\mu([a, 1[)},$$

and, in particular, $v_\mu([a, 1[) > 0$. Thus, with the help of (10), v_μ necessarily satisfies the relation:

$$\int^{1-} \frac{da}{v_\mu([a, 1[)} = +\infty.$$

(3) The function D_μ is characterized by its values on $]0, 1[$ (since $D_\mu(0) = 0$ and $D_\mu(1) = 1$). Hence, D_μ only depends on the values of ψ_μ on $]0, 1[$, and therefore, D_μ only depends on the restriction of μ to $]0, 1]$. The value of $\mu(\{0\})$ is irrelevant for the $(M \cdot Y)$ condition.

2.3.4 Characterizing the Measures v_μ

Theorem 2 invites to ask for the following question: given a probability measure v on $]0, 1[$, under which conditions on v does there exists a positive measure μ on $[0, 1]$ with finite total mass[1] such that μ satisfies $(M \cdot Y)$?

In particular, are the conditions given in Point (1) and (2) of the previous Remark 3 sufficient? In the following Theorem, we answer this question in the affirmative.

---

[1] Note that since D_μ remains unchanged if we replace μ by a multiple of μ, μ can always be chosen to be a probability.

Notation. We adopt the following notation:

- $\mathcal{P}_1$ denotes the set of all probabilities μ on $[0, 1]$, whose support contains 1, and which satisfy $(M \cdot Y)$.
- $\mathcal{P}_1^0 = \{\mu \in \mathcal{P}_1 ; \mu(\{0\}) = 0\}$.
- $\mathcal{P}_1'$ denotes the set of all probabilities ν on $]0, 1[$ such that:

  (i) $\nu([a, 1[) > 0$ for every $a \in]0, 1[$,

  (ii) $a \longmapsto \dfrac{\nu(]0, a[)}{a}$ is a decreasing function on $]0, 1]$ such that
  $$c_\nu := \lim_{a \to 0+} \frac{\nu(]0, a[)}{a} < \infty,$$

  (iii) $\displaystyle\int^{1^-} \frac{da}{\nu([a, 1[)} = +\infty.$

- We define a map Γ on $\mathcal{P}_1$ as follows: if $\mu \in \mathcal{P}_1$, then $\Gamma(\mu)$ is the measure ν on $]0, 1[$ such that
  $$D_\mu(a) = \nu(]0, a[), \quad a \in [0, 1].$$

In other words, $\Gamma(\mu) = \nu_\mu$ defined by (13).

With the help of the above notation, we can state:

Theorem 3.

(1) $\Gamma(\mathcal{P}_1^0) = \Gamma(\mathcal{P}_1) = \mathcal{P}_1'$.

(2) If $\mu \in \mathcal{P}_1$ and $\mu_0 \in \mathcal{P}_1^0$, then

$$\Gamma(\mu) = \Gamma(\mu_0) \quad \text{if and only if} \quad \mu = \mu(\{0\})\delta_0 + (1 - \mu(\{0\}))\mu_0$$

(where δ_0 denotes the Dirac measure at 0).

As a consequence of 1) and 2), Γ induces a bijection between $\mathcal{P}_1^0$ and $\mathcal{P}_1'$.

Proof (Proof of Theorem 3).

(a) Remark 3 entails that:
$$\Gamma(\mathcal{P}_1^0) \subset \Gamma(\mathcal{P}_1) \subset \mathcal{P}_1'.$$

(b) We now prove $\mathcal{P}_1' \subset \Gamma(\mathcal{P}_1^0)$. Let $\nu \in \mathcal{P}_1'$. We define $u^{(\nu)}$ by:

$$\begin{cases} u^{(\nu)}(x) := \dfrac{\nu(]0, x[)}{x(1 - \nu(]0, x[))} & \text{for } x \in]0, 1[, \\ u^{(\nu)}(0) := c_\nu = \lim_{x \to 0+} u^{(\nu)}(x), \end{cases}$$

and we set, for $x \in [0, 1[$:

$$m(x) = \frac{1}{c_\nu} u^{(\nu)}(x) \exp\left(-\int_0^x u^{(\nu)}(y)dy\right). \tag{14}$$

We remark that m is left-continuous on $]0, 1[$, right-continuous at 0 and $m(0) = 1$. To prove that m is decreasing on $[0, 1[$, it suffices to show that

m is decreasing on $]0, 1[$ or, equivalently (see Remark 2), that the distribution on $]0, 1[$: $\left(u^{(v)}\right)^2 - \left(u^{(v)}\right)'$, is a positive measure.

Now, from the definition of $u^{(v)}$, and setting:

$$v(a) := v(]0, a[),$$

we need to prove that (on $]0, 1[$):

$$v^2(a)da \geq a(1 - v(a))dv(a) - v(a)(1 - v(a))da + av(a)dv(a)$$
$$\Longleftrightarrow v^2(a)da \geq adv(a) - v(a)\bigl(1 - v(a)\bigr)da$$
$$\Longleftrightarrow 0 \geq adv(a) - v(a)da$$
$$\Longleftrightarrow 0 \geq d\left(\frac{v(a)}{a}\right).$$

The latter is ensured by Property (ii) in the definition of $\mathcal{P}_1'$. Hence, there exists a probability μ on $[0, 1]$ such that

$$\overline{\mu}(x) = m(x), \quad x \in [0, 1[.$$

In particular, since m is right-continuous at 0, $\mu(\{0\}) = 0$. Using Property (iii) in the definition of $\mathcal{P}_1'$, we obtain from (14), by integration:

$$\overline{\overline{\mu}}(0) = \frac{1}{c_v}.$$

Therefore, by Lemma 2, $u^{(v)} = u_\mu$, or:

$$u_\mu(a) = \frac{v(]0, a[)}{a(1 - v(]0, a[))}, \quad a \in]0, 1[.$$

Consequently,

$$D_\mu(a) = \frac{au_\mu(a)}{au_\mu(a) + 1} = v(]0, a[), \quad a \in]0, 1[,$$

and hence, $\mu \in \mathcal{P}_1^0$ and $\Gamma(\mu) = v$.

(c) We now prove Point (2). Suppose first that $\mu \in \mathcal{P}_1$, $\mu_0 \in \mathcal{P}_1^0$ and $\Gamma(\mu) = \Gamma(\mu_0)$. We then have:

$$u_\mu(a) = u_{\mu_0}(a), \quad a \in]0, 1[.$$

By Lemma 2, this entails that there exists $\lambda > 0$ such that:

$$\overline{\mu}(x) = \lambda\overline{\mu}_0(x), \quad x \in]0, 1[$$

and therefore, by differentiation, the restriction of μ to $]0, 1]$ is equal to $\lambda\mu_0$. Consequently, $\mu = \mu(\{0\})\delta_0 + \lambda\mu_0$ and, since μ is a probability, $\lambda = 1 - \mu(\{0\})$.

Conversely, suppose that $\mu = \mu(\{0\})\delta_0 + (1 - \mu(\{0\}))\mu_0$. Since 1 belongs to the support of μ, $\mu(\{0\}) < 1$. Therefore, $\psi_\mu(x) = \psi_{\mu_0}(x)$ for $x \in]0, 1]$, and hence, $D_\mu(a) = D_{\mu_0}(a)$ for $a \in]0, 1]$, which entails $\Gamma(\mu) = \Gamma(\mu_0)$.

$\square$

Example 1. If ν is a measure which admits a continuous density g which is decreasing on $]0, 1[$, and strictly positive in a neighborhood of 1, then $\nu \in \mathcal{P}'_1$. For example, let us take for $\beta \geq 2\alpha > 0$, $g(x) = \dfrac{\beta - 2\alpha x}{\beta - \alpha} 1_{]0,1[}(x)$. Then,

$\nu(]0, x[) = \dfrac{\beta x - \alpha x^2}{\beta - \alpha}$, and some easy computations show that:

$$\overline{\mu}(x) = \frac{\beta - \alpha x}{\beta}(1 - x)^{\frac{\alpha}{\beta - 2\alpha}}\left(1 - \frac{\alpha x}{\beta - \alpha}\right)^{-\frac{\beta - \alpha}{\beta - 2\alpha}}.$$

In particular, letting α tend to 0, we obtain: $\forall x \in [0, 1]$, $\overline{\mu}(x) = 1$, i.e. the correspondence:

$$\nu(dx) = 1_{]0,1[}(x)dx \longleftrightarrow \mu(dx) = \delta_1(dx)$$

where δ_1 denotes the Dirac measure at 1.

2.3.5   Examples of Elements of $\mathcal{P}'_1$

To a positive measure ρ on $]0, +\infty[$ such that $\int_0^{+\infty} y\rho(dy) < \infty$, we associate the measure:

$$\nu(]0, a[) = \frac{1}{Z}\int_0^{+\infty}(1 - e^{-ay})\rho(dy)$$

where $Z := \int_0^{+\infty}(1 - e^{-y})\rho(dy)$ is such that $\nu(]0, 1[) = 1$. Clearly, $a \longmapsto \frac{\nu(]0,a[)}{a} = \frac{1}{Z}\int_0^{+\infty} e^{-au}\overline{\rho}(u)du$, where $\overline{\rho}(u) = \rho(]u, +\infty[)$, is decreasing and $c_\nu = \frac{1}{Z}\int_0^{+\infty} y\rho(dy) < \infty$. Furthermore, $\lim\limits_{a\to 1^-} \frac{\nu([a,1[)}{1-a} = \frac{1}{Z}\int_0^{+\infty} ye^{-y}\rho(dy) > 0$, hence $\int^1 \frac{da}{\nu([a,1[)} = +\infty$, and Theorem 3 applies.

We now give some examples:

(i) For $\rho(dx) = e^{-\lambda x}dx$ $(\lambda > 0)$, we obtain: $\nu(]0, a[) = \dfrac{(\lambda + 1)a}{\lambda + a}$ $(a \in [0, 1[)$ and

$$\overline{\mu}(a) = \frac{1}{1 - a}\exp\left(-\int_0^a \frac{\lambda + 1}{\lambda}\frac{dx}{1 - x}\right) = (1 - a)^{1/\lambda}$$

(ii) For $\rho(dx) = \mathbb{P}(\Gamma \in dx)$ where Γ is a positive r.v. with finite expectation, we obtain

$$\nu(]0, a[) = \mathbb{P}\left(\frac{e}{\Gamma} \leq a \,\Big|\, \frac{e}{\Gamma} \leq 1\right)$$

where e is a standard exponential r.v. independent from Γ. In this case, we also note that:

$$\frac{1}{\psi_\mu(a)} = \frac{\nu(]0, a[)}{a} = K \int_0^{+\infty} e^{-ax} \mathbb{P}(\Gamma > x) dx$$

$$= K \mathbb{E}\left[\int_0^{\Gamma} e^{-ax} dx\right] = \frac{K}{a} \mathbb{E}\left[1 - e^{-a\Gamma}\right]$$

where $K = 1/\mathbb{E}\left[1 - e^{-\Gamma}\right]$. Consequently, the Madan-Yor function

$$D_\mu(a) = \frac{a}{\psi_\mu(a)} = K \mathbb{E}\left[1 - e^{-a\Gamma}\right]$$

is the Lévy exponent of a compound Poisson process.

(iii) For $\rho(dx) = \dfrac{e^{-\lambda x}}{x} dx$, we obtain: $\nu(]0, a[) = \dfrac{\log(1 + a)}{\log(2)}$.

2.3.6 Another Presentation of Theorem 2

In the previous section, we have parameterized the measure μ by its tail function $\overline{\mu}(x) := \int_{[x,1]} \mu(dy)$ and its tails ratio u_μ (cf. Lemma 2). Here is another parametrization of μ which provides an equivalent statement to that of Theorem 3.

Theorem 4. *Let μ be a finite positive measure on $[0, 1]$ whose support contains 1. Then, μ satisfies $(M \cdot Y)$ (i.e. D_μ is increasing on $[0, 1]$) if and only if there exists a function $\alpha_\mu :]0, 1[\longrightarrow \mathbb{R}_+$ such that:*

(i) α_μ is an increasing left-continuous function on $]0, 1[$,
(ii) $\left(\alpha_\mu^2(x) + \alpha_\mu(x)\right) dx - x d\alpha_\mu(x)$ is a positive measure on $]0, 1[$,
(iii) $\displaystyle\lim_{x \to 0^+} \frac{\alpha_\mu(x)}{x} < \infty$, and $\displaystyle\int^{1^-} \alpha_\mu(x) dx = +\infty$

and such that:

$$\overline{\mu}(x) = \overline{\mu}(0) \exp\left(-\int_0^x \frac{\alpha_\mu(y)}{y} dy\right). \tag{15}$$

Proof (Proof of Theorem 4). Properties (i), (ii) and (iii) are equivalent to the fact that the measure ν, defined on $]0, 1[$ by

$$\nu(]0, x[) = \frac{\alpha_\mu(x)}{\alpha_\mu(x) + 1}$$

belongs to $\mathcal{P}'_1$. By Theorem 3, this is equivalent to the existence of $\mu \in \mathcal{P}_1$ such that $\Gamma(\mu) = \nu$, which, in turn, is equivalent to

$$u_\mu(x) = \frac{\alpha_\mu(x)}{x}, \quad x \in]0, 1[,$$

and, finally, is equivalent to (15).   □

2.4 Some Sufficient Conditions for $(M \cdot Y)$

Throughout this section, we consider a positive finite measure μ on $\mathbb{R}_+$ which admits a density, denoted by h. Our aim is to give some sufficient conditions on h which ensure that $(M \cdot Y)$ holds. We start with a general lemma which takes up Madan-Yor condition as given in [MY02, Theorem 4] (this is Condition (iii) below):

Proposition 2. *Let h be a strictly positive function of $\mathcal{C}^1$ class on $]0, l[$ ($0 < l \leq +\infty$). The three following conditions are equivalent:*

(i) *For every $c \in]0, 1[$, $a \longmapsto \dfrac{h(a)}{h(ac)}$ is a decreasing function.*

(ii) *The function $\varepsilon(y) := -\dfrac{yh'(y)}{h(y)}$ is increasing.*

(iii) $h(a) = e^{-V(a)}$ *where $a \longmapsto aV'(a)$ is an increasing function.*

We denote this condition by (S_0).
Moreover, V and ε are related by, for any $a, b \in]0, l[$:

$$V(a) - V(b) = \int_b^a dy \frac{\varepsilon(y)}{y},$$

so that:

$$h(a) = h(b) \exp\left(-\int_b^a \frac{\varepsilon(y)}{y} dy\right).$$

Remark 4. Here are some general observations about condition (S_0):

- If both h_1 and h_2 satisfy condition (S_0), then so does $h_1 h_2$.
- If h satisfies condition (S_0), then, for every $\alpha \in \mathbb{R}$ and $\beta \geq 0$, so does $a \longmapsto a^\alpha h(a^\beta)$.
- As an example, we note that the Laplace transform $h(a) = \mathbb{E}\left[e^{-aX}\right]$ of a positive self-decomposable r.v. X satisfies condition (i). Indeed, by definition, for every $c \in [0, 1]$, there exists a positive r.v. $X^{(c)}$ independent from X such that:

$$X \overset{(\text{law})}{=} cX + X^{(c)}.$$

Taking Laplace transforms of both sides, we obtain:

$$h(a) := \mathbb{E}\left[e^{-aX}\right] = \mathbb{E}\left[e^{-acX}\right]\mathbb{E}\left[e^{-aX^{(c)}}\right],$$

which can be rewritten:

$$\frac{h(a)}{h(ac)} = \mathbb{E}\left[e^{-aX^{(c)}}\right].$$

– We note that in Theorem 5 of Madan-Yor [MY02], the second and third observations above are used jointly, as the authors remark that the function:
$$k(a) := \mathbb{E}\left[e^{-a^2 X}\right] = h(a^2) \text{ for } X \text{ positive and self-decomposable satisfies } (S_0).$$

Proof (Proof of Proposition 2).

(1) We prove that $(i) \Longleftrightarrow (ii)$
The implication $(ii) \Longrightarrow (i)$ is clear. Indeed, for $c \in]0, 1[$, we write:

$$\frac{h(a)}{h(ac)} = \exp\left(-\int_{ac}^{a}\frac{\varepsilon(y)}{y}dy\right) = \exp\left(-\int_{c}^{1}\frac{\varepsilon(ax)}{x}dx\right) \qquad (16)$$

which is a decreasing function of a since ε is increasing and $0 < c < 1$.
We now prove that $(i) \Longrightarrow (ii)$. From (16), we know that for every $c \in]0, 1[$,
$a \longmapsto \int_{ac}^{a}\frac{\varepsilon(x)}{x}dx$ is an increasing function. Therefore, by differentiation,

$$\forall a \in]0, l[, \ \forall c \in]0, 1[, \quad \varepsilon(a) - \varepsilon(ac) \geq 0$$

which proves that ε is an increasing function.
(2) We prove that $(ii) \Longleftrightarrow (iii)$
From the two representations of h, we deduce that $V(a) = \int_{b}^{a}\frac{\varepsilon(y)}{y}dy - \ln h(b)$, which gives, by differentiation:

$$aV'(a) = \varepsilon(a). \qquad (17)$$

This ends the proof of Proposition 2.

$\square$

In the following, we shall once again restrict our attention to probabilities μ on $[0, 1]$, and shall assume that they admit a density h which is strictly positive in a neighborhood of 1 (so that 1 belongs to the support of μ). We now give a first set of sufficient conditions (including (S_0)) which encompass most of the examples we shall deal with in the next section.

Theorem 5. *We assume that the density h of μ is continuous on $]0, 1[$. Then, the following conditions imply $(M \cdot Y)$:*

(S_0) *h is strictly positive on $]0, 1[$ and satisfies condition (i) of Proposition 2.*
(S_1) *for every $a \in]0, 1[$*

$$\overline{\mu}(a) := \int_a^1 h(x)dx \geq a(1-a)h(a).$$

(S_1') *the function $a \longmapsto a^2h(a)$ is increasing on $]0, 1[$.*
(S_2) *the function $a \longmapsto \log(a\overline{\mu}(a))$ is concave on $]0, 1[$ and*
$$\lim_{a \to 1^-} (1-a)h(a) = 0.$$

Proof (Proof of Theorem 5).

(1) We first prove: $(S_0) \Longrightarrow (M \cdot Y)$

We write for $a > 0$:

$$\frac{1}{D_\mu(a)} = \frac{\int_a^1 yh(y)dy}{a\overline{\mu}(a)} = \frac{[-y\overline{\mu}(y)]_a^1 + \int_a^1 \overline{\mu}(y)dy}{a\overline{\mu}(a)} = 1 + \int_1^{1/a} \frac{\overline{\mu}(ax)}{\overline{\mu}(a)}dx.$$

Clearly, $(M \cdot Y)$ is implied by the property: for all $x > 1$, $a \longmapsto \dfrac{\overline{\mu}(ax)}{\overline{\mu}(a)}$ is a decreasing function on $]0, \frac{1}{x}[$. Differentiating with respect to a, we obtain:

$$\frac{\partial}{\partial a}\left(\frac{\overline{\mu}(ax)}{\overline{\mu}(a)}\right) = \frac{-xh(ax)\overline{\mu}(a) + h(a)\overline{\mu}(ax)}{(\overline{\mu}(a))^2}.$$

We then rewrite the numerator as:

$$h(a)\int_{ax}^1 h(y)dy - xh(ax)\int_a^1 h(u)du$$

$$= xh(a)\int_a^{1/x} h(ux)du - xh(ax)\int_a^1 h(u)du$$

$$= xh(a)\int_a^{1/x} h(u)\left(\frac{h(ux)}{h(u)} - \frac{h(ax)}{h(a)}\right)du - xh(ax)\int_{1/x}^1 h(u)du \leq 0$$

from assertion (i) of Proposition 2, since for $x > 1$, the function $u \longmapsto$
$\dfrac{h(ux)}{h(u)} = \dfrac{h(ux)}{h\left(ux\frac{1}{x}\right)}$ is decreasing.

(2) We now prove: $(S_1) \Longrightarrow (M \cdot Y)$

We must prove that under (S_1), the function $D_\mu(a) := \dfrac{a\overline{\mu}(a)}{\int_a^1 xh(x)dx}$ is increasing. Elementary computations lead, for $a \in]0, 1[$, to:

$$\frac{D_\mu'(a)}{D_\mu(a)} = \frac{1}{a} - h(a)\frac{1 - D_\mu(a)}{\overline{\mu}(a)}. \tag{18}$$

From (S_1) and (12):

$$0 \leq \frac{h(a)}{\overline{\mu}(a)}(1 - D_\mu(a)) \leq \frac{1}{a(1-a)}(1-a) = \frac{1}{a}.$$

Hence, from (18):

$$\frac{D_\mu'(a)}{D_\mu(a)} \geq \frac{1}{a} - \frac{1}{a} = 0.$$

(3) We then prove: $(S_1') \Longrightarrow (S_1)$, hence $(M \cdot Y)$ holds

We have, for $a > 0$:

$$
\begin{aligned}
\overline{\mu}(a) := \int_a^1 h(x)dx &= \int_a^1 \frac{x^2 h(x)}{x^2}dx \\
&\geq a^2 h(a) \int_a^1 \frac{1}{x^2}dx \quad \text{(since } x \longmapsto x^2 h(x) \text{ is increasing.)} \\
&= a^2 h(a)\left(\frac{1}{a} - 1\right) = ah(a)(1-a).
\end{aligned}
$$

(4) We finally prove: $(S_2) \Longrightarrow (M \cdot Y)$

We set $\theta(a) = \log(a\overline{\mu}(a))$. Since

$$\int_a^1 th(t)dt = a\overline{\mu}(a) + \int_a^1 \overline{\mu}(t)dt$$

by integration by parts, we have, for $a \in]0, 1[$,

$$D_\mu(a) = \frac{e^{\theta(a)}}{e^{\theta(a)} + \int_a^1 \frac{1}{t}e^{\theta(t)}dt}.$$

Therefore, we must prove that the function $a \longmapsto e^{-\theta(a)}\int_a^1 \frac{1}{t}e^{\theta(t)}dt$ is decreasing. Differentiating this function, we need to prove:

$$l(a) := \theta'(a)\int_a^1 \frac{1}{t}e^{\theta(t)}dt + \frac{1}{a}e^{\theta(a)} \geq 0.$$

Now, since $\lim\limits_{a \to 1^-} \theta(a) = -\infty$, an integration by parts gives:

$$l(a) = \int_a^1 \frac{1}{t} e^{\theta(t)} (\theta'(a) - \theta'(t)) + \int_a^1 \frac{1}{t^2} e^{\theta(t)} dt,$$

and, θ' being a decreasing function, this last expression shows that l is also a decreasing function. Therefore, it remains to prove that:

$$\lim_{a \to 1^-} \frac{\overline{\mu}(a) - a h(a)}{\overline{\mu}(a)} \int_a^1 \overline{\mu}(t) dt \geq 0$$

or

$$\lim_{a \to 1^-} \frac{h(a)}{\overline{\mu}(a)} \int_a^1 \overline{\mu}(t) dt = 0.$$

Since $\int_a^1 \overline{\mu}(t) dt \leq (1-a)\overline{\mu}(a)$, the result follows from the assumption $\lim\limits_{a \to 1^-} (1-a) h(a) = 0$.

$\square$

Here are now some alternative conditions which ensure that $(M \cdot Y)$ is satisfied:

Proposition 3. *We assume that μ admits a density h of C^1 class on $]0, 1[$ which is strictly positive in a neighborhood of 1. The following conditions imply $(M \cdot Y)$:*

(S_3) $a \longmapsto a^3 h'(a)$ *is increasing on $]0, 1[$.*
(S_4) $a \longmapsto a^3 h'(a)$ *is decreasing on $]0, 1[$.*
(S_4') h *is decreasing and concave.*
    (Clearly, (S_4') implies (S_4)).
(S_5) h *is a decreasing function and $a \longmapsto \dfrac{ah(a)}{1-a}$ is increasing on $]0, 1[$.*
(S_5') $0 \geq h'(x) \geq -4h(x)$. *(In particular, h is decreasing).*

Proof (Proof of Proposition 3).

(1) We first prove: $(S_3) \implies (S_1')$
     We denote $\ell := \lim\limits_{a \to 0^+} a^3 h'(a) \geq -\infty$. If $\ell < 0$, then, there exists $A > 0$ and
     $\varepsilon \in]0, 1[$ such that for $x \in]0, \varepsilon[$, $h'(x) \leq -\dfrac{A}{x^3}$. This implies:

$$h(\varepsilon) - h(x) \leq \frac{A}{2} \left(\frac{1}{\varepsilon^2} - \frac{1}{x^2} \right) \qquad \text{i.e.} \quad h(x) \geq C + \frac{A}{2x^2},$$

which contradicts the fact that $\int_0^1 h(x)dx < \infty$. Therefore $\ell \geq 0$, h is positive and increasing and $h(0^+) := \lim_{x \to 0^+} h(x)$ exists. We then write:

$$a^2 h(a) = a^2 \left(h(0^+) + \int_0^a h'(x)dx \right) = a^2 h(0^+) + a^3 \int_0^1 h'(ay)dy$$

$$= a^2 h(0^+) + \int_0^1 \frac{dy}{y^3}(ay)^3 h'(ay),$$

which implies that $a \longmapsto a^2 h(a)$ is increasing as the sum of two increasing functions.

(2) We now prove: $(S_4) \Longrightarrow (M \cdot Y)$

We set $\widehat{\mu}(a) := \int_{[a,1]} x\mu(dx)$. Thus: $D_\mu(a) := \dfrac{a\overline{\mu}(a)}{\widehat{\mu}(a)}$ and, differentiation shows that $D'_\mu(a) \geq 0$ is equivalent to:

$$\gamma(a) := \overline{\mu}(a)\widehat{\mu}(a) + a^2 h(a)\overline{\mu}(a) - ah(a)\widehat{\mu}(a) \geq 0 \qquad a \in]0, 1] \quad (19)$$

We shall prove that, under (S_4), $\gamma(1^-) = 0$, $\gamma'(1^-) = 0$ and that γ is convex, which will of course imply that $\gamma \geq 0$ on $]0, 1]$. We denote $\ell := \lim_{a \to 1^-} a^3 h'(a) \geq -\infty$. Observe first that $h(1^-)$ is finite. Indeed, if ℓ is finite, then $h'(1^-)$ exists, and so does $h(1^-)$, while if $\ell = -\infty$, then $\lim_{a \to 1^-} h'(a) = -\infty$, hence h is decreasing in the neighborhood of 1 and h being positive, $h(1^-)$ also exists. Therefore, letting $a \to 1$ in (19), we obtain that $\gamma(1^-) = 0$. Now differentiating (19), we obtain:

$$\gamma'(a) = -2h(a)\widehat{\mu}(a) + ah(a)\overline{\mu}(a) + ah'(a)(a\overline{\mu}(a) - \widehat{\mu}(a)),$$

and to prove that $\gamma'(1^-) = 0$, we need to show, since $\widehat{\mu}(a) - a\overline{\mu}(a) = \overline{\overline{\mu}}(a)$, that:

$$\lim_{a \to 1^-} h'(a) \int_a^1 \overline{\overline{\mu}}(t)dt = 0.$$

If $h'(1^-)$ is finite, this property is clearly satisfied. Otherwise $\lim_{a \to 1^-} h'(a) = -\infty$. In this case, we write for a in the neighborhood of 1:

$$0 \leq -h'(a) \int_a^1 \overline{\overline{\mu}}(t)dt \leq -h'(a)(1 - a)\overline{\overline{\mu}}(a),$$

and it is sufficient to prove that:

$$\lim_{a \to 1^-} (1 - a)h'(a) = 0. \qquad (20)$$

Now, since $x \longmapsto x^3 h'(x)$ is decreasing:

$$h(1^-) - h(a) = \int_a^1 h'(x)dx \le a^3 h'(a) \left[-\frac{1}{2x^2} \right]_a^1 = \frac{a(1+a)}{2} h'(a)(1-a) \le 0$$

and (20) follows by passing to the limit as $a \to 1$.

Finally, denote by φ the decreasing continuous function: $a \longmapsto a^3 h'(a)$. Then:

$$\gamma'(a) = -\varphi(a) \frac{\overline{\overline{\mu}}(a)}{a^2} - h(a) \left(\overline{\mu}(a) + \widehat{\mu}(a) \right).$$

Consequently, γ' is a continuous function with locally finite variation, and we obtain by differentiation:

$$d\gamma'(a) = -\frac{\overline{\overline{\mu}}(a)}{a^2} d\varphi(a) + h(a) \left(ah(a) + \overline{\mu}(a) \right) da.$$

Hence, $d\gamma'$ is a positive measure on $]0, 1[$, which entails that γ is convex on $]0, 1[$.

(3) We then prove: $(S_5) \Longrightarrow (M \cdot Y)$

From (19), to prove that D_μ is increasing, we need to show that:

$$\rho(a) := \frac{\overline{\mu}(a) \widehat{\mu}(a)}{ah(a)} + a\overline{\mu}(a) - \widehat{\mu}(a) \ge 0.$$

Under (S_5), h is decreasing and hence, for $a \in]0, 1[$,

$$\overline{\mu}(a) \le h(a)(1-a). \tag{21}$$

Consequently, $\lim_{a \to 1} \rho(a) = 0$, and it is now sufficient to see that $\rho'(a) \le 0$ on $]0, 1[$.

$$\rho'(a) = -\frac{\overline{\mu}(a) \widehat{\mu}(a)}{a^2 h^2(a)} \left(h(a) + ah'(a) \right) - \frac{\widehat{\mu}(a)}{a}$$

hence, the assertion $\rho'(a) \le 0$ on $]0, 1[$ is equivalent to:

$$-\frac{1}{ah(a)} - \frac{h'(a)}{h^2(a)} \le \frac{1}{\overline{\mu}(a)}. \tag{22}$$

But, under (S_5), $a \longmapsto \frac{ah(a)}{1-a}$ is increasing, and therefore we have, for $a \in]0, 1[$,

$$\frac{1}{a(1-a)} + \frac{h'(a)}{h(a)} \ge 0. \tag{23}$$

Then, using (21) and (23), we obtain:

$$-\frac{1}{ah(a)} - \frac{h'(a)}{h^2(a)} \leq -\frac{1}{ah(a)} + \frac{1}{a(1-a)h(a)}$$

$$= \frac{1}{ah(a)}\left(\frac{1}{1-a} - 1\right) = \frac{1}{h(a)(1-a)} \leq \frac{1}{\overline{\mu}(a)}$$

which gives (22).

(4) We finally prove: $(S_5') \Longrightarrow (S_5)$

We must prove that $a \longmapsto \dfrac{ah(a)}{1-a}$ is increasing. Differentiating, we obtain:

$$\left(\frac{ah(a)}{1-a}\right)' = \frac{h(a)}{1-a}\left(\frac{1}{a(1-a)} + \frac{h'(a)}{h(a)}\right) = \frac{h(a)}{1-a}\left(\frac{1}{a(1-a)} - \left|\frac{h'(a)}{h(a)}\right|\right)$$

$$\geq \frac{h(a)}{1-a}\left(\frac{1}{a(1-a)} - 4\right) \geq 0$$

since, for $a \in [0, 1]$, $a(1-a) \leq \frac{1}{4}$.

$\square$

Remark 5. We observe that there exist some implications between these conditions. In particular:

- $(S_1') \Longrightarrow (S_2)$. Indeed, note first that since (S_1') implies (S_1), the relation $\overline{\mu}(a) \geq a(1-a)h(a)$ holds, and implies $\lim_{a \to 1^-}(1-a)h(a) = 0$. Then, for $a \in]0, 1[$, condition (S_1') is equivalent to $2h(a) + ah'(a) \geq 0$ and we can write:

$$-\left(\log\left(a\int_a^1 h(x)dx\right)\right)'' = \frac{1}{a^2} + \frac{h'(a)\overline{\mu}(a) + h^2(a)}{\overline{\mu}^2(a)}$$

$$= \frac{h(a)}{a\overline{\mu}(a)}\left(\frac{\overline{\mu}(a)}{ah(a)} + \frac{ah'(a)}{h(a)} + \frac{ah(a)}{\overline{\mu}(a)}\right)$$

$$\geq \frac{h(a)}{a\overline{\mu}(a)}\left(\frac{ah'(a)}{h(a)} + 2\right)$$

$$\left(\text{since for } x \geq 0,\ x + \frac{1}{x} \geq 2\right)$$

$$= \frac{1}{a\overline{\mu}(a)}\left(ah'(a) + 2h(a)\right) \geq 0. \qquad (24)$$

This is condition (S_2), i.e. $a \longmapsto \log(a\overline{\mu}(a))$ is a concave function.

- (S'_4) implies both (S_0) and (S_5).

    - (S_0) is satisfied since the function $y \longrightarrow \dfrac{yh'(y)}{h(y)}$ is clearly decreasing.
    - To prove that (S_5) is satisfied, we write:

$$h(1)-h(a) = \int_a^1 h'(x)dx = \int_a^1 \frac{x^2 h'(x)}{x^2}dx \le a^2 h'(a) \int_a^1 \frac{dx}{x^2} = ah'(a)(1-a),$$

hence:

$$\frac{h(a)}{1-a} + ah'(a) \ge \frac{h(1)}{1-a} \ge 0.$$

We sum up the implications between these different conditions in the following diagram:

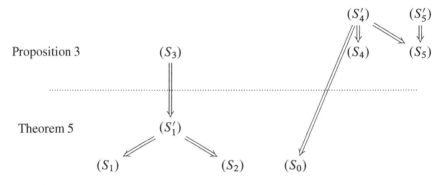

Remark 6. Let h be a decreasing function with bounded derivative h'. Then, for large enough c, the measure $\mu^{(c)}(dx) := (h(x) + c)dx$ satisfies condition (S'_5), hence $(M \cdot Y)$. Indeed, for $h^{(c)}(x) = h(x) + c$, we have:

$$\left| \frac{h^{(c)\prime}(x)}{h^{(c)}(x)} \right| = \frac{|h'(x)|}{h(x) + c} \xrightarrow[c \to +\infty]{} 0$$

This convergence being uniform, for large enough c, we obtain:

$$\sup_{x \in [0,1]} \left| \frac{h^{(c)\prime}(x)}{h^{(c)}(x)} \right| \le 4.$$

2.5 Case Where the Support of μ Is $\mathbb{R}_+$

In this section, we assume that $\mu(dx) = h(x)dx$ is a positive measure whose density h is strictly positive a.e. on $\mathbb{R}_+$. The following theorem gives sufficient conditions on h for the function D_μ to be increasing and converging to 1 when a tends to $+\infty$.

Theorem 6. *We assume that μ admits a density h on $\mathbb{R}_+$ which satisfies (S_0) (see Proposition 2).*

(1) *Then, there exists $\rho > 2$ (possibly $+\infty$) such that:*

$$\forall c \in]0, 1[, \quad \lim_{a \to +\infty} \frac{h(a)}{h(ac)} = c^\rho. \tag{25}$$

Furthermore:

$$\rho = \lim_{a \to +\infty} \varepsilon(a) = \lim_{a \to +\infty} a V'(a).$$

(2) *D_μ is an increasing function which converges towards ℓ with:*

- *If $\rho < +\infty$, then $\ell = \dfrac{\rho - 2}{\rho - 1}$*
- *If $\rho = +\infty$, then $\ell = 1$.*

In particular, if $\rho = +\infty$, then, there exists a probability measure ν_μ such that:

$$D_\mu(a) = \nu_\mu(]0, a[), \qquad a \geq 0.$$

Remark 7.

- Point (1) of Theorem 6 casts a new light on Proposition 2. Indeed, from (25), we see that h is a regularly varying function in the sense of Karamata, and Proposition 2 looks like a version of Karamata's representation Theorem (see [BGT89, Chap. 1, Theorems 1.3.1 and 1.4.1]).
- The property that the function $a \longmapsto \dfrac{h(a)}{h(ac)}$ is decreasing is not necessary to obtain the limit of D_μ, see [BGT89, Theorem 8.1.4].

Proof (Proof of Theorem 6).

(1) We first prove Point (1)

We assume that h satisfies (S_0) on $\mathbb{R}_+$. Therefore the decreasing limit $\gamma_c :=$ $\lim_{a \to +\infty} \dfrac{h(a)}{h(ac)}$ exists and belongs to $[0, 1]$. Then, for all $c, d \in]0, 1[$:

$$\gamma_{cd} = \lim_{a \to +\infty} \frac{h(a)}{h(acd)} = \lim_{a \to +\infty} \frac{h(a)}{h(ac)} \frac{h(ac)}{h(acd)} = \gamma_c \gamma_d.$$

This implies that $\gamma_c = c^\rho$ with $\rho \in \mathbb{R}_+$. Now, let $\eta(a) = \displaystyle\int_a^{+\infty} y h(y) dy$. For $A > 1$, we have

$$\eta(a) = \int_a^{+\infty} yh(y)dy = a^2 \int_1^{+\infty} zh(az)dz \geq a^2 \int_1^A z\frac{h(az)}{h(z)}h(z)dz$$

$$\geq a^2 \frac{h(aA)}{h(A)} \int_1^A zh(z)dz$$

$$\xrightarrow[A\to+\infty]{} a^{2-\rho} \int_1^{+\infty} zh(z)dz.$$

Letting a tend to $+\infty$, we obtain, since $\eta(a) \xrightarrow[a\to+\infty]{} 0$, that necessarily $\rho > 2$. Then, passing to the limit in (16), we obtain:

$$c^\rho = \exp\left(-\int_c^1 \frac{\varepsilon(+\infty)}{y}dy\right), \quad \text{i.e. } \varepsilon(+\infty) = \rho.$$

The last equality is a direct consequence of (17).

(2) We now prove that D_μ is increasing and converges towards ℓ

As in Theorem 1, we denote $\overline{\mu}(a) = \int_a^{+\infty} h(y)dy$. Then:

$$\frac{1}{D_\mu(a)} = \frac{[-y\overline{\mu}(y)]_a^{+\infty} + \int_a^{+\infty}\overline{\mu}(y)dy}{a\overline{\mu}(a)} = 1 + \int_1^{+\infty} \frac{\overline{\mu}(ax)}{\overline{\mu}(a)}dx. \quad (26)$$

Now, the proof of the increase of D_μ is exactly the same as that of the implication $S_0 \implies (M \cdot Y)$ (see Theorem 5). Then, D_μ being bounded by 1, it converges towards a limit ℓ, and it remains to identify ℓ. We write, for $x > 1$:

$$\frac{\overline{\mu}(a)}{\overline{\mu}(ax)} = \frac{\int_a^{+\infty} h(y)dy}{\int_{ax}^{+\infty} h(y)dy} = \frac{\int_1^{+\infty} h(au)du}{\int_x^{+\infty} h(au)du}$$

$$= \frac{\int_1^x h(au)du}{\int_x^{+\infty} h(au)du} + 1$$

$$= \frac{\int_1^x \frac{h(ax\frac{u}{x})}{h(ax)}du}{\int_x^{+\infty} \frac{h(au)}{h\left(au\frac{x}{u}\right)}du} + 1$$

$$\xrightarrow[a\to+\infty]{} \frac{\int_1^x \left(\frac{x}{u}\right)^\rho du}{\int_x^{+\infty} \left(\frac{x}{u}\right)^\rho du} + 1$$

from (25). Now, we must discuss different cases:

- If $\rho = +\infty$, then $\lim\limits_{a \to +\infty} \dfrac{\overline{\mu}(a)}{\overline{\mu}(ax)} = +\infty$, and plugging this limit into (26), we obtain $\ell = 1$.
- If $\rho < +\infty$, we obtain:

$$\lim_{a \to +\infty} \frac{\overline{\mu}(a)}{\overline{\mu}(ax)} = \frac{1}{x^{1-\rho}}.$$

Plugging this into (26), we obtain:

$$\frac{1}{\ell} = 1 + \int_1^{+\infty} \frac{dx}{x^{\rho-1}} = 1 - \frac{1}{2-\rho} = \frac{1-\rho}{2-\rho}.$$

$\square$

Remark 8. More generally, for $p \geq 1$, there is the equivalence:

$$\int^{+\infty} y^p h(y) dy < \infty \iff \rho > p + 1.$$

The implication $\implies$ can be proven in exactly the same way as Point (1). Conversely, since $\varepsilon(y)$ tends to ρ when y tends to $+\infty$, there exists $A > 0$ and $\theta > 0$ such that: $\forall y \geq A$, $\varepsilon(y) \geq p + 1 + \theta$. Then applying Proposition 2, we obtain:

$$h(a) = h(A) \exp\left(-\int_A^a \frac{\varepsilon(y)}{y} dy\right) \leq h(A) \exp\left(-(p+1+\theta)\int_A^a \frac{dy}{y}\right)$$

$$= h(A) \left(\frac{A}{a}\right)^{p+1+\theta}.$$

We note in particular that μ admits moments of all orders if and only if $\rho = +\infty$.

2.6 Examples

We take $\mu(dx) = h(x)dx$ and give some examples of functions h which enjoy the $(M \cdot Y)$ property. For some of them, we draw the graphs of h, D_μ, u_μ and $a \longmapsto \dfrac{v_\mu(]0, a[)}{a}$.

2.6.1 Beta Densities $h(x) = x^\alpha(1 - x)^\beta 1_{]0,1[}(x)$ $(\alpha, \beta > -1)$

(*i*) For $-1 < \beta \leq 0$ (and $\alpha > -1$), the function $x \longmapsto x^2 h(x)$ is increasing, hence from (S_1'), condition $(M \cdot Y)$ holds.
(*ii*) For $\beta \geq 0$:

$$\frac{h(a)}{h(ac)} = \frac{1}{c^\alpha} \left(1 - \frac{a(1-c)}{1-ac}\right)^\beta$$

which, for $0 < c < 1$, is a decreasing function of a, hence condition (S_0) is satisfied and $(M \cdot Y)$ also holds in that case. See Fig. 1.

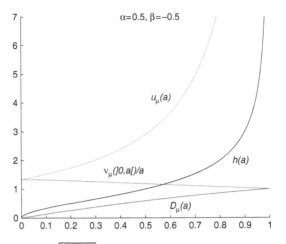

Fig. 1 Graphs for $h(x) = \sqrt{\dfrac{x}{(1-x)}}\,1_{[0,1[}(x)$

2.6.2 Further Examples

– The function $h(x) = \dfrac{x^\alpha}{(1+x^2)^\beta}\,1_{[0,1]}(x)$ $(\alpha > -1, \beta \in \mathbb{R})$ satisfies $(M \cdot Y)$.
  Indeed, for $\beta \leq 0$, $x \longmapsto x^2 h(x)$ is an increasing function on $[0,1]$, hence
  condition (S_1') holds, while, for $\beta \geq 0$, condition (S_0) is satisfied.
– The function $h(x) = \dfrac{x^\alpha}{(1-x^2)^\beta}\,1_{[0,1]}(x)$ $(\alpha > -1, \beta < 1)$ satisfies $(M \cdot Y)$. As
  in the previous example, for $0 \leq \beta \leq 1$, the function $x \longmapsto x^2 h(x)$ is increasing
  on $[0,1]$, and for $\beta \leq 0$, this results from condition (S_0).

2.6.3 $h(x) = |\cos(\pi x)|^m 1_{[0,1]}(x)$ $(m \in \mathbb{R}_+)$

We check that this example satisfies condition (S_1). Indeed, for $a \geq \frac{1}{2}$, $a \longmapsto h(a)$
is increasing, hence:

$$\int_a^1 |\cos(\pi x)|^m \, dx \geq |\cos(\pi a)|^m (1 - a) \geq a |\cos(\pi a)|^m (1 - a).$$

For $a \leq \frac{1}{2}$ we write by symmetry. See Fig. 2:

$$\int_a^1 |\cos(\pi x)|^m \, dx = \int_0^{1-a} |\cos(\pi x)|^m \, dx$$

$$\geq \int_0^a |\cos(\pi x)|^m \, dx$$

$$\geq a |\cos(\pi a)|^m \geq a |\cos(\pi a)|^m (1 - a).$$

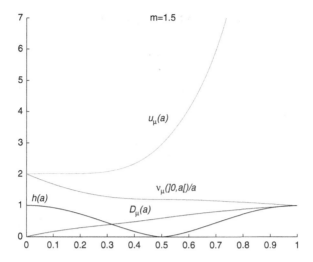

Fig. 2 Graphs for $h(x) = |\cos(\pi x)|^{3/2} 1_{[0,1]}(x)$

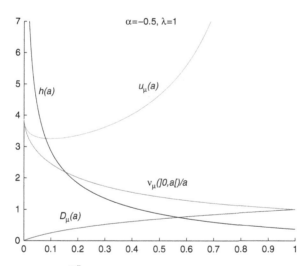

Fig. 3 Graphs for $h(x) = \dfrac{e^{-x}}{\sqrt{x}} 1_{]0,1]}(x)$

Remark 9. More generally, every function which is symmetric with respect to the axis $x = \frac{1}{2}$, and is first decreasing and then increasing, satisfies condition (S_1).

2.6.4 $h(x) = x^\alpha e^{-x^\lambda} 1_{]0,1]}(x)$ $(\alpha > -1, \lambda \in \mathbb{R})$

This is a direct consequence of (S_0). See Fig. 3.

2.6.5 An Example Where ($M \cdot Y$) is Not Satisfied

Let μ be the measure with density h defined by:

$$h(x) = c1_{[0,p[}(x) + e1_{[p,1]}(x) \qquad (c, e \geq 0, p \in]0, 1[).$$

For $a < p$, it holds:

$$D_\mu(a) := \frac{2a\left(c(p - a) + e(1 - p)\right)}{c(p^2 - a^2) + e(1 - p^2)}.$$

D_μ is C^∞ on $[0, p[$, and, for $a < p$, we have:

$$D'_\mu(a) = 2\frac{c^2 p(p - a)^2 + e^2(1 - p)^2(1 + p) + ec(1 - p)\left((p - a)^2 + p^2 + p - 2a\right)}{\left(c(p^2 - a^2) + e(1 - p^2)\right)^2}$$

and

$$D'_\mu(p^-) = 2\frac{e^2(1 - p)^2\left(1 + p\left(1 - \frac{c}{e}\right)\right)}{e^2(1 - p^2)^2} = 2\frac{1 + p\left(1 - \frac{c}{e}\right)}{(1 + p)^2}.$$

Therefore, it is clear that, for $\frac{c}{e}$ large enough, $D'_\mu(p^-) < 0$, hence D_μ is not increasing on $[0, 1]$. Note that, if $e \geq c$ (h is increasing), then $D'_\mu \geq 0$ (see condition (S'_1)), and that D_μ is increasing if and only if $D'_\mu(p^-) \geq 0$, i.e. $\frac{c}{e} \leq 1 + \frac{1}{p}$. See Figs. 4 and 5.

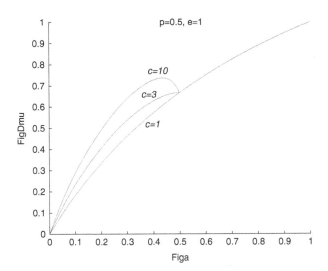

Fig. 4 Graph of D_μ for $h(x) = c1_{[0,1/2[}(x) + 1_{[1/2,1]}(x)$

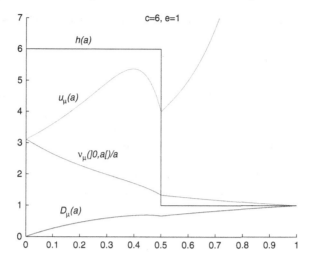

Fig. 5 Graphs for $h(x) = 6 1_{[0,1/2[}(x) + 1_{[1/2,1]}(x)$

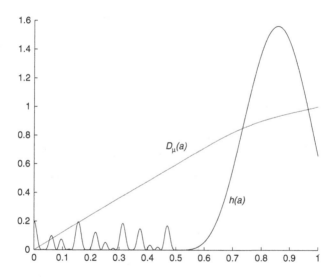

Fig. 6 Graphs for h satisfying neither condition $(S_i)_{i=0...5}$

2.6.6 A Situation Where Neither Condition $(S_i)_{i=0...5}$ Is Satisfied, But $(M \cdot Y)$ Is

Let h be a function such that, for $a \in [1/2, 1]$, $D'_\mu(a) > 0$. We define h on $[0, 1/2]$ such that $\int_0^{1/2} h(x)dx < \varepsilon$ and $\sup\limits_{x \in [0,1/2]} h(x) \le \eta$. Then, for $\varepsilon > 0$ and $\eta \ge 0$ small enough, the measure $\mu(dx) = h(x)dx$ satisfies $(M \cdot Y)$ and h may be chosen in such a way that none of the preceding conditions is satisfied. See Fig. 6.

3 Construction of Randomized Skorokhod Embeddings

3.1 Introduction

In this Sect. 3, our aim is still to construct martingales satisfying the properties (a) and (b), this time by a (seemingly) new Skorokhod embedding method, in the spirit of the original Skorokhod method, and of the so-called Hall method (see [Obl04] for comments and references; we also thank J. Obló́j (personal communication, 2009) for writing a short informal draft about this method).

Our method of randomized Skorokhod embedding will ensure directly that the family of stopping times $(\tau_t, t \geq 0)$ is increasing.

Here is the content of this Sect. 3:

- In Sect. 3.2, we consider a real valued, integrable and centered random variable X. We prove that there exist an $\mathbb{R}_+$-valued random variable V and an $\mathbb{R}_-^*$-valued random variable W, with V and W independent and independent of $(B_u, u \geq 0)$, such that, denoting:

$$\tau = \inf\{u \geq 0 \; ; \; B_u = V \; \text{ or } \; B_u = W\},$$

  Property (Sk1) is satisfied by this randomized stopping time τ, i.e: $B_\tau \overset{(\text{law})}{=} X$. To prove this result we use, as an essential tool, the Schauder-Tychonoff fixed point theorem (see [DS88]).

- In Sect. 3.3, we prove that the stopping time τ defined in Sect. 3.2 satisfies (Sk2), i.e: the martingale $B^\tau := (B_{u \wedge \tau}, u \geq 0)$ is uniformly integrable. Moreover, for every $p \geq 1$, we state conditions ensuring that B^τ is a martingale belonging to the space H^p consisting of all martingales $(M_t, t \geq 0)$ such that $\sup_{t \geq 0} |M_t| \in L^p$.

  Recall that, for $p > 1$, H^p consists of uniformly integrable martingales $(M_t, t \geq 0)$ such that $M_\infty \in L^p$, and, from Doob's L^p inequality:

$$\mathbb{E}\left[\left(\sup_{t \geq 0} |M_t| \right)^p \right] \leq \left(\frac{p}{p-1} \right)^p \mathbb{E}[|M_\infty|^p]. \tag{27}$$

- In Sect. 3.4, we follow the method presented in the general introduction, and construct an increasing family of randomized stopping times $(\tau_t, t \geq 0)$, such that $(B_{\tau_t}, t \geq 0)$ is a martingale satisfying properties (a) and (b).

3.2 Randomized Skorokhod Embedding

3.2.1 Notation

We denote by $\mathbb{R}_+$ (resp. $\mathbb{R}_-^*$) the interval $[0, +\infty[$ (resp. $]-\infty, 0[$), and by $\mathcal{M}_+$ (resp. $\mathcal{M}_-$) the set of positive finite measures on $\mathbb{R}_+$ (resp. $\mathbb{R}_-^*$), equipped with the weak topology:

$$\sigma(\mathcal{M}_+ , \mathcal{C}^0(\mathbb{R}_+)) \quad (\text{resp. } \sigma(\mathcal{M}_- , \mathcal{C}^0(\mathbb{R}_-^*)))$$

where $\mathcal{C}^0(\mathbb{R}_+)$ (resp. $\mathcal{C}^0(\mathbb{R}_-^*)$) denotes the space of continuous functions on $\mathbb{R}_+$ (resp. $\mathbb{R}_-^*$) tending to 0 at $+\infty$ (resp. at 0 and at $-\infty$).

$B = (B_u, u \geq 0)$ denotes a standard Brownian motion started from 0.

In the sequel we consider a real valued, integrable, centered random variable X, the law of which we denote by μ. The restrictions of μ to $\mathbb{R}_+$ and $\mathbb{R}_-^*$ are denoted respectively by μ_+ and μ_- .

3.2.2 Existence of a Randomized Stopping Time

This section is devoted to the proof of the following Skorokhod embedding method.

Theorem 7.

(i) *There exist an $\mathbb{R}_+$-valued random variable V and an $\mathbb{R}_-^*$-valued random variable W, V and W being independent and independent of $(B_u, u \geq 0)$, such that, setting*

$$\tau = \inf\{u \geq 0 ; \ B_u = V \ \text{or} \ B_u = W\} ,$$

one has: $B_\tau \overset{(law)}{=} X$.

(ii) *Denoting by γ_+ (resp. γ_-) the law of V (resp. W), then:*

$$\mu_+ \leq \gamma_+ \ll \mu_+ \quad \text{and} \quad \mu_- \leq \gamma_- \ll \mu_-.$$

Moreover,

$$\mathbb{E}[V \wedge (-W)] \leq \mathbb{E}[|X|] \leq 2\,\mathbb{E}[V \wedge (-W)] \tag{28}$$

and, for every $p > 1$,

$$\frac{1}{2}\mathbb{E}\left[(V \wedge (-W)) \left(V^{p-1} + (-W)^{p-1}\right)\right]$$
$$\leq \mathbb{E}[|X|^p] \leq \mathbb{E}\left[(V \wedge (-W)) \left(V^{p-1} + (-W)^{p-1}\right)\right] \tag{29}$$

Proof (Proof of Theorem 7). In the following, we exclude the case $\mu = \delta_0$, the Dirac measure at 0. Otherwise, it suffices to set: $V = 0$. Then, (i) is satisfied since $\tau = 0$, and (ii) is also satisfied except the property $\gamma_- \ll \mu_-$ (since $\mu_- = 0$).

1. We first recall the following classical result: Let $b < 0 \leq a$ and

$$T_{b,a} = \inf\{u \geq 0 \ ; \ B_u = a \ \text{ or } \ B_u = b\}.$$

Then,

$$\mathbb{P}(B_{T_{b,a}} = a) = \frac{-b}{a-b} \quad \text{and} \quad \mathbb{P}(B_{T_{b,a}} = b) = \frac{a}{a-b}.$$

2. Let V and W be respectively an $\mathbb{R}_+$-valued random variable and an $\mathbb{R}_-^*$-valued random variable, V and W being independent and independent of B, and let τ, γ_+, γ_- be defined as in the statement of the theorem. As a direct consequence of Point 1, we obtain that $B_\tau \overset{(\text{law})}{=} X$ if and only if:

$$\mu_+(dv) = \left(\int_{\mathbb{R}_-^*} \frac{-w}{v-w} \gamma_-(dw) \right) \gamma_+(dv) \quad \text{on } \mathbb{R}_+ \tag{30}$$

$$\mu_-(dw) = \left(\int_{\mathbb{R}_+} \frac{v}{v-w} \gamma_+(dv) \right) \gamma_-(dw) \quad \text{on } \mathbb{R}_-^* \tag{31}$$

As γ_+ and γ_- are probabilities, the above equations entail:

$$\gamma_+(dv) = \mu_+(dv) + \left(\int_{\mathbb{R}_-^*} \frac{v}{v-w} \gamma_-(dw) \right) \gamma_+(dv) \quad \text{on } \mathbb{R}_+ \tag{32}$$

$$\gamma_-(dw) = \mu_-(dw) + \left(\int_{\mathbb{R}_+} \frac{-w}{v-w} \gamma_+(dv) \right) \gamma_-(dw) \quad \text{on } \mathbb{R}_-^* \tag{33}$$

To prove Point i) of the theorem, we shall now solve this system of equations (32) and (33) by a fixed point method, and then we shall verify that the solution thus obtained is a pair of probabilities, which will entail (30) and (31).

3. We now introduce some further notation. If $(a,b) \in \mathcal{M}_+ \times \mathcal{M}_-$ and $\varepsilon > 0$, we set

$$a(\varepsilon) = \int 1_{]0,\varepsilon[}(v) \, a(dv) \quad \text{and} \quad b(\varepsilon) = \int 1_{]-\varepsilon,0[}(w) \, b(dw).$$

We also set: $m_+ = \int \mu_+(dv)$, $m_- = \int \mu_-(dw)$. We note that, since μ is centered and is not the Dirac measure at 0, then $m_+ > 0$ and $m_- > 0$. We then define:

$$\rho(\varepsilon) := 4 \sup \left(\mu_+(\varepsilon) \, m_+^{-1}, \mu_-(\varepsilon) \, m_-^{-1} \right)$$

and

$$\Theta := \{(a,b) \in \mathcal{M}_+ \times \mathcal{M}_- ; \ a \ge \mu_+, b \ge \mu_-, \int a(dv) + \int b(dw) \le 2$$

$$\text{and for every } \varepsilon \le \varepsilon_0, \ a(\varepsilon) \le \rho(\varepsilon) \text{ and } b(\varepsilon) \le \rho(\varepsilon)\}$$

where ε_0 will be defined subsequently.
Finally, we define $\Gamma = (\Gamma_+, \Gamma_-) : \mathcal{M}_+ \times \mathcal{M}_- \longrightarrow \mathcal{M}_+ \times \mathcal{M}_-$ by:

$$\Gamma_+(a,b)(dv) = \mu_+(dv) + \left(\int_{\mathbb{R}_*^-} \frac{v}{v-w} b(dw)\right) a(dv)$$

$$\Gamma_-(a,b)(dw) = \mu_-(dw) + \left(\int_{\mathbb{R}_+} \frac{-w}{v-w} a(dv)\right) b(dw)$$

Lemma 3. *Θ is a convex compact subset of $\mathcal{M}_+ \times \mathcal{M}_-$ (equipped with the product of the weak topologies), and $\Gamma(\Theta) \subset \Theta$.*

Proof (Proof of Lemma 3). The first part is clear. Suppose that $(a,b) \in \Theta$. By definition of Γ, we have:

$$\Gamma_+(a,b) \ge \mu_+ , \ \Gamma_-(a,b) \ge \mu_-$$

and

$$\int \Gamma_+(a,b)(dv) + \int \Gamma_-(a,b)(dw) = 1 + \left(\int a(dv)\right)\left(\int b(dw)\right). \quad (34)$$

Consequently,

$$\int \Gamma_+(a,b)(dv) + \int \Gamma_-(a,b)(dw) \le 2$$

and

$$\int \Gamma_+(a,b)(dv) \le 2 - m_- , \ \int \Gamma_-(a,b)(dw) \le 2 - m_+. \quad (35)$$

On the other hand,

$$\Gamma_+(a,b)(\varepsilon) = \mu_+(\varepsilon) + \int 1_{]0,\varepsilon[}(v) a(dv) \int 1_{]-v,0]}(w) \frac{v}{v-w} b(dw)$$

$$+ \int 1_{]0,\varepsilon[}(v) a(dv) \int 1_{]-\infty,-v]}(w) \frac{v}{v-w} b(dw).$$

Since $\dfrac{v}{v-w} \le 1$, and $\dfrac{v}{v-w} \le 1/2$ if $w \le -v$, taking into account (35) we obtain:

$$\Gamma_+(a,b)(\varepsilon) \leq \mu_+(\varepsilon) + a(\varepsilon)\, b(\varepsilon) + a(\varepsilon) \left(1 - \frac{m_+}{2}\right).$$

Hence,

$$\Gamma_+(a,b)(\varepsilon) \leq \rho^2(\varepsilon) + \rho(\varepsilon) \left(1 - \frac{m_+}{2}\right) + \mu_+(\varepsilon).$$

In order to deduce from the preceding that: $\Gamma_+(a,b)(\varepsilon) \leq \rho(\varepsilon)$, it suffices to prove:

$$\rho^2(\varepsilon) - \frac{m_+}{2}\, \rho(\varepsilon) + \mu_+(\varepsilon) \leq 0$$

or

$$\rho(\varepsilon) \in \left[\frac{1}{4}(m_+ - \sqrt{m_+^2 - 16\,\mu_+(\varepsilon)}),\, \frac{1}{4}(m_+ + \sqrt{m_+^2 - 16\,\mu_+(\varepsilon)})\right],$$

which is satisfied for $\varepsilon \leq \varepsilon_0$ for some choice of ε_0, by definition of ρ. The proof of $\Gamma_-(a,b)(\varepsilon) \leq \rho(\varepsilon)$ is similar. $\square$

Lemma 4. *The restriction of the map Γ to Θ is continuous.*

Proof (Proof of Lemma 4). We first prove the continuity of Γ_+. For $\varepsilon > 0$, we denote by h_ε a continuous function on $\mathbb{R}_-^*$ satisfying:

$$h_\varepsilon(w) = 0 \ \text{ for } \ -\varepsilon < w < 0 \ , \ \ h_\varepsilon(w) = 1 \text{ for } w < -2\,\varepsilon$$

and, for every $w < 0, 0 \leq h_\varepsilon(w) \leq 1$. We set: $\Gamma_+^\varepsilon(a,b) = \Gamma_+(a, h_\varepsilon b)$. Then, $\Gamma_+^\varepsilon(a,b) \leq \Gamma_+(a,b)$ and

$$0 \leq \int \Gamma_+(a,b)(dv) - \int \Gamma_+^\varepsilon(a,b)(dv) \leq 2\,\rho(2\,\varepsilon)\,,$$

which tends to 0 as ε tends to 0. Therefore, by uniform approximation, it suffices to prove the continuity of the map Γ_+^ε.

Let (a_n, b_n) be a sequence in Θ, weakly converging to (a,b), and let $\varphi \in C^0(\mathbb{R}_+)$. It is easy to see that the set:

$$\left\{\frac{v\,\varphi(v)}{v - \bullet}\, h_\varepsilon(\bullet) \, ; \, v \geq 0\right\}$$

is relatively compact in the Banach space $C^0(\mathbb{R}_-^*)$. Consequently,

$$\lim_{n \to \infty} \int \frac{v\,\varphi(v)}{v - w}\, h_\varepsilon(w)\, b_n(dw) = \int \frac{v\,\varphi(v)}{v - w}\, h_\varepsilon(w)\, b(dw) \tag{36}$$

uniformly with respect to v. Since

$$\left|\int \frac{v\,\varphi(v)}{v - w}\, h_\varepsilon(w)\, b_n(dw)\right| \leq 2\,|\varphi(v)|\,, \tag{37}$$

then

$$\left\{ \int \frac{v \, \varphi(v)}{v - w} \, h_\varepsilon(w) \, b_n(\mathrm{d}w) \; ; \; n \geq 0 \right\}$$

is relatively compact in the Banach space $C^0(\mathbb{R}_+)$. Therefore,

$$\lim_{n \to \infty} \int \varphi(v) \, \Gamma_+^\varepsilon(a_n, b_p)(\mathrm{d}v) = \int \varphi(v) \, \Gamma_+^\varepsilon(a, b_p)(\mathrm{d}v)$$

uniformly with respect to p, and, by (36) and (37):

$$\lim_{n \to \infty} \int \varphi(v) \, \Gamma_+^\varepsilon(a, b_n)(\mathrm{d}v) = \int \varphi(v) \, \Gamma_+^\varepsilon(a, b)(\mathrm{d}v).$$

Finally,

$$\lim_{n \to \infty} \int \varphi(v) \, \Gamma_+^\varepsilon(a_n, b_n)(\mathrm{d}v) = \int \varphi(v) \, \Gamma_+^\varepsilon(a, b)(\mathrm{d}v) \, ,$$

which proves the desired result.

The proof of the continuity of Γ_- is similar, but simpler since it does not need an approximation procedure. $\qquad \square$

As a consequence of Lemmas 3 and 4, we may apply the Schauder–Tychonoff fixed point theorem (see, for instance, [DS88, Theorem V.10.5]), which yields the existence of a pair $(\gamma_+, \gamma_-) \in \Theta$ satisfying (32) and (33). We set

$$\alpha_+ = \int \gamma_+(\mathrm{d}v) \, , \; \alpha_- = \int \gamma_-(\mathrm{d}w)$$

and we shall now prove that $\alpha_+ = \alpha_- = 1$.

4. By (34) applied to $(a, b) = (\gamma_+, \gamma_-)$, we obtain:

$$\alpha_+ + \alpha_- = 1 + \alpha_+ \alpha_-$$

and therefore, $\alpha_+ = 1$ or $\alpha_- = 1$. Suppose, for instance, $\alpha_+ = 1$. Since $\alpha_+ + \alpha_- \leq 2$, then $\alpha_- \leq 1$. We now suppose $\alpha_- < 1$. By (32), $\gamma_+ \leq \mu_+ + \alpha_- \gamma_+$, and hence, $\gamma_+ \leq (1 - \alpha_-)^{-1} \mu_+$. Consequently,

$$\int v \, \gamma_+(\mathrm{d}v) \leq (1 - \alpha_-)^{-1} \int v \, \mu_+(\mathrm{d}v) < \infty.$$

We deduce from (32) and (33) that, for every $r > 0$,

$$\int_0^\infty v \, \gamma_+(\mathrm{d}v) + \int_{-r}^0 w \, \gamma_-(\mathrm{d}w)$$

$$= \varepsilon_1(r) + \varepsilon_2(r) + \int_0^\infty \gamma_+(\mathrm{d}v) \int_{-r}^0 \gamma_-(\mathrm{d}w)(v + w) \qquad (38)$$

with

$$\varepsilon_1(r) = \int_{-r}^{+\infty} x \, \mu(dx) \qquad \text{and}$$

$$\varepsilon_2(r) = \int_0^\infty \gamma_+(dv) \int_{-\infty}^{-r} \gamma_-(dw) \frac{v^2}{v-w}.$$

Since X is centered, $\lim\limits_{r \to +\infty} \varepsilon_1(r) = 0$. On the other hand,

$$\varepsilon_2(r) \le \left(\int v \, \gamma_+(dv) \right) \left(\int_{-\infty}^{-r} \gamma_-(dw) \right)$$

and therefore, $\lim\limits_{r \to +\infty} \varepsilon_2(r) = 0$. Since $\alpha_+ = 1$, we deduce from (38):

$$\left(\int v \, \gamma_+(dv) \right) \left(1 - \int_{-r}^0 \gamma_-(dw) \right) = \varepsilon_1(r) + \varepsilon_2(r).$$

Since μ is not the Dirac measure at 0, then $\gamma_+(]0, +\infty[) > 0$. Therefore, letting r tend to ∞, we obtain $\alpha_- = 1$, which contradicts the assumption $\alpha_- < 1$. Thus, $\alpha_- = 1$ and $\alpha_+ = 1$.

5. We now prove Point ii). We have already seen: $\gamma_+ \ge \mu_+$ and $\gamma_- \ge \mu_-$. The property: $\gamma_+ \ll \mu_+$ follows directly from (30). More precisely, the Radon–Nikodym density of γ_+ with respect to μ_+ is given by:

$$\left(\int_{\mathbb{R}^*_-} \frac{-w}{v-w} \gamma_-(dw) \right)^{-1},$$

which is well defined since γ_- is a probability and $\dfrac{-w}{v-w}$ is > 0 for $w < 0$ and $v \ge 0$. On the other hand, since μ is not the Dirac measure at 0, then $\gamma_+(]0, +\infty[) > 0$. By (31), this easily entails the property: $\gamma_- \ll \mu_-$, the Radon–Nikodym density of γ_- with respect to μ_- being given by:

$$\left(\int_{\mathbb{R}_+} \frac{v}{v-w} \gamma_+(dv) \right)^{-1}.$$

On the other hand, we have for $v \ge 0$ and $w < 0$,

$$\frac{1}{2}(v \wedge (-w)) \le \frac{-vw}{v-w} \le v \wedge (-w). \tag{39}$$

Moreover, we deduce from (30) and (31)

$$\mathbb{E}\left[|X|^p\right] = \int \int \frac{-vw}{v-w} \left(v^{p-1} + (-w)^{p-1} \right) \gamma_+(dv)\gamma_-(dw) \tag{40}$$

for every $p \geq 1$. Then, (28) and (29) in Theorem 7 follow directly from (39) and (40).

□

We have obtained a theorem of existence, thanks to the application of the Schauder-Tychonoff fixed point theorem, which, of course, says nothing about the uniqueness of the pair (γ_+, γ_-) of probabilities satisfying the conditions (30) and (31). However, the following theorem states that this uniqueness holds.

Theorem 8. *Assume $\mu \neq \delta_0$. Then the laws of the r.v.'s V and W satisfying Point (i) in Theorem 7 are uniquely determined by μ.*

Proof (Proof of Theorem 8). Consider $\left(\gamma_+^{(j)}, \gamma_-^{(j)} \right)$, $j = 1, 2$, two pairs of probabilities in $\mathcal{M}_+ \times \mathcal{M}_+$ satisfying (30) and (31). We set, for $j = 1, 2$, $v \geq 0$ and $w < 0$,

$$a^{(j)}(v) = \int_{\mathbb{R}^*_-} \frac{-w}{v - w} \gamma_-^{(j)}(dw), \tag{41}$$

$$b^{(j)}(w) = \int_{\mathbb{R}_+} \frac{v}{v - w} \gamma_+^{(j)}(dv). \tag{42}$$

By (30) and (31), we have:

$$\gamma_+^{(j)} = \frac{1}{a^{(j)}} \mu_+ \quad \text{and} \quad \gamma_-^{(j)} = \frac{1}{b^{(j)}} \mu_-. \tag{43}$$

On the other hand, the following obvious equality holds:

$$\int \int_{\mathbb{R}_+ \times \mathbb{R}_-} \frac{v - w}{v - w} \left(\gamma_+^{(1)}(dv) + \gamma_+^{(2)}(dv) \right) \left(\gamma_-^{(1)}(dw) + \gamma_-^{(2)}(dw) \right) = 4. \tag{44}$$

Therefore, developing (44) and using (41), (42) and (43), we obtain:

$$\int_{\mathbb{R}_+} \left(a^{(1)}(v) + a^{(2)}(v) \right) \left(\frac{1}{a^{(1)}(v)} + \frac{1}{a^{(2)}(v)} \right) \mu_+(dv)$$

$$+ \int_{\mathbb{R}^*_-} \left(b^{(1)}(w) + b^{(2)}(w) \right) \left(\frac{1}{b^{(1)}(w)} + \frac{1}{b^{(2)}(w)} \right) \mu_-(dw) = 4 \tag{45}$$

Now, for $x > 0$, $x + \frac{1}{x} \geq 2$, and $x + \frac{1}{x} = 2$ if and only if $x = 1$. Therefore,

$$\left(a^{(1)}(v) + a^{(2)}(v) \right) \left(\frac{1}{a^{(1)}(v)} + \frac{1}{a^{(2)}(v)} \right) \geq 4$$

and $\left(a^{(1)}(v) + a^{(2)}(v) \right) \left(\dfrac{1}{a^{(1)}(v)} + \dfrac{1}{a^{(2)}(v)} \right) = 4$ if and only if $a^{(1)}(v) = a^{(2)}(v)$, and similarly with $b^{(1)}(w)$ and $b^{(2)}(w)$. Since μ is a probability, we deduce from (45) and the preceding that:

$$a^{(1)}(v) = a^{(2)}(v) \ \ \mu_+\text{-a.s.} \quad \text{and} \quad b^{(1)}(w) = b^{(2)}(w) \ \ \mu_-\text{-a.s.}$$

We then deduce from (43):

$$\gamma_+^{(1)} = \gamma_+^{(2)} \quad \text{and} \quad \gamma_-^{(1)} = \gamma_-^{(2)},$$

which is the desired result. $\square$

3.2.3 Remark

We have:

$$\forall v \geq 0, \ \forall w < 0, \qquad \frac{-w}{v-w} \geq \frac{1}{(v \vee 1)} \frac{-w}{1-w}.$$

Therefore, by (30), for $p > 1$:

$$\mathbb{E}[V^{p-1}] \leq \left(\int \frac{-w}{1-w} \gamma_-(\mathrm{d}w) \right)^{-1} \int (v \vee 1) v^{p-1} \mu_+(\mathrm{d}v) ,$$

and similarly for $\mathbb{E}[(-W)^{p-1}]$. Consequently,

$$\mathbb{E}[|X|^p] < \infty \Longrightarrow \mathbb{E}[V^{p-1}] < \infty \quad \text{and} \quad \mathbb{E}[(-W)^{p-1}] < \infty.$$

However, the converse generally does not hold (see Example 4 below), but it holds if $p \geq 2$ (see Remark 3.3.2).

3.2.4 Remark

If we no longer require the independence of the two r.v.'s V and W, then, easy computations show that Theorem 7 is still satisfied upon taking for the law of the couple (V, W):

$$2 \left(\mathbb{E}[|X|] \right)^{-1} (v - w) \, \mathrm{d}\mu_+(v) \mathrm{d}\mu_-(w). \tag{46}$$

This explicit formula, which results at once from [Bre68, 13.3, Problem 2], appears in [Hal68]. The results stated in the following Sects. 3.3 and 3.4 remain valid with the law of the couple (V, W) given by (46), except that, in Theorem 10, one must take care of replacing $\mathbb{E}[V]\mathbb{E}[-W]$ by $\mathbb{E}[-VW]$. Thus the difference between our embedding method and the one which relies on the Breiman-Hall formula is that we impose the independence of V and W. We then have the uniqueness of the laws of V and W (Theorem 8) but no general explicit formula.

3.2.5 Some Examples

In this section, we develop some explicit examples. We keep the previous notation. For $x \in \mathbb{R}$, δ_x denotes the Dirac measure at x.

Example 2. Let $0 < \alpha < 1$ and $x > 0$. We define $\mu_+ = \alpha \delta_x$ and we take for μ_- any measure in $\mathcal{M}_-$ such that

$$\int \mu_-(dw) = 1 - \alpha \qquad \text{and} \qquad \int w \, \mu_-(dw) = -\alpha \, x.$$

Then, the unique pair of probabilities (γ_+, γ_-) satisfying (30) and (31) is given by:

$$\gamma_+ = \delta_x \qquad \text{and} \qquad \gamma_-(dw) = \left(1 - \frac{w}{x}\right) \mu_-(dw).$$

Example 3. Let $0 < \alpha < 1$ and $0 < x < y$. We consider a symmetric measure μ such that:

$$\mu_+ = \frac{1}{2} \left(\alpha \, \delta_x + (1 - \alpha) \, \delta_y\right).$$

By an easy computation, we obtain that the unique pair of probabilities (γ_+, γ_-) satisfying (30) and (31) is given by:

$$\gamma_+ = \frac{y - \sqrt{(1 - \alpha) \, y^2 + \alpha \, x^2}}{y - x} \, \delta_x + \frac{-x + \sqrt{(1 - \alpha) \, y^2 + \alpha \, x^2}}{y - x} \, \delta_y$$

and $\gamma_-(dw) = \gamma_+(-dw)$.

Example 4. Let $0 < \alpha < 1$ and $0 < \beta < 1$ such that $\alpha + \beta > 1$. We define μ by:

$$\mu_+(dv) = \frac{\sin \alpha \pi}{\pi} \frac{v^{\alpha - 1}}{(1 + v^\beta) (1 + 2v^\alpha \cos \alpha \pi + v^{2\alpha})} \, dv$$

and

$$\mu_-(dw) = \frac{\sin \beta \pi}{\pi} \frac{(-w)^{\beta - 1}}{(1 + (-w)^\alpha) (1 + 2(-w)^\beta \cos \beta \pi + (-w)^{2\beta})} \, dw.$$

Then, the unique pair of probabilities (γ_+, γ_-) satisfying (30) and (31) is given by:

$$\gamma_+(dv) = \frac{\sin \alpha \pi}{\pi} \frac{v^{\alpha - 1}}{1 + 2v^\alpha \cos \alpha \pi + v^{2\alpha}} \, dv = (1 + v^\beta) \mu_+(dv)$$

and

$$\gamma_-(dw) = \frac{\sin \beta \pi}{\pi} \frac{(-w)^{\beta-1}}{1 + 2(-w)^\beta \cos \beta \pi + (-w)^{2\beta}} \, dw = (1 + (-w)^\alpha)\mu_-(dw).$$

This follows from the classical formula, which gives the Laplace transform of the resolvent of index 1 of a stable subordinator of index α (see Chaumont-Yor [CY03], Exercise 4.21]):

$$\frac{1}{1+v^\alpha} = \frac{\sin \alpha \pi}{\pi} \int_0^{+\infty} \frac{w^\alpha}{(v+w)(1 + 2w^\alpha \cos \alpha \pi + w^{2\alpha})} \, dw.$$

We note that, in this example, if $p > 1$, the condition: $\mathbb{E}[|X|^p] < \infty$ is satisfied if and only if $p < \alpha + \beta$, whereas the conditions: $\mathbb{E}[V^{p-1}] < \infty$ and $\mathbb{E}[(-W)^{p-1}] < \infty$ are satisfied if and only if $p < 1 + \alpha \wedge \beta$. Now, $\alpha + \beta < 1 + \alpha \wedge \beta$ since $\alpha \vee \beta < 1$. This illustrates Remark 3.2.3.

Example 5. We now consider a symmetric measure μ such that:

$$\mu_+(dv) = \frac{2}{\pi}(1+v^2)^{-2}\left(1 + \frac{2}{\pi} v \log v\right) dv.$$

By an easy computation, we obtain that the unique pair of probabilities (γ_+, γ_-) satisfying (30) and (31) is given by:

$$\gamma_+(dv) = \frac{2}{\pi}(1+v^2)^{-1} \, dv$$

and $\gamma_-(dw) = \gamma_+(-dw)$.

Example 6. Let μ be a symmetric measure such that:

$$\mu_+(dv) = \frac{1}{\pi}\left(\frac{1}{\sqrt{v(1-v)}} - \frac{1}{\sqrt{1-v^2}}\right) 1_{]0,1[}(v) \, dv.$$

Then, the unique pair of probabilities (γ_+, γ_-) satisfying (30) and (31) is given by:

$$\gamma_+(dv) = \frac{1}{\pi} \frac{1}{\sqrt{v(1-v)}} 1_{]0,1[}(v) \, dv$$

and $\gamma_-(dw) = \gamma_+(-dw)$. Thus, γ_+ is the Arcsine law.
This follows from the formula:

$$\frac{1}{\pi} \int_0^1 \frac{w}{v+w} \frac{1}{\sqrt{w(1-w)}} \, dw = 1 - \sqrt{\frac{v}{1+v}},$$

which can be found in [BFRY06, (1.18) and (1.23)].

3.3 Uniform Integrability

In this section, we consider again an integrable, centered, real-valued r.v. X, and we keep the notation of Theorem 7. We shall study the properties of uniform integrability of the martingale: $B^\tau := (B_{u \wedge \tau}, u \geq 0)$.

Theorem 9. *The martingale B^τ is uniformly integrable. Moreover, if $\mathbb{E}[\phi(X)] < \infty$ where $\phi : \mathbb{R} \to \mathbb{R}_+$ is defined by $\phi(x) = |x| \log^+(|x|)$, then, the martingale B^τ belongs to H^1, i.e. $\mathbb{E}\left[\sup_{u \geq 0} |B_u^\tau|\right] < \infty.$*

Proof (Proof of Theorem 9).

1. We first prove that B^τ is bounded in L^1. We denote by $\mathbb{E}_{W,V}$ the expectation with respect to the law of (W, V), and by $\mathbb{E}_B$ the expectation with respect to the law of Brownian motion B.

$$
\begin{aligned}
\sup_{u \geq 0} \mathbb{E}\left[|B_u^\tau|\right] &= \lim_{u \to +\infty} \uparrow \mathbb{E}\left[|B_u^\tau|\right] \\
&= \lim_{u \to +\infty} \uparrow \mathbb{E}_{W,V}\left[\mathbb{E}_B\left[|B_{u \wedge T_{W,V}}|\right]\right] \\
&= \mathbb{E}_{W,V}\left[\lim_{u \to +\infty} \uparrow \mathbb{E}_B\left[|B_{u \wedge T_{W,V}}|\right]\right] \\
&= \mathbb{E}_{W,V}\left[\mathbb{E}_B\left[|B_{T_{W,V}}|\right]\right] \\
&\qquad \text{(by the dominated convergence theorem,} \\
&\qquad \text{since } |B_{u \wedge T_{W,V}}| \leq V \vee (-W)) \\
&= \mathbb{E}\left[|B_\tau|\right] = \mathbb{E}[|X|].
\end{aligned}
$$

2. We have:

$$
\lambda \mathbb{P}\left(\sup_{u \geq 0} |B_u^\tau| \geq \lambda\right) = \mathbb{E}_{W,V}\left[\lambda \mathbb{P}_B\left(\sup_{u \geq 0} |B_{u \wedge T_{W,V}}| \geq \lambda\right)\right]. \tag{47}
$$

Now, since $\sup_{u \geq 0} |B_{u \wedge T_{W,V}}| \leq V \vee (-W)$,

$$
\lambda \mathbb{P}_B\left(\sup_{u \geq 0} |B_{u \wedge T_{W,V}}| \geq \lambda\right) \xrightarrow[\lambda \to +\infty]{} 0,
$$

and from Doob's maximal inequality and Point 1.:

$$
\lambda \mathbb{P}_B\left(\sup_{u \geq 0} |B_{u \wedge T_{W,V}}| \geq \lambda\right) \leq \sup_{u \geq 0} \mathbb{E}_B\left[|B_{u \wedge T_{W,V}}|\right] = \mathbb{E}_B\left[|B_{T_{W,V}}|\right],
$$

which is $\mathbb{P}_{W,V}$ integrable. Therefore, applying the dominated convergence theorem to the right hand side of (47), we obtain:

$$\lambda\mathbb{P}\left(\sup_{u\geq0}|B_u^\tau|\geq\lambda\right)\xrightarrow[\lambda\to+\infty]{}0.$$

Since B^τ is bounded in L^1, this proves, from Azéma-Gundy-Yor [AGY80, Théorème 1], the uniform integrability of B^τ.

3. We now suppose that $\mathbb{E}[\phi(X)] < \infty$. Applying the previous computation of Point 1. to the submartingale $(\phi(B_u^\tau), u \geq 0)$ (ϕ is convex), we obtain

$$\sup_{u\geq0}\mathbb{E}\left[\phi(B_u^\tau)\right] = \lim_{u\to+\infty}\uparrow\mathbb{E}\left[\phi(B_u^\tau)\right] = \mathbb{E}\left[\phi(B_\tau)\right]$$
$$= \mathbb{E}\left[\phi(X)\right] < \infty. \qquad (48)$$

Note that, under the hypothesis $\mathbb{E}[\phi(X)] < \infty$, (48) gives another proof of the fact that B^τ is a uniformly integrable martingale ([Mey66, Chap. 2, Théorème T22]).

On the other hand, from Doob's $L\log L$ inequality [RY99, p. 55],

$$\mathbb{E}\left[\sup_{u\geq0}|B_u^\tau|\right] \leq \frac{e}{e-1}\left(1 + \sup_{u\geq0}\mathbb{E}\left[\phi(B_u^\tau)\right]\right) = \frac{e}{e-1}\left(1 + \mathbb{E}\left[\phi(X)\right]\right) < \infty$$

from (48). Therefore, B^τ belongs to H^1. Actually, the martingale B^τ belongs to the $L\log L$ class (cf. [RY99, Exercise 1.16]).

$\square$

3.3.1 Remark

By Azéma-Gundy-Yor [AGY80, Théorème 1], we also deduce from the above Points 1. and 2. that:

$$\lim_{\lambda\to+\infty}\lambda\mathbb{P}\left(\sqrt{\tau}\geq\lambda\right) = 0.$$

We now complete Theorem 9 when the r.v. X admits moments of order $p > 1$. We start with $p = 2$.

Theorem 10. *The following properties are equivalent:*

(i) $\mathbb{E}[V] < \infty$ *and* $\mathbb{E}[-W] < \infty$.
(ii) $\mathbb{E}[X^2] < \infty$.
(iii) $\mathbb{E}[\tau] < \infty$.
(iv) *The martingale B^τ is in H^2.*

Moreover, if these properties are satisfied, then

$$\mathbb{E}[X^2] = \mathbb{E}[V]\,\mathbb{E}[-W] = \mathbb{E}[\tau].$$

Proof (Proof of Theorem 10). We deduce from (30) and (31) by addition:

$$\mathbb{E}[X^2] = \mathbb{E}[V]\,\mathbb{E}[-W].$$

This entails the equivalence of properties i) and ii).

On the other hand, if $b \geq 0$ and $a < 0$, the martingale $\left(B^2_{u \wedge T_{a,b}} - (u \wedge T_{a,b}), \right.$ $u \geq 0$) is uniformly integrable and hence, $\mathbb{E}[T_{a,b}] = \mathbb{E}\left[B^2_{T_{a,b}} \right] = -ab$. Consequently,

$$\mathbb{E}[\tau] = \mathbb{E}[T_{W,V}] = -\mathbb{E}[W\,V] = \mathbb{E}[V]\mathbb{E}[-W].$$

This shows that properties (i) and (iii) are equivalent.

By Doob's L^2 inequality,

$$\mathbb{E}\left[\left(\sup_{u \geq 0} |B^\tau_u| \right)^2 \right] \leq 4 \sup_{u \geq 0} \mathbb{E}\left[\left(B^\tau_u \right)^2 \right] = 4\mathbb{E}[\tau]$$

Hence, $(iii) \Longrightarrow (iv)$. The converse follows from:

$$\mathbb{E}\left[u \wedge \tau \right] = \mathbb{E}\left[\left(B^\tau_u \right)^2 \right] \leq \mathbb{E}\left[\left(\sup_{u \geq 0} |B^\tau_u| \right)^2 \right],$$

upon letting u tend to $+\infty$. Therefore:

$$\mathbb{E}[\tau] \leq \mathbb{E}\left[\left(\sup_{u \geq 0} |B^\tau_u| \right)^2 \right] \leq 4\mathbb{E}[\tau].$$

$\square$

We now replace the L^2 space by L^p for $p > 1$.

Theorem 11. *Let $p > 1$. The following properties are equivalent:*

(i) $\mathbb{E}[(V \wedge (-W))(V^{p-1} + (-W)^{p-1})] < \infty.$
(ii) $\mathbb{E}[|X|^p] < \infty.$
(iii) $\mathbb{E}[\tau^{p/2}] < \infty.$
(iv) The martingale B^τ is in H^p.

Proof (Proof of Theorem 11).

1. By (29), properties (i) and (ii) are equivalent.
2. Since we know from Theorem 9 that B^τ is uniformly integrable under the condition that X is integrable and centered, then properties (ii), (iii) and (iv) are well-known to be equivalent, see (27).

$\square$

3.3.2 Remark

If $p \geq 2$, the property $\mathbb{E}[|X|^p] < \infty$ is equivalent to: $\mathbb{E}[V^{p-1}] < \infty$ and $\mathbb{E}[(-W)^{p-1}] < \infty$. This is proven in Theorem 10 for $p = 2$.

Now, suppose $p > 2$. We saw in Remark 3.2.3 that:

$$\mathbb{E}[|X|^p] < \infty \implies \mathbb{E}[V^{p-1}] < \infty \text{ and } \mathbb{E}[(-W)^{p-1}] < \infty.$$

Conversely, suppose $\mathbb{E}[V^{p-1}] < \infty$ and $\mathbb{E}[(-W)^{p-1}] < \infty$. In particular, $\mathbb{E}[V] < \infty$ and $\mathbb{E}[(-W)] < \infty$. We deduce from (30) and (31):

$$\mathbb{E}[|X|^p] \leq \mathbb{E}[-W]\,\mathbb{E}[V^{p-1}] + \mathbb{E}[V]\,\mathbb{E}[(-W)^{p-1}]$$

which entails $\mathbb{E}[|X|^p] < \infty$.

3.4 Construction of Self-similar Martingales

In this section, we consider a real valued, centered, random variable X. Let V, W, be as in Theorem 7. We set:

$$\tau_t = \inf\{u \geq 0 \ ; \ B_u = \sqrt{t}\,V \ \text{ or } \ B_u = \sqrt{t}\,W\}.$$

Theorem 12.

(i) *The process $(B_{\tau_t}, t \geq 0)$ is a left-continuous martingale such that, for every fixed t, $B_{\tau_t} \overset{(law)}{=} \sqrt{t}\,X$.*

(ii) *For any $c > 0$,*

$$(B_{\tau_{c^2 t}}, t \geq 0) \overset{(law)}{=} (c\,B_{\tau_t}, t \geq 0).$$

(iii) *The process $(B_{\tau_t}, t \geq 0)$ is an inhomogeneous Markov process.*

In particular, $(B_{\tau_t}, t \geq 0)$ is a martingale associated to the peacock $(\sqrt{t}\,X, t \geq 0)$ (see Sect. 1.4 in the General Introduction).

Proof (Proof of Theorem 12).

1. By the definition of times τ_t and the continuity of B, one easily sees that the process $(\tau_t, t \geq 0)$ is a left-continuous increasing process. As a consequence, $(B_{\tau_t}, t \geq 0)$ is a left-continuous process.

2. Since, for a given $t \geq 0$, $(M_u := B_{u \wedge \tau_t}, u \geq 0)$ is a uniformly integrable martingale, and for $s < t$, $\tau_s \leq \tau_t$, then $(B_{\tau_t}, t \geq 0)$ is a martingale.

   Let, for $c > 0$, $(B_t^{(c)} := c\, B_{c^{-2}t}, t \geq 0)$, and denote by $(\tau_t^{(c)})$ the family of stopping times associated with the Brownian motion $B^{(c)}$. In other words,

   $$\tau_t^{(c)} = \inf\{u \geq 0 \; ; \; B_u^{(c)} = \sqrt{t}\, V \text{ or } B_u^{(c)} = \sqrt{t}\, W\}.$$

   We easily obtain, for every $t \geq 0$, $\tau_t^{(c)} = c^2 \tau_{c^{-2}t}$ and then, $B^{(c)}_{\tau_{c^2 t}^{(c)}} = c\, B_{\tau_t}$, which proves Point iii) since $(B_t^{(c)}, t \geq 0) \overset{(\text{law})}{=} (B_t, t \geq 0)$. Moreover, since $B_{\tau_1} \overset{(\text{law})}{=} X$, we also have, for every $t \geq 0$, $B_{\tau_t} \overset{(\text{law})}{=} \sqrt{t}\, X$.

3. We now consider the Brownian motion B as a strong Markov process in $\mathbb{R}$. We may define $\widetilde{\tau}_t$ by:

   $$\widetilde{\tau}_t = \inf\{u \geq 0 \; ; \; B_u \notin]\sqrt{t}\, W, \sqrt{t}\, V[\}.$$

(Note that $\widetilde{\tau}_t = \tau_t$ under $\mathbb{P}_0$, whereas, if $x \neq 0$, then $\widetilde{\tau}_t \neq \tau_t$ under $\mathbb{P}_x$.) For $s < t$, we have with the usual notation about time translation operators (θ_u),

$$\widetilde{\tau}_t = \widetilde{\tau}_s + \widetilde{\tau}_t \circ \theta_{\widetilde{\tau}_s}$$

and consequently: $B_{\widetilde{\tau}_t} = B_{\widetilde{\tau}_t} \circ \theta_{\widetilde{\tau}_s}$, which entails, for f a bounded Borel function,

$$\mathbb{E}[f(B_{\widetilde{\tau}_t}) \mid \mathcal{F}_{\widetilde{\tau}_s}] = \mathbb{E}_{B_{\widetilde{\tau}_s}}[f(B_{\widetilde{\tau}_t})],$$

which proves Point ii). More precisely, the transition semi group: $(P_{s,t}, 0 \leq s < t)$ is given by:

$$
\begin{aligned}
P_{s,t}\, f(x) &= \mathbb{E}\left[\left(f(\sqrt{t}\, V)\frac{x - \sqrt{t}\, W}{\sqrt{t}\,(V - W)} + f(\sqrt{t}\, W)\frac{-x + \sqrt{t}\, V}{\sqrt{t}\,(V - W)}\right) 1_{]\sqrt{t}\, W, \sqrt{t}\, V[}(x)\right] \\
&\quad + f(x)\,\mathbb{P}\left(x \notin]\sqrt{t}\, W, \sqrt{t}\, V[\right).
\end{aligned}
$$

Thus, $(P_{s,t}, 0 \leq s < t)$ is a transition semi group of a very special kind since, actually, $P_{s,t}$ does not depend on $s \in [0, t[$.

$\square$

References

Alb08. Albin, J.M.P.: A continuous non-Brownian motion martingale with Brownian motion marginal distributions. Stat. Probab. Lett. **78**(6), 682–686 (2008)

Azé85. Azéma, J.: Sur les fermés aléatoires. In: Séminaire de probabilités, XIX, 1983/84. Lecture Notes in Mathematics, vol. 1123, pp. 397–495. Springer, Berlin (1985)

AGY80. Azéma, J., Gundy, R.F., Yor, M.: Sur l'intégrabilité uniforme des martingales continues. In: Séminaire de Probabilités, XIV (Paris, 1978/1979) (French). Lecture Notes in Mathematics, vol. 784, pp. 53–61. Springer, Berlin (1980)

AY89. Azéma, J., Yor, M.: Étude d'une martingale remarquable. In: Séminaire de Probabilités, XXIII. Lecture Notes in Mathematics, vol. 1372, pp. 88–130. Springer, Berlin (1989)

AY79. Azéma, J., Yor, M.: Une solution simple au problème de Skorokhod. In: Séminaire de Probabilités, XIII (Univ. Strasbourg, Strasbourg, 1977/78). Lecture Notes in Mathematics, vol. 721, pp. 90–115. Springer, Berlin (1979)

BDMY10. Baker, D., Donati-Martin, C., Yor, M.: A sequence of Albin type continuous martingales, with Brownian marginals and scaling. In this volume (2010)

BFRY06. Bertoin, J., Fujita, T., Roynette, B., Yor, M.: On a particular class of self-decomposable random variables: the durations of Bessel excursions straddling independent exponential times. Probab. Math. Stat. **26**(2), 315–366 (2006)

BGT89. Bingham, N.H., Goldie, C.M., Teugels, J.L.: Regular variation. Encyclopedia of Mathematics and its Applications, vol. 27. Cambridge University Press, Cambridge (1989)

Bre68. Breiman, L.: Probability. Addison-Wesley, Reading, MA (1968)

CY03. Chaumont, L., Yor, M.: Exercises in probability. Cambridge Series in Statistical and Probabilistic Mathematics. A guided tour from measure theory to random processes, via conditioning, vol. 13. Cambridge University Press, Cambridge (2003)

DS88. Dunford, N., Schwartz, J.T.: Linear operators. Part I. Wiley Classics Library. Wiley, NY (1988). General theory, with the assistance of William G. Bade and Robert G. Bartle. Reprint of the 1958 original, A Wiley-Interscience Publication

Éme89. Émery, M.: On the Azéma martingales. In: Séminaire de Probabilités, XXIII. Lecture Notes in Mathematics, vol. 1372, pp. 66–87. Springer, Berlin (1989)

Éme96. Émery, M.: On the chaotic representation property for martingales. In: Probability theory and mathematical statistics (St. Petersburg, 1993), pp. 155–166. Gordon and Breach, Amsterdam (1996)

Éme90. Émery, M.: Sur les martingales d'Azéma (suite). In: Séminaire de Probabilités, XXIV, 1988/89. Lecture Notes in Mathematics, vol. 1426, pp. 442–447. Springer, Berlin (1990)

Hal68. Hall, W.J.: On the Skorokhod embedding theorem. In: Technical Report, vol. 33. Stanford University, Department of Statistics, (1968)

HK07. Hamza, K., Klebaner, F.C.: A family of non-Gaussian martingales with Gaussian marginals. J. Appl. Math. Stoch. Anal. ID 92723, 19 pp. (2007)

HPRY10. Hirsch, F., Profeta, C., Roynette, B., Yor, M.: Peacocks and associated martingales. (Submitted to Bacconi-Springer) (2010)

JY81. Jeulin, T., Yor, M.: Sur les distributions de certaines fonctionnelles du mouvement brownien. In: Séminaire de Probabilités, XV (Univ. Strasbourg, Strasbourg, 1979/1980) (French). Lecture Notes in Mathematics, vol. 850, pp. 210–226. Springer, Berlin (1981)

Kel72. Kellerer, H.G.: Markov-Komposition und eine Anwendung auf Martingale. Math. Ann. **198**, 99–122 (1972)

MY02. Madan, D.B., Yor, M.: Making Markov martingales meet marginals: with explicit constructions. Bernoulli **8**(4), 509–536 (2002)

Mey89a. Meyer, P.-A.: Construction de solutions d' "équations de structure". In: Séminaire de Probabilités, XXIII, Lecture Notes in Mathematics, vol. 1372, pp. 142–145. Springer, Berlin (1989)

Mey89b. Meyer, P.-A.: Équations de structure des martingales et probabilités quantiques. In: Séminaire de Probabilités, XXIII, Lecture Notes in Mathematics, vol. 1372, pp. 139–141. Springer, Berlin (1989)

Mey66. Meyer, P.-A.: Probabilités et potentiel. Publications de l'Institut de Mathématique de l'Université de Strasbourg, No. XIV. Actualités Scientifiques et Industrielles, No. 1318. Hermann, Paris (1966)

MYY09. Meziane, K., Yen, J.-Y., Yor, M.: Some examples of Skorokhod embeddings obtained from the Azéma-Yor algorithm. In preparation (2009)

Obl04. Obłój, J.: The Skorokhod embedding problem and its offspring. Probab. Surv. **1**, 321–390 (electronic) (2004)

Ole08. Oleszkiewicz, K.: On fake Brownian motions. Stat. Probab. Lett. **78**(11), 1251–1254 (2008)

Pie80. Pierre, M.: Le problème de Skorokhod: une remarque sur la démonstration d'Azéma-Yor. In: Séminaire de Probabilités, XIV (Paris, 1978/1979) (French). Lecture Notes in Mathematics, vol. 784, pp. 392–396. Springer, Berlin (1980)

RY99. Revuz, D., Yor, M.: Continuous martingales and Brownian motion. Grundlehren der Mathematischen Wissenschaften [Fundamental Principles of Mathematical Sciences], vol. 293, 3rd edn. Springer, Berlin (1999)

Rog81. Rogers, L.C.G.: Williams' characterisation of the Brownian excursion law: proof and applications. In: Séminaire de Probabilités, XV (Univ. Strasbourg, Strasbourg, 1979/1980) (French). Lecture Notes in Mathematics, vol. 850, pp. 227–250. Springer, Berlin (1981)

Yor97. Yor, M.: Some aspects of Brownian motion. Part II: Some recent martingale problems. Lecture Notes in Mathematics ETH Zürich. Birkhäuser, Basel (1997)

Lecture Notes in Mathematics

For information about earlier volumes
please contact your bookseller or Springer
LNM Online archive: springerlink.com

Vol. 1862: B. Helffer, F. Nier, Hypoelliptic Estimates and Spectral Theory for Fokker-Planck Operators and Witten Laplacians (2005)

Vol. 1863: H. Führ, Abstract Harmonic Analysis of Continuous Wavelet Transforms (2005)

Vol. 1864: K. Efstathiou, Metamorphoses of Hamiltonian Systems with Symmetries (2005)

Vol. 1865: D. Applebaum, B.V. R. Bhat, J. Kustermans, J. M. Lindsay, Quantum Independent Increment Processes I. From Classical Probability to Quantum Stochastic Calculus. Editors: M. Schürmann, U. Franz (2005)

Vol. 1866: O.E. Barndorff-Nielsen, U. Franz, R. Gohm, B. Kümmerer, S. Thorbjønsen, Quantum Independent Increment Processes II. Structure of Quantum Lévy Processes, Classical Probability, and Physics. Editors: M. Schürmann, U. Franz, (2005)

Vol. 1867: J. Sneyd (Ed.), Tutorials in Mathematical Biosciences II. Mathematical Modeling of Calcium Dynamics and Signal Transduction. (2005)

Vol. 1868: J. Jorgenson, S. Lang, $Pos_n(R)$ and Eisenstein Series. (2005)

Vol. 1869: A. Dembo, T. Funaki, Lectures on Probability Theory and Statistics. Ecole d'Eté de Probabilités de Saint-Flour XXXIII-2003. Editor: J. Picard (2005)

Vol. 1870: V.I. Gurariy, W. Lusky, Geometry of Müntz Spaces and Related Questions. (2005)

Vol. 1871: P. Constantin, G. Gallavotti, A.V. Kazhikhov, Y. Meyer, S. Ukai, Mathematical Foundation of Turbulent Viscous Flows, Martina Franca, Italy, 2003. Editors: M. Cannone, T. Miyakawa (2006)

Vol. 1872: A. Friedman (Ed.), Tutorials in Mathematical Biosciences III. Cell Cycle, Proliferation, and Cancer (2006)

Vol. 1873: R. Mansuy, M. Yor, Random Times and Enlargements of Filtrations in a Brownian Setting (2006)

Vol. 1874: M. Yor, M. Émery (Eds.), In Memoriam Paul-André Meyer - Séminaire de Probabilités XXXIX (2006)

Vol. 1875: J. Pitman, Combinatorial Stochastic Processes. Ecole d'Eté de Probabilités de Saint-Flour XXXII-2002. Editor: J. Picard (2006)

Vol. 1876: H. Herrlich, Axiom of Choice (2006)

Vol. 1877: J. Steuding, Value Distributions of L-Functions (2007)

Vol. 1878: R. Cerf, The Wulff Crystal in Ising and Percolation Models, Ecole d'Eté de Probabilités de Saint-Flour XXXIV-2004. Editor: Jean Picard (2006)

Vol. 1879: G. Slade, The Lace Expansion and its Applications, Ecole d'Eté de Probabilités de Saint-Flour XXXIV-2004. Editor: Jean Picard (2006)

Vol. 1880: S. Attal, A. Joye, C.-A. Pillet, Open Quantum Systems I, The Hamiltonian Approach (2006)

Vol. 1881: S. Attal, A. Joye, C.-A. Pillet, Open Quantum Systems II, The Markovian Approach (2006)

Vol. 1882: S. Attal, A. Joye, C.-A. Pillet, Open Quantum Systems III, Recent Developments (2006)

Vol. 1883: W. Van Assche, F. Marcellàn (Eds.), Orthogonal Polynomials and Special Functions, Computation and Application (2006)

Vol. 1884: N. Hayashi, E.I. Kaikina, P.I. Naumkin, I.A. Shishmarev, Asymptotics for Dissipative Nonlinear Equations (2006)

Vol. 1885: A. Telcs, The Art of Random Walks (2006)

Vol. 1886: S. Takamura, Splitting Deformations of Degenerations of Complex Curves (2006)

Vol. 1887: K. Habermann, L. Habermann, Introduction to Symplectic Dirac Operators (2006)

Vol. 1888: J. van der Hoeven, Transseries and Real Differential Algebra (2006)

Vol. 1889: G. Osipenko, Dynamical Systems, Graphs, and Algorithms (2006)

Vol. 1890: M. Bunge, J. Funk, Singular Coverings of Toposes (2006)

Vol. 1891: J.B. Friedlander, D.R. Heath-Brown, H. Iwaniec, J. Kaczorowski, Analytic Number Theory, Cetraro, Italy, 2002. Editors: A. Perelli, C. Viola (2006)

Vol. 1892: A. Baddeley, I. Bárány, R. Schneider, W. Weil, Stochastic Geometry, Martina Franca, Italy, 2004. Editor: W. Weil (2007)

Vol. 1893: H. Hanßmann, Local and Semi-Local Bifurcations in Hamiltonian Dynamical Systems, Results and Examples (2007)

Vol. 1894: C.W. Groetsch, Stable Approximate Evaluation of Unbounded Operators (2007)

Vol. 1895: L. Molnár, Selected Preserver Problems on Algebraic Structures of Linear Operators and on Function Spaces (2007)

Vol. 1896: P. Massart, Concentration Inequalities and Model Selection, Ecole d'Été de Probabilités de Saint-Flour XXXIII-2003. Editor: J. Picard (2007)

Vol. 1897: R. Doney, Fluctuation Theory for Lévy Processes, Ecole d'Été de Probabilités de Saint-Flour XXXV-2005. Editor: J. Picard (2007)

Vol. 1898: H.R. Beyer, Beyond Partial Differential Equations, On linear and Quasi-Linear Abstract Hyperbolic Evolution Equations (2007)

Vol. 1899: Séminaire de Probabilités XL. Editors: C. Donati-Martin, M. Émery, A. Rouault, C. Stricker (2007)

Vol. 1900: E. Bolthausen, A. Bovier (Eds.), Spin Glasses (2007)

Vol. 1901: O. Wittenberg, Intersections de deux quadriques et pinceaux de courbes de genre 1, Intersections of Two Quadrics and Pencils of Curves of Genus 1 (2007)

Vol. 1902: A. Isaev, Lectures on the Automorphism Groups of Kobayashi-Hyperbolic Manifolds (2007)

Vol. 1903: G. Kresin, V. Maz'ya, Sharp Real-Part Theorems (2007)

Vol. 1904: P. Giesl, Construction of Global Lyapunov Functions Using Radial Basis Functions (2007)

Vol. 1905: C. Prévôt, M. Röckner, A Concise Course on Stochastic Partial Differential Equations (2007)

Vol. 1906: T. Schuster, The Method of Approximate Inverse: Theory and Applications (2007)

Vol. 1907: M. Rasmussen, Attractivity and Bifurcation for Nonautonomous Dynamical Systems (2007)

Vol. 1908: T.J. Lyons, M. Caruana, T. Lévy, Differential Equations Driven by Rough Paths, Ecole d'Été de Probabilités de Saint-Flour XXXIV-2004 (2007)

Vol. 1909: H. Akiyoshi, M. Sakuma, M. Wada, Y. Yamashita, Punctured Torus Groups and 2-Bridge Knot Groups (I) (2007)

Vol. 1910: V.D. Milman, G. Schechtman (Eds.), Geometric Aspects of Functional Analysis. Israel Seminar 2004-2005 (2007)

Vol. 1911: A. Bressan, D. Serre, M. Williams, K. Zumbrun, Hyperbolic Systems of Balance Laws. Cetraro, Italy 2003. Editor: P. Marcati (2007)

Vol. 1912: V. Berinde, Iterative Approximation of Fixed Points (2007)

Vol. 1913: J.E. Marsden, G. Misiołek, J.-P. Ortega, M. Perlmutter, T.S. Ratiu, Hamiltonian Reduction by Stages (2007)

Vol. 1914: G. Kutyniok, Affine Density in Wavelet Analysis (2007)

Vol. 1915: T. Bıyıkoğlu, J. Leydold, P.F. Stadler, Laplacian Eigenvectors of Graphs. Perron-Frobenius and Faber-Krahn Type Theorems (2007)

Vol. 1916: C. Villani, F. Rezakhanlou, Entropy Methods for the Boltzmann Equation. Editors: F. Golse, S. Olla (2008)

Vol. 1917: I. Veselić, Existence and Regularity Properties of the Integrated Density of States of Random Schrödinger (2008)

Vol. 1918: B. Roberts, R. Schmidt, Local Newforms for GSp(4) (2007)

Vol. 1919: R.A. Carmona, I. Ekeland, A. Kohatsu-Higa, J.-M. Lasry, P.-L. Lions, H. Pham, E. Taflin, Paris-Princeton Lectures on Mathematical Finance 2004. Editors: R.A. Carmona, E. Çinlar, I. Ekeland, E. Jouini, J.A. Scheinkman, N. Touzi (2007)

Vol. 1920: S.N. Evans, Probability and Real Trees. Ecole d'Été de Probabilités de Saint-Flour XXXV-2005 (2008)

Vol. 1921: J.P. Tian, Evolution Algebras and their Applications (2008)

Vol. 1922: A. Friedman (Ed.), Tutorials in Mathematical BioSciences IV. Evolution and Ecology (2008)

Vol. 1923: J.P.N. Bishwal, Parameter Estimation in Stochastic Differential Equations (2008)

Vol. 1924: M. Wilson, Littlewood-Paley Theory and Exponential-Square Integrability (2008)

Vol. 1925: M. du Sautoy, L. Woodward, Zeta Functions of Groups and Rings (2008)

Vol. 1926: L. Barreira, V. Claudia, Stability of Nonautonomous Differential Equations (2008)

Vol. 1927: L. Ambrosio, L. Caffarelli, M.G. Crandall, L.C. Evans, N. Fusco, Calculus of Variations and Non-Linear Partial Differential Equations. Cetraro, Italy 2005. Editors: B. Dacorogna, P. Marcellini (2008)

Vol. 1928: J. Jonsson, Simplicial Complexes of Graphs (2008)

Vol. 1929: Y. Mishura, Stochastic Calculus for Fractional Brownian Motion and Related Processes (2008)

Vol. 1930: J.M. Urbano, The Method of Intrinsic Scaling. A Systematic Approach to Regularity for Degenerate and Singular PDEs (2008)

Vol. 1931: M. Cowling, E. Frenkel, M. Kashiwara, A. Valette, D.A. Vogan, Jr., N.R. Wallach, Representation Theory and Complex Analysis. Venice, Italy 2004. Editors: E.C. Tarabusi, A. D'Agnolo, M. Picardello (2008)

Vol. 1932: A.A. Agrachev, A.S. Morse, E.D. Sontag, H.J. Sussmann, V.I. Utkin, Nonlinear and Optimal Control Theory. Cetraro, Italy 2004. Editors: P. Nistri, G. Stefani (2008)

Vol. 1933: M. Petkovic, Point Estimation of Root Finding Methods (2008)

Vol. 1934: C. Donati-Martin, M. Émery, A. Rouault, C. Stricker (Eds.), Séminaire de Probabilités XLI (2008)

Vol. 1935: A. Unterberger, Alternative Pseudodifferential Analysis (2008)

Vol. 1936: P. Magal, S. Ruan (Eds.), Structured Population Models in Biology and Epidemiology (2008)

Vol. 1937: G. Capriz, P. Giovine, P.M. Mariano (Eds.), Mathematical Models of Granular Matter (2008)

Vol. 1938: D. Auroux, F. Catanese, M. Manetti, P. Seidel, B. Siebert, I. Smith, G. Tian, Symplectic 4-Manifolds and Algebraic Surfaces. Cetraro, Italy 2003. Editors: F. Catanese, G. Tian (2008)

Vol. 1939: D. Boffi, F. Brezzi, L. Demkowicz, R.G. Durán, R.S. Falk, M. Fortin, Mixed Finite Elements, Compatibility Conditions, and Applications. Cetraro, Italy 2006. Editors: D. Boffi, L. Gastaldi (2008)

Vol. 1940: J. Banasiak, V. Capasso, M.A.J. Chaplain, M. Lachowicz, J. Miękisz, Multiscale Problems in the Life Sciences. From Microscopic to Macroscopic. Będlewo, Poland 2006. Editors: V. Capasso, M. Lachowicz (2008)

Vol. 1941: S.M.J. Haran, Arithmetical Investigations. Representation Theory, Orthogonal Polynomials, and Quantum Interpolations (2008)

Vol. 1942: S. Albeverio, F. Flandoli, Y.G. Sinai, SPDE in Hydrodynamic. Recent Progress and Prospects. Cetraro, Italy 2005. Editors: G. Da Prato, M. Röckner (2008)

Vol. 1943: L.L. Bonilla (Ed.), Inverse Problems and Imaging. Martina Franca, Italy 2002 (2008)

Vol. 1944: A. Di Bartolo, G. Falcone, P. Plaumann, K. Strambach, Algebraic Groups and Lie Groups with Few Factors (2008)

Vol. 1945: F. Brauer, P. van den Driessche, J. Wu (Eds.), Mathematical Epidemiology (2008)

Vol. 1946: G. Allaire, A. Arnold, P. Degond, T.Y. Hou, Quantum Transport. Modelling, Analysis and Asymptotics. Cetraro, Italy 2006. Editors: N.B. Abdallah, G. Frosali (2008)

Vol. 1947: D. Abramovich, M. Mariño, M. Thaddeus, R. Vakil, Enumerative Invariants in Algebraic Geometry and String Theory. Cetraro, Italy 2005. Editors: K. Behrend, M. Manetti (2008)

Vol. 1948: F. Cao, J-L. Lisani, J-M. Morel, P. Musé, F. Sur, A Theory of Shape Identification (2008)

Vol. 1949: H.G. Feichtinger, B. Helffer, M.P. Lamoureux, N. Lerner, J. Toft, Pseudo-Differential Operators. Quantization and Signals. Cetraro, Italy 2006. Editors: L. Rodino, M.W. Wong (2008)

Vol. 1950: M. Bramson, Stability of Queueing Networks, Ecole d'Eté de Probabilités de Saint-Flour XXXVI-2006 (2008)

Vol. 1951: A. Moltó, J. Orihuela, S. Troyanski, M. Valdivia, A Non Linear Transfer Technique for Renorming (2009)

Vol. 1952: R. Mikhailov, I.B.S. Passi, Lower Central and Dimension Series of Groups (2009)

Vol. 1953: K. Arwini, C.T.J. Dodson, Information Geometry (2008)

Vol. 1954: P. Biane, L. Bouten, F. Cipriani, N. Konno, N. Privault, Q. Xu, Quantum Potential Theory. Editors: U. Franz, M. Schuermann (2008)

Vol. 1955: M. Bernot, V. Caselles, J.-M. Morel, Optimal Transportation Networks (2008)

Vol. 1956: C.H. Chu, Matrix Convolution Operators on Groups (2008)

Vol. 1957: A. Guionnet, On Random Matrices: Macroscopic Asymptotics, Ecole d'Eté de Probabilités de Saint-Flour XXXVI-2006 (2009)

Vol. 1958: M.C. Olsson, Compactifying Moduli Spaces for Abelian Varieties (2008)

Vol. 1959: Y. Nakkajima, A. Shiho, Weight Filtrations on Log Crystalline Cohomologies of Families of Open Smooth Varieties (2008)

Vol. 1960: J. Lipman, M. Hashimoto, Foundations of Grothendieck Duality for Diagrams of Schemes (2009)

Vol. 1961: G. Buttazzo, A. Pratelli, S. Solimini, E. Stepanov, Optimal Urban Networks via Mass Transportation (2009)

Vol. 1962: R. Dalang, D. Khoshnevisan, C. Mueller, D. Nualart, Y. Xiao, A Minicourse on Stochastic Partial Differential Equations (2009)

Vol. 1963: W. Siegert, Local Lyapunov Exponents (2009)

Vol. 1964: W. Roth, Operator-valued Measures and Integrals for Cone-valued Functions and Integrals for Cone-valued Functions (2009)

Vol. 1965: C. Chidume, Geometric Properties of Banach Spaces and Nonlinear Iterations (2009)

Vol. 1966: D. Deng, Y. Han, Harmonic Analysis on Spaces of Homogeneous Type (2009)

Vol. 1967: B. Fresse, Modules over Operads and Functors (2009)

Vol. 1968: R. Weissauer, Endoscopy for GSP(4) and the Cohomology of Siegel Modular Threefolds (2009)

Vol. 1969: B. Roynette, M. Yor, Penalising Brownian Paths (2009)

Vol. 1970: M. Biskup, A. Bovier, F. den Hollander, D. Ioffe, F. Martinelli, K. Netočný, F. Toninelli, Methods of Contemporary Mathematical Statistical Physics. Editor: R. Kotecký (2009)

Vol. 1971: L. Saint-Raymond, Hydrodynamic Limits of the Boltzmann Equation (2009)

Vol. 1972: T. Mochizuki, Donaldson Type Invariants for Algebraic Surfaces (2009)

Vol. 1973: M.A. Berger, L.H. Kauffmann, B. Khesin, H.K. Moffatt, R.L. Ricca, De W. Sumners, Lectures on Topological Fluid Mechanics. Cetraro, Italy 2001. Editor: R.L. Ricca (2009)

Vol. 1974: F. den Hollander, Random Polymers: École d'Été de Probabilités de Saint-Flour XXXVII – 2007 (2009)

Vol. 1975: J.C. Rohde, Cyclic Coverings, Calabi-Yau Manifolds and Complex Multiplication (2009)

Vol. 1976: N. Ginoux, The Dirac Spectrum (2009)

Vol. 1977: M.J. Gursky, E. Lanconelli, A. Malchiodi, G. Tarantello, X.-J. Wang, P.C. Yang, Geometric Analysis and PDEs. Cetraro, Italy 2001. Editors: A. Ambrosetti, S.-Y.A. Chang, A. Malchiodi (2009)

Vol. 1978: M. Qian, J.-S. Xie, S. Zhu, Smooth Ergodic Theory for Endomorphisms (2009)

Vol. 1979: C. Donati-Martin, M. Émery, A. Rouault, C. Stricker (Eds.), Séminaire de Probablitiés XLII (2009)

Vol. 1980: P. Graczyk, A. Stos (Eds.), Potential Analysis of Stable Processes and its Extensions (2009)

Vol. 1981: M. Chlouveraki, Blocks and Families for Cyclotomic Hecke Algebras (2009)

Vol. 1982: N. Privault, Stochastic Analysis in Discrete and Continuous Settings. With Normal Martingales (2009)

Vol. 1983: H. Ammari (Ed.), Mathematical Modeling in Biomedical Imaging I. Electrical and Ultrasound Tomographies, Anomaly Detection, and Brain Imaging (2009)

Vol. 1984: V. Caselles, P. Monasse, Geometric Description of Images as Topographic Maps (2010)

Vol. 1985: T. Linß, Layer-Adapted Meshes for Reaction-Convection-Diffusion Problems (2010)

Vol. 1986: J.-P. Antoine, C. Trapani, Partial Inner Product Spaces. Theory and Applications (2009)

Vol. 1987: J.-P. Brasselet, J. Seade, T. Suwa, Vector Fields on Singular Varieties (2010)

Vol. 1988: M. Broué, Introduction to Complex Reflection Groups and Their Braid Groups (2010)

Vol. 1989: I.M. Bomze, V. Demyanov, Nonlinear Optimization. Cetraro, Italy 2007. Editors: G. di Pillo, F. Schoen (2010)

Vol. 1990: S. Bouc, Biset Functors for Finite Groups (2010)

Vol. 1991: F. Gazzola, H.-C. Grunau, G. Sweers, Polyharmonic Boundary Value Problems (2010)

Vol. 1992: A. Parmeggiani, Spectral Theory of Non-Commutative Harmonic Oscillators: An Introduction (2010)

Vol. 1993: P. Dodos, Banach Spaces and Descriptive Set Theory: Selected Topics (2010)

Vol. 1994: A. Baricz, Generalized Bessel Functions of the First Kind (2010)

Vol. 1995: A.Y. Khapalov, Controllability of Partial Differential Equations Governed by Multiplicative Controls (2010)

Vol. 1996: T. Lorenz, Mutational Analysis. A Joint Framework for Cauchy Problems *In* and *Beyond* Vector Spaces (2010)

Vol. 1997: M. Banagl, Intersection Spaces, Spatial Homology Truncation, and String Theory (2010)

Vol. 1998: M. Abate, E. Bedford, M. Brunella, T.-C. Dinh, D. Schleicher, N. Sibony, Holomorphic Dynamical Systems. Editors: G. Gentili, J. Guenot, G. Patrizio (2010)

Vol. 1999: H. Schoutens, The Use of Ultraproducts in Commutative Algebra (2010)

Vol. 2000: H. Yserentant, Regularity and Approximability of Electronic Wave Functions (2010)

Vol. 2001: T. Duquesne, O. Reichmann, K.-I. Sato, C. Schwab, Lévy Matters I. Editors: O.E. Barndorff-Nielson, J. Bertoin, J. Jacod, C. Klüppelberg (2010)

Vol. 2002: C. Pötzsche, Geometric Theory of Discrete Nonautonomous Dynamical Systems (2010)

Vol. 2003: A. Cousin, S. Crépey, O. Guéant, D. Hobson, M. Jeanblanc, J.-M. Lasry, J.-P. Laurent, P.-L. Lions, P. Tankov, Paris-Princeton Lectures on Mathematical Finance 2010. Editors: R.A. Carmona, E. Cinlar, I. Ekeland, E. Jouini, J.A. Scheinkman, N. Touzi (2010)

Vol. 2004: K. Diethelm, The Analysis of Fractional Differential Equations (2010)

Vol. 2005: W. Yuan, W. Sickel, D. Yang, Morrey and Campanato Meet Besov, Lizorkin and Triebel (2011)

Vol. 2006: C. Donati-Martin, A. Lejay, A. Rouault (Eds.), Séminaire de Probabilités XLIII (2011)

Recent Reprints and New Editions

Vol. 1702: J. Ma, J. Yong, Forward-Backward Stochastic Differential Equations and their Applications. 1999 – Corr. 3rd printing (2007)

Vol. 830: J.A. Green, Polynomial Representations of GL_n, with an Appendix on Schensted Correspondence and Littelmann Paths by K. Erdmann, J.A. Green and M. Schoker 1980 – 2nd corr. and augmented edition (2007)

Vol. 1693: S. Simons, From Hahn-Banach to Monotonicity (Minimax and Monotonicity 1998) – 2nd exp. edition (2008)

Vol. 470: R.E. Bowen, Equilibrium States and the Ergodic Theory of Anosov Diffeomorphisms. With a preface by D. Ruelle. Edited by J.-R. Chazottes. 1975 – 2nd rev. edition (2008)

Vol. 523: S.A. Albeverio, R.J. Høegh-Krohn, S. Mazzucchi, Mathematical Theory of Feynman Path Integral. 1976 – 2nd corr. and enlarged edition (2008)

Vol. 1764: A. Cannas da Silva, Lectures on Symplectic Geometry 2001 – Corr. 2nd printing (2008)

LECTURE NOTES IN MATHEMATICS

🐎 Springer

Edited by J.-M. Morel, F. Takens, B. Teissier, P.K. Maini

Editorial Policy (for Multi-Author Publications: Summer Schools/Intensive Courses)

1. Lecture Notes aim to report new developments in all areas of mathematics and their applications - quickly, informally and at a high level. Mathematical texts analysing new developments in modelling and numerical simulation are welcome. Manuscripts should be reasonably self-contained and rounded off. Thus they may, and often will, present not only results of the author but also related work by other people. They should provide sufficient motivation, examples and applications. There should also be an introduction making the text comprehensible to a wider audience. This clearly distinguishes Lecture Notes from journal articles or technical reports which normally are very concise. Articles intended for a journal but too long to be accepted by most journals, usually do not have this "lecture notes" character.

2. In general SUMMER SCHOOLS and other similar INTENSIVE COURSES are held to present mathematical topics that are close to the frontiers of recent research to an audience at the beginning or intermediate graduate level, who may want to continue with this area of work, for a thesis or later. This makes demands on the didactic aspects of the presentation. Because the subjects of such schools are advanced, there often exists no textbook, and so ideally, the publication resulting from such a school could be a first approximation to such a textbook. Usually several authors are involved in the writing, so it is not always simple to obtain a unified approach to the presentation.

   For prospective publication in LNM, the resulting manuscript should not be just a collection of course notes, each of which has been developed by an individual author with little or no co-ordination with the others, and with little or no common concept. The subject matter should dictate the structure of the book, and the authorship of each part or chapter should take secondary importance. Of course the choice of authors is crucial to the quality of the material at the school and in the book, and the intention here is not to belittle their impact, but simply to say that the book should be planned to be written by these authors jointly, and not just assembled as a result of what these authors happen to submit.

   This represents considerable preparatory work (as it is imperative to ensure that the authors know these criteria before they invest work on a manuscript), and also considerable editing work afterwards, to get the book into final shape. Still it is the form that holds the most promise of a successful book that will be used by its intended audience, rather than yet another volume of proceedings for the library shelf.

3. Manuscripts should be submitted either online at www.editorialmanager.com/lnm/ to Springer's mathematics editorial, or to one of the series editors. Volume editors are expected to arrange for the refereeing, to the usual scientific standards, of the individual contributions. If the resulting reports can be forwarded to us (series editors or Springer) this is very helpful. If no reports are forwarded or if other questions remain unclear in respect of homogeneity etc, the series editors may wish to consult external referees for an overall evaluation of the volume. A final decision to publish can be made only on the basis of the complete manuscript; however a preliminary decision can be based on a pre-final or incomplete manuscript. The strict minimum amount of material that will be considered should include a detailed outline describing the planned contents of each chapter.

   Volume editors and authors should be aware that incomplete or insufficiently close to final manuscripts almost always result in longer evaluation times. They should also be aware that parallel submission of their manuscript to another publisher while under consideration for LNM will in general lead to immediate rejection.

4. Manuscripts should in general be submitted in English. Final manuscripts should contain at least 100 pages of mathematical text and should always include
   - a general table of contents;
   - an informative introduction, with adequate motivation and perhaps some historical remarks: it should be accessible to a reader not intimately familiar with the topic treated;
   - a global subject index: as a rule this is genuinely helpful for the reader.

   Lecture Notes volumes are, as a rule, printed digitally from the authors' files. We strongly recommend that all contributions in a volume be written in the same LaTeX version, preferably LaTeX2e. To ensure best results, authors are asked to use the LaTeX2e style files available from Springer's web-server at

   ftp://ftp.springer.de/pub/tex/latex/svmonot1/ (for monographs) and
   ftp://ftp.springer.de/pub/tex/latex/svmultt1/ (for summer schools/tutorials).

   Additional technical instructions are available on request from: lnm@springer.com.
5. Careful preparation of the manuscripts will help keep production time short besides ensuring satisfactory appearance of the finished book in print and online. After acceptance of the manuscript authors will be asked to prepare the final LaTeX source files and also the corresponding dvi-, pdf- or zipped ps-file. The LaTeX source files are essential for producing the full-text online version of the book. For the existing online volumes of LNM see: http://www.springerlink.com/openurl.asp?genre=journal&issn=0075-8434.

   The actual production of a Lecture Notes volume takes approximately 12 weeks.
6. Volume editors receive a total of 50 free copies of their volume to be shared with the authors, but no royalties. They and the authors are entitled to a discount of 33.3% on the price of Springer books purchased for their personal use, if ordering directly from Springer.
7. Commitment to publish is made by letter of intent rather than by signing a formal contract. Springer-Verlag secures the copyright for each volume. Authors are free to reuse material contained in their LNM volumes in later publications: a brief written (or e-mail) request for formal permission is sufficient.

Addresses:

Professor J.-M. Morel, CMLA,
École Normale Supérieure de Cachan,
61 Avenue du Président Wilson,
94235 Cachan Cedex, France
E-mail: Jean-Michel.Morel@cmla.ens-cachan.fr

Professor F. Takens, Mathematisch Instituut,
Rijksuniversiteit Groningen, Postbus 800,
9700 AV Groningen, The Netherlands
E-mail: F.Takens@math.rug.nl

Professor B. Teissier,
Institut Mathématique de Jussieu,
UMR 7586 du CNRS,
Équipe "Géométrie et Dynamique",
175 rue du Chevaleret
75013 Paris, France
E-mail: teissier@math.jussieu.fr

For the "Mathematical Biosciences Subseries" of LNM:

Professor P.K. Maini, Center for Mathematical Biology
Mathematical Institute, 24-29 St Giles,
Oxford OX1 3LP, UK
E-mail: maini@maths.ox.ac.uk

Springer, Mathematics Editorial I, Tiergartenstr. 17,
69121 Heidelberg, Germany,
Tel.: +49 (6221) 487-8259
Fax: +49 (6221) 4876-8259
E-mail: lnm@springer.com